FIELD THEORY OF NON-EQUILIBRIUM SYSTEMS

The physics of non-equilibrium many-body systems is a rapidly expanding area of theoretical physics. Traditionally employed in laser physics and superconducting kinetics, these techniques have more recently found applications in the dynamics of cold atomic gases, mesoscopic, and nano-mechanical systems, and quantum computation. This book provides a detailed presentation of modern non-equilibrium field-theoretical methods, applied to examples ranging from biophysics to the kinetics of superfluids and superconductors. A highly pedagogical and self-contained approach is adopted within the text, making it ideal as a reference for graduate students and researchers in condensed matter physics.

In this second edition, the text has been substantially updated to include recent developments in the field such as driven-dissipative quantum systems, kinetics of fermions with Berry curvature, and Floquet kinetics of periodically driven systems, among many other important new topics. Problems have been added throughout, structured as compact guided research projects that encourage independent exploration.

ALEX KAMENEV is a Professor at the William I. Fine Theoretical Physics Institute (FTPI) and at the School of Physics and Astronomy, University of Minnesota. His main areas of research include mesoscopic physics, low-dimensional and correlated fermionic and bosonic models, topological systems, and quantum computing. He is a Fellow of the American Physical Society.

FIELD THEORY OF NON-EQUILIBRIUM SYSTEMS

Second Edition

ALEX KAMENEV
University of Minnesota

CAMBRIDGE
UNIVERSITY PRESS

CAMBRIDGE
UNIVERSITY PRESS

Shaftesbury Road, Cambridge CB2 8EA, United Kingdom

One Liberty Plaza, 20th Floor, New York, NY 10006, USA

477 Williamstown Road, Port Melbourne, VIC 3207, Australia

314–321, 3rd Floor, Plot 3, Splendor Forum, Jasola District Centre, New Delhi – 110025, India

103 Penang Road, #05–06/07, Visioncrest Commercial, Singapore 238467

Cambridge University Press is part of Cambridge University Press & Assessment, a department of the University of Cambridge.
We share the University's mission to contribute to society through the pursuit of education, learning and research at the highest international levels of excellence.

www.cambridge.org
Information on this title: www.cambridge.org/9781108488259
DOI: 10.1017/9781108769266

First published 2023

A catalogue record for this publication is available from the British Library.

ISBN 978-1-108-48825-9 Hardback

To Julia and Andrei

Contents

Preface to the Second Edition

Since publication of the first edition more than 10 years ago, there have been significant developments in both experimental realization and theoretical understanding of non-equilibrium matter. The second edition is an attempt to catch up with some of them. It is also an opportunity to rectify certain omissions in the scope of the first book. Major new subjects that found their way into this edition include Lindblad dynamics of driven-dissipative quantum systems, kinetics of fermions with Berry curvature, Floquet kinetics of periodically driven systems, butterfly effect and out-of-time-order correlation functions, macroscopic fluctuation theory and counting statistics in classical stochastic models, and Boltzmann–Langevin theory. The book also received new chapters on hydrodynamics of Fermi liquids and electron–phonon interactions in disordered metals and superconductors. Needless to say, all these topics are presented within the unified framework of the functional treatment of evolution along the closed time contour. A notable addition to the book's structure is a set of problems at the end of each chapter. They are structured as compact guided research projects, which may be suggested to students for independent exploration. Most of them contain useful supplemental information, which is scattered around the periodic literature but has not yet been presented in textbooks. I hope you'll find it useful.

I am deeply indebted to numerous colleagues and current and former students and post-docs, who helped to shape my views and saved me from many pitfalls and omissions. Last but not least, I can't express enough appreciation to my family, whose love, support, and patience made this book possible. During my work on the second edition I was partially supported by National Science Foundation grant DMR-2037654.

Preface to the First Edition

The quantum field theory (QFT) is the universal common language of the condensed matter community. Like any living language it keeps evolving and changing. The change comes as a response to new problems and developments, trends from other branches of physics, and from internal pressure to optimize its own vocabulary to make it more flexible and powerful. There are many excellent books that document this evolution and give snapshots of "modern" QFT in condensed matter theory for almost half a century. In the beginning QFT was developed in the second quantization operator language. It produced such monumental books as Kadanoff and Baym [1], Abrikosov, Gor'kov, and Dzyaloshinski (AGD) [2], Fetter and Walecka [3], and Mahan [4]. The advent of renormalization group and Grassmann integrals stimulated development of functional methods of QFT. They were reflected in the next generation of books such as Itzykson and Zuber [5], Negele and Orland [6], and Fradkin [7]. The latest generation (e.g. Tsvelik [8], Altland and Simons [9]) is not only fully based on functional methods, but also deeply incorporates ideas of symmetry-based universality, geometry, and topology. (I do not mention here some excellent specialized texts devoted to applications of QFT in superconductivity, magnetism, phase transitions, mesoscopics, one-dimensional physics, etc.)

Following AGD authority, most of these books (with the notable exception of Kadanoff and Baym) employ imaginary time Matsubara formalism [11] of finite temperature equilibrium QFT. The irony is that a much more powerful non-equilibrium QFT pioneered by Schwinger [12], Konstantinov and Perel [13], Kadanoff and Baym [1], and Keldysh [14] was developed almost at the same time as the Matsubara technique. Being widely scattered across the periodic scientific literature, it has barely penetrated into the mainstream pedagogical texts. The very few books I am aware of are Kadanoff and Baym [1], Lifshitz and Pitaevskii [15], Smith and Jensen [16], Haug and Jauho [17], and Rammer [18]. (There are also a number of useful reviews [19–23].) In my personal opinion, the reasons for such a disparity are twofold: (i) There may be a perception that all subtle and interesting effects take place only in equilibrium; non-equilibrium systems are too "violent" to be treated by QFT. Instead, the kinetic equation approach is the best one can hope for; the latter may be obtained with the Golden rule and thus does not need QFT. (ii) The formalism is too involved, too complicated, to be a part of the "common" knowledge.

As far as the first reason is concerned, it was realized decades ago that even the "simple" kinetic equation may not actually be so simple. Time and again it was shown that the kinetic equation for superfluids, superconductors, fermion–boson mixtures, disordered normal metals, and so on can't be deduced from the Golden rule and has to be derived using the non-equilibrium QFT methods. Yet, most of the traditional experimental systems, such as liquid helium, bulk magnets, superconductors, or disordered normal metals, can hardly be driven substantially away from the equilibrium. This created the comforting impression that studying the equilibrium plus linear response properties is largely sufficient to describe experiment. The last two decades have changed this perception dramatically. First, mesoscopic normal metals and superconductors have demonstrated that non-equilibrium conditions may be achieved in controlled and reproducible ways and a number of unusual specifically non-equilibrium phenomena do emerge. Then came cold atomic gases in magnetic and optical traps. These systems are rarely truly at equilibrium, yet they exhibit a rich phenomenology, which calls for a theoretical description. Lately, the flourishing fields of nanomechanics and nanomagnetics emerged, which deal with stochastic mechanical and magnetic systems driven far away from the equilibrium. All these developments call for the systematic non-equilibrium theory to be a part of the "standard package" of a theoretical physicist.

As for technical complexity and lack of "aesthetic" appeal, there is some truth to this, especially when the story is told in the old operator formalism. (This is exactly how most of the currently existing books approach the subject.) I can see how one can be overwhelmed by the number of different Green functions, rules to follow, and the length of the calculations. Fortunately, the structure of the theory becomes much more transparent when it is presented in the functional formalism. Instead of keeping track of matrix Green functions and tensor vertices, one has to follow the scalar action, which is a functional of two fields. Yes, one still has to double the number of degrees of freedom. However, when taken in appropriate linear combinations (Keldysh rotation), they acquire a transparent physical meaning. Then the causality principle emerges as a simple and natural way to navigate through the calculations. The main goal of this book is to present a thorough and self-contained exposition of the non-equilibrium QFT *entirely in the functional formalism.*

I tried to pay special attention to specific peculiarities of non-equilibrium (i.e. closed time contour) QFT, which do not show up in the imaginary time or $T = 0$

formalism. In particular, presentation starts from the simplest possible systems and develops all minute technical details, exposes pitfalls, and explains the internal structure for such "trivial" situations. Then the systems are gradually taken to be more and more complex. I still tried my best to emphasize peculiarities of non-equilibrium calculations in comparison with the probably more familiar equilibrium ones. Although the book is meant to be entirely self-contained, some common subjects between equilibrium and non-equilibrium techniques (e.g. diagrammatic expansion, Dyson equation, renormalization group) are introduced in a rather compact way. In such cases I mostly focused on differences between the two approaches, possibly at the expense of the common themes. Therefore some prior familiarity with the imaginary time QFT is beneficial (although not compulsory) for a reader.

What are the benefits of learning non-equilibrium QFT? (i) It naturally provides a way to go beyond the linear response and derive consistent kinetic theory (e.g. quasiparticle kinetics coupled to the dynamics of the order parameter). As we have mentioned, there is a rapidly growing list of fields where such an approach is unavoidable. (ii) Even for linear response problems it allows one to circumvent analytic continuation to real time, which may be quite cumbersome. (iii) In its functional form it provides a natural and seamless connection to the huge field of classical stochastic systems, their universality classes, and phase transitions. In fact, a big part of this book is devoted to such classical problems. For this reason the word "quantum" does not appear in the title. Yet from the point of view of the formalism, non-equilibrium QFT and the theory of classical stochastic systems are virtually indistinguishable. (iv) Some subjects of great current interest (e.g. full counting statistics, or fluctuation relations) cannot even be approached without the formalism, presented here. (v) Non-equilibrium QFT (again in its functional form) appears to be extremely effective in dealing with systems with quenched disorder. Even if purely equilibrium or linear response properties are in question, closed time contour QFT is much more natural and efficient than the imaginary time one. All these items are the subject of the present book. I hope you'll find it useful.

This book is intended for advanced graduate students, post-docs, and faculty who want to enrich their understanding of non-equilibrium physics. It may be used as a guide for an upper-division graduate class on QFT methods in condensed matter physics. There is practically no discussion of relevant experimental results in the book. This omission is intentional, since the scope is rather broad and inclusion of experiments could easily increase the volume by a factor. Finally, the bibliography is not meant to be exhaustive or complete. In most cases the references are to the original works where the presented results were obtained, to their immediate extensions, and to review articles. I sincerely apologize to the many authors whose works I was not able to cover.

Finally, this is an opportunity to express my deep appreciation to all of my coauthors, colleagues, and students from whom I learned a great deal about the subjects of this book. During my work on the first edition I was partially supported by National Science Foundation grant DMR-0804266.

1
Introduction

1.1 Closed Time Contour

Consider a quantum many-body system governed by a time-dependent Hamiltonian $\hat{H}(t)$. Let us assume that in the distant past $t = -\infty$ the system was in a state specified by a many-body density matrix $\hat{\rho}(-\infty)$. The precise form of the latter is of no importance. It may be, for example, the equilibrium density matrix associated with the Hamiltonian $\hat{H}(-\infty)$. We shall also assume that the time-dependence of the Hamiltonian is such that at $t = -\infty$ the particles were noninteracting. The interactions are then adiabatically switched on to reach their actual physical strength sometime prior to the observation time. *In addition*, the Hamiltonian may contain *true* time dependence through, for example, external fields or boundary conditions. Due to such true time-dependent perturbations the density matrix is driven away from equilibrium.

The density matrix evolves according to the Von Neumann equation

$$\partial_t \hat{\rho}(t) = -\mathrm{i}\big[\hat{H}(t), \hat{\rho}(t)\big], \tag{1.1}$$

where we set $\hbar = 1$. It is formally solved with the help of the unitary evolution operator as $\hat{\rho}(t) = \hat{\mathcal{U}}_{t,-\infty}\hat{\rho}(-\infty)\big[\hat{\mathcal{U}}_{t,-\infty}\big]^{\dagger} = \hat{\mathcal{U}}_{t,-\infty}\hat{\rho}(-\infty)\hat{\mathcal{U}}_{-\infty,t}$, where the † denotes Hermitian conjugation. The evolution operator obeys

$$\partial_t \hat{\mathcal{U}}_{t,t'} = -\mathrm{i}\hat{H}(t)\hat{\mathcal{U}}_{t,t'}\,; \qquad \partial_{t'}\hat{\mathcal{U}}_{t,t'} = \mathrm{i}\hat{\mathcal{U}}_{t,t'}\,\hat{H}(t').$$

Notice that the Hamiltonian operators taken at different moments of time, in general, do not commute with each other. As a result, $\hat{\mathcal{U}}_{t,t'}$ must be understood as an infinite product of incremental evolution operators with instantaneous locally constant Hamiltonians

$$
\begin{aligned}
\hat{\mathcal{U}}_{t,t'} &= \lim_{N\to\infty} e^{-i\hat{H}(t-\delta_t)\delta_t} e^{-i\hat{H}(t-2\delta_t)\delta_t} \dots e^{-i\hat{H}(t-N\delta_t)\delta_t} e^{-i\hat{H}(t')\delta_t} \\
&= \mathbb{T} \exp\left(-i\int_{t'}^{t} \hat{H}(t)\, dt\right),
\end{aligned} \tag{1.2}
$$

where an infinitesimal time-step is $\delta_t = (t - t')/N$ and to shorten the notations the infinite product is abbreviated as the time-ordered, or $\mathbb{T}$-exponent.

One is usually interested to know an expectation value of some observable $\hat{\mathcal{O}}$ (say density or current operator) at a time t.[1] It is defined as

$$
\langle\hat{\mathcal{O}}\rangle(t) \equiv \frac{\mathrm{Tr}\{\hat{\mathcal{O}}\hat{\rho}(t)\}}{\mathrm{Tr}\{\hat{\rho}(t)\}} = \frac{1}{\mathrm{Tr}\{\hat{\rho}(t)\}}\, \mathrm{Tr}\big\{\hat{\mathcal{U}}_{-\infty,t}\hat{\mathcal{O}}\hat{\mathcal{U}}_{t,-\infty}\hat{\rho}(-\infty)\big\}, \tag{1.3}
$$

where the trace is performed over many-body Hilbert space and in the last equality we cyclically permuted the $\hat{\mathcal{U}}_{-\infty,t}$ operator under the trace sign. The expression under the last trace describes (read from right to left) evolution from $t = -\infty$, where the initial density matrix is specified, toward t, where the observable is calculated, and then back to $t = -\infty$. Therefore, calculation of an observable implies evolving the initial state both *forward and backward*.

Such forward–backward evolution is avoided in the equilibrium quantum field theory with a special trick. Let us recall how it works, for example, in the zero-temperature equilibrium formalism [2]. The latter deals with the ground state expectation values of the type $\langle \mathrm{GS}|\hat{\mathcal{O}}|\mathrm{GS}\rangle$, where $|\mathrm{GS}\rangle$ is a ground state of an *interacting* many-body system. It is obtained from the known and simple ground state of the corresponding *noninteracting* system $|0\rangle$ by acting on the latter with the evolution operator $|\mathrm{GS}\rangle = \hat{\mathcal{U}}_{t,-\infty}|0\rangle$. Since we are dealing with the equilibrium situation, the only time dependence allowed for the Hamiltonian is an adiabatic switching of the interactions on and off in the distant past and distant future, respectively. The evolution operator therefore describes the evolution of a simple noninteracting ground state $|0\rangle$ toward $|\mathrm{GS}\rangle$ upon adiabatic switching of the interactions and thus $\langle \mathrm{GS}|\hat{\mathcal{O}}|\mathrm{GS}\rangle = \langle 0|\hat{\mathcal{U}}_{-\infty,t}\hat{\mathcal{O}}\hat{\mathcal{U}}_{t,-\infty}|0\rangle$.

Now comes the trick: one argues that

$$
\hat{\mathcal{U}}_{+\infty,-\infty}|0\rangle = e^{iL}|0\rangle. \tag{1.4}
$$

That is, evolution of the noninteracting ground state upon adiabatic switching of the interactions on and subsequent adiabatic switching them off brings the system back into the state $|0\rangle$, up to a phase factor e^{iL}. This statement is based on the belief that the adiabatic perturbation keeps the system in its (evolving) ground state at all times. If so, in view of normalization $\langle 0|0\rangle = 1$, the only

[1] We work in the Schrödinger picture, where observables are t-independent operators, while the wavefunctions and the density matrix evolve.

possible change is the phase of the noninteracting ground state $e^{iL} = \langle 0|\hat{\mathcal{U}}_{+\infty,-\infty}|0\rangle$. Similarly $\langle 0|\hat{\mathcal{U}}_{+\infty,-\infty} = \langle 0|e^{iL}$. Accepting this, one proceeds as follows:

$$\begin{aligned}\langle \mathrm{GS}|\hat{\mathcal{O}}|\mathrm{GS}\rangle &= \langle 0|\hat{\mathcal{U}}_{-\infty,t}\hat{\mathcal{O}}\hat{\mathcal{U}}_{t,-\infty}|0\rangle = e^{-iL}\langle 0|e^{iL}\hat{\mathcal{U}}_{-\infty,t}\hat{\mathcal{O}}\hat{\mathcal{U}}_{t,-\infty}|0\rangle \\ &= e^{-iL}\langle 0|\hat{\mathcal{U}}_{+\infty,-\infty}\hat{\mathcal{U}}_{-\infty,t}\hat{\mathcal{O}}\hat{\mathcal{U}}_{t,-\infty}|0\rangle = \frac{\langle 0|\hat{\mathcal{U}}_{+\infty,t}\hat{\mathcal{O}}\hat{\mathcal{U}}_{t,-\infty}|0\rangle}{\langle 0|\hat{\mathcal{U}}_{+\infty,-\infty}|0\rangle}, \end{aligned} \quad (1.5)$$

where in the last equality we used $\hat{\mathcal{U}}_{+\infty,-\infty}\hat{\mathcal{U}}_{-\infty,t} = \hat{\mathcal{U}}_{+\infty,t}$, which is an immediate consequence of Eq. (1.2). The result of this procedure is that one needs to consider only the *forward* evolution. Indeed, the numerator in the last expression (being read from right to left) calls for evolving the noninteracting ground state $|0\rangle$ from the distant past to the observation time, where the observable operator acts, and then proceeding toward the distant future, where the overlap with the same known state $\langle 0|$ is evaluated.

The similar strategy works in the finite-temperature equilibrium formalism [11, 2, 4]. There, one treats the equilibrium density matrix $e^{-\beta\hat{H}}$, where $\beta = 1/T$ is the inverse temperature, as the evolution operator in the imaginary time τ. The latter is defined on a finite interval $0 \le \tau < \beta$. The observables (or correlation functions) are also evaluated at imaginary time points $\tau_1, \tau_2, \ldots$ and the result must be analytically continued back to the real-time axis. One may argue that, since the adiabatic switching of interactions does not drive the system out of equilibrium, a statement similar to Eq. (1.4) still holds. As a result one is again left to describe only the forward evolution, albeit along the finite time interval in the imaginary direction.

Let us mention that elimination of the backward evolution comes with a price: the normalization denominator in the last expression in Eq. (1.5). It offsets the phase accumulation e^{iL} of the noninteracting ground state $|0\rangle$. In diagrammatic language it amounts to subtracting the so-called disconnected or vacuum loop diagrams. This denominator is a serious liability in the theory of disordered systems. The reason is that the accumulated phase e^{iL} sensitively depends on a specific realization of the disorder (which may be thought of as being absent at $t \to \pm\infty$ and adiabatically switched on and off in the process of evolution). Therefore, the denominator absolutely must be included in any disorder averaging procedure, which complicates the treatment in a very substantial way.

The much more serious trouble with the outlined procedure is that Eq. (1.4) does *not* work in a non-equilibrium situation. If the Hamiltonian $\hat{H}(t)$ contains nonadiabatic time-dependent external fields, boundary conditions, and so on, the evolution drives the system away from equilibrium. Even if all such fields are eventually switched off in the distant future, there is no guarantee that the system returns to its ground (or equilibrium) state. Therefore, acting with the operator $\hat{\mathcal{U}}_{+\infty,-\infty}$ on

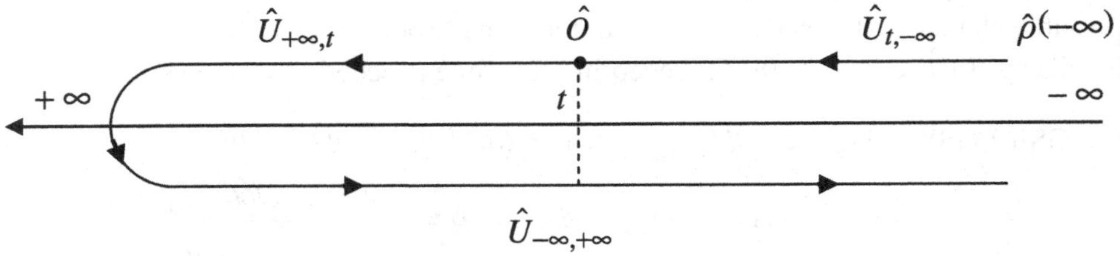

Figure 1.1 Closed time contour $\mathcal{C}$. Evolution along such a contour is described by Eq. (1.6).

the initial ground (or equilibrium) state results in an unknown superposition of excited states. As a result, the backward evolution, inherent to Eq. (1.3), can't be eliminated.

Nevertheless, it is still convenient to extend the evolution in Eq. (1.3) toward $t = +\infty$ and then back to t. This is achieved with the help of the trivial identity $\hat{\mathcal{U}}_{t,+\infty}\hat{\mathcal{U}}_{+\infty,t} = \hat{1}$. Inserting it into Eq. (1.3) and using $\hat{\mathcal{U}}_{-\infty,t}\hat{\mathcal{U}}_{t,+\infty} = \hat{\mathcal{U}}_{-\infty,+\infty}$, one finds

$$\langle\hat{O}\rangle(t) = \frac{1}{\mathrm{Tr}\{\hat{\rho}(-\infty)\}}\,\mathrm{Tr}\big\{\hat{\mathcal{U}}_{-\infty,+\infty}\hat{\mathcal{U}}_{+\infty,t}\hat{O}\hat{\mathcal{U}}_{t,-\infty}\hat{\rho}(-\infty)\big\}. \tag{1.6}$$

Here we also used the fact that, according to the Von Neumann equation (1.1), the trace of the initial density matrix is unchanged under the unitary evolution. Equation (1.6) describes evolution along the *closed time contour* $\mathcal{C}$ depicted in Fig. 1.1. The observable $\hat{O}$ is inserted at time t, somewhere along the forward branch of the contour. Notice that, by inserting the operator $\hat{\mathcal{U}}_{t,+\infty}\hat{\mathcal{U}}_{+\infty,t} = \hat{1}$ to the right of $\hat{O}$ in Eq. (1.3), one could equally well arrange to have the observable on the backward branch of the contour. As we shall see, the most convenient choice is to take a half sum of these two equivalent representations.

Evolution along the closed time contour $\mathcal{C}$ is the central subject of this book. The fact that the field theory can be constructed with the time ordering along such a contour was first realized by Schwinger [12] and further developed in [25, 26]. About the same time Konstantinov and Perel [13] developed a diagrammatic technique, based on the time contour containing forward and backward branches in the real-time direction along with the imaginary time portion of length β. The formalism was significantly advanced, in particular its utility to derive the kinetic theory, in the seminal book of Kadanoff and Baym [1]. Independently Keldysh [14] (for some of the historic context, see [27]) suggested a formulation that does not rely on imaginary time (and thus on the equilibrium density matrix). He also introduced a convenient choice of variables (Keldysh rotation), which made derivation of the kinetic theory particularly transparent. The time contour without the imaginary time piece, along with the Keldysh variables (which we call "classical" and "quantum"), appear to be by far the most convenient choices for the functional formulation of the theory presented in this book. For this reason we occasionally refer to the construction as the Keldysh technique (this should by no way diminish the credit deserved by the other authors). Reformulation of the theory for the case of fermions, given later by Larkin and Ovchinnikov [28], became universally accepted. In fact, other theories developed about the same time, while not using the time contour explicitly, appear to be close relatives of the Schwinger–Kadanoff–Baym–Keldysh construction. Among them are Feynman and Vernon [29], Wyld's [30] diagrammatic technique for fluid dynamics, and Martin, Siggia, and Rose [31] and DeDominicis' [32] calculus for classical stochastic systems.

The central object of the theory is the evolution operator along the closed contour $\hat{\mathcal{U}}_C = \hat{\mathcal{U}}_{-\infty,+\infty}\hat{\mathcal{U}}_{+\infty,-\infty}$. If the Hamiltonian is the same on the forward and backward branches, then the forward–backward evolution of *any* state brings it back exactly to the original state. (Not even a phase factor is accumulated; indeed, any phase gained on the forward branch is exactly "unwound" on the backward branch.) As a result $\hat{\mathcal{U}}_C = \hat{1}$ and the partition function, defined as $Z \equiv \mathrm{Tr}\{\hat{\mathcal{U}}_C\hat{\rho}(-\infty)\}/\mathrm{Tr}\{\hat{\rho}(-\infty)\}$, is identically equal to unity, $Z = 1$. Nevertheless, the partition function is a convenient object to develop the functional representation, and the normalization identity $Z = 1$ is a useful check of its consistency. For this reason we shall use it widely in what follows.

To insert an observable somewhere along the forward (as prescribed by Eq. (1.6)) or backward branches it is convenient to modify the Hamiltonian $\hat{H}(t)$ by adding the source term $\hat{H}_V^{\pm}(t) \equiv \hat{H}(t) \pm \hat{O}V(t)$, where the plus (minus) sign refers to the forward (backward) part of the contour. Now, since the Hamiltonian is different on the two branches, the evolution operator along the contour $\hat{\mathcal{U}}_C[V] \neq \hat{1}$ becomes nontrivial and so does the *generating function*

$$Z[V] \equiv \frac{\mathrm{Tr}\{\hat{\mathcal{U}}_C[V]\,\hat{\rho}(-\infty)\}}{\mathrm{Tr}\{\hat{\rho}(-\infty)\}}. \tag{1.7}$$

The expectation value of the observable $\hat{O}$, given by Eq. (1.6) (or rather by a half sum of the observable inserted along the forward and backward branches) may be found as $\langle\hat{O}\rangle(t) = (\mathrm{i}/2)\delta Z[V]/\delta V(t)\big|_{V=0}$. This expression should be compared with the equilibrium technique [2, 4], where the observables are given by variational derivatives of the *logarithm* of the generating (or partition) function. In our case, since $Z = Z[0] = 1$, the presence of the logarithm is optional.[2] Knowledge of

[2] It is worth mentioning that the denominators in Eqs. (1.5) and (1.7) have very different status. In the latter case $\mathrm{Tr}\{\hat{\rho}(-\infty)\}$ refers entirely to the distant past, when both interactions and disorder are switched off. It is therefore a simple constant, which may be easily evaluated. In the former case $\langle 0|\hat{\mathcal{U}}_{+\infty,-\infty}|0\rangle$ involves evolution of the ground state upon switching on and off the interactions and disorder. It thus depends on both disorder and interactions and requires a separate calculation. The absence of a disorder-dependent denominator makes the closed time contour formalism especially suitable to deal with the averaging over the quenched disorder. The fact that observables do not require the *logarithm* is another manifestation of the absence of the nontrivial denominator.

the generating function allows one thus to find observables of interest. Therefore, after developing the functional formalism for the partition function, we extend it to include the generating function as well.

1.2 Brief Outline of the Book

Part I is devoted to systems with one or few degrees of freedom. It is used as a pedagogical device to develop the functional integral treatment of the evolution along the closed time contour, introduced previously. There are several important steps that need to be dealt with within the time-discretized treatment, before the continuous notations turn into a practical theoretical device. We take extra care exposing those steps in Chapter 2 for Von Neumann unitary evolution and then again in Chapter 5 in Lindblad dissipative setting. To illustrate the utility of the developed framework we apply it to a number of quantum mechanical problems, which include treatment of systems in contact with classical or quantum bath. A particular focus is on non-perturbative phenomena, such as tunneling in a time-dependent potential, or quantum activation, where the functional integral treatment is truly indispensable.

Chapter 4 is used to demonstrate that the classical limit of the closed time contour theory yields the Martin–Siggia–Rose–DeDominicis framework for stochastic models. We use this observation as an opportunity to derive Langevin and Fokker–Planck equations, which are widely employed through the rest of the book. Once again, the developed machinery is deployed to treat a number of examples. Those include escape from a metastable state, full counting statistics, reaction models, fluctuation relations, and time-dependent perturbations in stochastic systems. These subjects resurface in subsequent chapters.

Part II extends the machinery to many-body quantum Bose systems as well as classical stochastic models. Chapter 6 serves to adopt the standard arsenal of the quantum field theory to the closed time contour setting. This includes the diagrammatic technique and the Dyson equation. Specifically we focus on the realization that a particular component of the Dyson equation yields the quantum kinetic equation. We present several pedagogical examples of how the kinetic equation is derived from the first principles and is used in practice. In Chapter 7, the classical collisionless plasma is used as a vehicle to expose some useful concepts: collective modes (plasmons), Landau damping, random phase approximation (RPA), Hubbard–Stratonovich transformation, and coupled kinetic equations for collective modes and quasiparticles degrees of freedom.

Chapter 8 employs the machinery to discuss kinetics of the Bose–Einstein condensation. Here we derive the mean-field Gross–Pitaevskii framework as the stationary point of the Keldysh action and Bogoliubov collective modes as its linearized excitations. We then develop the kinetic theory of coupled collective modes and quasiparticles. Chapter 9 is devoted to the dynamics of classical phase transitions. It includes nucleation of critical droplets in first-order transitions as well as dynamics of continuous phase transitions, starting from equilibrium transitions and their Hohenberg–Halperin classification. We then turn to essentially non-equilibrium phase transitions, which include absorbing state transitions in reaction-diffusion models and Kardar–Parisi–Zhang (KPZ) theory of the roughening transition on growing interfaces. The 1d KPZ equation is used to illustrate concepts of the macroscopic fluctuation theory (MFT), geared toward rare events statistics in stochastic models.

The rest of the book, Parts III and IV, is devoted to fermions. Chapter 10 extends treatment of Chapters 2 and 6 to fermionic systems, using Grassmann functional integrals on the closed time contour. Chapter 11 serves to derive the kinetic equation for interacting fermions and uses it to introduce a hydrodynamic description of the Fermi liquid. Chapter 12 deals with various aspects of the fermionic kinetic theory. Those include treatment of fermions in presence of a Berry curvature, kinetics of periodically driven Floquet systems, stochastic Boltzmann–Langevin terms in the kinetic equation. Finally we discuss the "butterfly effect" with the out-of-time-order correlation functions (OTOC), which requires four-branch time contour.

In Chapter 13 we use fermionic formalism to discuss non-equilibrium quantum transport. In particular we derive the Landauer formula for tunneling conductance, the Lesovik formula for shot-noise, the Levitov's result for the full counting statistics of transmitted charge, the Brouwer formula for adiabatic pumping of charge, and, following Nazarov and Tobiska, the exact fluctuation relation and its consequences. We also deal with the spin transport, deriving the Slonczewski–Berger spin-torque term in the Landau–Lifshitz equation along with the spin-torque noise and associated Gilbert damping.

Part IV of the book addresses fermions in the presence of a quenched disorder. The main technical tool of this part is Keldysh nonlinear sigma-model (NLSM), which is painstakingly developed in Chapter 14 and applied in various settings in this and subsequent chapters. Chapter 14 features weak-localization along with the scaling theory of Anderson localization. Chapter 15 deals with the mesoscopic phenomena ranging from spectral statistics and universal conductance fluctuations (UCF) to full counting statistics in tunneling, diffusive, and quantum chaotic systems.

We then include electron–electron interactions in disordered systems, which lead to singular behavior in the density of states (zero bias anomaly) and conductivity (Altshuler–Aronov corrections). Interactions also yield collision terms to the diffusive quantum kinetic equation. These phenomena are the subject of Chapter 16.

Chapter 17 is devoted to disordered superconductors. We generalize the nonlinear sigma-model to include superconducting correlations. It yields the Usadel equation as its stationary point, which includes equations for the quasiparticle spectrum as well as the quasiparticle kinetic equation for the quasiparticles distribution function. Together with the self-consistency condition they provide a complete framework to study non-equilibrium superconductivity. As examples, we work out the spectrum of the collective (Carlson–Goldman) mode of the superconductor, and derive the time-dependent Ginzburg–Landau theory and fluctuation corrections to the conductivity above the critical temperature. Finally, Chapter 18 deals with electron–phonon interactions in disordered normal metals and superconductors. We derive coupled kinetic equations for electron and phonon degrees of freedom and use them to analyze electron relaxation as well as ultrasound attenuation.

All chapters are equipped with several problems. They are structured as compact guided research projects, which may be suggested to students for an independent exploration. Most of them carry useful supplemental information, which has not yet found its way to textbooks.

Part I

Systems with Few Degrees of Freedom

2

Bosons

The aim of this chapter is to develop a functional integral representation for the evolution operator along the closed time contour. To this end we use an example of a single quantized level populated by bosonic particles. Notations and structures introduced in this chapter are used throughout the rest of the book.

2.1 Bosonic Coherent States

We start by considering a single quantum level occupied by bosonic particles. A *many-body* state with n bosons is denoted by $|n\rangle$. Such *pure number states* form a complete orthonormal basis, meaning $\langle n|n'\rangle = \delta_{nn'}$ and $\sum_n |n\rangle\langle n| = \hat{1}$. It is convenient to introduce bosonic annihilation and creation operators, $\hat{b}$ and $\hat{b}^\dagger$, which operate in the many-body Hilbert space of the system according to the following rules:

$$\hat{b}\,|n\rangle = \sqrt{n}\,|n-1\rangle\,; \qquad \hat{b}^\dagger\,|n\rangle = \sqrt{n+1}\,|n+1\rangle. \tag{2.1}$$

By acting on an arbitrary basis state, one may check the following relations:

$$\hat{b}^\dagger\hat{b}|n\rangle = n|n\rangle\,; \qquad \hat{b}\hat{b}^\dagger|n\rangle = (n+1)|n\rangle\,; \qquad [\hat{b},\hat{b}^\dagger] = \hat{1}. \tag{2.2}$$

An extremely useful tool for our purposes is the algebra of bosonic coherent states, which we summarize briefly in this section. A coherent state, parametrized by a complex number ϕ, is defined as a right eigenstate of the annihilation operator with the eigenvalue ϕ:

$$\hat{b}\,|\phi\rangle = \phi|\phi\rangle\,; \qquad \langle\phi|\,\hat{b}^\dagger = \bar{\phi}\langle\phi|, \tag{2.3}$$

where the bar denotes complex conjugation. As a result, the matrix elements in the coherent state basis of any *normally ordered* operator $\hat{H}(\hat{b}^\dagger,\hat{b})$ (i.e. such that all the creation operators are to the left of all the annihilation operators) are given by

$$\langle\phi|\hat{H}(\hat{b}^\dagger,\hat{b})|\phi'\rangle = H(\bar{\phi},\phi')\,\langle\phi|\phi'\rangle. \tag{2.4}$$

One may check by direct substitution using Eq. (2.1) that the following linear superposition of the pure number states is indeed the required right eigenstate of the operator $\hat{b}$:

$$|\phi\rangle = \sum_{n=0}^{\infty} \frac{\phi^n}{\sqrt{n!}} |n\rangle = \sum_{n=0}^{\infty} \frac{\phi^n}{n!} (\hat{b}^\dagger)^n |0\rangle = \mathrm{e}^{\phi \hat{b}^\dagger} |0\rangle, \tag{2.5}$$

where $|0\rangle$ is the vacuum state, $\hat{b}\,|0\rangle = 0$. Upon Hermitian conjugation, one finds $\langle\phi| = \langle 0|\,\mathrm{e}^{\bar{\phi}\hat{b}} = \sum_n \langle n|\bar{\phi}^n/\sqrt{n!}$. The coherent states are not mutually orthogonal and they form an over-complete basis. The overlap of two coherent states is given by

$$\langle\phi|\phi'\rangle = \sum_{n,n'=0}^{\infty} \frac{\bar{\phi}^n \phi'^{n'}}{\sqrt{n!n'!}} \langle n|n'\rangle = \sum_{n=0}^{\infty} \frac{(\bar{\phi}\phi')^n}{n!} = \mathrm{e}^{\bar{\phi}\phi'}, \tag{2.6}$$

where we employed the orthonormality of the pure number states.

One may express resolution of unity in the coherent states basis. It takes the following form:

$$\hat{1} = \int \mathrm{d}[\bar{\phi},\phi]\; \mathrm{e}^{-|\phi|^2}\, |\phi\rangle\langle\phi|, \tag{2.7}$$

where $\mathrm{d}[\bar{\phi},\phi] \equiv \mathrm{d}(\mathrm{Re}\,\phi)\,\mathrm{d}(\mathrm{Im}\,\phi)/\pi$. To prove this relation one may employ the Gaussian integral

$$Z[\bar{J},J] = \int \mathrm{d}[\bar{\phi},\phi]\,\mathrm{e}^{-\bar{\phi}\phi + \bar{\phi}J + \bar{J}\phi} = \mathrm{e}^{\bar{J}J}, \tag{2.8}$$

where J is an arbitrary complex number. As its consequence one obtains

$$\int \mathrm{d}[\bar{\phi},\phi]\,\mathrm{e}^{-|\phi|^2}\, \bar{\phi}^n \phi^{n'} = \frac{\partial^{n+n'}}{\partial J^n \partial \bar{J}^{n'}} Z[\bar{J},J]\Bigg|_{\bar{J}=J=0} = n!\,\delta_{n,n'}. \tag{2.9}$$

Substituting Eq. (2.5) and its conjugate into the right-hand side of Eq. (2.7) and employing Eq. (2.9) along with the resolution of unity in the number state basis $\hat{1} = \sum_n |n\rangle\langle n|$, one proves the identity (2.7).

The trace of an arbitrary operator $\hat{\mathcal{O}}$, acting in the space of the occupation numbers, is evaluated as

$$\begin{aligned} \mathrm{Tr}\{\hat{\mathcal{O}}\} &\equiv \sum_{n=0}^{\infty} \langle n|\hat{\mathcal{O}}|n\rangle = \sum_{n=0}^{\infty} \int \mathrm{d}[\bar{\phi},\phi]\,\mathrm{e}^{-|\phi|^2}\, \langle n|\hat{\mathcal{O}}|\phi\rangle\langle\phi|n\rangle \\ &= \int \mathrm{d}[\bar{\phi},\phi]\mathrm{e}^{-|\phi|^2} \sum_{n=0}^{\infty} \langle\phi|n\rangle\langle n|\hat{\mathcal{O}}|\phi\rangle = \int \mathrm{d}[\bar{\phi},\phi]\,\mathrm{e}^{-|\phi|^2}\, \langle\phi|\hat{\mathcal{O}}|\phi\rangle, \end{aligned} \tag{2.10}$$

where we have employed resolution of unity first in the coherent state basis and second in the number state basis.

Another useful identity is

$$f(\rho) \equiv \langle\phi|\rho^{\hat{b}^\dagger\hat{b}}|\phi'\rangle = \mathrm{e}^{\bar{\phi}\phi'\rho}. \tag{2.11}$$

The proof is based on the following operator relation: $g(\hat{b}^\dagger\hat{b})\,\hat{b} = \hat{b}\,g(\hat{b}^\dagger\hat{b}-1)$ valid for an arbitrary function $g(\hat{b}^\dagger\hat{b})$, which is verified by acting on an arbitrary basis vector $|n\rangle$. As a result,

$$\partial_\rho f(\rho) = \langle\phi|\hat{b}^\dagger\hat{b}\,\rho^{\hat{b}^\dagger\hat{b}-1}|\phi'\rangle = \langle\phi|\hat{b}^\dagger\,\rho^{\hat{b}^\dagger\hat{b}}\,\hat{b}|\phi'\rangle = \bar{\phi}\phi' f(\rho).$$

Integrating this differential equation with the initial condition $f(1) = \mathrm{e}^{\bar{\phi}\phi'}$, which follows from Eq. (2.6), one proves the identity (2.11).

2.2 Partition Function

Let us consider the simplest many-body system: bosonic particles occupying a single quantum state with the energy ω_0. Its secondary quantized Hamiltonian has the form

$$\hat{H}(\hat{b}^\dagger,\hat{b}) = \omega_0\,\hat{b}^\dagger\hat{b}, \tag{2.12}$$

where $\hat{b}^\dagger$ and $\hat{b}$ are bosonic creation and annihilation operators with the commutation relation $[\hat{b},\hat{b}^\dagger] = \hat{1}$. Let us define the partition function as

$$Z = \frac{\mathrm{Tr}\{\hat{\mathcal{U}}_C\hat{\rho}\}}{\mathrm{Tr}\{\hat{\rho}\}}. \tag{2.13}$$

If one assumes that all external fields are exactly the same on the forward and backward branches of the contour, then $\hat{\mathcal{U}}_C = \hat{1}$ and therefore $Z = 1$. The initial density matrix $\hat{\rho} = \hat{\rho}(\hat{H})$ is some operator-valued function of the Hamiltonian. To simplify the derivations one may choose it to be the equilibrium density matrix, $\hat{\rho}_0 = \exp\{-\beta(\hat{H}-\mu\hat{N})\} = \exp\{-\beta(\omega_0-\mu)\hat{b}^\dagger\hat{b}\}$, where $\beta = 1/T$ is the inverse temperature and μ is the chemical potential. Since arbitrary external perturbations may be switched on (and off) at a later time, the choice of the equilibrium initial density matrix does not prevent one from treating non-equilibrium dynamics. For the equilibrium initial density matrix one finds

$$\mathrm{Tr}\{\hat{\rho}_0\} = \sum_{n=0}^{\infty} \mathrm{e}^{-\beta(\omega_0-\mu)n} = [1-\rho(\omega_0)]^{-1}, \tag{2.14}$$

where $\rho(\omega_0) = \mathrm{e}^{-\beta(\omega_0-\mu)}$. An important observation is that, in general, $\mathrm{Tr}\{\hat{\rho}\}$ is an interaction- and disorder-independent constant. Indeed, both interactions and disorder are switched on (and off) on the forward (backward) parts of the contour sometime after (before) $t = -\infty$. This constant is therefore frequently omitted without causing confusion.

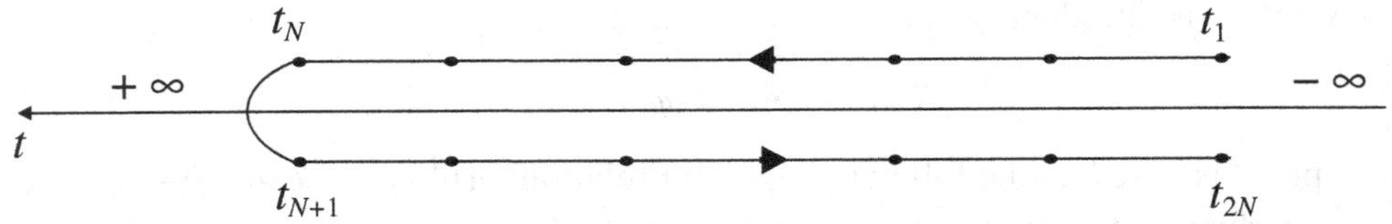

Figure 2.1 The closed time contour $\mathcal{C}$. Dots on the forward and backward branches of the contour denote discrete time points.

The next step is to divide the $\mathcal{C}$ contour into $(2N-2)$ time intervals of length δ_t, such that $t_1 = t_{2N} = -\infty$ and $t_N = t_{N+1} = +\infty$, as shown in Fig. 2.1. One then inserts the resolution of unity in the over-complete coherent state basis, Eq. (2.7),

$$\hat{1} = \int d[\bar{\phi}_j, \phi_j]\, e^{-|\phi_j|^2} |\phi_j\rangle\langle\phi_j|, \tag{2.15}$$

at each point $j = 1, 2, \ldots, 2N$ along the contour. For example, for $N = 3$ one obtains the following sequence in the expression for $\mathrm{Tr}\{\hat{\mathcal{U}}_\mathcal{C}\hat{\rho}_0\}$, Eq. (2.10) (read from right to left):

$$\langle\phi_6|\hat{\mathcal{U}}_{-\delta_t}|\phi_5\rangle\langle\phi_5|\hat{\mathcal{U}}_{-\delta_t}|\phi_4\rangle\langle\phi_4|\hat{1}|\phi_3\rangle\langle\phi_3|\hat{\mathcal{U}}_{+\delta_t}|\phi_2\rangle\langle\phi_2|\hat{\mathcal{U}}_{+\delta_t}|\phi_1\rangle\langle\phi_1|\hat{\rho}_0|\phi_6\rangle, \tag{2.16}$$

where $\hat{\mathcal{U}}_{\pm\delta_t}$ is the evolution operator (1.1) during the time interval δ_t in the positive (negative) time direction. Its matrix elements are given by

$$\begin{aligned}\left\langle\phi_j\left|\hat{\mathcal{U}}_{\pm\delta_t}\right|\phi_{j-1}\right\rangle &\equiv \left\langle\phi_j\left|e^{\mp i\hat{H}(b^\dagger,b)\delta_t}\right|\phi_{j-1}\right\rangle \approx \left\langle\phi_j\left|\left(1\mp i\hat{H}(b^\dagger,b)\delta_t\right)\right|\phi_{j-1}\right\rangle \\ &= \langle\phi_j|\phi_{j-1}\rangle\left(1\mp iH(\bar{\phi}_j,\phi_{j-1})\delta_t\right) \approx e^{\bar{\phi}_j\phi_{j-1}}\, e^{\mp iH(\bar{\phi}_j,\phi_{j-1})\delta_t},\end{aligned} \tag{2.17}$$

where the approximate equalities are valid up to the linear order in δ_t. Here we have employed expression (2.4) for the matrix elements of a normally ordered operator along with Eq. (2.6) for the overlap of the coherent states. For the toy example (2.12) one finds $H(\bar{\phi}_j, \phi_{j-1}) = \omega_0\bar{\phi}_j\phi_{j-1}$. However, Eq. (2.17) is not restricted to it, but holds for any *normally ordered* Hamiltonian. Notice that there is no evolution operator inserted between t_N and t_{N+1}. Indeed, these two points are physically indistinguishable and thus the system does not evolve during this time interval.

Employing the following property of the coherent states (see Eq. (2.11)): $\langle\phi_1|e^{-\beta(\omega_0-\mu)b^\dagger b}|\phi_{2N}\rangle = \exp\left\{\bar{\phi}_1\phi_{2N}\rho(\omega_0)\right\}$ and collecting all the exponential factors along the contour, one finds for the partition function, Eq. (2.13),

$$Z = \frac{1}{\mathrm{Tr}\{\hat{\rho}_0\}}\int\prod_{j=1}^{2N} d[\bar{\phi}_j,\phi_j]\,\exp\left(i\sum_{j,j'=1}^{2N}\bar{\phi}_j\, G^{-1}_{jj'}\,\phi_{j'}\right). \tag{2.18}$$

For $N = 3$ (see Eq. (2.16)), the $2N \times 2N$ matrix $\mathrm{i}G^{-1}_{jj'}$ takes the form

$$\mathrm{i}G^{-1}_{jj'} \equiv \left[\begin{array}{ccc|ccc} -1 & & & & & \rho(\omega_0) \\ h_- & -1 & & & & \\ & h_- & -1 & & & \\ \hline & & 1 & -1 & & \\ & & & h_+ & -1 & \\ & & & & h_+ & -1 \end{array}\right], \tag{2.19}$$

where $h_\mp \equiv 1 \mp \mathrm{i}\omega_0\delta_t$. The main diagonal of this matrix originates from the resolution of unity, Eq. (2.15), while the lower sub-diagonal comes from the matrix elements (2.17). Finally, the upper-right element comes from $\langle\phi_1|\hat{\rho}_0|\phi_{2N}\rangle$ in Eq. (2.16). This structure of the $\mathrm{i}\hat{G}^{-1}$ matrix is straightforwardly generalized to arbitrary N.

To proceed with the multiple integrals appearing in Eq. (2.18), we remind the reader of some properties of the Gaussian integrals.

2.3 Bosonic Gaussian Integrals

For any complex $N \times N$ matrix $\hat{A}_{ij}$, where $i,j = 1,\ldots,N$, such that all its eigenvalues, λ_i, have nonnegative real parts, $\mathrm{Re}\lambda_i \geq 0$, the following statement holds:

$$Z[\bar{J},J] = \int \prod_{j=1}^{N} \mathrm{d}[\bar{z}_j,z_j]\; \mathrm{e}^{-\sum\limits_{ij}^{N} \bar{z}_i\hat{A}_{ij}z_j + \sum\limits_{j}^{N}[\bar{z}_jJ_j+\bar{J}_jz_j]} = \frac{\mathrm{e}^{\sum\limits_{ij}^{N}\bar{J}_i(\hat{A}^{-1})_{ij}J_j}}{\det\hat{A}}, \tag{2.20}$$

where J_j is an arbitrary complex vector and $\mathrm{d}[\bar{z}_jz_j] = \mathrm{d}(\mathrm{Re}z_j)\mathrm{d}(\mathrm{Im}z_j)/\pi$. This equality is a generalization of the Gaussian integral (2.8). To prove it, one starts from a Hermitian matrix $\hat{A}$, which may be diagonalized by a unitary transformation $\hat{A} = \hat{U}^\dagger\hat{\Lambda}\hat{U}$, where $\hat{\Lambda} = \mathrm{diag}\{\lambda_j\}$. The identity is then proven by a change of variables with a unit Jacobian to $w_i = \sum_j \hat{U}_{ij}z_j$, which leads to

$$Z[\bar{J},J] = \prod_{j=1}^{N}\int \mathrm{d}[\bar{w}_j,w_j]\; \mathrm{e}^{-\bar{w}_j\lambda_jw_j+\bar{w}_jI_j+\bar{I}_jw_j} = \prod_{j=1}^{N}\frac{\mathrm{e}^{\bar{I}_j\lambda_j^{-1}I_j}}{\lambda_j},$$

where $I_i = \sum_j \hat{U}_{ij}J_j$. Using $\sum_j \bar{I}_j\lambda_j^{-1}I_j = \vec{\bar{J}}^{\mathrm{T}}\hat{U}^\dagger\hat{\Lambda}^{-1}\hat{U}\vec{J} = \vec{\bar{J}}^{\mathrm{T}}\hat{A}^{-1}\vec{J}$, along with $\det\hat{A} = \prod_j \lambda_j$, one obtains the right-hand side of Eq. (2.20). Finally, one notices that the right-hand side of Eq. (2.20) is an analytic function of both $\mathrm{Re}A_{ij}$ and $\mathrm{Im}A_{ij}$. Therefore, one may continue them analytically to the complex plane to reach an arbitrary complex matrix $\hat{A}_{ij}$. The identity (2.20) is thus valid as long as the integral is well defined, that is, all the eigenvalues of $\hat{A}$ have nonnegative real parts.

The Wick theorem deals with the average value of $z_{a_1} \dots z_{a_k} \bar{z}_{b_1} \dots \bar{z}_{b_k}$ weighted with the factor $\exp\left(-\sum_{ij} \bar{z}_i \hat{A}_{ij} z_j\right)$. The theorem states that this average is given by the sum of all possible products of pair-wise averages. For example, with the help of Eq. (2.20) one finds

$$\begin{aligned}\langle z_a \bar{z}_b \rangle &\equiv \frac{1}{Z[0,0]} \frac{\delta^2 Z[\bar{J},J]}{\delta \bar{J}_a \delta J_b}\bigg|_{J=0} = \hat{A}^{-1}_{ab}, \\ \langle z_a z_b \bar{z}_c \bar{z}_d \rangle &\equiv \frac{1}{Z[0,0]} \frac{\delta^4 Z[\bar{J},J]}{\delta \bar{J}_a \delta \bar{J}_b \delta J_c \delta J_d}\bigg|_{J=0} = \hat{A}^{-1}_{ac} \hat{A}^{-1}_{bd} + \hat{A}^{-1}_{ad} \hat{A}^{-1}_{bc},\end{aligned} \tag{2.21}$$

and so on.

The Gaussian identity for integration over real variables has the form

$$Z[J] = \int \prod_{j=1}^{N} \left(\frac{\mathrm{d}x_j}{\sqrt{2\pi}}\right) \mathrm{e}^{-\frac{1}{2}\sum\limits_{ij}^{N} x_i \hat{A}_{ij} x_j + \sum\limits_{j}^{N} x_j J_j} = \frac{\mathrm{e}^{\frac{1}{2}\sum\limits_{ij}^{N} J_i (\hat{A}^{-1})_{ij} J_j}}{\sqrt{\det \hat{A}}}, \tag{2.22}$$

where $\hat{A}$ is a *symmetric* complex matrix with all its eigenvalues having nonnegative real parts. The proof is similar to that in the case of complex variables: one starts from a real symmetric matrix, which may be diagonalized by an orthogonal transformation. The identity (2.22) is then easily proven by a change of variables. Finally, one may analytically continue the right-hand side (as long as the integral is well defined) from a real symmetric matrix $\hat{A}_{ij}$ to a *complex symmetric* one.

The corresponding Wick theorem for the average value of $x_{a_1} \dots x_{a_{2k}}$ weighted with the factor $\exp\left(-\frac{1}{2}\sum_{ij} x_i \hat{A}_{ij} x_j\right)$ takes the form

$$\begin{aligned}\langle x_a x_b \rangle &\equiv \frac{1}{Z[0]} \frac{\delta^2 Z[J]}{\delta J_a \delta J_b}\bigg|_{J=0} = \hat{A}^{-1}_{ab}, \\ \langle x_a x_b x_c x_d \rangle &\equiv \frac{1}{Z[0]} \frac{\delta^4 Z[J]}{\delta J_a \delta J_b \delta J_c \delta J_d}\bigg|_{J=0} = \hat{A}^{-1}_{ab} \hat{A}^{-1}_{cd} + \hat{A}^{-1}_{ac} \hat{A}^{-1}_{bd} + \hat{A}^{-1}_{ad} \hat{A}^{-1}_{bc},\end{aligned} \tag{2.23}$$

and so on. Notice the additional term in the second line in comparison with the corresponding complex result (2.21). The symmetry of $\hat{A}$ (and thus of $\hat{A}^{-1}$) is necessary to satisfy the obvious relation $\langle x_a x_b \rangle = \langle x_b x_a \rangle$.

2.4 Normalization and Continuum Notation

Having established the Gaussian identity (2.20), one can apply it to Eq. (2.18) to check the normalization factor. In this case $\hat{A} = -\mathrm{i}\hat{G}^{-1}$ and it is straightforward to evaluate the corresponding determinant employing Eq. (2.19):

$$\begin{aligned}\det\left[-\mathrm{i}\hat{G}^{-1}\right] &= 1 - \rho(\omega_0)(h_- h_+)^{N-1} = 1 - \rho(\omega_0)\left(1 + \omega_0^2 \delta_t^2\right)^{N-1} \\ &\approx 1 - \rho(\omega_0)\, \mathrm{e}^{\omega_0^2 \delta_t^2 (N-1)} \overset{N\to\infty}{\longrightarrow} 1 - \rho(\omega_0),\end{aligned} \tag{2.24}$$

where one used that $\delta_t^2 N \to 0$ if $N \to \infty$. Indeed, we divide the contour in a way to keep $\delta_t N = \text{const}$ (given by a full extent of the time axis) and as a result $\delta_t^2 \sim N^{-2}$. Employing the fact that the Gaussian integral in Eq. (2.18) is equal to the inverse determinant of the $-\mathrm{i}\hat{G}^{-1}$ matrix, Eq. (2.20), along with Eq. (2.14), one finds

$$Z = \frac{1}{\mathrm{Tr}\{\hat{\rho}_0\}} \frac{1}{\det\left[-\mathrm{i}\hat{G}^{-1}\right]} = 1, \tag{2.25}$$

as it should be, of course. Notice that keeping the upper-right element of the discrete matrix, Eq. (2.19), is crucial to maintain this normalization identity.

One may now take the limit $N \to \infty$ and formally write the partition function (2.18) in the continuum notation, $\phi_j \to \phi(t)$, as

$$Z = \int \mathbf{D}[\bar{\phi}(t), \phi(t)]\, \mathrm{e}^{\mathrm{i}S[\bar{\phi},\phi]}, \tag{2.26}$$

where the integration measure is the shorthand notation for $\mathbf{D}[\bar{\phi}(t), \phi(t)] = \prod_{j=1}^{2N} \mathrm{d}[\bar{\phi}_j, \phi_j] / \mathrm{Tr}\{\hat{\rho}_0\}$. According to Eqs. (2.18) and (2.19), the action is given by

$$S[\bar{\phi}, \phi] = \sum_{j=2}^{2N} \delta t_j \left[\mathrm{i}\bar{\phi}_j \frac{\phi_j - \phi_{j-1}}{\delta t_j} - \omega_0 \bar{\phi}_j \phi_{j-1} \right] + \mathrm{i}\,\bar{\phi}_1 \left[\phi_1 - \mathrm{i}\rho(\omega_0)\phi_{2N} \right], \tag{2.27}$$

where $\delta t_j \equiv t_j - t_{j-1} = \pm\delta_t$ on the forward and backward branches, correspondingly. In continuum notation, $\phi_j \to \phi(t)$, the action acquires the form

$$S[\bar{\phi}, \phi] = \int_C \mathrm{d}t\, \bar{\phi}(t) \hat{G}^{-1} \phi(t), \tag{2.28}$$

where the continuum form of the operator $\hat{G}^{-1}$ is (see the first square bracket on the right-hand side of Eq. (2.27))

$$\hat{G}^{-1} = \mathrm{i}\partial_t - \omega_0. \tag{2.29}$$

It is extremely important to remember that this continuum notation is only an abbreviation that represents the large discrete matrix, Eq. (2.19). In particular, the upper-right element of the matrix (the last term in Eq. (2.27)), which contains the information about the distribution function, is seemingly absent in the continuum notation, Eq. (2.29). The necessity of keeping the boundary terms originates from the fact that the continuum operator (2.29) possesses the zero mode $\mathrm{e}^{-\mathrm{i}\omega_0 t}$. Its inverse operator $\hat{G}$ is therefore not uniquely defined, unless the boundary terms are included.

To avoid integration along the closed time contour, it is convenient to split the bosonic field $\phi(t)$ into the two components $\phi^+(t)$ and $\phi^-(t)$, which reside on the forward and backward parts of the time contour, respectively. The continuum action then may be rewritten as

$$S[\bar{\phi},\phi] = \int_{-\infty}^{+\infty} \mathrm{d}t \left[\bar{\phi}^+(t)(\mathrm{i}\partial_t - \omega_0)\phi^+(t) - \bar{\phi}^-(t)(\mathrm{i}\partial_t - \omega_0)\phi^-(t)\right], \tag{2.30}$$

where the relative minus sign comes from the reversed direction of the time integration on the backward part of the contour. Once again, the continuum notation is somewhat misleading. Indeed, it creates an undue impression that the $\phi^+(t)$ and $\phi^-(t)$ fields are completely uncorrelated. In fact, they are connected due to the presence of the nonzero off-diagonal blocks in the discrete matrix, Eq. (2.19). It is therefore desirable to develop a continuum representation that automatically takes into account the proper regularization and mutual correlations. We shall achieve this in the following sections. First, the Green functions should be discussed.

2.5 Green Functions

According to the basic properties of the Gaussian integrals (see Section 2.3), the correlator of the two complex bosonic fields is given by

$$\langle \phi_j \bar{\phi}_{j'} \rangle \equiv \int \mathbf{D}[\bar{\phi},\phi]\; \phi_j \bar{\phi}_{j'} \exp\left(\mathrm{i} \sum_{k,k'=1}^{2N} \bar{\phi}_k\, G^{-1}_{kk'}\, \phi_{k'} \right) = \mathrm{i}G_{jj'}. \tag{2.31}$$

Notice the absence of the factor Z^{-1} in comparison with the analogous definition in the equilibrium theory [6]. Indeed, in the present construction $Z = 1$. This seemingly minor difference turns out to be the major issue in the theory of disordered systems. (See further discussion in Chapter 14, devoted to fermions with quenched disorder.) Inverting the $2N \times 2N$ matrix (2.19) with $N = 3$, one finds

$$\mathrm{i}G_{jj'} = \frac{1}{\det[-\mathrm{i}\hat{G}^{-1}]} \left[\begin{array}{ccc|ccc} 1 & \rho h_+^2 h_- & \rho h_+^2 & \rho h_+^2 & \rho h_+ & \rho \\ h_- & 1 & \rho h_+^2 h_- & \rho h_+^2 h_- & \rho h_+ h_- & \rho h_- \\ h_-^2 & h_- & 1 & \rho h_+^2 h_-^2 & \rho h_+ h_-^2 & \rho h_-^2 \\ \hline h_-^2 & h_- & 1 & 1 & \rho h_-^2 h_+ & \rho h_-^2 \\ h_-^2 h_+ & h_- h_+ & h_+ & h_+ & 1 & \rho h_-^2 h_+ \\ h_-^2 h_+^2 & h_- h_+^2 & h_+^2 & h_+^2 & h_+ & 1 \end{array} \right], \tag{2.32}$$

where $\rho \equiv \rho(\omega_0)$. Generalization of the $N = 3$ example to an arbitrary N is again straightforward. We switch now to the fields $\phi_j^\pm$ residing on the forward (backward) branches of the contour. Hereafter $j = 1,\ldots,N$ and thus the $2N \times 2N$ matrix we have just written is indexed as $1,2,\ldots,N,N,\ldots,2,1$. Then the following correlations may be read out of the matrix (2.32):

$$\langle\phi_j^+\bar{\phi}_{j'}^-\rangle \equiv \mathrm{i}G_{jj'}^< = \frac{\rho\, h_+^{j'-1}h_-^{j-1}}{\det[-\mathrm{i}\hat{G}^{-1}]}, \tag{2.33a}$$

$$\langle\phi_j^-\bar{\phi}_{j'}^+\rangle \equiv \mathrm{i}G_{jj'}^> = \frac{h_+^{N-j}h_-^{N-j'}}{\det[-\mathrm{i}\hat{G}^{-1}]} = \frac{(h_+h_-)^{N-1}h_+^{1-j}h_-^{1-j'}}{\det[-\mathrm{i}\hat{G}^{-1}]}, \tag{2.33b}$$

$$\langle\phi_j^+\bar{\phi}_{j'}^+\rangle \equiv \mathrm{i}G_{jj'}^{\mathbb{T}} = \frac{h_-^{j-j'}}{\det[-i\hat{G}^{-1}]} \times \begin{cases} 1, & j \geq j' \\ \rho(h_+h_-)^{N-1}, & j < j' \end{cases}, \tag{2.33c}$$

$$\langle\phi_j^-\bar{\phi}_{j'}^-\rangle \equiv \mathrm{i}G_{jj'}^{\widetilde{\mathbb{T}}} = \frac{h_+^{j'-j}}{\det[-i\hat{G}^{-1}]} \times \begin{cases} \rho(h_+h_-)^{N-1}, & j > j' \\ 1, & j \leq j' \end{cases}. \tag{2.33d}$$

Here the symbols $\mathbb{T}$ and $\widetilde{\mathbb{T}}$ stand for time ordering and anti-ordering correspondingly, while $<$ ($>$) is a convenient notation indicating that the first time argument is taken before (after) the second one on the Keldysh contour. Since $h_+^* = h_-$, one notices that

$$\left[G^{<(>)}\right]^\dagger = -G^{<(>)}; \qquad \left[G^{\mathbb{T}}\right]^\dagger = -G^{\widetilde{\mathbb{T}}}, \tag{2.34}$$

where the Hermitian conjugation involves interchange of the two time arguments along with complex conjugation.

Recalling that $h_\mp = 1 \mp \mathrm{i}\omega_0\delta_t$, one can take the $N \to \infty$ limit, keeping $N\delta_t$ a constant. To this end notice that $(h_+h_-)^N = (1 + \omega_0^2\delta_t^2)^N \overset{N\to\infty}{\longrightarrow} 1$, while $h_\mp^j \overset{N\to\infty}{\longrightarrow} \mathrm{e}^{\mp\mathrm{i}\omega_0\delta_t j} = \mathrm{e}^{\mp\mathrm{i}\omega_0 t}$, where we denoted $t = \delta_t j$ and correspondingly $t' = \delta_t j'$. Employing also the evaluation of the determinant given by Eq. (2.24), one obtains for the correlation functions in the continuum limit

$$\langle\phi^+(t)\bar{\phi}^-(t')\rangle = \mathrm{i}G^<(t,t') = n_\mathrm{B}\,\mathrm{e}^{-\mathrm{i}\omega_0(t-t')}, \tag{2.35a}$$

$$\langle\phi^-(t)\bar{\phi}^+(t')\rangle = \mathrm{i}G^>(t,t') = (n_\mathrm{B}+1)\,\mathrm{e}^{-\mathrm{i}\omega_0(t-t')}, \tag{2.35b}$$

$$\langle\phi^+(t)\bar{\phi}^+(t')\rangle = \mathrm{i}G^{\mathbb{T}}(t,t') = \theta(t-t')\mathrm{i}G^>(t,t') + \theta(t'-t)\mathrm{i}G^<(t,t'), \tag{2.35c}$$

$$\langle\phi^-(t)\bar{\phi}^-(t')\rangle = \mathrm{i}G^{\widetilde{\mathbb{T}}}(t,t') = \theta(t'-t)\mathrm{i}G^>(t,t') + \theta(t-t')\mathrm{i}G^<(t,t'), \tag{2.35d}$$

where we introduced the *bosonic occupation number* n_B as

$$n_\mathrm{B}(\omega_0) = \frac{\rho(\omega_0)}{1-\rho(\omega_0)}. \tag{2.36}$$

Indeed, to calculate the number of bosons at a certain point in time one needs to insert the operator $\hat{b}^\dagger\hat{b}$ into the corresponding point along the forward or backward branches of the contour. This leads to the correlation function $\langle\phi_{j-1}\bar{\phi}_j\rangle$, or in terms of $\phi^\pm$ fields to either $\langle\phi_{j-1}^+\bar{\phi}_j^+\rangle$ or $\langle\phi_j^-\bar{\phi}_{j-1}^-\rangle$ (notice the reversed indexing along the backward branch). According to Eqs. (2.33c, d), in the $N \to \infty$ limit both of them equal n_B.

The step function $\theta(t)$ in Eqs. (2.35c, d) is defined as $\theta(t-t') = 1$ if $t > t'$ and $\theta(t-t') = 0$ if $t < t'$. There is an ambiguity about equal times. Consulting with the discrete version of the correlation functions, Eqs. (2.33a–d), one notices that in both Eqs. (2.35c) and (2.35d) the first step function should be understood as having $\theta(0) = 1$, and the second as having $\theta(0) = 0$. Although slightly inconvenient, this ambiguity will disappear in the formalism that follows.

By analogy with the definition of the discrete correlation functions as a $2N$-fold integral, Eq. (2.31), it is convenient to write their continuum limit, Eq. (2.35), formally as a *functional* integral,

$$\langle \phi^{\pm}(t)\, \bar{\phi}^{\pm}(t') \rangle = \int \mathbf{D}[\bar{\phi}, \phi]\, \phi^{\pm}(t)\bar{\phi}^{\pm}(t')\, \mathrm{e}^{\,\mathrm{i}S[\bar{\phi},\phi]}, \tag{2.37}$$

where the action $S[\bar{\phi}, \phi]$ is given by Eq. (2.30). Notice that, despite the impression that the integrals over $\phi^{+}(t)$ and $\phi^{-}(t)$ may be split from each other and performed separately, there are nonvanishing cross-correlations between these fields, Eqs. (2.35a,b). The reason, of course, is that the continuum notation (2.37) is nothing but a shorthand abbreviation for the $N \to \infty$ limit of the discrete integral (2.31). The latter contains the matrix (2.19) with nonzero off-diagonal blocks, which are the sole reason for the existence of the cross-correlations. It is highly desirable to develop a continuum formalism, which automatically accounts for the proper cross-correlations without the need to resort to the discrete notations.

This task is facilitated by the observation that not all four Green functions we have defined are independent. Indeed, direct inspection shows that

$$G^{\mathbb{T}}(t,t') + G^{\tilde{\mathbb{T}}}(t,t') - G^{>}(t,t') - G^{<}(t,t') = 0. \tag{2.38}$$

This suggests that one may benefit explicitly from this relation by performing a linear transformation. The Keldysh rotation achieves just that. Notice that, due to the regularization of $\theta(0)$ discussed above Eq. (2.37), the identity does *not* hold for $t = t'$. Indeed at $t = t'$ the left-hand side of Eq. (2.38) is one rather than zero. However, since the $t = t'$ line is a manifold of measure zero, the violation of Eq. (2.38) for most purposes is inconsequential. (Notice that the left-hand side of Eq. (2.38) is not a delta-function $\delta(t - t')$. It is rather a Kronecker delta $\delta_{jj'}$ in the discrete version, which disappears in the continuum limit.)

2.6 Keldysh Rotation

Let us introduce a new pair of fields according to

$$\phi^{\mathrm{cl}}(t) = \frac{1}{\sqrt{2}}\big(\phi^{+}(t) + \phi^{-}(t)\big), \qquad \phi^{\mathrm{q}}(t) = \frac{1}{\sqrt{2}}\big(\phi^{+}(t) - \phi^{-}(t)\big), \tag{2.39}$$

with the analogous transformation for the conjugated fields. The superscripts "cl" and "q" stand for the *classical* and the *quantum* components of the fields, respectively. The rationale for this notation will become clear shortly. First, a simple algebraic manipulation with Eqs. (2.33a)–(2.33d) shows that

$$\langle \phi^{\alpha}(t)\bar{\phi}^{\beta}(t')\rangle \equiv \mathrm{i}G^{\alpha\beta}(t,t') = \begin{pmatrix} \mathrm{i}G^{\mathrm{K}}(t,t') & \mathrm{i}G^{\mathrm{R}}(t,t') \\ \mathrm{i}G^{\mathrm{A}}(t,t') & 0 \end{pmatrix}, \tag{2.40}$$

where hereafter $\alpha, \beta = (\mathrm{cl}, \mathrm{q})$. The fact that the (q, q) element of this matrix is zero is a manifestation of the identity (2.38). Superscripts R, A, and K stand for the *retarded, advanced*, and *Keldysh* components of the Green function, respectively. These three Green functions are the fundamental objects of the Keldysh technique. They are defined as

$$G^{\mathrm{R}}(t,t') = G^{\mathrm{cl,q}}(t,t') = \frac{1}{2}\left(G^{\mathbb{T}} - G^{\tilde{\mathbb{T}}} + G^{>} - G^{<}\right) = \theta(t-t')\left(G^{>} - G^{<}\right), \tag{2.41a}$$

$$G^{\mathrm{A}}(t,t') = G^{\mathrm{q,cl}}(t,t') = \frac{1}{2}\left(G^{\mathbb{T}} - G^{\tilde{\mathbb{T}}} - G^{>} + G^{<}\right) = \theta(t'-t)\left(G^{<} - G^{>}\right), \tag{2.41b}$$

$$G^{\mathrm{K}}(t,t') = G^{\mathrm{cl,cl}}(t,t') = \frac{1}{2}\left(G^{\mathbb{T}} + G^{\tilde{\mathbb{T}}} + G^{>} + G^{<}\right) = G^{>} + G^{<}. \tag{2.41c}$$

As mentioned after Eq. (2.38), the last equality in each line here holds for $t \neq t'$ only, while the diagonal $t = t'$ is discussed in what follows. Employing Eq. (2.34), one notices that

$$G^{\mathrm{A}} = \left[G^{\mathrm{R}}\right]^{\dagger}, \qquad G^{\mathrm{K}} = -\left[G^{\mathrm{K}}\right]^{\dagger}, \tag{2.42}$$

where the Green functions are understood as matrices in the time domain. Hermitian conjugation therefore includes complex conjugation along with interchanging of the two time arguments.

The retarded (advanced) Green function is a lower (upper) triangular matrix in the time domain. Since a product of any number of triangular matrices is again a triangular matrix, one obtains the simple rule that the convolution of any number of retarded (advanced) Green functions is also a retarded (advanced) Green function:

$$G_1^{\mathrm{R}} \circ G_2^{\mathrm{R}} \circ \cdots \circ G_l^{\mathrm{R}} = G^{\mathrm{R}}, \tag{2.43a}$$

$$G_1^{\mathrm{A}} \circ G_2^{\mathrm{A}} \circ \cdots \circ G_l^{\mathrm{A}} = G^{\mathrm{A}}, \tag{2.43b}$$

where the circular multiplication sign stands for the convolution operation, that is, multiplication of matrices in the time domain, and subscripts denote all other indices apart from the time.

Both retarded and advanced matrices have nonzero main diagonals, that is, $t = t'$. The important observation, however, is that

$$G^{\mathrm{R}}(t,t) + G^{\mathrm{A}}(t,t) = 0; \tag{2.44}$$

see Eqs. (2.35c,d) and the discussion of $\theta(0)$ regularization below them. It may be traced back to the fact that $G^{\mathrm{R}} + G^{\mathrm{A}} = G^{\mathbb{T}} - G^{\tilde{\mathbb{T}}}$, and since at the coinciding times the time ordering and anti-ordering are equivalent, the result is zero. This consideration shows that Eq. (2.44) is not restricted to our toy model, but is completely general. In the energy representation Eq. (2.44) takes the form

$$\int \frac{\mathrm{d}\epsilon}{2\pi} \left[G^{\mathrm{R}}(\epsilon) + G^{\mathrm{A}}(\epsilon)\right] = 0, \tag{2.45}$$

and it is tempting to attribute it to the fact that energy integral of a function analytic in the upper (lower) complex half-plane is zero. One should be aware, however, that, according to Eqs. (2.41a,b) and (2.35a,b),

$$G^{\mathrm{R}}(t,t) - G^{\mathrm{A}}(t,t) = -\mathrm{i}. \tag{2.46}$$

Once again this expression is not restricted to the toy model, but is very general. Indeed, $G^{\mathrm{R}} - G^{\mathrm{A}} = G^{>} - G^{<} = -\mathrm{i}\langle\langle bb^\dagger - b^\dagger b\rangle\rangle = -\mathrm{i}$ since the commutation relation (2.2) at the coinciding times is a generic property of any bosonic system. As a result, $\int \mathrm{d}\epsilon \left[G^{\mathrm{R}}(\epsilon) - G^{\mathrm{A}}(\epsilon)\right] = -2\pi\mathrm{i}$ and therefore $\int \mathrm{d}\epsilon\, G^{\mathrm{R(A)}}(\epsilon) = \mp\pi\mathrm{i}$, which is coming from the integration along the large semicircle closing the integration contour in the upper (lower) complex half-plane. In practical calculations the difference $G^{\mathrm{R}} - G^{\mathrm{A}}$ always comes with the distribution function (see what follows). The latter usually exhibits poles or branch cuts in both upper and lower energy half-planes and therefore the contour integration is not helpful anyway. Wherever G^{R} or G^{A} show up without the distribution function, they *always* appear in the combination $G^{\mathrm{R}} + G^{\mathrm{A}}$, calling for the contour integration (2.45). We shall thus frequently quote (as a rule of thumb) that $\int \mathrm{d}\epsilon\, G^{\mathrm{R(A)}}(\epsilon) = 0$, or equivalently $G^{\mathrm{R(A)}}(t,t) = 0$, understanding that it is always the sum of the two that matters. This never leads to confusion and therefore there is no danger in extending Eqs. (2.41a,b) to the diagonal $t = t'$ (with the understanding $\theta(0) = 0$) in the continuum formalism.

It is useful to introduce graphic representations for the three Green functions. To this end, let us denote the classical component of the field by a full line and the quantum component by a dashed line. Then the retarded Green function is represented by a full arrow and dashed line, the advanced by a dashed arrow and full line, and the Keldysh by a full arrow and full line; see Fig. 2.2. Notice that the dashed arrow and dashed line that would represent the $\langle\phi^{\mathrm{q}}\bar{\phi}^{\mathrm{q}}\rangle$ Green function is absent. The arrow shows the direction from ϕ^{α} toward $\bar{\phi}^{\beta}$.

Employing Eqs. (2.35a–d), one finds for our toy example of the single boson level

$$G^{\mathrm{R}} = -\mathrm{i}\theta(t-t')\,\mathrm{e}^{-\mathrm{i}\omega_0(t-t')} \overset{\mathrm{FT}}{\to} (\epsilon - \omega_0 + \mathrm{i}0)^{-1}, \tag{2.47a}$$

$$G^{\mathrm{A}} = \mathrm{i}\theta(t'-t)\,\mathrm{e}^{-\mathrm{i}\omega_0(t-t')} \overset{\mathrm{FT}}{\to} (\epsilon - \omega_0 - \mathrm{i}0)^{-1}, \tag{2.47b}$$

$$G^{\mathrm{K}} = -\mathrm{i}\,[2n_{\mathrm{B}}(\omega_0) + 1]\,\mathrm{e}^{-\mathrm{i}\omega_0(t-t')} \overset{\mathrm{FT}}{\to} -2\pi\mathrm{i}[2n_{\mathrm{B}}(\epsilon) + 1]\delta(\epsilon - \omega_0). \tag{2.47c}$$

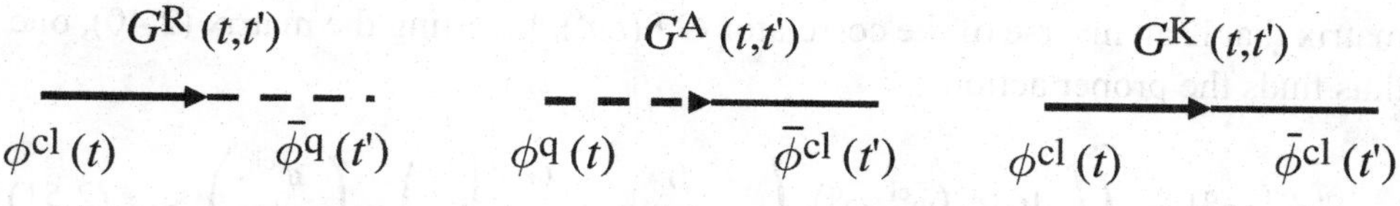

Figure 2.2 Graphic representation of G^{R}, G^{A}, and G^{K}. The full line represents the classical field component ϕ^{cl}, while the dashed line represents the quantum component ϕ^{q}. The arrows are directed from the annihilation operator toward the creation one, that is, from ϕ^{α} to $\bar{\phi}^{\beta}$.

The Fourier transforms (FT) with respect to $t - t'$ are given for each of the three Green functions. Notice that the retarded and advanced components contain information only about the spectrum and are independent of the occupation number, whereas the Keldysh component depends on it. In thermal equilibrium $\rho = e^{-(\omega_0-\mu)/T}$, while $n_B = (e^{(\omega_0-\mu)/T} - 1)^{-1}$ and therefore

$$G^{K}(\epsilon) = \coth\frac{\epsilon - \mu}{2T}\left[G^{R}(\epsilon) - G^{A}(\epsilon)\right], \tag{2.48}$$

where $T = \beta^{-1}$ is the system's temperature, expressed in units of energy.

The previous equation constitutes the statement of the *fluctuation–dissipation theorem* (FDT). As we shall see, the FDT is a general property of thermal equilibrium that is not restricted to the toy example considered here. It implies a rigid relation between the response functions and the correlation functions in equilibrium.

In general, it is convenient to parametrize the anti-Hermitian Keldysh Green function, Eq. (2.42), with the help of a Hermitian matrix $F = F^{\dagger}$, as follows:

$$G^{K} = G^{R} \circ F - F \circ G^{A}, \tag{2.49}$$

where $F = F(t,t')$. The Wigner transform (see Section 6.6), $F(t,\epsilon)$, of the matrix F is referred to as the *distribution function*. In thermal equilibrium $F(\epsilon) = \coth((\epsilon - \mu)/2T)$, Eq. (2.48).

2.7 Keldysh Action and Its Structure

One would like to have a continuum action, written in terms of ϕ^{cl}, ϕ^{q}, that properly reproduces the correlators Eqs. (2.40) and (2.47), namely

$$\langle \phi^{\alpha}(t)\bar{\phi}^{\beta}(t')\rangle = iG^{\alpha\beta}(t,t') = \int \mathbf{D}[\phi^{cl},\phi^{q}]\ \phi^{\alpha}(t)\bar{\phi}^{\beta}(t')\ e^{iS[\phi^{cl},\phi^{q}]}, \tag{2.50}$$

where the conjugated fields are not listed in the action arguments or the integration measure for brevity. According to the basic properties of Gaussian integrals, Section 2.3, the action should be taken as a quadratic form of the fields with the matrix that is an inverse of the correlator $G^{\alpha\beta}(t,t')$. Inverting the matrix (2.40), one thus finds the proper action:

$$S[\phi^{\rm cl},\phi^{\rm q}] = \iint\limits_{-\infty}^{+\infty} {\rm d}t\,{\rm d}t'\,(\bar\phi^{\rm cl},\bar\phi^{\rm q})_t \begin{pmatrix} 0 & \left[G^{-1}\right]^{\rm A} \\ \left[G^{-1}\right]^{\rm R} & \left[G^{-1}\right]^{\rm K} \end{pmatrix}_{t,t'} \begin{pmatrix} \phi^{\rm cl} \\ \phi^{\rm q} \end{pmatrix}_{t'} . \tag{2.51}$$

The off-diagonal elements are found from the condition $\left[G^{-1}\right]^{\rm R}\circ G^{\rm R} = 1$ and the similar one for the advanced component. The right-hand side here is the unit matrix, which in the continuum time representation is $\delta(t-t')$. As a result, the off-diagonal components are obtained by the matrix inversion of the corresponding components of the Green functions $\left[G^{-1}\right]^{\rm R(A)} = \left[G^{\rm R(A)}\right]^{-1}$. Such an inversion is most convenient in the energy representation

$$\left[G^{-1}\right]^{\rm R(A)} = \epsilon - \omega_0 \pm {\rm i}0 \to \delta(t-t')\,({\rm i}\partial_{t'} - \omega_0 \pm {\rm i}0), \tag{2.52a}$$

where in the last step we performed the inverse Fourier transform back to the time representation, employing the fact that the Fourier transform of ϵ is $\delta(t-t'){\rm i}\partial_{t'}$.

Although in the continuum limit these matrices look diagonal, it is important to remember that in the discrete regularization $\left[G^{\rm R(A)}\right]^{-1}$ contains $\mp{\rm i}$ along the main diagonal and $\pm{\rm i}-\omega_0\delta_t$ along the lower (upper) sub-diagonal. The determinants of the corresponding matrices are given by the product of all diagonal elements, $\det\left[G^{-1}\right]^{\rm R}\det\left[G^{-1}\right]^{\rm A} = \prod_{j=1}^{N}{\rm i}(-{\rm i}) = 1$. To obtain this statement without resorting to discretization, one notices that in the energy representation the Green functions are diagonal and therefore $\det\left[G^{-1}\right]^{\rm R}\left[G^{-1}\right]^{\rm A} = \prod_\epsilon\left[G^{\rm R}(\epsilon)G^{\rm A}(\epsilon)\right]^{-1} = \exp\left\{-\int\frac{{\rm d}\epsilon}{2\pi}[\ln G^{\rm R}+\ln G^{\rm A}]\right\} = 1$. Here we used the fact that Eq. (2.45) holds not only for the Green functions themselves but, thanks to Eqs. (2.43), also for any function of them. This property is important for maintaining the normalization identity $Z = \int \mathbf{D}[\phi^{\rm cl},\phi^{\rm q}]\,{\rm e}^{{\rm i}S} = 1$. Indeed, the integral is equal to minus (due to the factor of i in the exponent) the determinant of the quadratic form, while the latter is (-1) times the product of the determinants of the off-diagonal elements in the quadratic form (2.51).

The diagonal Keldysh component, $[G^{-1}]^{\rm K}$, of the quadratic form (2.51) is found from the condition $G^{\rm K}\circ[G^{\rm A}]^{-1} + G^{\rm R}\circ[G^{-1}]^{\rm K} = 0$. Employing the parametrization (2.49), one finds

$$\left[G^{-1}\right]^{\rm K} = -\left[G^{\rm R}\right]^{-1}\circ G^{\rm K}\circ[G^{\rm A}]^{-1} = \left[G^{\rm R}\right]^{-1}\circ F - F\circ\left[G^{\rm A}\right]^{-1}. \tag{2.52b}$$

The action (2.51) should be viewed as a construction devised to reproduce the proper continuum limit of the correlation functions according to the rules of Gaussian integration. It is fully self-consistent in the following senses: (i) it does not need to appeal to the discrete representation for regularization and (ii) its general structure is intact upon renormalization or "dressing" of its components by the interaction corrections (see Chapter 6).

Here we summarize the main features of the action (2.51), which, for lack of better terminology, we call the *causality structure*.

- The cl − cl component of the quadratic form is zero. It reflects the fact that for a pure classical field configuration ($\phi^{\mathrm{q}} = 0$) the action is zero. Indeed, in this case $\phi^+ = \phi^-$ and the action on the forward part of the contour is canceled by that on the backward part (except for the boundary terms, which are implicit in the continuum limit). The very general statement is, therefore, that

$$S[\phi^{\mathrm{cl}}, 0] = 0. \tag{2.53}$$

  Obviously this statement is not restricted to the Gaussian action of the form given by Eq. (2.51), but holds for any generic action (see Section 2.8 and Chapter 6).
- The cl − q and q − cl components are mutually Hermitian conjugated upper and lower (advanced and retarded) triangular matrices in the time domain. This property is responsible for the causality of the response functions as well as for protecting the cl − cl component from a perturbative renormalization (see Section 2.8). Relations (2.44) and (2.45) are crucial for this last purpose and necessary for the consistency of the theory.
- The q − q component is an anti-Hermitian matrix [see Eq. (2.42)]. It is responsible for the convergence of the functional integral and keeps information about the distribution function. In our simple example $\left[G^{\mathrm{K}}\right]^{-1} = 2i0F$, where F is a Hermitian matrix. The fact that it is infinitesimally small is a peculiarity of the noninteracting model. We shall see in the following chapters that it acquires a finite value, once interactions with other degrees of freedom are included.

2.8 Nonlinear Interactions

The discussion so far is mostly content with the linear harmonic system. Let us discuss now how nonlinear interactions enter the theory. In fact, this is exactly the subject of the rest of the book. It is worth the time, however, to have a brief preliminary exposition already at this juncture.

To this end consider a *normally ordered* bosonic Hamiltonian of the form

$$\hat{H}(\hat{b}^\dagger, \hat{b}) = \omega_0\, \hat{b}^\dagger \hat{b} + \frac{g}{2}\, \hat{b}^\dagger \hat{b}^\dagger \hat{b}\hat{b}, \tag{2.54}$$

where g is the two-particle interaction energy.[1] We now proceed to writing the partition function (2.13), where $\hat{\mathcal{U}}_{\mathcal{C}}$ is the closed time contour evolution operator

[1] Since $\hat{b}^\dagger\hat{b}^\dagger\hat{b}\hat{b} = (\hat{b}^\dagger\hat{b})^2 - \hat{b}^\dagger\hat{b}$, the eigenfunctions of the Hamiltonian (2.54) are those of the harmonic oscillator, $|n\rangle$, while the corresponding energies are $E_n = \omega_0 n + (g/2)(n^2 - n)$.

for the Hamiltonian (2.54). To this end we discretize the time contour and insert coherent state resolutions of identity. As mentioned after Eq. (2.17), calculation of the corresponding matrix elements are applicable for any normally ordered Hamiltonian. As a result, we arrive at a generalization of the continuum action (2.30), written as

$$S[\bar{\phi},\phi] = \int_{-\infty}^{+\infty} \mathrm{d}t \left[\bar{\phi}^{+}\mathrm{i}\partial_t\phi^{+} - H(\bar{\phi}^{+},\phi^{+}) - \bar{\phi}^{-}\mathrm{i}\partial_t\phi^{-} + H(\bar{\phi}^{-},\phi^{-})\right]. \quad (2.55)$$

As a next step we perform the Keldysh rotation (2.39) and rewrite the action in terms of classical and quantum field components. The harmonic part of the Hamiltonian, $\omega_0\,\hat{b}^\dagger\hat{b}$, brings the already familiar quadratic action (2.51), which is discussed in detail in Section 2.7. Here we focus on the interaction part of the action, which after a simple algebra acquires the form

$$S_{\mathrm{int}} = -\frac{g}{2}\int_{-\infty}^{+\infty} \mathrm{d}t \left[\bar{\phi}^{\mathrm{cl}}\bar{\phi}^{\mathrm{q}}\phi^{\mathrm{cl}}\phi^{\mathrm{cl}} + \bar{\phi}^{\mathrm{cl}}\bar{\phi}^{\mathrm{q}}\phi^{\mathrm{q}}\phi^{\mathrm{q}} + \bar{\phi}^{\mathrm{cl}}\bar{\phi}^{\mathrm{cl}}\phi^{\mathrm{cl}}\phi^{\mathrm{q}} + \bar{\phi}^{\mathrm{q}}\bar{\phi}^{\mathrm{q}}\phi^{\mathrm{cl}}\phi^{\mathrm{q}}\right]. \quad (2.56)$$

The corresponding vertices are represented diagramatically in Fig. 2.3. One immediately notices that all vertices contain at least one dashed line (i.e. quantum field). This assures that Eq. (2.53) is indeed satisfied. One can now expand $\mathrm{e}^{\mathrm{i}S}$ in (2.50) in powers of S_{int} and employ the Wick theorem (2.21) to calculate perturbative interaction corrections to the Green functions. One can also evaluate such corrections to the partition function. For the consistency of the theory it is important that the partition function remains to be normalized, $Z = 1$ (i.e. it does not acquire any interaction corrections), while the Green functions retain their causality structure. This is indeed the case; see Problem 2.10.3 and Section 6.2.

One may also investigate stationary configurations of the functional integral (2.50), given by the variations $\delta S/\delta\bar{\phi}^{\mathrm{q}} = \delta S/\delta\bar{\phi}^{\mathrm{cl}} = 0$ and their complex conjugates. This way, one arrives at the equations of motion:

$$\mathrm{i}\partial_t\phi^{\mathrm{cl}} = \omega_0\phi^{\mathrm{cl}} + \frac{g}{2}\left[\bar{\phi}^{\mathrm{cl}}\phi^{\mathrm{cl}}\phi^{\mathrm{cl}} + \bar{\phi}^{\mathrm{cl}}\phi^{\mathrm{q}}\phi^{\mathrm{q}} + 2\bar{\phi}^{\mathrm{q}}\phi^{\mathrm{q}}\phi^{\mathrm{cl}}\right]; \quad (2.57\mathrm{a})$$

$$\mathrm{i}\partial_t\phi^{\mathrm{q}} = \omega_0\phi^{\mathrm{q}} + \frac{g}{2}\left[\bar{\phi}^{\mathrm{q}}\phi^{\mathrm{cl}}\phi^{\mathrm{cl}} + \bar{\phi}^{\mathrm{q}}\phi^{\mathrm{q}}\phi^{\mathrm{q}} + 2\bar{\phi}^{\mathrm{cl}}\phi^{\mathrm{cl}}\phi^{\mathrm{q}}\right], \quad (2.57\mathrm{b})$$

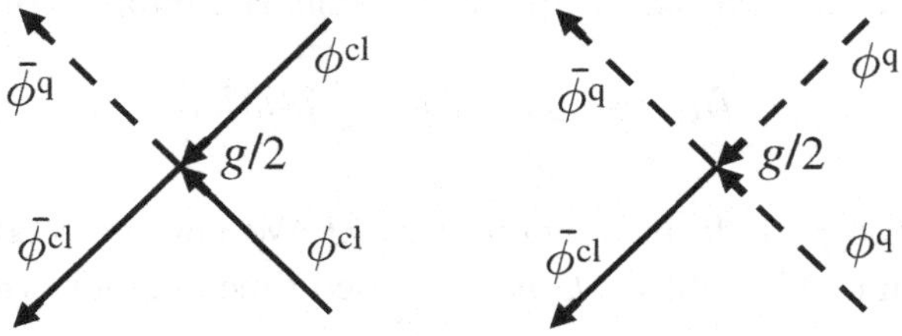

Figure 2.3 Interaction vertices for the Hamiltonian (2.54). There are also two complex conjugated vertices with a reversed direction of all arrows.

and their conjugated counterparts. In view of the causality constraint (2.53), the second of such equations, $\delta S/\delta\bar{\phi}^{\rm cl} = 0$, can always be solved by a particular solution $\phi^{\rm q} = 0$. In our example this leaves us with the two equations for the classical anharmonic oscillator:

$$i\partial_t\phi = \omega_0\phi + \frac{g}{2}|\phi|^2\phi; \qquad i\partial_t\bar{\phi} = -\omega_0\bar{\phi} - \frac{g}{2}|\phi|^2\bar{\phi}, \tag{2.58}$$

where the superscript cl is suppressed for brevity. These equations are solved by harmonic functions (cf. footnote 1) $\phi(t) = \phi_n e^{-i\omega_n t}$, where $\omega_n = \omega_0 + g|\phi_n|^2/2$. This should be compared with the transition frequencies $\omega_n = E_{n+1} - E_n = \omega_0 + gn$, calling for the identification $|\phi_n|^2 = 2n$, that is, twice the number of particles (the unfortunate factor of 2 is a result of our definition of the classical field as $\phi^{\rm cl} = (\phi^+ + \phi^-)/\sqrt{2}$). Generalization of Eq. (2.58) appears in the theory of Bose condensation as the Gross–Pitaevskii equation, Section 8.1.

2.9 External Sources

So far we have been content with the representation of the partition function. The latter does not carry any information in the Keldysh technique, since $Z = 1$. To make the entire construction meaningful one should introduce source fields, which enable one to compute various observables. As an example, let us introduce an external time-dependent potential $V(t)$. It interacts with the bosons through the Hamiltonian $\hat{H}_V = V(t)\hat{b}^\dagger\hat{b}$. One can now introduce the generating function $Z[V]$ defined similarly to the partition function (2.13), $Z[V] = \mathrm{Tr}\{\hat{\mathcal{U}}_C[V]\hat{\rho}\}/\mathrm{Tr}\{\hat{\rho}\}$, where the evolution operator $\hat{\mathcal{U}}_C[V]$ includes the source Hamiltonian $\hat{H}_V$ along with the bare one, Eq. (2.12). While any classical external field is the same on both branches of the contour, it is convenient to allow $V^+(t)$ and $V^-(t)$ to be distinct and put them equal only at the very end. Repeating the construction of the coherent state functional integral of Section 2.2, one obtains for the generating function

$$Z_{\rm d}[V] = \frac{1}{\mathrm{Tr}\{\hat{\rho}_0\}} \int \prod_{j=1}^{2N} d[\bar{\phi}_j, \phi_j] \exp\left(i \sum_{j,j'=1}^{2N} \bar{\phi}_j \, G_{jj'}^{-1}[V] \, \phi_{j'} \right), \tag{2.59}$$

where the subscript d stands for the discrete representation. The $2N \times 2N$ matrix $iG_{jj'}^{-1}[V]$ is similar to Eq. (2.19) with $h_\mp \to h_\mp[V] = 1 \mp i(\omega_0 + V_j)\delta_t$, where $V_j = V(t_j)$. According to Eq. (2.20) the generating function is proportional to the inverse determinant of the $-iG_{jj'}^{-1}[V]$ matrix. The latter is calculated in a way very similar to Eq. (2.24), leading to

$$Z_{\rm d}[V] = \frac{1}{\mathrm{Tr}\{\hat{\rho}_0\}}\,\frac{1}{\det[\,-\mathrm{i}\hat{G}^{-1}[V]]} = \frac{1-\rho(\omega_0)}{1-\rho(\omega_0)\mathrm{e}^{-\mathrm{i}\int_C \mathrm{d}t\,V(t)}}. \tag{2.60}$$

It is convenient to introduce classical and quantum components of the source potential $V(t)$ as

$$V^{\rm cl}(t) = \frac{1}{2}\left[V^+(t) + V^-(t)\right]; \qquad V^{\rm q}(t) = \frac{1}{2}\left[V^+(t) - V^-(t)\right], \tag{2.61}$$

where $V^{\pm}(t)$ is the source potential on the forward (backward) branch of the contour. With this notation along with Eq. (2.36) the generating function takes the form

$$Z_{\rm d}[V^{\rm cl}, V^{\rm q}] = \left[1 - n_{\rm B}(\omega_0)\left(\mathrm{e}^{-2\mathrm{i}\int \mathrm{d}t\,V^{\rm q}(t)} - 1\right)\right]^{-1}. \tag{2.62}$$

The fact that the generating function depends only on the integral of the quantum component of the source and does not depend on its classical component is a peculiarity of our toy model. (Indeed, since $[\hat{H}, b^\dagger b] = 0$, the number of particles is conserved, making the generating function independent of the classical external potential $V^{\rm cl}$). The very general statement, though, is

$$Z[V^{\rm cl}, 0] = 1. \tag{2.63}$$

Indeed, if $V^{\rm q} = 0$ the source potential is the same on the two branches, $V^+(t) = V^-(t)$, and thus the evolution operator brings the system exactly to its initial state, namely $\hat{\mathcal{U}}_C[V^{\rm cl}] = \hat{1}$. One crucially needs therefore a fictitious potential $V^{\rm q}(t)$ to generate observables.

Since the source potential is coupled to the number of particles operator $\hat{n} = \hat{b}^\dagger\hat{b}$, differentiation over $V^{\rm q}(t)$ generates an expectation value of $-2\mathrm{i}\langle\hat{n}(t)\rangle$ (the factor of two here is due to the fact that we insert $\hat{b}^\dagger(t)\hat{b}(t)$ on both branches): $\langle\hat{n}(t)\rangle = (\mathrm{i}/2)\delta Z_{\rm d}[V^{\rm q}]/\delta V^{\rm q}(t)|_{V^{\rm q}=0} = n_{\rm B}(\omega_0)$, as was established in Section 2.5. The higher-order correlation functions may be obtained by repetitive differentiation of the generating function. To generate *irreducible* correlators (i.e. cumulants) $\langle\langle\hat{n}^k(t)\rangle\rangle \equiv \langle(\hat{n}(t) - n_{\rm B})^k\rangle$ one needs to differentiate the *logarithm* of the generating function, Eq. (2.62), for example,

$$\begin{aligned}
\langle\langle\hat{n}^2(t)\rangle\rangle &= \left(\frac{\mathrm{i}}{2}\right)^2 \frac{\delta^2 \ln Z_{\rm d}}{\delta[V^{\rm q}(t)]^2}\bigg|_{V=0} = n_{\rm B}^2 + n_{\rm B};\\
\langle\langle\hat{n}^3(t)\rangle\rangle &= \left(\frac{\mathrm{i}}{2}\right)^3 \frac{\delta^3 \ln Z_{\rm d}}{\delta[V^{\rm q}(t)]^3}\bigg|_{V=0} = 2n_{\rm B}^3 + 3n_{\rm B}^2 + n_{\rm B};\\
\langle\langle\hat{n}^4(t)\rangle\rangle &= \left(\frac{\mathrm{i}}{2}\right)^4 \frac{\delta^4 \ln Z_{\rm d}}{\delta[V^{\rm q}(t)]^4}\bigg|_{V=0} = 6n_{\rm B}^4 + 12n_{\rm B}^3 + 7n_{\rm B}^2 + n_{\rm B};
\end{aligned} \tag{2.64}$$

and so on.

Let us see now how these results can be reproduced in the continuum technique, without resorting to discretization. The continuum generating function is defined as

$$Z_c[V] = \int \mathbf{D}[\bar{\phi}, \phi]\, e^{iS[\bar{\phi},\phi]+iS_V[\bar{\phi},\phi]}, \tag{2.65}$$

where the bare action $S[\bar{\phi}, \phi]$ is given by Eq. (2.30) and

$$S_V[\bar{\phi}, \phi] = -\int_{\mathcal{C}} dt\, V(t)\bar{\phi}(t)\phi(t) = -\int_{-\infty}^{+\infty} dt \left[V^+\bar{\phi}^+\phi^+ - V^-\bar{\phi}^-\phi^-\right] \tag{2.66}$$
$$= -\int_{-\infty}^{+\infty} dt \left[V^{\mathrm{cl}}(\bar{\phi}^+\phi^+ - \bar{\phi}^-\phi^-) + V^{\mathrm{q}}(\bar{\phi}^+\phi^+ + \bar{\phi}^-\phi^-)\right] = -\int_{-\infty}^{+\infty} dt\, \vec{\bar{\phi}}^{\mathrm{T}}\hat{V}\vec{\phi},$$

where $\vec{\phi} = (\phi^{\mathrm{cl}}, \phi^{\mathrm{q}})^{\mathrm{T}}$ and

$$\hat{V}(t) = \begin{pmatrix} V^{\mathrm{q}}(t) & V^{\mathrm{cl}}(t) \\ V^{\mathrm{cl}}(t) & V^{\mathrm{q}}(t) \end{pmatrix}. \tag{2.67}$$

As a result, for our example of the single bosonic level the continuum generating function is given by

$$Z_c[V^{\mathrm{cl}}, V^{\mathrm{q}}] = \int \mathbf{D}[\bar{\phi}, \phi]\, e^{i\int dt\, \vec{\bar{\phi}}^{\mathrm{T}}\left(\hat{G}^{-1} - \hat{V}(t)\right)\vec{\phi}} = \frac{1}{\mathrm{Tr}\{\hat{\rho}_0\}}\, \frac{1}{\det\left[-i\hat{G}^{-1} + i\hat{V}\right]}$$
$$= \frac{1}{\det\left[1 - \hat{G}\hat{V}\right]} = e^{-\mathrm{Tr}\ln\left[1 - \hat{G}\hat{V}\right]}, \tag{2.68}$$

where we have used Eq. (2.25) along with the identity $\ln\det\hat{A} = \mathrm{Tr}\ln\hat{A}$. According to Eqs. (2.40) and (2.47), the matrix Green function is

$$\hat{G}(t, t') = -i\, e^{-i\omega_0(t-t')} \begin{pmatrix} F(\omega_0) & \theta(t - t') \\ -\theta(t' - t) & 0 \end{pmatrix} \tag{2.69}$$

and $F(\omega_0) = 2n_{\mathrm{B}}(\omega_0) + 1$.

The continuum generating function Z_c is *not* identical to the discrete one, Z_d. However, as we shall show, it possesses the same general properties and generates exactly the same statistics as the number operator. First, let us verify Eq. (2.63) by expanding the logarithm in Eq. (2.68). To first order in $\hat{V}$ one finds $-\mathrm{Tr}\ln\left[1 - \hat{G}\hat{V}\right] \approx \mathrm{Tr}\hat{G}\hat{V} = \int dt\, [G^{\mathrm{R}}(t,t) + G^{\mathrm{A}}(t,t)]V^{\mathrm{cl}}(t) = 0$, where we put $V^{\mathrm{q}} = 0$ and employed Eq. (2.44). To second order one encounters $\int dt dt'\, G^{\mathrm{R}}(t,t')V^{\mathrm{cl}}(t')G^{\mathrm{R}}(t',t)V^{\mathrm{cl}}(t)$ and similarly for G^{A}. Since $G^{\mathrm{R}}(t,t') = 0$ if $t < t'$, while $G^{\mathrm{R}}(t',t) = 0$ if $t > t'$, the expression under the integral is non-zero only if $t = t'$. In the continuum limit ($N \to \infty$) this is the manifold of zero measure, making the integral zero. Clearly the same holds in all orders in V^{cl}. This illustrates how the generic feature of the Keldysh technique, Eq. (2.63), works in our simple example.

Consider now $\mathrm{i}\delta Z_c[V]/\delta V^{\mathrm{q}}(t)|_{V=0} = \langle \bar\phi^+(t)\phi^+(t) + \bar\phi^-(t)\phi^-(t)\rangle$; we refer to Eqs. (2.65) and (2.66) to see this relation. The expectation value of which operator is calculated this way? The naive answer is that $\bar\phi(t)\phi(t)$ is generated by $\langle \hat b^\dagger(t)\hat b(t)\rangle$ and we deal with the sum of this operator inserted on the forward and backward branches. If this were the case, $\bar\phi$ would be taken one time step ahead of the ϕ field, as is indeed the case in the discrete representation. However, our continuum expression indiscriminately places both $\bar\phi^\pm$ and $\phi^\pm$ at the same time t. One can check that such a "democratic" choice of the time arguments corresponds to the expectation value of the symmetric combination $\hat F(t) \equiv \hat b^\dagger(t)\hat b(t) + \hat b(t)\hat b^\dagger(t)$. Employing the equal time commutation relation $[\hat b(t), \hat b^\dagger(t)] = \hat 1$, one finds $\hat F(t) = 2\hat n(t) + 1$ and $\langle \hat F(t)\rangle = \mathrm{i}\, \delta Z_c[V^{\mathrm{cl}}, V^{\mathrm{q}}]/\delta V^{\mathrm{q}}(t)\big|_{V=0} = \mathrm{i}G^{\mathrm{K}}(t,t) = F(\omega_0)$, as it should be, of course. For higher-order irreducible correlators one obtains

$$\begin{aligned}
\langle\langle \hat F^2(t)\rangle\rangle &= \mathrm{i}^2 \left.\frac{\delta^2 \ln Z_c}{\delta[V^{\mathrm{q}}(t)]^2}\right|_{V=0} = F^2 - 1;\\
\langle\langle \hat F^3(t)\rangle\rangle &= \mathrm{i}^3 \left.\frac{\delta^3 \ln Z_c}{\delta[V^{\mathrm{q}}(t)]^3}\right|_{V=0} = 2F^3 - 2F;\\
\langle\langle \hat F^4(t)\rangle\rangle &= \mathrm{i}^4 \left.\frac{\delta^4 \ln Z_c}{\delta[V^{\mathrm{q}}(t)]^4}\right|_{V=0} = 6F^4 - 8F^2 + 2;
\end{aligned} \tag{2.70}$$

and so on. To see how it works, consider, for example, the third-order term in the expansion of $\ln Z_c = -\mathrm{Tr}\ln[1 - \hat G\hat V]$ in Eq. (2.68) in powers of $V^{\mathrm{q}}(t)$ at $V^{\mathrm{cl}} = 0$:

$$\begin{aligned}
\frac{1}{3}\mathrm{Tr}\{(\hat G\hat V)^3\} &= \frac{1}{3}\int \mathrm{d}t\mathrm{d}t'\mathrm{d}t''\mathrm{Tr}\left\{\hat G(t,t')V^{\mathrm{q}}(t')\hat G(t',t'')V^{\mathrm{q}}(t'')\hat G(t'',t)V^{\mathrm{q}}(t)\right\}\\
&= \mathrm{i}\frac{F^3}{3}\left(\int \mathrm{d}tV^{\mathrm{q}}(t)\right)^3 - \mathrm{i}F\int \mathrm{d}tV^{\mathrm{q}}(t)\left(\int_t \mathrm{d}t'V^{\mathrm{q}}(t')\right)^2\\
&= \mathrm{i}\frac{F^3 - F}{3}\left(\int \mathrm{d}tV^{\mathrm{q}}(t)\right)^3.
\end{aligned}$$

To calculate the last integral in the intermediate expression here, one introduces $W(t) = \int_t V^{\mathrm{q}}(t)$ and therefore $V^{\mathrm{q}} = -\dot W$, and the integral in question is thus $-\int \mathrm{d}t\dot W W^2 = -\int \mathrm{d}W W^2 = -(1/3)W^3(t)|_{-\infty}^{\infty} = (1/3)(\int \mathrm{d}tV^{\mathrm{q}})^3$. Differentiating over V^{q} three times, one arrives at Eq. (2.70).

Substituting $\hat F = 2\hat n + 1$ and $F = 2n_{\mathrm{B}} + 1$, it is easy to check that the respective moments (2.64) and (2.70) are exactly equivalent! Therefore, although the generating functions Z_d and Z_c generate slightly different sets of cumulants, their statistical content is equivalent. From now on we shall always deal with the continuum version, circumventing the tedious discretization procedure.

The generating function $Z[V^q]$ gives access not only to the moments, but to a *full counting statistics* of the operator $\hat{n}(t_0)$, or $\hat{F}(t_0)$. Let us define the probability of measuring n bosons at a time t_0 as $\mathcal{P}(n)$. Then $\langle \hat{n}^k(t_0)\rangle = \int \mathrm{d}n\, n^k \mathcal{P}(n)$. The generating function $Z[\eta] \equiv \int \mathrm{d}n\, \mathrm{e}^{i\eta n}\mathcal{P}(n) = \sum_k (i\eta)^k \langle \hat{n}^k(t_0)\rangle / k!$, where η is called the counting "field." Comparing this with $Z_d[V^q]$, one notices that $Z[\eta]$ may be obtained from it by the substitution $V^q(t) = -(\eta/2)\delta(t-t_0)$. Employing Eq. (2.60), one finds

$$Z[\eta] = \frac{1-\rho(\omega_0)}{1-\rho(\omega_0)\, e^{i\eta}} = \big(1-\rho(\omega_0)\big)\sum_{k=0}^{\infty}[\rho(\omega_0)]^k\, \mathrm{e}^{ik\eta}. \tag{2.71}$$

Performing the inverse Fourier transform and recalling that $\rho(\omega_0) = \mathrm{e}^{-\beta(\omega_0-\mu)}$, one finds

$$\mathcal{P}(n) = \sum_{k=0}^{\infty} \delta(n-k)\left(1-\mathrm{e}^{-\beta(\omega_0-\mu)}\right)\mathrm{e}^{-\beta(\omega_0-\mu)k}. \tag{2.72}$$

That is, one can measure only an integer number of bosons, and the corresponding probability is proportional to $\mathrm{e}^{-\beta(E_n-\mu n)}$, where the energy $E_n = n\omega_0$. This is, of course, a trivial result, which we have already de facto employed in Eq. (2.14). The important message, however, is that the counting field η is nothing but a particular realization of the quantum source field $V^q(t)$, tailored to generate an appropriate statistic. As opposed to the calculation of the moments (2.64) and (2.70), one should *not* put the quantum source to zero when the *full statistics* are evaluated. We shall employ this lesson in Sections 4.10, 13.3, and 15.4 to discuss less obvious examples of the full counting statistics.

2.10 Problems

2.10.1 Uncertainty Principle

Define two Hermitian operators $X = (\hat{b}^\dagger + \hat{b})/\sqrt{2}$ and $P = i(\hat{b}^\dagger - \hat{b})/\sqrt{2}$. Verify that they satisfy the canonical commutation relation $[X, P] = i$. Show that the Heisenberg uncertainty principle is saturated by the coherent states, namely

$$\sqrt{\left\langle (X-\langle X\rangle)^2\right\rangle}\sqrt{\left\langle (P-\langle P\rangle)^2\right\rangle} = \frac{1}{2}, \tag{2.73}$$

where $\langle \ldots \rangle = \mathrm{e}^{-|\phi|^2}\langle\phi|\ldots|\phi\rangle$ stands for the expectation value in the *normalized* coherent state.

2.10.2 Bogoliubov Transformation

Consider a quadratic Hamiltonian of the following form:

$$\hat{H} = \Delta \hat{b}^\dagger \hat{b} + \lambda(\hat{b}^\dagger \hat{b}^\dagger + \hat{b}\hat{b}) = (\hat{b}^\dagger, \hat{b}) \begin{pmatrix} \Delta/2 & \lambda \\ \lambda & \Delta/2 \end{pmatrix} \begin{pmatrix} \hat{b} \\ \hat{b}^\dagger \end{pmatrix} - \frac{\Delta}{2}. \tag{2.74}$$

Such Hamiltonians appear as a rotating wave description of a parametrically driven oscillator, Section 5.5, and in the theory of Bose condensation, Chapter 8. In the former case Δ represents a frequency detuning and λ is the driving amplitude. Show that with the help of the canonical Bogoliubov transformation

$$\hat{b} = \cosh(\alpha)\, \hat{\mathfrak{b}} + \sinh(\alpha)\, \hat{\mathfrak{b}}^\dagger; \qquad \hat{b}^\dagger = \cosh(\alpha)\, \hat{\mathfrak{b}}^\dagger + \sinh(\alpha)\, \hat{\mathfrak{b}}, \tag{2.75}$$

where $[\hat{\mathfrak{b}}, \hat{\mathfrak{b}}^\dagger] = 1$ and $\tanh(2\alpha) = -2\lambda/\Delta$, the Hamiltonian (2.74) may be brought to the diagonal form (up to a constant) $\hat{H} = \Omega\, \hat{\mathfrak{b}}^\dagger \hat{\mathfrak{b}}$, and $\Omega^2 = \Delta^2 - 4\lambda^2$. Show that its normalized ground state, $|\mathfrak{o}\rangle$, found from the condition $\hat{\mathfrak{b}}|\mathfrak{o}\rangle = 0$, is given by the coherent state of pairs

$$|\mathfrak{o}\rangle = \frac{1}{\sqrt{\cosh(\alpha)}}\, \mathrm{e}^{-\tanh(\alpha)\, \hat{b}^\dagger \hat{b}^\dagger/2} |0\rangle, \tag{2.76}$$

where $|0\rangle$ is the ground state of the bare oscillator, $\hat{b}|0\rangle = 0$. Notice that it is a superposition of only even-particle number states. All the other eigenstates $|\mathfrak{n}\rangle = (\hat{\mathfrak{b}}^\dagger)^n |\mathfrak{o}\rangle/\sqrt{n!}$ also have a definite parity, which alternates between even and odd. Show that the expected number of quanta of the bare oscillator in the ground state of (2.74) is $\langle \mathfrak{o}|\hat{b}^\dagger \hat{b}|\mathfrak{o}\rangle = \sinh^2(\alpha)$. It diverges if $\lambda \to \Delta/2$, signaling an instability of the driven oscillator; see Section 5.5.

Using the first quantization representation, where $\hat{b} = (\partial_X + X)/\sqrt{2}$ and $\hat{b}^\dagger = (-\partial_X + X)/\sqrt{2}$, which satisfy $[\hat{b}, \hat{b}^\dagger] = 1$, rewrite the Hamiltonian (2.74) as a differential operator. Discuss the meaning of Eq. (2.76) in terms of the oscillator eigenstates in the coordinate representation, $\psi_n(X)$.

2.10.3 Keldysh Perturbation Theory

Expand the exponent $\mathrm{e}^{\mathrm{i}S}$ in (2.50) up to the second order in S_{int}, (2.56), and employ the Wick theorem (2.21) to calculate perturbative interaction corrections to the partition function and components of the matrix Green function. Represent them diagrammatically using notations of Figs. 2.2 and 2.3. Show that (i) the partition function does not acquire any perturbative corrections and thus remains $Z = 1$; (ii) retarded and advanced components of the perturbed Green function remain mutually conjugated triangular matrices in the time space; (iii) corrections to the Keldysh component are anti-Hermitian matrices in the time space. You'll need to employ that $G^{\mathrm{R}}(t,t) + G^{\mathrm{A}}(t,t) = 0$, Eq. (2.44), and $G^{\mathrm{R}}(t,t')G^{\mathrm{A}}(t,t') = 0$ due to causality. Some help may be found in Section 6.2.

3

Single-Particle Quantum Mechanics

In this chapter we discuss quantum mechanics formulated on the closed time contour. We also derive a real-time version of the Caldeira–Leggett model for a quantum particle interacting with a bath of harmonic oscillators. A semiclassical treatment of quantum tunneling on the closed time contour is developed and used to evaluate the tunneling rate through a time-dependent potential barrier with and without coupling to the bath.

3.1 Harmonic Oscillator

The simplest many-body system of a single bosonic state, considered previously, is equivalent to a quantum harmonic oscillator. To make this connection explicit, consider the Keldysh contour action Eq. (2.28) with the correlator Eq. (2.29) written in terms of the complex field $\phi(t)$. The latter may be parametrized by its real and imaginary parts as

$$\phi(t) = \frac{1}{\sqrt{2\omega_0}}\big(\omega_0 X(t) + \mathrm{i}\, P(t)\big), \qquad \bar{\phi}(t) = \frac{1}{\sqrt{2\omega_0}}\big(\omega_0 X(t) - \mathrm{i}\, P(t)\big). \tag{3.1}$$

In terms of the real fields $P(t)$ and $X(t)$ the action, Eq. (2.28), takes the form

$$S[X,P] = \int_C \mathrm{d}t \left[P\dot{X} - \frac{1}{2}P^2 - \frac{\omega_0^2}{2}X^2 \right], \tag{3.2}$$

where the full time derivatives of P^2, X^2, and PX were omitted, since they contribute only to the boundary terms, implicit in the continuum notations. (They have to be kept for the proper regularization, though.) Equation (3.2) is nothing but the action of the quantum harmonic oscillator in the Hamiltonian form. One may perform the Gaussian integration over the real field $P(t)$, with the help of Eq. (2.22), to obtain

$$S[X] = \int_C \mathrm{d}t \left[\frac{1}{2}\dot{X}^2 - \frac{\omega_0^2}{2}X^2 \right]. \tag{3.3}$$

This is the Feynman Lagrangian action of the harmonic oscillator [33], written on the closed time contour. It may be generalized for an arbitrary single-particle potential $V(X)$:

$$S[X] = \int_C \mathrm{d}t \left[\frac{1}{2} \dot{X}^2 - V(X) \right]. \tag{3.4}$$

One may split the $X(t)$ field into two components, $X^+(t)$ and $X^-(t)$, residing on the forward and backward branches of the contour. The Keldysh rotation for real fields is conveniently defined as

$$X^{\mathrm{cl}}(t) = \frac{1}{2}\left[X^+(t) + X^-(t)\right]; \qquad X^{\mathrm{q}}(t) = \frac{1}{2}\left[X^+(t) - X^-(t)\right]. \tag{3.5}$$

In terms of these fields the action takes the form

$$S[X^{\mathrm{cl}}, X^{\mathrm{q}}] = \int_{-\infty}^{+\infty} \mathrm{d}t \left[-2X^{\mathrm{q}}\ddot{X}^{\mathrm{cl}} - V\big(X^{\mathrm{cl}} + X^{\mathrm{q}}\big) + V\big(X^{\mathrm{cl}} - X^{\mathrm{q}}\big) \right], \tag{3.6}$$

where the integration by parts was performed on the term $\dot{X}^{\mathrm{q}}\dot{X}^{\mathrm{cl}}$. This is the Keldysh form of the Feynman path integral. The omitted boundary terms provide a convergence factor of the form $\sim \mathrm{i}0(X^q)^2$.

If the fluctuations of the quantum component $X^{\mathrm{q}}(t)$ are regarded as small, one may expand the potential to first order and find for the action

$$S[X^{\mathrm{cl}}, X^{\mathrm{q}}] = -\int_{-\infty}^{+\infty} \mathrm{d}t \left[2X^{\mathrm{q}} \left(\ddot{X}^{\mathrm{cl}} + V'(X^{\mathrm{cl}}) \right) + O[(X^{\mathrm{q}})^3] \right], \tag{3.7}$$

where $V'(X) = \partial V(X)/\partial X$. In this approximation the integration over the quantum component, X^{q}, may be explicitly performed, leading to the functional delta-function of the expression in the round brackets. This delta-function enforces the classical Newtonian dynamics of X^{cl}:

$$\ddot{X}^{\mathrm{cl}} = -V'\big(X^{\mathrm{cl}}\big). \tag{3.8}$$

This is the reason the symmetric (over the forward and backward branches) part of the field is called the classical component. One should be careful with this name, though. If the higher-order terms in X^{q} are kept in the action, *both* X^{q} and X^{cl} are subject to quantum fluctuations.

Returning to the harmonic oscillator, $V(X) = \omega_0^2 X^2/2$, one may rewrite its Feynman–Keldysh action (3.3) in the matrix form

$$S[\vec{X}] = \frac{1}{2} \int_{-\infty}^{+\infty} \mathrm{d}t\, \vec{X}^{\mathrm{T}} \hat{D}^{-1} \vec{X}, \tag{3.9}$$

where in analogy with the complex field, Eq. (2.51), we introduced

$$\vec{X}(t) = \begin{pmatrix} X^{\mathrm{cl}}(t) \\ X^{\mathrm{q}}(t) \end{pmatrix}; \qquad \hat{D}^{-1} = \begin{pmatrix} 0 & [D^{-1}]^{\mathrm{A}} \\ [D^{-1}]^{\mathrm{R}} & [D^{-1}]^{\mathrm{K}} \end{pmatrix} \tag{3.10}$$

and the superscript T stands for matrix transposition. Here the retarded and advanced components of the quadratic form in the action are given by $\frac{1}{2}[D^{-1}]^{\mathrm{R(A)}} = (\mathrm{i}\partial_t \pm \mathrm{i}0)^2 - \omega_0^2$. As before, one should understand that this expression is simply a continuous abbreviation for the large lower (upper) triangular matrices with $-\delta_t^{-1}$ along the main diagonal, $2\delta_t^{-1} - \omega_0^2\delta_t$ along the lower (upper) sub-diagonal, and $-\delta_t^{-1}$ along the second lower (upper) sub-diagonal. This makes the $\hat{D}^{-1}$ matrix symmetric, since its $[D^{-1}]^{\mathrm{K}}$ component must be symmetric by construction (its antisymmetric part does not enter the action). In continuous notation the Keldysh component $[D^{-1}]^{\mathrm{K}}$ is only a regularization. It is convenient to keep it explicitly, since it suggests the way the matrix $\hat{D}^{-1}$ should be inverted to find the Green function:

$$\left\langle X^\alpha(t) X^\beta(t')\right\rangle = \int \mathbf{D}[\vec{X}]\, X^\alpha(t) X^\beta(t')\, \mathrm{e}^{\mathrm{i}S[\vec{X}]} = \mathrm{i}\hat{D}^{\alpha\beta}(t,t'), \tag{3.11}$$

where $\alpha, \beta = (\mathrm{cl}, \mathrm{q})$ and the matrix inverse of Eq. (3.10) is given by

$$\hat{D}^{\alpha\beta}(t,t') = \begin{pmatrix} D^{\mathrm{K}}(t,t') & D^{\mathrm{R}}(t,t') \\ D^{\mathrm{A}}(t,t') & 0 \end{pmatrix}. \tag{3.12}$$

To apply the rules of Gaussian integration for real variables (see Section 2.3), it is crucial that the matrix $\hat{D}^{-1}$ is symmetric. In the Fourier representation components of the equilibrium correlation matrix are given by

$$D^{\mathrm{R(A)}}(\epsilon) = \frac{1}{2}\frac{1}{(\epsilon \pm \mathrm{i}0)^2 - \omega_0^2}, \tag{3.13a}$$

$$D^{\mathrm{K}}(\epsilon) = \coth\frac{\epsilon}{2T}\left[D^{\mathrm{R}}(\epsilon) - D^{\mathrm{A}}(\epsilon)\right], \tag{3.13b}$$

where we have assumed an equilibrium thermal distribution with zero chemical potential. One way to check the consistency of the expression for the Keldysh component is to express X^α through $\bar{\phi}^\alpha$ and ϕ^α and employ the correlation functions for the complex fields, derived in Chapter 2. The fact that the chemical potential of a real field *must be zero* follows directly from the symmetry of $D^{\mathrm{K}}(t,t')$ (making $D^{\mathrm{K}}(\epsilon)$ an even function) and the identity $D^{\mathrm{R}}(-\epsilon) = D^{\mathrm{A}}(\epsilon)$.

The normalization identity, $\int \mathbf{D}[\vec{X}]\,\mathrm{e}^{\mathrm{i}S[\vec{X}]} = 1$, is maintained in the following way: (i) first, due to the structure of the $\hat{D}^{-1}$ matrix, explained previously, $\det[\frac{1}{\mathrm{i}}\hat{D}^{-1}] = -\det[\frac{1}{\mathrm{i}}D^{-1}]^{\mathrm{R}}\det[\frac{1}{\mathrm{i}}D^{-1}]^{\mathrm{A}} = (2/\delta_t)^{2N}$; (ii) the integration measure is understood as $\mathbf{D}[\vec{X}] = \prod_{j=1}^{N} 2\left(\mathrm{d}X_j^{\mathrm{cl}}/\sqrt{2\pi\delta_t}\right)\left(\mathrm{d}X_j^{\mathrm{q}}/\sqrt{2\pi\delta_t}\right)$ (in comparison with Eq. (2.22) there is an additional factor of 2, which originates from the Jacobian of the transformation (3.5), and factor δ_t^{-1} at each time slice, coming from the integrations over $P_j = P(t_j)$). According to the real Gaussian identity (2.22) this leads exactly to the proper normalization. One can also understand the normalization in the way discussed after Eq. (2.52a), without resorting to the discrete representation.

3.2 Quantum Particle in Contact with an Environment

Consider a quantum particle with coordinate $X(t)$, placed in a potential $V(X)$ and brought into contact with a bath of harmonic oscillators. The bath oscillators are labeled by an index s and their coordinates are denoted by φ_s. They possess a set of frequencies ω_s. The Keldysh action of such a system is given by the three terms $S = S_{\mathrm{p}} + S_{\mathrm{bath}} + S_{\mathrm{int}}$, where

$$S_{\mathrm{p}}[X] = \int_{-\infty}^{+\infty} \mathrm{d}t \left[-2X^{\mathrm{q}}\ddot{X}^{\mathrm{cl}} - V\left(X^{\mathrm{cl}} + X^{\mathrm{q}}\right) + V\left(X^{\mathrm{cl}} - X^{\mathrm{q}}\right) \right], \tag{3.14a}$$

$$S_{\mathrm{bath}}[\varphi_s] = \frac{1}{2}\sum_s \int_{-\infty}^{+\infty} \mathrm{d}t\, \vec{\varphi}_s^{\,\mathrm{T}} \hat{D}_s^{-1} \vec{\varphi}_s, \tag{3.14b}$$

$$S_{\mathrm{int}}[X, \varphi_s] = \sum_s g_s \int_{-\infty}^{+\infty} \mathrm{d}t\, \vec{X}^{\mathrm{T}}\, \hat{\sigma}_1\, \vec{\varphi}_s, \tag{3.14c}$$

where the symmetric quadratic form $\hat{D}_s^{-1}$ is given by Eq. (3.10) with the frequency ω_s. The interaction term between the particle and the bath oscillators is taken as $\sum_s g_s \int_{\mathcal{C}} \mathrm{d}t\, X(t)\varphi_s(t) = \sum_s g_s \int \mathrm{d}t (X^+\varphi_s^+ - X^-\varphi_s^-)$. Performing the Keldysh rotation according to Eq. (3.5), one arrives at Eq. (3.14c), where $\hat{\sigma}_1$ is the first Pauli matrix in the Keldysh (cl, q) space. The corresponding coupling constants are denoted by g_s.

One may now integrate out the degrees of freedom of the bath to reduce the problem to the particle coordinate only. Employing Eq. (2.22) for the Gaussian integration over the real variables, one arrives at the so-called dissipative action for the particle:

$$S_{\mathrm{diss}} = \frac{1}{2} \iint_{-\infty}^{+\infty} \mathrm{d}t\, \mathrm{d}t'\, \vec{X}^{\mathrm{T}}(t)\, \hat{\mathfrak{D}}^{-1}(t - t')\, \vec{X}(t'), \tag{3.15a}$$

$$\hat{\mathfrak{D}}^{-1}(t - t') = -\hat{\sigma}_1 \left[\sum_s g_s^2 \hat{D}_s(t - t') \right] \hat{\sigma}_1. \tag{3.15b}$$

Straightforward matrix multiplication shows that the dissipative quadratic form $\hat{\mathfrak{D}}^{-1}$ possesses the causality structure as, for example, Eq. (3.10). For the Fourier transform of its retarded (advanced) components, one finds

$$\left[\mathfrak{D}^{-1}(\epsilon)\right]^{\mathrm{R(A)}} = -\frac{1}{2}\sum_s \frac{g_s^2}{(\epsilon \pm \mathrm{i}0)^2 - \omega_s^2} = \int \frac{\mathrm{d}\omega}{2\pi} \frac{\omega J(\omega)}{\omega^2 - (\epsilon \pm \mathrm{i}0)^2}, \tag{3.16}$$

where $J(\omega) = \pi \sum_s (g_s^2/\omega_s)\delta(\omega - \omega_s)$ is the bath spectral density.

We shall assume now that the spectral density behaves as $J(\omega) = 4\gamma\omega$, where γ is a constant at small frequencies. This is the so-called *Ohmic* bath, which is frequently found in more realistic models of the environment (see, e.g. Section 8.8). Substituting it into Eq. (3.16), one finds

$$[\mathfrak{D}^{-1}(\epsilon)]^{\mathrm{R(A)}} = 4\gamma \int \frac{d\omega}{2\pi} \frac{\omega^2}{\omega^2 - (\epsilon \pm i0)^2} = \text{const} \pm 2i\gamma\epsilon, \qquad (3.17)$$

where the ϵ-independent real positive constant (the same for R and A components) may be absorbed into the redefinition of the harmonic part of the particle's potential $V(X) = \text{const} \times X^2 + \cdots$ and, thus, may be omitted. If the bath is in equilibrium, the Keldysh component of the correlator is set by FDT,

$$[\mathfrak{D}^{-1}(\epsilon)]^{\mathrm{K}} = \left([\mathfrak{D}^{\mathrm{R}}]^{-1} - [\mathfrak{D}^{\mathrm{A}}]^{-1}\right) \coth\frac{\epsilon}{2T} = 4i\gamma\epsilon \coth\frac{\epsilon}{2T}, \qquad (3.18)$$

where we assumed that the bath is at temperature T and, as explained after Eqs. (3.13), the chemical potential of the real bath oscillators must be zero. Notice that the validity of this expression does *not* rely on the particle being at equilibrium, but only the bath. The Keldysh component is an anti-Hermitian operator with a positive-definite imaginary part, rendering convergence of the functional integral over $\vec{X}(t)$.

In the time representation the retarded (advanced) component of the correlator takes a time-local form: $[\mathfrak{D}^{\mathrm{R(A)}}]^{-1} = \mp 2\gamma\, \delta(t-t')\, \partial_{t'}$. On the other hand, the Keldysh component is a nonlocal function that may be found by the inverse Fourier transform of Eq. (3.18):

$$[\mathfrak{D}^{-1}(t-t')]^{\mathrm{K}} = 4i\gamma \left[(2T + C)\delta(t-t') - \frac{\pi T^2}{\sinh^2[\pi T(t-t')]}\right], \qquad (3.19)$$

where the infinite constant $C = \pi T^2 \int dt/\sinh^2(\pi T t)$ serves to satisfy the condition $\int dt [\mathfrak{D}^{-1}(t)]^{\mathrm{K}} = [\mathfrak{D}^{-1}(\epsilon = 0)]^{\mathrm{K}} = 8i\gamma T$. Finally, one obtains for the Keldysh action of the particle connected to the ohmic bath

$$S[\vec{X}] = \int_{-\infty}^{+\infty} dt \left[-2X^{\mathrm{q}}(\ddot{X}^{\mathrm{cl}} + \gamma\dot{X}^{\mathrm{cl}}) - V(X^{\mathrm{cl}} + X^{\mathrm{q}}) + V(X^{\mathrm{cl}} - X^{\mathrm{q}})\right]$$
$$+ 2i\gamma \int_{-\infty}^{+\infty} dt \left[2T\left(X^{\mathrm{q}}(t)\right)^2 + \frac{\pi T^2}{2}\int_{-\infty}^{+\infty} dt' \frac{\left(X^{\mathrm{q}}(t) - X^{\mathrm{q}}(t')\right)^2}{\sinh^2[\pi T(t-t')]}\right], \qquad (3.20)$$

where the infinite constant C is absorbed into the two diagonal terms $\sim \left(X^{\mathrm{q}}(t)\right)^2$. This action satisfies all the causality criteria listed in Section 2.7. Notice that in the present case the Keldysh $\mathrm{q}-\mathrm{q}$ component is not just a regularization, but a finite term, originating from the coupling to the bath and serving to limit fluctuations. This term breaks the symmetry $S[X^{\mathrm{cl}}, -X^{\mathrm{q}}] = -S[X^{\mathrm{cl}}, X^{\mathrm{q}}]$ present in the initial action (3.14). Such a symmetry of the action is a direct consequence of the time reversal symmetry of the problem. Thus the appearance of a finite $\mathrm{q}-\mathrm{q}$ component of the action is a manifestation of the breaking of the time-reversal symmetry. The latter takes place due to integrating out the continuum of the bath degrees of freedom.

The other manifestation of the bath is the presence of the friction term $\sim \gamma\partial_t$ in the R and the A components. In equilibrium the friction coefficient and fluctuation amplitude are rigidly connected by the FDT. The quantum dissipative action, Eq. (3.20), is a convenient playground to demonstrate various approximations and connections to other approaches. We shall discuss it in detail in Chapter 4. If only linear terms in X^{q} are kept in the action (3.20), the integration over $X^{\mathrm{q}}(t)$ results in the functional delta-function, which enforces the following relation:

$$\ddot{X}^{\mathrm{cl}} = -V'\big(X^{\mathrm{cl}}\big) - \gamma\dot{X}^{\mathrm{cl}}. \tag{3.21}$$

This is the classical Newtonian equation with the viscous friction force. Remarkably, we have obtained the $\dot{X}^{\mathrm{cl}}$ term in the equation of motion from the action principle. It would not be possible, if not for the doubling of the number of fields X^{cl} and X^{q}. Indeed, in any action depending on X^{cl} only, terms linear in the first time derivative may be written as a full time derivative and integrated out, not affecting the equation of motion.

3.3 From Matsubara to Keldysh

Most of the texts dealing with equilibrium systems at finite temperature employ the Matsubara technique [11, 2, 4, 6]. This method is designed to treat the equilibrium density matrix $\mathrm{e}^{-\beta\hat{H}}$ as the evolution operator. To this end one considers an imaginary time quantum mechanics, with the imaginary time τ restricted to the interval $0 \le \tau < \beta$. When calculating an expectation value of an observable $\hat{\mathcal{O}}(\tau)$, one evaluates a trace of the form $\langle\hat{\mathcal{O}}\rangle = \mathrm{Tr}\{\hat{\mathcal{O}}(\tau)\mathrm{e}^{-\beta\hat{H}}\}$. To this end one divides the imaginary time interval $[0,\beta]$ into N infinitesimal segments and inserts the resolution of unity in the coherent state basis at each segment, similar to our procedure in Section 2.2. As a result, one ends up with fields, say with coordinate $X(\tau)$, which, in view of the fact that one evaluates the trace, obeys the periodic boundary conditions $X(0) = X(\beta)$. In the Fourier representation it is represented by a discrete set of components $X_m = \int_0^\beta d\tau X(\tau)\,\mathrm{e}^{\mathrm{i}\epsilon_m\tau}$, where $\epsilon_m = 2\pi mT$ is a set of Matsubara frequencies and m is an integer.

We shall discuss now how to convert an action written with the Matsubara technique into the Keldysh representation. This may be useful, if one wishes to extend treatment of the problem to non-equilibrium or time-dependent conditions. As an example consider the following bosonic Matsubara action:

$$S[X_m] = \frac{\mathrm{i}}{2}\gamma T \sum_{m=-\infty}^{\infty} |\epsilon_m||X_m|^2. \tag{3.22}$$

Due to the absolute value sign, $|\epsilon_m| \neq i\partial_\tau$. In fact, in the imaginary time representation the kernel $K_m = |\epsilon_m|$ acquires the form $K(\tau) = \sum_m |\epsilon_m| e^{-i\epsilon_m \tau} = C\delta(\tau) - \pi T \sin^{-2}(\pi T \tau)$, where the infinite constant C is chosen to satisfy normalization, $\int_0^\beta d\tau K(\tau) = K_0 = 0$. As a result, in the imaginary time representation the action (3.22) obtains the following nonlocal form:

$$S[X] = \frac{i}{2}\gamma T \iint_0^\beta d\tau\, d\tau' X(\tau) K(\tau - \tau') X(\tau')$$
$$= \frac{i}{4\pi}\gamma \iint_0^\beta d\tau\, d\tau' \frac{\pi^2 T^2}{\sin^2[\pi T(\tau - \tau')]} \big(X(\tau) - X(\tau')\big)^2. \qquad (3.23)$$

This action is frequently named after Caldeira and Leggett [34], who used it to investigate the influence of dissipation on quantum tunneling.

To transform to the Keldysh representation one proceeds along the following steps: (i) double the number of degrees of freedom, correspondingly doubling the action, $X \to \vec{X} = (X^{\mathrm{cl}}, X^{\mathrm{q}})^{\mathrm{T}}$ and consider the latter as functions of the real time t or real frequency ϵ; (ii) according to the causality structure, Section 2.7, the general form of the quadratic time translationally invariant Keldysh action is:

$$S[\vec{X}] = \gamma \int \frac{d\epsilon}{2\pi} (X_\epsilon^{\mathrm{cl}}, X_\epsilon^{\mathrm{q}}) \begin{pmatrix} 0 & K^{\mathrm{A}}(\epsilon) \\ K^{\mathrm{R}}(\epsilon) & K^{\mathrm{K}}(\epsilon) \end{pmatrix} \begin{pmatrix} X_\epsilon^{\mathrm{cl}} \\ X_\epsilon^{\mathrm{q}} \end{pmatrix}; \qquad (3.24)$$

(iii) the retarded (advanced) component $K^{\mathrm{R(A)}}(\epsilon)$ is the analytic continuation of the Matsubara correlator $K(\epsilon_m) = |\epsilon_m|$ from the *upper* (*lower*) half-plane of the complex variable ϵ_m to the real axis: $\mp i\epsilon_m \to \epsilon$, see [2]. This leads to $K^{\mathrm{R(A)}}(\epsilon) = \pm i\epsilon$; (iv) in equilibrium the Keldysh component follows from FDT: $K^{\mathrm{K}}(\epsilon) = \big(K^{\mathrm{R}}(\epsilon) - K^{\mathrm{A}}(\epsilon)\big) \coth(\epsilon/2T) = 2i\epsilon \coth(\epsilon/2T)$; see Eqs. (3.17) and (3.18). We found thus that $\gamma \hat{K}(\epsilon) = \frac{1}{2}\hat{\mathfrak{D}}^{-1}(\epsilon)$ and therefore the Keldysh counterpart of the Matsubara action, Eq. (3.22) or (3.23), is the already familiar dissipative action (3.20) (without the potential and inertial terms, of course). One may now include external fields and allow the system to deviate from equilibrium.

3.4 Quantum Tunneling in a Time-Dependent Potential

We shall discuss here the quasi-classical description of quantum mechanical tunneling. The quasi-classical approach originates from evaluating the Feynman path integral in the stationary path approximation. Taking the variation of the Feynman action, one arrives at the classical Newtonian equation of motion. At first glance, the latter fails to describe motion in the classically forbidden under-barrier region. Let us look at it more closely, however. In particular, for a particle with unit mass and energy $E = P^2/2 + V(X)$ one finds

$$P(t) = \frac{dX}{dt} = \sqrt{2(E - V(X))}. \tag{3.25}$$

Integrating this equation, one finds, for the time t needed to reach infinity starting from a point X,

$$t(X) = \int_X^\infty \frac{dX'}{\sqrt{2(E - V(X'))}}. \tag{3.26}$$

As long as $X > X_2$, such that $E > V(X)$, the corresponding time is real; see Fig. 3.1. For $X_1 < X < X_2$, where $E < V(X)$, the time changes along the imaginary direction. Finally, for $X < X_1$ the time is complex, $t + i\tau_0$, with a constant imaginary part $\tau_0 = \int_{X_1}^{X_2} dX/\sqrt{2(V(X) - E)}$. Therefore, for a tunneling trajectory, with X going from negative to positive infinity, the time evolves along the C^+ contour depicted in Fig. 3.2.

If one wants the tunneling trajectory to be a solution of the stationary path equation, one has to consider the evolution operator along the contour C^+ in the complex time plane; see Fig. 3.2. The semiclassical approximation for the tunneling amplitude is given by the exponentiated action along the C^+ contour. To calculate the tunneling probability one has to supplement the latter with the conjugated backward contour C^-; see Fig. 3.2. Actually, the locations in time of the vertical parts of the contour are not necessarily the same on the forward and backward branches. This freedom is important for the treatment of multiple tunneling events. Since our immediate goal is to find the probability of a single tunneling event, we can restrict ourselves to the particular contour drawn in Fig. 3.2.

With exponential accuracy, the tunneling probability for a particle with energy E is given by

$$P(E) \sim e^{\,i\int_{C^+ + C^-} dt \left[\frac{1}{2}\dot{X}^2 - V(X) + E\right]}, \tag{3.27}$$

where $X(t)$ is a solution of the classical equation of motion $\ddot{X} = -V'(X)$ along the contour. The last term in the action, $\int dt\, E = E(t_f - t_i)$, where t_f

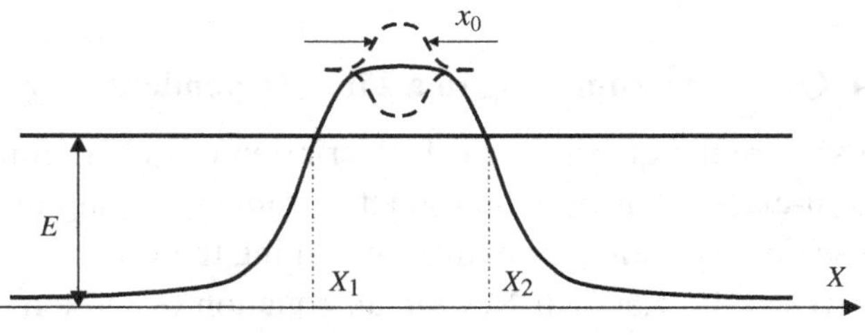

Figure 3.1 Tunneling potential $V(X)$ and classical turning points X_1 and X_2. Time goes in the imaginary direction for $X_1 < X < X_2$. The dashed lines show a small time-dependent part of the potential, Eq. (3.29).

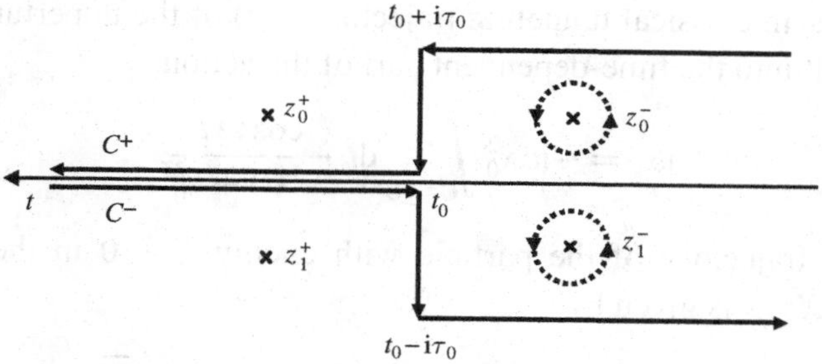

Figure 3.2 The time contour for semiclassical evaluation of the tunneling probability. The poles at $t = z_n^{\pm}$ appear upon perturbative treatment of a time-dependent potential. After deforming the contour (dotted lines) only the poles at z_0^- and z_1^- contribute to the action.

is the final point on C^- and t_i is the initial point on C^+, serves to fix the energy of the particle. Indeed, demanding stationarity with respect to variations over $t_{f,i}$ and using [35] $\delta S/\delta t_{f,i} = \mp H$, one finds that the energy is fixed on both branches of the contour $\dot{X}^2/2 + V(X) = E$. An alternative way of looking at this term is to view it as a Fourier transform from the time to the energy representation.

It is easy to see that the action along the horizontal parts of the contour sums up to zero. Indeed, the action of the backward branch cancels exactly that of the forward one. It is therefore the action along the two vertical segments that remains. With the help of the classical equation of motion (3.25), the latter is given by

$$\mathrm{i}S_0 = \mathrm{i}\int_{\mathrm{i}\tau_0}^{-\mathrm{i}\tau_0} \mathrm{d}t\,\dot{X}^2 = 2\mathrm{i}\int_{X_1}^{X_2} \mathrm{d}X\dot{X} = -2\int_{X_1}^{X_2} \mathrm{d}X\sqrt{2(V(X)-E)}. \tag{3.28}$$

For the tunneling probability one finds $P(E) \sim e^{\mathrm{i}S_0}$. This is the well-known WKB result [36], which is in fact obtained by the usual trick [36] of considering the imaginary time Schrödinger equation. (Indeed, the horizontal parts of the contour were inconsequential so far.)

However, doing the problem in the way we have presented allows one to consider tunneling in the presence of a time-dependent potential [37]. To be specific, let us consider a potential of the form (see Fig.3.1)

$$V(X,t) = -\frac{1}{2}\,\omega_0^2 X^2 + \frac{\varepsilon\, x_0^2}{X^2 + x_0^2}\,\cos\Omega\, t, \tag{3.29}$$

which consists of a parabolic barrier, along with the localized time-dependent potential, oscillating with a frequency Ω. We shall assume that the amplitude of the latter is small and consider a correction to the action linear in ε. To this end we need to find a semiclassical tunneling trajectory $X(t)$ of the unperturbed potential and substitute it into the time-dependent part of the action:

$$\mathrm{i}S_1 = -\mathrm{i}\varepsilon\, x_0^2 \int_{C^+ + C^-} \mathrm{d}t\, \frac{\cos\Omega\, t}{X^2(t) + x_0^2}. \tag{3.30}$$

The tunneling trajectory of the particle with energy $E < 0$ in the unperturbed potential $-\omega_0^2 X^2/2$ is given by

$$X(t) = X_2 \cosh\big(\omega_0(t - t_0)\big), \qquad X_2 = \sqrt{-2E}/\omega_0. \tag{3.31}$$

At time $t = -\infty + \mathrm{i}\tau_0$ the particle starts at $X = -\infty$ and reaches the point $X_1 = -X_2$ at time $t = t_0 + \mathrm{i}\tau_0$. Then it spends an imaginary time $\mathrm{i}\tau_0 = \mathrm{i}\pi/\omega_0$ under the barrier where $X(t_0 + \mathrm{i}\tau) = X_2 \cos\omega_0\tau$ and finally continues to move in real time from $X = X_2$ toward $X = \infty$. The integral in Eq. (3.30) has poles in the complex time plane at $t = z_n^{\pm}$, where $z_n^{\pm} = t_0 \pm \omega_0^{-1}\mathrm{arcsinh}(x_0/X_2) + \mathrm{i}\tau_0(1/2 - n)$ and n is an integer. Two of these poles are located inside the contour, z_0^- and z_1^-; see Fig. 3.2. Deforming the contour to run around the poles and evaluating the integral in Eq. (3.30) with the help of the residue theorem, one finds

$$\mathrm{i}S_1 = \frac{2\,\varepsilon\tau_0\, \cos\alpha}{\sqrt{(X_2/x_0)^2 + 1}}\, \cosh\frac{\Omega\,\tau_0}{2}, \tag{3.32}$$

where $\alpha = \Omega\, t_0 - (\Omega/\omega_0)\,\mathrm{arcsinh}(x_0/X_2)$. The tunneling probability is given by $P(E) \sim \mathrm{e}^{\mathrm{i}S_0 + iS_1}$, where in the present case $\mathrm{i}S_0 = 2E\tau_0 = 2\pi E/\omega_0$ (remember that $E < 0$). The correction $\mathrm{i}S_1$ has a random sign, dictated by $\cos\alpha$, which depends on t_0 – the free parameter of the tunneling trajectory (3.31). One should now fix t_0 by maximizing the tunneling probability, that is demanding $\cos\alpha = 1$. This way one obtains for the tunneling probability in the presence of the oscillating field

$$P(E) \sim \mathrm{e}^{-2|E|\tau_0} \exp\left\{ \frac{2\,\varepsilon\tau_0}{\sqrt{(X_2/x_0)^2 + 1}}\, \cosh\frac{\Omega\tau_0}{2} \right\}. \tag{3.33}$$

Therefore, the tunneling probability is exponentially enhanced! The most surprising feature of this result is that at high frequency $\Omega\tau_0 \gg 1$ the enhancement parameter is itself exponentially large $\sim \varepsilon\tau_0 \mathrm{e}^{\Omega\tau_0/2}$ [37]. This does not mean that a weak high-frequency field can make the barrier completely transparent. It rather means that there is a surprisingly small scale of the ac modulation amplitude $\varepsilon \sim \omega_0 \mathrm{e}^{-\Omega\tau_0/2}$, beyond which the linear correction to the action is not sufficient.

To understand this behavior qualitatively, consider absorption of n quanta of energy Ω. It elevates the energy of an incoming particle to $E + \Omega n$ and therefore changes its tunneling action to $\mathrm{i}S_0(E + \Omega n) \approx \mathrm{i}S_0 + 2\tau_0\Omega n$, since quite generally $\partial(\mathrm{i}S_0)/\partial E = 2\tau_0$, where τ_0 is the (imaginary) time the particle spends under the barrier. The amplitude of the n-quanta absorption process may be estimated as $(\varepsilon/\Omega)^n/n!$ As a result, the probability of tunneling upon absorption of n quanta from the ac field is $e^{iS_0+2\tau_0\Omega n}(\varepsilon/\Omega n)^{2n}$. Optimizing over n, one finds $iS_1 \sim (\varepsilon/\Omega)e^{\Omega\tau_0}$, similarly to what we found earlier. The difference from the actual result, Eq. (3.33), originates in the overestimated absorption amplitude (the actual one is probably reduced by another factor of $n!$). This consideration shows that the results are applicable as long as $\Omega \ll |E|$.

3.5 Dissipative Quantum Tunneling

Consider a particle with unit mass moving in a potential

$$V(X) = V_0\left[\delta\left(\frac{X}{a}\right)^2 - \left(\frac{X}{a}\right)^3\right], \tag{3.34}$$

where δ is a dimensionless bifurcation parameter, which governs the shape of the potential. For $\delta > 0$ the potential exhibits a metastable minimum at $X = 0$; see Fig. 3.3(a). If $V_0 \gg 1/(\delta^5 a^2)$ one may disregard energy quantization in the metastable well and consider escape of the particle with zero energy, $E = 0$, initially trapped in the metastable minimum. The semiclassical escape trajectory according to Eq. (3.26) is given by

$$X(t) = \frac{\delta a}{\cos^2\frac{\omega_0(t-t_0)}{2}}, \tag{3.35}$$

where $\omega_0 = \sqrt{2V_0\delta/a^2}$. The contour C^+, see Fig. 3.3(b), proceeds along $t + i\tau_0$, where $t \in]-\infty, t_0]$, then goes along the imaginary axis from $t = t_0 + i\tau_0$ to $t = t_0$ and finally goes along the real-time axis from $t = t_0$ to $t = t_0 + \pi/\omega_0$; see Fig. 3.3(b). In the coordinate space the three pieces of the C^+ contour correspond to (i) the particle staying at $X = 0$; (ii) the particle moving under the barrier, where $X(t_0 + i\tau) = \delta a/\cosh^2(\omega_0\tau/2)$; (iii) the classical motion from the "resurfacing" point $X = \delta a$ toward $X = \infty$. The imaginary time spent under the barrier is infinite in the present case, $\tau_0 = \infty$. The presence of the poles on the real axis in Eq. (3.35) is an artefact of the too-steep potential drop at $X > \delta a$, allowing the particle to reach $X = \infty$ in a finite time. According to Eq. (3.28), the under-the-barrier action is given by

$$iS_0 = -2\int_0^{\delta a} dX\sqrt{2V(X)} = -\frac{8}{15}\sqrt{2V_0}\,a\,\delta^{5/2}. \tag{3.36}$$

As long as $|iS_0| \gg 1$, the escape rate from the metastable well is $W \propto \omega_0 e^{iS_0}$.

We now consider how the coupling to the ohmic bath affects the tunneling escape rate. In this paragraph we restrict ourselves to the purely quantum, that is, zero-temperature scenario, $T = 0$ (the high-temperature case is discussed in Chapter 4).

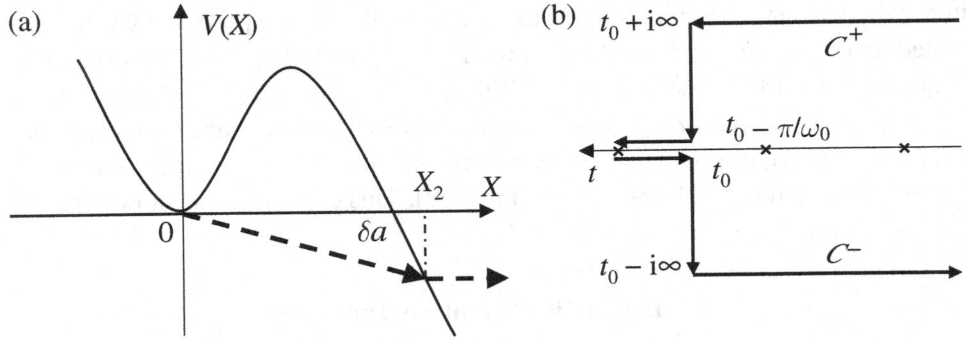

Figure 3.3 (a) Potential (3.34) with the metastable minimum at $X = 0$. If a particle loses energy by exciting the bath, the tunneling trajectory is plotted schematically by the dashed line, with the "resurfacing" point X_2. (b) Contour in the complex time plane. The particle reaches $X = +\infty$ at $t = t_0 + \pi/\omega_0$. There is an infinite set of poles of Eq. (3.35) at $t = t_0 - \pi(1 + 2n)/\omega_0$.

Since all the bath oscillators are in their ground states, they cannot transfer energy to the particle. Therefore, one does not expect any activation-like acceleration of the escape. On the other hand, the particle is very far from its ground state and may excite the bath oscillators during its escape. Such processes lead to the particle losing its energy and "sinking" deeper into the barrier; see Fig. 3.3(a). As a result, one expects that the particle emerges from under the barrier somewhere at $X = X_2 > \delta a$. (In fact, we'll see that in the limit of very strong coupling to the bath there is a universal result for such a "resurfacing" point $X_2 = \frac{4}{3}\delta a$.) At $X > X_2$ the particle moves in real time, and its action is real and cancels between the forward and backward branches of the contour. The finite imaginary part of the action is accumulated during the motion along the imaginary time direction $t = \mathrm{i}\tau$. Taking the limit $T \to 0$ in Eq. (3.23), one finds for the imaginary time action

$$S[X] = \mathrm{i}\int d\tau \left\{ \frac{1}{2}(\partial_\tau X)^2 + V(X) + \frac{\gamma}{4\pi}\int d\tau' \frac{\left(X(t_0 + \mathrm{i}\tau) - X(t_0 + \mathrm{i}\tau')\right)^2}{(\tau - \tau')^2} \right\}. \tag{3.37}$$

Variation of this action with respect to $X(t_0 + \mathrm{i}\tau)$ leads to the semiclassical equation for under the barrier motion:

$$\partial_\tau^2 X = V'(X) + \frac{\gamma}{\pi} ⨍ d\tau' \frac{X(t_0 + \mathrm{i}\tau) - X(t_0 + \mathrm{i}\tau')}{(\tau - \tau')^2}, \tag{3.38}$$

where the integral is understood as a principal value. A general solution of this equation is not known. In the limit of weak dissipation, $\gamma \ll \omega_0$, one may find a correction to the tunneling action using perturbation theory. To this end one needs to substitute the imaginary time tunneling trajectory $X(t_0 + \mathrm{i}\tau)$, given by Eq. (3.35), into the last term in Eq. (3.37). This way one finds for the small dissipative correction to the bare tunneling action (3.36) that $\mathrm{i}\delta S_{\mathrm{diss}} = -(12\zeta(3)/\pi^3)\gamma a^2\delta^2$.[1] Notice that this correction, being smaller than the bare tunneling action (3.36), may still result in the exponential suppression of the tunneling rate by the dissipation.

As was first realized by Caldeira and Leggett [34], one may also find a solution of Eq. (3.38) in the opposite limit of strong dissipation $\gamma \gg \omega_0$. In this case the inertia term $\partial_\tau^2 X$ on the left-hand side of the equation of motion (3.38) may be neglected. One can check then by direct substitution that the following trajectory is indeed the desired solution:[2]

$$X(t) = \frac{\frac{4}{3}\,\delta a}{1 - \omega_1^2(t - t_0)^2}, \tag{3.39}$$

where $t = t_0 + \mathrm{i}\tau$ and $\omega_1 = 2V_0\delta/(\gamma a^2) = \omega_0^2/\gamma$. At $t = t_0 + \mathrm{i}\infty$ it starts in the metastable minimum $X = 0$ and reaches $X_2 = 4\delta a/3$ at $t = t_0 + \mathrm{i}0$. Here the particle emerges from under the barrier and continues its motion in real time. The action (3.37) on this trajectory is given by (the inertia term $(\partial_\tau X)^2/2$ is neglected)

$$\mathrm{i}S_{\mathrm{diss}} = -\frac{2\pi}{9}\,\gamma a^2\delta^2. \tag{3.40}$$

Notice that $S_{\mathrm{diss}}/S_0 \sim \gamma/\omega_0 \gg 1$ in the limit of strong dissipation. As a result, the coupling to the bath leads to the exponential suppression of the escape rate $W \sim \mathrm{e}^{\mathrm{i}S_{\mathrm{diss}}}$. Remarkably, the scaling of the action with the bifurcation parameter changes from $\delta^{5/2}$ to δ^2. One expects that the Caldeira–Leggett scaling, δ^2, always wins in the immediate vicinity of the bifurcation point, that is, for $\delta \ll 1$. Indeed, since $\omega_0 \sim \delta^{1/2}$, for $\delta \ll 1$ the dissipation is always strong, that is, $\omega_0 \ll \gamma$. It may seem paradoxical that the escape rate is independent of the barrier height V_0. In fact, taking the inertia term $(\partial_\tau X)^2/2$ in Eq. (3.37) as a perturbation, that is, substituting in it the inertia-less solution (3.39), one finds a correction to the action (3.40)

[1] The calculations are easier in the Fourier representation, where the $T = 0$ Matsubara components of the trajectory (3.35) are given by $X_m = 4\pi\delta a\omega_0^{-2}\epsilon_m/\sinh(\pi\epsilon_m/\omega_0)$. Employing the dissipative action in the form of Eq. (3.22) and substituting summation by integration, one finds

$$\mathrm{i}\delta S_{\mathrm{diss}} = -(\gamma/2)(4\pi\delta a\omega_0^{-2})^2\int(\mathrm{d}\epsilon/2\pi)\epsilon^3/\sinh^2(\pi\epsilon/\omega_0) = -(12\zeta(3)/\pi^3)\gamma a^2\delta^2.$$

[2] Indeed, putting $t_0 = 0$ for simplicity,

$$\fint \mathrm{d}\tau'\,\frac{X(\mathrm{i}\tau)-X(\mathrm{i}\tau')}{(\tau-\tau')^2} = \frac{4}{3}\,\delta a\omega_1\fint\frac{\mathrm{d}z'}{(z'-z)^2}\left[\frac{1}{1+z^2}-\frac{1}{1+z'^2}\right] = \frac{4}{3}\frac{\delta a\omega_1}{1+z^2}\mathrm{Re}\int\frac{\mathrm{d}z'}{z'-z-\mathrm{i}0}\,\frac{z'+z}{1+z'^2},$$

where $z = \omega_1\tau$. Evaluating the integral with the help of the residue theorem, one finds

$$\frac{4\pi}{3}\,\frac{\delta a\omega_1}{1+(\omega_1\tau)^2}\,\mathrm{Re}\,\frac{\mathrm{i}+\omega_1\tau}{\mathrm{i}-\omega_1\tau} = \frac{4\pi}{3}\,\delta a\omega_1\frac{(\omega_1\tau)^2-1}{[1+(\omega_1\tau)^2]^2} = \frac{-\pi V_0}{\gamma}\left[\frac{2\delta X(\tau)}{a^2}-\frac{3X^2(\tau)}{a^3}\right] = \frac{-\pi}{\gamma}V'(X).$$

$\mathrm{i}\delta S_{\mathrm{inert}} = -4\pi V_0\delta^3/(9\gamma)$. Although smaller than the dissipative action (3.40), this correction still leads to the exponential dependence of the tunneling rate on V_0.

One may notice that the real-time Caldeira–Leggett solution (3.39) does not satisfy Newton's equation with viscous friction, Eq. (3.21). This is because during the imaginary-time part of the trajectory the particle has excited the bath oscillators. The latter also continue to evolve in real time, exerting an additional force on the particle. As discussed in Section 3.3, the real-time counterpart of the dissipative action (3.37) is the Keldysh action (3.20). The corresponding semiclassical equation of motion is given by Eq. (3.21). Combining Eqs. (3.21) and (3.38), one finds the equation of motion for $X(t)$ on the real-time part of the contour, Fig. 3.3b,

$$\ddot{X} = -V'(X) - \gamma\dot{X} - \frac{\gamma}{\pi}\int \mathrm{d}\tau\, \frac{X(t_0 + \mathrm{i}\tau)}{(t - t_0 - \mathrm{i}\tau)^2}, \tag{3.41}$$

where the τ-integration runs along the vertical part of the contour in Fig. 3.3b. It is easy to check that, neglecting the inertia term $\ddot{X}$, the Caldeira–Leggett solution (3.39) satisfies this equation too. Therefore, the trajectory (3.39) solves the semiclassical equations of motion along the entire contour! Notice that the tunneling event completed at $t = t_0$ exerts a slowly decaying $\sim (t-t_0)^{-2}$ (for $t-t_0 \gg \omega_1^{-1}$) force, altering the subsequent motion of the particle in real time. This fact may qualitatively change the picture of tunneling between two resonant wells, where multiple tunneling events are important. We shall not develop this theory here, referring the reader to a review [39].

One can now use the analytic form of the tunneling trajectories, Eqs. (3.35) and (3.39), to investigate the influence of an external time-dependent signal on the escape rates. To this end let us consider a weak spatially uniform oscillatory force by adding the following term to the potential (3.34):

$$V(X,t) = \varepsilon\frac{X}{a}\,\mathrm{e}^{\nu t}\cos\Omega t, \tag{3.42}$$

where ν is an infinitesimal energy scale, which describes an adiabatic switching on of the external time-dependent force. To the first order in ε the change in the tunneling action is evaluated by substituting the trajectory (3.35) or (3.39) into the action $S_1 = -\mathrm{i}\int_{C^+ + C^-}\mathrm{d}t\, V(X(t),t)$. Due to the factor $\mathrm{e}^{\nu t}$ (omitted from now on) the $t = -\infty$ part of the contour does not contribute to the action. Deforming the contour, one finds that the integral is reduced to the contribution of the poles of the $X(t)$ function along the $t < t_0$ part of the real-time axis. For the bare tunneling trajectory, Eq. (3.35), the relevant poles are at $t = z_n = t_0 - (\pi + 2\pi n)/\omega_0$, where $n = 0, 1, \ldots$. Summing over all of them and maximizing over the free parameter t_0, one finds [37]

$$\mathrm{i}S_1 = \frac{\varepsilon\Omega}{\omega_0^2}\frac{4\pi\delta}{\sin(\pi\Omega/\omega_0)}. \tag{3.43}$$

For $\Omega = \omega_0, 2\omega_0, \ldots$ the external field is in resonance with the small oscillations in the metastable minimum, and the linear response approach fails. In the limit $\Omega \to 0$ one finds $\mathrm{i}S_1 = 4\delta\varepsilon/\omega_0$. This may be directly obtained from Eq. (3.36) by changing δ^2 to $\delta^2 - 3\varepsilon/V_0$. Indeed, the potential (3.34) is equivalent, up to a trivial shift, to $V(X) = V_0[\delta^2 X/3a - (X/a)^3]$. Therefore, adding to it a static linear term $-\varepsilon X/a$ leads to the aforementioned redefinition of δ^2.

For the case of strong dissipation, the only relevant pole of the trajectory (3.39) is at $t = t_0 - 1/\omega_1$, which leads to the following correction to the action:

$$\mathrm{i}S_1 = \frac{4\pi}{3}\frac{\varepsilon\delta}{\omega_1} = \frac{2\pi}{9}\gamma a^2 \frac{3\epsilon}{V_0}. \tag{3.44}$$

This is nothing but the adiabatic change of the time-independent result (3.40) by the static reduction of $\delta^2 \to \delta^2 - 3\varepsilon/V_0$. Therefore, for not-too-large frequencies the effect of the ac force on the overdamped tunneling decay is the same as the dc one. At higher frequencies the fact that the $-X^3$ tail of the potential must flatten somewhere becomes important. In this case, the particle does not reach infinity in a finite time. This fact translates into the splitting of the poles and moving them away from the real axis by a small imaginary time $\mathrm{i}\tau_s \sim \mathrm{i}a/\sqrt{V_0}$. Similarly to Section 3.4, it leads to the exponential enhancement of the ac correction at very high frequencies $\mathrm{i}S_1 \propto \varepsilon e^{\Omega\tau_s}$ [37].

3.6 Problems

3.6.1 Feynman–Keldysh Action for Symmetric Potentials

Write the Feynman–Keldysh action (3.6) for some symmetric potentials, for example: (i) $V(X) = -X^2 + gX^4$, and (ii) $V(X) = \cosh X$. Explain the apparent symmetry $X^{\mathrm{cl}} \leftrightarrow X^{\mathrm{q}}$. This symmetry is implicitly broken by regularization and explicitly broken by coupling to a bath. The latter generates terms $\propto (X^{\mathrm{q}})^2$, restricting fluctuation of the quantum component. No terms like $\propto (X^{\mathrm{cl}})^2$ are generated, of course. Take variations of the action and derive coupled classical equations of motion for X^{cl} and X^{q}. Investigate them in the limit $X^{\mathrm{q}} = 0$.

3.6.2 Korshunov Instantons

Consider a dissipative quantum rotator, described by a 2π-periodic coordinate $\Phi(\tau)$. The imaginary time boundary conditions need to be generalized to allow for integer winding numbers W, such that $\Phi(\beta) = \Phi(0) + 2\pi W$. The dissipative action, known as Ambegoakar–Eckern–Schön (AES) action [40], is a generalization of the Caldeira–Leggett action (3.23) for the periodic field:

$$S[\Phi] = \mathrm{i}\,\frac{gT^2}{4}\iint_0^\beta \mathrm{d}\tau\,\mathrm{d}\tau'\,\frac{\sin^2\left[\frac{\Phi(\tau)-\Phi(\tau')}{2}\right]}{\sin^2[\pi T(\tau-\tau')]} + q\int_0^\beta \mathrm{d}\tau\,\partial_\tau\Phi(\tau), \qquad (3.45)$$

where q is a conjugate variable and the last term constitutes the weight $\mathrm{e}^{\mathrm{i}q2\pi W}$ of different winding numbers in the imaginary time partition sum.

Perform variation of the action over the field $\Phi(\tau)$ and find the imaginary time, stationary point equation of motion for the field. By doing contour integration in the z-plane, verify that the $W = 1$ realization [41],

$$\mathrm{e}^{\mathrm{i}\Phi(\tau)} = \frac{z - z_1}{1 - z\bar{z}_1}, \qquad (3.46)$$

satisfies the equation of motion. Here $z = \mathrm{e}^{2\pi\mathrm{i}T\tau}$ and z_1 is an arbitrary complex number inside the unit circle, $|z_1| < 1$. Plot $\Phi(\tau)$ for various z_1. Calculate the action (3.45) and show that it is given by $S = \mathrm{i}g/4 + 2\pi q$, independent of z_1. Therefore, z_1 is a zero mode of the saddle-point solution. This solution may be generalized for an arbitrary $W > 0$ as

$$\mathrm{e}^{\mathrm{i}\Phi(\tau)} = \prod_{a=1}^{W} \frac{z - z_a}{1 - z\bar{z}_a}, \qquad (3.47)$$

while $W < 0$ is obtained from here by $z \to 1/z$. Verify that the corresponding action is $S_W = \mathrm{i}g|W|/4 + 2\pi qW$, and thus z_a with $a = 1, 2, \ldots, W$ are all zero modes of this solution. Calculate the partition function as $Z(q) = \sum_W \mathrm{e}^{\mathrm{i}S_W}$. (Though such calculation misses W-dependent fluctuation factors [42, 44], it gives a qualitatively acceptable result.)

Show that the kinetic part of the action removes the degeneracy of the zero modes, since it exhibits an explicit dependence on z_a parameters

$$S_{\mathrm{kin}} = \frac{1}{4E_c}\int_0^\beta \mathrm{d}\tau\,(\partial_\tau\Phi)^2 = \frac{\pi^2 T}{E_c}\sum_{a,b=1}^{W}\frac{1 - |z_a|^2|z_b|^2}{|1 - z_a\bar{z}_b|^2}. \qquad (3.48)$$

The AES action and its Korshunov's stationary solutions are widely used in the theory of the Coulomb blockade [42, 44, 43], where E_c plays the role of the charging energy of a quantum dot, g is dimensionless conductance between a metallic lead and the dot, and q is a background charge on the dot. Periodicity of the partition function $Z(q)$ with the period one reflects the charge quantization. Notice that for $g \gg 1$ the modulation of the free energy $F(q) = -T \log Z(q)$ is exponentially small. This phenomenon is often referred to as a *weak Coulomb blockade*. Generalizations of this action to the real-time non-equilibrium situation were discussed in [45, 46].

4

Classical Stochastic Systems

This chapter is devoted to the classical limit of the quantum dissipative action obtained in Chapter 3. We show how it yields Langevin, Fokker–Planck, and optimal path descriptions of classical stochastic systems. These approaches are used to discuss activation escape, fluctuation relation, reaction models, and other examples.

4.1 Classical Dissipative Action

In Section 3.2 we derived the Keldysh action for a quantum particle coupled to an Ohmic environment, Eq. (3.20). If only linear terms in the quantum coordinate $X^{\rm q}(t)$ are kept in this action, it leads to a classical Newtonian equation with a viscous friction force, Eq. (3.21). Such an approximation completely disregards any fluctuations, both quantum and *classical*. Our goal now is to do better than that and to keep classical thermal fluctuations, while still neglecting quantum effects.

To this end it is convenient to restore the Planck constant $\hbar$ in the action and then take the limit $\hbar \to 0$. For dimensional reasons, the factor $\hbar^{-1}$ should stay in front of the entire action. To keep the part of the action responsible for the classical equation of motion (3.21) free from the Planck constant it is convenient to rescale the quantum component as $X^{\rm q} \to \hbar X^{\rm q}$. Indeed, when this is done all terms linear in $X^{\rm q}$ do not contain $\hbar$. Finally, to have the temperature in energy units, one needs to substitute T with $T/\hbar$. As a result, the term $\sim \gamma T\,(X^{\rm q}(t))^2$ does not contain the Planck constant either. The limit $\hbar \to 0$ is now straightforward: (i) one has to expand $\mp V(X^{\rm cl} \pm \hbar X^{\rm q})$ to first order in $\hbar X^{\rm q}$ and neglect all higher-order terms; (ii) in the last nonlocal term in Eq. (3.20) the $\hbar \to 0$ limit is taken with the help of the identity

$$\frac{\pi T^2/(2\hbar)}{\sinh^2(\pi T(t-t')/\hbar)} \stackrel{\hbar\to 0}{\longrightarrow} T\delta(t-t'). \tag{4.1}$$

Consequently the nonlocal term becomes local in the $\hbar \to 0$ limit. Finally, the classical limit of the dissipative action (3.20) is

$$S[\vec{X}] = \int_{-\infty}^{+\infty} \mathrm{d}t \left\{ -2X^{\mathrm{q}} \left[\ddot{X}^{\mathrm{cl}} + \gamma \dot{X}^{\mathrm{cl}} + V'(X^{\mathrm{cl}}) \right] + 4\mathrm{i}\gamma T \left(X^{\mathrm{q}} \right)^2 \right\}. \tag{4.2}$$

Notice that this action is local in time. Also, despite its name, the quantum component X^{q} still has a role to play in the classical setting.

Physically the limit $\hbar \to 0$ means that $\hbar\gamma$ and $\hbar\Omega \ll T$, where Ω is a characteristic frequency of the particle's classical motion. These conditions are sufficient for us to neglect both the time nonlocal term and the higher-order expansion of $V(X^{\mathrm{cl}} \pm \hbar X^{\mathrm{q}})$ in Eq. (3.20). Correspondingly, an alternative way to look at the classical expression (4.2) is to view it as a high-temperature limit of the full quantum action (3.20). On the technical level it amounts to substituting $\coth \epsilon/(2T)$ by $2T/\epsilon$ in, for example, Eq. (3.24). In this chapter we consider some implications of the classical dissipative action (4.2) as well as some of its generalizations.

4.2 Langevin Equation

One way to proceed with the classical action (4.2) is to notice that the exponent of its last term (times i) may be rewritten in the following way:

$$\mathrm{e}^{-4\gamma T \int \mathrm{d}t \left(X^{\mathrm{q}}(t) \right)^2} = \int \mathbf{D}[\xi(t)]\, \mathrm{e}^{-\int \mathrm{d}t \left[\frac{1}{4\gamma T} \xi^2(t) - 2\mathrm{i}\xi(t) X^{\mathrm{q}}(t) \right]}. \tag{4.3}$$

This identity is called the Hubbard–Stratonovich transformation, where $\xi(t)$ is an auxiliary Hubbard–Stratonovich field. With the integration measure $\mathbf{D}[\xi(t)]$ normalized such that $\int \mathbf{D}[\xi(t)]\, \mathrm{e}^{-\int \mathrm{d}t\, \xi^2/4\gamma T} = 1$, the identity (4.3) is an immediate consequence of the real Gaussian integral (2.22).

Any observable $\mathcal{O}[X^{\mathrm{cl}}]$ formulated in terms of the classical coordinate (possibly taken in more than one instance of time) may be written as follows (recall that $Z = 1$ and thus no normalization factor is needed):

$$\begin{aligned} \langle \mathcal{O}[X^{\mathrm{cl}}] \rangle &= \int \mathbf{D}[X^{\mathrm{cl}}, X^{\mathrm{q}}]\, \mathcal{O}[X^{\mathrm{cl}}]\, \mathrm{e}^{\mathrm{i}S[\vec{X}]} \\ &= \int \mathbf{D}[\xi]\, \mathrm{e}^{-\frac{1}{4\gamma T}\int \mathrm{d}t\, \xi^2} \int \mathbf{D}[X^{\mathrm{cl}}] \mathcal{O}[X^{\mathrm{cl}}] \int \mathbf{D}[X^{\mathrm{q}}]\, \mathrm{e}^{-2\mathrm{i}\int \mathrm{d}t\, X^{\mathrm{q}} (\ddot{X}^{\mathrm{cl}} + \gamma \dot{X}^{\mathrm{cl}} + V'(X^{\mathrm{cl}}) - \xi)} \\ &= \int \mathbf{D}[\xi]\, \mathrm{e}^{-\frac{1}{4\gamma T}\int \mathrm{d}t\, \xi^2} \int \mathbf{D}[X^{\mathrm{cl}}] \mathcal{O}[X^{\mathrm{cl}}]\, \delta\left(\ddot{X}^{\mathrm{cl}} + \gamma \dot{X}^{\mathrm{cl}} + V'(X^{\mathrm{cl}}) - \xi \right), \end{aligned} \tag{4.4}$$

where the last line includes the functional delta-function of the expression in the round brackets. This functional delta-function enforces its argument to be zero at every moment of time. Therefore, among all possible trajectories $X^{cl}(t)$ only those contribute to the observable that satisfy

$$\ddot{X}^{cl} = -\gamma \dot{X}^{cl} - V'(X^{cl}) + \xi(t). \tag{4.5}$$

This is the Newton equation with a friction force $-\gamma \dot{X}$ and a time-dependent external force $\xi(t)$, known also as the Langevin equation.

Equation (4.4) implies the following strategy for finding the expectation value $\langle \mathcal{O}[X^{cl}]\rangle$: (i) choose a particular realization of the force $\xi(t)$; (ii) solve Eq. (4.5) (e.g. numerically); (iii) having its solution, $X^{cl}(t)$, calculate the observable $\mathcal{O}[X^{cl}]$; (iv) average the result over an ensemble of realizations of the random force $\xi(t)$ with the Gaussian weight $\exp\{-\int dt\, \xi^2(t)/4\gamma T\}$. The Gaussian statistics of the random force $\xi(t)$ means that only its first and second *irreducible* moments must be specified. In our example $\langle \xi(t)\rangle = 0$ (if the first moment is not zero, it may always be viewed as a part of the deterministic force $-V'$). This means that the Langevin equation (4.5) must be supplemented only with the second moment of the random force, given by

$$\langle \xi(t)\xi(t')\rangle = \int \mathbf{D}[\xi]\, \xi(t)\xi(t')\, e^{-\frac{1}{4\gamma T}\int dt\, \xi^2} = 2\gamma T \delta(t-t'), \tag{4.6}$$

where we employed the Wick theorem, Eq. (2.23). Since in the frequency representation the right-hand side of this equation is a constant, the corresponding random force is often referred to as a *white* noise. It originates from the classical thermal fluctuations of bath oscillators. The fact that the noise amplitude is proportional to the friction coefficient, γ, and temperature T is a manifestation of FDT in its classical limit (i.e. $\coth \epsilon/2T \to 2T/\epsilon$). The latter holds because we assumed the bath to be in thermal equilibrium.

4.3 Multiplicative Noise and Martin–Siggia–Rose Method

The Langevin equation (4.5) with the white noise force (4.6) provides a convenient way for a numerical treatment of the classical dissipative action (4.2). It is not very useful, though, for analytical approaches. In fact, many problems may be initially formulated as Langevin equations with certain random forces, and one would like to have a way to convert them into a proper classical action. Such a procedure, which is in essence an inversion of what was done in the previous section, was formulated by Martin, Siggia, and Rose (MSR) [31]. It is presented here in the form suggested by DeDominicis [32] and Janssen [47].

Consider a Langevin equation

$$\dot{X} = A(X) + b(X)\xi(t). \tag{4.7}$$

We have restricted ourselves to the first-order differential operator ∂_t. It may be viewed as an overdamped limit (i.e. $\gamma \gg \Omega$, where Ω is a characteristic classical frequency) of the Newton equation (4.5). We shall generalize, however, Eq. (4.7) to more than one variable. This will allow us to treat an arbitrary order operator by representing it as a higher-dimensional first-order one (see what follows). The most important difference between Eqs. (4.7) and (4.5) is the fact that the noisy force $\xi(t)$ is modulated in a coordinate-dependent way. This is achieved by multiplying it by a coordinate-dependent function $b(X)$, hence the name *multiplicative* noise. The Gaussian white noise $\xi(t)$ is fully specified by its second moment, which without loss of generality may be normalized as

$$\langle \xi(t)\xi(t') \rangle = 2\delta(t - t'). \tag{4.8}$$

In fact the multiplicative Langevin equation (4.7) is ill-defined unless the regularization of the differential operator is explicitly specified. We shall choose such a regularization in a way to be consistent with the field theoretical treatment of the previous chapters. To this end consider the "partition function"

$$Z[\xi] = \int \mathbf{D}[X(t)]\, \mathcal{J}[X]\, \delta\big(\partial_t X - A(X) - b(X)\xi\big) \equiv 1. \tag{4.9}$$

It is identically equal to unity by virtue of the integration of the delta-function, provided $\mathcal{J}[X]$ is the Jacobian of the operator $\hat{\mathcal{N}}[X] = \partial_t X - A(X) - b(X)\xi$. The way to interpret Eq. (4.9) is to discretize the time axis, introducing N-dimensional vectors $X_j = X(t_j)$ and $\xi_j = \xi(t_j)$, where $j = 1, \ldots, N$. The operator takes the form $\mathcal{N}_j = \mathcal{N}^{(0)}_j + \mathcal{N}^{(1)}_{jl} X_l + \frac{1}{2}\mathcal{N}^{(2)}_{jlk} X_l X_k + \cdots$, where summation is understood over repeated indices. The Jacobian $\mathcal{J}[X]$ in the partition function (4.9) is given by the absolute value of the determinant of the following $N \times N$ matrix: $\mathcal{J}_{jl} \equiv \partial \mathcal{N}_j / \partial X_l = \mathcal{N}^{(1)}_{jl} + \mathcal{N}^{(2)}_{jlk} X_k + \cdots$. It is possible to choose a proper (*retarded*) regularization, where $\mathcal{J}_{jl}$ is the lower triangular matrix with unit main diagonal (coming entirely from the $\mathcal{N}^{(1)}_{jj} = 1$ term). Clearly, in this case $\mathcal{J} = 1$. To this end let us choose the discrete version of the operator as

$$\mathcal{N}_j = X_j - X_{j-1} - \delta_t\big[A(X_{j-1}) + b(X_{j-1})\xi_{j-1}\big]. \tag{4.10}$$

Clearly, in this case $\mathcal{J}_{jj} = 1$ and $\mathcal{J}_{j,j-1} = -1 - \delta_t[A'(X_{j-1}) + b'(X_{j-1})\xi_{j-1}]$, while all other matrix elements $\mathcal{J}_{jl} = 0$. As a result $\mathcal{J}[X] = 1$ for any realization X_j and ξ_j. The regularization (4.10) of the differential operator (4.7) is retarded since the right-hand side of Eq. (4.7) is always taken in the "preceding" moment of time $j-1$.

Such an understanding of the Langevin equation (4.7) is called *Ito regularization* [48, 49] and it is the most convenient one for field-theoretical treatment.

Although the partition function (4.9) is trivial, it is clear that all meaningful observables and correlation functions may be obtained by inserting a factor $\mathcal{O}[X]$ in the functional integral (4.9). Having this in mind along with the fact that $\mathcal{J}[X] = 1$ due to Ito regularization, let us proceed with the partition function. Employing the integral representation of the delta-function with the help of an auxiliary field $X^{\mathrm{q}}(t)$, one obtains

$$Z[\xi] = \int \mathbf{D}[X] \int \mathbf{D}[X^{\mathrm{q}}]\ \mathrm{e}^{-2\mathrm{i}\int dt\, X^{\mathrm{q}}(t)\left(\partial_t^{\mathrm{R}} X - A(X) - b(X)\xi(t)\right)}, \tag{4.11}$$

where ∂_t^{R} stays for the retarded (Ito) regularization of the operator. One may average now the partition function over the white noise, Eq. (4.8), by performing the Gaussian integration over $\xi(t)$:

$$Z = \int \mathbf{D}[\xi]\ \mathrm{e}^{-\frac{1}{4}\int dt\, \xi^2} Z[\xi] = \int \mathbf{D}[X, X^{\mathrm{q}}]\, \mathrm{e}^{\int dt \left[-2\mathrm{i}X^{\mathrm{q}}\left(\partial_t^{\mathrm{R}} X - A(X)\right) - 4(X^{\mathrm{q}})^2 D(X)\right]}, \tag{4.12}$$

where $D(X) \equiv b^2(X) \geq 0$. The exponent on the right-hand side is (i times) the MSR action for the Ito–Langevin process (4.7), (4.8). The main difference from the classical limit of the Keldysh action (4.2) is the X-dependent coefficient $D(X)$ in the Keldysh component $\sim (X^{\mathrm{q}})^2$. It clearly originates from the multiplicative nature of the noise term. Notice also that the retarded derivative $\sim X^{\mathrm{q}}\partial_t^{\mathrm{R}} X$ has a correct regularization of the lower triangular matrix with the unit main diagonal. This shows that taking Ito regularization (4.10) of the Langevin process (4.7), is indeed crucial to establishing correspondence with the Keldysh formalism. Let us reiterate thus the discrete form of the MSR action:

$$S[\vec{X}] = \sum_{j=1}^{N} \left[-2X_j^{\mathrm{q}}(X_j - X_{j-1} - \delta_t A(X_{j-1})) + 4\mathrm{i}\delta_t (X_j^{\mathrm{q}})^2 D(X_{j-1})\right], \tag{4.13}$$

which appears to be *normally ordered* (in the sense that the auxiliary variable X^{q} is taken one time step ahead of the physical variable X, apart from the diagonal term $-2X_j^{\mathrm{q}} X_j$). The MSR method provides a way to go from a classical stochastic problem to its proper functional representation. The latter is useful for analytical analysis. Some examples are discussed below.

One can generalize the above consideration for an M-component vector variable $X_\alpha(t)$, where $\alpha = 1, \ldots, M$. The corresponding Ito–Langevin process reads as

$$\dot{X}_\alpha = A_\alpha(X) + b_{\alpha\beta}(X)\xi_\beta(t); \tag{4.14}$$

$$\langle \xi_\beta(t)\xi_\gamma(t')\rangle = 2\delta_{\beta\gamma}\delta(t - t'), \tag{4.15}$$

where summation over repeated indices is understood. Introducing the corresponding vector of auxiliary fields X^{q}_α, one obtains the following MSR action (in the continuous notation):

$$S[\vec{X}] = \int \mathrm{d}t \left[-2X^{\mathrm{q}}_\alpha(\dot{X}_\alpha - A_\alpha(X)) + 4\mathrm{i}X^{\mathrm{q}}_\alpha X^{\mathrm{q}}_\beta D_{\alpha\beta}(X) \right], \tag{4.16}$$

where $D_{\alpha\beta}(X) = \sum_{\gamma=1}^{M} b_{\alpha\gamma}(X)\, b_{\beta\gamma}(X)$ is a symmetric nonnegative-definite matrix.[1]

As an example, consider the second-order Langevin equation (4.5). Renaming the variables as $X_1 = X^{\mathrm{cl}}$ and $X_2 = \dot{X}^{\mathrm{cl}}$, Eq. (4.5) may be brought to the form of Eq. (4.14) with $A_1(X) = X_2$, $A_2(X) = -\gamma X_2 - V'(X_1)$, and $b_{22} = \sqrt{\gamma T}$, while all other components of $b_{\alpha\beta}$ are zero. One may then write the MSR action (4.16) and notice that X_1^{q} enters the action only linearly. Integrating over X_1^{q}, one thus obtains $\delta(\dot{X}_1 - X_2)$, which allows one now to perform integration over X_2. The resulting action written in terms of $X_1 = X^{\mathrm{cl}}$ and $X_2^{\mathrm{q}} = X^{\mathrm{q}}$ is exactly the classical dissipative action (4.2). This illustrates that considering the first-order Langevin equations is not a real limitation. It also shows that, since the equation $\dot{X}_1 = X_2$ should be understood in the Ito way, namely $X_{1,j} - X_{1,j-1} = \delta_t X_{2,j-1}$, the proper regularization of Eq. (4.5) is $X_j - 2X_{j-1} + X_{j-2} = -\delta_t\gamma(X_{j-1} - X_{j-2}) - \delta_t^2 V'(X_{j-2})$. That is, the corresponding quadratic action again has a lower triangular structure with unit diagonal.

4.4 Optimal Path Approximation

For some applications (most notably associated with rare events) the functional integral in Eq. (4.12) may be evaluated in the stationary path approximation. The corresponding equations are obtained by the variation of the action with respect to $X^{\mathrm{q}}(t)$ and $X(t)$ and have the form

$$\begin{aligned} \dot{X} &= A(X) + 4\mathrm{i}X^{\mathrm{q}}D(X), \\ \mathrm{i}\dot{X}^{\mathrm{q}} &= -\mathrm{i}X^{\mathrm{q}}A'(X) + 2(X^{\mathrm{q}})^2 D'(X). \end{aligned} \tag{4.17}$$

One possible solution of these equations is $X^{\mathrm{q}} = 0$, while $\dot{X} = A(X)$. Clearly this solution corresponds to the noiseless evolution of $X(t)$. Such a noiseless trajectory is by no means the only solution of the stationary path equations (4.17). There are other solutions, which ought to be considered. Since $X(t)$ as well as $A(X)$ and $D(X)$ are all real, one expects that stationary trajectories of the variable X^{q} are purely imaginary. This does not contradict, of course, the fact that $\mathbf{D}[X^{\mathrm{q}}] = \prod_j \mathrm{d}X_j^{\mathrm{q}}$ integrations run along the real axis. What we observed is that the stationary points are located away from the initial integration contour and therefore the latter must be

[1] Indeed, the eigenvalue equation is $D_{\alpha\beta}s_\beta = b_{\alpha\gamma}b_{\beta\gamma}s_\beta = \lambda s_\alpha$. Multiplying by s_α, one finds $\lambda = (b_{\beta\gamma}s_\beta)^2/(s_\alpha)^2 \geq 0$. The zero eigenvalue is possible if the matrix $b_{\alpha\beta}$ possesses a left zero mode, i.e. if $s_\alpha b_{\alpha\beta} = 0$.

deformed in the complex planes of $X_j^{\rm q}$ to pass through purely imaginary stationary points (unless $X^{\rm q} = 0$).

To avoid complex notation it is convenient to rename a *stationary* trajectory $X^{\rm q}(t)$ as $X^{\rm q}(t) = P(t)/(2{\rm i})$, where $P(t)$ is real on the stationary trajectories. With this notation, Eqs. (4.17) acquire the Hamiltonian structure

$$\dot{X} = \partial_P H(P,X), \qquad \dot{P} = -\partial_X H(P,X); \tag{4.18}$$

$$H(P,X) = PA(X) + P^2 D(X). \tag{4.19}$$

Notice that P is not the physical momentum (indeed we deal with the overdamped motion (4.7)). It is rather an auxiliary variable that encodes the noise. Nevertheless, it is useful to view it as the canonical pair of the physical variable X. Due to their Hamiltonian nature, the stationary path equations possess the integral of motion: the "energy" $H(P,X) = \text{const}$. The corresponding MSR action, acquired along an *optimal trajectory* (i.e. the one satisfying the equations of motion (4.18)), takes the standard form [35]

$${\rm i}S[X,P] = -\int {\rm d}t\, \left[P\dot{X} - H(P,X)\right], \tag{4.20}$$

where $H(P,X)$ is a constant along the trajectory. The statistical weight of the corresponding path is given by $\exp\{{\rm i}S\}$.

One may visualize solutions of Eqs. (4.18) by plotting the phase portrait, namely the curves of constant energy on the phase plane (P,X). The special role is played by the curves of zero energy $H = 0$. Generally (i.e. if $D(X) \neq 0$) there are two of them: $P = 0$ and $P = -A(X)/D(X)$. The first one corresponds to the noiseless relaxation according to $\dot{X} = A(X)$, while the second one is responsible for fluctuations. These two intersect at the points where $A(X) = 0$, namely at the fixed points of the noiseless dynamics. Along the fluctuation curve $P = -A(X)/D(X)$ the equation of motion reads $\dot{X} = A(X) + 2PD(X) = -A(X)$, namely it describes the evolution, which is time reversed compared to that along the noiseless $P = 0$ line. The fact that the fluctuations are time-reversed partners of the relaxation is *not* generic. It is a consequence of the potential nature of the force; see Section 4.13.

As an example, consider an overdamped thermal motion in a potential $V(X)$. In this case $A(X) = -V'(X)$ and $D(X) = T$ (we put $\gamma = 1$ for brevity). The fluctuation zero energy curve takes the form $P = V'(X)/T$. Figure 4.1(a) depicts the phase portrait for a potential with a single stable minimum at $X = 0$. The noiseless relaxation drives the system toward the origin $X = 0$ along the $P = 0$ line. If we are interested in a relative weight for finding the system at some $X_0 \neq 0$, we need to identify an optimal trajectory, which brings the system to X_0 in a given time. If no time limitations are imposed (i.e. the observation time is unlimited), the proper optimal trajectory is the zero energy curve $P = V'(X)/T$. Indeed, it takes

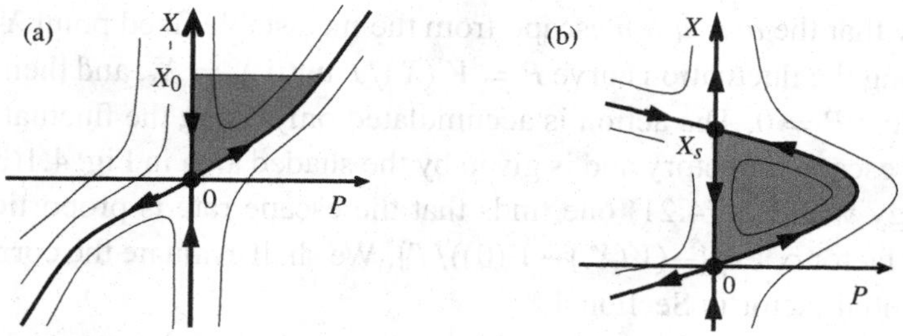

Figure 4.1 Phase portraits of the Fokker–Planck Hamiltonians: (a) for a potential with a single minimum at $X = 0$; (b) for a potential with a metastable minimum at $X = 0$ and unstable maximum at $X = X_s$. Bold lines are curves of zero energy $P = 0$ and $P = V'(X)/T$. The shaded areas give actions of the optimal paths, reaching points X_0 and X_s, respectively.

an infinite time to depart from the fixed point $X = 0$. Since along the optimal path $H(P,X) = 0$, the accumulated action (4.20) may be written as

$$\mathrm{i}S(X_0) = -\int \mathrm{d}t\, P\dot{X} = -\int_0^{X_0} P\,\mathrm{d}X,$$

that is, it is given by the geometric area shaded in Fig. 4.1a. Employing the fact that $P = V'(X)/T$, one further obtains

$$\mathrm{i}S(X_0) = -\int_0^{X_0} P\,\mathrm{d}X = -\frac{1}{T}\int_0^{X_0} V'(X)\mathrm{d}X = -\frac{V(X_0) - V(0)}{T}. \tag{4.21}$$

As a result, the relative statistical weight for finding the system at $X = X_0$ is $\propto \exp\{-V(X_0)/T\}$. This is, of course, nothing but the Boltzmann distribution. So far we have found it with exponential accuracy only, namely without a pre-exponential factor, which, in principle, could be X_0-dependent. In the next section we'll prove that this is not the case.

Consider now a potential that has a *metastable* minimum at $X = 0$ and an unstable maximum at $X = X_s$; see Fig. 3.2. The corresponding phase portrait is depicted in Fig. 4.1b. The fluctuation curve $P = V'(X)/T$ now has two intersections with the relaxation line $P = 0$. The relaxation dynamics in a local vicinity of $X = 0$ is stable (attractive), while at $X = X_s$ it is unstable (repulsive). According to the Liouville theorem of classical mechanics [35], the Hamiltonian motion conserves the area of the phase space. This implies that both fixed points must be hyperbolic, that is, have one attractive and one repulsive direction. As a result, the stability of the two fixed points along the fluctuation curve is opposite to that along the relaxation line, that is, $X = 0$ is repulsive, while $X = X_s$ is attractive; see Fig. 4.1(b). It is clear now that the *activation* escape from the metastable fixed point $X = 0$ must proceed along the fluctuation curve $P = V'(X)/T$ until $X = X_s$, and then follow the relaxation line $P = 0$. The action is accumulated only along the fluctuation part of the optimal escape trajectory and is given by the shaded area in Fig. 4.1(b). In complete analogy with Eq. (4.21) one finds that the escape rate is proportional to the Boltzmann factor $\propto \exp\{-(V(X_s) - V(0))/T\}$. We shall evaluate the corresponding pre-exponential factor in Section 4.8.

Let us discuss now an overdamped particle in a harmonic potential subject to a multiplicative noise (in the Ito sense), proportional to a certain positive power of $|X|$:

$$\dot{X} = -\kappa X + |X|^{\nu}\xi(t), \tag{4.22}$$

where the Gaussian white noise $\xi(t)$ is specified by Eq. (4.8). The question is whether the particle sticks to the bottom of the well and does not ever leave it, because the noise near the bottom is too weak. The corresponding Hamiltonian reads as $H(P,X) = -\kappa PX + P^2|X|^{2\nu}$ and its phase portrait for the case $\nu > 1/2$ is plotted in Fig. 4.2. Again, the relative weight of reaching a point $X_0 \neq 0$ is given by an exponentiated (negative) area enclosed by the curves of zero energy, that is, $\exp\{-\kappa X_0^{2-2\nu}/(2-2\nu)\}$ for $\nu < 1$. On the other hand, for $\nu \geq 1$ the corresponding area diverges, nullifying the long-time probability of finding the particle away from $X_0 = 0$. As a result, for $\nu \geq 1$ the particle eventually sticks to the bottom and the only steady-state distribution is $\delta(X_0)$.

The message of this section is that the stationary path dynamics of *dissipative* stochastic models may be described by the effective Hamiltonian system. The role of momentum is played by the auxiliary MSR variable (times i), which is nothing but the classical limit of the Keldysh "quantum" component. A lot of insight into the behavior of the corresponding stochastic model may be gained

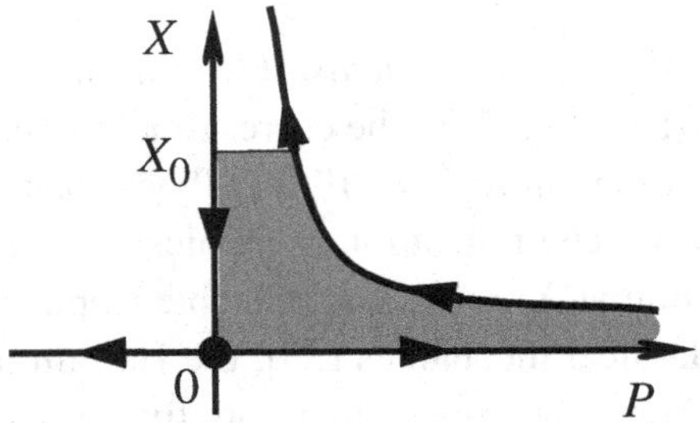

Figure 4.2 Zero energy lines of the Fokker–Planck Hamiltonian corresponding to Eq. (4.22): $P = 0$, $X = 0$, and $P = \kappa/X^{2\nu-1}$ with $\nu > 1/2$. The action (i.e. the shaded area) diverges for $\nu \geq 1$.

by an inspection of the phase portrait of the corresponding Hamiltonian. The action (4.20) was written earlier only for stationary trajectories (i.e. satisfying the equations of motion (4.18)). However, it may be equally well extended to any trajectory $X(t)$ and $P(t)$ and used as a weight in the functional integral, much as the action (4.13) or (4.16). The only thing to remember is that the $\mathbf{D}[P(t)] = \prod_j dP_j$ integrations run along the *imaginary* axis. Notice that quantities such as the escape rate are determined by the optimal trajectories with some nonzero $X^{\mathrm{q}}(t)$. This means, in turn, that such trajectories are different along the forward and backward branches of the time contour. The latter is a consequence of the time-reversal invariance being broken by the integration over the bath with a continuous spectrum.

4.5 Fokker–Planck Equation

The consideration of the previous section closely resembles the WKB approximation in quantum mechanics. One may take one step forward toward the analogy between the theory of classical stochastic models and quantum mechanics and look for a corresponding Schrödinger equation. The latter is derived through the transfer-matrix treatment of the Feynman path integral [33]. To this end one integrates over trajectories, which at time $t = t_j$ arrive at the point $X = X_j$ with an arbitrary momentum (i.e. arbitrary X^{q}). From the definition (4.9) it is clear that the corresponding *restricted* partition function $\mathcal{P}(X,t) = Z|_{X(t)=X}$ is proportional to the *probability* (not the amplitude!) of finding the system at the point X at time t. The fact that $\mathcal{P}(X,t)$ is real follows immediately from the form of the action (4.12) and the symmetry $X^{\mathrm{q}} \to -X^{\mathrm{q}}$, while $Z \to Z^*$. In other words, the Keldysh contour provides the product of an amplitude (forward branch) and its complex conjugate (backward branch), resulting in the probability.

We shall formally derive the equation for $\mathcal{P}(X,t)$ in what follows. The result, however, may be anticipated from the Hamiltonian formulation, Eqs. (4.19), (4.20), and analogy with quantum mechanics. The latter states that the required equation has the form $\partial_t \mathcal{P} = \hat{H}\mathcal{P}$. Here the Hamiltonian *operator* $\hat{H}$ is obtained from the *normally ordered* classical Hamiltonian $H(P,X)$ by the substitution of $P \to \hat{P}$, which satisfies the canonical commutation relation $[X,\hat{P}] = 1$ (in our case P runs along the imaginary axis), that is, $\hat{P} = -\partial_X$. Using Eq. (4.19) for the Hamiltonian, one obtains

$$\partial_t \mathcal{P}(X,t) = -\partial_X\Big[A(X)\mathcal{P}(X,t) - \partial_X\big[D(X)\mathcal{P}(X,t)\big]\Big]. \tag{4.23}$$

This is the Fokker–Planck equation [50] for the evolution of the probability distribution function of the stochastic system (4.7). The normal ordering of the action (4.13) is crucial to employing the quantum mechanical analogy. Therefore, this form of the Fokker–Planck equation is specific to the Ito regularization.

The Fokker–Planck equation has the structure of the continuity relation $\partial_t \mathcal{P} + \partial_X J = 0$, where the probability current is $J = A\mathcal{P} - \partial_X[D\mathcal{P}]$. This fact is responsible for the conservation of probability $\partial_t \int dX \mathcal{P} = 0$. On the classical level it may be traced back to the observation that $H(P,X) \sim P$, that is, there are no terms with the zero power of momentum P in the Hamiltonian. Therefore the property of the Hamiltonian

$$H(0,X) = 0 \tag{4.24}$$

is crucial to the conservation of probability. On the other hand, this relation along with expression (4.20) for the action are completely equivalent to the basic Keldysh symmetry $S[X,0] = 0$, Eq. (2.53) (recall that $P \sim X^{\mathrm{q}}$).

In the case of the additive noise $D(X) = T$ the Fokker–Planck Hamiltonian (4.19) may be transformed into the conventional Schrödinger form. This is achieved by the canonical transformation $x = X$ and $\hat{p} = \hat{P} - V'(X)/(2T)$, which preserves the commutation relation $[x, \hat{p}] = 1$ and thus $\hat{p} = -\partial_x$. With these new variables the Fokker–Planck equation acquires the form of the imaginary-time Schrödinger equation:

$$\partial_t \Psi(x,t) = -T\left[-\partial_x^2 + W(x)\right]\Psi(x,t), \tag{4.25}$$

where the effective potential, $W(x)$, and the "wave function", $\Psi(x,t)$, are given by

$$W(x) = \frac{[V'(x)]^2}{4T^2} - \frac{V''(x)}{2T}; \qquad \Psi(x,t) = \mathrm{e}^{V(x)/(2T)}\mathcal{P}(x,t). \tag{4.26}$$

It is worth mentioning that the effective quantum mechanics (4.25) is *supersymmetric* [51, 52], which is due to the fact that the initial Hamiltonian (4.19) satisfies the normalization identity (4.24).[2]

We turn now to the transfer matrix derivation of the Fokker–Planck equation (4.23). Consider $\mathcal{P}(X_{j-1}, t_{j-1})$, which is obtained from Eq. (4.12) by integration over all X_i with $i = 1, \ldots, j-2$ and all X_i^{q} with $i = 1, \ldots, j-1$. Notice that the X^{q} integration runs one step ahead of the X integration. To find $\mathcal{P} = \mathcal{P}(X,t) = \mathcal{P}(X_j, t_j)$

[2] Indeed the "Schrödinger equation" (4.25) may be written as

$$T^{-1}\partial_t \Psi = -\hat{B}^\dagger \hat{B}\, \Psi; \qquad \hat{B} = \mathrm{i}\partial_x + \frac{\mathrm{i}}{2T} V'(x). \tag{4.27}$$

This equation admits a stationary zero mode annihilated by the operator $\hat{B}$, i.e. $\hat{B}\Psi_0 = 0$. This results in $\Psi_0(x) \propto \mathrm{e}^{-V(x)/(2T)}$ and thus $\mathcal{P}(x) \propto \mathrm{e}^{-V(x)/T} \propto \Psi_0^2$, which is, of course, the Boltzmann distribution. This is indeed a solution as long as $\Psi_0(x)$ is a normalizable "wave-function", i.e. $V(x) \to +\infty$, when $x \to \pm\infty$. In this case the supersymmetry is said to be *unbroken* [52]. In the opposite case of the *broken* supersymmetry (e.g. $V(x) = \delta x^2 - x^3$) there is no zero eigenvalue of the Hamiltonian $\hat{B}^\dagger \hat{B}$, but instead there is an exponentially small one, considered in Section 4.8 (notice that the super-potential $W(x)$ is still confining, even if the potential $V(x)$ is not).

one needs to perform two more integrations over $dX_{j-1}dX_j^q$ with the weight specified by Eq. (4.12):

$$\mathcal{P}=\int dX_{j-1}dX_j^q\, e^{-2iX_j^q\left(X_j-X_{j-1}-\delta_t A(X_{j-1})\right)-4\delta_t(X_j^q)^2 D(X_{j-1})}\mathcal{P}(X_{j-1},t_{j-1}). \quad (4.28)$$

We now rename the integration variables as $X_j^q = X^q$ and $X_{j-1} = X_j - \delta_X$ and expand the exponent to second order in the small fluctuations δ_X and X^q. This leads to the already familiar Keldysh structure

$$\exp\left\{-(\delta_X,X^q)\begin{pmatrix} 0 & i \\ i & 4\delta_t D\end{pmatrix}\begin{pmatrix}\delta_X \\ X^q\end{pmatrix}\right\}. \quad (4.29)$$

From here one concludes that as $\delta_t \to 0$ the fluctuations scale as $X^q \sim \delta_t^{-1/2}$ and $\delta_X \sim \delta_t^{1/2}$. We then approximate $A(X_{j-1}) \approx A - \delta_X A'$, while $D(X_{j-1}) \approx D - \delta_X D' + \delta_X^2 D''/2$ and $\mathcal{P}(X_{j-1}, t_{j-1}) \approx \mathcal{P} - \delta_X \mathcal{P}' + \delta_X^2 \mathcal{P}''/2 - \partial_t \mathcal{P}\delta_t$, where $A = A(X_j)$, $D = D(X_j)$, and $\mathcal{P} = \mathcal{P}(X_j, t_j)$ and primes denote derivatives with respect to X_j. Expanding the exponent up to second order in terms $\delta_X \sim \delta_t^{1/2}$ and up to first order in terms $\delta_X^2 \sim \delta_t$, we find

$$\partial_t\mathcal{P} = -(A'\mathcal{P}+A\mathcal{P}')2i\langle\delta_X X^q\rangle - (D''\mathcal{P}+2D'\mathcal{P}')2\langle\delta_X^2(X^q)^2\rangle + \mathcal{P}''\frac{\langle\delta_X^2\rangle}{2\delta_t},$$

where the angular brackets stand for averaging with the Gaussian weight (4.29) and we took into account that, as always, $\langle(X^q)^2\rangle = 0$ and also $\langle\delta_X(X^q)^3\rangle = \langle\delta_X^2(X^q)^4\rangle = 0$. The remaining nonzero averages are given by $\langle\delta_X X^q\rangle = -i/2$, $\langle\delta_X^2(X^q)^2\rangle = -1/2$, and $\langle\delta_X^2\rangle = 2\delta_t D$. As a result one obtains the Fokker–Planck equation (4.23), as expected.

The derivation may be straightforwardly extended to the multivariable Ito–Langevin process (4.14), yielding

$$\partial_t\mathcal{P}(X,t) = -\partial_\alpha\Big[A_\alpha(X)\mathcal{P}(X,t) - \partial_\beta\big[D_{\alpha\beta}(X)\mathcal{P}(X,t)\big]\Big], \quad (4.30)$$

where $\partial_\alpha = \partial_{X_\alpha}$ and summation over repeated indices is understood. Again the equation has the structure of the continuity relation $\partial_t\mathcal{P} + \mathrm{div}J = 0$, where the probability current vector $J_\alpha = A_\alpha\mathcal{P} - \partial_\beta[D_{\alpha\beta}(X)\mathcal{P}]$ consists of the drift part and the diffusive part.

For a particular case where the drift is provided by a potential force, that is, $A_\alpha(X) = -\partial_\alpha V(X)$ and the noise is isotropic and additive, namely $D_{\alpha\beta} = \delta_{\alpha\beta}T$, one may look for a stationary solution of Eq. (4.30) by demanding that the current vector is zero: $\partial_\alpha V\mathcal{P} = -T\partial_\alpha\mathcal{P}$. Solving this first-order equation, one finds

$$\mathcal{P}(X) = Z^{-1}\, e^{-V(X)/T}, \quad (4.31)$$

which is a proper stationary probability distribution as long as it can be normalized. This means the normalization constant, also known as the partition function, $Z = \int \prod_\alpha dX_\alpha \, e^{-V(X)/T}$ exists.[3] This is, of course, the Boltzmann distribution, which we have already found with exponential accuracy using the optimal path method; see Eq. (4.21). Here we proved that the pre-exponential factor is an X-independent constant. Notice that if the drift force is not a potential one, the stationary distribution (if it exists) implies, in general, a nonzero divergenceless current, $\mathrm{div}J = 0$, whereas $J_\alpha \neq 0$. The distribution (4.31) is thus not applicable.

The fact that the Fokker–Planck equation has a stationary solution (for the class of normalizable potentials) may be formulated as the presence of a *zero eigenvalue* of the corresponding Hamiltonian operator $\hat{H}(\hat{P}, X)$. Equation (4.31) provides the corresponding eigenfunction, or *zero mode*. In the transformed Schrödinger variables (4.25) the presence of the zero mode follows from the *supersymmetric* nature of the effective potential $W(x)$ [51, 52]; see footnote 2.

If there is an inertia term $\ddot{X}$ in the Langevin equation (4.5), one needs to consider particle momentum as just another coordinate $X_1 = X$ and $X_2 = \dot{X} = K$. Employing that $A_1(K) = K$, $A_2(X,K) = -\gamma K - V'(X)$ and $D_{22} = \gamma T$ is the only nonzero component of $D_{\alpha\beta}$, one may rewrite Eq. (4.30) for the probability distribution function $\mathcal{P} = \mathcal{P}(X,K,t)$ as

$$\partial_t \mathcal{P} + K\partial_X \mathcal{P} - V'(X)\partial_K \mathcal{P} = \gamma \partial_K \left(K\mathcal{P} + T\partial_K \mathcal{P} \right). \qquad (4.32)$$

The left-hand side, called the *kinetic* term, may be written as $\partial_t \mathcal{P} - \{E, \mathcal{P}\}$, where the classical Hamiltonian function is $E(K,X) = K^2/2 + V(X)$ and we used standard Poisson brackets. It describes evolution of the distribution function due to the drift of position in the presence of the velocity $K = v_K = \partial_K(K^2/2)$ and the drift of momentum in the presence of the force $-\partial_X V(X)$. The right-hand side, also known as the *collision* term, originates from the interaction with the thermal bath. It describes random diffusion in the momentum space superimposed on the drift toward $K = 0$ in the effective "potential" $K^2/2$. The latter is responsible for the particle losing energy and cooling down, if the temperature T is too low.

One may look for a stationary solution of the Fokker–Planck equation (4.32), which separately nullifies the kinetic and the collision terms. From the latter condition one finds that $\mathcal{P}(X,K) = \mathcal{P}(X)\, e^{-K^2/2T}$. Substituting it into the kinetic term, one finally finds for the corresponding zero mode

$$\mathcal{P}(X,K) = Z^{-1}\, e^{-V(X)/T}\, e^{-K^2/2T} = Z^{-1}\, e^{-E(K,X)/T}, \qquad (4.33)$$

[3] Here Z is not the Keldysh "partition function" normalized to one, but a usual equilibrium statistical mechanics partition function.

where the normalization constant Z, namely the partition function, is given by $Z = \int dX dK\, e^{-V(X)/T}\, e^{-K^2/2T}$. This is the Maxwell–Boltzmann distribution for the particle's potential and kinetic energy in thermal equilibrium.

If the system is out of equilibrium, but all characteristic time scales are much longer than the relaxation time γ^{-1}, one may look for a solution of the Fokker–Planck equation (4.32) in the form

$$\mathcal{P}(X,K,t) = (2\pi T)^{-1/2}\, e^{-K^2/2T}\big[\mathcal{P}(X,t) + K\mathcal{N}(X,t)\big],$$

where the exponential factor is chosen to nullify the right-hand side of Eq. (4.32) and thus to compensate for the large factor γ. We now substitute this trial solution in Eq. (4.32) and (i) integrate over K; (ii) multiply by K and then integrate over K. This way we obtain two coupled equations:

$$\partial_t\mathcal{P} + T\partial_X\mathcal{N} = 0\,; \qquad T\partial_X\mathcal{P} + V'(X)\,\mathcal{P} = -\gamma T\mathcal{N},$$

where in the second equation we neglected the term $T\partial_t\mathcal{N}$ as being much smaller than its right-hand side. Substituting $\mathcal{N}$ from the second equation into the first one, one finds a closed equation for $\mathcal{P}(X,t)$:

$$\gamma\,\partial_t\mathcal{P} = \partial_X\big[V'(X)\,\mathcal{P} + T\,\partial_X\mathcal{P}\big]. \tag{4.34}$$

This is, of course, the already familiar overdamped Fokker–Planck equation (4.23). The fact that the diffusion coefficient in the coordinate space is $D = T/\gamma$ is known as the *Einstein relation*. Notice that the diffusion coefficient in momentum space, according to Eq. (4.32), is $D_K = \gamma T$. Both of these facts are manifestations of FDT.

Since we are dealing with a classical particle, there is no problem in exactly specifying its coordinate X and momentum K simultaneously. This should be contrasted with the fictitious momentum $P = 2iX^{q}$, introduced in Section 4.4. The latter is conjugated to the coordinate X in the sense of the functional integral. It thus obeys the uncertainty principle $\Delta X\Delta P \geq 1$ even in a purely classical setting. One may still discuss trajectories in the phase space (P, X) in the semiclassical (i.e. weak fluctuations, or low temperature) approximation. If the inertia and thus the physical momentum K are taken into account, the semiclassical phase space is four-dimensional: (P, P_2, X, K), where P_2 is conjugate to $K = \dot{X}_2$.

4.6 Ito versus Stratonovich

Although Ito regularization of stochastic processes, discussed in Section 4.3, is the most convenient for the field-theoretical representation, one must be aware that there are other regularizations. The one frequently found in physics literature is known as Stratonovich regularization. It appears upon changing variables in stochastic evolution equations. Consider, for example, the Langevin equation (4.7), (4.8) with the additive noise, namely $b = 1$. The corresponding Fokker–Planck equation for the probability distribution function $\mathcal{P}(X,t)$ is given by Eq. (4.23) with $D = b^2 = 1$,

$$\dot{X} = A(X) + \xi(t)\,; \qquad \partial_t \mathcal{P} = -\partial_X\big[A\mathcal{P} - \partial_X \mathcal{P}\big]. \tag{4.35}$$

Suppose now we want to change the coordinate $X(t)$ to another coordinate $Y(t)$, such that

$$X = f(Y), \tag{4.36}$$

where f is a monotonic function, which provides a one-to-one correspondence between X and Y. We shall assume thus that $f'(Y) = \mathrm{d}X/\mathrm{d}Y > 0$, for convenience. Substituting it in the Langevin equation (4.35), one notices that the corresponding stochastic equation for the new variable $Y(t)$ formally acquires the multiplicative form

$$\dot{Y} = \tilde{A}(Y) + \tilde{b}(Y)\,\xi(t), \tag{4.37}$$

where $\tilde{b}(Y) = 1/f'(Y)$ and $\tilde{A}(Y) = A(f(Y))/f'(Y)$.

Naively one may think that the corresponding Fokker–Planck equation is given by Eq. (4.23) with $D(Y) = \tilde{b}^2(Y) = [1/f'(Y)]^2$. Let us, however, perform the change of variables (4.36) directly in the Fokker–Planck equation (4.35). To maintain normalization of the probability distribution function one has to demand that $\tilde{\mathcal{P}}(Y,t)\,\mathrm{d}Y = \mathcal{P}(X,t)\,\mathrm{d}X$, and thus the proper distribution of the Y variable is $\tilde{\mathcal{P}}(Y,t) = \mathcal{P}(f(Y),t)f'(Y)$. Notice also that $\partial_X = (1/f'(Y))\,\partial_Y$. As a result the Fokker–Planck equation (4.35) transforms into

$$\partial_t \tilde{\mathcal{P}}(Y,t) = -\partial_Y\Big[\tilde{A}(Y)\,\tilde{\mathcal{P}}(Y,t) - \tilde{b}(Y)\,\partial_Y\big[\tilde{b}(Y)\tilde{\mathcal{P}}(Y,t)\big]\Big]. \tag{4.38}$$

As before, $\tilde{b}(Y) = 1/f'(Y)$. While the drift current is what we expect from the Langevin equation (4.37), the diffusive current is different from that in the Ito–Fokker–Planck equation (4.23). The latter has the form $\partial_Y[\tilde{b}^2\tilde{\mathcal{P}}]$.

The reason for this difference is that the multiplicative noise in Eq. (4.37) does *not* have the Ito retarded regularization. Indeed, the discrete form of the Langevin equation (4.35) is $X_j - X_{j-1} = \delta_t A(X_{j-1}) + \delta_t \xi_{j-1}$. Upon the change of variables given by Eq. (4.36) the left-hand side takes the form $f(Y_j) - f(Y_{j-1}) = f(\bar{Y} + \delta_Y/2) - f(\bar{Y} - \delta_Y/2) = f'(\bar{Y})(Y_j - Y_{j-1}) + O(\delta_Y^3)$, where $\delta_Y = Y_j - Y_{j-1}$ and $\bar{Y} = (Y_j + Y_{j-1})/2$. As a result the discrete version of the Langevin equation (4.37) is

$$Y_j - Y_{j-1} = \delta_t \tilde{b}\left(\frac{Y_j + Y_{j-1}}{2}\right) A(Y_{j-1}) + \delta_t \tilde{b}\left(\frac{Y_j + Y_{j-1}}{2}\right)\xi_{j-1}. \tag{4.39}$$

The first term on the right-hand side is already $\sim \delta_t$ and therefore may be substituted by $\delta_t \tilde{A}_{j-1} = \delta_t \tilde{b}(Y_{j-1})A(Y_{j-1})$. This is not so with the second term: since $\xi_j \sim \delta_t^{-1/2}$ (indeed the corresponding statistical weight is $\mathrm{e}^{-\delta_t \xi_j^2/4}$), one finds $\delta_t \xi_{j-1} \sim \delta_t^{1/2}$. Therefore it is important to keep the argument of the noise modulation function as $\tilde{b}\big((Y_j + Y_{j-1})/2\big)$. This is different from the Ito retarded regularization, which assumes $\tilde{b}(Y_{j-1})$ instead. The symmetric regularization of the multiplicative noise as in Eq. (4.39) is known as Stratonovich regularization. It leads to the different form of the diffusion term in the Fokker–Planck equation (4.38).

One may formally bring the Stratonovich diffusion term to the Ito form, at the expense of adding $(\partial_Y \tilde{b})\,\tilde{b}\,\tilde{\mathcal{P}}$ to the drift term. All the considerations may be straightforwardly generalized to an arbitrary dimension M. As a result the Stratonovich–Langevin equation $\dot{X} = A^{(\mathrm{S})} + b\xi$ is equivalent to the Ito–Langevin one $\dot{X} = A^{(\mathrm{I})} + b\xi$ with

$$A_\alpha^{(\mathrm{I})}(X) = A_\alpha^{(\mathrm{S})}(X) + \sum_{\beta,\gamma=1}^{M} \big[\partial_\beta b_{\alpha\gamma}(X)\big]\, b_{\beta\gamma}(X). \tag{4.40}$$

The Ito process may be then used for the field-theoretical treatment via the MSR procedure of Section 4.3.

4.7 Noise with a Finite Correlation Time

Another context where the Stratonovich interpretation appears naturally is stochastic systems with noise that has a short, but finite correlation time τ. Consider Gaussian "colored" noise with the correlation function

$$\langle \eta(t)\eta(t')\rangle = \frac{1}{\tau}\,\mathrm{e}^{-|t-t'|/\tau}, \tag{4.41}$$

known also as the Ornstein–Uhlenbeck process. The white noise (4.8) is obtained in the limit $\tau \to 0$. Such a random function is a result of "filtering" the white noise force $\xi(t)$, Eq. (4.8), with an overdamped harmonic oscillator (e.g. an RC circuit), having the time constant τ. This means that $\eta(t)$ satisfies

$$\dot{\eta} = \frac{1}{\tau}\big[-\eta + \xi(t)\big]. \tag{4.42}$$

Indeed, the solution of this equation is $\eta(t) = \frac{1}{\tau}\int^t \mathrm{d}t_1\,\xi(t_1)\,\mathrm{e}^{(t_1-t)/\tau}$. Employing Eq. (4.8), one readily establishes the correlation function (4.41).

One may consider the Ornstein–Uhlenbeck process (4.41), (4.42) as a random force term in a multiplicative Langevin equation:

$$\dot{X} = A(X) + b(X)\,\eta(t). \tag{4.43}$$

Then in the limit $\tau \ll \Omega^{-1}$ (but still $\tau \gg \delta_t$), where Ω is a characteristic frequency of the deterministic process $\dot{X} = A(X)$, the probability distribution function $\mathcal{P}(X,t)$ obeys the Stratonovich–Fokker–Planck equation (4.38).

To prove this statement let us consider the two evolutionary equations (4.43) and (4.42) as a two-dimensional Langevin process (4.14). In this case $A_X = A(X) + b(X)\,\eta$ and $A_\eta = -\eta/\tau$, while the noise $\xi(t)$ is non-multiplicative with the coupling constant $b_{\eta,1} = 1/\tau$. Because the noise is non-multiplicative, there is no need to specify the regularization. The Fokker–Planck equation for the joint probability distribution $\mathcal{P}(X,\eta,t)$ acquires the form (cf. Eq. (4.30))

$$\partial_t \mathcal{P} = -\partial_X\big[(A(X) + b(X)\,\eta)\mathcal{P}\big] + \partial_\eta\left[\frac{1}{\tau}\,\eta\mathcal{P} + \frac{1}{\tau^2}\,\partial_\eta \mathcal{P}\right]. \tag{4.44}$$

In the absence of coupling, $b = 0$, the η-variable quickly equilibrates to a symmetric Gaussian distribution $\sim e^{-\eta^2\tau/2}$. One can thus proceed in a way analogous to the one that led from Eq. (4.32) to Eq. (4.34). To this end we look for a solution in the form $\mathcal{P}(X,\eta,t) \sim e^{-\eta^2\tau/2}\big[\mathcal{P}(X,t) + \eta\mathcal{N}(X,t)\big]$. Substituting it into Eq. (4.44) and then (i) integrating over η and (ii) multiplying by η and then integrating over it, one obtains two equations, $\partial_t\mathcal{P} = -\partial_X[A\mathcal{P}] - \partial_X[b\mathcal{N}]/\tau$ along with $\mathcal{N}/\tau = -\partial_X[b\mathcal{P}]$, where in the last equation we have neglected $\partial_t\mathcal{N}$ and $\partial_x[A\mathcal{N}]$, as being much less than $\mathcal{N}/\tau$. Substituting the resulting $\mathcal{N}$ into the equation for $\mathcal{P}$, one finds the Stratonovich–Fokker–Planck equation (4.38).

4.8 Kramers Problem

We return now to the problem of activation escape of an overdamped particle from a metastable potential minimum. It was briefly considered in Section 4.4 in the stationary path approximation. Here we address it employing the Ito–Fokker–Planck equation (4.23). To modify the result for the Stratonovich stochastic process, one can take advantage of the correspondence rule (4.40). The potential $V(X)$ is similar to the one plotted in Fig. 3.3a. Since the Boltzmann distribution (4.31) with such a potential is not normalizable, there is no stationary solution (i.e. zero mode) of the Fokker–Planck equation. One expects, however, that there is a long-lived solution localized in the vicinity of the metastable minimum. We shall look for such a solution in the form

$$\mathcal{P}(X,t) = \mathcal{P}(X)\,e^{-t/\tau_{es}}, \tag{4.45}$$

where τ_{es} is the escape time, which is expected to be exponentially long. The total probability is not conserved, because there is a probability current J toward $X = \infty$. Substituting this form into Eq. (4.23), one obtains the stationary Fokker–Planck equation for $\mathcal{P}(X)$:

$$\frac{1}{\tau_{\mathrm{es}}}\mathcal{P} = \partial_X\big[-V'\mathcal{P} - \partial_X[D\mathcal{P}]\big] = \partial_X J. \tag{4.46}$$

Since the escape rate $1/\tau_{\mathrm{es}}$ is expected to be exponentially small, so is $\partial_X J$ everywhere, except near the narrow peak of the metastable distribution $\mathcal{P}(X)$ around $X = 0$. As we show here, the latter has the characteristic width $l_0 = \sqrt{D(0)/V''(0)} \ll X_{\mathrm{s}}$, if $D(0)$ is small enough (see Fig. 4.1(b)). Therefore the current J out of the metastable state is practically a constant for $|X| \gg l_0$. Obviously the current is zero in the negative direction: $J(-\infty) = 0$, while at large positive X it approaches a constant value, which we denote $J(\infty)$. This observation leads to the linear first-order differential equation

$$-V'(X)\mathcal{P}(X) - \partial_X[D(X)\mathcal{P}(X)] = J(X), \tag{4.47}$$

which may be easily solved to express $\mathcal{P}$ through $J(X)$. The result is

$$\mathcal{P}(X) = \frac{1}{D(X)}\,\mathrm{e}^{-S(X)}\int_X^\infty \mathrm{d}Y\, J(Y)\,\mathrm{e}^{S(Y)}, \tag{4.48}$$

where S satisfies $S' = V'/D$, that is,

$$S(X) = \int_{-\infty}^X \mathrm{d}Y\,\frac{V'(Y)}{D(Y)}. \tag{4.49}$$

The upper limit of the Y-integration in Eq. (4.48) is basically arbitrary (with exponential accuracy) as long as it is well to the right of the point $Y = X_{\mathrm{s}}$. We put it at infinity for brevity. One can now integrate the stationary Fokker–Planck equation (4.46) from minus infinity, where the current is zero, to plus infinity (in the same sense as previously), where the current is $J(\infty)$. This leads to

$$\tau_{\mathrm{es}} = \frac{1}{J(\infty)}\int_{-\infty}^\infty \mathrm{d}X\,\mathcal{P}(X) = \int_{-\infty}^\infty \frac{\mathrm{d}X}{D(X)}\,\mathrm{e}^{-S(X)}\int_X^\infty \mathrm{d}Y\,\frac{J(Y)}{J(\infty)}\,\mathrm{e}^{S(Y)}. \tag{4.50}$$

Notice that adding a constant to S does not change the result. This means that the lower limit of integration in Eq. (4.49) is of no importance.

The X-integral is dominated by the vicinity of the potential minimum, that is, $|X| \lesssim l_0$. On the other hand, the Y-integral is coming from the vicinity of the maximum, that is, $|X - X_{\mathrm{s}}| \lesssim l_{\mathrm{s}}$, where the characteristic width of the maximum is $l_{\mathrm{s}} = \sqrt{D(X_{\mathrm{s}})/|V''(X_{\mathrm{s}})|}$. If $l_0 + l_{\mathrm{s}} \ll X_{\mathrm{s}}$, which is the case for sufficiently small D, one may extend the Y-integral to minus infinity and perform both integrals in the stationary point approximation. The crucial observation is that under the same condition, $J(X_{\mathrm{s}})/J(\infty) = 1$ with exponential accuracy. As a result one obtains for the escape time, including the pre-exponential factor [53],

$$\tau_{\mathrm{es}} = \frac{2\pi}{D(0)}\,\frac{\mathrm{e}^{\Delta S}}{\sqrt{S''(0)|S''(X_{\mathrm{s}})|}} = \sqrt{\frac{D(X_{\mathrm{s}})}{D(0)}}\,\frac{2\pi}{\sqrt{V''(0)|V''(X_{\mathrm{s}})|}}\,\mathrm{e}^{\Delta S}, \tag{4.51}$$

where we took into account that $S'' = (V'/D)' = V''/D$, since $V'(0) = V'(X_s) = 0$, and

$$\Delta S = S(X_s) - S(0) = \int_0^{X_s} dX \, \frac{V'(X)}{D(X)}. \tag{4.52}$$

This is exactly the action along the zero energy trajectory $P = V'/D$ of the Hamiltonian (4.19), introduced in Section 4.4. The Fokker–Planck equation allowed us to determine the pre-exponential factor. In the case of non-multiplicative noise $D = T = \text{const}$, the exponent is the Boltzmann one, $(V(X_s) - V(0))/T$, while the prefactor is temperature independent, $2\pi[V''(0)|V''(X_s)|]^{-1/2}$. This is the celebrated Kramers result [54]. In the case where the inertia term may not be neglected, the pre-exponential factor was evaluated in [55].

In a vicinity of the bifurcation point, a wide class of problems may be modeled by the cubic potential (3.34) and the additive noise with variance D. According to Kramers formula the corresponding escape time is

$$\tau_{es} = \frac{\pi a^2}{V_0 \delta} \, e^{\frac{4}{27} \frac{V_0 \delta^3}{D}}. \tag{4.53}$$

The scaling of the action with the bifurcation parameter δ is rather different from both the pure tunneling exponent (3.36) and the dissipative tunneling exponent (3.40).

4.9 Counting Statistics: Donsker–Varadhan Theory and Beyond

Here we focus on an equilibrium stochastic process:

$$\dot{X} = -\partial_X V(X) + \sqrt{T}\, \xi(t), \tag{4.54}$$

with the white noise, $\xi(t)$, normalized according to Eq. (4.8). We will be concerned with the long time *counting statistics* of a certain observable, $Q(X)$, accumulated during a long time interval, t_0. Specifically, we will define a counter:

$$q = \frac{1}{t_0} \int_0^{t_0} dt \, Q(X(t)). \tag{4.55}$$

Since $X(t)$ depends on a specific realization of the noise $\xi(t)$, so does the counter, q. Therefore q is a stochastic variable, and we will be interested in its probability distribution function, $\mathcal{P}(q)$. Specifically we will look for it in the long time limit, $t_0 \to \infty$. In this limit, X is distributed according to the Boltzmann distribution, $\mathcal{P}_0(X) = Z^{-1} e^{-V(X)/T}$, and therefore the expected value of the counter is $\langle q \rangle = \int dX \, Q(X) \mathcal{P}_0(X)$. The question is how probable are deviations of the counter q

from its expectation value $\langle q \rangle$. We will show that in many cases the long time limit of $\mathcal{P}(q)$ takes the form

$$\mathcal{P}(q) \propto \mathrm{e}^{-t_0\Lambda(q)}, \tag{4.56}$$

where $\Lambda(q) = -\lim_{t_0\to\infty} \log \mathcal{P}(q)/t_0$ is called the *large deviation function*. The normalization requires

$$\Lambda(\langle q \rangle) = 0. \tag{4.57}$$

The counting statistics is calculated as

$$\begin{aligned} \mathcal{P}(q) &= \int \mathbf{D}[\xi(t)]\, \mathrm{e}^{-\frac{1}{4}\int \mathrm{d}t\, \xi^2(t)}\, \delta\left(q - \frac{1}{t_0}\int_0^{t_0} \mathrm{d}t\, Q(X(t))\right) \\ &= \int \mathrm{d}\lambda\, \mathrm{e}^{t_0\lambda q} \int \mathbf{D}[X(t), P(t)]\, \mathrm{e}^{-\int_0^{t_0} \mathrm{d}t [P\dot{X} - TP^2 + P\partial_X V + \lambda Q(X)]}, \end{aligned} \tag{4.58}$$

where we used integral representation of the delta-function $\delta(\ldots) = \int_{-\mathrm{i}\infty}^{-\mathrm{i}\infty} \mathrm{d}\lambda\, \mathrm{e}^{t_0\lambda \ldots}$ and the Martin–Siggia–Rose procedure of Section 4.3 with $P = 2\mathrm{i}X^{\mathrm{q}}$. We notice now that the functional integral over X and P represents the Green function, $G(X_0, X_{t_0}; t_0)$, of the imaginary time quantum mechanics with the Hamiltonian $H_\lambda(P, X) = TP^2 - P\partial_X V - \lambda Q(X)$. One can write such a Green function as $G(X_0, X_{t_0}; t_0) = \sum_n \Psi_n(X_0)\Psi_n(X_{t_0})\, \mathrm{e}^{-t_0 E_n(\lambda)}$, where $\Psi_n(X)$ are eigenfunctions of the Hamiltonian H_λ and $E_n(\lambda)$ are the corresponding eigenvalues. In the long time limit only the ground-state, $E_0(\lambda)$, should be retained, resulting in

$$\mathcal{P}(q) \propto \int_{-\mathrm{i}\infty}^{\mathrm{i}\infty} \mathrm{d}\lambda\; \mathrm{e}^{-t_0[E_0(\lambda) - \lambda q]}. \tag{4.59}$$

Since t_0 is large, this integral may be evaluated in the saddle-point approximation, which leads to the extremum condition $\partial_\lambda E_0(\lambda) = q$.

To proceed further, we employ the quantum mechanical variational principle, which states that the ground-state energy $E_0(\lambda) = \min\langle \Psi | \hat{H}_\lambda | \Psi \rangle$, where the variation is taken over the space of normalized functions $\int \mathrm{d}x\, \Psi^2(x) = 1$. Here the Hamiltonian operator should be taken in the conventional Schrödinger form (4.25): $\hat{H}_\lambda = -T\partial_x^2 + [V'(x)]^2/(4T) - V''(x)/2 + \lambda Q(x)$. The extremum condition takes the form $\partial_\lambda E_0(\lambda) = \langle \Psi | Q(x) | \Psi \rangle = \int \mathrm{d}x\, Q(x)\Psi^2(x) = q$. As a result, the large deviation function is found as

$$\Lambda(q) = E_0(\lambda) - \lambda q = \min\langle \Psi | \hat{H}_0 | \Psi \rangle, \tag{4.60}$$

where the wave function is constrained by the normalization $\int \mathrm{d}x\, \Psi^2(x) = 1$ *and* $\int \mathrm{d}x\, Q(x)\Psi^2(x) = q$ conditions, and $\hat{H}_0$ is the supersymmetric Hamiltonian

$\hat{H}_0 = T\left(\mathrm{i}\partial_x + \frac{\mathrm{i}}{2T}V'(x)\right)^{\dagger}\left(\mathrm{i}\partial_x + \frac{\mathrm{i}}{2T}V'(x)\right)$; see footnote 2. This shows that the large deviation function is manifestly nonnegative:

$$\Lambda(q) = \min\left\{T\int \mathrm{d}x\left[\partial_x\Psi + \frac{1}{2T}V'(x)\Psi\right]^2\right\} \geq 0. \tag{4.61}$$

Notice that $\Lambda(q) = 0$ if and only if $[\partial_x + V'(x)/(2T)]\Psi(x) = 0$. This leads to $\Psi(x) \propto \mathrm{e}^{-V(x)/(2T)} \propto \sqrt{\mathcal{P}_0(x)}$ and therefore $q = \langle q\rangle$. This agrees with the condition (4.57).

One may reformulate the variational principle in terms of a nonnegative probability density, $\mathcal{P}(x) = \Psi^2(x)$. Substituting $\Psi = \sqrt{\mathcal{P}}$ into Eq. (4.61), one finds

$$\Lambda(q) = \min\left\{\frac{T}{4}\int \mathrm{d}x\,\mathcal{P}(x)\left[\partial_x \log\frac{\mathcal{P}(x)}{\mathcal{P}_0(x)}\right]^2\right\}, \tag{4.62}$$

with the constraints $\int \mathrm{d}x\,\mathcal{P}(x) = 1$ and $\int \mathrm{d}x\,Q(x)\mathcal{P}(x) = q$. Here we have used the fact that $V(x) = -T\log\mathcal{P}_0(x)$. This is the form of the large deviation function found by Donsker and Varadhan [56]. For a generalization to the case where a stationary distribution admits a nonzero steady-state current, see Ref. [60].

As an example, let's take the harmonic potential $V(x) = \kappa x^2/2$ and set the counter q as a fraction of time t_0 the stochastic particles spend at a positive x. To this end we choose $Q(x) = \theta(x)$ – the Heaviside step-function. The corresponding Schrödinger operator is $\hat{H}_\lambda = -T\partial_x^2 + (\kappa x)^2/(4T) - \kappa/2 + \lambda\theta(x)$. Its ground state energy may be shown to have the form

$$E_0(\lambda) = \kappa\, e_0(\lambda/\kappa); \qquad e_0(y) \approx \begin{cases} 1 & y \to +\infty; \\ y/2 - cy^2 & |y| \ll 1; \\ 1 + y & y \to -\infty. \end{cases} \tag{4.63}$$

Here $e_0(\lambda/\kappa)$ is a monotonously increasing function and $c = \log 2/4$ is found from the second-order perturbation theory. The extremum condition, $e_0'(\lambda/\kappa) = q$, dictates that $0 < q < 1$, as it should be according to the definition of q. One also notices that $e_0'(0) = \langle q\rangle = 1/2$, also as expected from the reflection symmetry of the harmonic potential. Finally, one finds the probability distribution function from Eq. (4.59) as $\mathcal{P}(q) \propto \exp\{-f(q)\,t_0/\tau_{\mathrm{rel}}\}$, where $\tau_{\mathrm{rel}} = \gamma/\kappa$ is the relaxation time (we have restored the friction constant γ from Eq. (4.5)) and $f(q)$ is a dimensionless function with $f(0) = f(1) = 1$ and $f(q) \approx (q - 1/2)^2/\log 2$ for $|q - 1/2| \ll 1$. This result holds for $t_0 \gg \tau_{\mathrm{rel}}$. Notice that the distribution function is temperature independent.

As another example, let us take the harmonic potential $V(x) = \kappa x^2/2$ and $Q(x) = x$, so the counter, $q = \langle x(t)\rangle_{t_0}$, is a long time average value of the coordinate. From the reflection symmetry one expects a symmetric distribution $\mathcal{P}(q)$

with the zero expectation value, $\langle q\rangle = 0$. The Schrödinger operator is $\hat{H}_\lambda = -T\partial_x^2 + (\kappa x)^2/(4T) - \kappa/2 + \lambda x$. Since the potential is still harmonic, the ground state wave function is Gaussian $\Psi = (2\pi T/\kappa)^{1/4}\exp\{-\kappa(x-q)^2/4T\}$, where the center is chosen to satisfy $\langle\Psi|x|\Psi\rangle = q$. Substituting this minimizer into Eq. (4.61), one finds $\Lambda(q) = (\kappa q)^2/4T$. Finally the distribution function of the average coordinate is a Gaussian, $\mathcal{P}(q) \propto \exp\{-(t_0/\tau_{\rm rel})(\kappa q^2/4T)\}$, with the width that scales as $\sqrt{T/t_0}$.

Let us now turn to a case where the Donsker–Varadhan theory is not applicable. To this end we again consider the harmonic potential $V(x) = \kappa x^2/2$ and take $Q(x) = x^3$. The counter, $q = \langle x^3(t)\rangle_{t_0}$, represents an average value of the cube of the coordinate. Due to the reflection symmetry, its distribution function, $\mathcal{P}(q)$, is expected to be symmetric with zero mean value. The corresponding Schrödinger operator contains potential $(\kappa x)^2/(4T) + \lambda x^3$, which does *not* admit a ground state. This indicates a departure of the long time distribution from the Donsker–Varadhan form (4.56). The physical reason for such a departure is that typical trajectories, $x(t)$, which result in a fixed value of $q = t_0^{-1}\int_0^{t_0} x^3(t)\mathrm{d}t$, are short bursts with a large amplitude (whereas in the Donsker–Varadhan case they are given by small deviations sustained for a long time $\sim t_0$). To illustrate this point,[4] imagine a meander-like trajectory $x(t)$, which stays at $x(t) = x_q$ for a time $\tau \le t_0$ and at $x = 0$ for the rest of the time (one may assume sharp fronts of the meander with duration $\tau_{\rm rel}$). The two variational parameters x_q and τ are related through $q = x_q^3\tau/t_0$. The statistical weight, e^{-S}, of such a trajectory is given by

$$S = \frac{\kappa}{4T\tau_{\rm rel}}\int \mathrm{d}t\left[(\dot{x}\tau_{\rm rel})^2 + x^2\right] \approx \frac{\kappa}{4T}\left[x_q^2 + x_q^2\frac{\tau}{\tau_{\rm rel}}\right] = \frac{\kappa}{4T}\left[x_q^2 + \frac{qt_0}{x_q\tau_{\rm rel}}\right]. \tag{4.64}$$

Optimizing this over x_q, one finds $x_q \propto (qt_0/\tau_{\rm rel})^{1/3}$. From here and $q = x_q^3\tau/t_0$ one finds $\tau \propto \tau_{\rm rel}$, that is, the entire burst has a short duration, $\sim \tau_{\rm rel}$, while its amplitude scales as $t_0^{1/3}$. The corresponding statistical weight is $S \sim (qt_0)^{2/3}/T$. Therefore, the Donsker–Varadhan scaling (4.56) gives way[5] to a very different scaling of the probability with the counting time, t_0,

$$\mathcal{P}(q) \propto \frac{t_0}{\tau_{\rm rel}}\exp\left\{-\alpha\frac{\kappa(qt_0/\tau_{\rm rel})^{2/3}}{2T}\right\}, \tag{4.65}$$

with yet undetermined numerical constant α. The pre-factor, $t_0/\tau_{\rm rel}$, is due to the fact that the short burst may happen at any time during the counting interval, t_0.

[4] I am grateful to Ofer Zeitouni for providing these arguments.

[5] At smallest q the Donsker–Varadhan scaling still holds. This can be seen from a brute force evaluation of low moments, $\langle q^2\rangle, \langle q^4\rangle, \ldots$. The corresponding large deviation function is found as $\Lambda(q) \propto \kappa^3 q^2/(T^3\tau_{\rm rel})$. Comparing Eqs. (4.56) and (4.65), one finds that the Donsker–Varadhan scaling holds for $q < q_c$, where $q_c \propto (T/\kappa)^{3/2}(\tau_{\rm rel}/t_0)^{1/4} \to 0$, as $t_0 \to \infty$. At $q = q_c$ there is a first-order transition to the new counting statistics, which is given by Eq. (4.65) for $q \gg q_c$ [61]. I am indebted to Baruch Meerson and Naftali Smith for clarifications of this issue.

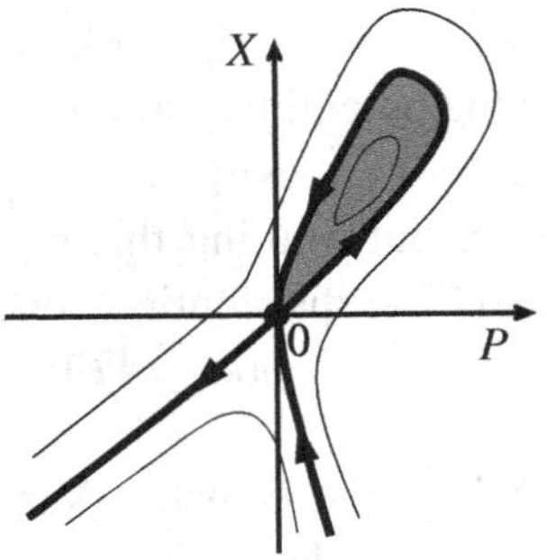

Figure 4.3 Lines of the constant energy for the counting Hamiltonian $H_\lambda(P, X) = TP^2 - \kappa PX - \lambda X^3$ with $\lambda < 0$. The zero energy line is in bold. The corresponding action, $S(\lambda)$, is given by the area of the shaded area.

To make these qualitative arguments more quantitative, we follow Nickelsen and Touchette [62]. They argued that the optimal path $X(t)$ is given by a special stationary trajectory of the action in the second line of Eq. (4.58). Such trajectories are given by lines of constant energy $H_\lambda(P, X) = TP^2 - \kappa PX - \lambda X^3$, depicted in Fig. 4.3. The trajectory of interest, shown in bold, starts and ends at the stationary point $X = P = 0$. It thus has zero energy and describes a short, large amplitude burst reaching $X = -\kappa^2/(4\lambda T)$. The corresponding action, $S(\lambda) = \int dt\, P\dot{X} = \int P dX$, is given by the area shaded in Fig. 4.3:

$$S(\lambda) = \frac{1}{T} \int_0^{-\kappa^2/(4\lambda T)} \sqrt{\kappa^2 X^2 + 4\lambda T X^3}\, dX = \frac{1}{60} \frac{\kappa^5}{T^3} \frac{1}{\lambda^2}. \tag{4.66}$$

One can now optimize $S = -t_0 \lambda q + S(\lambda)$ (cf. Eq. (4.58)) over λ and substitute the optimal λ back into this expression. This way we recover Eq. (4.65) with $\alpha = (9/10)^{1/3}$, in agreement with Ref. [62].

4.10 Fluctuation Relation

Consider an overdamped Langevin dynamics in a time-dependent potential

$$\dot{X} = -\partial_X V(X, t) + \sqrt{T}\, \xi(t), \tag{4.67}$$

where the white noise is normalized according to Eq. (4.8). The time dependence of the potential is limited to a time window $t_i < t < t_f$. Moreover, we shall assume that initially at $t = t_i$ the system is in equilibrium with a bath maintained at temperature T. During the time interval $[t_i, t_f]$ the potential is changing by the action of an external device. Such a device (e.g. a piston) is performing work W on the system, which may be written as

$$W[X] = \int_{t_i}^{t_f} \mathrm{d}t\, \partial_t V(X(t), t). \tag{4.68}$$

The work is a functional of the stochastic trajectory $X(t)$, which the system follows upon a given realization of the random noise $\xi(t)$. As a result, the work W is itself a random quantity, dependent on the noise. One may ask about statistics of the work, for example, the work distribution function $\mathcal{P}(W)$. This question probably can't be answered for an arbitrary potential. There is, however, a particular function of the work, $\mathrm{e}^{-W/T}$, whose average value may be found in a very general form [63, 65].

Employing the MSR method of Section 4.3, one finds for the corresponding average value (cf. Eq. (4.12)),

$$\begin{aligned}\left\langle \mathrm{e}^{-W/T} \right\rangle &= \int \mathbf{D}[X, X^{\mathrm{q}}]\, \mathrm{e}^{\int \mathrm{d}t \left[-2\mathrm{i}X^{\mathrm{q}}\left(\dot{X} + \partial_X V(X,t)\right) - 4T(X^{\mathrm{q}})^2\right]}\, \mathrm{e}^{-W[X]/T} \\ &= \int \mathbf{D}[X, X^{\mathrm{q}}]\, \mathrm{e}^{\int \mathrm{d}t \left[-2\mathrm{i}X^{\mathrm{q}}\left(\dot{X} + \partial_X V(X,t)\right) - 4T(X^{\mathrm{q}})^2 - \frac{1}{T}\partial_t V(X,t)\right]}. \end{aligned} \tag{4.69}$$

Let us first analyze this expression within the stationary path approximation. Following the procedure of Section 4.4, it is convenient to rename the auxiliary variable as $X^{\mathrm{q}} = P/(2\mathrm{i})$. The action acquires the Hamiltonian form (4.20), where

$$H(P, X, t) = -P\, \partial_X V(X, t) + TP^2 - \frac{1}{T}\, \partial_t V(X, t). \tag{4.70}$$

Notice that this Hamiltonian does not satisfy the probability conservation condition (4.24), because it includes the specific observable $-W/T$. Since the Hamiltonian is explicitly time dependent, the energy is not conserved and the solution of the stationary path equations

$$\dot{X} = \frac{\partial H}{\partial P} = -\partial_X V + 2TP, \qquad \dot{P} = -\frac{\partial H}{\partial X} = P\, \partial_X^2 V + \frac{1}{T}\, \partial_X \partial_t V \tag{4.71}$$

is not immediately obvious. Remarkably, the activation trajectory of the time-independent problem, that is, the time-reversed path of the noiseless relaxation $\dot{X} = +\partial_X V$ and $P = \partial_X V/T$, still solves the equations of motion. This fact may be checked by direct substitution of this solution into the equations (4.71).[6] Notice

[6] This solution may be traced back to the existence of the canonical transformation $(P, X, H) \to (p, x, h)$, with the generating function [35] $\Phi = \Phi(x, P, t) = -xP + V(x, t)/T$. Then the following relations hold:

$$X = -\frac{\partial \Phi}{\partial P} = x, \qquad p = -\frac{\partial \Phi}{\partial x} = P - \frac{1}{T}\, \partial_X V(x, t).$$

The transformed Hamiltonian is

$$\begin{aligned} h = H + \frac{\partial \Phi}{\partial t} &= -\left(p + \frac{1}{T}\, \partial_X V(x,t)\right) \partial_X V(x,t) + T\left(p + \frac{1}{T}\, \partial_X V(x,t)\right)^2 \\ &\quad - \frac{1}{T}\, \partial_t V(x,t) + \frac{1}{T}\, \partial_t V(X,t) = p\, \partial_X V(x,t) + Tp^2. \end{aligned}$$

that to have such a solution it is crucial to average $e^{-\eta W}$ with $\eta = 1/T$. It would not work for any other η. The action along this trajectory is given by

$$iS = -\int dt\,[P\dot{X} - H] = -\frac{1}{T}\int dt\,[\partial_X V\dot{X} + \partial_t V] = -\frac{1}{T}\int dV = \frac{V_i - V_f}{T},$$

where $V_{i/f} = V(X_{i/f}, t_{i/f})$ and e^{iS} gives the relative weight of a particle moving from X_i to X_f under the action of the time-dependent potential. Since at $t = t_i$ the system is assumed to be in thermal equilibrium, the initial coordinate is to be weighted with the Boltzmann distribution $e^{-V_i/T}/Z(t_i)$. On the other hand, there is no control over the final coordinate X_f and therefore it should be integrated over with the plane measure, resulting in $Z(t_f)$. One thus obtains

$$\left\langle e^{-W/T}\right\rangle = \frac{e^{-V_i/T}}{Z(t_i)}\int dX_f\, e^{(V_i - V_f)/T} = \frac{Z(t_f)}{Z(t_i)} = e^{-(F(t_f)-F(t_i))/T}, \tag{4.72}$$

where $Z(t_{i/f}) = \int dX e^{-V(X,t_{i/f})/T} \equiv e^{-F(t_{i/f})/T}$ are the *equilibrium* partition functions in the potentials $V(X, t_i)$ and $V(X, t_f)$, respectively. Therefore this particular average value of the non-equilibrium work may be expressed through the equilibrium free energies of the system allowed to equilibrate in the initial and final potential configurations. This remarkable statement is known as the Jarzynski fluctuation relation [63, 65]. As a matter of principle, it allows one to measure the equilibrium free energy of the final state, without waiting for the system to equilibrate. To this end one has to accumulate statistics of the work performed to bring the system into final (yet non-equilibrium) states, and average $e^{-W/T}$.

So far we have derived the fluctuation relation in the stationary path approximation. Let us show now that Eq. (4.72) is actually exact, that is, there is no pre-exponential factor on its right-hand side. To this end we need to derive the Fokker–Planck equation corresponding to the functional integral (4.69) [67]. As explained in Section 4.5 the "quantization" procedure is $\partial_t \mathcal{P}(X,t) = \hat{H}\mathcal{P}(X,t)$, where $\hat{H}$ is obtained from Eq. (4.70) by the substitution $P \to -\partial_X$:

$$\partial_t \mathcal{P} = \partial_X[\partial_X V\, \mathcal{P}] + T\partial_X^2 \mathcal{P} - \frac{1}{T}\,\partial_t V\, \mathcal{P}. \tag{4.73}$$

The initial condition is $\mathcal{P}(X, t_i) = e^{-V(X,t_i)/T}/Z(t_i)$. Motivated by the stationary path result, we look for the solution of this equation in the following form: $\mathcal{P}(X,t) = e^{-V(X,t)/T}/Z(t_i)$. It is easy to check that it is indeed the solution, satisfying the initial condition. Notice that having the coefficient $1/T$ in the last term (the observable) is vital to find such a simple solution. By construction of the functional integral (4.69), $\left\langle e^{-W/T}\right\rangle = \int dX\mathcal{P}(X, t_f)$, leading directly to Eq. (4.72). This proves that the

It is conserved (despite being time dependent), $h = 0$, along the following obvious solution of the equations of motion: $p = 0$, while $\dot{x} = \partial_X V$. Being transformed back to the original variables, it yields the required solution.

fluctuation relation (4.72) is not restricted to the stationary path approximation, but is actually exact.

4.11 Reaction Models

Another important class of classical stochastic models is provided by reaction systems. These models are formulated in terms of reaction rules that are followed by certain agents. The latter are typically denoted as A, B, and so on and can be atoms, molecules, viruses, organisms, or other things. An example of such a reaction rule is $A + A \xrightarrow{\lambda} B$, which states that two agents A may coagulate to form an agent B. A probability for this to happen per unit time (in other words, a reaction rate) is denoted as λ. One would like to have a description that would be able to predict an outcome of many such reactions, provided some initial conditions are specified. It is clear that such a description has to be probabilistic, since there is no way to say with absolute certainty how many and what reactions will happen in a long time span. The other important thing to remember is that the number of agents at any time is always an integer. Therefore a state of the system may be characterized by a time-dependent probability $\mathcal{P}(n, m, \ldots, t)$ of finding n agents A, m agents B, and so on at time t, where $n, m, \ldots$ are integers. Such a probability is normalized as

$$\sum_{n,m,\ldots} \mathcal{P}(n, m, \ldots, t) = 1. \tag{4.74}$$

For any given set of reaction rules one may formulate an evolution equation, also known as a *Master* equation, for probabilities $\mathcal{P}(n, m, \ldots, t)$. For example, for a single species reaction model, the Master equation is

$$\partial_t \mathcal{P}(n, t) = \sum_{n'} \left[W_{n' \to n} \mathcal{P}(n', t) - W_{n \to n'} \mathcal{P}(n, t) \right], \tag{4.75}$$

where $W_{n \to n'}$ is the rate of going from a state with n agents to a state with n' ones. The first term on the right-hand side is the rate of *in* processes, that is, those which lead into the state n from any other state, while the second term is the rate of *out* processes, namely, those which lead out of the state n into any other state. If, for example, the reaction rules are $A + A \xrightarrow{\lambda} \emptyset$, $A \xrightarrow{\mu} \emptyset$, and $A \xrightarrow{\sigma} 2A$, the corresponding rates are

$$W_{n \to n'} = \lambda\, \delta_{n', n-2}\, n(n-1)/2 + \mu\, \delta_{n', n-1}\, n + \sigma\, \delta_{n', n+1}\, n, \tag{4.76}$$

where $n(n-1)/2$ is the number of pairs that can enter the coagulation reaction, and n is the number of agents amenable to annihilation or branching.

Reaction models may have a stationary state, that is a time-independent solution of the Master equation $\mathcal{P}(n)$. If, in such a stationary state, every term on the right-hand side of Eq. (4.75) (i.e. for any integer n') is zero, $W_{n' \to n} \mathcal{P}(n') = W_{n \to n'} \mathcal{P}(n)$, it is said that the reaction scheme satisfies the *detailed balance condition*. If there are no reactions creating agents out of the empty state Ø, in many cases the only stationary solution is the complete *extinction* $\mathcal{P}(n) = \delta_{n,0}$. In this case the detailed balance condition is clearly absent.

The Master equation (4.75) may be written as a differential equation. To this end let us formally extend the integer variable n onto the entire real axis. The "in" term of the Master equation includes the shift operation of the function $W_{n\to n+r}\mathcal{P}(n,t)$, where r is an integer, on r units to bring it to the form $W_{n-r\to n}\mathcal{P}(n-r,t)$. Such a shift operation may be written as $e^{-r\partial_n}$.[7] As a result, the Master equation (4.75) acquires the form

$$\partial_t \mathcal{P}(n,t) = \sum_r \left[e^{-r\partial_n} - 1\right] W_{n\to n+r}\mathcal{P}(n,t). \tag{4.77}$$

It may thus be written as $\partial_t \mathcal{P} = \hat{H}(\hat{p}, n)\,\mathcal{P}$, where the "momentum" operator stands for $\hat{p} = -\partial_n$ and the reaction Hamiltonian is given by [68]

$$H(\hat{p}, n) = \sum_r \left[e^{r\hat{p}} - 1\right] W_{n\to n+r}. \tag{4.78}$$

The reaction Hamiltonian is normally ordered, meaning that all $\hat{p}$ operators stay on the left of the n-dependent functions. It also satisfies the identity (4.24), $H(0,n) = 0$, which is necessary to maintain the conservation of probability (4.74). This way of writing the Master equation brings it into the same category as the Ito–Fokker–Planck equation. The only difference is that the latter has only terms of the first and second powers of the $\hat{p}$-operator. For some problems one may expand the exponent in Eq. (4.78) up to second power in rp, reducing the Master equation to the Fokker–Planck equation (4.23). In this case the drift term is $A(n) = \sum_r rW_{n\to n+r}$, while the diffusion coefficient $D(n) = \sum_r r^2 W_{n\to n+r}/2$. Other problems do not allow for such an expansion, nevertheless all the tools developed for the treatment of the Ito–Fokker–Planck dynamics may be directly transferred to the reaction models.

In particular, a solution to the Master equation may be formally written as $\mathcal{P}(n,t) = \int dn_i\, \hat{\mathcal{U}}(n,t;n_i,t_i)\,\mathcal{P}(n_i,t_i)$, where the evolution operator acts on an initial distribution function. The evolution operator $\hat{\mathcal{U}}$ may be represented by the Hamiltonian path integral; see Eq. (4.12):

$$\hat{\mathcal{U}}(n,t;n_i,t_i) = \int \mathbf{D}[n,p]\, e^{-\int dt[p\dot{n} - H(p,n)]}, \tag{4.79}$$

where, as explained at the end of Section 4.4, the $\mathbf{D}[p]$ integration runs along the imaginary axis. The trajectories satisfy $n(t_i) = n_i$ and $n(t) = n$. We shall first

[7] Indeed, $e^{-r\partial_n}f(n) = f(n) - rf'(n) + r^2 f''(n)/2 - \cdots = f(n-r)$.

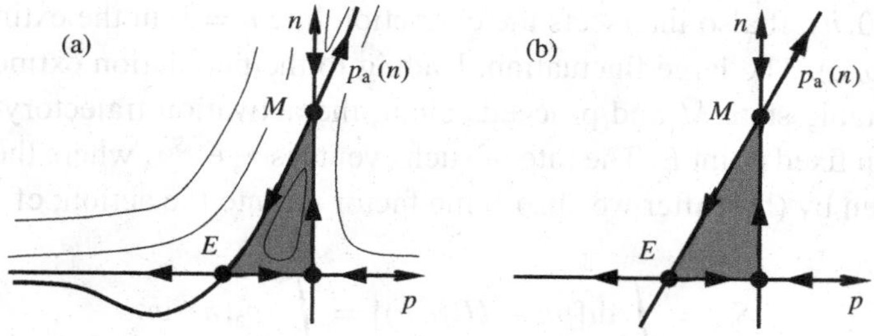

Figure 4.4 Phase portraits of the reaction Hamiltonians: (a) (4.81), which corresponds to the reaction scheme (4.76); (b) the universal Hamiltonian (4.83) close to the bifurcation point. Bold lines are curves of zero energy, M is the metastable point, and E the extinction fixed point.

analyze this expression in the stationary path approximation. The corresponding equations of motion are the Hamilton equations $\dot{n} = \partial_p H$ and $\dot{p} = -\partial_n H$. Since they conserve the energy H, one may visualize the solutions by plotting curves of constant energy on the phase plane (p, n), Fig. 4.4(a). As discussed in Section 4.4, the long-time behavior is described by the curves of zero energy $H = 0$. Due to the conservation of probability (4.74), one such line is always $p = 0$. The corresponding stationary path equation is nothing but the *rate* equation

$$\dot{n} = \partial_p H(p, n)\big|_{p=0} = \sum_r r\, W_{n\to n+r} = A(n). \tag{4.80}$$

It provides the mean-field description, which disregards fluctuations and discreteness of the agents. For the reaction scheme of Eq. (4.76) it predicts the stable fixed point at $n = \bar{n} \approx (\sigma - \mu)/\lambda$ and the unstable fixed point at $n = 0$. According to the rate equation (4.80), the population stabilizes at $n \approx \bar{n}$. This is indeed the case at the intermediate time scale (provided $\bar{n} \gg 1$). However, in the long-time limit the only stationary solution of the corresponding Master equation is the extinct state. In the stationary path approximation this fact is reflected in the presence of the $n = 0$ line of constant zero energy, which is thus the invariant line of the Hamiltonian dynamics. This is always the case if $n = 0$ is the *absorbing* state, namely $W_{0\to n'} = 0$. As a result all $W_{n\to n'} \sim n$ and therefore $H(p, 0) = 0$.

For a scheme exemplified by Eq. (4.76) one finds, with the help of Eq. (4.78), the following reaction Hamiltonian:

$$H(p, n) = \frac{\lambda}{2}(e^{-2p} - 1)n(n-1) + \mu(e^{-p} - 1)n + \sigma(e^{p} - 1)n. \tag{4.81}$$

Its inspection shows that in addition to $p = 0$ and $n = 0$, there is the third curve of zero energy $p = p_a(n)$, which we call the *activation trajectory*; see Fig. 4.4(a). The activation trajectory intersects the rate equation line $p = 0$ in the metastable fixed point $M = (0, \bar{n})$. It also intersects the extinction line $n = 0$ in the extinction fixed point $E = (\bar{p}, 0)$. The large fluctuation, leading to the population extinction, starts at the metastable state M and proceeds along the activation trajectory $p_a(n)$ until the extinction fixed point E. The rate of such events is $\sim e^{-S_{ex}}$, where the extinction action is given by (hereafter we absorb the factor $-i$ into the action; cf. Eq. (4.20))

$$S_{ex} = \int dt[p\dot{n} - H(p,n)] = \int_{\bar{n}}^{0} p_a(n)\, dn. \tag{4.82}$$

Here we took into account that $H = 0$ along the activation trajectory. The extinction time is thus proportional to the exponentiated area of the shaded triangle in Fig. 4.4(a) [68, 69].

If the two fixed points of the rate equation $n = \bar{n}$ and $n = 0$ are relatively close to each other, the problem may be substantially simplified. For our example (4.76), (4.81) this is the case when $0 < \sigma - \mu \ll \sigma$. One may then disregard the curvature of the activation trajectory between the fixed points M and E and substitute it by a straight line. This leads to the universal reaction Hamiltonian of the form

$$H(p,n) = p\left(\delta - \frac{n}{N} + p\right) n. \tag{4.83}$$

Its three zero-energy lines form the right triangle, Fig. 4.4(b). In terms of our example (4.76) we put $\mu + \sigma = 1$, which fixes units of time, and introduced notations $\delta = (\sigma - \mu)/(\sigma + \mu) \ll 1$ for the so-called bifurcation parameter and $N = (\sigma + \mu)/\lambda \gg 1$ for the effective system size. A large class of models in the vicinity of the bifurcation point may be described by this Hamiltonian. The activation trajectory is given by $p_a(n) = n/N - \delta$ and therefore the extinction action (4.82) is $S_{ex} = N\delta^2/2$.

Substituting $p \to -\partial_n$ and keeping the normal ordering in Eq. (4.83), one obtains the universal limit of the Master equation for $\mathcal{P}(n,t)$. Its only true stationary solution is the extinct state $\mathcal{P}(n) = \delta_{n,0}$. There is, however, a long-lived metastable solution, which we shall look for in the form $\mathcal{P}(n)\, e^{-t/\tau_{ex}}$. With the help of the Master equation with the universal Hamiltonian (4.83), one finds $\mathcal{P}/\tau_{ex} = \partial_n\big[(-p_a(n) - \partial_n)n\mathcal{P}\big] = \partial_n J$. Since the extinction time τ_{ex} is expected to be exponentially long, the probability current $J(n)$ is practically a constant away from the narrow peak of the metastable distribution around $\bar{n} = N\delta \gg 1$. This constant current is obviously zero for $n \gg \bar{n}$ and is finite, $J(0)$, in the direction of the absorbing boundary at $n = 0$. Integrating the expression for the current, one finds $n\mathcal{P}(n) = -e^{-S(n)} \int_0^n dl\, J(l)\, e^{S(l)}$, where $S(n) = \int^n p_a(n)dn$ and we demanded that $n\mathcal{P} \overset{n\to 0}{\to} 0$. Integrating the equality $\mathcal{P} = \tau_{ex}\partial_n J$ over the entire range of n, one finds

$$\tau_{\rm ex} = -\frac{1}{J(0)}\int_0^\infty {\rm d}n\,\mathcal{P}(n) = \int_0^\infty \frac{{\rm d}n}{n}\,{\rm e}^{-S(n)}\int_0^n {\rm d}l\,\frac{J(l)}{J(0)}\,{\rm e}^{S(l)}. \tag{4.84}$$

The dn integral is dominated by the minimum of the action $S(n)$, that is, by $n \approx N\delta = \bar{n}$. Under these conditions the dl integral is given by the boundary region $l \approx 0$, where $J(l) \approx J(0)$ with exponential precision. Evaluating the integrals in the stationary/boundary point approximation, one finds

$$\tau_{\rm ex} = \sqrt{\frac{2\pi}{S''(\bar{n})}}\,\frac{1}{\bar{n}}\,\frac{{\rm e}^{S_{\rm ex}}}{|S'(0)|} = \sqrt{\frac{2\pi}{p_{\rm a}'(\bar{n})}}\,\frac{{\rm e}^{S_{\rm ex}}}{\bar{n}|p_{\rm a}(0)|} = \sqrt{\frac{2\pi}{N}}\,\frac{1}{\delta^2}\,{\rm e}^{N\delta^2/2}. \tag{4.85}$$

The result is valid for $N^{-1/2} < \delta \ll 1$. Notice that scaling of the action and the pre-exponential factor with the bifurcation parameter δ is very different from the Kramers activation (4.53). This result, along with pre-exponential factors in more general situations, was found in [70, 71, 72].

4.12 Time-Dependent Problems

Imagine that the particle's potential or reaction rates are modulated in time. In Sections 3.4 and 3.5 we discussed how such a modulation affects the quantum tunneling. Here we consider its influence on the activation escape time, or the extinction time. In the language of optimal paths these rare events correspond to *instanton* trajectories, which bring the system from, for example, the metastable fixed point M to the extinction fixed point E; see Fig. 4.4. In a time-independent setting, such an instanton trajectory may be written $n = n_0(t - t_0)$ and $p = p_0(t - t_0)$, where t_0 is an arbitrary constant that specifies the time of the extinction event. The action does not depend on t_0 and it is therefore said to be a "zero mode." If the Hamiltonian is an explicit function of time, the independence of the action on t_0 is lifted. Indeed, there are more and less preferable instances of undertaking the fluctuation that leads to the extinction. The probabilities of these fluctuations differ exponentially and therefore are largely dominated by the "best chance" t_0, when the extinction is most likely to occur [73].

One can analytically access such an optimal t_0 and the corresponding extinction probability in some limiting cases. The first such case is a *weak* time-dependent modulation of the system's parameters. It leads to a time-dependent reaction (or Fokker–Planck) Hamiltonian:

$$H(p, n, t) = H_0(p, n) + \varepsilon H_1(p, n, t), \tag{4.86}$$

where ε is a small parameter. According to the Melnikov theorem of classical mechanics [75, 76], the perturbed Hamiltonian still allows for the optimal trajectory. This is the case if ε is small enough and another condition, explained in what follows, is satisfied. Such a deformed optimal trajectory may be written as

$$n(t, t_0) = n_0(t - t_0) + \varepsilon n_1(t, t_0), \qquad p(t, t_0) = p_0(t - t_0) + \varepsilon p_1(t, t_0). \tag{4.87}$$

The corresponding action to first order in ε is given by the integral of

$$(p_0 + \varepsilon p_1)(\dot{n}_0 + \varepsilon \dot{n}_1) - H_0 - \partial_n H_0\, \varepsilon n_1 - \partial_p H_0\, \varepsilon p_1 - \varepsilon H_1 = p_0 \dot{n}_0 - \varepsilon H_1,$$

where we employed the fact that $H_0(n_0, p_0) = 0$ along with the equations of motion $\dot{n}_0 = \partial_p H_0$ and $\dot{p}_0 = -\partial_n H_0$. We have also disregarded the full time derivative $p_0 \dot{n}_1 + \dot{p}_0 n_1$. The first term on the right-hand side is the unperturbed action S_{ex}. Therefore, to first order in ε the change of the action (4.20) is

$$S_1(t_0) = -\varepsilon \int \mathrm{d}t\, H_1\big(p_0(t - t_0), n_0(t - t_0), t\big). \tag{4.88}$$

To maximize the extinction (escape) probability $\sim \mathrm{e}^{-(S_{\text{ex}} + S_1(t_0))}$, one needs to find minima of $S_1(t_0)$ with respect to the center of the bare instanton t_0. That is, find a t_0 such that $\partial_{t_0} S_1(t_0) = 0$, while the second derivative is positive. This leads to the condition

$$\int \mathrm{d}t \left(\partial_p H_1\, \dot{p}_0 + \partial_n H_1\, \dot{n}_0\right) = \int \mathrm{d}t \left\{H_1, H_0\right\} = 0, \tag{4.89}$$

where $\{\,,\,\}$ denotes classical Poisson brackets [35] and we again employed equations of motion $\dot{n}_0 = \partial_p H_0$ and $\dot{p}_0 = -\partial_n H_0$. Existence of simple zeros of this function is the condition of the Melnikov theorem [75, 73, 77].

As an example, consider H_0 given by the universal extinction Hamiltonian (4.83). Its bare instanton trajectory may be easily obtained by putting $p = p_{\text{a}}(n) = n/N - \delta$ into the Hamilton equation of motion $\dot{n} = \partial_p H(p, n)$ and integrating this first-order differential equation. The result is

$$n_0(t - t_0) = \frac{N\delta}{1 + N\,\mathrm{e}^{\delta(t - t_0)}}, \qquad p_0(t - t_0) = -\frac{N\delta\,\mathrm{e}^{\delta(t - t_0)}}{1 + N\,\mathrm{e}^{\delta(t - t_0)}}. \tag{4.90}$$

For the time-dependent part we take a weak harmonic modulation of the bifurcation parameter $\delta(t) = \delta(1 + \varepsilon \cos \Omega t)$, leading to $H_1(p, n, t) = pn\delta \cos \Omega t$. The resulting correction (4.88) to the action is

$$S_1(t_0) = \varepsilon N^2 \delta^3 \int \mathrm{d}t\, \frac{\mathrm{e}^{\delta(t - t_0)} \cos \Omega t}{\left(1 + N\,\mathrm{e}^{\delta(t - t_0)}\right)^2} = \frac{\varepsilon\, \pi N \delta \Omega}{\sinh \pi \Omega/\delta} \cos\left[\Omega(t_0 + \delta^{-1} \ln N)\right].$$

Once every period there is the "best chance" t_0, rendering the last cosine to be -1. For such optimal trajectories one finds the negative correction to the extinction action

$$S_1 = -\varepsilon S_{\text{ex}} \frac{2\pi \Omega/\delta}{\sinh\left(\pi \Omega/\delta\right)}, \tag{4.91}$$

where, as before, $S_{ex} = N\delta^2/2$. The modulation thus leads to the *exponential* reduction of the extinction time (4.84) by the factor $e^{-|S_1|}$. The result is valid as long as $1 < |S_1| \ll S_{ex}$. In the limit $\Omega \ll \delta$ the correction is $-2\varepsilon S_{ex}$, which may be immediately found as the minimizing adiabatic form of the extinction action $S_{ex} = N\delta^2(t)/2$. At large frequency $\Omega > \delta$ the linear correction decays exponentially.[8] This should be compared with the very different frequency dependence for the underdamped, Eq. (3.32), and overdamped, Eq. (3.43), quantum tunneling.

Another example where the optimal path may be explicitly constructed is a sudden temporary change in the system's parameters [79]. We call it a "catastrophic" event. Consider, for example, the reaction scheme (4.76) and assume that during the time window $-t_c < t < t_c$ the branching rate σ suddenly drops to zero. There is a chance that, after it recovers back to its pre-catastrophic value at $t = t_c$, the population does not recover (provided it was not extinct at the time the catastrophe struck at $t = -t_c$). Our goal is to evaluate the probability that the population goes extinct during the catastrophe or in its immediate aftermath. The corresponding optimal trajectory starts at the metastable fixed point M sometime before the catastrophe arrives and ends up in the extinction fixed point E after it ends. Therefore the initial and final pieces of the optimal path follow the *zero energy* activation trajectory $p_a(n)$ of the pre-catastrophic Hamiltonian $H_0(p,n)$. During the time window $|t| < t_c$ the Hamiltonian acquires a different form $H_c(p,n)$, and the optimal path follows one of its *finite energy* trajectories, Fig. 4.5. The latter is selected in such a way that the time elapsed between its two intersections with $p_a(n)$ is exactly $2t_c$.

To be specific, let us model the pre-catastrophic $H_0(p,n)$ by Eq. (4.83), while during the catastrophe $H_c(p,n) = -\mu pn$. The latter corresponds to the pure annihilation reaction $A \xrightarrow{\mu} \emptyset$, where we took into account that $|p| < \delta \ll 1$. Its constant energy H trajectory $p_H(n) = -H/(\mu n)$ intersects the activation trajectory $p_a(n) = n/N - \delta$ in points $n_\pm = (1 \pm \epsilon)N\delta/2$, where $\epsilon = \sqrt{1 - 4H/\mu N\delta^2}$. Since $n(t) \sim e^{-t\mu}$ along $p_H(n)$, the time elapsed between points n_+ and n_- is found to be $e^{-2t_c\mu} = n_-/n_+ = (1-\epsilon)/(1+\epsilon)$. From here one finds the proper energy to be $H = \mu N\delta^2/4\cosh^2(t_c\mu)$. The corresponding extinction action is given by $S(t_c) = \int dt[p\dot{n} - H]$, where the first term is the area shaded in Fig. 4.5, while the second one is $-2t_c H$. Straightforward calculation yields

$$S(t_c) = S_{ex}\left[1 - \tanh(t_c\mu)\right] = N\delta^2/(1 + e^{2t_c\mu}). \qquad (4.92)$$

The extinction probability in the aftermath of the catastrophe is $\sim e^{-S(t_c)}$. It is of order one if $N\delta^2 e^{-2t_c\mu} \approx 1$. However, according to the rate equation $\dot{n} = -\mu n$, by the time such a catastrophe ends $n(2t_c) = N\delta\, e^{-2t_c\mu} \approx 1/\delta \gg 1$, and one could

8 There is, however, the second-order correction to the action $\sim -(\varepsilon\delta/\Omega)^2 S_{ex}$, which decays only as a power law of frequency [78]. The same type of correction is responsible for the Kapitsa pendulum effect. Therefore, at $\Omega \gtrsim (\delta/\pi)\ln(1/\varepsilon)$ the linear correction (4.91) may be disregarded.

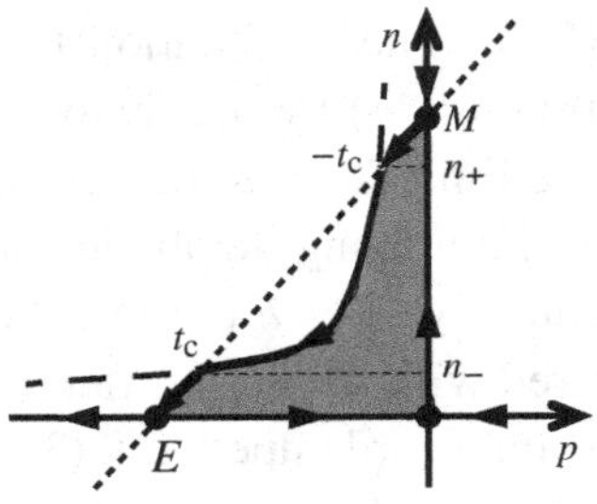

Figure 4.5 Optimal path to extinction facilitated by the catastrophe – bold line. The dotted line is the activation trajectory of the pre-catastrophic Hamiltonian $p_a(n)$. The dashed line is a finite energy trajectory of the Hamiltonian during the catastrophe, $p_H(n)$. The switches between the two occur at $t = \mp t_c$.

expect that the population is still in no immediate danger of extinction. The message is that the population may be much less catastrophe-tolerant than a naive expectation based on the rate equations.

4.13 Large Deviations in Multivariable Systems

We discuss now applications of the optimal path approach of Section 4.4 for systems with several degrees of freedom. The ideas touched upon here were introduced in seminal works of Graham and Tél [80], Dykman and Smelyanskiy [84], and Maier and Stein [86]. Consider, for example, an overdamped stochastic system with two degrees of freedom X_1 and X_2, that is, $\dot{X}_\alpha = A_\alpha(X) + \sqrt{T}\,\xi_\alpha(t)$, with the white noise (4.15). Its Fokker–Planck Hamiltonian (4.19) is given by

$$H(P_1, P_2, X_1, X_2) = P_1 A_1(X) + P_2 A_2(X) + TP_1^2 + TP_2^2. \tag{4.93}$$

The corresponding noiseless motion is described by the zero-energy invariant plane of this Hamiltonian $P_1 = P_2 = 0$, indeed $\dot{X}_\alpha = A_\alpha(X) = \partial_{P_\alpha} H\big|_{P=0}$. Let us assume for simplicity that such a noiseless dynamics admits a fixed point at $X_1 = X_2 = 0$, namely it is in the origin of the *four*-dimensional phase space. Being the fixed point means $A_\alpha(0) = 0$. Linearizing noiseless equations of motion in the vicinity of this point, one finds $\dot{X}_\alpha = A_{\alpha\beta} X_\beta$, where $A_{\alpha\beta} = \partial_\beta A_\alpha(X)\big|_{X=0}$. The matrix $A_{\alpha\beta}$ may have either two real eigenvalues $\lambda_{1,2}$, or two complex conjugated eigenvalues $\lambda_{1,2} = \kappa \pm i\omega$. We consider the latter case and assume that $\kappa < 0$, that is the fixed point $X = 0$ is a locally stable *focus*. This means that the noiseless relaxation tends to bring the system to the fixed point along a spiral trajectory. In a near vicinity of $X = 0$ the spirals may be characterized in terms of the two right eigenvectors of $A_{\alpha\beta}$, which both belong to the (X_1, X_2) plane of the four-dimensional phase space (analog of the $P = 0$ line in the one degree of freedom example of Fig. 4.1).

Thermal fluctuations take the system out of the fixed point and lead to a certain probability of finding the system at $X_0 \neq 0$. As explained in Section 4.4, such a probability is given by the exponentiated action of an *activation trajectory*, which goes from the origin to a point of the phase space with the coordinates X_0. In case of the single degree of freedom there is only one possible trajectory that departs from the fixed point: the curve of zero energy $P = -A(X)/T$, Fig. 4.1. The situation is much more interesting now. Linearizing the Hamiltonian equations of motion determined by Eq. (4.93) near the origin, one finds

$$\begin{pmatrix} \dot{X}_\alpha \\ \dot{P}_\alpha \end{pmatrix} = \begin{pmatrix} A_{\alpha\beta} & 2T\delta_{\alpha\beta} \\ 0 & -A^T_{\alpha\beta} \end{pmatrix} \begin{pmatrix} X_\beta \\ P_\beta \end{pmatrix}. \tag{4.94}$$

This 4×4 matrix possesses four eigenvalues. Two of them are already familiar eigenvalues of $A_{\alpha\beta}$ denoted as $\lambda_{1,2}$. The corresponding two right eigenvectors have zero components in the P directions. They thus give rise to the relaxation trajectories, which stay entirely within the invariant hyperplane $P = 0$. Two additional eigenvalues $\lambda_{3,4} = -\lambda_{1,2}$ have *positive* real parts and thus describe the activation trajectories that depart from the fixed point. The corresponding right eigenvectors have, in general, nonzero components in all four directions of the phase space. All trajectories that depart from the fixed point along an arbitrary linear superposition of these *two* eigenvectors form the *two-dimensional* Lagrangian manifold of activation trajectories. The energy is still a conserved quantity and therefore all the trajectories forming the Lagrangian manifold have the fixed energy, which is *zero*. (Indeed, all these trajectories depart from the origin, which has zero energy). The Lagrangian manifold is a generalization of the activation zero energy trajectory of Fig. 4.1. Since for a non-potential force A_α the Lagrangian manifold is *not* given by $P_\alpha(X) = -A_\alpha(X)/T$,[9] the activation trajectories are *not* time-reversed counterparts of the relaxation ones.

This observation may have dramatic consequences. As shown in [80, 84, 86], for non-potential forces the Lagrangian manifold develops cusp singularities some distance away from the origin, Fig. 4.6(a). The projection of the manifold onto the physical (X_1, X_2) plane exhibits two caustics, emanating from the cusp, Fig. 4.6(b). In between them the projection is three-valued. There are thus three distinct trajectories, whose projections pass through the same point $X_0 = (X_{01}, X_{02})$. Two of them, 1 and 2, reach the point X_0 before being reflected by one of the caustics, that is, they meet X_0 while being on the top and bottom sheets of the manifold.

[9] Indeed, according to the Hamilton–Jacobi equation [35] the Lagrangian manifold is characterized by $P_\alpha(X) = -\partial_\alpha S(X)$, where S is the action along a trajectory leading to X. Therefore, for $P_\alpha(X) = -A_\alpha(X)/T$ to be true one needs to have $\partial_2 A_1 = \partial_1 A_2$, which implies $A_\alpha = -\partial_\alpha V(X)$, that the force is potential (no vorticity). In the language of the Fokker–Planck equation the condition $P_\alpha(X) = -A_\alpha(X)/T$ means that in a stationary state all components of the current vector are zero. Again, this is not the case in presence of vorticity – a stationary state does support a finite divergenceless current. The author is indebted to M. Dykman for clarifying this point.

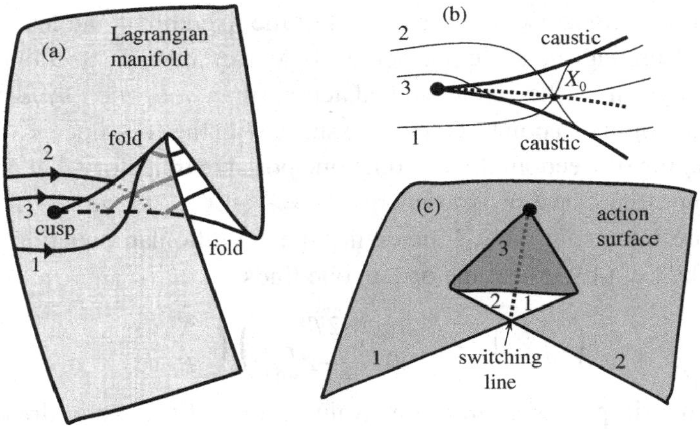

Figure 4.6 (a) Lagrangian manifold with a cusp singularity and three characteristic trajectories coming from the origin; (b) Projection of the Lagrangian manifold onto the (X_1, X_2) plane. The two folds project onto caustics, while projections of the three trajectories intersect at the point X_0; (c) The action $S(X_0)$ is a three-valued function in between the caustics. Two of its lower branches intersect along the switching line. After [84].

The projection of 3 passes through X_0 after being reflected once by a caustic, that is, the corresponding trajectory meets X_0 being on the middle sheet of the Lagrangian manifold. The action $S(X_0)$ calculated along the trajectories is therefore a three-valued function of the physical coordinate in between the two caustics, Fig. 4.6(c). The biggest action is due to trajectories of type 3, which underwent reflection before arriving at the point X_0. The two smaller action branches intersect each other along the *switching line*, which emanates from the projection of the cusp and stays in between the two caustics.

The stationary state probability $\mathcal{P}(X_0)$ of finding the system at point X_0 is given by the exponentiated action; see Section 4.4. If the action is multivalued, one can observe only its smallest branch, which gives rise to the largest probability. Therefore the large deviation function: $-\lim_{T\to 0}[T \ln \mathcal{P}(X_0)] = T\ \min\{S(X_0)\}$ is a *nonanalytic* function of the coordinates along the switching line. Notice that for the potential forces $A_\alpha = -\partial_\alpha V$, this function is simply the potential $V(X_0)$ and thus is perfectly smooth. As a result, the stationary state of non-equilibrium systems (e.g. with non-potential forces) is qualitatively different from that of equilibrium ones. Strictly speaking, this is only true in the limit of weak noise $T \to 0$, while for a finite noise the singularities are smeared. However, since the action is in the exponent, the change in the derivative across the switching line may be extremely sharp. There is a large mathematical literature devoted to this phenomenon [89, 90].

4.14 Problems

4.14.1 Noise-Induced Stabilization

Consider the following stochastic system:

$$\dot{X} = X^2 - Y^2 + \xi_X; \qquad \dot{Y} = -2Y + \xi_Y, \tag{4.95}$$

where the two uncorrelated Gaussian noises are $\langle \xi_X(t)\xi_X(t')\rangle = 2T_X\delta(t-t')$ and $\langle \xi_Y(t)\xi_Y(t')\rangle = 2T_Y\delta(t-t')$. Clearly the Y-degree of freedom equilibrates toward the stationary Boltzmann distribution $\mathcal{P}(Y) \propto \mathrm{e}^{-Y^2/T_Y}$. If in the X-equation one substitutes $Y^2 \to \langle Y^2\rangle = T_Y/2$, it creates an effective potential $V(X) = -X^3/3 + X\langle Y^2\rangle$, which has a metastable minimum at $X_{\min} = -\sqrt{\langle Y^2\rangle}$ and a local maximum at $X_{\max} = \sqrt{\langle Y^2\rangle}$. Consequently, X may be trapped near $X_{\min}$ for the exponentially long time $\tau_{\mathrm{es}} \propto \mathrm{e}^{(V(X_{\max})-V(X_{\min}))/T_X} = \mathrm{e}^{\sqrt{2}T_Y^{3/2}/3T_X}$. Notice that the trapping time *increases* exponentially with the increase of the Y-noise amplitude.

Show that for $T_Y \gg \sqrt{T_X}$ the actual trapping time, though exponentially long, is much shorter than suggested by the preceding estimate. This is due to rare periods of anomalously quiet Y-noise, during which X has a chance to escape. To estimate this effect: (i) calculate the probability that for the time t_0 the Y degree of freedom is confined to $|Y| < Y_0$ (to this end one needs to solve the Fokker–Planck equation with the absorbing boundary conditions at $Y = \pm Y_0$); (ii) calculate the probability that in time t_0 the X coordinate diffuses the distance $X_{\max} - X_{\min} \approx 2Y_0$ and thus escapes the trapping (the corresponding diffusion coefficient is T_X). One can now optimize the product of these two probabilities over t_0 and Y_0 (it actually only depends on the ratio t_0/Y_0^2) to find the escape rate. This brings the estimate $\log \tau_{\mathrm{es}} \propto \sqrt{T_Y/T_X}$. The numerical factor is calculated in [91].

4.14.2 Anomalous Counting Statistics

Following Section 4.9 and reference [62], derive long time counting statistics of $q_n = t_0^{-1}\int_0^{t_0} \mathrm{d}t\, x^n(t)$, where $x(t)$ are trajectories of a stochastic harmonic oscillator and $n \geq 3$ is an integer. Show that $-\log \mathcal{P}(q_n) \propto (q_n t_0)^{2/n}$ at large t_0.

4.14.3 Lotka–Volterra Model

Consider a population dynamics model of interacting fox (F) and rabbit (R) populations:

$$F \xrightarrow{\mu} \emptyset; \qquad R \xrightarrow{\mu} 2R; \qquad F + R \xrightarrow{\lambda} 2F, \tag{4.96}$$

where we put fox death and rabbit birth rates to be equal. (This assumption does not change any of the qualitative features of the model.) Derive the corresponding mean-field equations for $n_F(t)$ and $n_R(t)$, the corresponding reaction Hamiltonian (4.78), $H(p_F, p_R; n_F, n_R)$, and the Fokker–Planck equation for $\mathcal{P}(n_F, n_R, t)$. The latter may be obtained by expanding the normally ordered reaction Hamiltonian up to the second order in p_F and p_R and "quantizing" it with the substitution $\hat{p}_F = -\partial_{n_F}$ and $\hat{p}_R = -\partial_{n_R}$.

Show that with the change of variables $n_F, n_R \to \mathcal{F}, \mathcal{R}$ as $(N = \mu/\lambda)$

$$\mathcal{F} = \log(n_F/N); \qquad \mathcal{R} = \log(n_R/N) \tag{4.97}$$

the mean-field equations acquire the Hamiltonian form $\dot{\mathcal{F}} = \partial_{\mathcal{R}} G$; $\dot{\mathcal{R}} = -\partial_{\mathcal{F}} G$, where the "Hamiltonian" $G(\mathcal{F}, \mathcal{R}) = \mathrm{e}^{\mathcal{F}} - 1 - \mathcal{F} + \mathrm{e}^{\mathcal{R}} - 1 - \mathcal{R} \geq 0$, and the time is measured in units of $1/\mu$. This implies a presence of the mean-field integral of motion: the "energy," G, and therefore the motion along closed orbits of constant G in the $(\mathcal{F}, \mathcal{R})$ phase space. This way $G = 0$, namely $\mathcal{F} = \mathcal{R} = 0$, corresponds to stationary populations of size $n_F = n_R = N$. The $G > 0$ orbits describe periodic population cycles, while $G \to \infty$ signals either the total extinction, or the extinction of the fox with the unlimited growth of the rabbit population.

The Fokker–Planck description does *not* conserve G and leads to a slow growth of G, resulting in an inevitable extinction. To show this, transform the Fokker–Planck equation for $\mathcal{P}(\mathcal{F}, \mathcal{R}, t) = n_F n_R \mathcal{P}(n_F, n_R, t)$, where factor $n_F n_R$ is the Jacobean of (4.97), into the form of the continuity relation

$$\partial_t \mathcal{P} = -\vec{\nabla} \cdot \left(\vec{J}_{\mathrm{MF}} + \frac{\vec{J}_{\mathrm{D}}}{N} \right); \qquad \vec{\nabla} = (\partial_{\mathcal{F}}, \partial_{\mathcal{R}}); \qquad \vec{J}_{\mathrm{MF}} = (\mathcal{P}\partial_{\mathcal{R}} G, \, -\mathcal{P}\partial_{\mathcal{F}} G).$$

In the limit $N \to \infty$ where diffusive current component, $\vec{J}_{\mathrm{D}}$, may be disregarded, any function $\mathcal{P}(G(\mathcal{F}, \mathcal{R}))$ is a stationary solution of the Fokker–Planck equation. This corresponds to the mean-field notion of never-ending population cycles. Corrections to this picture are small in $1/N \ll 1$, yet they lead to a drift of G toward infinity and therefore to extinction in a finite time $\tau_{\mathrm{ex}} \sim N$. A quantitative theory may be derived by averaging over the fast cycles and deducing an effective radial Fokker–Planck equation for $\mathcal{P}(G, t)$, [92].

5
Driven-Dissipative Systems

This chapter considers driven oscillatory systems weakly coupled to a bath. To this end we derive the Lindblad equation for the density matrix as a limiting case of the effective Keldysh action. The formalism is then applied to a number of examples where the combined effect of the drive and the dissipation leads to qualitatively new phenomena.

5.1 Rotating-Wave Approximation

We again consider a quantum system in contact with a harmonic bath, very similar to the one presented in Section 3.2. There we have shown that already the simplest linear coupling to the environment generates an effective quadratic action, with nonlocal in time Keldysh component. The reason for this behavior is that for an equilibrium bath with temperature T the aforementioned quantum-quantum component of the bath-induced self-energy[1] is given by Eqs. (3.17), (3.18):

$$\left[\Sigma(\epsilon)\right]^{\mathrm{K}} = -\mathrm{i}J(\epsilon)\coth\frac{\epsilon}{2T}, \tag{5.1}$$

where $J(\epsilon)$ is a spectral density of the bath, defined after Eqs. (3.17). Since at small temperatures $\coth\epsilon/2T \to \operatorname{sign}\epsilon$ is a nonanalytic function of ϵ, this leads to a nonlocal in time $\left[\Sigma(t-t')\right]^{\mathrm{K}}$. For example, an ohmic bath with $J(\epsilon) \sim \epsilon$ leads to $\Sigma^{\mathrm{K}}(\epsilon) \sim |\epsilon|$. In the time domain this results in the nonlocal Caldeira–Leggett type action (3.20), [34], with $\Sigma^{\mathrm{K}} \sim (t-t')^{-2}$ kernel.

This predicament may be avoided, however, if the system filters out the low frequency part of the bath spectrum and self-selects some relatively narrow high-frequency band. To illustrate this point, we consider a weakly nonlinear bosonic oscillator, which may represent, for example, a particular mode of a resonant cavity.

[1] Here we use a more appropriate self-energy notation for the inverse bath correlator of Section 3.2: $\hat{\Sigma} = -\hat{\mathfrak{D}}^{-1}$.

We will also assume that it is subject to an external monochromatic drive, close to its resonance frequency. In the secondary quantized representation the corresponding Hamiltonian of the oscillator linearly coupled to the bath is (cf. Eqs. (3.14)) $H = H_{\text{p}} + H_{\text{bath}} + H_{\text{int}}$, with[2]

$$\hat{H}_{\text{p}} = \omega_0 \hat{b}^\dagger \hat{b} + \frac{g}{12}(\hat{b}^\dagger + \hat{b})^4 + 2\mathrm{i}\eta(\hat{b}^\dagger - \hat{b})\cos\omega_p t; \tag{5.2a}$$

$$\hat{H}_{\text{bath}} = \sum_s \omega_s \hat{a}_s^\dagger \hat{a}_s; \qquad \hat{H}_{\text{int}} = \sum_s \frac{g_s}{\sqrt{2\omega_s}}(\hat{a}_s^\dagger \hat{b} + \hat{b}^\dagger \hat{a}_s), \tag{5.2b}$$

where η is an amplitude and ω_p a frequency of the external drive. We now perform a gauge transformation to pass to the rotating frame $\hat{\hat{b}} = \hat{b}e^{\mathrm{i}\omega_p t}$ and $\hat{\hat{b}}^\dagger = \hat{b}^\dagger e^{-\mathrm{i}\omega_p t}$. In the limit $|\omega_0 - \omega_p| \ll \omega_p$, one may disregard all terms rotating with frequency $2\omega_p$ and its higher harmonics. The resulting Hamiltonian in the *rotating-wave* approximation takes the form:

$$\hat{H}_{\text{p}} = \Delta \hat{\hat{b}}^\dagger \hat{\hat{b}} + \frac{g}{2}\, \hat{\hat{b}}^\dagger \hat{\hat{b}}^\dagger \hat{\hat{b}}\hat{\hat{b}} + \mathrm{i}\eta(\hat{\hat{b}}^\dagger - \hat{\hat{b}}); \tag{5.3a}$$

$$\hat{H}_{\text{bath}} = \sum_s \omega_s \hat{a}_s^\dagger \hat{a}_s; \qquad \hat{H}_{\text{int}} = \sum_s \frac{g_s}{\sqrt{2\omega_s}} \left(\hat{a}_s^\dagger \hat{\hat{b}} e^{-\mathrm{i}\omega_p t} + \hat{\hat{b}}^\dagger \hat{a}_s e^{\mathrm{i}\omega_p t}\right), \tag{5.3b}$$

where $\Delta \equiv \omega_0 - \omega_p + g \ll \omega_p$ is the detuning frequency. Notice that the time-dependence enters now only in the interaction term between the system and the bath. One could, in principle, absorb it into a gauge transformation of the bath degrees of freedom. However this would obscure the fact that the original bath oscillators are assumed to be in thermal equilibrium. We thus prefer to keep their time-dependence explicitly.

The evolution operator along the closed time contour with the *normally ordered* Hamiltonian (5.3) may be written using the coherent state functional representation, using fields $\bar{\phi}(t), \phi(t)$ for the system and $\bar{\varphi}_s(t), \varphi_s(t)$ for the bath degrees of freedom. The time-dependent coupling terms acquire the following form: $\bar{\varphi}_s^+(t)\phi^+(t)e^{\mathrm{i}\omega_p t} - \bar{\varphi}_s^-(t)\phi^-(t)e^{\mathrm{i}\omega_p t} = [\bar{\varphi}_s^{\text{q}}(t)\phi^{\text{cl}}(t) + \bar{\varphi}_s^{\text{cl}}(t)\phi^{\text{q}}(t)]e^{\mathrm{i}\omega_p t}$ and similarly for the conjugated term. Evaluating Gaussian integrals over $\vec{\varphi}_s = (\varphi_s^{\text{cl}}, \varphi_s^{\text{q}})^{\text{T}}$, one finds the dissipative action for the oscillator (cf. Eq. (3.15)):

$$S_{\text{diss}} = -\iint_{-\infty}^{+\infty} \mathrm{d}t\, \mathrm{d}t'\, \vec{\bar{\phi}}^{\,\text{T}}(t)\, \hat{\Sigma}(t-t')\, \vec{\phi}(t'), \tag{5.4a}$$

$$\hat{\Sigma}(t-t') = \hat{\sigma}_1 \left[\sum_s \frac{g_s^2}{2\omega_s}\, e^{\mathrm{i}\omega_p t}\hat{G}_s(t-t')e^{-\mathrm{i}\omega_p t'}\right] \hat{\sigma}_1, \tag{5.4b}$$

[2] We use $g_s/\sqrt{2\omega_s}$ as a coupling constant to keep the same notations as in the first quantized treatment of Section 3.2. Indeed, connection between the field variables and first quantized coordinates involves the square root of the oscillator frequency, Eq. (3.1).

where $\hat{G}_s$ is the Green function of the *equilibrium* bath oscillators, given by Eqs. (2.40), (2.47). Using bath spectral density, $J(\omega) = \pi \sum_s (g_s^2/\omega_s)\delta(\omega - \omega_s)$, one obtains for the components of the dissipative self-energy:

$$\mathrm{Im}\big[\Sigma(\epsilon)\big]^{\mathrm{R(A)}} = \mp\frac{\mathrm{i}}{2}J(\epsilon + \omega_p); \qquad \big[\Sigma(\epsilon)\big]^{\mathrm{K}} = -\mathrm{i}J(\epsilon + \omega_p)\coth\frac{\epsilon + \omega_p}{2T}. \quad (5.5)$$

The characteristic frequencies of the weakly nonlinear oscillator (5.3a) are centered in the narrow band around $\epsilon \approx \Delta$. In this case one is justified in approximating $\epsilon+\omega_p \approx \Delta+\omega_p \approx \omega_0$ on the right-hand sides of these equations and approximating them by constants:

$$\mathrm{Im}\big[\Sigma(\epsilon)\big]^{\mathrm{R(A)}} \approx \mp\mathrm{i}\kappa; \qquad \big[\Sigma(\epsilon)\big]^{\mathrm{K}} \approx -2\mathrm{i}\kappa(2n_B + 1) = -2\mathrm{i}\kappa_1, \qquad (5.6)$$

where $\kappa = J(\omega_0)/2$ is the spectral density of the bath at the resonance frequency and

$$\frac{\kappa_1 + \kappa}{\kappa_1 - \kappa} = \frac{n_B + 1}{n_B} = e^{\omega_0/T}, \qquad (5.7)$$

independent of the details of the bath. Here $n_B = n_B(\omega_0) = (e^{\omega_0/T} - 1)^{-1}$ is the occupation number of the bath oscillators at the oscillator frequency. Notice that $\mathrm{Re}\,\Sigma^{\mathrm{R}} = \mathrm{Re}\,\Sigma^{\mathrm{A}}$ may also be approximated by a constant, which may then be absorbed in the redefinition of the detuning frequency, Δ, in Eq. (5.3a). The approximation is valid for weak dissipation $\kappa \ll \omega_0$ and weak nonlinearity $gT \ll \omega_0^2$.

Comparing Eqs. (5.5) and (5.1), one observes that the nonanalytic behavior of $\Sigma^{\mathrm{K}}(\epsilon)$ at $\epsilon = 0$ is avoided by shifting the nonanalytic point far away from the range of relevant frequencies. The ϵ-independent form of the dissipative self-energy (5.6) results in the *time-local* effective action, which includes the dissipative part (5.4) along with the unitary evolution with the time-independent Hamiltonian (5.3a):

$$S[\phi^{\mathrm{cl}},\phi^{\mathrm{q}}] = \int_{-\infty}^{+\infty} \mathrm{d}t\,(\bar{\phi}^{\mathrm{cl}},\bar{\phi}^{\mathrm{q}})\begin{pmatrix} 0 & \mathrm{i}\partial_t - \Delta - \frac{g}{2}|\phi^{\mathrm{cl}}|^2 - \mathrm{i}\kappa \\ \mathrm{i}\partial_t - \Delta - \frac{g}{2}|\phi^{\mathrm{cl}}|^2 + \mathrm{i}\kappa & 2\mathrm{i}\kappa_1 \end{pmatrix}\begin{pmatrix} \phi^{\mathrm{cl}} \\ \phi^{\mathrm{q}} \end{pmatrix}$$
$$-\int_{-\infty}^{+\infty} \mathrm{d}t\left[\frac{g}{2}\left(\bar{\phi}^{\mathrm{q}}\bar{\phi}^{\mathrm{q}}\phi^{\mathrm{q}}\phi^{\mathrm{cl}} + \bar{\phi}^{\mathrm{cl}}\bar{\phi}^{\mathrm{q}}\phi^{\mathrm{q}}\phi^{\mathrm{q}}\right) + \sqrt{2}\mathrm{i}\,\eta\,(\bar{\phi}^{\mathrm{q}} - \phi^{\mathrm{q}})\right]. \qquad (5.8)$$

The driving term, $\sim \eta$, may be eliminated by a shift of the classical field variable, $\phi^{\mathrm{cl}} \to \phi^{\mathrm{cl}} - \phi_\eta$, by a complex constant ϕ_η, which is found as a solution of the nonlinear equation $(\Delta - i\kappa)\phi_\eta + \frac{g}{2}|\phi_\eta|^2\phi_\eta = \sqrt{2}i\eta$. Due to nonlinearity such a shift generates Bogoliubov terms $\bar{\phi}^{\mathrm{q}}\bar{\phi}^{\mathrm{cl}}$ and $\phi^{\mathrm{q}}\phi^{\mathrm{cl}}$ as well as extra cubic nonlinear terms of the form $\bar{\phi}^{\mathrm{q}}\phi^{\mathrm{cl}}\phi^{\mathrm{cl}}$ and $\bar{\phi}^{\mathrm{q}}\phi^{\mathrm{q}}\phi^{\mathrm{q}}$.

The time-local driven-dissipative Keldysh action (5.8), in the rotating-wave approximation, is the central result of this section. It describes an inherently non-equilibrium setup and possesses the causality structure. To some extent it resembles the high temperature limit of the Caldeira–Leggett action (4.2), which leads to classical Langevin and Fokker–Planck equations, Sections 4.2 and 4.5. Nevertheless, there are important distinctions between the local actions (4.2) and (5.8). Most notably, while the classical action (4.2) is limited to terms linear and quadratic in the quantum field, the driven-dissipative action (5.8) contains higher order (cubic) terms in the quantum component. As a result, the latter contains the full quantum information, similar to Eq. (3.6), while the former is restricted to the classical limit, cf. Eqs. (3.7) and (3.8).

One may expect that a time-local action is equivalent to some evolution equation. In the case of the quantum mechanical action (3.6), this is the Von Neumann equation (1.1), $\partial_t\hat{\rho}(t) = -\mathrm{i}\big[\hat{H}(t),\hat{\rho}(t)\big]$, for the density matrix *operator*, $\hat{\rho}$, governed by some hermitian Hamiltonian, $\hat{H}$. In the case of the classical action (4.2), this is the Fokker–Planck equation (4.23) for the classical probability distribution *function*. One may ask if there is an evolution equation, corresponding to the local driven-dissipative action (5.8). Since the latter contains the quantum physics, the corresponding equation should be written in terms of a density matrix. However, unlike the Von Neumann equation, it can't describe a unitary evolution. Indeed, since the bath degrees of freedom were integrated out, the unitarity is expected to be lost. On the formal level this observation manifests itself in the fact that the driven-dissipative action (5.8) does *not* have the form of $S[\phi^+] - S[\phi^-]$ (while Eq. (3.6) does), because of κ and κ_1 terms. Nevertheless, the corresponding non-unitary evolution equation for the *reduced* density matrix indeed exists. It is known [93, 94, 95, 96] as the Lindblad equation.

5.2 Lindblad Equation

The full quantum system, which includes the nonlinear oscillator and the bath, is described by the density matrix, $\hat{\rho}$. The latter undergoes the unitary evolution according to the Von Neumann equation (1.1) with the Hamiltonian, $\hat{H}$, (5.2). We have already tacitly used this fact, when we wrote the coherent state representation for the corresponding evolution operator. Since one is not interested in the bath, one may define the *reduced* density matrix by tracing out degrees of freedom belonging to the bath (i.e. $\hat{a}_s$ oscillators):

$$\hat{\rho}_{\mathrm{p}} = \mathrm{Tr}_{\mathrm{bath}}\big\{\hat{\rho}\big\}, \tag{5.9}$$

where $\mathrm{Tr}_{\mathrm{bath}}\{\dots\}$ stays for the trace over the bath coordinates, as opposed to the trace over all degrees of freedom, $\mathrm{Tr}\{\dots\} = \mathrm{Tr}_{\mathrm{p}}\big\{\mathrm{Tr}_{\mathrm{bath}}\{\dots\}\big\}$. Since $\mathrm{Tr}\{\hat{\rho}\} = 1$, one concludes that

$$\mathrm{Tr}_{\mathrm{p}}\big\{\hat{\rho}_{\mathrm{p}}\big\} = 1. \tag{5.10}$$

One can also show that $\hat{\rho}_{\mathrm{p}}$ is Hermitian and positive semi-definite (since $\hat{\rho}$ is). It therefore possesses all the properties of the density matrix of an isolated p-subsystem.

This does not mean, of course, that $\hat{\rho}_{\mathrm{p}}$ evolves according to the Von Neumann equation. Moreover, since coupling to the bath may introduce long-time memory effects within the p-subsystem, $\hat{\rho}_{\mathrm{p}}(t)$ may not satisfy any local in time evolution equation. Assuming that such memory effects are not present – the assumption known as the *Markovian approximation* – one may show [94, 95, 96] that the reduced density matrix evolves according to the Lindblad equation:

$$\partial_t \hat{\rho}_{\mathrm{p}}(t) = -\mathrm{i}[\hat{H}'_{\mathrm{p}}, \hat{\rho}_{\mathrm{p}}(t)] + \sum_a \gamma_a \left(\hat{L}_a \hat{\rho}_{\mathrm{p}}(t) \hat{L}_a^\dagger - \frac{1}{2} \left\{ \hat{L}_a^\dagger \hat{L}_a, \hat{\rho}_{\mathrm{p}}(t) \right\} \right), \tag{5.11}$$

where $\gamma_a \geq 0$ is a set of real nonnegative constants and $\{.,.\}$ is the anti-commutator of two operators. In general non-Hermitian operators $\hat{L}_a$ are called *quantum jump operators*. The Hamiltonian, $\hat{H}'_{\mathrm{p}}$, is that of an isolated p-subsystem, possibly renormalized[3] by the coupling to the bath – thus the prime mark.

The Lindblad equation is trace-preserving, making Eq. (5.10) valid at all times. It also preserves Hermiticity and positive semi-definiteness of the reduced density matrix. In fact, it may be shown [94, 96] that the Lindblad form is the most general Markovian equation, which possesses these properties. Its evolution operator realizes quantum dynamical semigroup. That is, the product of any two evolution operators is again an evolution operator, but inverse operators may not be defined. Notice that the inverse operators would represent backward in time evolution. The fact that they may not exist reflects irreversible dynamics of the subsystem and appearance of the time arrow.

In the next section we will show, following Refs. [97, 98], that the Markovian approximation is another name for a *time-local* form of the effective Keldysh action, obtained after integration over the bath degrees of freedom. In particular, the driven-dissipative action (5.8) provides the evolution operator for the Lindblad equation with the Hamiltonian (5.3a) and two jump operators (hereafter we omit the tilde symbol for the rotating frame operators):

$$\hat{L}_1 = \hat{b}, \qquad \gamma_1 = \kappa_1 + \kappa; \qquad \hat{L}_2 = \hat{b}^\dagger, \qquad \gamma_2 = \kappa_1 - \kappa. \tag{5.12}$$

One may consider the right-hand side of the Lindblad equation (5.11) as a linear super-operator, $\hat{\mathcal{L}}$ – the Lindbladian, acting on a "vectorized" density matrix ρ. For an $\mathcal{N}$–dimensional Hilbert space, the reduced density matrix is $\mathcal{N} \times \mathcal{N}$ and therefore its vectorized form is a vector with $\mathcal{N}^2$ components. Thus, the super-operator,

[3] Such renormalization is provided by Re Σ^{R} = Re Σ^{A} in the previous section. It is absorbed into a redefinition of the detuning frequency, Δ, in the Hamiltonian (5.3a). A celebrated example is the Lamb shift of S and P atomic hydrogen levels due to their interactions with the quantized electromagnetic field.

$\hat{\mathcal{L}}$, is $\mathcal{N}^2 \times \mathcal{N}^2$, in general non-Hermitian operator. In these notations the Lindblad equation (5.11) acquires the following form:

$$\partial_t \rho = \hat{\mathcal{L}}[\rho], \tag{5.13}$$

where from now on we omit subscript p, indicating reduced density matrix, for brevity. The spectrum of the super-operator consists of $\mathcal{N}^2$, in general complex, eigenvalues. In a particular case of a pure Hamiltonian evolution (i.e. all $\gamma_a = 0$) all eigenvalues are imaginary, given by $\mathrm{i}(E_n - E_m)$, where E_n are real eigenvalues of the Hamiltonian and $n, m = 1, \ldots, \mathcal{N}$. Therefore there are at least $\mathcal{N}$ zero eigenvalues with $n = m$, corresponding to pure eigenstates of the Hamiltonian, $\rho_n = |n\rangle\langle n|$. In the presence of a bath ($\gamma_a \neq 0$) some of the eigenvalues of $\hat{\mathcal{L}}$ acquire real parts. For systems that admit a long-time stationary state (or possibly states) such real parts ought to be nonpositive. At least one eigenvalue must remain zero. This can be seen from the presence of at least one conserved quantity – the trace of the reduced density matrix, Eq. (5.10).

In some cases (see Section 5.7 for an example) there is more than one zero eigenvalue of the Lindbladian. The corresponding eigenvectors form a subspace, which is stationary in the process of evolution. Such subspace is called *dark space*, or *dissipation-free subspace*. The simplest realization is the case of a pure dissipative evolution, namely $\hat{H}'_p = 0$, and all quantum jump operators annihilating a set of states $\hat{L}_a|\mu\rangle = 0 = \langle\mu|\hat{L}_a^\dagger$, for *all* a, and $\mu, \nu = 1, \ldots, \mathcal{M}$. Then there is $\mathcal{M}^2$-dimensional dark space, spanned by the zero eigenstates of the form $|\mu\rangle\langle\nu|$.

5.3 From Keldysh to Lindblad

Following references [97, 98] we now develop coherent state representation for the Lindbladian evolution. The formal solution of equation (5.13) may be written as the time-ordered-exponent of the super-operator (cf. Eq. (1.2)),

$$\rho(t) = \mathbb{T}\, e^{\int_{t'}^{t} \mathrm{d}t\, \hat{\mathcal{L}}(t)} \rho(t'). \tag{5.14}$$

Here we allowed for an explicit time dependence of the Lindbladian super-operator, either through time-dependent Hamiltonian, or quantum jump operators, or both. Notice that, since this is the super-operator, which incorporates acting on the density matrix from the left and from the right, one does not need a closed time contour. Therefore the time here runs along the usual real axis – the field doubling and the corresponding two branches appearing automatically. One now discretizes the time axis onto N infinitesimal steps, labelled by $j = 1, \ldots N$, and writes one infinitesimal evolution step from t_j to t_{j+1} as

$$\rho_{j+1} = e^{\delta_t \hat{\mathcal{L}}_j} \rho_j \approx \big(1 + \delta_t \hat{\mathcal{L}}_j\big)\rho_j, \tag{5.15}$$

where $\delta_t = \int_{t_j}^{t_{j+1}} \mathrm{d}t$ is a duration of one time step (cf. Eq. (2.17)).

We restrict ourselves to the case of one bosonic degree of freedom, that is, assume that all operators $\hat{H}'_{\rm p}$ and $\hat{L}_a$ are expressed in terms of (normally ordered) operators $\hat{b}$ and $\hat{b}^\dagger$. To proceed one needs to introduce *two sets* [98] of coherent state over-complete basis, $|\phi_j^+\rangle$ and $|\phi_j^-\rangle$, such that $\hat{b}|\phi_j^\pm\rangle = \phi_j^\pm|\phi_j^\pm\rangle$, at each time discretization point. With their help and employing the resolution of unity (2.15) one represents the instantaneous density matrix operator as

$$\hat{\rho}_j = \iint d[\bar{\phi}_j^+, \phi_j^+]\, d[\bar{\phi}_j^-, \phi_j^-]\, e^{-|\phi_j^+|^2 - |\phi_j^-|^2}\, |\phi_j^+\rangle\langle\phi_j^+|\hat{\rho}_j|\phi_j^-\rangle\langle\phi_j^-|. \tag{5.16}$$

One can now use the explicit form of the Lindbladian (5.11) to write the action of the super-operator $\hat{\mathcal{L}}_j$ on the vectorized form of each operator component

$$\begin{aligned}\hat{\mathcal{L}}_j[|\phi_j^+\rangle\langle\phi_j^-|] &= -i\hat{H}'_{{\rm p},j}|\phi_j^+\rangle\langle\phi_j^-| + i|\phi_j^+\rangle\langle\phi_j^-|\hat{H}'_{{\rm p},j} \\ &\quad + \sum_a \gamma_a\Big[\hat{L}_{a,j}|\phi_j^+\rangle\langle\phi_j^-|\hat{L}^\dagger_{a,j} - \frac{1}{2}\hat{L}^\dagger_{a,j}\hat{L}_{a,j}|\phi_j^+\rangle\langle\phi_j^-| \\ &\quad - \frac{1}{2}|\phi_j^+\rangle\langle\phi_j^-|\hat{L}^\dagger_{a,j}\hat{L}_{a,j}\Big].\end{aligned} \tag{5.17}$$

We are now in a position to write down matrix elements of the density matrix in the subsequent instant of time, $\langle\phi_{j+1}^+|\hat{\rho}_{j+1}|\phi_{j+1}^-\rangle$. They include

$$\begin{aligned}\langle\phi_{j+1}^+|\hat{\mathcal{L}}_j[|\phi_j^+\rangle\langle\phi_j^-|]|\phi_{j+1}^-\rangle &= -\,i\langle\phi_{j+1}^+|\hat{H}'_{\rm p}|\phi_j^+\rangle\langle\phi_j^-|\phi_{j+1}^-\rangle + i\langle\phi_{j+1}^+|\phi_j^+\rangle\langle\phi_j^-|\hat{H}'_{\rm p}|\phi_{j+1}^-\rangle \\ &\quad + \sum_a \gamma_a\Big[\langle\phi_{j+1}^+|\hat{L}_a|\phi_j^+\rangle\langle\phi_j^-|\hat{L}_a^\dagger|\phi_{j+1}^-\rangle \\ &\quad - \frac{1}{2}\langle\phi_{j+1}^+|\hat{L}_a^\dagger\hat{L}_a|\phi_j^+\rangle\langle\phi_j^-|\phi_{j+1}^-\rangle \\ &\quad - \frac{1}{2}\langle\phi_{j+1}^+|\phi_j^+\rangle\langle\phi_j^-|\hat{L}_a^\dagger\hat{L}_a|\phi_{j+1}^-\rangle\Big],\end{aligned} \tag{5.18}$$

where possible time dependence of the operators is suppressed for brevity. To make progress one needs to assume that *all* operators, $\hat{H}'_{\rm p}$ as well as $\hat{L}_a$ and $\hat{L}_a^\dagger\hat{L}_a$, are *normally ordered*. For the reasons articulated in what follows, we shall not be concerned about any subtleties that may arise upon normal ordering. With this understanding all matrix elements in the last lengthy expression may be immediately evaluated by simple substitution of operators by complex variables: $\hat{b} \to \phi^\pm$ and $\hat{b}^\dagger \to \bar{\phi}^\pm$. Notice also that in the case of (+)-fields the time ordering is $\bar{\phi}_{j+1}^+\phi_j^+$, while in the case of (−)-fields it is opposite: $\bar{\phi}_j^-\phi_{j+1}^-$. This is precisely what happens along the closed time contour, where the (+)-fields are understood as being on the forward branch, while the (−)-fields are on the backward one. As a result one finds

$$\langle\phi_{j+1}^+|\hat{\mathcal{L}}_j[|\phi_j^+\rangle\langle\phi_j^-|]|\phi_{j+1}^-\rangle = \langle\phi_{j+1}^+|\phi_j^+\rangle\langle\phi_j^-|\phi_{j+1}^-\rangle\, \mathcal{L}_j(\bar{\phi}_{j+1}^+, \bar{\phi}_j^-, \phi_j^+, \phi_{j+1}^-), \tag{5.19}$$

where the *function* $\mathcal{L}(\ldots)$ is given by

$$\mathcal{L}(\bar{\phi}^+,\bar{\phi}^-,\phi^+,\phi^-) = -\mathrm{i}\left(H_{\mathrm{p}}'^{+} - H_{\mathrm{p}}'^{-}\right) + \sum_a \gamma_a\Big(L_a^+\bar{L}_a^- - \frac{1}{2}\bar{L}_a^+L_a^+ - \frac{1}{2}\bar{L}_a^-L_a^-\Big). \tag{5.20}$$

Here all the functions are obtained by the aforementioned substitution: for example, $H_{\mathrm{p}}'^{\pm} = \hat{H}_{\mathrm{p}}'(\bar{\phi}^{\pm},\phi^{\pm})$, $\bar{L}_a^{\pm} = \hat{L}^{\dagger}(\bar{\phi}^{\pm},\phi^{\pm})$, and so on.

We now approximate $(1+\delta_t\mathcal{L}) \approx e^{\delta_t\mathcal{L}}$ – valid to the first order in δ_t, and use coherent states overlaps, Eq. (2.6), to write for the matrix element of the reduced density matrix

$$\langle\phi_{j+1}^+|\hat{\rho}_{j+1}|\phi_{j+1}^-\rangle = \iint \mathrm{d}[\bar{\phi}_j^+,\phi_j^+,\bar{\phi}_j^-,\phi_j^-]\,\mathrm{e}^{\delta_t\left[\partial_t\bar{\phi}_j^+\phi_j^+ + \bar{\phi}_j^-\partial_t\phi_j^- + \mathcal{L}_j(\ldots)\right]}\,\langle\phi_j^+|\hat{\rho}_j|\phi_j^-\rangle, \tag{5.21}$$

where we use the notation $\partial_t\phi_j^{\pm} = (\phi_{j+1}^{\pm} - \phi_j^{\pm})/\delta_t$. One can now iterate this expression all the way from an initial time, t', to a final time, t, to obtain the evolution super-operator, Eq. (5.14), in the form of the functional integral. To streamline the notations we take trace of the reduced density matrix at the final time to obtain the "partition function," $Z = \mathrm{Tr}\hat{\rho}(t) = 1$. This entails integrating over $\phi_N^{\pm}$ with the Gaussian weight, Eq. (2.10). We also introduce obvious continuum notations for the discrete sum, accumulated in the exponent upon iteration of the time steps. This way, one finds (cf. Eqs. (2.26) and (2.30)),

$$Z = \int \mathbf{D}[\bar{\phi}(t),\phi(t)]\;\mathrm{e}^{\,\mathrm{i}S[\bar{\phi},\phi]}, \tag{5.22}$$

where

$$\begin{aligned} S[\bar{\phi},\phi] = \int \mathrm{d}t\Big[&\bar{\phi}^+\mathrm{i}\partial_t\phi^+ - \bar{\phi}^-\mathrm{i}\partial_t\phi^- - H_{\mathrm{p}}'^{+} + H_{\mathrm{p}}'^{-} \\ &-\mathrm{i}\sum_a \gamma_a\Big(L_a^+\bar{L}_a^- - \frac{1}{2}\bar{L}_a^+L_a^+ - \frac{1}{2}\bar{L}_a^-L_a^-\Big)\Big]. \end{aligned} \tag{5.23}$$

One recognizes the structure similar to the Keldysh path integral of Chapter 2. In fact, without dissipative terms, $\gamma_a = 0$, this is exactly the same object, with the structure $S = S[\phi^+] - S[\phi^-]$, derived in a slightly different way. The dissipative terms violate this structure. Most notably, the $\hat{L}_a\hat{\rho}\hat{L}_a^{\dagger}$ term in the Lindblad equation (5.11) results in the $L_a^+\bar{L}_a^-$ part of the action, which mixes variables on the two branches of the contours. This is not unexpected – integration of the bath degrees of freedom, illustrated in Section 5.1, indeed generates such terms. Therefore, the structure of the action (5.23) is not inconsistent with that of an *effective* Keldysh action, where some degrees of freedom were integrated out.

Indeed, one notices a characteristic feature of the causality structure, Section 2.7, namely the action (5.23) is nullified if $\phi^+(t) = \phi^-(t)$. This property may be traced back to the trace-preserving nature of the Lindblad equation. This motivates to introduce classical and quantum fields in the usual way, Eq. (2.39):

$$\phi^{\mathrm{cl}}(t) = \frac{1}{\sqrt{2}}(\phi^+(t) + \phi^-(t)), \qquad \phi^{\mathrm{q}}(t) = \frac{1}{\sqrt{2}}(\phi^+(t) - \phi^-(t)), \tag{5.24}$$

and to notice that $S[\phi^{\mathrm{cl}}, \phi^{\mathrm{q}} = 0] = 0$ (cf. Eq. (2.53)). To verify the full causality structure of the action (5.23) one needs to prove retarded/advanced character of the quantum–classical correlations. This can be done by carefully following the time discretization and operator normal ordering [97]. We will not follow this root here. Instead, we take a point of view that the Keldysh action, derived along the lines of Section 5.1, is the primary object, while the Lindblad equation is the secondary. In this approach the Keldysh action, for example, (5.8), is guaranteed to possess the causality structure. The role of Eq. (5.23), derived here, is to help identifying quantum jump operators, $\hat{L}_a$, in a Lindblad equation, corresponding to an approximately time-local *effective* Keldysh action, such as (5.8).

Following this line of reasoning, let us examine the action (5.23) with the simplest jump operators $\hat{L}_1 = \hat{b}$ and $\hat{L}_2 = \hat{b}^\dagger$. Upon substitution $\hat{b} \to \phi^\pm$ and $\hat{b}^\dagger \to \bar{\phi}^\pm$, the dissipative parts of the action (5.23) acquire the following form:

$$-\mathrm{i}\gamma_1\left(\phi^+\bar{\phi}^- - \frac{1}{2}\bar{\phi}^+\phi^+ - \frac{1}{2}\bar{\phi}^-\phi^-\right) = \frac{1}{2}(\bar{\phi}^{\mathrm{cl}}, \bar{\phi}^{\mathrm{q}})\begin{pmatrix} 0 & -\mathrm{i}\gamma_1 \\ \mathrm{i}\gamma_1 & 2\mathrm{i}\gamma_1 \end{pmatrix}\begin{pmatrix} \phi^{\mathrm{cl}} \\ \phi^{\mathrm{q}} \end{pmatrix}; \tag{5.25a}$$

$$-\mathrm{i}\gamma_2\left(\bar{\phi}^+\phi^- - \frac{1}{2}\bar{\phi}^+\phi^+ - \frac{1}{2}\bar{\phi}^-\phi^-\right) = \frac{1}{2}(\bar{\phi}^{\mathrm{cl}}, \bar{\phi}^{\mathrm{q}})\begin{pmatrix} 0 & \mathrm{i}\gamma_2 \\ -\mathrm{i}\gamma_2 & 2\mathrm{i}\gamma_2 \end{pmatrix}\begin{pmatrix} \phi^{\mathrm{cl}} \\ \phi^{\mathrm{q}} \end{pmatrix}. \tag{5.25b}$$

Comparing this to Eq. (5.8), one finds $\kappa = (\gamma_1 - \gamma_2)/2$ and $\kappa_1 = (\gamma_1 + \gamma_2)/2$. This establishes equivalence between the time-local action (5.8) and the Lindblad equation (5.11) with the quantum jump operators (5.12).

One can immediately generalize this construction by taking a more general form of the interactions between the oscillator and the bath. To this end we modify Eq. (5.2b) as

$$\hat{H}_{\mathrm{int}} = \sum_s \frac{g_s}{\sqrt{2\omega_s}}(\hat{a}_s^\dagger \hat{L} + \hat{L}^\dagger \hat{a}_s), \tag{5.26}$$

where $\hat{L} = \hat{b}^n$. Integrating over the bath variables and comparing the resulting action with Eq. (5.23), one again arrives at the Lindblad equation with the two competing jump operators: "down" jumps with $\hat{L}_1 = \hat{L}$ and "up" jumps with $\hat{L}_2 = \hat{L}^\dagger$. According to Eq. (5.7) the ratio of the corresponding "down" to "up" rates is

$$\frac{\gamma_1}{\gamma_2} = \frac{n_B + 1}{n_B}; \qquad n_B = \frac{1}{e^{n\omega_0/T} - 1}, \tag{5.27}$$

independent of the details of the coupling mechanism and the bath. This statement may be dubbed as *rotating-wave FDT*. Notice that n_B here is the equilibrium occupation number of the bath oscillators at integer multiples of the oscillator frequency. The rotating-wave FDT relies thus on the bath being in equilibrium at a temperature T. For $T \ll n\omega_0$ only "down" jumps remain.

5.4 Linear Systems

Here we focus on driven-dissipative harmonic systems. To this end we disregard nonlinearity by putting $g = 0$ in Eqs. (5.2) and (5.8). We then perform a constant shift of the classical field variables by a complex constant $\phi_\eta = \sqrt{2}\mathrm{i}\eta/(\Delta - i\kappa)$, as $\phi^{\mathrm{cl}} \to \phi^{\mathrm{cl}} - \phi_\eta$ and $\bar{\phi}^{\mathrm{cl}} \to \bar{\phi}^{\mathrm{cl}} - \bar{\phi}_\eta$ to eliminate the drive, $\sim \eta$, term from the action (5.8). The physical meaning of this shift is that the original driven oscillator follows the driving force by rotating in the phase space with the frequency ω_p, the amplitude $|\phi_\eta|$, and the phase shift $\arg(\phi_\eta)$. This is exactly what one expects from a steady-state *classical* damped harmonic oscillator, driven by a periodic external force. The residual Gaussian action (cf. Eq. (5.8)),

$$S[\phi^{\mathrm{cl}}, \phi^{\mathrm{q}}] = \int_{-\infty}^{+\infty} \mathrm{d}t\, (\bar{\phi}^{\mathrm{cl}}, \bar{\phi}^{\mathrm{q}}) \begin{pmatrix} 0 & \mathrm{i}\partial_t - \Delta - \mathrm{i}\kappa \\ \mathrm{i}\partial_t - \Delta + \mathrm{i}\kappa & 2\mathrm{i}\kappa(2n_B + 1) \end{pmatrix} \begin{pmatrix} \phi^{\mathrm{cl}} \\ \phi^{\mathrm{q}} \end{pmatrix}, \quad (5.28)$$

describes quantum and thermal fluctuations on top of the aforementioned classical rotation in the phase space. Here $\Delta = \omega_0 - \omega_p + \mathrm{Re}[\Sigma(\omega_0)]^{\mathrm{R}}$ is a detuning frequency, renormalized by a coupling to the bath.

According to the previous section this action is equivalent to the following Lindblad equation:

$$\partial_t \hat{\rho} = -\mathrm{i}\big[\hat{H}'_{\mathrm{p}}, \hat{\rho}\big] + 2\kappa(n_B + 1)\Big(\hat{b}\hat{\rho}\,\hat{b}^\dagger - \frac{1}{2}\big\{\hat{b}^\dagger\hat{b}, \hat{\rho}\big\}\Big) + 2\kappa n_B\Big(\hat{b}^\dagger\hat{\rho}\,\hat{b} - \frac{1}{2}\big\{\hat{b}\hat{b}^\dagger, \hat{\rho}\big\}\Big), \quad (5.29)$$

where $\hat{H}'_{\mathrm{p}} = \Delta\hat{b}^\dagger\hat{b}$. Here $n_B = n_B(\omega_0)$ is the occupation number at the resonance frequency. Notice that one could equally well use the harmonic Hamiltonian with the driving force $H'_{\mathrm{p}} = \Delta\hat{b}^\dagger\hat{b} + \mathrm{i}\eta(\hat{b}^\dagger - \hat{b})$ and perform operator canonical transformation $\hat{b} \to \hat{b} + \phi_\eta/\sqrt{2}$ and $\hat{b}^\dagger \to \hat{b}^\dagger + \bar{\phi}_\eta/\sqrt{2}$ in the corresponding Lindblad equation to arrive at Eq. (5.29).

One way to solve Eq. (5.29) is to employ the Glauber–Sudarshan coherent state representation [99] of the density matrix,

$$\hat{\rho}(t) = \int \mathrm{d}[\bar{\phi}, \phi]\, \mathrm{e}^{-|\phi|^2}\, \mathcal{P}_{\mathrm{GS}}(\bar{\phi}, \phi, t)\, |\phi\rangle\langle\phi|. \quad (5.30)$$

In the present example $\mathcal{P}_{\mathrm{GS}}(\bar{\phi}, \phi, t)$ happens to be positive everywhere and thus may be associated with the probability distribution. We now recall properties of the coherent states

$$\hat{b}|\phi\rangle = \phi|\phi\rangle; \qquad \hat{b}^\dagger|\phi\rangle = \partial_\phi|\phi\rangle; \qquad \langle\phi|\hat{b}^\dagger = \bar{\phi}\langle\phi|; \qquad \langle\phi|\hat{b} = \partial_{\bar{\phi}}\langle\phi|, \quad (5.31)$$

which follow from the fact that $|\phi\rangle = \mathrm{e}^{\phi\hat{b}^\dagger}|0\rangle$ and $\langle\phi| = \langle 0|\mathrm{e}^{\hat{b}\bar{\phi}}$. Substituting Eq. (5.30) into Eq. (5.29) and performing integrations by parts one finds

$$\partial_t \mathcal{P}_{\mathrm{GS}} = \Big[(\kappa + \mathrm{i}\Delta)\,\partial_\phi\phi + (\kappa - \mathrm{i}\Delta)\,\partial_{\bar{\phi}}\bar{\phi}\Big]\mathcal{P}_{\mathrm{GS}} + 2\kappa n_B\,\partial^2_{\bar{\phi}\phi}\mathcal{P}_{\mathrm{GS}}. \qquad (5.32)$$

This may be viewed as a Fokker–Planck equation for the probability distribution in the phase space. Notice that the dissipative terms, $\propto \kappa$, are of potential nature $\kappa[\partial_\phi V_{\bar{\phi}} + \partial_{\bar{\phi}} V_\phi]\mathcal{P}$, where the potential is $V(\phi,\bar{\phi}) = \bar{\phi}\phi$. On the other hand, the dynamical terms, $\sim \mathrm{i}\Delta$, generates rotation in the phase space. Anticipating that the *stationary* distribution depends only on the radius (i.e. action) in the phase space, but not on the angle, one finds Boltzmann-like distribution with n_B playing the role of temperature

$$\mathcal{P}_{\mathrm{GS}}(\bar{\phi},\phi) = \frac{1}{n_B}\,\exp\left\{-\frac{|\phi|^2}{n_B}\right\}, \qquad (5.33)$$

where we have enforced normalization $\mathrm{Tr}\,\hat{\rho} = \int \mathrm{d}[\bar{\phi},\phi]\,\mathcal{P}_{\mathrm{GS}}(\bar{\phi},\phi) = 1$. One thus finds for the stationary density matrix in the coherent state basis

$$\hat{\rho} = \frac{1}{n_B}\int \mathrm{d}[\bar{\phi},\phi]\,\exp\left\{-\mathrm{e}^{\omega_0/T}|\phi|^2\right\}\,|\phi\rangle\langle\phi|. \qquad (5.34)$$

This expression can be translated into the number state basis with the help of the identity $|\phi\rangle = \sum_{n=0}^{\infty}\phi^n|n\rangle/\sqrt{n!}$, Eq. (2.5). Performing straightforward integration over $\mathrm{d}[\bar{\phi},\phi]$ using Eq. (2.9), one finds (see problem 5.11.1 for an alternative derivation of this result)

$$\hat{\rho} = \sum_{n=0}^{\infty}\mathcal{P}_n\,|n\rangle\langle n|; \qquad \mathcal{P}_n = \frac{e^{-n\omega_0/T}}{n_B+1}. \qquad (5.35)$$

This is the equilibrium thermal density matrix for the oscillator, describing fluctuations around the driven classical path in the phase space. The linear drive of a harmonic oscillator does not result in a non-equilibrium distribution. In the next section we will see that this is not the case for a parametric drive. One can employ Eq. (5.35) to evaluate moments of the occupation number $\hat{n} = \hat{b}^\dagger\hat{b}$, for example, $\langle\hat{n}\rangle = \mathrm{Tr}\{\hat{\rho}\,\hat{n}\} = n_B$, while the higher moments are given by Eq. (2.64).

In the unshifted frame the Glauber–Sudarshan stationary function takes the form $\mathcal{P}_{\mathrm{GS}} \propto \exp\left\{|\phi - \phi_\eta|^2/n_B\right\}$. For a zero temperature bath it reduces to the two-dimensional delta-function, $\mathcal{P}_{\mathrm{GS}} = \delta(\phi - \phi_\eta)\delta(\bar{\phi} - \bar{\phi}_\eta)$. As a result, the stationary density matrix acquires a pure coherent state form:

$$\hat{\rho} = e^{-|\phi_\eta|^2}\,|\phi_\eta\rangle\langle\phi_\eta|. \qquad (5.36)$$

An alternative approach to deal with the local quadratic action (5.28) is to perform the Hubbard–Stratonovich transformation[4] of the $e^{-2\kappa(2n_B+1)\int dt|\phi^q|^2}$ term with a complex field $\xi(t)$. Subsequent integration over $\phi^{\rm q}$ leads to the Langevin equations for the classical field components:

$$\partial_t\phi^{\rm cl} = -(\kappa + {\rm i}\Delta)\phi^{\rm cl} - {\rm i}\xi(t); \qquad \partial_t\bar{\phi}^{\rm cl} = -(\kappa - {\rm i}\Delta)\bar{\phi}^{\rm cl} + {\rm i}\bar{\xi}(t). \tag{5.37}$$

The noise correlator is given by

$$\langle \xi(t)\bar{\xi}(t')\rangle = 2\kappa(2n_B + 1)\delta(t - t'). \tag{5.38}$$

The corresponding Fokker–Planck equation, Section 4.5, for the probability distribution function, $\mathcal{P}(\bar{\phi}^{\rm cl}, \phi^{\rm cl}, t)$, takes the form:[5]

$$\partial_t\mathcal{P} = \Big[(\kappa + {\rm i}\Delta)\,\partial_{\phi^{\rm cl}}\phi^{\rm cl} + (\kappa - {\rm i}\Delta)\,\partial_{\bar{\phi}^{\rm cl}}\bar{\phi}^{\rm cl}\Big]\mathcal{P} + 2\kappa(2n_B + 1)\,\partial^2_{\bar{\phi}^{\rm cl}\phi^{\rm cl}}\mathcal{P}. \tag{5.39}$$

Its normalized stationary solution is

$$\mathcal{P}(\bar{\phi}^{\rm cl}, \phi^{\rm cl}) = \frac{1}{2n_B + 1}\exp\left\{-\frac{|\phi^{\rm cl}|^2}{2n_B + 1}\right\}, \tag{5.40}$$

which implies, for example, $\langle\phi^{\rm cl}(t)\bar{\phi}^{\rm cl}(t)\rangle = 2n_B + 1$.

Comparing the Fokker–Planck equation (5.39) and its stationary solution (5.40) to the Glauber–Sudarshan, (5.32), (5.33), one observes that they coincide exactly[6] in the high-temperature limit, where $n_B \gg 1$. Yet they are extremely different in the opposite limit, $T \ll \omega_0$, where n_B is exponentially small. Does it mean that at least one of these approaches is wrong at low temperatures? Fortunately both are right. They simply address statistics of different observables. The Glauber–Sudarshan stationary solutions (5.33), (5.35) of the Lindblad equation provide statistics of the number operator $\hat{n} = \hat{b}^\dagger\hat{b}$, while the Langevin–Fokker–Planck approach deals with a different operator, $\hat{F} = \hat{b}^\dagger\hat{b} + \hat{b}\hat{b}^\dagger$. This situation is discussed in detail in

[4] To this end one writes Gaussian integral

$$e^{-2\kappa(2n_B+1)\int dt|\phi^{\rm q}|^2} = \int \mathbf{D}[\bar{\xi}(t), \xi(t)]\; e^{-\int dt\left[|\xi|^2/2\kappa(2n_B+1) + {\rm i}\bar{\xi}\phi^{\rm q} + {\rm i}\bar{\phi}^{\rm q}\xi\right]}.$$

The action becomes linear in $\phi^{\rm q}, \bar{\phi}^{\rm q}$ and the corresponding integrations enforce delta-functions of classical equations for $\bar{\phi}^{\rm cl}, \phi^{\rm cl}$.

[5] A quick and instructive way to write it down is to follow the recipe formulated earlier in Eq. (4.23). To this end one introduces notations $p_\phi = \bar{\phi}^{\rm q}$ and $p_{\bar{\phi}} = -\phi^{\rm q}$ and rewrites the action (5.28) in the Hamiltonian way: $S = {\rm i}\int dt\big[p_\phi\partial_t\phi^{\rm cl} + p_{\bar{\phi}}\partial_t\bar{\phi}^{\rm cl} - H(p_\phi, p_{\bar{\phi}}, \phi^{\rm cl}, \bar{\phi}^{\rm cl})\big]$. One can then "quantize" this action by first *ordering* the Hamiltonian, i.e. moving quantum components to the left of classical (if there are multiple quantum components, they ought to be symmetrized) and turning momenta into operators as $p_\phi \to \hat{p}_\phi = -\partial_{\phi^{\rm cl}}$ and $p_{\bar{\phi}} \to \hat{p}_{\bar{\phi}} = -\partial_{\bar{\phi}^{\rm cl}}$. The Fokker–Planck equation is the corresponding imaginary time "Schrödinger" equation: $\partial_t\mathcal{P} = \hat{H}(\hat{p}_\phi, \hat{p}_{\bar{\phi}}, \phi^{\rm cl}, \bar{\phi}^{\rm cl})\mathcal{P}$.

[6] The factor of two differences in front of the second derivative term may be absorbed by a trivial rescaling of the $\phi^{\rm cl}$ field, which amounts to adopting the definition $\phi^{\rm cl} = (\phi^+ + \phi^-)/2$, instead of Eq. (5.24).

Section 2.9, where it is shown that the two correspond to discrete and continuous regularizations of the generating function. They generate distinct, but *equivalent* sets of statistical moments[7] (cf. Eqs. (2.64) and (2.70)).

The stationary solutions (5.33) or (5.40) are unique. They are given by a mixed thermal state (unless $T = 0$, i.e. $n_B = 0$ and thus $\gamma_2 = 0$, leading to a pure stationary state (5.36)). To access truly non-equilibrium dynamics one needs to follow the time-dependent evolution of either Eq. (5.32), or (5.39). Alternatively (and frequently more informative) one may look for multiple-time correlation functions in the stationary state. To execute this program in the Lindblad formalism, one has to go from the Schrödinger picture, Eq. (5.29), to the Heisenberg one, which provides for the Lindbladian evolution of *operators*. This is achieved through the *adjoint* Lindbladian super-operator, $\mathcal{L}^\dagger$, and the corresponding adjoint evolution super-operator [96]. Due to the irreversibility of the Lindbladian evolution, this procedure is less straightforward than the more familiar one, leading from the Von Neumann equation to the Heisenberg operator evolution. We will not take this route here.

Instead, we notice that in the equivalent formulation in terms of the Keldysh action, the multi-time correlation functions are readily available. In particular, for the harmonic system, described by the action (5.28), the steady-state two-point correlation functions are given by

$$\langle \phi^{\text{cl}}(t)\,\bar\phi^{\text{q}}(t')\rangle = \theta(t-t')\langle[\hat{b}(t),\hat{b}^\dagger(t')]\rangle = \mathrm{i}\tilde{G}^{\text{R}}(t,t') = \theta(t-t')\mathrm{e}^{(-\kappa-\mathrm{i}\Delta)(t-t')}, \tag{5.41a}$$

$$\langle \phi^{\text{q}}(t)\,\bar\phi^{\text{cl}}(t')\rangle = -\theta(t'-t)\langle[\hat{b}(t),\hat{b}^\dagger(t')]\rangle = \mathrm{i}\tilde{G}^{\text{A}}(t,t') = -\theta(t'-t)\mathrm{e}^{(\kappa-\mathrm{i}\Delta)(t-t')}, \tag{5.41b}$$

$$\langle \phi^{\text{cl}}(t)\,\bar\phi^{\text{cl}}(t')\rangle = \langle\{\hat{b}(t),\hat{b}^\dagger(t')\}\rangle = \mathrm{i}\tilde{G}^{\text{K}}(t,t') = F\,\mathrm{e}^{-\mathrm{i}\Delta(t-t')-\kappa|t-t'|}, \tag{5.41c}$$

where $F = (2n_B+1) = \coth\omega_0/2T$ and, as always, $\langle\phi^{\text{q}}(t)\,\bar\phi^{\text{q}}(t')\rangle = 0$. The time-dependent operators, $\hat{b}(t), \hat{b}^\dagger(t)$, here are defined in the standard Heisenberg picture, which assumes unitary evolution with the *full* Hamiltonian, namely the one that explicitly includes the bath degrees of freedom. The averaging, $\langle\ldots\rangle$, is over the stationary density matrix (5.35).

[7] In fact (5.33) and (5.40) representations of the thermal state are two points in the continuous family of the *quasi-probability distributions*,

$$W(\bar\phi,\phi,p) = \int \mathrm{d}[\bar\xi,\xi]\,\mathrm{Tr}\left\{\hat\rho\,\mathrm{e}^{(\hat{b}^\dagger-\bar\phi)\xi-\bar\xi(\hat{b}-\phi)}\right\}\,\mathrm{e}^{p|\xi|^2/2},$$

labeled by $p \le 1$, [99]. Glauber–Sudarshan (5.33) corresponds to $p = 1$, known to be a generator of moments of normally ordered operators. The Fokker–Planck (5.40) – $p = 0$, which generates moments of the symmetrized products. The anti-normal ordered moments are generated by $p = -1$ Husimi function; see problem 5.11.2. Generic quasi-probability representation of the thermal state is a Gaussian with the dispersion $2n_B + 1 - p$. I am indebted to Foster Thompson and Sebastian Diehl for discussing this point.

The tilde on top of the Green functions is a reminder that those are correlation functions of the rotating wave operators. The original Heisenberg operators $\hat{b}(t)$ and $\hat{b}^\dagger(t)$ take the following form:

$$\hat{b}(t) = e^{-i\omega_p t}\,\phi_0/\sqrt{2} + \delta\hat{b}(t); \qquad \hat{b}^\dagger(t) = e^{i\omega_p t}\,\bar{\phi}_0/\sqrt{2} + \delta\hat{b}^\dagger(t), \tag{5.42}$$

where the Green functions of $\delta\hat{b}$ operators in the stationary state are

$$iG^R(t-t') = \theta(t-t')\langle[\delta\hat{b}(t), \delta\hat{b}^\dagger(t')]\rangle = \theta(t-t')\,e^{(-\kappa - i\omega_0)(t-t')}; \tag{5.43a}$$

$$iG^K(t-t') = \langle\{\delta\hat{b}(t), \delta\hat{b}^\dagger(t')\}\rangle = F\,e^{-i\omega_0(t-t') - \kappa|t-t'|}. \tag{5.43b}$$

Their Fourier transforms are

$$G^{R(A)}(\epsilon) = \frac{1}{\epsilon - \omega_0 \pm i\kappa}, \qquad G^K(\epsilon) = F\left(G^R - G^A\right) = \frac{-2i\kappa F}{(\epsilon-\omega_0)^2 + \kappa^2}. \tag{5.44}$$

The multi-point functions are given by a suitable application of the Wick theorem. Unlike their unitary counterparts, Eq. (2.47), the correlation functions decay exponentially in time with the inverse decay time κ. This behavior reflects the real parts of the complex eigenvalues of the Lindbladian super-operator.

One may also discuss the evolution of the distribution function $F(\epsilon, t)$ (which is the Wigner transform of the two-time-point function $F(t, t')$, defined through $G^K = G^R \circ F - F \circ G^A$) of a driven dissipative oscillator, which is away from its steady state. To this end we write the kinetic equation in the form of Eqs. (6.45), (6.47):

$$\partial_t F(t) = i\Sigma^K + F(t)\Big(\text{Im}\,\Sigma^R - \text{Im}\,\Sigma^A\Big) = 2\kappa\big(2n_B + 1 - F(t)\big), \tag{5.45}$$

where we used the rotating-wave approximation, Eq. (5.6), for the components of the bath-induced self-energy, $\Sigma^K = -2i\kappa(2n_B+1)$ and $\text{Im}\,\Sigma^{R(A)} = \mp\kappa$. According to the kinetic equation (5.45), any initial distribution function, $F(\epsilon, t')$, exponentially approaches the steady-state value $2n_B+1$ with the rate 2κ. Notice that this rate is *twice larger* than the rate of correlations decay in the steady state, Eqs. (5.41). An alternative way to derive kinetic equation (5.45) is discussed in problem 5.11.3.

5.5 Quantum Heating

The previous section dealt with a stable linear system. It exhibits a unique stationary state, localized in the phase space. Any initial state exponentially approaches this stationary state. This is not the only scenario, however. Linear systems may be driven out of stability toward an exponentially runaway behavior. Eventually the instability is curtailed by nonlinearities, which will be considered in the subsequent sections. Here we demonstrate that, upon approaching such classical instability, there are strong quantum fluctuations, known as *quantum heating*.

To be specific we consider a *parametrically* driven harmonic oscillator with a modulated resonance frequency. The corresponding classical Hamiltonian is $H = P^2/2 + (\omega_0^2 + 8\lambda\omega_0 \cos 2\omega_p t)X^2/2$. In the secondary quantized representation it takes the form (up to inessential terms)

$$\hat{H}_{2p} = \omega_0 \hat{b}^\dagger \hat{b} + 2\lambda(\hat{b}^\dagger \hat{b}^\dagger + \hat{b}\hat{b}) \cos 2\omega_p t. \tag{5.46}$$

We will refer to it as the *two-photon* drive, as opposed to the one-photon drive of Eq. (5.2a). This Hamiltonian is supplemented with the coupling to the bath, exactly as in Eq. (5.2b).

Closely following Section 5.1, we now (i) adopt rotating-wave approximation; (ii) pass to the coherent state representation, and (iii) integrate out the bath degrees of freedom. This leads to the time-local action (5.28), supplemented with the *quadratic* drive term

$$S = \int dt\, (\bar{\phi}^{\mathrm{cl}}, \bar{\phi}^{\mathrm{q}}) \begin{pmatrix} 0 & i\partial_t - \Delta - i\kappa \\ i\partial_t - \Delta + i\kappa & 2i\kappa(2n_B + 1) \end{pmatrix} \begin{pmatrix} \phi^{\mathrm{cl}} \\ \phi^{\mathrm{q}} \end{pmatrix} - 2\lambda\, (\bar{\phi}^{\mathrm{q}}\bar{\phi}^{\mathrm{cl}} + \phi^{\mathrm{q}}\phi^{\mathrm{cl}}). \tag{5.47}$$

The corresponding Lindblad equation is of the form of Eq. (5.29) with the quadratic rotating-wave two-photon Hamiltonian:

$$\hat{H}'_{2p} = \Delta \hat{b}^\dagger \hat{b} + \lambda(\hat{b}^\dagger \hat{b}^\dagger + \hat{b}\hat{b}), \tag{5.48}$$

with $\Delta = \omega_0 - \omega_p$.

What is the physics behind the local in time dissipative action (5.47), or equivalently the corresponding Lindblad equation? The Hamiltonian (5.48) may be diagonalized by the canonical Bogoliubov transformation (problem 2.10.2),

$$\hat{b} = \cosh\alpha\, \hat{\mathfrak{b}} + \sinh\alpha\, \hat{\mathfrak{b}}^\dagger; \qquad \hat{b}^\dagger = \cosh\alpha\, \hat{\mathfrak{b}}^\dagger + \sinh\alpha\, \hat{\mathfrak{b}}, \tag{5.49}$$

where $[\hat{\mathfrak{b}}, \hat{\mathfrak{b}}^\dagger] = 1$ and $\tanh 2\alpha = -2\lambda/\Delta$. It brings the Hamiltonian to the form (up to a constant)

$$\hat{H}'_{2p} = \Omega\, \hat{\mathfrak{b}}^\dagger \hat{\mathfrak{b}}; \qquad \Omega^2 = \Delta^2 - 4\lambda^2. \tag{5.50}$$

Its ground state, $|\mathfrak{o}\rangle$, is found from the condition $\hat{\mathfrak{b}}|\mathfrak{o}\rangle = 0$ and is given by the coherent state of pairs $|\mathfrak{o}\rangle \propto \exp\{-\tanh\alpha\, \hat{b}^\dagger \hat{b}^\dagger/2\}|0\rangle$. Notice that it is a superposition of only even-particle number states. All the other eigenstates $|\mathfrak{n}\rangle \propto (\hat{\mathfrak{b}}^\dagger)^n|\mathfrak{o}\rangle$ also have a definite parity, which alternates between even and odd.

One notices, however, that this procedure makes sense only for not-too-strong drive amplitude, $2|\lambda| < \Delta$, when $\Omega^2 > 0$. We will show in what follows that for a stronger drive the harmonic system is classically unstable and thus a nonlinearity must be added to stabilize it. Here we focus on the effect of a weakly coupled bath *below* the instability threshold. To this end we write the dissipative part of the Lindblad operator (5.29) in the Bogoliubov rotated representation:

$$
\begin{aligned}
&2\kappa(n_B+1)\Big(\hat{b}\hat{\rho}\,\hat{b}^\dagger-\frac{1}{2}\{\hat{b}^\dagger\hat{b},\hat{\rho}\}\Big)+2\kappa n_B\Big(\hat{b}^\dagger\hat{\rho}\,\hat{b}-\frac{1}{2}\{\hat{b}\hat{b}^\dagger,\hat{\rho}\}\Big) \qquad (5.51)\\
&=2\kappa\Big[(n_B+1)\cosh^2\alpha+n_B\sinh^2\alpha\Big]\Big(\hat{\mathfrak{b}}\hat{\rho}\,\hat{\mathfrak{b}}^\dagger-\frac{1}{2}\{\hat{\mathfrak{b}}^\dagger\hat{\mathfrak{b}},\hat{\rho}\}\Big)\\
&+2\kappa\Big[(n_B+1)\sinh^2\alpha+n_B\cosh^2\alpha\Big]\Big(\hat{\mathfrak{b}}^\dagger\hat{\rho}\,\hat{\mathfrak{b}}-\frac{1}{2}\{\hat{\mathfrak{b}}\hat{\mathfrak{b}}^\dagger,\hat{\rho}\}\Big)+\dots,
\end{aligned}
$$

where ... stands for terms with two $\hat{\mathfrak{b}}$, or two $\hat{\mathfrak{b}}^\dagger$ operators. These latter terms are fast rotated as $e^{\pm i2\Omega t}$ and thus average out to zero. Comparing this expression with the Lindbladian (5.29), one can define an *effective* occupation number for the $\Omega\,\hat{\mathfrak{b}}^\dagger\hat{\mathfrak{b}}$ oscillator:

$$
n_B^{\text{eff}}+1=(n_B+1)\cosh^2\alpha+n_B\sinh^2\alpha;\qquad n_B^{\text{eff}}=(n_B+1)\sinh^2\alpha+n_B\cosh^2\alpha. \tag{5.52}
$$

Therefore, one can describe the Bogoliubov oscillator by its *effective* temperature

$$
e^{\hbar\Omega/T^{\text{eff}}}=\frac{n_B^{\text{eff}}+1}{n_B^{\text{eff}}}=\frac{(2n_B+1)\cosh 2\alpha+1}{(2n_B+1)\cosh 2\alpha-1}, \tag{5.53}
$$

where $\cosh 2\alpha=\Delta/\Omega$. Close to the instability, where $\Omega\ll\Delta$, one finds

$$
T^{\text{eff}}\approx\frac{\hbar\Delta}{2}(2n_B+1). \tag{5.54}
$$

This shows that even at strictly zero bath temperature, $T=0$ and thus $n_B=0$, there is a finite effective temperature for the $\Omega\,\hat{\mathfrak{b}}^\dagger\hat{\mathfrak{b}}$ Bogoliubov oscillator. It therefore never resides in its ground state $|\mathfrak{o}\rangle$, but reaches a quasi-equilibrium state with the relative occupation of the excited states $|\mathfrak{n}\rangle$ given by $e^{-\hbar n\Omega/T^{\text{eff}}}$. The effective temperature, being proportional to $\hbar$, is of the pure quantum nature. This remarkable observation was first made by M. Dykman and colleagues [100, 101, 102] and called *quantum heating*. The physics of this phenomenon [100, 101, 102] is that the quasi-energy states of the periodically driven system are superpositions of the Fock states of the non-driven one. The $T=0$ dissipation induces only downward transitions among the latter. However, in terms of the former there are both downward and upward transitions. Therefore, the non-equilibrium driven system exhibits an effective temperature, $T^{\text{eff}}>0$, even for $T=0$. Notice that, although the coupling to the bath plays a crucial role, the coupling constant, κ, does not affect T^{eff} (for $\kappa\ll\Delta$). However, it determines the time scale of equilibration toward the effective thermal state.

Before addressing nonlinearities, let us briefly discuss classical stability of the parametrically driven oscillator. To this end we look at the semiclassical equations of motion for the Keldysh action (5.47). They are obtained from $\delta S/\delta\bar{\phi}^q|_{\phi^q=0}=0$ and its complex conjugated partner:

$$i\partial_t \begin{pmatrix} \phi^{\rm cl} \\ \bar{\phi}^{\rm cl} \end{pmatrix} = \begin{pmatrix} \Delta - i\kappa & 2\lambda \\ -2\lambda & -\Delta - i\kappa \end{pmatrix} \begin{pmatrix} \phi^{\rm cl} \\ \bar{\phi}^{\rm cl} \end{pmatrix}. \tag{5.55}$$

Diagonalizing the 2×2 matrix of the right-hand side, one finds that as long as

$$4\lambda^2 < \Delta^2 + \kappa^2 \tag{5.56}$$

is satisfied, $\phi^{\rm cl} = 0$ is the stable point of these equations. In the opposite limit, $\phi^{\rm cl} = 0$ is exponentially unstable and $\phi^{\rm cl}(t)$ runs to infinity. We have found that, although a finite dissipation strength shifts the stability boundary, there is always the instability, if the driving force is strong enough.

Treatment of linear systems may be extended to multiple degrees of freedom, that is, $\hat{b} \to \hat{b}_a$, where $a = 1, \ldots, M$. In this case the parameters become $M \times M$ matrices: Hermitian $\hat{\Delta}_{ab}$, positive-definite $\hat{\kappa}_{ab}$, and complex symmetric $\hat{\lambda}_{ab}$. The right-hand side of Eq. (5.55) turns to a $2M \times 2M$ block-matrix, and the stability criteria is given by zero of its determinant.

5.6 Quantum Activation

The instability of the parametrically driven harmonic oscillator, discussed in Section 5.5, is stabilized by either oscillator's nonlinearity (e.g. quartic), or nonlinear dissipation. Both of these scenarios contain some important lessons to be learned. Here we focus on the former one, while the latter is presented in Sections 5.7 and 5.9. The rotated wave Hamiltonian of the nonlinear parametrically driven oscillator, linearly coupled to the bath, is given by (cf. Eqs. (5.3), (5.48))

$$\hat{H}_{\rm 2p} = \Delta \hat{b}^\dagger \hat{b} + \frac{g}{2} \hat{b}^\dagger \hat{b}^\dagger \hat{b}\hat{b} + \lambda(\hat{b}^\dagger \hat{b}^\dagger + \hat{b}\hat{b}); \tag{5.57a}$$

$$\hat{H}_{\rm bath} = \sum_s \omega_s \hat{a}_s^\dagger \hat{a}_s; \qquad \hat{H}_{\rm int} = \sum_s \frac{g_s}{\sqrt{2\omega_s}} \left(\hat{a}_s^\dagger \hat{b} e^{-i\omega_p t} + \hat{b}^\dagger \hat{a}_s e^{i\omega_p t} \right). \tag{5.57b}$$

Upon integrating out the bath, the corresponding local action is given by Eq. (5.47) with the nonlinear terms identical to those in Eq. (5.8). It is equivalent to the Lindblad equation (5.29) with the Hamiltonian $\hat{H}_{\rm 2p}$, Eq. (5.57a).

To analyze this setup we start from the semiclassical equations of motion: $\delta S/\delta \bar{\phi}^{\rm q}|_{\phi^{\rm q}=0} = 0$ and $\delta S/\delta \phi^{\rm q}|_{\phi^{\rm q}=0} = 0$, where the action is given by Eq. (5.47):

$$i\partial_t \phi = \partial_{\bar{\phi}} V(\phi, \bar{\phi}) - i\kappa\phi; \qquad i\partial_t \bar{\phi} = -\partial_\phi V(\phi, \bar{\phi}) - i\kappa\bar{\phi}, \tag{5.58}$$

where we have omitted the superscript "cl" for brevity and the effective potential, $V(\phi, \bar{\phi})$, is given by

$$V(\phi, \bar{\phi}) = \Delta \bar{\phi}\phi + \frac{g}{4}(\bar{\phi}\phi)^2 + \lambda(\phi^2 + \bar{\phi}^2). \tag{5.59}$$

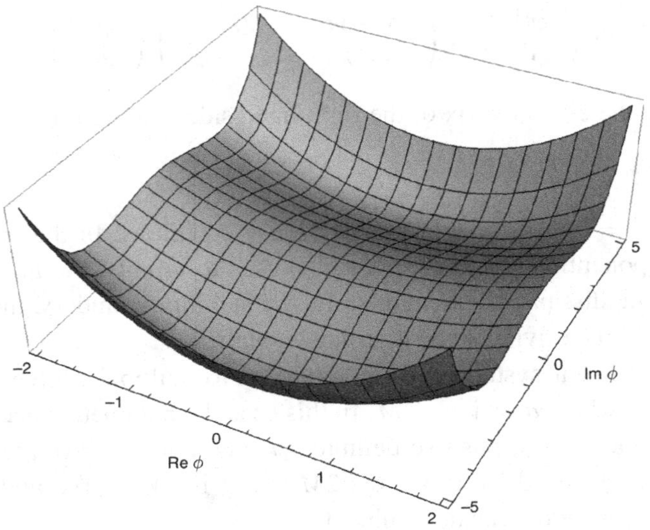

Figure 5.1 Effective potential of the parametric-driven oscillator, $V(\phi, \bar{\phi})$, Eq. (5.59), for $2\lambda > \Delta$.

For $2\lambda < \Delta$ it has a single minimum at $\phi = \bar{\phi} = 0$, corresponding to the stable regime of the parametric harmonic oscillator. For $2\lambda > \Delta$ (and small κ) the linear stability is lost and the effective potential develops two symmetric minima at $\phi = \pm\phi_0$, where $\phi_0 = \mathrm{i}\sqrt{2(2\lambda - \Delta)/g}$, Fig. 5.1. These two minima correspond to parametrically induced oscillations with the amplitude $|\phi_0|$ and two opposite phases relative to the driving force.

Classically both these two minima are stable at $T = 0$. At $T > 0$ there are activated transitions between the two states, restoring the symmetry of the stationary distribution function. Physically such transitions are the phase slips of the parametric oscillations. Quantum mechanically the symmetry is restored even at $T = 0$. That is, the stationary state involves a superposition of oscillations with the two phases. In the absence of coupling to the bath it is achieved by quantum tunneling. The corresponding tunneling rate may be evaluated by the analytical continuation to the imaginary time [103]. However, in the presence of any (non-exponentially weak) coupling the dominant switching mechanism is *not* the quantum tunneling. Instead, the *quantum heating* mechanism, Section 5.5, leads to activated transitions between the two wells of Fig. 5.1. Dykman and colleagues called it *quantum activation* [100, 101, 102] and showed that it always dominates over the tunneling.

To highlight relations with the quantum heating we begin with the linearized problem. To this end we focus on a local vicinity of one of the two potential minima, say $\phi = \phi_0$, write $\phi(t) = \phi_0 + \delta\phi(t)$, and expand the potential to the second order in $\delta\phi$. This way we find:

$$V(\delta\phi,\delta\bar{\phi}) \approx V(\phi_0) + \frac{1}{2}\,(\delta\bar{\phi},\delta\phi)\begin{pmatrix} 4\lambda-\Delta & \Delta \\ \Delta & 4\lambda-\Delta \end{pmatrix}\begin{pmatrix} \delta\phi \\ \delta\bar{\phi} \end{pmatrix}. \tag{5.60}$$

The quadratic form here may be diagonalized by the Bogoliubov rotation (cf. Eq. (5.49))

$$\delta\phi(t) = \cosh\alpha\,\zeta(t) + \sinh\alpha\,\bar{\zeta}(t); \qquad \delta\bar{\phi}(t) = \cosh\alpha\,\bar{\zeta}(t) + \sinh\alpha\,\zeta(t), \tag{5.61}$$

where $\tanh 2\alpha = -\Delta/(4\lambda-\Delta)$. This brings the local potential to the canonical form of the harmonic oscillator, $V(\zeta,\bar{\zeta}) \approx V(\phi_0) + \Omega\,\bar{\zeta}\zeta$, where $\Omega^2 = 8\lambda(2\lambda-\Delta)$. The corresponding linearized non-dissipative action in terms of $(\zeta^{\rm cl},\zeta^{\rm q})$ takes the simple form

$$S = \int dt\,(\bar{\zeta}^{\rm cl},\bar{\zeta}^{\rm q})\begin{pmatrix} 0 & i\partial_t - \Omega \\ i\partial_t - \Omega & 0 \end{pmatrix}\begin{pmatrix} \zeta^{\rm cl} \\ \zeta^{\rm q} \end{pmatrix}, \tag{5.62}$$

where $\phi^{\rm q}(t) = \cosh\alpha\,\zeta^{\rm q}(t) + \sinh\alpha\,\bar{\zeta}^{\rm q}(t)$ and the same for the bar-fields. (Indeed the Bogoliubov transformation (5.61) may be performed separately on the forward and backward branches and thus the classical and quantum components ought to transform in the same way)).

We now examine how the Bogoliubov transformation (5.61) acts on the dissipative terms in the action (5.47). Upon the rotation of the classical and quantum components, one arrives at

$$S_{\rm diss} \approx \int dt\,(\bar{\zeta}^{\rm cl},\bar{\zeta}^{\rm q})\begin{pmatrix} 0 & -i\kappa \\ i\kappa & 2i\kappa(2n_B^{\rm eff}+1) \end{pmatrix}\begin{pmatrix} \zeta^{\rm cl} \\ \zeta^{\rm q} \end{pmatrix}, \tag{5.63}$$

where we have disregarded terms with two bar-fields and two non-bar-fields, since they are fast rotated as $e^{\pm i2\Omega t}$ and average out to zero. The effective occupation number and the effective temperature are defined in exactly the same way[8] as in Eqs. (5.52), (5.53) with the only difference that $\cosh 2\alpha = (4\lambda-\Delta)/\Omega$. Introducing a dimensionless parameter $\mu \equiv \Delta/2\lambda < 1$, one thus finds [101]

$$\frac{2\lambda}{T^{\rm eff}} = \frac{1}{2\sqrt{1-\mu}}\,\log\frac{(2n_B+1)(2-\mu)+2\sqrt{1-\mu}}{(2n_B+1)(2-\mu)-2\sqrt{1-\mu}}. \tag{5.64}$$

Close to the bifurcation point, where $1-\mu \ll 1$, one recovers Eq. (5.54). This shows that the effective quantum temperature is continuous across the bifurcation transition.

To make a quantitative progress beyond the harmonic approximation, one may take advantage of the relative weakness of the dissipation, $\kappa \ll \lambda$. This ensures that solutions of the equations of motion (5.58) are close to the dissipationless Hamiltonian trajectories given by the lines of constant effective potential

[8] A straightforward way to establish it is to use forward and backward fields $\phi^{\pm}$ and employ Eq. (5.23).

$V(\phi, \bar{\phi}) = \text{const}$. This motivates the *canonical* transformation to the action-angle coordinates $(\phi, \bar{\phi}) \to (I, \theta)$.[9] In these coordinates the effective potential is a function of the action variable only, $V(\phi, \bar{\phi}) \to V(I)$, and the dissipation-less equations of motion acquire the simple form: $\partial_t I = 0$ and $\partial_t \theta = \partial_I V(I) = \omega(I)$. One can thus take advantage of the timescale separation between the fast angular motion (which is only weakly affected by the dissipation) and the slow motion of the action variable. The latter is entirely due to the coupling to the bath.

To derive the effective slow dynamics of the action variable it is convenient to return to the forward–backward Keldysh action (5.23), where the Hamiltonian part is $H^{\pm} = V(\phi^{\pm}, \bar{\phi}^{\pm})$ and the Lindbladian part is given by Eqs. (5.25). One passes now to the action-angle coordinates separately on the forward and backward branches of the contour, $(\phi^{\pm}, \bar{\phi}^{\pm}) \to (I^{\pm}, \theta^{\pm})$ and performs the Keldysh rotation $I^{\text{cl}} = (I^{+} + I^{-})/2$; $I^{\text{q}} = I^{+} - I^{-}$ and $\theta^{\text{cl}} = (\theta^{+} + \theta^{-})/2$; $\theta^{\text{q}} = \theta^{+} - \theta^{-}$. The Hamiltonian part of the action (5.23) acquires the following form (neglecting full time derivatives):

$$\begin{aligned} &\bar{\phi}^{+} \mathrm{i}\partial_t \phi^{+} - \bar{\phi}^{-} \mathrm{i}\partial_t \phi^{-} - V^{+} + V^{-} = I^{+}\partial_t \theta^{+} - I^{-}\partial_t \theta^{-} - V(I^{+}) + V(I^{-}) \qquad (5.65) \\ &= I^{\text{cl}}\partial_t \theta^{\text{q}} + I^{\text{q}}\partial_t \theta^{\text{cl}} - V\left(I^{\text{cl}} + \frac{I^{\text{q}}}{2}\right) + V\left(I^{\text{cl}} - \frac{I^{\text{q}}}{2}\right) \approx I^{\text{cl}}\partial_t \theta^{\text{q}} + I^{\text{q}}\left[\partial_t \theta^{\text{cl}} - \omega(I^{\text{cl}})\right], \end{aligned}$$

where $\omega(I^{\text{cl}}) = \partial_I V(I^{\text{cl}})$ and we restricted ourselves to the first order in I^{q}. Integration over I^{q} enforces the classical equation of motion, $\partial_t \theta^{\text{cl}} = \omega(I^{\text{cl}})$, resulting in the fast angular rotation $\theta^{\text{cl}}(t) = \omega(I^{\text{cl}})t$. Since this is the fast degree of freedom, we shall disregard both quantum (coming from the higher orders in I^{q}) and bath-induced fluctuations of θ^{cl} (we will fully keep, though, fluctuations of the slow degree of freedom, I^{cl}). This amounts to putting $I^{\text{q}} = 0$ in all subsequent calculations.

We turn now to the dissipative terms in the action (5.23), given by Eqs. (5.25). To this end we utilize periodicity of the angular degree of freedom to write (cf. footnote 9)

$$\phi^{\pm} = \sum_{r=-\infty}^{\infty} c_r(I^{\pm})\, e^{-\mathrm{i}r\theta^{\pm}} \approx \sum_{r=-\infty}^{\infty} c_r(I)\, e^{-\mathrm{i}r(\theta^{\text{cl}} \pm \theta^{\text{q}}/2)}, \qquad (5.66)$$

where we put $I^{\text{q}} = 0$ and use $I^{\text{cl}} = I$ for brevity hereafter. Complex functions $c_r(I)$ are Fourier coefficients of the time-periodic trajectory $\phi(t)$, which is a solution of $\mathrm{i}\partial_t \phi = \partial_{\bar{\phi}} V(\phi, \bar{\phi})$ with the constant $V(\phi, \bar{\phi}) = V(I)$. We now substitute this expression, along with its complex conjugated $\bar{\phi}^{\pm}$, into Eqs. (5.25) and average over the fast rotating angle θ^{cl}. This way, one finds for

[9] For, e.g. harmonic oscillator this transformation is accomplished by $\phi = \sqrt{I}\, e^{-\mathrm{i}\theta}$ and $\bar{\phi} = \sqrt{I}\, e^{\mathrm{i}\theta}$, while $V(\phi, \bar{\phi}) = \omega_0 \bar{\phi}\phi = \omega_0 I$. For a generic potential, $V(\phi, \bar{\phi})$, the transformation takes the form of Eq. (5.66).

$$\phi^+\bar{\phi}^- = \sum_{r,r'} c_r(I)\bar{c}_{r'}(I)\, e^{\mathrm{i}(r'-r)\theta^{\mathrm{cl}}-\mathrm{i}(r'+r)\theta^{\mathrm{q}}/2} \to \sum_r |c_r(I)|^2\ e^{-\mathrm{i}r\theta^{\mathrm{q}}};$$

$$\phi^+\bar{\phi}^+ = \sum_{r,r'} c_r(I)\bar{c}_{r'}(I)\, e^{\mathrm{i}(r'-r)\theta^{\mathrm{cl}}+\mathrm{i}(r'-r)\theta^{\mathrm{q}}/2} \to \sum_r |c_r(I)|^2$$

$$\bar{\phi}^+\phi^- \to \sum_r |c_r(I)|^2\ e^{\mathrm{i}r\theta^{\mathrm{q}}}; \qquad \phi^-\bar{\phi}^- \to \sum_r |c_r(I)|^2.$$

Finally it is convenient to rename the quantum angle as $p_I = \mathrm{i}\theta^{\mathrm{q}}$. This allows one to have the effective equations of motion in purely real notations and is very close in spirit to the change of variables in Eq. (4.18).

Putting everything together, we have succeeded to reduce the microscopic action (5.23) to the following effective action:

$$\mathrm{i}S_{\mathrm{eff}}[p_I, I] = -\int dt \left[p_I \partial_t I - H_{\mathrm{eff}}(p_I, I)\right]. \tag{5.67}$$

The kinetic term $p_I\partial_t I$ comes from Eq. (5.65), and $H_{\mathrm{eff}}(p_I, I)$ comes from the Lindblad quantum jump operators (5.25):

$$H_{\mathrm{eff}}(p_I, I) = \gamma_1 h(-p_I, I) + \gamma_2 h(p_I, I); \qquad h(p_I, I) = \sum_{r=-\infty}^{\infty} [\mathrm{e}^{rp_I} - 1]\, |c_r(I)|^2. \tag{5.68}$$

As before, $\gamma_{1(2)}$ are the rates of the down(up) transitions with the relative magnitude $\gamma_1/\gamma_2 = \mathrm{e}^{\omega_0/T}$. We have arrived thus at the action of a reaction (chemical kinetics) model [103], discussed in Section 4.11; compare Eqs. (4.78) and (4.79). The action variable, I, plays the role of a number of excitation quanta. Its canonical pair, p_I, upon quantization takes the form $p_I \to -\partial_I$ and therefore e^{rp_I} downshifts the action by an integer number r. The integerness of the action, I, is, of course, a manifestation of the Bohr–Sommerfeld quantization. The r-step up-down transitions between the quantum states are induced by the bath. Their rates are proportional to the Lindbladian constants, γ_a, and weighted by the Fourier power spectrum of the classical trajectory, $|c_r(I)|^2$.

As an example let's consider an equilibrium oscillator with $V(\phi, \bar{\phi}) = V(|\phi|^2)$. Similarly to the harmonic oscillator (see footnote 9), in this case $c_r(I) = \sqrt{I}\,\delta_{r,1}$, and thus

$$H_{\mathrm{eff}} = \left[\gamma_1(\mathrm{e}^{-p_I} - 1) + \gamma_2(\mathrm{e}^{p_I} - 1)\right] I = [\gamma_1 - \gamma_2\,\mathrm{e}^{p_I}](\mathrm{e}^{-p_I} - 1)I. \tag{5.69}$$

Equations (5.67) and (5.69) are completely equivalent to the Master equation (5.93) for the diagonal elements of the density matrix. The Hamiltonian (5.69) admits three zero-energy lines, $I = 0$, $p_I = 0$, and $p_I = \log(\gamma_1/\gamma_2) = \omega(I)/T$. The corresponding phase portrait is depicted in Fig. 5.2. The stationary probability to reach an action I is given by $\mathcal{P}(I) \propto \mathrm{e}^{S(I)}$, where $S(I)$ is the shaded area,

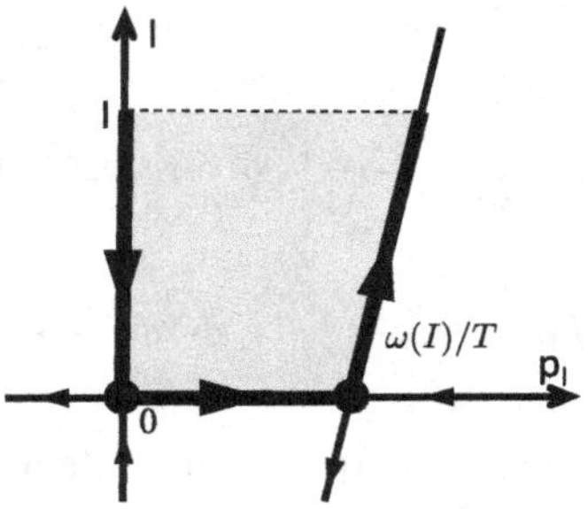

Figure 5.2 Phase portrait of the effective Hamiltonian (5.69) for an equilibrium oscillator with $V(\phi,\bar{\phi}) = \Delta\bar{\phi}\phi + \frac{g}{4}(\bar{\phi}\phi)^2$. The probability to reach an action I is given by the exponentiated area of the shaded region.

$S(I) = -\int_0^I p_I \mathrm{d}I = -\int_0^I(\omega(I)/T)\mathrm{d}I = -(V(I) - V(0))/T$, where we used the fact that $\omega(I) = \partial_I V(I)$. This is, of course, the equilibrium Boltzmann distribution.

The semiclassical evolution along the vertical line, $p_I = 0$, describes the mean-field relaxation of the action. It is given by

$$\partial_t I = \left.\frac{\partial H_{\text{eff}}}{\partial p_I}\right|_{p_I=0} = (\gamma_2 - \gamma_1)\sum_{r=-\infty}^{\infty} r\,|c_r(I)|^2 = (\gamma_2 - \gamma_1)I = -2\kappa I. \tag{5.70}$$

Thus the action relaxation, $I(t) \propto \mathrm{e}^{-2\kappa t}$, is consistent with the equilibration rate, Eq. (5.45), of the distribution in linear systems. Equation (5.70) is more general, however, since it is not limited to the linear models. Indeed quite generally, $\sum_r r\,|c_r(I)|^2 = \int \mathrm{d}t\,\bar{\phi}\,\mathrm{i}\partial_t\phi = \mathrm{i}\oint\bar{\phi}\,\mathrm{d}\phi = I$.

For a parametrically driven oscillator, Eq. (5.59), the classical trajectories $\phi(t)$ are given by the elliptic functions. Their Fourier coefficients for $|r| \gg 1$ decay as $c_r(I) \propto \exp\{-rp_>(I)/2\}$ for $r > 0$ and $c_r(I) \propto \exp\{-|r|p_<(I)/2\}$ for $r < 0$, where $p_{>(<)}(I)$ are related to the poles of the elliptic function in the upper (lower) half-plane of the complex time. In the present problem they are related as $p_<(I) = 3p_>(I)$, which may be traced to the times of reaching infinite ϕ, moving in the imaginary time direction according to Eqs. (5.58).[10] As a result, the function $h(-p_I, I)$ is only convergent within the interval $-p_> < p_I < p_<$ and reaches positive infinity at both of its boundaries. Since $h(-p_I, I) = 0$ for $p_I = 0$, there must be at least one other value of p_I within the convergence interval where $h(-p_I, I) = 0$. For example, for $\Delta = 0$, it happens to be $p_I = 2p_>$, Fig. 5.3(a).

It is convenient to plot the phase portrait of the effective Hamiltonian (5.68) in the rescaled coordinates $(p_I/p_>(I), I)$. At zero temperature, $\gamma_2 = 2\kappa n_B = 0$, and

[10] It is worth noticing that, being proportional to the imaginary part of the second period of the elliptic function, $p_>(I)$ diverges logarithmically upon approaching the bottom of the potential, $p_>(I) \propto -\log I$. On the other hand, the first period, $\omega^{-1}(I)$, diverges at the saddle point of the potential, $\omega^{-1}(I) \propto -\log(I_0 - I)$, where I_0 is the area enclosed by the separatrix line, $V(\phi,\bar{\phi}) = 0$.

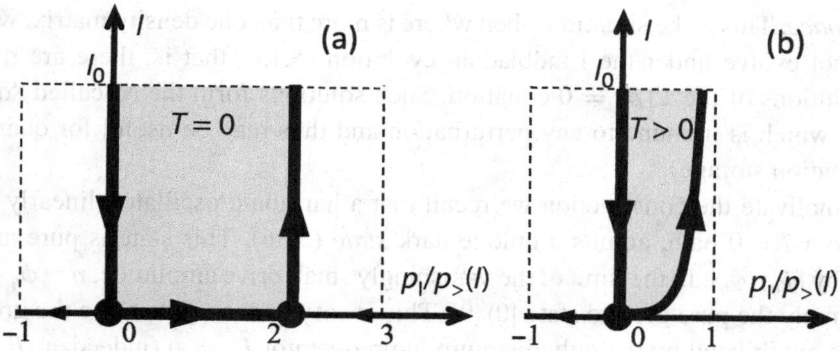

Figure 5.3 Phase portraits of the effective Hamiltonian (5.68) for (a) $T = 0$ (i.e. $\gamma_2 = 0$); (b) $T > 0$. Bold lines are zero energy trajectories. Vertical dashed lines are boundaries of convergence of the sum in Eq. (5.68). The line $I = 0$ corresponds to the bottom of the two wells in Fig. 5.1, the top line $I = I_0$ corresponds to the saddle point $V = 0$ – this is the place where a trajectory can switch between the two wells. The probability of the switching is given by the exponentiated action (5.67) on the trajectory encircling the shaded regions.

the effective Hamiltonian, $H_{\text{eff}}(p_I, I) = 2\kappa h(-p_I, I)$, is defined within the interval $-p_> < p_I < 3p_>$; see Fig. 5.3(a). For a small, but finite temperature both $h(-p_I, I)$ and $h(p_I, I)$ should be considered and thus the interval of convergence immediately shrinks to $-p_> < p_I < p_>$. One may see that the new zero-energy line appears in this case, Fig. 5.3(b), and the activation probability is given by the exponentiated action (5.67) calculated along this line. This leads to a peculiar phenomenon of the *zero temperature fragility* [101], when the result of the $T = 0$ calculation is different from the $T \to 0$ one. This signifies the emergence of a new (extremely small) temperature scale where the semiclassical approximation breaks down. The phenomenon may be traced back to the divergence of $h(p_I, I)$ at $p_I = p_>$. This divergence is due to the upward quantum transitions with $r \to \infty$. Therefore, the "long-ranged" (in the quantization number) transitions are responsible for the low-temperature quantum activation of the phase slips in the parametrically driven oscillator [101]. Such long-ranged transitions occur due to the nonlinearity of the potential (5.59). As a consequence, the zero temperature fragility of the quantum activation is absent in the harmonic approximation.

5.7 Dark Space

Here we discuss an alternative mechanism of limiting the instability of the parametrically driven oscillator. Instead of the Hamiltonian quartic nonlinearity, considered in the previous section, here we focus on the nonlinear dissipation. The latter results from the two-photon coupling to the bath. It allows us to illustrate a concept of the *dark space*. This is the situation when where is more than one density matrix, which does not evolve under the Lindbladian evolution (5.13), that is, there are multiple solutions of the $\mathcal{L}[\hat{\rho}] = 0$ equation. Such solutions form the so-called "dark" space, which is immune to any perturbation and thus may be useful for quantum information storage.

To motivate the construction we recall that a harmonic oscillator, linearly coupled to a $T=0$ bath, admits a unique dark *state* (5.36). This state is pure and is given by $|\phi_\eta\rangle\langle\phi_\eta|$. In the limit of the vanishingly small drive amplitude, $\eta \sim \phi_\eta \to 0$, it is simply the pure ground state $|0\rangle\langle 0|$. This is easy to anticipate since the ground state is annihilated by the only quantum jump operator $\hat{L}_1 = \hat{b}$ (indeed, at $T = 0$, the rate of $\hat{L}_2 = \hat{b}^\dagger$ is zero). This observation motivates choosing a quantum jump operator, which annihilates more than one state. The simplest example is provided by $\hat{L} = \hat{b}^2$, which annihilates both the ground state, $|0\rangle$ and the first excited state, $|1\rangle$.

With this in mind we investigate a model, where both the drive and the dissipation involve only *two*-photon processes (cf. Eq. (5.2)):

$$\hat{H}_{2\mathrm{p}} = \omega_0 \hat{b}^\dagger \hat{b} + 2\lambda(\hat{b}^\dagger \hat{b}^\dagger + \hat{b}\hat{b}) \cos 2\omega_p t; \tag{5.71a}$$

$$\hat{H}_{\mathrm{bath}} = \sum_s \omega_s \hat{a}_s^\dagger \hat{a}_s; \qquad \hat{H}_{\mathrm{int}} = \sum_s \frac{g_s}{\sqrt{2\omega_s}} \left(\hat{a}_s^\dagger \hat{b}\hat{b} + \hat{b}^\dagger \hat{b}^\dagger \hat{a}_s\right). \tag{5.71b}$$

We have omitted a nonlinearity for brevity, since it is anyways generated by the interactions with the bath. Adopting the rotating wave approximation and integrating out the bath degrees of freedom for the $T = 0$ bath, one arrives at the following action:

$$= \int dt\, (\bar{\phi}^{\mathrm{cl}}, \bar{\phi}^{\mathrm{q}}) \begin{pmatrix} 0 & i\partial_t - \Delta \\ i\partial_t - \Delta & 0 \end{pmatrix} \begin{pmatrix} \phi^{\mathrm{cl}} \\ \phi^{\mathrm{q}} \end{pmatrix} - 2\lambda\, (\bar{\phi}^{\mathrm{q}}\bar{\phi}^{\mathrm{cl}} + \phi^{\mathrm{q}}\phi^{\mathrm{cl}}) \tag{5.72}$$

$$+ i\kappa_2 \int dt [\bar{\phi}^{\mathrm{q}}\bar{\phi}^{\mathrm{cl}}(\phi^{\mathrm{cl}}\phi^{\mathrm{cl}} + \phi^{\mathrm{q}}\phi^{\mathrm{q}}) - \phi^{\mathrm{q}}\phi^{\mathrm{cl}}(\bar{\phi}^{\mathrm{cl}}\bar{\phi}^{\mathrm{cl}} + \bar{\phi}^{\mathrm{q}}\bar{\phi}^{\mathrm{q}}) + 2\bar{\phi}^{\mathrm{q}}\bar{\phi}^{\mathrm{cl}}\phi^{\mathrm{q}}\phi^{\mathrm{cl}} + 2\phi^{\mathrm{q}}\phi^{\mathrm{cl}}\bar{\phi}^{\mathrm{q}}\bar{\phi}^{\mathrm{cl}}].$$

The first line here represents the unitary evolution with the quadratic rotating-wave Hamiltonian, $\hat{H}'_{2\mathrm{p}}$, (5.48). The second line in Eq. (5.72) represents dissipation with the friction coefficient $\kappa_2 = J(2\omega_0)/2$, given by the spectral function of the bath. It originates[11] from $i\kappa_2 \int dt\, [\bar{L}^{\mathrm{q}} L^{\mathrm{cl}} - L^{\mathrm{q}}\bar{L}^{\mathrm{cl}} + \bar{L}^{\mathrm{q}} L^{\mathrm{q}} + L^{\mathrm{q}}\bar{L}^{\mathrm{q}}]$, where $L^{\mathrm{q}} = \sqrt{2}\phi^{\mathrm{q}}\phi^{\mathrm{cl}}$ and $L^{\mathrm{cl}} = (\phi^{\mathrm{cl}}\phi^{\mathrm{cl}} + \phi^{\mathrm{q}}\phi^{\mathrm{q}})/\sqrt{2}$. Comparing this line with the Keldysh–Lindblad action (5.23), one notices that it encodes Lindbladian evolution with the single quantum jump operator, $\hat{L} = \hat{b}\hat{b}$, and $\gamma = 2\kappa_2$. We thus conclude that the action (5.72) is the functional representation of the following Lindblad equation:

[11] This way of writing the dissipative action conforms with the ordering prescription of footnote 5. Substituting now $\phi^{\mathrm{q}} \to \partial_{\bar{\phi}}$ and $\bar{\phi}^{\mathrm{q}} \to -\partial_\phi$, one obtains the equation for the Wigner function.

$$\partial_t \hat{\rho} = -\mathrm{i}[\hat{H}'_{2\mathrm{p}}, \hat{\rho}] + 2\kappa_2 \left(\hat{b}\hat{b}\,\hat{\rho}\,\hat{b}^\dagger\hat{b}^\dagger - \frac{1}{2}\left\{\hat{b}^\dagger\hat{b}^\dagger\hat{b}\hat{b}, \hat{\rho}\right\}\right) \tag{5.73}$$

with the rotated wave Hamiltonian (5.48).

We already know from Section 5.5 that the unitary evolution exhibits the instability for the drive amplitude exceeding the critical value, $\lambda > \Delta/2$. However, the instability is arrested by the two-photon bath coupling, Eq. (5.71b). To understand how it works consider a static classical stationary point of the action (5.72) given by $\delta S/\delta\bar{\phi}^{\mathrm{q}}|_{\phi^{\mathrm{q}}=0} = 0$ and its complex conjugated partner:

$$\left(-\Delta + \mathrm{i}\kappa_2|\phi^{\mathrm{cl}}|^2\right)\phi^{\mathrm{cl}} = 2\lambda\bar{\phi}^{\mathrm{cl}}; \qquad \left(-\Delta - \mathrm{i}\kappa_2|\phi^{\mathrm{cl}}|^2\right)\bar{\phi}^{\mathrm{cl}} = 2\lambda\phi^{\mathrm{cl}}. \tag{5.74}$$

Multiplying the two equations, one finds

$$\left(\Delta^2 - 4\lambda^2 + \kappa_2^2|\phi^{\mathrm{cl}}|^4\right)|\phi^{\mathrm{cl}}|^2 = 0. \tag{5.75}$$

It shows that for $\lambda < \Delta/2$, the only static semiclassical solution is (trivially symmetric) $\phi^{\mathrm{cl}} = 0$. This is the regime, which quantized description is closely related to Eq. (5.50). In the opposite limit, $\lambda > \Delta/2$, the symmetric solution $\phi^{\mathrm{cl}} = 0$ is unstable. Instead, the symmetry-broken solutions are stabilized with $|\phi_0^{\mathrm{cl}}|^2 = \sqrt{4\lambda^2 - \Delta^2}/\kappa_2$. An inspection of Eqs. (5.74) shows that there are exactly two such solutions with $\phi_0^{\mathrm{cl}} = \pm\mathrm{i}|\phi_0^{\mathrm{cl}}|\mathrm{e}^{-\mathrm{i}\alpha/2}$, where $\tan\alpha = \sqrt{4\lambda^2 - \Delta^2}/\Delta$. Notice that in the limit $\Delta \to 0$ the two solutions approach $\phi_0^{\mathrm{cl}} = \pm\sqrt{2\mathrm{i}\lambda/\kappa_2}$.

This indicates that in the presence of dissipation, there are *two* distinct stationary solutions of the Lindblad equation (5.73). In the limit $\lambda = 0$ those are nothing but $|0\rangle\langle 0|$ and $|1\rangle\langle 1|$, which may be directly checked by substitution into Eq. (5.73) (notice that $|0\rangle\langle 1|$ and $|1\rangle\langle 0|$ are also stationary only if $\Delta = 0$). As functions of λ, the two stationary solutions evolve, maintaining their even and odd character in the particle number. At $\lambda > \Delta/2$ they sharply (for small κ_2) cross over to being close to the coherent states with $\phi = \pm\phi_0^{\mathrm{cl}}/\sqrt{2}$ (see footnote 6), where $\pm\phi_0^{\mathrm{cl}}$ are the two solutions of the saddle-point equations (5.74). Yet, since there are no true phase transitions in 0D, the actual stationary solutions are rather symmetric and antisymmetric combinations of these two coherent states.[12]

[12] To actually see this dynamics one may transform the Lindblad equation (5.73) into a differential form, for example, Glauber–Sudarshan or Husimi, and investigate its stationary solutions. The most convenient one is probably the Wigner symmetric form, which may be directly read out from the Keldysh action (5.72) using the prescription in footnote 5. This way one finds $\partial_t\mathcal{P} = \partial_\phi J_\phi + \partial_{\bar{\phi}} J_{\bar{\phi}}$, where

$$J_\phi = \kappa_2\left[\bar{\phi}(\phi^2 + \partial^2_{\bar{\phi}\bar{\phi}}) + 2\bar{\phi}\phi\partial_{\bar{\phi}}\right]\mathcal{P} + \mathrm{i}(2\lambda\bar{\phi} + \Delta\phi)\mathcal{P}$$

and we omitted superscript "cl" for brevity. The resulting Wigner equation is of the third order (Glauber–Sudarshan is second order, while Husimi is fourth order). They all ought to have the same number of stationary solutions, though. For $\lambda = 0$ those are $\mathcal{P} = \mathrm{e}^{-|\phi|^2}$ and $\mathcal{P} = |\phi|^2\,\mathrm{e}^{-|\phi|^2}$, corresponding to $|0\rangle\langle 0|$ and $|1\rangle\langle 1|$ states. If also $\Delta = 0$, then there are two extra stationary solutions $\mathcal{P} = \bar{\phi}\,\mathrm{e}^{-|\phi|^2}$ and $\mathcal{P} = \phi\,\mathrm{e}^{-|\phi|^2}$, corresponding to $|0\rangle\langle 1|$ and $|1\rangle\langle 0|$ states.

We have shown that the model generically exhibits two stationary states with the opposite particle number parity. This is not good news from the point of view of quantum information. Indeed, the two stationary states may act as a classical, but not a quantum bit. The latter requires the full space of a quantum two-level system, namely $|\uparrow\rangle\langle\uparrow|, |\downarrow\rangle\langle\downarrow|, |\downarrow\rangle\langle\uparrow|$ and $|\uparrow\rangle\langle\downarrow|$ – the *four*fold degenerate dark space. In our driven-dissipative system, this is only realized at strictly zero detuning, $\Delta = 0$. In this particular case the evolution equation (5.73) may be rewritten[13] as a purely dissipative one with the new jump operator $\hat{L} = \hat{b}\hat{b} - \phi_0^2$ and no Hamiltonian part:

$$\partial_t \hat{\rho} = 2\kappa_2 \left((\hat{b}\hat{b} - \phi_0^2)\, \hat{\rho}\, (\hat{b}^\dagger \hat{b}^\dagger - \bar{\phi}_0^2) - \frac{1}{2} \left\{ (\hat{b}^\dagger \hat{b}^\dagger - \bar{\phi}_0^2)(\hat{b}\hat{b} - \phi_0^2), \hat{\rho} \right\} \right), \quad (5.76)$$

where $\phi_0^2 = \mathrm{i}\lambda/\kappa_2$. Since $(\hat{b}\hat{b} - \phi_0^2)|\pm\phi_0\rangle = 0$, where $|\pm\phi_0\rangle$ are the two coherent states, we conclude that there are *four* stationary states of this Lindbladian: $|\phi_0\rangle\langle\phi_0|$, $|-\phi_0\rangle\langle-\phi_0|$, as well as $|-\phi_0\rangle\langle\phi_0|$ and $|\phi_0\rangle\langle-\phi_0|$. The first two of them are precisely the ones we found in the semiclassical treatment in the limit $\Delta \to 0$. They are descendents of the two generic stationary states. The latter two exist only for $\Delta = 0$; the corresponding eigenvalues of the Lindbladian super-operator, $\hat{\mathcal{L}}$, acquire nonzero values as soon as $\Delta \neq 0$. They thus lead to a finite dephasing rate, which is, however, exponentially small for a weak dissipation:

$$\frac{\hbar}{\tau_{\text{dephasing}}} \propto \Delta\, \mathrm{e}^{-2\lambda/\kappa_2}. \quad (5.77)$$

We derive this result in Section 5.9 using a semiclassical approach. Therefore, the model does realize a qubit dark space, but it is not protected from a dephasing (albeit exponentially weak) due to a nonzero detuning, Δ. Ideas of how to utilize topology to achieve better protection from the decoherence were put forward in Refs. [104, 105].

5.8 Symmetries of Dissipative Evolution

To systematically understand the degeneracy of dark spaces of a dissipative evolution, one needs to discuss the underlying symmetries. In the unitary case the

[13] This is a manifestation of the general property of the Lindblad equation that the jump operators are defined up to a constant. Namely the transformations

$$\hat{L}_a \to \hat{L}_a - C_a; \qquad \hat{H} \to \hat{H} - \frac{\mathrm{i}}{2} \sum_a \gamma_a \left(C_a \hat{L}_a^\dagger - \bar{C}_a \hat{L}_a \right)$$

leave the Lindblad equation (5.11) invariant. Here C_a are complex constants.

Table 5.1 *Symmetries of 0D systems with one-photon (1-ph) and two-photon (2-ph) drive and coupling to the bath.*

	1-ph drive (5.2a)	2-ph drive (5.71a)
1-ph coupling (5.2b)	none: $[\hat{H}'_{\rm p}, \hat{P}] \neq 0$	weak: $[\hat{H}'_{\rm 2p}, \hat{P}] = \{\hat{L}_a, \hat{P}\} = 0$
2-ph coupling (5.71b)	none: $[\hat{H}'_{\rm p}, \hat{P}] \neq 0$	strong: $[\hat{H}'_{\rm 2p}, \hat{P}] = [\hat{L}_a, \hat{P}] = 0$

symmetries are associated with operators that commute with the system's Hamiltonian. For a dissipative evolution of a system with a bath integrated out, the situation is more subtle. One needs to distinguish between *strong* and *weak* symmetries [106]. A case in point is bosonic parity, which in the unitary case is associated with the parity operator, $\hat{P} = {\rm e}^{i\pi\hat{b}^\dagger\hat{b}}$. The latter has an eigenvalue $+1$ for even states and -1 for odd states: $\hat{P}|\pm\rangle = \pm|\pm\rangle$, where $(\pm)$ denote even and odd states correspondingly.

A *strong* symmetry is the one that is respected on the microscopic level. That is, the elementary processes of the system–bath interactions preserve parity. Hamiltonian (5.71) is an example of the strong parity symmetry, (5.2b) is not. An immediate consequence of this (and thus a sufficient condition of a strong symmetry) is that both systems' Hamiltonian and *all* jump operators commute with the symmetry operator:

$$[\hat{H}'_{\rm 2p}, \hat{P}] = [\hat{L}_a, \hat{P}] = 0. \tag{5.78}$$

A *weak* symmetry is absent on the microscopic level, but appears in an effective description after the bath is integrated. Hamiltonians (5.2b) and (5.46) provide an example of the weak parity symmetry. In this case the individual jump operators do not respect parity, $[\hat{L}_a, \hat{P}] \neq 0$, but within the Lindblad equation, where there are always two jump operators acting on the density matrix, the parity is conserved. A sufficient condition for this to happen is

$$[\hat{H}'_{\rm 2p}, \hat{P}] = \{\hat{L}_a, \hat{P}\} = 0. \tag{5.79}$$

This is indeed what happens in our examples: $\{\hat{b}, {\rm e}^{i\pi\hat{b}^\dagger\hat{b}}\} = \{\hat{b}^\dagger, {\rm e}^{i\pi\hat{b}^\dagger\hat{b}}\} = 0$. Hamiltonian (5.2a) does not obey neither weak nor strong symmetry, since it does not commute with the parity operator. This information is collected in Table 5.1.

To appreciate the consequences of strong and weak symmetries one needs to understand how they manifest themselves on the level of the super-operator. Here we follow the discussion and notations of Refs. [107, 108]. The super-operator version of the parity operator is denoted as $\hat{\mathcal{P}}$ (do not confuse it with a probability distribution, denoted by the same letter). It acts as $\hat{\mathcal{P}}[\hat{\rho}] = \hat{P}\hat{\rho}\hat{P}^\dagger$. In the case of a *weak* symmetry, the super-parity commutes with the Lindbladian,

$[\hat{\mathcal{L}}, \hat{\mathcal{P}}] = 0$. Therefore, all density matrices may be classified as having $+1$ or -1 super-parity. The super-parity $+1$ have the structure $|\pm\rangle\langle\pm|$, while the super-parity -1 have the structure $|\pm\rangle\langle\mp|$. Notice that all density matrices with the super-parity -1 must have zero trace. Due to $[\hat{\mathcal{L}}, \hat{\mathcal{P}}] = 0$, the Lindbladian super-operator acquires a block-diagonal structure in the super-parity subspaces $\hat{\mathcal{L}} = \mathrm{diag}\{\hat{\mathcal{L}}_+, \hat{\mathcal{L}}_-\}$. Only $\hat{\mathcal{L}}_+$ operates in the sub-space of trace-one operators, while $\hat{\mathcal{L}}_-$ operates in the sub-space of trace-zero operators. As a result, only $\hat{\mathcal{L}}_+$ admits a zero eigenvalue (i.e. stationary) density matrix eigen-operator. Thus in the case of a *weak* symmetry one generically expects a unique stationary *state* – not a dark *space*.

In case of a *strong* symmetry, there are two symmetry super-operators: $\hat{\mathcal{P}}_l$ and $\hat{\mathcal{P}}_r$. They act from the left and right correspondingly: $\hat{\mathcal{P}}_l[\hat{\rho}] = \hat{P}\hat{\rho}$ and $\hat{\mathcal{P}}_r[\hat{\rho}] = \hat{\rho}\hat{P}^\dagger$. They both commute with the Lindbladian super-operator, $[\hat{\mathcal{L}}, \hat{\mathcal{P}}_l] = [\hat{\mathcal{L}}, \hat{\mathcal{P}}_r] = 0$, and thus their eigenvalues resolve $|+\rangle\langle+|$ from $|-\rangle\langle-|$ and similarly $|+\rangle\langle-|$ from $|-\rangle\langle+|$. As a result, the Lindbladian super-operator is split further onto four sub-blocks, $\hat{\mathcal{L}} = \mathrm{diag}\{\hat{\mathcal{L}}_{++}, \hat{\mathcal{L}}_{--}, \hat{\mathcal{L}}_{+-}, \hat{\mathcal{L}}_{-+}\}$. Two of these four blocks act on trace-one operators. As a result, there are at least two distinct stationary density matrices: the even one $|+\rangle\langle+|$ and the odd one $|-\rangle\langle-|$. Those form the two-dimensional dark space, encountered in the previous section.

The dimensionality of the dark space may be further increased if a *weak* or a *strong* symmetry is spontaneously broken. As discussed in Section 5.5, a sufficiently strong two-photon excitation drives the system unstable. The instability is arrested by a two-photon bath coupling, Section 5.7, with the "friction" coefficient κ_2. For any finite κ_2 the 0D system does not allow for a broken symmetry. Yet, in the limit $\kappa_2 \to 0$ (and the driving amplitude above the threshold (5.56)), the symmetry is indeed broken (as apparent e.g. from Eq. (5.77)). In this limit *weakly* symmetric systems exhibit a twofold degenerate dark space, while *strongly* symmetric ones exhibit a fourfold degenerate one. In the special case $\kappa = \Delta = 0$, the strongly symmetric system (accidentally) breaks the symmetry and gains the fourfold qubit degenerate space even for a finite κ_2. For a small nonzero detuning Δ the symmetry is restored by instantons, Section 5.9, and the qubit is decoherent with the exponentially small rate (5.77).

One may wonder if it is possible to engineer dark spaces with the degeneracy other than two or four. The answer is affirmative [109], if there is a *set* of *strong* symmetry operators, which realize a *projective* representation of a certain discrete group. The simplest example is provided by the Abelian group $\mathbb{Z}_m \times \mathbb{Z}_m$. Its projective representation is given by two *non-commuting* unitary operators, $\hat{P}_1$ and $\hat{P}_2$, such that

$$\hat{P}_1\hat{P}_2 = \mathrm{e}^{-2\pi \mathrm{i}/m}\, \hat{P}_2\hat{P}_1 \tag{5.80}$$

and $\hat{P}_1^m = \hat{P}_2^m = \hat{1}$. Here $e^{-2\pi i/m}$ is an example of 2-cocycle with the period m. If a Hamiltonian and all jump operators commute with both $\hat{P}_1$ and $\hat{P}_2$, there is a strong symmetry,[14] which guarantees an m-dimensional dark space. If, in addition, the Hamiltonian is absent (analog of $\Delta = 0$ case in Section 5.7), the dark space is m^2-fold degenerate. Reference [109] provides an example of such a construction with a fractional quantum Hall liquid on a surface of a torus, at a filling fraction $\nu = 1/m$. The operators $\hat{P}_1$ and $\hat{P}_2$ insert one flux quanta through the two principal periods of the torus.

5.9 Lindbladian Instantons

Here we address the issue of symmetry restoration and tunneling in the Lindbladian dynamics [110]. We take the two-photon drive with the two-photon dissipation model (5.71) of Section 5.7 to illustrate the construction. In the rotating-wave approximation the model is reduced to the Lindblad equation (5.73), which admits the strong parity symmetry. As a result, generically there are two dark states with the definite parity $|+\rangle\langle+|$ and $|-\rangle\langle-|$. The symmetry is broken at a special point of zero detuning, $\Delta = 0$, where the evolution may be written as a purely dissipative one, given by Eq. (5.76). Here there are four steady states, which can be written as pure coherent states: $|\phi_0\rangle\langle\phi_0|$, $|-\phi_0\rangle\langle-\phi_0|$, and $|-\phi_0\rangle\langle\phi_0|$, $|\phi_0\rangle\langle-\phi_0|$, where $\phi_0^2 = i\lambda/\kappa_2$. None of these pure states, taken separately, has a definite parity. One can form, however, even/odd states as $|\pm\rangle = |\phi_0\rangle \pm |-\phi_0\rangle$ and then the two stationary states with super-parity $+1$ take the form $|\pm\rangle\langle\pm|$. These states are adiabatically connected with two stationary states at $\Delta \neq 0$. The two states with super-parity -1, $|\pm\rangle\langle\mp|$, are only stationary at $\Delta = 0$. The unitary perturbation in the form $\hat{\mathcal{H}}[\hat{\rho}] = -i[\Delta\hat{b}^\dagger\hat{b}, \hat{\rho}]$ introduces a super-matrix element between these two super-parity -1 states (but *not* between states with opposite super-parity, since the perturbation commutes with the super-parity operator, $[\hat{\mathcal{H}}, \hat{\mathcal{P}}] = 0$). This matrix element leads to repulsion between the corresponding eigenvalues of the Lindbladian super-operator making them nonzero (in fact, imaginary in the first order in $\hat{\mathcal{H}}$). Being a close analog of the usual quantum mechanical tunneling, this mechanism *restores* the strong parity symmetry.

Our goal is to calculate the corresponding matrix element and thus nonzero eigenvalues of the Lindbladian. The latter provide the dephasing rate of the qubit. The problem is that, since the super-operator, $\hat{\mathcal{L}}$, is non-Hermitian, its left eigenfunctions are not complex conjugated of the right ones. While we know the latter,

14 This condition is actually too strong. If there are multiple jump operators, $\hat{L}_a$, it is enough to require that they mutually transform as $\hat{P}_{1,2}\hat{L}_a = \mathcal{U}_{ab}^{(1,2)}\hat{L}_b\hat{P}_{1,2}$, where $\mathcal{U}^{(1,2)}$ are two unitary matrices (not operators) rotating between different jump operators.

to calculate the former one needs to deal with the adjoint super-operator, $\hat{\mathcal{L}}^\dagger$. To overcome this, we employ the field-theory formulation and argue that the corresponding matrix element is given by the functional integral with fixed initial and final points in the phase-space [100]. In the limit of small dissipation, $\kappa_2 \ll \lambda$ (and thus $|\phi_0| \gg 1$) the functional integral is dominated by stationary – *instanton* – field configurations, which may be found by solving appropriate semiclassical equations of motion.

To execute this program we start from the action (5.72), which we reproduce here in the $\Delta = 0$ limit for convenience:

$$S = \int dt\,[\bar{\phi}^{\rm q} i\partial_t \phi^{\rm cl} + \bar{\phi}^{\rm cl} i\partial_t \phi^{\rm q} - 2\lambda\,(\bar{\phi}^{\rm q}\bar{\phi}^{\rm cl} + \phi^{\rm q}\phi^{\rm cl})] \\ + i\kappa_2 \int dt\,\big[\bar{\phi}^{\rm q}\bar{\phi}^{\rm cl}(\phi^{\rm cl}\phi^{\rm cl} + \phi^{\rm q}\phi^{\rm q}) - (\bar{\phi}^{\rm cl}\bar{\phi}^{\rm cl} + \bar{\phi}^{\rm q}\bar{\phi}^{\rm q})\phi^{\rm q}\phi^{\rm cl} + 4\,\bar{\phi}^{\rm q}\bar{\phi}^{\rm cl}\phi^{\rm q}\phi^{\rm cl}\big]. \tag{5.81}$$

Notice that the action possesses the two symmetries:

$$\phi^{\rm cl,q} \to -\phi^{\rm cl,q}; \qquad\qquad \phi^{\rm cl} \leftrightarrow \phi^{\rm q}. \tag{5.82}$$

The first one *simultaneously* reverses the sign of the classical and the quantum components, while the second one exchanges the two. These are the consequences of the presence of the *strong* parity symmetry, which dictates invariance of the action with respect to two independent symmetry transformations $\phi^+ \to -\phi^+$ and $\phi^- \to -\phi^-$. Upon Keldysh rotation, $\phi^{\rm cl,q} = (\phi^+ \pm \phi^-)/\sqrt{2}$; this leads to the symmetries (5.82). The second of them, $\phi^{\rm cl} \leftrightarrow \phi^{\rm q}$, shows that one can't expect solutions with the quantum component being smaller than the classical one.

The corresponding stationary field configurations are obtained by taking independent variations with respect to $\bar{\phi}^{\rm q}$, $\phi^{\rm q}$ and $\bar{\phi}^{\rm cl}$, $\phi^{\rm cl}$. This way, one obtains

$$\begin{aligned}
i\partial_t\phi &= 2\lambda\bar{\phi} - i\kappa_2\big[\bar{\phi}\phi\phi + \bar{\phi}\phi^{\rm q}\phi^{\rm q} - 2\bar{\phi}^{\rm q}\phi^{\rm q}\phi + 4\,\bar{\phi}\phi\phi^{\rm q}\big];\\
-i\partial_t\bar{\phi} &= 2\lambda\phi + i\kappa_2\big[\bar{\phi}\bar{\phi}\phi + \bar{\phi}^{\rm q}\bar{\phi}^{\rm q}\phi - 2\bar{\phi}\bar{\phi}^{\rm q}\phi^{\rm q} - 4\,\bar{\phi}^{\rm q}\bar{\phi}\phi\big];\\
i\partial_t\phi^{\rm q} &= 2\lambda\bar{\phi}^{\rm q} - i\kappa_2\big[\bar{\phi}^{\rm q}\phi^{\rm q}\phi^{\rm q} + \bar{\phi}^{\rm q}\phi\phi - 2\bar{\phi}\phi\phi^{\rm q} + 4\,\bar{\phi}^{\rm q}\phi^{\rm q}\phi\big];\\
-i\partial_t\bar{\phi}^{\rm q} &= 2\lambda\phi^{\rm q} + i\kappa_2\big[\bar{\phi}^{\rm q}\bar{\phi}^{\rm q}\phi^{\rm q} + \bar{\phi}\bar{\phi}\phi^{\rm q} - 2\bar{\phi}^{\rm q}\bar{\phi}\phi - 4\,\bar{\phi}\bar{\phi}^{\rm q}\phi^{\rm q}\big],
\end{aligned}$$

where we have omitted superscript "cl" but kept "q". Nominally $\bar{\phi}^{\rm q}$ denotes a complex conjugated of $\phi^{\rm q}$, namely $\bar{\phi}^{\rm q} = \phi^{{\rm q}*}$. However, the saddle-point equations do *not* respect this structure (because of the last terms on their right-hand sides). This means that to reach the saddle point the integration contours should be deformed in the complex planes of $\phi^{\rm q}$ and $\bar{\phi}^{\rm q}$, understood as two *independent* complex variables. In other words: both ${\rm Re}\,\phi^{\rm q}$ and ${\rm Im}\,\phi^{\rm q}$ integration contours are independently deformed into respective complex planes. This unpleasant feature may be repaired by renaming $\phi^{\rm q} = \bar{\chi}$ and $\bar{\phi}^{\rm q} = -\chi$ with the understanding that $\bar{\chi} = \chi^*$ – a complex conjugated of χ. With this substitution equations for $\bar{\phi}$ and $\bar{\chi}$ indeed become complex conjugated of those for ϕ and χ, correspondingly.[15] The latter equations take the following form:

$$i\partial_t\phi = 2\lambda\,\bar{\phi} - i\kappa_2\left[\bar{\phi}\phi\phi + \bar{\phi}\bar{\chi}\bar{\chi} + 2\bar{\chi}\chi\phi + 4\,\bar{\chi}\bar{\phi}\phi\right]; \tag{5.83a}$$

$$i\partial_t\chi = 2\lambda\,\bar{\chi} + i\kappa_2\left[\bar{\chi}\chi\chi + \bar{\chi}\bar{\phi}\bar{\phi} + 2\bar{\phi}\phi\chi + 4\,\bar{\phi}\bar{\chi}\chi\right]. \tag{5.83b}$$

We now look for stationary points of these equations. Some of them we already know; see Eqs. (5.74), (5.75) and the discussion following them. Those are found in the classical limit, $\phi^q = \bar{\chi} = 0$, where Eq. (5.83a) takes the form $2\lambda\bar{\phi} - i\kappa_2\bar{\phi}\phi\phi = 0$. It has two solutions: $\phi = \pm\sqrt{2\lambda/i\kappa_2}$. Similarly in the pure quantum subspace, where $\phi = 0$, Eq. (5.83b) takes the form $2\lambda\bar{\chi} + i\kappa_2\bar{\chi}\chi\chi = 0$, which has two solutions: $\chi = \pm\sqrt{2i\lambda/\kappa_2}$. This motivates one to introduce new variables (P,X) as

$$\chi = \sqrt{\frac{2i\lambda}{\kappa_2}}\,P; \qquad \phi = \sqrt{\frac{2\lambda}{i\kappa_2}}\,X, \tag{5.84}$$

where the classical and quantum stationary points take the form $(0,\pm 1)$ and $(\pm 1, 0)$, correspondingly, in (P,X) notations. Since all the stationary points belong now to the plane of *real* (P,X), one may try to look for instanton solutions of Eqs. (5.83) within this invariant subspace. Substituting (5.84) with real (P,X) into Eq. (5.83), one finds

$$\dot{X} = X - X^3 - 3P^2X - 4PX^2 = \partial_P H(P,X); \tag{5.85a}$$

$$\dot{P} = -P + P^3 + 3X^2P + 4XP^2 = -\partial_X H(P,X), \tag{5.85b}$$

where we have rescaled the (real) time as $2\lambda t \to t$ and denoted $\dot{X} = \partial_t X$ and $\dot{P} = \partial_t P$. Here the effective Hamiltonian is

$$H(P,X) = XP - X^3P - P^3X - 2X^2P^2 = XP(1 - X - P)(1 + X + P). \tag{5.86}$$

The system's dynamics may be read out from the phase portrait of the Hamiltonian (5.86), depicted in Fig. 5.4. It exhibits two classically stable points $(0, \pm 1)$, which are quasi-classical representations of $|\pm\phi_0\rangle$ stationary states. The matrix element, which provides for repulsion of the super-parity -1 states, is associated with the path leading from one classically stable point to the other. According to Fig. 5.4, such a path starts at $(0, 1)$, proceeds through the quantum stationary point $(1, 0)$, then returns to the classical manifold $P = 0$ at $(0, 0)$ and finally follows classically allowed motion toward $(0, -1)$. Since the equations of motion have the Hamiltonian form, they conserve the "energy," $H(P,X) = 0$, along the instanton trajectory. The only contribution to the action is coming from the $\int dt P\partial_t X = \int P dX$ term. It is proportional to the area of the shaded triangle in Fig. 5.4. Employing Eq. (5.84)

15 This is very similar to the analytical continuation $X^q = P/2i$, done in classical stochastic dynamics to arrive at real-valued Hamiltonian dynamics in the phase space (P, X^{cl}); see Eq. (4.18).

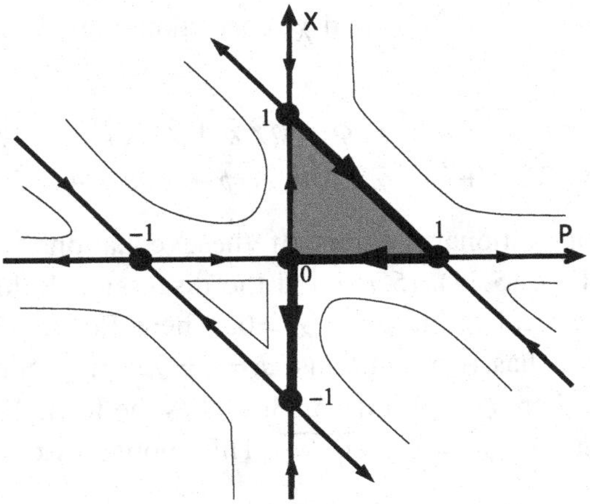

Figure 5.4 Phase portrait of the effective Hamiltonian (5.86) in the (P, X) plane. Notice its symmetries: $(P, X) \to (-P, -X)$ and $X \leftrightarrow P$; compare Eq. (5.82). Bold lines are lines of zero energy $P = 0$, $X = 0$, and $X = \pm 1 - P$. The instanton trajectory starts at $(0, 1)$ and goes through $(1, 0)$, $(0, 0)$ and finally arrives at $(0, -1)$. It thus accomplishes a switch between the two classically stable points $(0, \pm 1)$. The shaded area gives the instanton action.

and using that along the instanton path $P = 1 - X$, one finds the following for the action (5.81):

$$S_{\text{inst}} = \mathrm{i}\int [\bar{\phi}^{\mathrm{q}} \mathrm{d}\phi^{\mathrm{cl}} - \phi^{\mathrm{q}} \mathrm{d}\bar{\phi}^{\mathrm{cl}}] = -\mathrm{i}\int [\chi\, \mathrm{d}\phi + \bar{\chi}\, \mathrm{d}\bar{\phi}] = -2\mathrm{i}\frac{2\lambda}{\kappa_2}\int_1^0 (1 - X)\mathrm{d}X = \mathrm{i}\frac{2\lambda}{\kappa_2}. \tag{5.87}$$

We thus conclude that the matrix element of $\hat{\mathcal{H}}$ between the two states with the super-parity -1 is given by $\mathrm{i}\Delta\, \mathrm{e}^{\mathrm{i}S_{\text{inst}}}$ [110]. This leads to the qubit dephasing rate quoted in Eq. (5.77).

5.10 Optomechanical Cooling

Here we consider an effect of a strong drive, which is far detuned from the system's resonance frequency. The case in point is a mechanical oscillator with frequency ω_M, driven by an optical laser with the frequency $\omega_p \gg \omega_M$ [111, 112, 113, 114]. A schematic example is shown in Fig. 5.5a, where the laser is pumping an optical cavity, close to its eigenmode, ω_0, while one of the cavity's mirrors may oscillate mechanically with the frequency ω_M. The mechanical vibrations modulate the cavity's resonance frequency $\omega_0 \to \omega_0(1 + X(t)/L)$, where $X(t)$ is the mirror's displacement and L is the cavity length. The corresponding quantum Hamiltonian of coupled optical, $\hat{b}, \hat{b}^\dagger$, and mechanical, $\hat{B}, \hat{B}^\dagger$, oscillators takes the form

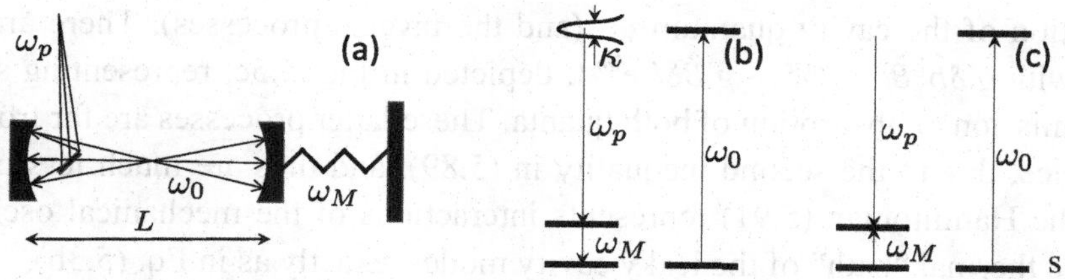

Figure 5.5 (a) A schematic representation of an optical cavity with an eigenmode frequency ω_0, coupled to a mechanical oscillator with the frequency ω_M. (b) Cooling transitions described by the interaction Hamiltonian (5.91). The pumping frequency ω_p is chosen to fall within the spectral width, κ, of the optical cavity mode. (c) Heating transitions with two absorbed (or emitted) quanta, not included in Eq. (5.91). Probability of these transitions is suppressed by the relative factor $(\kappa/2\omega_M)^2 \ll 1$.

$$\hat{H}_{\text{om}} = \omega_0 \hat{b}^\dagger \hat{b} + \omega_M \hat{B}^\dagger \hat{B} + \mathrm{i}g\, \hat{b}^\dagger \hat{b} \left(\hat{B}^\dagger - \hat{B}\right) + 2\mathrm{i}\eta \left(\hat{b}^\dagger - \hat{b}\right) \cos \omega_p t, \tag{5.88}$$

where $g \sim L^{-1}$ and the last term represent the driving laser with the amplitude $\propto \eta$. We also assume that the optical cavity is coupled (leakage) to a continuum of bulk electromagnetic modes, $\hat{a}_s, \hat{a}_s^\dagger$, exactly as in Eq. (5.2b). Such coupling results in a finite spectral width, κ, of the cavity mode, discussed in detail in Sections 5.1 and 5.4. For the purpose of cooling the mechanical oscillator we consider the case where $\kappa \ll \omega_M \ll \omega_p \lesssim \omega_0$ and

$$|\omega_0 - \omega_M - \omega_p| \lesssim \kappa; \qquad \omega_0 + \omega_M - \omega_p \gg \kappa, \tag{5.89}$$

which will be explained shortly.

After a time $\sim \kappa^{-1}$ the resonator reaches a steady-state balancing between the pumping and the leakage. Neglecting for a moment the nonlinear coupling term, the dynamics of the resonator mode in the steady state is fully described by Eqs. (5.42)–(5.44). Namely,

$$\hat{b}(t) = \frac{\mathrm{i}\eta\, \mathrm{e}^{-\mathrm{i}\omega_p t}}{\omega_0 - \omega_p - \mathrm{i}\kappa} + \delta\hat{b}(t); \qquad \hat{b}^\dagger(t) = -\frac{\mathrm{i}\eta\, \mathrm{e}^{\mathrm{i}\omega_p t}}{\omega_0 - \omega_p + \mathrm{i}\kappa} + \delta\hat{b}^\dagger(t), \tag{5.90}$$

where Gaussian correlations of $\delta\hat{b}(t)$ and $\delta\hat{b}^\dagger(t)$ are given by Eqs. (5.43) and (5.44). We now substitute this into the nonlinear interaction term in Eq. (5.88) and focus on terms linear in the laser amplitude η (other contributions are of a lesser interest). This way we arrive at the effective interaction Hamiltonian

$$\hat{H}_{\text{int}} = g'\, \delta\hat{b}^\dagger \hat{B}\, \mathrm{e}^{-\mathrm{i}\omega_p t} + \bar{g}' \hat{B}^\dagger \delta\hat{b}\, \mathrm{e}^{\mathrm{i}\omega_p t}, \tag{5.91}$$

where $g' = g\eta/(\omega_0 - \omega_p - \mathrm{i}\kappa) \approx g\eta/\omega_M$ (cf. Eq. (5.89)). This Hamiltonian represents processes depicted in Fig. 5.5b, where the external drive stimulates emission of the mechanical oscillation quantum ω_M with the simultaneous absorption of the cavity quantum ω_0 (and the inverse processes). There are also terms with $g'\delta\hat{b}^\dagger\hat{B}^\dagger\,\mathrm{e}^{-\mathrm{i}\omega_p t} + \bar{g}'\hat{B}\delta\hat{b}\,\mathrm{e}^{\mathrm{i}\omega_p t}$, depicted in Fig. 5.5c, representing stimulated emission or absorption of both quanta. These latter processes are far from the resonance, due to the second inequality in (5.89), and thus are much less important. The Hamiltonian (5.91) represents interactions of the mechanical oscillator with the thermal "bath" of the leaky cavity modes, exactly as in Eq. (5.3b).

One can now integrate out the cavity modes, $\delta\hat{b}^\dagger$ and $\delta\hat{b}$, using Eq. (5.44). This way, one arrives at the dissipative action for the mechanical degree of freedom of the form of Eq. (5.4). It is characterized by the dissipation constant $2\kappa^{\mathrm{eff}} = |g'|^2\mathrm{Im}\,G^{\mathrm{A}}(\omega_M+\omega_p) = |g'|^2\kappa/[(\omega_M+\omega_p-\omega_0)^2+\kappa^2] \approx g^2\eta^2/\omega_M^2\kappa$ and the effective occupation number $n^{\mathrm{eff}} = (\mathrm{e}^{(\omega_M+\omega_p)/T} - 1)^{-1}$. The last observation is particularly important. It shows that the effective temperature, T^{eff}, of the mechanical oscillator with frequency ω_M is

$$\mathrm{e}^{\omega_M/T^{\mathrm{eff}}} = \frac{n^{\mathrm{eff}}+1}{n^{\mathrm{eff}}} \approx \mathrm{e}^{\omega_0/T}; \qquad T^{\mathrm{eff}} = T\,\frac{\omega_M}{\omega_0} \ll T. \tag{5.92}$$

This temperature is reached after a time $(\kappa^{\mathrm{eff}})^{-1}$. To avoid overheating of the cavity, this time should be longer than the cavity equilibration time, κ^{-1}, dictating $\kappa > g\eta/\omega_M$.

The physics behind the optomechanical cooling is apparent from Fig. 5.5b. The resonance condition requires emission of the mechanical quantum and absorption of the optical one. For $T \ll \omega_0$ the optical fluctuations are rare: $\langle\delta\hat{b}^\dagger\delta\hat{b}\rangle = n_B \ll 1$. Thus the optical absorption is much more likely than the emission, with the ratio $(n_B+1)/n_B \gg 1$. Therefore, the mechanical mode undergoes many more stimulated emission processes than absorption ones, with the same ratio. As a result, there is a decrease in energy drift for the mechanical mode, even though $\omega_M < T$. The corresponding Lindblad equation takes the form of Eq. (5.29) with $\Delta \to \omega_M$, $\kappa \to \kappa^{\mathrm{eff}}$, and $n_B \to n^{\mathrm{eff}}$. It leads to the steady-state thermal population of the mechanical modes with $T^{\mathrm{eff}} \ll T$. For the cooling to be effective, it is important that the simultaneous mechanical and optical absorptions, Fig. 5.5c, are off-resonance and thus occur very rarely. This is insured by the relatively narrow width of the optical "bath" spectral density $\kappa \ll \omega_M$, which allows for the two inequalities (5.89) to be consistent.

5.11 Problems

5.11.1 Master Equation in the Number Basis

Consider the Lindblad equation (5.29) in the number state basis. Show that diagonal matrix elements $\mathcal{P}_n(t) = \langle n|\hat{\rho}(t)|n\rangle$ satisfy a closed Master equation

$$\partial_t \mathcal{P}_n = 2\kappa(n_B+1)\big[(n+1)\mathcal{P}_{n+1} - n\mathcal{P}_n\big] + 2\kappa n_B\big[n\mathcal{P}_{n-1} - (n+1)\mathcal{P}_n\big], \quad (5.93)$$

where $n_B = (e^{\omega_0/T}-1)^{-1}$. Show that Eq. (5.35) provides a stationary solution for this equation. This is the Fokker–Planck equation for the stochastic action (5.67) with the effective Hamiltonian (5.69), which we have derived using the action-angle representation.

5.11.2 Husimi Function

The Husimi function is defined as

$$\mathcal{Q}(\bar\phi,\phi,t) = e^{-|\phi|^2}\langle\phi|\hat\rho|\phi\rangle. \quad (5.94)$$

It is normalized as $\int d[\bar\phi,\phi]\mathcal{Q}(\bar\phi,\phi,t) = \operatorname{Tr}\hat\rho = 1$. Starting from the Lindblad equation (5.29) and employing Eq. (5.31), derive the evolution equation for the Husimi function. Show that its stationary solution is

$$\mathcal{Q}(\bar\phi,\phi) = \frac{1}{n_B+1}\exp\left\{-\frac{|\phi|^2}{n_B+1}\right\}. \quad (5.95)$$

Calculate the Husimi function for the density matrix (5.35) and compare it to the preceding stationary solution.

5.11.3 Kinetic Equation

According to Eq. (5.41c), $\langle\phi^{\rm cl}(t)\bar\phi^{\rm cl}(t)\rangle = F(t)$ is the equal time expectation value of the classical fields. It may be understood as $F(t) = \int d[\bar\phi,\phi]\,\phi\bar\phi\,\mathcal{P}(\bar\phi,\phi,t)$, where the probability distribution, $\mathcal{P}(\bar\phi,\phi,t)$, satisfies the Fokker–Planck equation (5.39). Derive a closed evolution equation for $F(t)$.

Solution: Multiplying the Fokker–Planck equation (5.39) by $\phi^{\rm cl}\bar\phi^{\rm cl}$ and integrating over the phase plane, one finds (show that rotational terms, $\sim i\Delta$, drop out)

$$\partial_t F = \kappa\int d[\bar\phi,\phi]\phi\bar\phi\Big[\partial_\phi(\phi\mathcal{P}) + \partial_{\bar\phi}(\bar\phi\mathcal{P}) + 2(2n_B+1)\partial^2_{\bar\phi\phi}\mathcal{P}\Big] = 2\kappa(2n_B+1-F),$$

where in the last equality we employed integration by parts and normalization $\int d[\bar\phi,\phi]\mathcal{P} = 1$. We have found the kinetic equation (5.45) for the distribution function.

5.11.4 Bloch Equations

Consider a *two-level system* (e.g. spin 1/2) with a 2×2 density matrix satisfying the Lindblad equation

$$\partial_t \hat{\rho} = \mathrm{i}\big[\hat{\rho}, \vec{B} \cdot \hat{\vec{\sigma}}\big] + 2\kappa(n_B + 1)\Big(\hat{\sigma}_- \hat{\rho} \hat{\sigma}_+ - \frac{1}{2}\{\hat{\sigma}_+ \hat{\sigma}_-, \hat{\rho}\}\Big)$$
$$+ 2\kappa n_B \Big(\hat{\sigma}_+ \hat{\rho} \hat{\sigma}_- - \frac{1}{2}\{\hat{\sigma}_- \hat{\sigma}_+, \hat{\rho}\}\Big),$$

where $\hat{\vec{\sigma}} = (\hat{\sigma}_1, \hat{\sigma}_2, \hat{\sigma}_3)$ are the Pauli matrices and $\hat{\sigma}_\pm = (\hat{\sigma}_1 \pm \mathrm{i}\hat{\sigma}_2)/2$. The density matrix may be written as $\hat{\rho}(t) = \frac{1}{2}[\hat{\sigma}_0 + \vec{m}(t) \cdot \hat{\vec{\sigma}}]$, where $\vec{m}(t) = \mathrm{Tr}\{\hat{\rho}(t)\hat{\vec{\sigma}}\}$ is the magnetization vector with $\vec{m} \cdot \vec{m} \le 1$. Rewrite the Lindblad equation as equations for components of the magnetization vector $\vec{m}(t)$. These equations are known as Bloch equations.

Solution:

$$\partial_t m_{1,2} = 2[\vec{B} \times \vec{m}]_{1,2} - \frac{m_{1,2}}{T_2}; \qquad \partial_t m_3 = 2[\vec{B} \times \vec{m}]_3 - \frac{m_3 - \langle m_3 \rangle}{T_1},$$

where $T_2^{-1} = \kappa(2n_B + 1)$, $T_1^{-1} = 2\kappa(2n_B + 1)$ and $\langle m_3 \rangle = -(2n_B + 1)^{-1}$.

Part II

Bosonic and Classical Fields

6

Bosonic Fields

In this chapter we generalize the formalism of Chapter 2 for the case of complex and real interacting bosonic fields. We then develop a perturbative diagrammatic technique and use it to derive the quantum kinetic equation.

6.1 Complex Bosonic Fields

Consider a box of size L filled with bosonic particles of mass m. The single-particle states within the box are labeled by the wavenumber vector $\mathbf{k} = (2\pi/L)\mathbf{n}$, where the vector $\mathbf{n} = (n_x, n_y, n_z)$ has integer components $n_\mu = 0, \pm 1, \pm 2, \ldots$ (we have assumed periodic boundary conditions in all directions). The corresponding energies are given by $\omega_{\mathbf{k}} = \mathbf{k}^2/(2m)$. One may associate bosonic creation and annihilation operators $\hat{b}^\dagger_{\mathbf{k}}$ and $\hat{b}_{\mathbf{k}}$, obeying the commutation relations $[\hat{b}_{\mathbf{k}}, \hat{b}^\dagger_{\mathbf{k}'}] = \delta_{\mathbf{k},\mathbf{k}'}$, with each of these single-particle states. The kinetic energy part of the Hamiltonian written in terms of such operators takes the form

$$\hat{H}_0 = \sum_{\mathbf{k}} \omega_{\mathbf{k}} \hat{b}^\dagger_{\mathbf{k}} \hat{b}_{\mathbf{k}}. \tag{6.1}$$

Assuming some initial density matrix, for example, $\hat{\rho}_0 = \exp\{-\beta(\hat{H}_0 - \mu \hat{N})\}$, where the number operator is $\hat{N} = \sum_{\mathbf{k}} \hat{b}^\dagger_{\mathbf{k}} \hat{b}_{\mathbf{k}}$, one may write the Keldysh partition function $Z = 1$ (see Eq. (2.13)), as a functional integral over the closed time contour (Fig. 2.1). The coherent states are parametrized by a set of complex numbers $\phi_j(\mathbf{k})$, labeled by the discrete time index j along with the state index $\mathbf{k}$. Transforming to continuum notation and performing the Keldysh rotation according to Eq. (2.39) for each state $\mathbf{k}$, one obtains the two sets of complex fields $\phi^{\mathrm{cl}}(\mathbf{k}, t)$ and $\phi^{\mathrm{q}}(\mathbf{k}, t)$. The partition function acquires the form

$$Z = \int \mathbf{D}[\phi^{\mathrm{cl}}, \phi^{\mathrm{q}}]\, e^{iS_0[\phi^{\mathrm{cl}}, \phi^{\mathrm{q}}]}, \tag{6.2}$$

where the integration measure is given by

$$\mathbf{D}[\phi^{\mathrm{cl}},\phi^{\mathrm{q}}] = \frac{1}{\mathrm{Tr}\{\hat\rho_0\}}\prod_{\mathbf{k}}\prod_{j=1}^{N}\frac{\mathrm{d}(\mathrm{Re}\phi_j^{\mathrm{cl}}(\mathbf{k}))\mathrm{d}(\mathrm{Im}\phi_j^{\mathrm{cl}}(\mathbf{k}))}{\pi}\,\frac{\mathrm{d}(\mathrm{Re}\phi_j^{\mathrm{q}}(\mathbf{k}))\mathrm{d}(\mathrm{Im}\phi_j^{\mathrm{q}}(\mathbf{k}))}{\pi}, \tag{6.3}$$

and the limit $N \to \infty$ is understood. The Keldysh action of the free complex Bose field is written, employing Eqs. (2.51) and (2.52), as

$$S_0[\phi^{\mathrm{cl}},\phi^{\mathrm{q}}] = \sum_{\mathbf{k}}\int_{-\infty}^{\infty}\mathrm{d}t\,(\bar\phi^{\mathrm{cl}},\bar\phi^{\mathrm{q}})\begin{pmatrix} 0 & \mathrm{i}\partial_t - \omega_{\mathbf{k}} - \mathrm{i}0 \\ \mathrm{i}\partial_t - \omega_{\mathbf{k}} + \mathrm{i}0 & 2\mathrm{i}0F(\omega_{\mathbf{k}})\end{pmatrix}\begin{pmatrix}\phi^{\mathrm{cl}}\\ \phi^{\mathrm{q}}\end{pmatrix}, \tag{6.4}$$

where $\phi^{\mathrm{cl,q}} = \phi^{\mathrm{cl,q}}(\mathbf{k},t)$. The $\pm\mathrm{i}0$ indicates the retarded/advanced nature of the off-diagonal operators and specifies how the corresponding inverted operators are to be understood. The q − q Keldysh component is a pure regularization for the free field. Unlike the cl − cl component, it becomes finite (and in general non-local with respect to time and state indices) once the interactions between the particles are included. We kept it explicitly here to remind us that it determines the way the quadratic form in the action is inverted.

The corresponding free (bare) Green function is defined as

$$G_0^{\alpha\beta}(\mathbf{k},\mathbf{k}',t,t') = -\mathrm{i}\int\mathbf{D}[\phi^{\mathrm{cl}},\phi^{\mathrm{q}}]\;\phi^{\alpha}(\mathbf{k},t)\,\bar\phi^{\beta}(\,\mathbf{k}',t')\;\mathrm{e}^{\mathrm{i}S_0[\phi^{\mathrm{cl}},\phi^{\mathrm{q}}]} \tag{6.5}$$

and according to the rules of the Gaussian integration is given by the inverse of the quadratic form in the action

$$G_0^{\alpha\beta}(\mathbf{k},\mathbf{k}',t,t') = \delta_{\mathbf{k},\mathbf{k}'}\begin{pmatrix} G_0^{\mathrm{K}}(\mathbf{k},t-t') & G_0^{\mathrm{R}}(\mathbf{k},t-t') \\ G_0^{\mathrm{A}}(\mathbf{k},t-t') & 0\end{pmatrix}. \tag{6.6}$$

The three nonzero components of the Green function are

$$G_0^{\mathrm{R}}(\mathbf{k},t) = -\mathrm{i}\theta(t)\,\mathrm{e}^{-\mathrm{i}\omega_{\mathbf{k}}t} \overset{\mathrm{FT}}{\to} (\epsilon-\omega_{\mathbf{k}}+\mathrm{i}0)^{-1}; \tag{6.7a}$$

$$G_0^{\mathrm{A}}(\mathbf{k},t) = \mathrm{i}\theta(-t)\,\mathrm{e}^{-\mathrm{i}\omega_{\mathbf{k}}t} \overset{\mathrm{FT}}{\to} (\epsilon-\omega_{\mathbf{k}}-\mathrm{i}0)^{-1}; \tag{6.7b}$$

$$G_0^{\mathrm{K}}(\mathbf{k},t) = -\mathrm{i}F(\omega_{\mathbf{k}})\,\mathrm{e}^{-\mathrm{i}\omega_{\mathbf{k}}t} \overset{\mathrm{FT}}{\to} -2\pi\mathrm{i}F(\epsilon)\,\delta(\epsilon-\omega_{\mathbf{k}}). \tag{6.7c}$$

In equilibrium the distribution function is $F(\epsilon) = \coth(\epsilon-\mu)/(2T)$. Here we also quoted the Fourier transforms with respect to the time argument for all three components.

It is sometimes convenient to perform the linear change of variables in the functional integral to introduce the coordinate space representation for the two complex bosonic fields

$$\phi^\alpha(\mathbf{r},t) = \sum_{\mathbf{k}} \phi^\alpha(\mathbf{k},t)\, e^{i\mathbf{kr}}. \tag{6.8}$$

In terms of these fields the bare bosonic action (6.4) takes the form

$$S_0 = \int d\mathbf{r} \int_{-\infty}^{\infty} dt\, (\bar\phi^{cl}, \bar\phi^{q}) \begin{pmatrix} 0 & i\partial_t + \frac{\nabla_{\mathbf{r}}^2}{2m} - V^{cl} \\ i\partial_t + \frac{\nabla_{\mathbf{r}}^2}{2m} - V^{cl} & 2i0F \end{pmatrix} \begin{pmatrix} \phi^{cl} \\ \phi^{q} \end{pmatrix}, \tag{6.9}$$

where we have added an external classical potential $V^{cl} = V^{cl}(\mathbf{r},t)$ in accordance with Eq. (2.66). In the absence of such an external potential, the correlators of the coordinate space bosonic fields are given by the Fourier transform of the Green function (6.6), (6.7)

$$\langle \phi^\alpha(\mathbf{r},t)\bar\phi^\beta(\mathbf{r}',t')\rangle = iG_0^{\alpha\beta}(\mathbf{r}-\mathbf{r}',t-t') = i\sum_{\mathbf{k}} G_0^{\alpha\beta}(\mathbf{k},t-t')e^{i\mathbf{k}(\mathbf{r}-\mathbf{r}')}. \tag{6.10}$$

6.2 Interactions

Let us now include interactions between bosonic particles through a pairwise interaction potential $U(\mathbf{r}-\mathbf{r}')$. The corresponding *normally ordered* Hamiltonian takes the form

$$\hat H_{\text{int}} = \frac{1}{2}\sum_{\mathbf{q},\mathbf{k},\mathbf{k}'} U(\mathbf{q})\, \hat b^\dagger_{\mathbf{k}} \hat b^\dagger_{\mathbf{k}'} \hat b_{\mathbf{k}'+\mathbf{q}} \hat b_{\mathbf{k}-\mathbf{q}}, \tag{6.11}$$

where $U(\mathbf{q})$ is the Fourier transform of the interaction potential. In the case of dilute atomic gases the interaction potential may be thought of as being short-ranged, that is, momentum-independent, $U(\mathbf{q}) = g$, where the interaction constant may be expressed through the s-wave scattering length a_s as $g = 4\pi a_s/m$ [115]. The corresponding term in the action takes the form

$$S_{\text{int}} = -\frac{g}{2}\sum_{\mathbf{q},\mathbf{k},\mathbf{k}'} \int_{\mathcal{C}} dt\; \bar\phi(\mathbf{k},t)\bar\phi(\mathbf{k}',t)\phi(\mathbf{k}'+\mathbf{q},t)\phi(\mathbf{k}-\mathbf{q},t).$$

Going to the coordinate space representation, one finds

$$S_{\text{int}} = -\frac{g}{2}\int d\mathbf{r} \int_{\mathcal{C}} dt\,(\bar\phi\phi)^2 = -\frac{g}{2}\int d\mathbf{r}\int_{-\infty}^{+\infty} dt\,\left[(\bar\phi^+\phi^+)^2 - (\bar\phi^-\phi^-)^2\right]. \tag{6.12}$$

It is important to remember that there are no interactions in the distant past, $t = -\infty$ (while they are present in the future, $t = +\infty$). The interactions are supposed to be adiabatically switched on and off on the forward and backward branches correspondingly. Therefore, the interactions modify only those matrix elements of the evolution operator, Eq. (2.17), that are away from $t = -\infty$. It

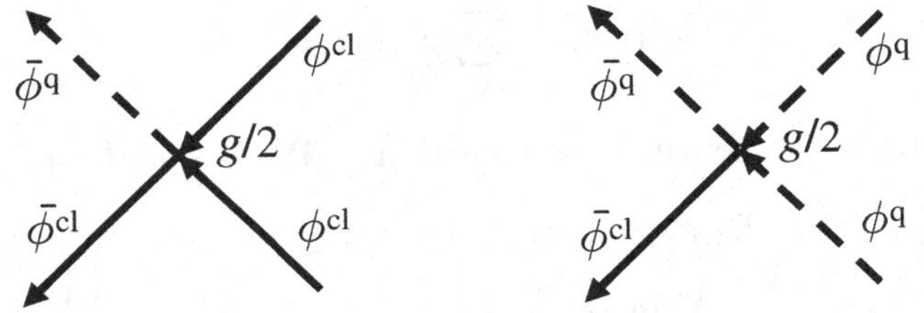

Figure 6.1 Graphic representation of the two interaction vertices of the $|\phi|^4$ theory. There are also two complex conjugated vertices with a reversed direction of all arrows.

is also worth remembering that in the discrete time form the $\bar\phi$ fields are taken one time step δ_t *after* the ϕ fields along the contour $\mathcal{C}$. Performing the Keldysh rotation, Eq. (2.39), one finds

$$S_{\rm int}[\phi^{\rm cl},\phi^{\rm q}] = -\frac{g}{2}\int d\mathbf{r}\int_{-\infty}^{+\infty} dt\left[\bar\phi^{\rm cl}\bar\phi^{\rm q}\phi^{\rm cl}\phi^{\rm cl} + \bar\phi^{\rm cl}\bar\phi^{\rm q}\phi^{\rm q}\phi^{\rm q} + \text{c.c.}\right], \qquad (6.13)$$

where c.c. stands for the complex conjugate of the first two terms. The interaction action, Eq. (6.13), obviously satisfies the normalization condition, Eq. (2.53). Diagrammatically, the action (6.13) generates two types of vertex depicted in Fig. 6.1: one with three classical fields (full lines) and one quantum field (dashed line) and the other with one classical field and three quantum fields (as well as two complex conjugated vertices, obtained by reversing the direction of the arrows).

Let us demonstrate that the addition of the interaction term to the action does not violate the normalization identity, $Z = 1$. To this end, one may expand $\exp(iS_{\rm int})$ in powers of g and then average term by term with the help of the Gaussian action (6.9). To show that the normalization, $Z = 1$, is intact, one needs to show that $\langle S_{\rm int}\rangle = \langle S_{\rm int}^2\rangle = \cdots = 0$. Applying the Wick theorem, Eq. (2.21), one finds for the term linear in g

$$\langle S_{\rm int}\rangle = -\frac{g}{2}\int d\mathbf{r}\, dt\left\langle \bar\phi^{\rm cl}\bar\phi^{\rm q}\phi^{\rm cl}\phi^{\rm cl} + \bar\phi^{\rm cl}\bar\phi^{\rm cl}\phi^{\rm cl}\phi^{\rm q} + \bar\phi^{\rm q}\bar\phi^{\rm cl}\phi^{\rm q}\phi^{\rm q} + \bar\phi^{\rm q}\bar\phi^{\rm q}\phi^{\rm q}\phi^{\rm cl}\right\rangle.$$

The first two terms upon application of the Wick theorem lead to diagrams of the type of Fig. 6.2(a):

$$\left\langle \bar\phi^{\rm cl}\bar\phi^{\rm q}\phi^{\rm cl}\phi^{\rm cl} + \bar\phi^{\rm cl}\bar\phi^{\rm cl}\phi^{\rm cl}\phi^{\rm q}\right\rangle = -2\left[G_0^{\rm R}(t,t) + G_0^{\rm A}(t,t)\right]G_0^{\rm K}(t,t) = 0,$$

where we have suppressed the space arguments and focused only on the time ones and the factor of two originates from the two combinatorial possibilities to make Wick's contractions. This expression vanishes due to the identity (2.44). The last two terms in $\langle S_{\rm int}\rangle$ trivially vanish because $\langle \phi^{\rm q}\bar\phi^{\rm q}\rangle = 0$.

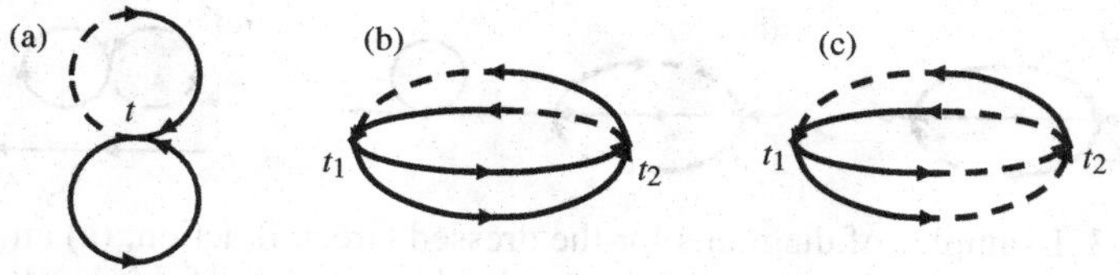

Figure 6.2 Diagrams for the first- (a) and second- (b), (c) order interaction corrections to the partition function Z. As explained in the text, they do not change the normalization identity $Z = 1$.

There are two families of terms that are second order in g and contain not more than four quantum fields (terms with six quantum fields unavoidably lead to q − q contractions and therefore vanish). They contain

$$\langle \bar{\phi}_1^{\rm q} \bar{\phi}_1^{\rm cl} \phi_1^{\rm cl} \phi_1^{\rm cl} \times \phi_2^{\rm q} \phi_2^{\rm cl} \bar{\phi}_2^{\rm cl} \phi_2^{\rm cl} \rangle = 2G_0^{\rm R}(t_2,t_1) G_0^{\rm A}(t_2,t_1) [G_0^{\rm K}(t_1,t_2)]^2,$$

Fig. 6.2(b), and

$$\langle \bar{\phi}_1^{\rm q} \bar{\phi}_1^{\rm cl} \phi_1^{\rm cl} \phi_1^{\rm cl} \times \phi_2^{\rm q} \phi_2^{\rm cl} \bar{\phi}_2^{\rm q} \phi_2^{\rm q} \rangle = 2G_0^{\rm R}(t_2,t_1) G_0^{\rm A}(t_2,t_1) [G_0^{\rm R}(t_1,t_2)]^2,$$

Fig. 6.2(c), where $\phi_{1,2}^{\alpha} = \phi^{\alpha}(\mathbf{r}_{1,2}, t_{1,2})$. Both of these terms are zero, because $G_0^{\rm R}(t_2,t_1) \sim \theta(t_2 - t_1)$, while $G_0^{\rm A}(t_2,t_1) \sim G_0^{\rm R}(t_1,t_2)^* \sim \theta(t_1 - t_2)$, and thus their product has no support in the time domain. One may be concerned that $G_0^{\rm R}(t_2,t_1)$ and $G_0^{\rm A}(t_2,t_1)$ are simultaneously nonzero on the diagonal $t_1 = t_2$. The contribution of the diagonal to the double integral over $dt_1 dt_2$, however, is of the order $\sim \delta_t^2 N \to 0$, when $N \to \infty$. It is easy to see that, for exactly the same reasons, all higher-order terms in g vanish and thus the fundamental normalization is indeed intact (at least in the perturbative expansion). However, the observables and correlation functions are affected by the interactions. We demonstrate it in what follows on the example of the Green functions.

6.3 Dyson Equation

We define the full or *dressed* Green function as the correlator of the fields averaged with the weight, which includes both the bare action S_0 and the interaction action:

$$G^{\alpha\beta}(\mathbf{r},\mathbf{r}',t,t') = -\mathrm{i} \int \mathbf{D}[\bar{\phi}\phi]\, \phi^{\alpha}(\mathbf{r},t)\, \bar{\phi}^{\beta}(\mathbf{r}',t')\, e^{\mathrm{i}(S_0+S_{\rm int})}, \tag{6.14}$$

here $\alpha, \beta = ({\rm cl}, {\rm q})$ and the action is given by Eqs. (6.9) and (6.13). To evaluate the full Green function one may expand the exponent in powers of $S_{\rm int}$. The functional integration with the remaining Gaussian action S_0 is then performed using the Wick theorem. This procedure leads to an infinite series of terms which are convenient to represent by Feynman diagrams. Each of these diagrams has two external “legs”: an

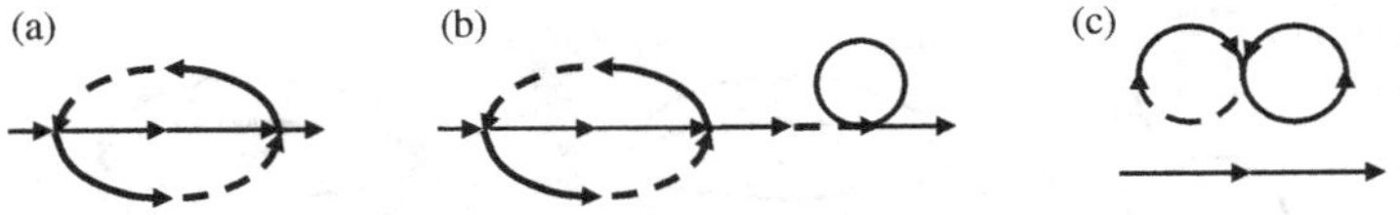

Figure 6.3 Examples of diagrams for the dressed Green function: (a) an irreducible diagram of second order in g; (b) a reducible diagram of third order, which contains two irreducible blocks; (c) a disconnected diagram of first order.

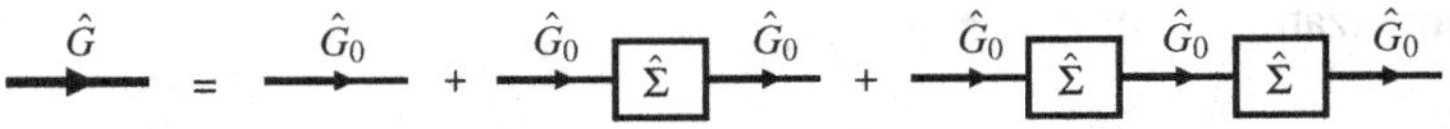

Figure 6.4 Diagrammatic series for the dressed Green function $\hat{G}$, rearranged into the Dyson series. The self-energy blocks contain the sum of all *irreducible* diagrams.

incoming, staying for the contraction $\langle \phi^\alpha(x)\bar{\phi}^\gamma(x_1)\rangle = iG_0^{\alpha\gamma}(x,x_1)$, and an outgoing, representing $\langle \phi^\delta(x_2)\bar{\phi}^\beta(x')\rangle = iG_0^{\delta\beta}(x_2,x')$, where we introduced a combined notation $x = \mathbf{r},t$. The interior of a diagram, which is a matrix in Keldysh indices γ,δ as well as in space-time coordinates x_1,x_2, contains a number of internal four-leg vertices, each carrying a factor of $g/2$. Integration over space-time coordinates of all internal vertices as well as summation over Keldysh indices is assumed. Examples of the diagrams are given in Fig. 6.3.

One can now define *irreducible* diagrams as those which cannot be cut into two disconnected parts by cutting a single line in the interior of the diagram. The diagram in Fig. 6.3(a) is irreducible, while the one in Fig. 6.3(b) is reducible. The diagram in Fig. 6.3(c) is a disconnected one. The disconnected diagrams contain all the same building blocks as in Fig. 6.2 and thus are zero, as explained previously.[1] Rearranging the order of terms in the perturbative expansion, one may formally sum up the inner parts of all irreducible diagrams and call the resulting object the *self-energy* $\Sigma^{\gamma\delta}(x_1,x_2)$. The full series may be written then (see Fig. 6.4) as

$$\hat{G} = \hat{G}_0 + \hat{G}_0 \circ \hat{\Sigma} \circ \hat{G}_0 + \hat{G}_0 \circ \hat{\Sigma} \circ \hat{G}_0 \circ \hat{\Sigma} \circ \hat{G}_0 + \cdots = \hat{G}_0 + \hat{G}_0 \circ \hat{\Sigma} \circ \hat{G}, \quad (6.15)$$

where the circular multiplication sign implies convolution of the space-time coordinates as well as a 2×2 Keldysh matrix multiplication. The only difference compared with the standard diagrammatic expansion [2, 4, 6] is the presence of the 2×2 matrix structure. The fact that the series is arranged as a sequence of matrix products is no surprise. Indeed, the Keldysh index, $\alpha = (\mathrm{cl}, \mathrm{q})$, is just one more index in addition to time, space, spin, and others. Therefore, as with any other

[1] Cancellation of disconnected diagrams is a direct consequence of the normalization identity $Z = 1$. Notice that in equilibrium theory the disconnected diagrams are *not* zero and serve to compensate for the denominator e^{iL}, Eq. (1.5), [2, 4].

index, there is a summation over all of its intermediate values, hence the matrix multiplication. The concrete form of the self-energy matrix, $\hat{\Sigma}$, is specific to the Keldysh technique and is discussed in what follows in some detail.

Multiplying both sides of Eq. (6.15) by $\hat{G}_0^{-1}$ from the left, one obtains an equation for the *exact* dressed Green function, $\hat{G}$,

$$\left(\hat{G}_0^{-1} - \hat{\Sigma}\right) \circ \hat{G} = \hat{1}, \tag{6.16}$$

where $\hat{1}$ is the unit matrix. This equation is named after Dyson. The very nontrivial feature of the Keldysh technique is that the self-energy matrix, $\hat{\Sigma}$, possesses the same causality structure as $\hat{G}_0^{-1}$, Eq. (2.51), namely

$$\hat{\Sigma} = \begin{pmatrix} 0 & \Sigma^{\mathrm{A}} \\ \Sigma^{\mathrm{R}} & \Sigma^{\mathrm{K}} \end{pmatrix}, \tag{6.17}$$

where $\Sigma^{\mathrm{R(A)}}$ are mutually Hermitian conjugated lower (upper) triangular matrices with respect to the two time indices, while Σ^{K} is an anti-Hermitian matrix

$$\Sigma^{\mathrm{R}}(x_1,x_2) = [\Sigma^{\mathrm{A}}(x_2,x_1)]^* \sim \theta(t_1 - t_2); \quad \Sigma^{\mathrm{K}}(x_1,x_2) = -[\Sigma^{\mathrm{K}}(x_2,x_1)]^*. \tag{6.18}$$

This fact will be explicitly demonstrated in what follows. Since both $\hat{G}_0^{-1}$ and $\hat{\Sigma}$ have the same causality structure, one concludes that the dressed Green function, $\hat{G}$, also possesses the causality structure, like Eq. (2.40). As a result, the Dyson equation acquires the form

$$\begin{pmatrix} 0 & \left[G_0^{\mathrm{A}}\right]^{-1} - \Sigma^{\mathrm{A}} \\ \left[G^{\mathrm{R}}\right]_0^{-1} - \Sigma^{\mathrm{R}} & -\Sigma^{\mathrm{K}} \end{pmatrix} \circ \begin{pmatrix} G^{\mathrm{K}} & G^{\mathrm{R}} \\ G^{\mathrm{A}} & 0 \end{pmatrix} = \hat{1}, \tag{6.19}$$

where one took into account that $\left[G_0^{-1}\right]^{\mathrm{K}}$ is a pure regularization ($\sim i0F$) and thus may be omitted in the presence of a nonzero self-energy component Σ^{K}. Employing the specific form of $\left[G_0^{\mathrm{R(A)}}\right]^{-1}$, Eq. (6.9), one obtains for the retarded (advanced) component

$$\left(i\partial_t + \frac{1}{2m}\nabla_{\mathbf{r}}^2 - V^{\mathrm{cl}}(\mathbf{r},t) - \Sigma^{\mathrm{R(A)}}\right) \circ G^{\mathrm{R(A)}}(x,x') = \delta(t-t')\delta(\mathbf{r}-\mathbf{r}'). \tag{6.20}$$

Provided the self-energy component $\Sigma^{\mathrm{R(A)}}$ is known (in some approximation), Eq. (6.20) constitutes a closed equation for the retarded (advanced) component of the dressed Green function.

For the space-time translationally invariant system, $V^{\mathrm{cl}} = 0$, such that $\hat{G}(x,x') = \hat{G}(x-x')$, this equation may be solved explicitly with the help of the Fourier transform, leading to

$$G^{\mathrm{R(A)}}(\mathbf{k},\epsilon) = \left(\epsilon - \frac{\mathbf{k}^2}{2m} - \Sigma^{\mathrm{R(A)}}(\mathbf{k},\epsilon)\right)^{-1}. \tag{6.21}$$

Employing Eq. (6.18), one observes that $\mathrm{Re}\Sigma^{\mathrm{R}}(\mathbf{k},\epsilon) = \mathrm{Re}\Sigma^{\mathrm{A}}(\mathbf{k},\epsilon)$ and $\mathrm{Im}\Sigma^{\mathrm{R}}(\mathbf{k},\epsilon) = -\mathrm{Im}\Sigma^{\mathrm{A}}(\mathbf{k},\epsilon) \le 0$. The real part of the retarded (advanced) self-energy provides renormalization of the particle's dispersion relation. That is, the relation $\epsilon = \mathbf{k}^2/(2m)$ should be substituted by the solution of the equation $\epsilon - \mathbf{k}^2/(2m) - \mathrm{Re}\Sigma^{\mathrm{R(A)}}(\mathbf{k},\epsilon) = 0$. On the other hand, the imaginary part of the self-energy has the meaning of the inverse lifetime a particle spends in a given (renormalized) eigenstate $\mathbf{k}$ of the noninteracting system.

We turn now to the Keldysh component of the Dyson equation. As before, it is convenient to parametrize the Keldysh component of the Green function as

$$G^{\mathrm{K}} = G^{\mathrm{R}} \circ F - F \circ G^{\mathrm{A}} \tag{6.22}$$

(compare with Eq. (2.49)), where $F(x,x')$ is a Hermitian matrix in the space-time domain. The Keldysh, namely (2, 1) matrix component, of the Dyson equation (6.19) then takes the form $\left(\left[G_0^{\mathrm{R}}\right]^{-1} - \Sigma^{\mathrm{R}}\right) \circ \left(G^{\mathrm{R}} \circ F - F \circ G^{\mathrm{A}}\right) = \Sigma^{\mathrm{K}} \circ G^{\mathrm{A}}$. Multiplying it from the right by $\left(\left[G_0^{\mathrm{A}}\right]^{-1} - \Sigma^{\mathrm{A}}\right)$ and employing Eq. (6.20), one finds $F \circ \left(\left[G_0^{\mathrm{A}}\right]^{-1} - \Sigma^{\mathrm{A}}\right) - \left(\left[G_0^{\mathrm{R}}\right]^{-1} - \Sigma^{\mathrm{R}}\right) \circ F = \Sigma^{\mathrm{K}}$. This may be written as

$$F \circ \left[G_0^{\mathrm{A}}\right]^{-1} - \left[G_0^{\mathrm{R}}\right]^{-1} \circ F = \Sigma^{\mathrm{K}} - \left(\Sigma^{\mathrm{R}} \circ F - F \circ \Sigma^{\mathrm{A}}\right). \tag{6.23}$$

Since $\left[G_0^{\mathrm{R}}\right]^{-1}(x',x) = \left[G_0^{\mathrm{A}}\right]^{-1}(x',x) = \delta(x'-x)\left(\mathrm{i}\partial_t + \nabla_{\mathbf{r}}^2/(2m) - V^{\mathrm{cl}}(x)\right)$, where the regularization $\pm\mathrm{i}0$ may be omitted in this context, one finally finds

$$-\left[\left(\mathrm{i}\partial_t + \frac{1}{2m}\nabla_{\mathbf{r}}^2 - V^{\mathrm{cl}}(x)\right) \overset{\circ}{,} F\right] = \Sigma^{\mathrm{K}} - \left(\Sigma^{\mathrm{R}} \circ F - F \circ \Sigma^{\mathrm{A}}\right), \tag{6.24}$$

where the symbol $[\,\overset{\circ}{,}\,]$ stands for the commutator. With the help of integration by parts, it may be understood as $[\partial_t \overset{\circ}{,} F] = (\partial_t + \partial_{t'})F(x,x')$, on the other hand $[\nabla_{\mathbf{r}}^2 \overset{\circ}{,} F] = (\nabla_{\mathbf{r}}^2 - \nabla_{\mathbf{r}'}^2)F(x,x')$ and $[V \overset{\circ}{,} F] = (V(x) - V(x'))F(x,x')$. This equation is the quantum kinetic equation for the distribution matrix $F(x,x')$. Schematically, its left-hand side forms the *kinetic* term, while the right-hand side is the *collision term* or the *collision integral*. In equilibrium the kinetic term vanishes. This implies, in turn, that the self-energy possesses the same structure as the Green function: $\Sigma^{\mathrm{K}} = \Sigma^{\mathrm{R}} \circ F - F \circ \Sigma^{\mathrm{A}}$. The latter is not the case, however, away from equilibrium.

6.4 Real Bosonic Fields

We briefly repeat now the construction of the interacting field theory for the case of real boson fields, such as, for example, elastic phonons. To this end we consider a toy model of a d-dimensional "quantum membrane." It is formed by a lattice of quantum particles, where a displacement of an $\mathbf{i}$th particle from the corresponding lattice point is denoted as $\varphi_{\mathbf{i}}$. For simplicity we consider φ as a scalar, which may be, for example, a transversal deformation of the membrane. We assume that the particles interact through some short-ranged potential, which leads to a bending rigidity κ of the membrane. The corresponding contribution to the energy is $U = (\kappa/2a^4)\sum_{\mathrm{i}}(\nabla^2\varphi_{\mathrm{i}})^2$, where ∇ is the lattice gradient operation and a is the lattice constant. In essence, the energy is paid for the *curvature* of the membrane. Furthermore, we also assume that each site experiences a static potential $V(\varphi_{\mathrm{i}})$, created by, for example, an external substrate. Being expanded around its minimum, such a potential may be written as

$$V(\varphi) = \frac{\omega_0^2}{2}\varphi^2 + \frac{\gamma}{6}\varphi^3 + \cdots . \tag{6.25}$$

The Keldysh action of an individual quantum particle is given by Eq. (3.4). Generalizing it for the lattice, one finds

$$S[\varphi] = \int_C \mathrm{d}t \sum_{\mathrm{i}} \left(\frac{1}{2}\dot{\varphi}_{\mathrm{i}}^2 - \frac{\kappa}{2a^4}(\nabla^2\varphi_{\mathrm{i}})^2 - V(\varphi_{\mathrm{i}}) \right), \tag{6.26}$$

where $\varphi_{\mathrm{i}} = \varphi_{\mathrm{i}}(t)$ with t running along the closed time contour. We take now the continuum limit by introducing the displacement density field $\varphi(\mathbf{r}, t)$, where $\mathbf{r}$ is the coordinate in the d-dimensional space of the membrane. In terms of this scalar *real field* the action takes the form

$$S[\varphi] = \int_C \mathrm{d}t \int \mathrm{d}\mathbf{r} \left[\frac{1}{2}\left(\dot{\varphi}^2 - \kappa(\nabla_{\mathbf{r}}^2\varphi)^2 - \omega_0^2\varphi^2 \right) - \frac{\gamma}{6}\varphi^3 \right]. \tag{6.27}$$

Performing the Keldysh rotation according to $\varphi^{\mathrm{cl,q}} = (\varphi^+ \pm \varphi^-)/2$, one finds for the quadratic part of the action (6.27)

$$S_0 = \frac{1}{2}\int \mathrm{d}\mathbf{r} \int_{-\infty}^{\infty} \mathrm{d}t\, (\varphi^{\mathrm{cl}}, \varphi^{\mathrm{q}}) \begin{pmatrix} 0 & -2(\partial_t^2 + \kappa\nabla_{\mathbf{r}}^4 + \omega_0^2) \\ -2(\partial_t^2 + \kappa\nabla_{\mathbf{r}}^4 + \omega_0^2) & -0[\partial_t, F] \end{pmatrix} \begin{pmatrix} \varphi^{\mathrm{cl}} \\ \varphi^{\mathrm{q}} \end{pmatrix}. \tag{6.28}$$

The matrix in the action is the inverse bare Green function $\hat{D}_0^{-1}$. As before, its Keldysh q – q component is a pure regularization, showing the way the matrix is to be inverted. Neglecting the cubic nonlinearity, the correlator of the real fields is given by the bare Green function

$$D_0^{\alpha\beta}(\mathbf{r}, \mathbf{r}', t, t') = -\mathrm{i} \int \mathbf{D}[\varphi^{\mathrm{cl}}, \varphi^{\mathrm{q}}]\, \varphi^{\alpha}(\mathbf{r}, t)\, \varphi^{\beta}(\mathbf{r}', t')\, \mathrm{e}^{\mathrm{i}S_0[\varphi^{\mathrm{cl}}, \varphi^{\mathrm{q}}]}, \tag{6.29}$$

which possesses the standard causality structure

$$D_0^{\alpha\beta}(\mathbf{r}, \mathbf{r}', t, t') = \begin{pmatrix} D_0^{\mathrm{K}}(\mathbf{r} - \mathbf{r}', t - t') & D_0^{\mathrm{R}}(\mathbf{r} - \mathbf{r}', t - t') \\ D_0^{\mathrm{A}}(\mathbf{r} - \mathbf{r}', t - t') & 0 \end{pmatrix}. \tag{6.30}$$

As discussed in Section 3.1, the matrix $\hat{D}$ is symmetric (unlike the case of the complex field). The Fourier transforms of the three nonzero components of the bare Green function, according to Eq. (3.13), are

$$D_0^{\mathrm{R(A)}}(\mathbf{k},\epsilon) = \frac{1}{2}\frac{1}{(\epsilon \pm i0)^2 - \omega_{\mathbf{k}}^2}, \tag{6.31a}$$

$$D_0^{\mathrm{K}}(\mathbf{k},\epsilon) = F(\epsilon)\left[D^{\mathrm{R}}(\mathbf{k},\epsilon) - D^{\mathrm{A}}(\mathbf{k},\epsilon)\right], \tag{6.31b}$$

where the dispersion relation for our model is given by $\omega_{\mathbf{k}}^2 = \kappa\mathbf{k}^4 + \omega_0^2$, but the construction may be generalized to accommodate an arbitrary dispersion $\omega_{\mathbf{k}}$. Due to the symmetry of the Green function the distribution function $F(\epsilon)$ must be an *odd* function of energy ϵ. It takes the form $F^{\mathrm{eq}}(\epsilon) = \coth(\epsilon/2T)$ in equilibrium.

The cubic unharmonicity of the action (6.27) after the Keldysh rotation leads to the following nonlinear term in the action:

$$S_{\mathrm{int}} = -\int d\mathbf{r}\int_{-\infty}^{+\infty} dt\,\left[\gamma\left(\varphi^{\mathrm{cl}}\right)^2\varphi^q + \frac{\gamma}{3}\left(\varphi^{\mathrm{q}}\right)^3\right]. \tag{6.32}$$

The normalization condition (2.53) is again satisfied. Diagrammatically, the cubic nonlinearity generates two types of vertex (Fig. 6.5): one with two classical fields (full lines) and one quantum field (dashed line), and the other with three quantum fields. The former vertex carries the factor γ, while the latter has $\gamma/3$. Note that for the real field the lines do not have a direction.

Similarly to the case of the complex field, one may check that addition of the interaction action does not affect the normalization identity $Z = 1$. This property is based on the identity $D_0^{\mathrm{R}}(t,t) + D_0^{\mathrm{A}}(t,t) = 0$ and the rule of thumb $D_0^{\mathrm{R}}(t,t')D_0^{\mathrm{A}}(t,t') = 0$, explained in Section 6.2. The Green function, on the other hand, is affected. The effect of nonlinearity on the dressed Green function $\hat{D}(x,x')$ is described by the Dyson equation

$$\left(\hat{D}_0^{-1} - \hat{\Sigma}\right)\circ\hat{D} = \hat{1}, \tag{6.33}$$

where $\hat{\Sigma}(x,x')$ is the self-energy of real bosons, possessing the causality structure, Eq. (6.17), and calculated in the next paragraph to second order in γ. The retarded and advanced components of the Dyson equation take the form

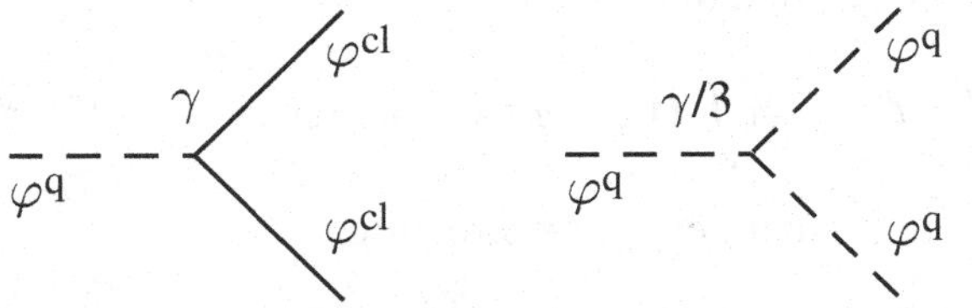

Figure 6.5 Graphic representation of the two interaction vertices of the φ^3 theory. Note the relative factor of one-third between them.

$$-\left(2\partial_t^2 + 2\kappa\,\nabla_{\mathbf{r}}^4 + 2\omega_0^2 + \Sigma^{\mathrm{R(A)}}\right) \circ D^{\mathrm{R(A)}}(x,x') = \delta(t-t')\delta(\mathbf{r}-\mathbf{r}'). \tag{6.34}$$

The Keldysh component of the Green function is again convenient to parametrize as $D^{\mathrm{K}} = D^{\mathrm{R}} \circ F - F \circ D^{\mathrm{A}}$ (see Eq. (2.49)), where $F(x,x')$ is a Hermitian matrix in the space-time domain. The Dyson equation for the Keldysh component then takes the form of the kinetic equation for the two-point distribution function $F(x,x')$:

$$\left[\left(2\partial_t^2 + 2\kappa\,\nabla_{\mathbf{r}}^4 + 2\omega_0^2\right) \stackrel{\circ}{,} F\right] = \Sigma^{\mathrm{K}} - \left(\Sigma^{\mathrm{R}} \circ F - F \circ \Sigma^{\mathrm{A}}\right). \tag{6.35}$$

The commutators involved in the kinetic term read $[\partial_t^2 \stackrel{\circ}{,} F] = (\partial_t^2 - \partial_{t'}^2)F(x,x')$ and $[\nabla_{\mathbf{r}}^4 \stackrel{\circ}{,} F] = (\nabla_{\mathbf{r}}^4 - \nabla_{\mathbf{r}'}^4)F(x,x')$. Finally $[\omega_0^2 \stackrel{\circ}{,} F] = 0$; one may allow, though, for some space- and/or time-dependent function $\omega_0(x)$, in the latter case $[\omega_0^2 \stackrel{\circ}{,} F] = (\omega_0^2(x) - \omega_0^2(x'))F(x,x')$. The self-energy components on the right-hand side of Eq. (6.35) are calculated in the following sections.

6.5 Self-Energy

Let us demonstrate that the self-energy matrix, $\hat{\Sigma}$, indeed possesses the causality structure, Eq. (6.17). To this end, we consider the real boson field with the $\gamma\varphi^3$ non-linearity, Eq. (6.32), and perform calculations up to second order in the nonlinearity γ. Employing the two vertices of Fig. 6.5, one finds the following.

(i) The cl – cl component of the self-energy (i.e. the diagram having two classical external legs) is given by the single diagram, depicted in Fig. 6.6(a). The corresponding analytic expression is

$$\Sigma^{\mathrm{cl-cl}}(x,x') = 4\mathrm{i}\gamma^2 D_0^{\mathrm{R}}(x,x')D_0^{\mathrm{A}}(x,x') = 0.$$

Indeed, the product $D_0^{\mathrm{R}}(t,t')D_0^{\mathrm{A}}(t,t')$ has no support in the time domain (see the discussion in Section 6.2).

(ii) The cl – q (advanced) component is given by the single diagram Fig. 6.6(b). The corresponding expression is

$$\Sigma^{\mathrm{A}}(x,x') = 4\mathrm{i}\gamma^2 D_0^{\mathrm{A}}(x,x')D_0^{\mathrm{K}}(x,x'). \tag{6.36}$$

Since $\Sigma^{\mathrm{A}}(t,t') \sim D_0^{\mathrm{A}}(t,t') \sim \theta(t'-t)$, it is, indeed, an advanced (upper triangular) matrix in the time domain. There is a combinatoric factor of 4, associated with the diagram (four ways of choosing external legs $\times$ 2 internal permutations $\times$ 1/(2!) for having two identical vertices).

(iii) The q – cl (retarded) component is given by the diagram of Fig. 6.6(c):

$$\Sigma^{\mathrm{R}}(x,x') = 4\mathrm{i}\gamma^2 D_0^{\mathrm{R}}(x,x')D_0^{\mathrm{K}}(x,x'), \tag{6.37}$$

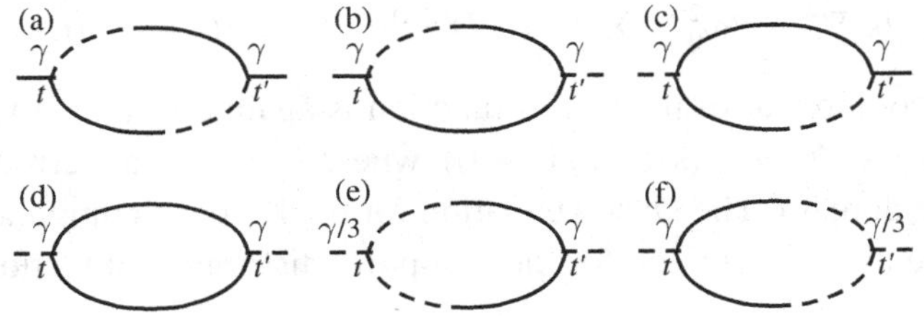

Figure 6.6 Self-energy diagrams for the φ^3 theory.

which is, in fact, the Hermitian conjugation of Eq. (6.36): $\Sigma^R = [\Sigma^A]^\dagger$. Since $\Sigma^R(t,t') \sim D_0^R(t,t') \sim \theta(t-t')$, it is indeed a retarded (lower triangular) matrix.

(iv) The q – q (Keldysh) component is given by the three diagrams, Fig. 6.6(d)–(f). The corresponding expression (sum of these diagrams) is

$$\begin{aligned}\Sigma^K(x,x') &= 2i\gamma^2\big[D_0^K(x,x')\big]^2 + 6i\left(\frac{\gamma}{3}\right)\gamma\big[D_0^A(x,x')\big]^2 + 6i\gamma\left(\frac{\gamma}{3}\right)\big[D_0^R(x,x')\big]^2 \\ &= 2i\gamma^2\left(\big[D_0^K(x,x')\big]^2 + \big[D_0^R(x,x') - D_0^A(x,x')\big]^2\right). \end{aligned} \tag{6.38}$$

The combinatoric factors are 2 for diagram (d) and 6 for (e) and (f). In the last equality the fact that $D_0^R(t,t')D_0^A(t,t') = 0$, due to the absence of support in the time domain, has been used again. Employing the symmetry properties of the Green functions, one finds $\Sigma^K = -[\Sigma^K]^\dagger$. This demonstrates that the self-energy $\hat{\Sigma}$ possesses the same structure as $\hat{D}_0^{-1}$. One may check that this statement is not restricted to second order in γ, but holds in higher orders as well.

6.6 Wigner Transformation

The distribution matrix $F(x_1,x_2) = F(\mathbf{r}_1,t_1,\mathbf{r}_2,t_2)$ is a function of the two space-time points. It is usually difficult to solve the kinetic equations (6.24) or (6.35) in full generality. One may often take advantage of scale separation between intrinsic microscopic space and time scales and the extrinsic ones, dictated by external perturbations and/or a measurement apparatus. In many instances the latter scales are macroscopic, or at least mesoscopic, and thus are much greater than the former ones. If this is indeed the case, the kinetic theory may be greatly simplified. Most elegantly, it is achieved with the help of the Wigner transformation (WT).

We employ combined notation for space-time $x = \mathbf{r}, t$ and momentum-energy $p = \mathbf{k}, \epsilon$, with $px = \mathbf{kr} - \epsilon t$. For a two-point function $A(x_1,x_2)$ one may change the variables to the central point coordinate $x = (x_1 + x_2)/2$ and the relative coordinate

$x' = x_1 - x_2$, such that $x_{1,2} = x \pm x'/2$. One then performs a Fourier transform, going from the relative coordinate x' to its Fourier image p. As a result, the Wigner transform of the two-point function $A(x_1, x_2)$ is a function of the central coordinate x and the relative momentum p, namely $A(x, p)$, defined as

$$A(x,p) = \int dx'\, e^{-ipx'} A\left(x + \frac{x'}{2}, x - \frac{x'}{2}\right). \tag{6.39}$$

The WT of $A^\dagger(x_1, x_2) = [A(x_2, x_1)]^*$ is simply $[A(x, p)]^*$. The inverse WT takes the form

$$A(x_1, x_2) = \sum_p e^{ip(x_1 - x_2)} A\left(\frac{x_1 + x_2}{2}, p\right), \tag{6.40}$$

where $\sum_p = \sum_{\mathbf{k}} \int d\epsilon/(2\pi)$.

Let us consider now a two-point function $C = A \circ B$, which means $C(x_1, x_2) = \int dx_3\, A(x_1, x_3)B(x_3, x_2)$. According to the preceding definitions, its WT is given by

$$C(x,p) = \int dx' e^{-ipx'} \int dx_3 \sum_{p_1, p_2} e^{ip_1(x + x'/2 - x_3) + ip_2(x_3 - x + x'/2)}$$
$$A\left(\frac{x + x'/2 + x_3}{2}, p_1\right) B\left(\frac{x_3 + x - x'/2}{2}, p_2\right).$$

We change coordinate variables from x_3, x' to $x_{a,b} = x_3 - x \pm x'/2$ and shift momenta as $p_{a,b} = p_{1,2} - p$ to obtain

$$C(x,p) = \iint dx_a dx_b \sum_{p_a, p_b} e^{i(p_b x_a - p_a x_b)} A\left(x + \frac{x_a}{2}, p + p_a\right) B\left(x + \frac{x_b}{2}, p + p_b\right).$$

We now formally expand the A and B functions in Taylor series in momenta $p_{a,b}$. The corresponding integrals over momenta may be evaluated by employing $\sum_p e^{\pm ipx} p^n = (\mp i)^n \delta^{(n)}(x)$, where $\delta^{(n)}$ denotes the nth derivative of the delta-function. Subsequently, the integrals over coordinates $x_{a,b}$ also may be evaluated, leading to the formally exact expression

$$C(x,p) = A(x,p)\, e^{\frac{i}{2}(\overleftarrow{\partial}_x \overrightarrow{\partial}_p - \overleftarrow{\partial}_p \overrightarrow{\partial}_x)} B(x,p), \tag{6.41}$$

where the arrows show the direction of the differentiation, and the scalar products in the exponent are $\partial_x \partial_p = \nabla_{\mathbf{r}} \nabla_{\mathbf{k}} - \partial_t \partial_\epsilon$.

This formally exact result is most useful when the exponential operator on its right-hand side may be expanded and only the few lowest-order terms kept. It is a legitimate procedure when the operator $\partial_x \partial_p$ may be regarded as small, that is, if $(\delta x)(\delta p) \gg 1$, where δx and δp are characteristic scales at which the x and p arguments of the WT functions change. This in turn implies that the two-point functions of interest, say $A(x_1, x_2)$, are relatively slow functions of the central coordinate

$x = (x_1 + x_2)/2$ and relatively fast functions of the distance between the two points $x' = x_1 - x_2$. The ultimate example is translationally invariant functions, for example, $A(x_1 - x_2) = A(x')$, for which $\partial_x = 0$ and therefore only the zeroth-order term in the expansion of the exponent in Eq. (6.41) survives, leading to $C(p) = A(p)B(p)$, which is, of course, the well-known convolution theorem of the Fourier analysis. For the case where dependence on the central coordinate x is slow one finds

$$C = AB + \frac{\mathrm{i}}{2}\big(\partial_x A \partial_p B - \partial_p A \partial_x B\big) + \cdots, \tag{6.42}$$

where the arguments of all the functions are (x, p). As a result, WT is a tool to approximately substitute *convolutions* of two-point functions by *algebraic* products of the Wigner transforms and their derivatives. In the same approximation one finds for the commutator of two-point functions

$$[A \overset{\circ}{,} B] \overset{\text{WT}}{\rightarrow} \mathrm{i}\big(\partial_x A \partial_p B - \partial_p A \partial_x B\big) + \cdots, \tag{6.43}$$

namely the classical Poisson bracket.

For an algebraic product of two-point functions, as, for example, in Eqs. (6.36)–(6.38), $C(x_1, x_2) = A(x_1, x_2)B(x_1, x_2)$, one finds after WT

$$C(x, p) = \sum_q A\,(x, p - q)\, B\,(x, q)\,. \tag{6.44}$$

6.7 Kinetic Term

A one-point function, such as, for example, $V^{\text{cl}}(x)$, should be considered as its own WT, which is momentum p independent. We find thus for the commutator in the kinetic term of Eq. (6.24)

$$[V^{\text{cl}} \overset{\circ}{,} F] \overset{\text{WT}}{\rightarrow} \mathrm{i}\partial_x V^{\text{cl}}(x)\partial_p F(x, p) = \mathrm{i}\nabla_{\mathbf{r}} V^{\text{cl}} \nabla_{\mathbf{k}} F - \mathrm{i}\partial_t V^{\text{cl}} \partial_\epsilon F,$$

where $F(x, p) = F(\mathbf{r}, t, \mathbf{k}, \epsilon)$ is the WT of the two-point function $F(\mathbf{r}_1, t_1, \mathbf{r}_2, t_2)$. We turn now to the other commutators in the kinetic terms on the right-hand sides of the kinetic equations (6.24) and (6.35). The WT of the translationally invariant operator $\mathrm{i}\partial_t$ is ϵ; as a result $[\mathrm{i}\partial_t \overset{\circ}{,} F] \overset{\text{WT}}{\rightarrow} \mathrm{i}\partial_\epsilon \epsilon \partial_t F = \mathrm{i}\partial_t F$, and in a similar way $[-\partial_t^2 \overset{\circ}{,} F] \overset{\text{WT}}{\rightarrow} \mathrm{i}\partial_\epsilon \epsilon^2 \partial_t F = 2\mathrm{i}\epsilon\partial_t F$. Finally, the WT of the operator $-\nabla_{\mathbf{r}}^2$ is $\mathbf{k}^2$ and thus $[-\nabla_{\mathbf{r}}^2 \overset{\circ}{,} F] \overset{\text{WT}}{\rightarrow} -\mathrm{i}\nabla_{\mathbf{k}} \mathbf{k}^2 \nabla_{\mathbf{r}} F = -2\mathrm{i}\mathbf{k}\nabla_{\mathbf{r}} F$. For a generic dispersion relation $\omega_{\mathbf{k}}$ one finds $[\omega_{\mathbf{k}}^2 \overset{\circ}{,} F] \overset{\text{WT}}{\rightarrow} -2\mathrm{i}\omega_{\mathbf{k}} \mathbf{v}_{\mathbf{k}} \nabla_{\mathbf{r}} F$, where we introduced the group velocity as $\mathbf{v}_{\mathbf{k}} = \nabla_{\mathbf{k}} \omega_{\mathbf{k}}$. As for the right-hand side of Eqs. (6.24) and (6.35), one finds for its WT

$$\Sigma^{\text{K}} - F\Big(\Sigma^{\text{R}} - \Sigma^{\text{A}}\Big) - \mathrm{i}\partial_x\big(\text{Re}\Sigma^{\text{R}}\big)\partial_p F + \mathrm{i}\partial_p\big(\text{Re}\Sigma^{\text{R}}\big)\partial_x F,$$

where we took into account that $\Sigma^{\text{A}}(x, p) = [\Sigma^{\text{R}}(x, p)]^*$.

Combining all the pieces together, one finds for the WT of the kinetic equation (6.24) for the complex boson field

$$\Big[\big(1-\partial_\epsilon \mathrm{Re}\Sigma^{\mathrm{R}}\big)\partial_t+(\partial_t\tilde{V})\partial_\epsilon+\tilde{\mathbf{v}}_{\mathbf{k}}\nabla_{\mathbf{r}}-(\nabla_{\mathbf{r}}\tilde{V})\nabla_{\mathbf{k}}\Big]F=I^{\mathrm{coll}}[F], \tag{6.45}$$

where

$$\tilde{V}(x,p)=V^{\mathrm{cl}}(x)+\mathrm{Re}\big[\Sigma^{\mathrm{R}}(x,p)\big]\,;\quad \tilde{\mathbf{v}}_{\mathbf{k}}=\nabla_{\mathbf{k}}(\omega_{\mathbf{k}}+\mathrm{Re}\Sigma^{\mathrm{R}}) \tag{6.46}$$

and the right-hand side, known as the *collision integral*, is

$$I^{\mathrm{coll}}[F]=\mathrm{i}\Sigma^{\mathrm{K}}(x,p)+2F(x,p)\,\mathrm{Im}\big[\Sigma^{\mathrm{R}}(x,p)\big]. \tag{6.47}$$

Notice that in a static (i.e. $\partial_t=0$, including spatially nonuniform) situation *any* function $F(\epsilon)$ that depends only on the energy argument nullifies the left-hand side of the kinetic equation (6.45). As we shall see in the next section, there is one such function, $F^{\mathrm{eq}}=\coth(\epsilon-\mu)/2T$, which also nullifies its right-hand side. This is the equilibrium solution.

To make progress away from equilibrium, one changes the energy argument of the distribution function as

$$F(\mathbf{r},t,\mathbf{k},\epsilon)=\tilde{F}(\mathbf{r},t,\mathbf{k},\epsilon-\omega_{\mathbf{k}}-\tilde{V}). \tag{6.48}$$

One may check that the distribution function $\tilde{F}$, defined this way, satisfies the equation, which differs from Eq. (6.45) only by the absence of the $(\partial_t\tilde{V})\partial_\epsilon$ term on the left-hand side. Thus, there is no derivative over the last argument in the equation for $\tilde{F}$. *Should* the collision integral depend only on the same local value of the renormalized energy $\tilde{\epsilon}=\epsilon-\omega_{\mathbf{k}}-\tilde{V}$, the kinetic equations for different $\tilde{\epsilon}$s would split and would not talk to each other. Strictly speaking, this is never the case. That is, the collision integral is actually a nonlocal function of $\tilde{\epsilon}=\tilde{\epsilon}(x,p)$ in both the energy and space-time directions. However, in many cases the distribution function $\tilde{F}(\mathbf{r},t,\mathbf{k},\tilde{\epsilon})$ is a much slower function of $\tilde{\epsilon}$ than $G^{\mathrm{R}}-G^{\mathrm{A}}$. The latter is a sharply peaked function at $\tilde{\epsilon}=0$ with the width given by the inverse quasiparticle lifetime $1/\tau_{\mathrm{qp}}$. As long as the characteristic energy scale $\delta\tilde{\epsilon}$ of the distribution function $\tilde{F}(\mathbf{r},t,\mathbf{k},\tilde{\epsilon})$ is much larger than it, $\delta\tilde{\epsilon}\gg 1/\tau_{\mathrm{qp}}$, one may approximately disregard the $\tilde{\epsilon}$ dependence of $\tilde{F}$ in the collision integral. Indeed, the distribution function, by its definition Eq. (2.49), always shows up in a product with WT of $G^{\mathrm{R}}-G^{\mathrm{A}}$. Since the latter is a sharp function of the renormalized energy at $\tilde{\epsilon}=0$, one may approximately put that

$$\tilde{F}(\mathbf{r},t,\mathbf{k},\tilde{\epsilon})\approx\tilde{F}(\mathbf{r},t,\mathbf{k},0)\equiv\tilde{F}(\mathbf{r},t,\mathbf{k}) \tag{6.49}$$

and write down a closed kinetic equation for the *three*-argument, or the *mass-shell restricted* distribution function $\tilde{F}(\mathbf{r}, t, \mathbf{k})$.[2] As long as quasiparticles are well defined, that is, $\delta\tilde{\epsilon}\tau_{\text{qp}} \gg 1$, the WT of $G^{\text{R}} - G^{\text{A}}$ remains a sharply peaked function at $\epsilon = \epsilon(\mathbf{r}, t, \mathbf{k})$ satisfying

$$\epsilon - \omega_{\mathbf{k}} - V^{\text{cl}}(\mathbf{r}, t) - \text{Re}\Sigma^{\text{R}}(\mathbf{r}, t, \ \mathbf{k}, \epsilon) = 0; \tag{6.50}$$

compare Eq. (6.21). As a result, all observables are *approximately* (in the leading order in $(\delta\tilde{\epsilon}\tau_{\text{qp}})^{-1}$) determined by $\tilde{F}(\mathbf{r}, t, \mathbf{k})$. Such a "mass-shell" distribution function obeys the following closed kinetic equation:

$$\left[\tilde{Z}^{-1}\,\partial_t + \tilde{\mathbf{v}}_{\mathbf{k}}\nabla_{\mathbf{r}} - (\nabla_{\mathbf{r}}\tilde{V})\nabla_{\mathbf{k}}\right]\tilde{F}(\mathbf{r}, t, \mathbf{k}) = I^{\text{coll}}[\tilde{F}], \tag{6.51}$$

where $\tilde{Z}^{-1}(\mathbf{r}, t, \mathbf{k}) = 1 - \partial_\epsilon \text{Re}\Sigma^{\text{R}}$. It is important that velocity and external potential are renormalized according to Eq. (6.46) and the energy argument $\epsilon = \epsilon(\mathbf{r}, t, \mathbf{k})$ of all functions is taken as the solution of Eq. (6.50).

The "mass-shell" distribution function $\tilde{F}(\mathbf{r}, t, \mathbf{k})$ is essentially a classical object. It may be considered as a time-dependent probability of finding a particle at a given point of the classical phase space $(\mathbf{r}, \mathbf{k})$. The quantum mechanics modifies the dispersion relation along with the effective potential and the *quasiparticle weight* $\tilde{Z}$ as well as (possibly) the collision integral. The kinetic equation (6.51) provides thus a semiclassical approximation of the full quantum description. It is instructive to compare the kinetic term (i.e. the left-hand side) of Eq. (6.51) with that of the Fokker–Planck equation (4.32). Provided $\tilde{Z} = 1$, both may be written as $\partial_t \ldots - \{E, \ldots\}$, where the curly brackets stand for the classical Poisson brackets and the classical Hamiltonian is $E(\mathbf{k}, \mathbf{r}) = \omega_{\mathbf{k}} + \tilde{V}(\mathbf{r}, \ \mathbf{k})$. One observes, therefore, that the mass-shell distribution function $\tilde{F}(\mathbf{r}, t, \mathbf{k})$ has basically the same meaning as the classical probability distribution function $\mathcal{P}(\mathbf{r}, \mathbf{k}, t)$. The right-hand side of the Fokker–Planck equation (4.32), being linear in $\mathcal{P}$, is different from the collision integral (see Section 6.8). The latter is a nonlinear functional of the distribution

[2] The mass-shell distribution function may be defined as $\tilde{F}(\mathbf{r}, t, \mathbf{k}) = \int d\epsilon\, F(\mathbf{r}, t, \mathbf{k}, \epsilon)\delta(\epsilon - \omega_{\mathbf{k}} - \tilde{V})$. For free *noninteracting* particles $G^{\text{R}} - G^{\text{A}} = -2\pi i\delta(\epsilon - \omega_{\mathbf{k}})$, while $G^{\text{K}} = F(G^{\text{R}} - G^{\text{A}})$ and thus this definition is equivalent to $\tilde{F}(\mathbf{r}, t, \mathbf{k}) = i\int (d\epsilon/2\pi)\, G^{\text{K}}(\mathbf{r}, t, \ \mathbf{k}, \epsilon) = iG^{\text{K}}(\mathbf{r}, \mathbf{k}, t, t)$. It is therefore frequently stated that the mass-shell distribution function is equivalent to the Keldysh Green function at the coinciding time arguments. As explained in Section 2.9, the latter is given by $2n_{\text{B}}(\mathbf{k}) + 1$, where $n_{\text{B}}(\mathbf{k})$ is the occupation number of the state $\mathbf{k}$. This latter relation between the equal-time Keldysh function and the occupation number is generic and remains true even in the interacting case. However, the relation between the equal-time Keldysh function and the mass-shell distribution function $\tilde{F}$ is *not*. It is restricted to the noninteracting case, where $G^{\text{R}} - G^{\text{A}} = -2\pi i\delta(\epsilon - \omega_{\mathbf{k}})$. It is therefore important to stress that the kinetic equation is written for the mass-shell distribution function $\tilde{F}(\mathbf{r}, t, \mathbf{k})$ and *not* for the equal-time Keldysh Green function $G^{\text{K}}(\mathbf{r}, \mathbf{k}, t, t)$. In particular, in equilibrium $\tilde{F} = \coth(\omega_{\mathbf{k}} - \mu)/2T$ is always a solution of the kinetic equation. On the other hand, the occupation number $n_{\text{B}}(\omega_{\mathbf{k}})$ in equilibrium may be very different from the Bose (or Fermi) distribution. The most famous example probably comes from the fermionic 1d Luttinger model [116], where the occupation number at $T = 0$ is not a Fermi step-function, but rather a power-law nonanalytic function. This function is *not* a solution of the kinetic equation; in equilibrium the latter is solved by the Fermi distribution.

function $\tilde{F}$. This difference originates from the fact that in the classical problems of Chapter 4 the bath was assumed to be passive and independent of the state of the system. In the present context the many-body system serves as a "bath" for itself. This latter "bath," however, is not passive and depends on the local state of the system. Hence the nonlinear character of the collision term. Such a nonlinearity is still a classical phenomenon (though specific transition rates may, of course, incorporate quantum mechanics in an essential way).

Finally, let us formulate the kinetic equation for real boson quasiparticles, such as, for example, elastic phonons. The Wigner transform of the kinetic term of the real boson Dyson equation (6.35) takes the form

$$\Big[\partial_\epsilon(2\epsilon^2 - \mathrm{Re}\Sigma^{\mathrm{R}})\partial_t + \partial_t\mathrm{Re}\Sigma^{\mathrm{R}}\partial_\epsilon + \nabla_{\mathbf{k}}(2\tilde{\omega}_{\mathbf{k}}^2)\nabla_{\mathbf{r}} - \nabla_{\mathbf{r}}(2\tilde{\omega}_{\mathbf{k}}^2)\nabla_{\mathbf{k}}\Big]F,$$

where $\tilde{\omega}_{\mathbf{k}}^2 = \omega_{\mathbf{k}}^2 + \mathrm{Re}\Sigma^{\mathrm{R}}/2$. Due to the symmetries of the real boson Green functions, the distribution function F obeys

$$F(\mathbf{r}, t, \mathbf{k}, \epsilon) = -F(\mathbf{r}, t, -\mathbf{k}, -\epsilon). \tag{6.52}$$

Changing the energy argument of the distribution function and acknowledging that $\epsilon \approx \pm\tilde{\omega}_{\mathbf{k}}$, one arrives at the three-argument "mass-shell" distribution function

$$F(\mathbf{r}, t, \mathbf{k}, \epsilon) = s\tilde{F}(\mathbf{r}, t, s\mathbf{k}, \epsilon^2 - \tilde{\omega}_{\mathbf{k}}^2) \to s\tilde{F}(\mathbf{r}, t, s\mathbf{k}, 0), \tag{6.53}$$

where $s = \mathrm{sign}(\epsilon)$. Such a "mass-shell" distribution function obeys the closed kinetic equation

$$\Big[\partial_t + \mathbf{v}_{\mathbf{k}}\nabla_{\mathbf{r}}\Big]\tilde{F}(\mathbf{r}, t, \mathbf{k}) = I^{\mathrm{coll}}[\tilde{F}], \tag{6.54}$$

where for simplicity we disregarded the dispersion renormalization by the real part of the self-energy. We also took $\epsilon = \omega_{\mathbf{k}} > 0$, which brings the collision integral to the following form:

$$I^{\mathrm{coll}}[\tilde{F}] = \frac{1}{4\omega_{\mathbf{k}}}\Big(i\Sigma^{\mathrm{K}}(x,p) + 2F(x,p)\,\mathrm{Im}\big[\Sigma^{\mathrm{R}}(x,p)\big]\Big)\Big|_{\epsilon=\omega_{\mathbf{k}}}. \tag{6.55}$$

Notice that taking $\epsilon = -\omega_{\mathbf{k}} < 0$ is equivalent to making a $\mathbf{k} \to -\mathbf{k}$ substitution in the kinetic equation (6.54).

6.8 Collision Integral

We discuss now the collision integral, using real bosons with cubic nonlinearity, Section 6.5, as an example. The collision integral for complex bosons is considered in Section 8.7. The proper collision integral is given by Eq. (6.55). To be consistent with the approximations adopted in the derivation of the kinetic term presented earlier, we need to restrict ourselves to products of WT only. In particular,

$D_0^K(x,p) \approx F(x,p)[D_0^R(x,p) - D_0^A(x,p)]$. Even though the Green functions here are the bare ones, the distribution function F is not determined by the dynamics of the free bosons. We should allow F to be self-consistently determined by the kinetic part of the Dyson equation. Employing Eqs. (6.36)–(6.38) along with Eq. (6.44), one finds for the corresponding parts of the collision integral

$$\mathrm{i}\Sigma^K(x,p) = 8\pi^2\gamma^2 \sum_q \Delta(x,p-q)\Delta(x,q)\Big[F(x,p-q)F(x,q)+1\Big], \quad (6.56a)$$

$$2\mathrm{Im}[\Sigma^R(x,p)] = -8\pi^2\gamma^2 \sum_q \Delta(x,p-q)\Delta(x,q)\Big[F(x,p-q)+F(x,q)\Big], \quad (6.56b)$$

where the right-hand side of the last equation is symmetrized with respect to arguments $p-q = \mathbf{k}-\mathbf{q}, \epsilon-\omega$, and $q = \mathbf{q}, \omega$. Here we defined

$$\Delta(x,p) = \frac{\mathrm{i}}{2\pi}\left[D_0^R(x,p) - D_0^A(x,p)\right] = \frac{1}{4\omega_{\mathbf{k}}}\Big(\delta(\epsilon-\omega_{\mathbf{k}}) - \delta(\epsilon+\omega_{\mathbf{k}})\Big). \quad (6.57)$$

To include the renormalization of the dispersion relation $\omega_{\mathbf{k}}$ by the real part of the self-energy, one may use here the dressed Green functions D. This corresponds to the so-called self-consistent Born approximation, where the self-energy diagram is evaluated using self-consistently defined Green functions. Such an approximation neglects vertex corrections, which may lead to a renormalization of the interaction parameter $\gamma \to \Gamma^{\alpha,\alpha',\alpha''}(q,p-q)$, where $\alpha = \mathrm{cl}, \mathrm{q}$. In some cases the full Γ may be found from independent considerations; in general, one should write an additional equation for the vertex tensor and solve it in an approximation consistent with that for the self-energy.

Employing Eqs. (6.55) and (6.56), one finds for the collision integral

$$I^{\mathrm{coll}}[F] = \frac{2\pi^2\gamma^2}{\omega_{\mathbf{k}}} \sum_q \Delta(x,p-q)\Delta(x,q)$$
$$\times \Big[F(x,p-q)F(x,q) + 1 - F(x,p)\big(F(x,p-q)+F(x,q)\big)\Big]. \quad (6.58)$$

The combination of the distribution functions in the square brackets is a very general construction, which repeats itself in higher orders in γ. Thanks to the energy delta-functions incorporated in the $\Delta(x,p)$ symbols and the "magic" identity:

$$\coth(a)\,\coth(b) + 1 = \coth(a+b)\big(\coth(a)+\coth(b)\big), \quad (6.59)$$

the collision integral is identically nullified by the equilibrium Bose distribution $F(\mathbf{r},t,\ \mathbf{k},\epsilon) = F^{\mathrm{eq}}(\epsilon) = \coth(\epsilon-\mu)/2T$, where T and μ are yet unspecified temperature and chemical potential. For real bosons, due to the requirement that F is an odd function of energy, one has to choose $\mu = 0$. As explained after Eq. (6.47), any function $F = F(\epsilon)$ also nullifies the kinetic term in a stationary situation. As a result, the thermal equilibrium distribution function $F^{\text{eq}}(\epsilon)$ solves the kinetic equation. Such a solution is (locally) stable for any temperature (the latter is determined either by an external reservoir, or, for a closed system, from the conservation of total energy). Since the equilibrium distribution obviously nullifies both left- and right-hand sides of Eq. (6.35) the *exact* equilibrium self-energy satisfies $\Sigma^{\text{K}} = \coth(\epsilon/2T)\big[\Sigma^{\text{R}} - \Sigma^{\text{A}}\big]$. Since the bare Green functions obey the same relation, Eq. (2.48), one concludes that in thermal equilibrium the *exact* dressed Green function satisfies

$$D^{\text{K}}(\mathbf{r},\mathbf{k},\epsilon) = \coth\frac{\epsilon}{2T}\Big(D^{\text{R}}(\mathbf{r},\mathbf{k},\epsilon) - D^{\text{A}}(\mathbf{r},\mathbf{k},\epsilon)\Big). \tag{6.60}$$

This is the statement of the *fluctuation–dissipation theorem* (FDT). Its consequence is that in equilibrium the Keldysh component does not contain any additional information with respect to the retarded one. Therefore, the Keldysh technique may be, in principle, substituted by a more compact construction – the Matsubara formalism. The latter does not work, of course, away from equilibrium. Notice that the Green functions may still be space dependent, since the equilibrium implies only stationarity, but *not* translational invariance in space.

To make progress away from equilibrium, one needs to restrict the two-point function F to the mass-shell function $\tilde{F}$ according to Eq. (6.53). This is possible due to the fact that $\Delta(x,p-q)$ and $\Delta(x,q)$ are sharply peaked at $\epsilon - \omega = \pm\omega_{\mathbf{k}-\mathbf{q}}$ and $\omega = \pm\omega_{\mathbf{q}}$, while the external argument is to put $\epsilon = \omega_{\mathbf{k}} > 0$. Once the distribution functions are restricted to the "mass-shell," the energy dependence in Eq. (6.58) is explicitly specified by Eqs. (6.53) and (6.57). Thus, one can perform the ω-integration explicitly with the help of the delta-functions and find for the collision integral

$$\begin{aligned} I^{\text{coll}}[\tilde{F}(\mathbf{k})] = \frac{\pi\gamma^2}{16\omega_{\mathbf{k}}}\sum_{\mathbf{q}}\frac{1}{\omega_{\mathbf{k}-\mathbf{q}}\omega_{\mathbf{q}}} &\times\Big\{\delta(\omega_{\mathbf{k}}-\omega_{\mathbf{q}}-\omega_{\mathbf{k}-\mathbf{q}})\big[\tilde{F}(\mathbf{k}-\mathbf{q})\tilde{F}(\mathbf{q})+1-\tilde{F}(\mathbf{k})\big(\tilde{F}(\mathbf{k}-\mathbf{q})+\tilde{F}(\mathbf{q})\big)\big] \\ &+\delta(\omega_{\mathbf{k}}+\omega_{\mathbf{q}}-\omega_{\mathbf{k}-\mathbf{q}})\big[\tilde{F}(\mathbf{k}-\mathbf{q})\tilde{F}(-\mathbf{q})-1+\tilde{F}(\mathbf{k})\big(\tilde{F}(\mathbf{k}-\mathbf{q})-\tilde{F}(-\mathbf{q})\big)\big] \\ &+\delta(\omega_{\mathbf{k}}-\omega_{\mathbf{q}}+\omega_{\mathbf{q}-\mathbf{k}})\big[\tilde{F}(\mathbf{q}-\mathbf{k})\tilde{F}(\mathbf{q})-1-\tilde{F}(\mathbf{k})\big(\tilde{F}(\mathbf{q}-\mathbf{k})-\tilde{F}(\mathbf{q})\big)\big]\Big\}, \end{aligned} \tag{6.61}$$

where we have suppressed slow space-time argument $\mathbf{r}, t$ in the distribution functions $\tilde{F}(\mathbf{r},t,\mathbf{k}) \to \tilde{F}(\mathbf{k})$. There are three types of process allowed by energy conservation. To appreciate the structure of the corresponding terms it is convenient to express their rates through the boson occupation number $n_{\mathbf{q}}$ related to the distribution function as $\tilde{F}(\mathbf{q}) \approx 2n_{\mathbf{q}} + 1$.[3] Then the rate of the first process is proportional to $[n_{\mathbf{k}-\mathbf{q}} n_{\mathbf{q}} - n_{\mathbf{k}}(n_{\mathbf{k}-\mathbf{q}} + n_{\mathbf{q}} + 1)]$. It states that the state $\mathbf{k}$ may be populated due to the merging of particles from states $\mathbf{q}$ and $\mathbf{k}-\mathbf{q}$ and depopulated due to stimulated emission of $\mathbf{k}-\mathbf{q}$ and $\mathbf{q}$ phonons, or spontaneous emission. The rate of the second process is proportional to $[n_{\mathbf{k}-\mathbf{q}}(n_{-\mathbf{q}} + n_{\mathbf{k}} + 1) - n_{\mathbf{k}} n_{-\mathbf{q}}]$. Here the state $\mathbf{k}$ may be populated due to stimulated or spontaneous decay of a higher energy state $\mathbf{k}-\mathbf{q}$ and depopulated by merging with a particle in a state $-\mathbf{q}$. Finally, the rate of the third process is proportional to $[n_{\mathbf{q}}(n_{\mathbf{q}-\mathbf{k}} + n_{\mathbf{k}} + 1) - n_{\mathbf{k}} n_{\mathbf{q}-\mathbf{k}}]$ and the physics is the same as in the second process with states $\mathbf{q}$ and $\mathbf{k}-\mathbf{q}$ interchanged.

Within the mass-shell approximation the equilibrium solution of the kinetic equation (6.61) takes the form $\tilde{F}^{\mathrm{eq}}(\mathbf{r},\mathbf{k}) = \coth \omega_{\mathbf{k}}(\mathbf{r})/2T$, which is, in general, a function of coordinates and momenta. Since on the mass-shell $\epsilon = \omega_{\mathbf{k}}(\mathbf{r})$, this is consistent with $F^{\mathrm{eq}}(\mathbf{r},\mathbf{k},\epsilon) = \coth \epsilon/2T$. The latter statement is exact, while the mass-shell one is only an approximation valid for well-defined quasiparticles.

For the dispersion relation of the quantum membrane $\omega_{\mathbf{k}} = \sqrt{\omega_0^2 + \kappa \mathbf{k}^4}$, the energy conservation law $\omega_{\mathbf{k}} = \omega_{\mathbf{q}} + \omega_{\mathbf{k}-\mathbf{q}}$ may be satisfied for $k > k_{\mathrm{c}}$, where $k_{\mathrm{c}}^4 = 4\omega_0^2/\kappa$. For smaller momenta $k < k_{\mathrm{c}}$ the cubic nonlinearity alone does not provide relaxation of the distribution function. Therefore, it does not lead to thermalization if the resulting temperature is too small ($T \lesssim 2\omega_0$). On the other hand, if a high-energy $k \gg k_{\mathrm{c}}$ mode is excited, it decays into $\mathbf{q}$ and $\mathbf{k}-\mathbf{q}$ modes with almost perpendicular momenta. To find the corresponding relaxation time we restrict the collision integral to the spontaneous emission part in the first term in Eq. (6.61) and find $\partial_t n_{\mathbf{k}} = -n_{\mathbf{k}}/\tau_{\mathbf{k}}$, where

$$\frac{1}{\tau_{\mathbf{k}}} = \frac{\pi\gamma^2}{8\omega_{\mathbf{k}}} \sum_{\mathbf{q}} \frac{\delta(\omega_{\mathbf{k}} - \omega_{\mathbf{q}} - \omega_{\mathbf{k}-\mathbf{q}})}{\omega_{\mathbf{k}-\mathbf{q}}\omega_{\mathbf{q}}} = \frac{3}{128\pi} \frac{\gamma^2}{\kappa^2 k^5} \ln\left(\frac{k}{k_{\mathrm{c}}}\right), \tag{6.62}$$

for $d = 3$.

Finally, let us discuss approximations involved in the Wigner transformations. It is a justified procedure as long as $\delta\mathbf{k}\,\delta\mathbf{r} \gg 1$, where $\delta\mathbf{k}$ is a characteristic microscopic scale of the momentum dependence of the distribution function, while $\delta\mathbf{r}$ is a characteristic scale of its spatial variations. One may ask if there is a similar requirement in the time domain: $\delta\epsilon\,\delta t \gg 1$, with $\delta\epsilon$ and δt being the characteristic energy and the time scale, respectively. Such a requirement is very demanding, since typically $\delta\epsilon \approx T$ and at low temperature it would allow us to treat only very slow processes with $\delta t \gg 1/T$. Fortunately, this is not the case. Because of the peaked structure of $\Delta(\mathbf{k},\epsilon)$, the energy argument ϵ is locked to $\tilde{\omega}_{\mathbf{k}}$ (i.e. to the "mass-shell") and does not have its own dynamics as long as the peak is sharp.

[3] As discussed in the footnote after Eq. (6.49), this relation is only approximate, valid to the leading order in $(\delta\tilde{\epsilon}\tau_{qp})^{-1}$.

The actual criterion is therefore that $\delta\epsilon$ is much larger than the width of the peak in $\Delta(\mathbf{k},\epsilon)$. The latter is the inverse quasiparticle lifetime, $1/\tau_{\rm qp}$, and therefore the actual condition is $\tau_{\rm qp} \gg 1/T$. This condition is indeed satisfied in systems with well-defined quasiparticles. Notice that this is exactly the same condition that we employed to justify the restriction of the distribution function to the mass-shell; see Eq. (6.49) and the discussion following it. This is thus a necessary condition for the applicability of the quasi-classical kinetic equation.

6.9 Problems

6.9.1 Diffusion in Energy Space

Consider complex bosons under the influence of some (classical) high-frequency perturbation. For example, one may think about charged particles subject to a spatially uniform ac electric field $\mathbf{E}^{\rm cl}(t) = \partial_t \mathbf{A}^{\rm cl}(t)$. The corresponding interaction vertex with the complex bosons is $\sum_{\mathbf{k}}\big[\bar{\phi}_{\mathbf{k}}^{\rm cl}(\mathbf{k}\cdot\mathbf{A}^{\rm cl})\phi_{\mathbf{k}}^{\rm q}+{\rm c.c}\big]$. Let us also assume that the distribution function remains isotropic in the momentum space (possibly due to some elastic scattering mechanism leading to a fast isotropization of momentum directions, but not altering particles' energy). By this reason we'll suppress all momentum indexes and focus exclusively on time and energy arguments.

Calculate the self-energy components in the second-order perturbation theory in the ac field. Show that $\Sigma^{\rm R,A,K}_{t,t'} \propto \mathbf{A}(t) G^{\rm R,A,K}_{t,t'} \mathbf{A}(t')$ (index cl is suppressed hereafter). Employ that $G^{\rm K} = G^{\rm R}\circ F - F\circ G^{\rm A}$ and calculate the Wigner transform of the collision integral $I^{\rm coll}[F] \propto \Sigma^{\rm K} - \big(\Sigma^{\rm R}\circ F - F\circ\Sigma^{\rm A}\big)$ with the understanding that $\mathbf{A} = \mathbf{A}(t)$, $G^{\rm R,A} = G^{\rm R,A}(\epsilon)$, and $F = F(t,\epsilon)$.

One can now average the resulting collision integral over the fast variations of the applied ac field. Use the fact that $\langle \mathbf{A}\partial_t\mathbf{A}\rangle = 0$ and $\langle \partial_t\mathbf{A}\partial_t\mathbf{A}\rangle = -\langle \mathbf{A}\partial_t^2\mathbf{A}\rangle = \langle \mathbf{E}^2\rangle$, where angular brackets denote time averaging, to show that the distribution function obeys the diffusion equation in the energy direction

$$\partial_t F(t,\epsilon) = \partial_\epsilon\big[\mathcal{D}(\epsilon)\,\partial_\epsilon F(t,\epsilon)\big], \tag{6.63}$$

where the energy-dependent diffusion coefficient $\mathcal{D}(\epsilon) \propto \langle \mathbf{E}^2\rangle \,{\rm Im}\, G^{\rm A}(\epsilon)$. Verify that $\mathcal{D}(\epsilon) > 0$. The current in energy space is given by $\mathcal{D}\,\partial_\epsilon F$, and the only stationary state with no current is the infinite temperature distribution with $\partial_\epsilon F = 0$. Equation (6.63) is useful for describing transient regimes and energy absorption rate; see Section 14.9 for more details.

7

Dynamics of Collisionless Plasma

In this chapter we consider a two-component gas of charged particles interacting through long-ranged Coulomb interactions. In most cases those are electrons and ions. For the sake of illustration, however, we shall treat both of them as being bosons. Since we focus on the high-temperature classical limit, the underlying quantum statistics is of almost no importance. We use this model to introduce collective modes, collisionless Landau damping, random phase approximation, and kinetics of particles coupled to the collective modes.

7.1 Plasma Action

Consider a two-component gas of oppositely charged bosonic particles with masses m_{e} and m_{i}. We shall call them "electrons" (e) and "ions" (i), with the understanding that the fermionic nature of one or both of them may be easily incorporated into the same scheme; see Chapter 10 and especially a remark at the end of Section 10.5. First, we focus not on the kinetics of the plasma itself, but rather on the dynamics of electric field waves propagating through it. To this end we start from the Keldysh action for the gas of interacting charged bosons:

$$S = \int_C \mathrm{d}t \left[\int \mathrm{d}\mathbf{r} \sum_{\nu=\mathrm{e,i}} \bar{\phi}_\nu \left(\mathrm{i}\partial_t + \frac{\nabla_{\mathbf{r}}^2}{2m_\nu} \right) \phi_\nu - \frac{1}{2} \iint \mathrm{d}\mathbf{r}\mathrm{d}\mathbf{r}' \rho(\mathbf{r},t) U(\mathbf{r}-\mathbf{r}') \rho(\mathbf{r}',t) \right], \tag{7.1}$$

where $\rho(\mathbf{r},t) = \bar{\phi}_{\mathrm{e}}(\mathbf{r},t)\phi_{\mathrm{e}}(\mathbf{r},t) - \bar{\phi}_{\mathrm{i}}(\mathbf{r},t)\phi_{\mathrm{i}}(\mathbf{r},t)$ is the density of electrons minus that of the neutralizing ions. The Coulomb interaction potential and its Fourier transform are

$$U(\mathbf{r}-\mathbf{r}') = \frac{e^2}{|\mathbf{r}-\mathbf{r}'|}; \qquad U(\mathbf{q}) = \frac{4\pi e^2}{\mathbf{q}^2}. \tag{7.2}$$

To single out the dynamics of the electric potential, we perform now the Hubbard–Stratonovich transformation of the interaction term:

$$\begin{aligned}
\exp\left\{-\frac{\mathrm{i}}{2}\int_C \mathrm{d}t \iint \mathrm{d}\mathbf{r}\mathrm{d}\mathbf{r}'\rho(\mathbf{r},t)U(\mathbf{r}-\mathbf{r}')\rho(\mathbf{r}',t)\right\} &= \mathrm{e}^{-\frac{\mathrm{i}}{2}\int_C \mathrm{d}t\sum_{\mathbf{q}}U(\mathbf{q})|\rho(\mathbf{q},t)|^2} \\
&= \int \mathbf{D}[V]\,\mathrm{e}^{\frac{\mathrm{i}}{2}\int_C \mathrm{d}t\sum_{\mathbf{q}}\left([U(\mathbf{q})]^{-1}|V(\mathbf{q},t)|^2-\rho^*(\mathbf{q},t)V(\mathbf{q},t)-V^*(\mathbf{q},t)\rho(\mathbf{q},t)\right)} \qquad (7.3)\\
&= \int \mathbf{D}[V]\exp\left\{\mathrm{i}\int_C \mathrm{d}t\int \mathrm{d}\mathbf{r}\,\left((8\pi e^2)^{-1}[\nabla_{\mathbf{r}}V(\mathbf{r},t)]^2-\rho(\mathbf{r},t)V(\mathbf{r},t)\right)\right\},
\end{aligned}$$

where we have used the fact that the operator $\mathbf{q}^2$ takes the form $-\nabla_{\mathbf{r}}^2$ in real space. This transformation is nothing but an implementation of the real variable Gaussian integral, discussed in Section 2.3. Here $V(\mathbf{r},t)$ is a real fluctuating scalar potential defined along the closed time contour. We have introduced thus an additional auxiliary functional integral. The great advantage of this procedure is that the boson fields $\bar{\phi}_\nu$ and ϕ_ν, where $\nu = \mathrm{e},\mathrm{i}$, enter the transformed action only quadratically. As a result, the functional integrals over $\bar{\phi}_\nu$ and ϕ_ν may be calculated explicitly(!), leaving us with the effective action for the scalar potential $S[V]$. The latter contains much of the information we need about dynamics of the plasma modes.

To perform the Gaussian integration over $\bar{\phi}_\nu$ and ϕ_ν, respecting the regularization, we need to go through the procedure described in Chapters 2 and 6. To this end we first split the fields into components residing on the forward and backward branches of the contour and then perform the Keldysh rotation according to Eq. (2.39) for the complex fields $\bar{\phi}_\nu$ and ϕ_ν and according to Eq. (2.61) for the real field V. This brings the Gaussian ϕ_ν-action to the form $\sum_\nu \left(S_0[\bar{\phi}_\nu,\phi_\nu] \pm S_V[\bar{\phi}_\nu,\phi_\nu]\right)$, given by Eqs. (6.4) and (2.66) correspondingly. (Here the plus sign corresponds to electrons and the minus sign to the ions.) The functional integrals may now be performed exactly. This leads to the inverse determinants of the corresponding quadratic forms; compare Eq. (2.68). Rewriting each determinant as the trace of the logarithm, one finds for the effective action of the scalar potential $V(\mathbf{r},t)$

$$S[V] = \int \mathrm{d}x\left[-\frac{1}{4\pi e^2}(V^{\mathrm{cl}},V^{\mathrm{q}})\begin{pmatrix}0 & \nabla_{\mathbf{r}}^2\\ \nabla_{\mathbf{r}}^2 & 0\end{pmatrix}\begin{pmatrix}V^{\mathrm{cl}}\\ V^{\mathrm{q}}\end{pmatrix} + \mathrm{i}\sum_{\nu=\mathrm{i},\mathrm{e}}\mathrm{Tr}\ln\left(\hat{G}_{0\nu}^{-1}\mp\hat{V}\right)\right], \qquad (7.4)$$

where the $\hat{V}(\mathbf{r},t)$ matrix is given by Eq. (2.67). For the fermionic plasma, we would obtain a minus sign in front of the trace logarithm, since the corresponding fermionic integral is given by the determinant of the quadratic form; see Chapter 10. It is important to notice that both species of particles interact with the same collective fields $V^{\mathrm{cl,q}}(x)$.

We shall first focus on the classical *fluctuationless* dynamics of the scalar potential. To this end one needs to expand the action $S[V]$ up to first order in the quantum component of the potential $V^{\mathrm{q}}(\mathbf{r},t)$. The classical component V^{cl} is not assumed to be small at this stage. Notice that there is no zeroth-order term in the expansion of

$\mathrm{Tr}\ln\left(\hat{G}_{0\nu}^{-1} \mp \hat{V}\right)$. Indeed, in the absence of V^{q} the properly regularized ϕ-integral is unity, due to the normalization identity (2.63), even in the presence of an arbitrary V^{cl}. The first-order term in V^{q} takes the form

$$S[V] = -2\int \mathrm{d}x\, V^{\mathrm{q}}(x)\left(\frac{\nabla_{\mathbf{r}}^2 V^{\mathrm{cl}}(x)}{4\pi e^2} + \langle\rho(x)\rangle\right), \tag{7.5}$$

where

$$\langle\rho(x)\rangle = \frac{\mathrm{i}}{2}\,\mathrm{Tr}\left\{\hat{G}_{\mathrm{e}}(x,x;V^{\mathrm{cl}}) - \hat{G}_{\mathrm{i}}(x,x;V^{\mathrm{cl}})\right\}.$$

Here $\hat{G}_\nu(x,x';V^{\mathrm{cl}})$ are the operators inverse to $\hat{G}_{0\nu}^{-1} \mp V^{\mathrm{cl}}\hat{\sigma}_1$. They are therefore solutions of the Dyson equations

$$\left[\hat{G}_{0\nu}^{-1} \mp V^{\mathrm{cl}}\hat{\sigma}_1\right] \circ \hat{G}_\nu = \hat{1}, \tag{7.6}$$

which possess the causality structure, Eq. (2.51). Rewriting the Dyson equation for the components of the Green functions, one finds for its retarded parts (cf. Eq. (6.20))

$$\left(\mathrm{i}\partial_t^{\mathrm{R}} + \frac{1}{2m_\nu}\nabla_{\mathbf{r}}^2 \mp V^{\mathrm{cl}}(x)\right) \circ G_\nu^{\mathrm{R}}(x,x';V^{\mathrm{cl}}) = \delta(x-x'). \tag{7.7}$$

The advanced components obey the same equations with advanced regularization of the time derivative $\partial_t^{\mathrm{R}} \to \partial_t^{\mathrm{A}}$. We shall assume that the classical electric potential $V^{\mathrm{cl}}(x)$ is a slow function of coordinates on the scale $1/k_T$, where k_T is the characteristic thermal momentum of particles. Under this assumption the Dyson equation is solved using WT, leading to $G_\nu^{\mathrm{R}}(x,p;V^{\mathrm{cl}}) = [\epsilon - \mathbf{k}^2/2m_\nu \mp V^{\mathrm{cl}}(x) + \mathrm{i}0]^{-1}$. Parametrizing the Keldysh components of the Green functions, by the two-point distribution functions $F_\nu(x,x')$, going to their WT $F_\nu(x,p)$, and restricting the latter to the "mass-shells" $\epsilon_\nu(\mathbf{k},x) = \mathbf{k}^2/2m_\nu \pm V^{\mathrm{cl}}(x)$, one arrives at the two kinetic equations for electrons and ions (cf. Eq. (6.51)):

$$\left[\partial_t^{\mathrm{R}} + \mathbf{v}_{\mathbf{k}\nu}\nabla_{\mathbf{r}} \mp \nabla_{\mathbf{r}}V^{\mathrm{cl}}(\mathbf{r},t)\nabla_{\mathbf{k}}\right]\tilde{F}_\nu(\mathbf{r},t,\mathbf{k}) = 0, \tag{7.8}$$

where $\mathbf{v}_{\mathbf{k}\nu} = \mathbf{k}/m_\nu$. We indicated the retarded regularization of the time derivative in the kinetic term here. Indeed, the way we derived the kinetic equation (cf. Eq. (6.23)) was by acting with the retarded (advanced) operator on the distribution function from the left (right). One can show that both of these operations imply the ∂_t^{R} operator acting from the left. If the distribution function approaches a time-independent constant at the boundary of the phase-space $\mathbf{r},\mathbf{k} \to \infty$, the total number of particles of each sort is conserved, $\partial_t \int \mathrm{d}\mathbf{r} \sum_{\mathbf{k}} \tilde{F}_\nu(\mathbf{r},t,\mathbf{k}) = 0$. It must be fixed by the condition of the global charge neutrality

$$\int \mathrm{d}\mathbf{r} \sum_{\mathbf{k}} \left[\tilde{F}_{\mathrm{e}}(\mathbf{r},t,\mathbf{k}) - \tilde{F}_{\mathrm{i}}(\mathbf{r},t,\mathbf{k})\right] = 0. \tag{7.9}$$

The plasma is called collisionless because of the absence of the collision integrals on the right-hand sides of Eq. (7.8). This is a consequence of our assumption that the $V^{\rm cl}(x)$ is a slow field. In principle, it also contains fast components, which need to be integrated out. Then the corresponding Dyson equations acquire self-energy parts, originating from such fast components of the potential. The self-energy in turn leads to the collision integral. The effect of such collisions is small, as long as the plasma is sufficiently dilute, $e^2\rho_0^{1/3} \ll T$, where T is a characteristic kinetic energy of the particles.

Provided solutions of Eqs. (7.7) and (7.8) are known, one may express $\langle\rho(x)\rangle$ through them to obtain a closed expression for the $S[V]$ effective action. This way, one finds

$$\langle\rho(x)\rangle = \frac{\mathrm{i}}{2}\big[G^{\rm K}_{\rm e}(x,x;V^{\rm cl}) - G^{\rm K}_{\rm i}(x,x;V^{\rm cl})\big] = \frac{1}{2}\sum_{\mathbf{k}}\big[\tilde{F}_{\rm e}(x,\mathbf{k}) - \tilde{F}_{\rm i}(x,\mathbf{k})\big], \tag{7.10}$$

where we have used the fact that after WT, $G^{\rm K}_{\nu} = (G^{\rm R}_{\nu} - G^{\rm A}_{\nu})F_{\nu}$ and also $G^{\rm R}_{\nu} - G^{\rm A}_{\nu} = -2\pi\mathrm{i}\delta(\epsilon - \mathbf{k}^2/2m_\nu \mp V^{\rm cl})$ to perform the energy integration.

Once we know to express all the terms in the $S[V]$ action (7.5) through the solutions of the collisionless kinetic equations (7.8), we can close the loop and write down the equation of motion for the scalar potential $V^{\rm cl}(x)$. To this end notice that the integral over $V^{\rm q}(x)$ of the exponentiated action (7.5) enforces the functional delta-function of the expression in the round brackets on the right-hand side of Eq. (7.5). This is nothing but the Poisson equation for the classical component of the scalar potential, which in view of Eq. (7.10) takes the form

$$\nabla^2_{\mathbf{r}} V^{\rm cl}(x) = -2\pi e^2 \sum_{\mathbf{k}}\big[\tilde{F}_{\rm e}(x,\mathbf{k}) - \tilde{F}_{\rm i}(x,\mathbf{k})\big]. \tag{7.11}$$

(In our notation V has the dimensionality of energy, i.e. it is e times the electric potential.) The collisionless kinetic equations (7.8), which are coupled through the common potential $V^{\rm cl}$, obeying the Poisson equation (7.11), are called Vlasov [117] equations. As we have shown here, these equations correspond to the approximation where one keeps only the linear term in $V^{\rm q}$. In the following sections we shall also retain terms quadratic in $V^{\rm q}$ to incorporate fluctuation effects. First, let us investigate Vlasov equations.

7.2 Plasmons and Landau Damping

We shall focus first on the linearized Vlasov equations, assuming that deviations from the equilibrium distribution $\tilde{F}^{(1)}_{\nu}(x,\mathbf{k}) = \tilde{F}_{\nu}(x,\mathbf{k}) - \tilde{F}^{\rm eq}(\omega_{\mathbf{k}\nu})$ are small at all times. Here $\omega_{\mathbf{k}\nu} = \mathbf{k}^2/2m_\nu$. According to the Poisson equation (7.11), this also implies smallness of the potential $V^{\rm cl}$. Therefore, in the last term in the kinetic equations (7.8) one may disregard deviations of the distribution function from the equilibrium one. As a result, one obtains the homogeneous linear system of equations

$$[\partial_t^R + \mathbf{v}_{\mathbf{k}\nu}\nabla_{\mathbf{r}}]\tilde{F}_\nu^{(1)}(x,\, \mathbf{k}) = \pm\nabla_{\mathbf{k}}\tilde{F}^{\mathrm{eq}}(\omega_{\mathbf{k}\nu})\nabla_{\mathbf{r}}V^{\mathrm{cl}}(x);$$
$$\nabla_{\mathbf{r}}^2 V^{\mathrm{cl}}(x) = -2\pi e^2 \sum_{\mathbf{k}} \big[\tilde{F}_{\mathrm{e}}^{(1)}(x,\mathbf{k}) - \tilde{F}_{\mathrm{i}}^{(1)}(x,\mathbf{k})\big].$$

We perform now the Fourier transformation from $x = \mathbf{r}, t$ to $q = \mathbf{q}, \omega$, find $\tilde{F}_\nu^{(1)}(\mathbf{q},\omega,\mathbf{k}) = \mp V^{\mathrm{cl}}(q)\mathbf{q}\nabla_{\mathbf{k}}\tilde{F}^{\mathrm{eq}}/(\omega + \mathrm{i}0 - \mathbf{v}_{\mathbf{k}\nu}\mathbf{q})$ from the kinetic equation, and substitute it into the Poisson equation. A nontrivial solution of the latter exists only if the following condition is satisfied:

$$\frac{\mathbf{q}^2}{4\pi e^2} + \sum_{\nu=\mathrm{e,i}} \Pi_\nu^R(\mathbf{q},\omega) = 0, \tag{7.12}$$

where

$$\Pi_\nu^R(\mathbf{q},\omega) = \frac{1}{2}\sum_{\mathbf{k}} \frac{\mathbf{q}\nabla_{\mathbf{k}}F^{\mathrm{eq}}(\omega_{\mathbf{k}\nu})}{\omega + \mathrm{i}0 - \mathbf{v}_{\mathbf{k}\nu}\mathbf{q}} = \sum_{\mathbf{k}} \frac{\mathbf{q}\nabla_{\mathbf{k}}n_{\mathrm{B}}(\omega_{\mathbf{k}\nu})}{\omega + \mathrm{i}0 - \mathbf{v}_{\mathbf{k}\nu}\mathbf{q}} \tag{7.13}$$

is the retarded component of the polarization matrix. Here we took into account that $\tilde{F}^{\mathrm{eq}}(\omega_{\mathbf{k}}) = 2n_{\mathrm{B}}(\omega_{\mathbf{k}}) + 1$ and indicated the retarded regularization of the pole, originating from the nature of the ∂_t operator in the kinetic equations. Solutions of Eq. (7.12), $\omega = \omega(\mathbf{q})$, are dispersion relations of the plasma modes, namely combined oscillations of the electric potential and the two distribution functions.

Let us first disregard dynamics of the ions, assuming them to be much heavier than the electrons, $m_{\mathrm{i}} \gg m_{\mathrm{e}}$. In this limit one may put $\Pi_{\mathrm{i}}^R \to 0$. Focusing on the high-frequency dynamics, one may expand the denominator of Π_{e}^R in powers of $\mathbf{v}_{\mathbf{k}}\mathbf{q}/\omega$. Performing integration by parts and employing $\nabla_{\mathbf{k}}(\mathbf{v}_{\mathbf{k}\mathrm{e}}\mathbf{q}) = \mathbf{q}/m_{\mathrm{e}}$, one finds

$$\Pi_{\mathrm{e}}^R(\mathbf{q},\omega) = -\frac{\mathbf{q}^2}{m_{\mathrm{e}}\omega^2}\sum_{\mathbf{k}} n_{\mathrm{B}}(\omega_{\mathbf{k}\mathrm{e}})\left[1 + \frac{3(\mathbf{v}_{\mathbf{k}\mathrm{e}}\mathbf{q})^2}{\omega^2} + \cdots\right].$$

Employing the fact that the average electron density is given by $\rho_0 = \sum_{\mathbf{k}} n_{\mathrm{B}}(\omega_{\mathbf{k}\mathrm{e}})$ and defining the plasma frequency as $\omega_{\mathrm{p}} = \sqrt{4\pi e^2\rho_0/m_{\mathrm{e}}}$, one finds from Eq. (7.12)

$$\omega^2 = \omega_{\mathrm{p}}^2\left[1 + \frac{\langle \mathbf{v}_{\mathbf{k}\mathrm{e}}^2\rangle \mathbf{q}^2}{\omega^2} + \cdots\right],$$

where $\langle \mathbf{v}_{\mathbf{k}\mathrm{e}}^2\rangle = \sum_{\mathbf{k}} n_{\mathrm{B}}(\omega_{\mathbf{k}^{\mathrm{e}}})\mathbf{v}_{\mathbf{k}\mathrm{e}}^2/\rho_0$ is the electron mean square thermal velocity. Solving this equation yields the dispersion relation of the longitudinal Langmuir mode also known as the *plasmon*:

$$\omega_{\rm p}(\mathbf{q}) = \omega_{\rm p}\left(1 + \frac{\langle \mathbf{v}_{\mathbf{k}\rm e}^2\rangle \mathbf{q}^2}{2\,\omega_{\rm p}^2} + \cdots\right). \tag{7.14}$$

At first sight, the expansion over $\mathbf{v}_{\mathbf{k}\rm e}\mathbf{q}/\omega$ seems to be perfectly justified at small enough momenta, $q \ll \omega_{\rm p}/\sqrt{\langle \mathbf{v}_{\mathbf{k}\rm e}^2\rangle}$. However, as was realized by Landau [119], the actual dispersion contains a small negative imaginary part, implying damping of the plasma oscillations. It originates from the retarded regularization of the pole in Eq. (7.13) and the relation

$$\mathrm{Im}\,\frac{1}{\omega + \mathrm{i}0 - \mathbf{v}_{\mathbf{k}\rm e}\mathbf{q}} = -\pi\,\delta(\omega - \mathbf{v}_{\mathbf{k}\rm e}\mathbf{q}). \tag{7.15}$$

For the case of the classical plasma, where $n_{\rm B}(\omega_{\,\mathbf{k}\rm e}) = \mathrm{e}^{-(\omega_{\mathbf{k}\rm e}-\mu_{\rm e})/T_{\rm e}}$ is the Maxwell distribution, one finds

$$\Pi_{\rm e}^{\rm R}(\mathbf{q},\omega) = -\frac{\mathbf{q}^2\rho_0}{m_{\rm e}\omega^2}\,[1+\cdots] - \mathrm{i}\pi\rho_0 q\,\frac{\sum_{k_x}\delta(\omega - k_x q/m_{\rm e})\,\partial_{k_x}\mathrm{e}^{-k_x^2/2m_{\rm e}T_{\rm e}}}{\sum_{k_x}\mathrm{e}^{-k_x^2/2m_{\rm e}T_{\rm e}}}, \tag{7.16}$$

where k_x is the direction along the $\mathbf{q}$ vector. Performing the integrations one finds for $q\kappa \ll 1$, where $\kappa = \sqrt{T_{\rm e}/4\pi e^2\rho_0}$ is the Debye–Huckel screening length,

$$\tilde{\omega}_{\rm p}(\mathbf{q}) = \omega_{\rm p}\left(1 + \frac{3}{2}\,(q\kappa)^2 + \cdots\right) - \mathrm{i}\Gamma(\mathbf{q}); \qquad \Gamma(\mathbf{q}) = \sqrt{\frac{\pi}{8}}\frac{\omega_p}{(q\kappa)^3}\,\mathrm{e}^{-\frac{1}{2(q\kappa)^2}}. \tag{7.17}$$

The Landau damping rate Γ is exponentially small at small q. It is not surprising, thus, that the expansion in powers of $\mathbf{q}^2$, Eq. (7.14), missed the effect. As apparent from the derivation, the fact that the plasmon is damped (rather than pumped) is due to the fact that $\partial_{k_x} n_{\rm B} <, 0$ at $v_x = \omega_{\rm p}/q$. That is, there are more electrons moving slower than the phase velocity of the plasma wave than there are moving faster. As a result, more electrons absorb energy from the plasmon than supply it. If an inverted population is created with $\partial_{k_x} n_{\rm B} > 0$, it may lead to an instability of the plasma oscillations.

7.3 Acoustic Modes in Plasma

A closer look at Eq. (7.12) shows that it has more than one solution. In Fig. 7.1 we plotted $\mathrm{Re}\Pi_{\rm e}^{\rm R}(\mathbf{q},\omega)$ as a function of ω at some fixed $\mathbf{q} \ll \kappa^{-1}$. As we saw earlier, at large frequencies it is negative and is given by $-\mathbf{q}^2\rho_0/(m_{\rm e}\omega^2)$. Yet at $\omega = 0$ it is positive and given by $-m_{\rm e}\sum_{\mathbf{k}}\partial_{k_x} n_{\rm B}/k_x = \rho_0/T_{\rm e}$. There is a point where it changes sign, and for small q there is a solution of the dispersion equation (7.12) close to this point. It is called the *electron acoustic* mode, since at small q its dispersion relation is linear, $\omega_{\rm ea}(\mathbf{q}) = v_{\rm ea}q$. The corresponding velocity $v_{\rm ea}$ is found from the condition $\mathrm{Re}\Pi_{\rm e}^{\rm R}(\mathbf{q}, v_{\rm ea}q) = 0$, namely $⨍\, dk\,\partial_{k_x} n_{\rm B}/(m_{\rm e}v_{\rm ea} - k_x) = 0$,

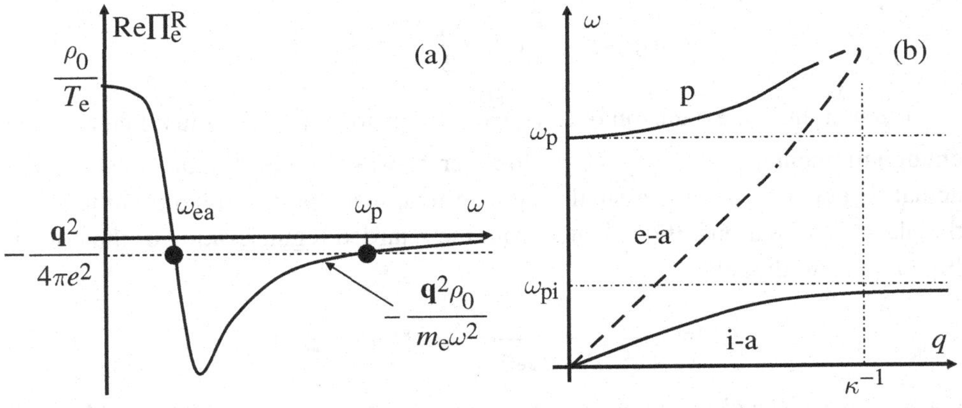

Figure 7.1 (a) Real part of the retarded polarization operator $\Pi_e^R(\mathbf{q},\omega)$ versus energy ω. Two solutions of Eq. (7.12) correspond to the plasma mode ω_p and the electron acoustic mode ω_{ea}. (b) Collective modes of plasma oscillations: p, plasmons; e-a, electron acoustic; i-a, ion acoustic. Dashed lines show overdamped modes.

where the integral is understood as the principal value. For the Maxwell distribution this yields $v_{ea} \approx 1.3\sqrt{T_e/m_e}$, that is, it is of the order of the electron thermal velocity [120]. At $\omega \approx \omega_{ea}$ one estimates $\mathrm{Re}\Pi_e^R(\mathbf{q},\omega) \sim (1 - \omega/v_{ea}q)\rho_0/T$; see Fig. 7.1. Let us look, however, at the imaginary part. With the help of Eq. (7.15) one finds $\mathrm{Im}\Pi_e^R(\mathbf{q}, v_{ea}q) \sim m\rho_0\partial_{k_x}n_B|_{k_x=mv_{ea}}/\sqrt{mT} \sim \rho_0/T$. Demanding $\Pi_e^R(\mathbf{q},\omega) = -q^2/(4\pi e^2) \approx 0$, one finds that the Landau damping of the electron acoustic mode is of the same order as its frequency. The mode is therefore overdamped and the corresponding oscillations can't be excited.

One can have an underdamped acoustic mode in the plasma, if one allows for the heavy ion motion $m_i \gg m_e$ *and* assumes that the electronic temperature is much higher than the ionic one, $T_e \gg T_i$. Then in the frequency range $q\sqrt{T_i/m_i} \ll \omega \ll q\sqrt{T_e/m_e}$ the electron motion is adiabatic and one may approximate $\mathrm{Re}\Pi_e^R(\mathbf{q},\omega) \approx \rho_0/T_e$. From the point of view of the ions, this is still a very high frequency and the ionic polarization operator is given by $\mathrm{Re}\Pi_i^R(\mathbf{q},\omega) \approx -\mathbf{q}^2\rho_0/(m_i\omega^2)$. For the real part of the dispersion equation (7.12) one thus finds

$$\frac{\mathbf{q}^2}{4\pi e^2} + \frac{\rho_0}{T_e} - \frac{\mathbf{q}^2\rho_0}{m_i\omega^2} = 0.$$

As a result one obtains the following dispersion relation of the *ion acoustic* mode:

$$\omega_{ia}^2(\mathbf{q}) = \omega_{pi}^2 \frac{\mathbf{q}^2}{\mathbf{q}^2 + \kappa^{-2}}, \tag{7.18}$$

where $\omega_{pi} = \sqrt{4\pi e^2\rho_0/m_i} \ll \omega_p$ is the ionic plasma frequency. At small wavenumbers $\kappa q \ll 1$ the dispersion relation is acoustic, $\omega_{ia}(\mathbf{q}) = v_{ia}q$, where the ion-acoustic velocity is $v_{\mathrm{ia}} = \sqrt{T_{\mathrm{e}}/m_{\mathrm{i}}}$. The sound velocity is determined by the electron temperature and the ionic mass. The Landau damping rate of the ion acoustic mode is evaluated in problem 7.7.1. There it is shown that the mode is underdamped if electronic temperature greatly exceeds the ionic one, $T_{\mathrm{e}} \gg T_{\mathrm{i}}$. This is indeed frequently the case in non-equilibrium plasmas, where electrons are quickly heated by an external field, while ions are slow to adjust. In equilibrium, $T_{\mathrm{e}} = T_{\mathrm{i}}$ the damping rate is basically the same as the mode frequency and it is thus overdamped. The modes of the two-component plasma are schematically summarized in Fig. 7.1(b).

7.4 Random Phase Approximation

We shall develop now an alternative approach to the linearized dynamics of the plasma modes, which will allow us to discuss their fluctuations as well. To this end we go back to the effective action (7.4) and expand it to second order in *both* V^{cl} and V^{q} field components. To this end it is useful to notice that, due to the normalization identity $Z = 1$, one has $\det[\hat{G}_{0\nu}^{-1}] = 1$ and therefore $\mathrm{Tr}\ln\big(\hat{G}_{0\nu}^{-1} \mp \hat{V}\big) = \mathrm{Tr}\ln\big(1 \mp \hat{G}_{0\nu}\hat{V}\big) = \mathrm{Tr}\ln\big(1 \mp \hat{G}_{0\nu}(\hat{\sigma}_1 V^{\mathrm{cl}} + V^{\mathrm{q}})\big)$. One expects that the cross-term $\sim V^{\mathrm{q}}V^{\mathrm{cl}}$ in the expansion contains the physics discussed in Sections 7.2 and 7.3. On the other hand, the quadratic term in V^{q} is going to be responsible for the thermal fluctuations of the plasma modes. Action (7.4) expanded to second order may be written as a quadratic form:

$$S[V] = \iint \mathrm{d}x\mathrm{d}x'\, \vec{V}(x)^{\mathrm{T}} \big[\hat{U}_{\mathrm{RPA}}(x,x')\big]^{-1} \vec{V}(x'). \tag{7.19}$$

Its kernel is known as the (inverse) random phase approximation (RPA) interaction potential and is given by

$$\big[\hat{U}_{\mathrm{RPA}}(x,x')\big]^{-1} = \hat{U}^{-1} + \hat{\Pi}_{\mathrm{e}} = \frac{\delta(x-x')}{4\pi e^2}\begin{pmatrix} 0 & -\nabla^2_{\mathbf{r}'} \\ -\nabla^2_{\mathbf{r}'} & 0 \end{pmatrix} + \begin{pmatrix} 0 & \Pi^{\mathrm{A}}_{\mathrm{e}}(x,x') \\ \Pi^{\mathrm{R}}_{\mathrm{e}}(x,x') & \Pi^{\mathrm{K}}_{\mathrm{e}}(x,x') \end{pmatrix}. \tag{7.20}$$

From now on we assume that ions are infinitely heavy and disregard their dynamics (this implies that $\hat{\Pi}_{\mathrm{i}} \to 0$), focusing exclusively on electrons. The electron polarization matrix $\hat{\Pi}_{\mathrm{e}}(x,x')$ has the standard causality structure of a bosonic self-energy; see Eq. (6.17). Its retarded component is given by

$$\Pi^{\mathrm{R}}_{\mathrm{e}}(x,x') = -\left.\frac{\delta\langle\rho(x)\rangle}{\delta V^{\mathrm{cl}}(x')}\right|_{V^{\mathrm{cl}}=0} = \frac{1}{2\mathrm{i}}\mathrm{Tr}\Big\{\hat{G}_0(x,x')\hat{\sigma}_1\hat{G}_0(x',x)\Big\}, \tag{7.21}$$

where the first equality here follows from the comparison of Eqs. (7.20) and (7.5), while the second is a result of the expansion of $\mathrm{Tr}\ln(1-\hat{G}\hat{V})$ to second order in $\hat{V}$. This is a response function of the average density at a space-time point x on a perturbation of the classical external potential at a point x'. On physical grounds one expects such a response to be retarded, that is, $\Pi_e^R(x,x') \sim \theta(t-t')$. This is the reason for the superscript R. Indeed, straightforward matrix algebra leads to

$$\Pi_e^R(x,x') = \frac{1}{2\mathrm{i}}\Big(G_0^K(x,x')G_0^A(x',x) + G_0^R(x,x')G_0^K(x',x)\Big), \tag{7.22}$$

where the retarded nature of $\Pi_e^R(x,x')$ is apparent from the causality properties of the Green functions. To make the structure of the action (7.19) symmetric, we have also introduced $\Pi_e^A(x,x') \sim \theta(t'-t)$. It is given by

$$\Pi_e^A(x,x') = \frac{1}{2\mathrm{i}}\,\mathrm{Tr}\Big\{\hat{G}_0\hat{G}_0\hat{\sigma}_1\Big\} = \frac{1}{2\mathrm{i}}\Big(G_0^K(x,x')G_0^R(x',x) + G_0^A(x,x')G_0^K(x',x)\Big). \tag{7.23}$$

Finally, the Keldysh component, obtained by the expansion of the action (7.4) to second order in V^q, is

$$\begin{aligned}\Pi_e^K(x,x') &= \frac{1}{2\mathrm{i}}\,\mathrm{Tr}\Big\{\hat{G}_0(x,x')\hat{G}_0(x',x)\Big\} = \frac{1}{2\mathrm{i}}\Big[G_0^K(x,x')G_0^K(x',x)\\ &\quad - (G_0^R(x,x')-G_0^A(x,x'))(G_0^R(x',x)-G_0^A(x',x))\Big],\end{aligned} \tag{7.24}$$

where we have used the fact that due to causality $G_0^R(x,x')G_0^R(x',x) = 0$. (The same is true for the advanced Green functions.)

For a translationally invariant system the components of the polarization matrix $\hat{\Pi}$ can be evaluated using the Fourier transformation and the explicit form of the bare Green functions, Eq. (6.7). This way one finds

$$\Pi_e^{R(A)}(\mathbf{q},\omega) = \frac{1}{2}\sum_{\mathbf{k}} \frac{\tilde{F}_e(\mathbf{k}+\mathbf{q}) - \tilde{F}_e(\mathbf{k})}{\omega \pm \mathrm{i}0 + \omega_{\mathbf{k}} - \omega_{\mathbf{k}+\mathbf{q}}}; \tag{7.25a}$$

$$\Pi_e^K(\mathbf{q},\omega) = \mathrm{i}\pi\sum_{\mathbf{k}} \delta(\omega + \omega_{\mathbf{k}} - \omega_{\mathbf{k}+\mathbf{q}})\big[\tilde{F}_e(\mathbf{k})\tilde{F}_e(\mathbf{k}+\mathbf{q}) - 1\big], \tag{7.25b}$$

where we employed that $G_0^K = \tilde{F}_e(G_0^R - G_0^A)$ and $G_0^R - G_0^A = -2\pi\mathrm{i}\delta(\epsilon - \omega_{\mathbf{k}})$ to perform the energy integrations with the help of the delta-function. For an equilibrium plasma $\tilde{F}_e(\mathbf{k}) = F^{\mathrm{eq}}(\omega_{\mathbf{k}}) = \coth(\omega_{\mathbf{k}} - \mu_e)/2T_e$ and one finds FDT relation between the components of the polarization matrix:

$$\Pi_e^K(\mathbf{q},\omega) = \coth\frac{\omega}{2T_e}\Big[\Pi_e^R(\mathbf{q},\omega) - \Pi_e^A(\mathbf{q},\omega)\Big]. \tag{7.26}$$

Notice that, though the particle distribution function contains a finite chemical potential, μ_e, the distribution function for the real boson field V turns out to have zero chemical potential. This must be the case, of course, for a real field; see Section 6.4.

Going back to the RPA action (7.19), one notices that the fluctuations of the classical component V^{cl} of the scalar potential are especially strong if $[U_{\text{RPA}}^{-1}]^{\text{R}} = U^{-1} + \Pi_{\text{e}}^{\text{R}} \approx 0$. In the Fourier representation, this condition boils down to

$$\frac{\mathbf{q}^2}{4\pi e^2} + \sum_{\mathbf{k}} \frac{n_{\text{B}}(\omega_{\mathbf{k}+\mathbf{q}}) - n_{\text{B}}(\omega_{\mathbf{k}})}{\omega + \text{i}0 + \omega_{\mathbf{k}} - \omega_{\mathbf{k}+\mathbf{q}}} = 0, \tag{7.27}$$

where we substituted $F^{\text{eq}}(\omega_{\mathbf{k}}) = 2n_{\text{B}}(\omega_{\mathbf{k}}) + 1$. If $q \ll k_{\text{T}}$, where $k_{\text{T}} = \sqrt{\text{T}_{\text{e}} m_{\text{e}}}$ is a typical thermal momentum, one may expand $\omega_{\mathbf{k}+\mathbf{q}} \approx \omega_{\mathbf{k}} + \mathbf{v}_{\mathbf{k}}\mathbf{q}$, while $n_{\text{B}}(\omega_{\mathbf{k}+\mathbf{q}}) \approx n_{\text{B}}(\omega_{\mathbf{k}}) + \mathbf{q}\nabla_{\mathbf{k}} n_{\text{B}}(\omega_{\mathbf{k}})$ and arrive exactly at the dispersion equation (7.12) of the plasma mode. The unexpanded version (7.27) probably offers a clearer explanation of Landau damping. Indeed, the imaginary part of the retarded polarization operator corresponds to the energy conservation delta-function $\omega_{\mathbf{k}+\mathbf{q}} = \omega_{\mathbf{k}} + \omega$. Therefore it describes a real (i.e. energy-conserving) transition of a particle with an initial momentum $\mathbf{k}$, which absorbs a quantum of plasma oscillation with momentum $\mathbf{q}$ and energy $\omega = \omega_{\text{p}}(\mathbf{q})$ and ends up in the state $\mathbf{k}+\mathbf{q}$. For small wavevectors $\mathbf{q}$ this is only possible if the initial momentum $\mathbf{k}$ is sufficiently large, so that $\mathbf{v}_{\mathbf{k}} \approx \omega_{\text{p}}/q$, and thus the corresponding initial energy $\mathbf{v}_{\mathbf{k}}^2/2m_{\text{e}} \approx \omega_{\text{p}}^2/(2m_{\text{e}}q^2) = T_{\text{e}}/2(q\kappa)^2$; see Fig. 7.2. The occupation numbers of such high energy states are exponentially small, and therefore so is the number of real transitions. Of course, there are also stimulated transitions, where particles emit energy ω_{p} into the plasma wave. Those, however, require even higher initial particle energy and their number is correspondingly smaller (in equilibrium). The difference between stimulated emissions and absorptions is described by the factor $n_{\text{B}}(\omega_{\mathbf{k}+\mathbf{q}}) - n_{\text{B}}(\omega_{\mathbf{k}})$ in Eq. (7.27).

In a vicinity of the plasma frequency $\omega \approx \omega_{\text{p}}(\mathbf{q})$ one may use the analysis of Section 7.2 to approximate

$$\left[U_{\text{RPA}}^{\text{R(A)}}(\mathbf{q},\omega)\right]^{-1} \approx \frac{\mathbf{q}^2}{4\pi e^2} \frac{\omega^2 - \omega_{\text{p}}^2(\mathbf{q}) \pm 2\text{i}\Gamma(\mathbf{q})\,\omega}{\omega_{\text{p}}^2}; \tag{7.28}$$

$$\left[U_{\text{RPA}}^{\text{K}}(\mathbf{q},\omega)\right]^{-1} \approx \coth\frac{\omega}{2T_{\text{e}}} \frac{\mathbf{q}^2}{4\pi e^2} \frac{4\text{i}\Gamma(\mathbf{q})\,\omega}{\omega_{\text{p}}^2} \approx \frac{\mathbf{q}^2}{4\pi e^2} \frac{8\text{i}T_{\text{e}}\Gamma(\mathbf{q})}{\omega_{\text{p}}^2},$$

where we employed the fact that in the classical plasma $T_{\text{e}} \gg \omega_{\text{p}}$. One can now discuss the Gaussian RPA action (7.19). To this end let us introduce the longitudinal electric field $\mathbf{E} = -\nabla_{\mathbf{r}} V/e$ and rewrite the action (7.19) in terms of its complex spatial Fourier components $\mathbf{E}_{\mathbf{q}}(t) = \text{i}\mathbf{q}V(\mathbf{q},t)/e$:

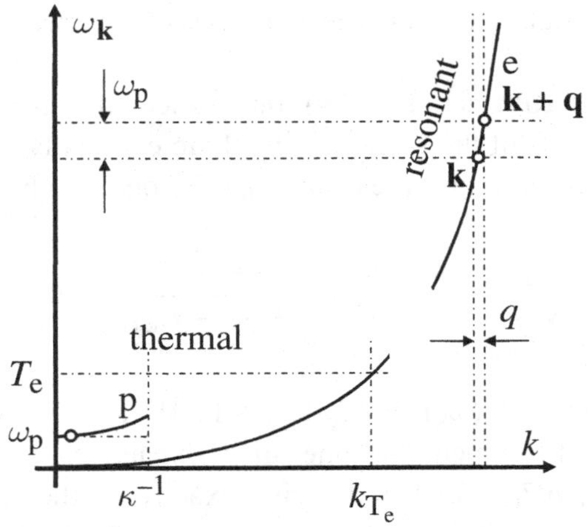

Figure 7.2 Landau damping of a plasmon with momentum $\mathbf{q}$ by the resonant transition of an electron from state $\mathbf{k}$ to state $\mathbf{k}+\mathbf{q}$. The thermal electrons determine the plasmon spectrum $\omega_{\mathrm{p}}(\mathbf{q})$ but do not participate in damping.

$$S[\mathbf{E}] = \sum_{\mathbf{q}} \int \frac{\mathrm{d}t}{4\pi\omega_{\mathrm{p}}^2} \left(\mathbf{E}_{\mathbf{q}}^{\mathrm{cl}}, \mathbf{E}_{\mathbf{q}}^{\mathrm{q}}\right)^* \begin{pmatrix} 0 & -\partial_t^2 - \omega_{\mathrm{p}}^2 + 2\Gamma\partial_t \\ -\partial_t^2 - \omega_{\mathrm{p}}^2 - 2\Gamma\partial_t & 8\mathrm{i}T_{\mathrm{e}}\Gamma \end{pmatrix} \begin{pmatrix} \mathbf{E}_{\mathbf{q}}^{\mathrm{cl}} \\ \mathbf{E}_{\mathbf{q}}^{\mathrm{q}} \end{pmatrix}. \tag{7.29}$$

Following Section 4.2, it is convenient to split the term quadratic in the quantum component, $\mathbf{E}_{\mathbf{q}}^{\mathrm{q}}$, with the help of the Hubbard–Stratonovich transformation, using a longitudinal auxiliary field $\mathbf{f}_{\mathbf{q}}(t)$:

$$\mathrm{e}^{-\sum_{\mathbf{q}}\int \mathrm{d}t \frac{8T_{\mathrm{e}}\Gamma}{4\pi\omega_{\mathrm{p}}^2}|\mathbf{E}_{\mathbf{q}}^{\mathrm{q}}|^2} = \int \mathbf{D}[\mathbf{f}]\, \mathrm{e}^{-(4\pi\omega_{\mathrm{p}}^2)^{-1}\sum_{\mathbf{q}}\left[(8T_{\mathrm{e}}\Gamma)^{-1}|\mathbf{f}_{\mathbf{q}}|^2 - \mathrm{i}(\mathbf{f}_{\mathbf{q}}^*\mathbf{E}_{\mathbf{q}} + \mathbf{E}_{\mathbf{q}}^*\mathbf{f}_{\mathbf{q}})\right]}. \tag{7.30}$$

The remaining action is linear in $\mathbf{E}_{\mathbf{q}}^{\mathrm{q}}$ and $(\mathbf{E}_{\mathbf{q}}^{\mathrm{q}})^*$. The corresponding functional integrals result in the delta-function of the following stochastic equation of motion:

$$\ddot{\mathbf{E}}_{\mathbf{q}}^{\mathrm{cl}} = -\omega_{\mathrm{p}}^2(\mathbf{q})\mathbf{E}_{\mathbf{q}}^{\mathrm{cl}} - 2\Gamma(\mathbf{q})\dot{\mathbf{E}}_{\mathbf{q}}^{\mathrm{cl}} + \mathbf{f}_{\mathbf{q}}(t) \tag{7.31}$$

as well as its complex conjugate. The stochastic force on the right-hand side is Gaussian white noise with zero mean and a second moment given by

$$\left\langle \mathbf{f}_{\mathbf{q}}^*(t)\mathbf{f}_{\mathbf{q}'}(t')\right\rangle = 16\pi\omega_{\mathrm{p}}^2 T_{\mathrm{e}}\Gamma(\mathbf{q})\delta_{\mathbf{q},\mathbf{q}'}\delta(t-t'). \tag{7.32}$$

The fact that the variance is proportional to the damping is a manifestation of the FDT. According to Eq. (7.31) the electric field exhibits an oscillatory motion at the plasma frequency $\omega_{\mathrm{p}}(\mathbf{q})$ with viscous friction given by the Landau damping. There is also a corresponding stochastic force exciting the oscillations. In the long-time limit the fluctuations and the dissipation combined lead to a stationary probability distribution of finding a certain electric filed amplitude $|\mathbf{E}^{\rm cl}_{\mathbf{q}}|$ and its time derivative $|\dot{\mathbf{E}}^{\rm cl}_{\mathbf{q}}|$. Such a distribution obeys the Fokker–Planck equation, which is discussed in Chapter 4. As shown there (see Eq. (4.33)), the stationary solution of the Fokker–Planck equation, corresponding to Eqs. (7.31) and (7.32), is

$$\mathcal{P}(|\mathbf{E}^{\rm cl}_{\mathbf{q}}|, |\dot{\mathbf{E}}^{\rm cl}_{\mathbf{q}}|) \propto \exp\left\{-\frac{|\dot{\mathbf{E}}^{\rm cl}_{\mathbf{q}}|^2 + \omega_{\rm p}^2 |\mathbf{E}^{\rm cl}_{\mathbf{q}}|^2}{8\pi\omega_{\rm p}^2 T_{\rm e}}\right\}. \tag{7.33}$$

This is nothing but the classical Maxwell–Boltzmann distribution function, describing the electric field noise. The latter is induced by the thermal fluctuations of the plasma close to the plasma frequency.

7.5 Beyond Linearized Dynamics

The long-wavelength fluctuations of the electronic plasma are dominated by the plasmons, which behave as quasiparticles with the dispersion relation $\omega_{\rm p}(\mathbf{q})$. Their linearized dynamics is governed by the Keldysh RPA action (7.29). A (moderately) excited plasma may be described by a non-equilibrium distribution function of the *plasmons* $\tilde{F}_{\rm p}(x, \mathbf{q})$, which is a slow function of the space-time location x. Our goal is to derive a kinetic equation that describes evolution of such a distribution function. To this end we need to discuss what is the main source of nonlinearity in the plasmon dynamic. One candidate is three- or four-plasmon collisions. The corresponding processes could be obtained by expanding the action (7.4) to the third or fourth powers in the potential V. The three-plasmon collisions cannot lead to real (energy-conserving) processes, due to the energy gap $\omega_{\rm p}$. Indeed, to split one plasmon into two new ones, one needs at least the energy $2\omega_{\rm p}$, which an initial long-wavelength plasmon does not have. As a result, three-plasmon collisions are virtual processes, which only lead to a renormalization of the dispersion relation. The four-plasmon collisions can be energy conserving. One may show, though, that the corresponding cross section is small.

Instead, the most important process is interaction of the plasmons with the resonant electrons. On the level of linearized dynamics those are exactly the processes which lead to Landau damping; see Fig. 7.2. At first glance, considering both plasmons and electrons as independent entities looks like double-counting. Indeed, we have derived the plasmon dynamics, for example, Eq. (7.4), by completely integrating out the electronic (and ionic) degrees of freedom. It is useful, however, to step back and acknowledge that different groups of electrons played very different roles in this process. The real part of the polarization operator that leads to the plasmon dispersion relation, namely to the inertial terms $\partial_t^2 + \omega_{\rm p}^2(\mathbf{q})$ in the plasmon action (7.29), is primarily given by thermal electrons with $k \lesssim k_{T{\rm e}} = \sqrt{T_{\rm e} m_{\rm e}}$. On the other hand, the Landau damping is entirely coming from the fast resonant electrons with $k \sim \sqrt{T_e m_e}/(\kappa q) \gg k_{T_e}$. Thus our program is to (i) split the electronic degrees of freedom into slow (thermal) and fast (resonant) ones (see Fig. 7.2); (ii) integrate out slow electrons and arrive at the undamped plasmon dynamics; (iii) treat the resulting long wavelength plasmons *and* remaining fast electrons as independent degrees of freedom; and (iv) derive coupled kinetic equations for their respective distribution functions $\tilde{F}_p(x, \mathbf{q})$ and $\tilde{F}_e(x, \mathbf{k})$, where $|\mathbf{q}| < \kappa^{-1} \ll k_{T_e} < |\mathbf{k}|$.

Kinetic equations for plasmons may be directly read out from the action (7.29). To this end we compare it with the real boson action (6.28) (in the present case $\omega_p^2(\mathbf{q}) \approx \omega_p^2 + \langle \mathbf{v}_\mathbf{k}^2 \rangle\, \mathbf{q}^2 \to \omega_p^2 - \langle \mathbf{v}_\mathbf{k}^2 \rangle \nabla_\mathbf{r}^2$) and treat the *imaginary* part of the polarization operator $\hat{\Pi}_e$ as the self-energy of the bosonic field (more precisely, $\hat{\Sigma}_p = -(2e^2/\mathbf{q}^2)\,\mathrm{Im}\,\hat{\Pi}_e$). The real part of the polarization operator went into the definition of the plasmon dispersion. The corresponding kinetic equation is given by Eq. (6.54) and the collision integral is Eq. (6.55). To write down the latter we need to employ expressions (7.25) for the components of the polarization matrix with the understanding that the electronic distribution function $\tilde{F}_e$ may be a non-equilibrium one. This way, we find

$$\Big[\partial_t + \mathbf{v}_{\mathbf{q}p}\nabla_\mathbf{r} - \nabla_\mathbf{r}\omega_p(x)\nabla_\mathbf{q}\Big]\tilde{F}_p(x,\mathbf{q}) = \frac{2\pi^2 e^2 \omega_p(\mathbf{q})}{\mathbf{q}^2}\sum_{\mathbf{k}>k_{T_e}} \delta(\omega_p(\mathbf{q}) + \omega_\mathbf{k} - \omega_{\mathbf{k}+\mathbf{q}})$$
$$\times\Big[\tilde{F}_e(x,\mathbf{k}+\mathbf{q})\tilde{F}_e(x,\mathbf{k}) - 1 + \tilde{F}_p(x,\mathbf{q})\big(\tilde{F}_e(x,\mathbf{k}+\mathbf{q}) - \tilde{F}_e(x,\,\mathbf{k})\big)\Big], \tag{7.34}$$

where $\omega_p^2(x) = (2\pi e^2/m_e)\sum_\mathbf{k}[\tilde{F}_e(x,\mathbf{k}) - 1]$ and $\mathbf{v}_{\mathbf{q}p} = \nabla_\mathbf{q}\omega_p(\mathbf{q}) = \mathbf{q}\langle \mathbf{v}_\mathbf{k}^2\rangle/\omega_p$.

We turn now to the kinetic equation for the fast electron distribution function $\tilde{F}_e$. It is given by Eq. (6.51), while the collision integral is Eq. (6.47). To find the proper self-energy one notices that the interaction of the fast resonant electrons with the plasmon field is described by the action $S_V = \int dx\, \vec{\bar{\phi}}_e \hat{V} \vec{\phi}_e$, Eq. (2.66). There is no significant vertex renormalization due to the integrated-out slow electrons. On the other hand, the propagator of the V-field is renormalized and is given by the RPA interaction potential (without Landau damping). Diagrams for the self-energy are plotted in Fig. 7.3, and the corresponding expressions are

$$\Sigma_e^{R(A)}(x,x') = \frac{i}{2}\left[G_0^K(x,x')U_{RPA}^{A(R)}(x',x) + G_0^{R(A)}(x,x')U_{RPA}^K(x',x)\right];$$
$$\Sigma_e^K(x,x') = \frac{i}{2}\left[G_0^K(x,x')U_{RPA}^K(x',x) - (G_0^R - G_0^A)(U_{RPA}^R - U_{RPA}^A)\right],$$

where $U_{RPA}^{R(A)} = (4\pi e^2\omega_p^2/\mathbf{q}^2)[(\omega \pm i0)^2 - \omega_p^2]^{-1}$; compare (7.28), while the Keldysh component is $U_{RPA}^K = U_{RPA}^R \circ F_p - F_p \circ U_{RPA}^A$. Once again we have used the fact that $G_0^{R(A)}(x,x')U_{RPA}^{R(A)}(x',x) = 0$, due to causality. Performing WT, putting the

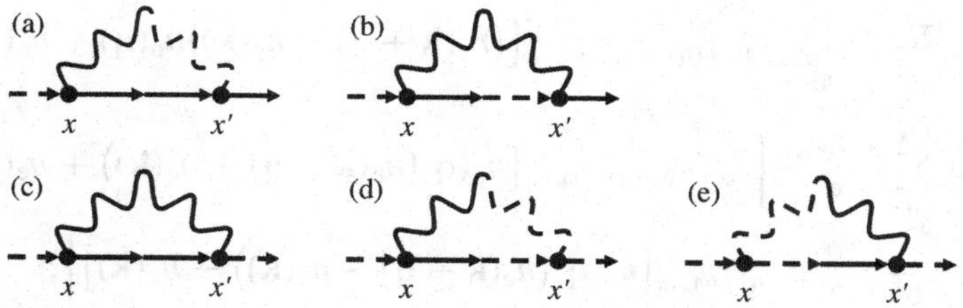

Figure 7.3 Self-energy of the resonant electrons due to interactions with the plasmons: $\Sigma_e^R(x,x')$ is given by diagrams (a) and (b); $\Sigma_e^K(x,x')$ – by diagrams (c)–(e). The straight lines are electronic Green functions G_0, the wavy lines are RPA plasmon propagators U_{RPA}. Full lines denote classical, and dashed lines quantum "legs."

distribution functions on "mass-shell" according to Eqs. (6.48) and (6.53) for F_e and F_p, respectively, and integrating over the plasmon energy ω with the help of the delta-function, one obtains the kinetic equation

$$\left[\tilde{Z}^{-1}\partial_t + \tilde{\mathbf{v}}_{\mathbf{k}e}\nabla_{\mathbf{r}} - \nabla_{\mathbf{r}}\mathrm{Re}\Sigma_e^R(x,\mathbf{k})\nabla_{\mathbf{k}}\right]\tilde{F}_e(x,\mathbf{k}) = \sum_{\mathbf{q}}^{1/\kappa}\frac{2\pi^2e^2\omega_p(\mathbf{q})}{\mathbf{q}^2}$$
$$\times\Big\{\delta(\omega_{\mathbf{k}}+\omega_p(\mathbf{q})-\omega_{\mathbf{k}+\mathbf{q}})\Big[\tilde{F}_e(\mathbf{k}+\mathbf{q})\tilde{F}_p(\mathbf{q})-1+\tilde{F}_e(\mathbf{k})\big(\tilde{F}_e(\mathbf{k}+\mathbf{q})-\tilde{F}_p(\mathbf{q})\big)\Big]$$
$$+\,\delta(\omega_{\mathbf{k}}-\omega_p(\mathbf{q})-\omega_{\mathbf{k}-\mathbf{q}})\Big[\tilde{F}_e(\mathbf{k}-\mathbf{q})\tilde{F}_p(\mathbf{q})+1-\tilde{F}_e(\mathbf{k})\big(\tilde{F}_e(\mathbf{k}-\mathbf{q})+\tilde{F}_p(\mathbf{q})\big)\Big]\Big\}, \tag{7.35}$$

where we have omitted space-time arguments x on the right-hand side and changed the summation variable $\mathbf{q} \to -\mathbf{q}$ in the last term. Here $\tilde{Z}^{-1} = 1 - \partial_\epsilon\mathrm{Re}\Sigma_e^R$ and $\tilde{\mathbf{v}}_{\mathbf{k}e} = \mathbf{k}^2/(2m_e) + \nabla_{\mathbf{k}}\mathrm{Re}\Sigma_e^R$. Equations (7.34) and (7.35) provide the kinetic theory of the coupled system of plasmons and resonant high-energy electrons [121, 122, 123, 15]. Despite having collision integrals on the right-hand sides, these equations still describe *collisionless* plasma. Indeed, their right-hand sides describe only interactions between fast electrons and slow fluctuations of the longitudinal electric field – plasmons. Collisions between the electrons are not included. Unlike the Vlasov equations, which treat the electric field as a deterministic one, Eqs. (7.34) and (7.35) take into account fluctuations of both electron and plasmon degrees of freedom.

To simplify these equations let us introduce non-equilibrium occupation numbers $\tilde{F}_p(x,\mathbf{q}) = 2n_p(x,\mathbf{q})+1$ for plasmons and $\tilde{F}_e(x,\mathbf{k}) = 2n_e(x,\mathbf{k})+1$ for electrons. Due to the classical nature of the problem $n_p \gg 1$ and $n_e \ll 1$. One may thus neglect terms quadratic in n_e. Restricting ourselves to a spatially uniform situation, we find for the system of coupled kinetic equations

$$\partial_t n_p(\mathbf{q}) = \frac{4\pi^2 e^2 \omega_p}{\mathbf{q}^2} \sum_{\mathbf{k}>k_{T_e}} \delta_{\omega_p(\mathbf{q})+\omega_\mathbf{k}-\omega_{\mathbf{k}+\mathbf{q}}} \left[\left(n_e(\mathbf{k}+\mathbf{q}) - n_e(\mathbf{k}) \right) n_p(\mathbf{q}) + n_e(\mathbf{k}+\mathbf{q}) \right];$$

$$\partial_t n_e(\mathbf{k}) = \sum_{\mathbf{q}}^{1/\kappa} \frac{4\pi^2 e^2 \omega_p}{\mathbf{q}^2} \Big\{ \delta_{\omega_\mathbf{k}+\omega_p(\mathbf{q})-\omega_{\mathbf{k}+\mathbf{q}}} \left[n_p(\mathbf{q}) \left(n_e(\mathbf{k}+\mathbf{q}) - n_e(\mathbf{k}) \right) + n_e(\mathbf{k}+\mathbf{q}) \right]$$
$$+ \delta_{\omega_\mathbf{k}-\omega_p(\mathbf{q})-\omega_{\mathbf{k}-\mathbf{q}}} \left[n_p(\mathbf{q}) \left(n_e(\mathbf{k}-\mathbf{q}) - n_e(\mathbf{k}) \right) - n_e(\mathbf{k}) \right] \Big\},$$

where we put $\tilde{Z} = 1$ for simplicity. Notice that the Maxwell distribution for electrons, $n_e(\mathbf{k}) = e^{-(\omega_\mathbf{k} - \mu_e)/T_e}$, and the Planck distribution for plasmons, $n_p(\mathbf{q}) = n_B(\omega_p(\mathbf{q}))$, are equilibrium solutions of the coupled system. One can now use the scale separation between long-wavelength plasmons and fast resonant electrons, that is, $q \lesssim \kappa^{-1} \ll k_{T_e} \lesssim k$ to expand in powers of q. While in the equation for n_p first order is sufficient, in the equation for n_e one should expand up to the second one. This procedure leads to

$$\partial_t n_p(t, \mathbf{q}) = -2\left[\Gamma(\mathbf{q}; n_e)\, n_p(t, \mathbf{q}) - L(\mathbf{q}; n_e) \right]; \tag{7.36}$$

$$\partial_t n_e(t, \mathbf{k}) = \nabla_{\mathbf{k}_\mu} \left[D_{\mu\nu}(\mathbf{k}; n_p) \nabla_{\mathbf{k}_\nu} n_e(t, \mathbf{k}) - \mathbf{j}_\mu(\mathbf{k})\, n_e(t,\ \mathbf{k}) \right], \tag{7.37}$$

where $\mu, \nu = x, y, z$. The plasmon damping coefficient and the fast electron diffusion tensor are given by

$$\Gamma(\mathbf{q}; n_e) = -\frac{2\pi^2 e^2 \omega_p^{\prime}(\mathbf{q})}{\mathbf{q}^2} \sum_{\mathbf{k}>k_{T_e}} \delta(\omega_p(\mathbf{q}) - \mathbf{v}_{\mathbf{k}e}\mathbf{q})\, \mathbf{q}_\mu \nabla_{\mathbf{k}_\mu} n_e(t, \mathbf{k}); \tag{7.38}$$

$$D_{\mu\nu}(\mathbf{k}; n_p) = 4\pi^2 e^2 \sum_{\mathbf{q}}^{1/\kappa} \delta(\omega_p(\mathbf{q}) - \mathbf{v}_{\mathbf{k}e}\mathbf{q}) \frac{\mathbf{q}_\mu \mathbf{q}_\nu}{\mathbf{q}^2}\, \omega_p(\mathbf{q})\, n_p(t, \mathbf{q}). \tag{7.39}$$

These quantities originate from the *stimulated* emission and absorption of plasmons by the resonant electrons. The *spontaneous* emission processes give rise to the second terms on the right-hand sides of Eqs. (7.36) and (7.37). They read as

$$L(\mathbf{q}; n_e) = \frac{2\pi^2 e^2 \omega_p(\mathbf{q})}{\mathbf{q}^2} \sum_{\mathbf{k}>k_{T_e}} \delta(\omega_p(\mathbf{q}) - \mathbf{v}_{\mathbf{k}e}\mathbf{q})\, n_e(t, \mathbf{k}); \tag{7.40}$$

$$\mathbf{j}_\mu(\mathbf{k}) = -4\pi^2 e^2 \sum_{\mathbf{q}}^{1/\kappa} \delta(\omega_p(\mathbf{q}) - \mathbf{v}_{\mathbf{k}e}\mathbf{q})\, \omega_p(\mathbf{q})\, \mathbf{q}_\mu / \mathbf{q}^2. \tag{7.41}$$

These terms are necessary for the equilibrium distributions to be solutions of the kinetic equations. Under the adopted approximations the Maxwell distribution for the fast electrons and the classical equipartition for plasmons $n_p^{eq}(\mathbf{q}) = T_e/\omega_p(\mathbf{q})$ are such equilibrium solutions of Eqs. (7.36) and (7.37). In equilibrium $\Gamma(\mathbf{q}) = \Gamma(\mathbf{q}; n_e^{eq})$ is the already familiar linear Landau damping (7.17). It may be modified if the slope of the electronic distribution $\nabla_{\mathbf{k}} n_{\rm e}(t,\mathbf{k})$ differs from the equilibrium one (see Section 7.6). The fact that equation (7.37) for the fast electron distribution has the form of the continuity relation reflects conservation of the number of electrons inside the resonant region, $\partial_t \sum_{\mathbf{k}} n_{\rm e}(t,\mathbf{k}) = 0$ (as opposed to the number of plasmons, which is not conserved). To prove this relation one has to perform integration by parts in Eq. (7.37) and assume that the particle current vanishes in the region of thermal electrons, $\mathbf{J}_\mu(\mathbf{k}) = \mathbf{j}_\mu n_{\rm e} - D_{\mu\nu}\nabla_{\mathbf{k}_\nu} n_{\rm e} = 0$ for $k < k_{T_{\rm e}}$. This is indeed the case if the nonresonant part of the electron distribution is kept in thermal equilibrium, for example, due to collisions. Under this assumption Eqs. (7.36) and (7.37) also conserve the total momentum as well as the total energy of the combined system of plasmons and the resonant electrons, that is, $\partial_t\big[\sum_{\mathbf{q}} \mathbf{q}\, n_{\rm p}(t,\mathbf{q}) + \sum_{\mathbf{k}} \mathbf{k}\, n_{\rm e}(t,\mathbf{k})\big] = 0$ and $\partial_t\big[\sum_{\mathbf{q}} \omega_{\rm p}(\,\mathbf{q})\, n_{\rm p}(t,\mathbf{q}) + \sum_{\mathbf{k}} \mathbf{k}^2/(2m)\, n_{\rm e}(t,\mathbf{k})\big] = 0$.

For distributions that are isotropic in momentum space one may write

$$D_{\mu\nu}(\mathbf{k}; n_{\rm p}) = \frac{\mathbf{k}_\mu \mathbf{k}_\nu}{\mathbf{k}^2} D_\parallel(k) + \left(\delta_{\mu\nu} - \frac{\mathbf{k}_\mu \mathbf{k}_\nu}{\mathbf{k}^2}\right) D_\perp(k); \quad \mathbf{j}(\mathbf{k}) = j_\parallel \frac{\mathbf{k}}{k}. \tag{7.42}$$

Performing the angular integrations, one finds for the corresponding kinetic coefficients

$$\begin{aligned}
D_\parallel(k) &= \frac{e^2\omega_{\rm p}^3 m_{\rm e}^3}{k^3} \int_{\omega_{\rm p} m_{\rm e}/k}^{1/\kappa} \frac{dq}{q}\, n_{\rm p}(t,q); \qquad \Gamma(q) = \frac{e^2\omega_{\rm p}^2 m_{\rm e}^2}{2q^3}\, n_{\rm e}\left(t, \frac{\omega_{\rm p} m_{\rm e}}{q}\right); \\
D_\perp(k) &= \frac{e^2\omega_{\rm p} m_{\rm e}}{k} \int_{\omega_{\rm p} m_{\rm e}/k}^{1/\kappa} q\, dq\, n_{\rm p}(t,q) \left(1 - \frac{\omega_{\rm p}^2 m_{\rm e}^2}{k^2 q^2}\right); \\
L(q) &= \frac{e^2\omega_{\rm p} m_{\rm e}}{2q^3} \int_{\omega_{\rm p} m_{\rm e}/q}^{\infty} k\, dk\, n_{\rm e}(t,k)\,; \qquad j_\parallel(k) = -\frac{e^2\omega_{\rm p}^2 m_{\rm e}^2}{k^2} \ln \frac{k}{\kappa\omega_{\rm p} m_{\rm e}}.
\end{aligned} \tag{7.43}$$

Notice that if the plasmon occupation number decays slower than $1/q^2$ there is a strong inequality $D_\perp \gg D_\parallel$, with the former being dependent on details of the cutoff at $q \sim 1/\kappa$. As a result, the fast diffusive relaxation first takes place in the direction perpendicular to $\mathbf{k}$. Diffusive relaxation in the parallel direction is much slower and therefore it is $D_\parallel$ which is relevant for kinetics. In equilibrium the parallel diffusion constant is $D_\parallel(k) = (e^2\omega_{\rm p}^2\, T_{\rm e} m_{\rm e}^3/k^3)\ln(k/\kappa\omega_{\rm p} m_{\rm e})$, that is, it depends on the cutoff only logarithmically and is therefore universal. One may check then that $D_\parallel(k)\partial_k n_{\rm e}^{\rm eq}(k) = j_\parallel(k) n_{\rm e}^{\rm eq}(k)$ and $\Gamma(q) n_{\rm p}^{\rm eq}(q) = L(q)$, as it should be according to Eqs. (7.36) and (7.37).

7.6 Nonlinear Landau Damping

Consider a spatially uniform and momentum isotropic plasma, where the plasmon modes are initially ($t = 0$) excited above the thermal population, $n_{\rm p}(0) \gg n_{\rm p}^{\rm eq}$

for the range of the wavenumbers $q \lesssim \kappa^{-1}$. It is convenient to choose $\sqrt{\omega_{\mathrm{p}} m_{\mathrm{e}}} = \sqrt{k_{T_{\mathrm{e}}}/\kappa}$ as a unit of momentum and $e^2\sqrt{\omega_{\mathrm{p}} m_{\mathrm{e}}} = \sqrt{2\mathrm{Ry}\,\omega_{\mathrm{p}}}$ as a unit of energy, where Ry is the Rydberg constant. In these units $q \ll 1 \ll k$, and the radial part of Eqs. (7.36) and (7.37) with the kinetic coefficients (7.43) takes the form

$$\dot{n}_{\mathrm{p}}(t,q) = -2\Gamma(q;n_{\mathrm{e}})\,n_{\mathrm{p}}(t,q); \qquad \dot{n}_{\mathrm{e}}(t,k) = k^{-2}\partial_k\left[k^2 D_{\parallel}(k;n_{\mathrm{p}})\,\partial_k n_{\mathrm{e}}(t,k)\right],$$
$$\Gamma(q;n_{\mathrm{e}}) = \frac{1}{2q^3}\,n_{\mathrm{e}}(t,q^{-1}); \qquad D_{\parallel}(k;n_{\mathrm{p}}) = \frac{1}{k^3}\int_{1/k}^{1/\kappa}\frac{\mathrm{d}q}{q}\,n_{\mathrm{p}}(t,q), \tag{7.44}$$

where we have neglected spontaneous emission terms, since they are of no importance in the strongly non-equilibrium case. One may look for scaling solutions of these equations in the following form: $n_{\mathrm{p}}(t,q) = u_{\mathrm{p}}(t) f_{\mathrm{p}}(q)$ and $n_{\mathrm{e}}(t,k) = u_{\mathrm{e}}(t) f_{\mathrm{e}}(k)$. It is then easy to see that $u_{\mathrm{p}}(t) = u_{\mathrm{e}}(t) = t^{-1}$ and $f_{\mathrm{e}}(k) = k^{-3}$. One then notices that $k\partial_k[3k^{-5}\int_{1/k}^{1/\kappa}(\mathrm{d}q/q) f_{\mathrm{p}}(q)] = 1$, from which it follows that $f_{\mathrm{p}}(q) = (5/3)\,q^{-5}\ln(1/\kappa q)$. Transforming back to physical units, one finds for the corresponding scaling solution of Eqs. (7.44)

$$n_{\mathrm{p}}(t,q) = \frac{5}{3\,t}\,\frac{4\pi\rho_0 m}{q^5}\,\ln\frac{1/\kappa}{q}; \qquad n_{\mathrm{e}}(t,k) = \frac{1}{\omega_{\mathrm{p}}\,t}\,\frac{4\pi\rho_0}{k^3}. \tag{7.45}$$

Such a solution is only possible once $n_{\mathrm{p}}(t,\kappa^{-1}) < n_{\mathrm{p}}(0)$, that is, $t \gtrsim t_0$, where $t_0 = \rho_0 m\kappa^5/n_{\mathrm{p}}(0)$. The interval of plasmon momenta where the scaling solution holds is $q_0(t) < q < \kappa^{-1}$, where the boundary $q_0(t)$ is determined by $n_{\mathrm{p}}(t, q_0(t)) = n_{\mathrm{p}}(0)$, that is, $q_0(t) \propto (t\,n_{\mathrm{p}}(0))^{-1/5}$. The corresponding scaling interval for the resonant electrons is $k_{T_{\mathrm{e}}} < k < \omega_{\mathrm{p}} m_{\mathrm{e}}/q_0(t)$. The plasmon occupation numbers for $q < q_0(t)$ stay approximately constant, given by $n_{\mathrm{p}}(0)$, until the time when $q = q_0(t)$. At a later time they decay as $1/t$ according to Eq. (7.45). As time increases past t_0 the scaling interval for plasmons grows toward smaller momenta ("red" shift), while for the resonant electrons it grows toward the larger momenta ("blue" shift). The energy of the plasmon modes in the scaling interval $E_{\mathrm{p}} = \int_{q_0(t)}^{1/\kappa} q^2\mathrm{d}q\,\omega_{\mathrm{p}}\,n_{\mathrm{p}}(t,q) \sim t^{-1}q_0^{-2}(t)$, that is, it decreases with time as $E_{\mathrm{p}} \sim t^{-3/5}$ at $t > t_0$. The same is true regarding the energy of the resonant electrons in their respective scaling interval $E_{\mathrm{e}} = \int_{k_{T_{\mathrm{e}}}}^{\omega_{\mathrm{p}} m_{\mathrm{e}}/q_0} k^2\mathrm{d}k\,(k^2/2m)\,n_{\mathrm{e}}(t,k) \sim t^{-3/5}$. The "missing" energy flows through the upper boundary of the electron scaling interval $\omega_{\mathrm{p}} m_{\mathrm{e}}/q_0(t)$ toward larger momenta. There is a corresponding particle flow through this boundary. Indeed, the total number of resonant electrons in the scaling interval is $N_{\mathrm{e}} = \int_{k_{T_{\mathrm{e}}}}^{\omega_{\mathrm{p}} m_{\mathrm{e}}/q_0} k^2\mathrm{d}k\,n_{\mathrm{e}}(t,k) \propto \rho_0(\omega_{\mathrm{p}} t)^{-1}\ln(1/\kappa q_0(t))$, that is, it decays as $N_{\mathrm{e}} \sim t^{-1}\ln(t/t_0)$.

One may show that the flux of particles and heat through the upper boundary of the scaling interval precisely accounts for the "missing" electrons and energy. If these high-energy electrons leave the system due to some mechanism of escape, then the relaxation of an initially non-equilibrium plasmon distribution $n_p(0)$ results in a slow *cooling* of the plasma. Indeed, the plasmon energy is channeled to the hot resonant electrons. The latter then leave the system, taking away the energy of the non-equilibrium plasmons as well as their own kinetic energy. Thus, after eventual equilibration (e.g. due to collisions) the remaining particles have a *lower* temperature than that before the plasmon excitation.

The exponents of the scaling solution (7.45) depend on the details of interactions between plasmons and resonant electrons, as well as on the system dimensionality and the dispersion relation of the participating quasiparticles. However, the existence of the scaling interval of momenta and a power-law distribution function within this interval is a very general phenomenon. It was called *wave turbulence* and found in a great variety of systems [124].

The observation that the plasmon occupation decays as $n_p \sim 1/t$ at long times is rather robust. It is based on the fact that $\dot{n}_p \sim -\nabla_{\mathbf{k}} n_e \, n_p$ and $\dot{n}_e \sim \nabla_{\mathbf{k}}(n_p \nabla_{\mathbf{k}} n_e)$. Then the solution $n_p \propto n_e \sim 1/t$ is to be expected in a broad range of parameters. It is one of the manifestations of nonlinear Landau damping. In its essence, deviations of the resonant electron distribution function from the equilibrium one invalidate the exponential plasmon relaxation $n_p \sim e^{-2\Gamma t}$ and transform it to a much slower, power-law decay.

7.7 Problems

7.7.1 Landau Damping of the Ion Acoustic Mode

Consider the ion acoustic plasma mode of Section 7.3. For $\kappa q \ll 1$ its dispersion relation is linear $\omega_{ia}(\mathbf{q}) = v_{ia} q$ with $v_{ia} = \sqrt{T_e/m_i}$. Calculate Landau damping of this mode by electrons, $\Gamma^{(e)}_{ia}(\mathbf{q})$, and by the ions $\Gamma^{(i)}_{ia}(\mathbf{q})$. Show that the electron contribution is always small, $\Gamma^{(e)}_{ia}(\mathbf{q})/v_{ia}q \propto \sqrt{m_e/m_i} \ll 1$. Show that the ion damping is given by

$$\frac{\Gamma^{(i)}_{ia}(\mathbf{q})}{v_{ia}q} \approx \sqrt{\frac{\pi}{8}} \left(\frac{T_e}{T_i}\right)^{3/2} e^{-T_e/2T_i}. \tag{7.46}$$

For $T_e > 10T_i$ the ratio is about $\Gamma^{(i)}_{ia}/v_{ia}q \lesssim 0.1$ and therefore the mode is reasonably underdamped.

To show this evaluate $\operatorname{Im} \Pi^R_e(\mathbf{q}, v_{ia}q)$ and $\operatorname{Im} \Pi^R_i(\mathbf{q}, v_{ia}q)$, using Eqs. (7.15) and (7.16) and solve Eq. (7.12) for the complex valued $\omega = \omega(\mathbf{q})$. The real parts of the polarization operators are explained in Section 7.3. Employ the fact that the occupation number $n_B(k) = e^{\mu_{e,(i)}/T_{e,(i)}} \prod_{\nu=x,y,x} e^{-k_\nu^2/2m_{e(i)}T_{e,(i)}}$, along with the normalization $\rho_0 = \sum_{\mathbf{k}} n_B(k)$ for both electrons and ions. This way the x-axis may be chosen along the $\mathbf{q}$ vector and the corresponding x-integral calculated with the help of the delta-function (7.15).

8
Kinetics of Bose Condensates

This chapter is devoted to the dynamics of Bose gases in the presence of a condensate. We derive a system of coupled equations for the condensate wavefunction and the distribution function of above-the-condensate quasiparticles. We also discuss quasiparticle fluctuations, which lead to stochastic Langevin forces acting on the condensate.

8.1 Gross–Pitaevskii Equation

The phenomenon of Bose–Einstein condensation is associated with the fact that the largest eigenvalue N_0 of the density matrix operator $\hat{\rho}(\mathbf{r}, \mathbf{r}')$ acquires a macroscopically large value [115] $\int d\mathbf{r}' \hat{\rho}(\mathbf{r}, \mathbf{r}')\Phi_0(\mathbf{r}') = N_0\Phi_0(\mathbf{r})$. We first consider the case where almost all particles belong to the condensate, meaning that $N - N_0 \ll N$, where N is the total number of bosons. This is the case if the gas is sufficiently dilute, $a_s^3\rho \ll 1$, where a_s is the scattering length, ρ is the density *and* the system is at equilibrium at a temperature that is much less than the critical temperature of Bose–Einstein condensation, $T \ll T_c$. Under these conditions the condensate wavefunction may be found as a stationary field configuration of the bosonic action $S = S_0 + S_{\text{int}}$ given by Eqs. (6.9) and (6.13):

$$S_0 = \int d\mathbf{r}\, dt\, (\bar{\phi}^{\text{cl}}, \bar{\phi}^{\text{q}}) \begin{pmatrix} 0 & i\partial_t + \frac{\nabla_\mathbf{r}^2}{2m} - V^{\text{cl}} \\ i\partial_t + \frac{\nabla_\mathbf{r}^2}{2m} - V^{\text{cl}} & 0 \end{pmatrix} \begin{pmatrix} \phi^{\text{cl}} \\ \phi^{\text{q}} \end{pmatrix}; \tag{8.1}$$

$$S_{\text{int}} = -\frac{g}{2} \int d\mathbf{r}\, dt \left[\bar{\phi}^{\text{q}}\bar{\phi}^{\text{cl}}\phi^{\text{cl}}\phi^{\text{cl}} + \bar{\phi}^{\text{cl}}\bar{\phi}^{\text{q}}\phi^{\text{q}}\phi^{\text{q}} + \text{c.c.} \right], \tag{8.2}$$

where $V^{\text{cl}}(\mathbf{r}, t)$ is an external potential, for example, of an optical or magnetic trap. The interaction parameter g is related to the s-wave scattering length a_s as $g = 4\pi a_s/m$ [115]. To find a stationary configuration of the action we notice that, according to (2.53), there are no terms in the action that have zero power of both $\bar{\phi}^{\text{q}}$ and ϕ^{q}. The same is obviously true regarding $\delta S/\delta\bar{\phi}^{\text{cl}}$ and $\delta S/\delta\phi^{\text{cl}}$. As a result, two of the saddle-point equations,

$$\frac{\delta S}{\delta \bar{\phi}^{\rm cl}} = 0; \qquad \frac{\delta S}{\delta \phi^{\rm cl}} = 0, \tag{8.3}$$

may always be satisfied by

$$\Phi^{\rm q} = \bar{\Phi}^{\rm q} = 0, \tag{8.4}$$

irrespective of what the classical component, $\Phi^{\rm cl} = \Phi_0$, is. By the capital $\Phi^{\rm cl(q)}$ we denote solutions of the stationary point equations. Under the condition (8.4) the second pair of stationary point equations takes the form

$$\begin{aligned} \frac{\delta S}{\delta \bar{\phi}^{\rm q}} &= \left(\mathrm{i}\partial_t + \frac{1}{2m}\nabla_{\mathbf{r}}^2 - V^{\rm cl}(\mathbf{r},t) - \frac{g}{2}\,|\Phi_0|^2 \right) \Phi_0 = 0; \\ \frac{\delta S}{\delta \phi^{\rm q}} &= \left(-\mathrm{i}\partial_t + \frac{1}{2m}\nabla_{\mathbf{r}}^2 - V^{\rm cl}(\mathbf{r},t) - \frac{g}{2}\,|\Phi_0|^2 \right) \bar{\Phi}_0 = 0. \end{aligned} \tag{8.5}$$

This nonlinear equation, called the time-dependent Gross–Pitaevskii equation, determines the condensate wavefunction $\Phi_0(\mathbf{r},t)$, provided some initial and boundary conditions are specified. The resulting macroscopic wavefunction must accommodate (almost) all particles, which dictates the normalization condition

$$\int d\mathbf{r}\, |\Phi_0(\mathbf{r},t)|^2 = 2N. \tag{8.6}$$

The factor of two on the right-hand side is associated with our definition of the Keldysh rotations, Eq. (2.39), and the resulting fact that the expectation value of the two classical fields is *twice* that of the occupation number, $\langle \phi^{\rm cl}\bar{\phi}^{\rm cl}\rangle \sim 2n_{\rm B} + 1$.

In the case of a static external potential $V^{\rm cl}(\mathbf{r})$ one may look for a solution of Eq. (8.5) in the form $\Phi_0(\mathbf{r},t) = \mathrm{e}^{-\mathrm{i}\mu t}\Phi_0(\mathbf{r})$, where $\Phi_0(\mathbf{r})$ satisfies the time-independent Gross–Pitaevskii equation

$$\left(\frac{1}{2m}\nabla_{\mathbf{r}}^2 - V^{\rm cl}(\mathbf{r},t) + \mu - \frac{g}{2}\,|\Phi_0|^2 \right) \Phi_0 = 0 \tag{8.7}$$

and the chemical potential μ is chosen to satisfy the normalization condition (8.6). In a particular case of an adiabatically smooth potential, one may disregard spatial gradients, finding $|\Phi_0(\mathbf{r})|^2 = 2(\mu - V^{\rm cl}(\mathbf{r}))/g$, if the right-hand side of this expression is positive, and $\Phi_0 = 0$ otherwise. This is the so-called Thomas–Fermi approximation for the condensate solution in a trap. Once again, the chemical potential μ is determined from Eq. (8.6), where the integral runs over the region where $V^{\rm cl}(\mathbf{r}) < \mu$. In the absence of an external potential $V^{\rm cl} = 0$ one finds $|\Phi_0|^2 = 2\mu/g$ and thus $\mu = \rho g$, where ρ is the density of the Bose gas.

Let us investigate now small dynamical fluctuations around the uniform condensate solution. To this end we write $\Phi_0(\mathbf{r},t) = \mathrm{e}^{-\mathrm{i}\mu t}(\Phi_0 + \varphi(\mathbf{r},t))$, where

$|\Phi_0|^2 = 2\mu/g = 2\rho$. We then substitute it into the time-dependent Gross–Pitaevskii equation (8.5) and linearize it with respect to φ. As a result one finds a pair of linear homogeneous differential equations for φ and $\bar{\varphi}$:

$$\begin{aligned} &\mathrm{i}\partial_t\varphi + \frac{1}{2m}\nabla_{\mathbf{r}}^2\varphi - \frac{g}{2}|\Phi_0|^2\varphi - \frac{g}{2}\Phi_0^2\bar{\varphi} = 0; \\ &-\mathrm{i}\partial_t\bar{\varphi} + \frac{1}{2m}\nabla_{\mathbf{r}}^2\bar{\varphi} - \frac{g}{2}|\Phi_0|^2\bar{\varphi} - \frac{g}{2}\bar{\Phi}_0^2\varphi = 0. \end{aligned} \tag{8.8}$$

Performing a Fourier transformation, one finds the algebraic relation

$$\begin{pmatrix} \omega - \mathbf{q}^2/(2m) - \frac{g}{2}|\Phi_0|^2 & -\frac{g}{2}\Phi_0^2 \\ -\frac{g}{2}\bar{\Phi}_0^2 & -\omega - \mathbf{q}^2/(2m) - \frac{g}{2}|\Phi_0|^2 \end{pmatrix} \begin{pmatrix} \varphi(\mathbf{q},\omega) \\ \bar{\varphi}(\mathbf{q},\omega) \end{pmatrix} = 0, \tag{8.9}$$

which may be satisfied only if the determinant of the matrix on its left-hand side is zero. This way, one finds the spectrum of the small fluctuations of the condensate:

$$\omega^2 = \left(\frac{\mathbf{q}^2}{2m} + \frac{g}{2}|\Phi_0|^2\right)^2 - \left(\frac{g}{2}\right)^2|\Phi_0|^4 = \left(\frac{\mathbf{q}^2}{2m}\right)^2 + \frac{\mathbf{q}^2}{m}\frac{g}{2}|\Phi_0|^2. \tag{8.10}$$

This is the celebrated Bogoliubov dispersion relation, which may be written as

$$\omega_{\mathrm{B}}(\mathbf{q}) = \sqrt{c^2\mathbf{q}^2 + \left(\frac{\mathbf{q}^2}{2m}\right)^2}. \tag{8.11}$$

For small wavenumbers the excitations are of an acoustic nature, $\omega_{\mathrm{B}}(\mathbf{q}) = cq$, where the speed of sound c is given by $c^2 = g|\Phi_0|^2/2m = g\rho/m$. At large wavenumbers the excitation spectrum approaches that of the usual particles, $\omega_{\mathrm{B}}(\mathbf{q}) = \mathbf{q}^2/(2m) + mc^2 + O(\mathbf{q}^{-2})$. Notice that the linearity of the spectrum at small wavenumbers originates from the presence of the off-diagonal terms in the matrix (8.9), which assumes a fixed phase of the condensate wavefunction. The latter assumption is not true for low-dimensional systems, where the phase fluctuations destroy the phase coherence at long distances and times. Nevertheless the $T = 0$ spectrum of the long wavelength excitations is still linear [125] even in these cases.

It is convenient to parametrize the condensate wavefunction with its amplitude and phase as $\Phi_0 = \sqrt{2\rho_0}\,\mathrm{e}^{\mathrm{i}\theta}$. Then the imaginary and real parts of the Gross–Pitaevskii equation (8.5) acquire the form

$$\partial_t\rho_0 + \nabla_{\mathbf{r}}\big(\rho_0\nabla_{\mathbf{r}}\theta/m\big) = 0; \tag{8.12}$$

$$\partial_t\theta = \frac{\nabla_{\mathbf{r}}^2\sqrt{\rho_0}}{2m\sqrt{\rho_0}} - \frac{(\nabla_{\mathbf{r}}\theta)^2}{2m} - V^{\mathrm{cl}} - g\rho_0. \tag{8.13}$$

The first equation here may be identified as the continuity relation for the condensate density and current, $\partial_t\rho_0 + \operatorname{div}\mathbf{j}_0 = 0$. Acknowledging that the condensate current is given by the product of its density and the superfluid velocity $\mathbf{j}_0 = \rho_0 \mathbf{v}_{\mathrm{sf}}$, one finds for the latter $\mathbf{v}_{\mathrm{sf}} = \nabla_{\mathbf{r}}\theta/m$. Neglecting then the density gradients, Eq. (8.13) is recognized as the Euler equation for the superfluid velocity of the condensate $\partial_t \mathbf{v}_{\mathrm{sf}} + \nabla_{\mathbf{r}}\big(\mathbf{v}_{\mathrm{sf}}^2/2 + \mu/m\big) = 0$, where $\mu = V^{\mathrm{cl}} + g\rho_0$ is the local chemical potential.

8.2 Quasiparticles

The Gross–Pitaevskii approach is adequate when most of the particles belong to the condensate and the relative number of above-the-condensate excitations is small. This is the case in equilibrium at temperatures much below the critical temperature of the Bose condensation, $T_{\mathrm{c}} \propto \rho^{2/3}/m$. At temperatures of the order of the critical one, or in a non-equilibrium situation, one can't neglect above-the-condensate quasiparticles. The quasiparticles are characterized by their distribution function $\tilde{F}(\mathbf{r}, t, \mathbf{k})$, which evolves according to a kinetic equation. Our goal thus is to describe the system with the complex wavefunction of the condensate $\Phi_0(\mathbf{r}, t)$ along with the quasiparticle distribution function $\tilde{F}(\mathbf{r}, t, \mathbf{k})$. The condensate wavefunction obeys a generalized Gross–Pitaevskii equation, which includes condensate interactions with above-the-condensate quasiparticles. The distribution function obeys the kinetic equation, which is coupled to the condensate dynamics [127]. This system of equations is ideologically similar to the Vlasov equations for plasma; see Section 7.1. Here, instead of the Poisson equation for the real scalar potential $V(\mathbf{r}, t)$ we have the Gross–Pitaevskii equation for the complex wavefunction $\Phi_0(\mathbf{r}, t)$. As in the case of a plasma, there is a lot of information in these equations even with the collisionless form of the kinetic equation. As a next step (similar to what is done in Section 7.4) one may include stochastic terms in the generalized Gross–Pitaevskii equation, which are due to the quasiparticle fluctuations.

To implement this program we split the complex field $\phi(\mathbf{r}, t)$ into slow $\Phi(\mathbf{r}, t)$ and fast $\varphi(\mathbf{r}, t)$ components. The former describes the condensate, while the latter describes above-the-condensate quasiparticles. Since away from equilibrium the condensate is a dynamic variable, the boundary between the two is somewhat arbitrary and itself may be a slow function of time. We thus write

$$\phi^{\mathrm{cl}}(x) = \mathrm{e}^{-\mathrm{i}\mathcal{K}(x)}\big[\Phi_0(x) + \varphi(x)\big]\,; \qquad \phi^{\mathrm{q}}(x) = \mathrm{e}^{-\mathrm{i}\mathcal{K}(x)}\big[\Phi^{\mathrm{q}}(x) + \varphi^{\mathrm{q}}(x)\big], \quad (8.14)$$

where $x = \mathbf{r}, t$. The yet unspecified local phase $\mathcal{K}(x)$ serves to fix a particular gauge. We shall choose $\mathcal{K}(x)$ to make the semiclassical condensate wavefunction $\Phi_0(x)$ as slow as possible. For example, in equilibrium $\mathcal{K} = \mu t$ and Φ_0 is static. Away from equilibrium the complex function $\Phi_0(x)$ can't be brought to a static form by a choice of phase. Nevertheless, proper choices of $\mathcal{K}(x)$ make the scale separation of the degrees of freedom well justified. We shall return to this issue in Section 8.6.

Substituting Eq. (8.14) into the action (8.1), (8.2), one obtains the action in terms of the condensate variables Φ_0 and Φ^{q} as well as the quasiparticle fields φ and φ^{q}. To derive semiclassical equations governing the condensate dynamics it is sufficient to keep only the terms with zero or first power of Φ^{q}. The interaction action (8.2) brings a host of terms, which may be classified as follows.

(i) Those containing four condensate fields $\bar{\Phi}^{\mathrm{q}}\bar{\Phi}_0\Phi_0\Phi_0$ – they are part of the Gross–Pitaevskii action.

(ii) There are no terms with three condensate fields. Indeed, such terms are necessarily linear in the fast fields φ or φ^{q}, and therefore cannot satisfy energy and momentum conservation.

(iii) Those containing two condensate fields. There are two kinds of them: (a) with only classical condensate fields, for example, $|\Phi_0|^2\bar{\varphi}^{\mathrm{q}}\varphi$ or $\Phi_0^2\bar{\varphi}^{\mathrm{q}}\bar{\varphi}$ – they are part of the quadratic action for the quasiparticles; (b) with one quantum and one classical condensate field. We shall treat them in the Popov approximation [128]

$$\bar{\Phi}^{\mathrm{q}}\Phi_0\bar{\varphi}\varphi \approx \bar{\Phi}^{\mathrm{q}}\Phi_0\langle\bar{\varphi}\varphi\rangle + \cdots ,$$

where the angular brackets stand for the expectation value of two classical fields (Keldysh Green function at the coinciding points). The remaining term $\bar{\Phi}^{\mathrm{q}}\Phi_0(\bar{\varphi}\varphi - \langle\bar{\varphi}\varphi\rangle)$ is a part of the collision action. Following Popov, we shall not single out the anomalous Bogoliubov expectation value $\langle\varphi\varphi\rangle$. This is justified for the temperature range $mc^2 < T < T_{\mathrm{c}}$ (see Section 8.3), where the anomalous averages are indeed much smaller than the normal ones. Those terms as well as $\bar{\Phi}^{\mathrm{q}}\bar{\Phi}_0\varphi^{\mathrm{q}}\varphi^{\mathrm{q}}$ do not influence the semiclassical dynamics.

(iv) Those containing one condensate field. Again there are two kinds of them: (a) with a classical condensate field, for example, $\bar{\Phi}_0\bar{\varphi}^{\mathrm{q}}\varphi\varphi$ or $\bar{\Phi}_0\bar{\varphi}^{\mathrm{q}}\varphi^{\mathrm{q}}\varphi^{\mathrm{q}}$ – they contribute to the self-energy and thus to the collision terms in the quasiparticle kinetic equation. This part of the quasiparticle collision integral is about particle exchange with the condensate; (b) with a quantum condensate field, for example, $\bar{\Phi}^{\mathrm{q}}\bar{\varphi}\varphi\varphi$ – they, being paired with one of the (iv (a)) terms, provide a part of the Gross–Pitaevskii equation, which describes particle exchange with the quasiparticle cloud.

(v) Those with no condensate fields. In the Popov approximation they ought to be treated as

$$\bar{\varphi}^{\mathrm{q}}\bar{\varphi}\varphi\varphi = 2\bar{\varphi}^{\mathrm{q}}\varphi\langle\bar{\varphi}\varphi\rangle + \cdots .$$

Along with (iii (a)) they are a part of the quadratic quasiparticle action. The remaining terms, being taken to second order, contribute to the quasiparticle collision integral. This part of the collision integral conserves the number of quasiparticles.

As a result the action may be split into the Gross–Pitaevskii part, the quadratic quasiparticle action, and the collision part. The Gross–Pitaevskii part takes the form

$$S_{\mathrm{GP}} = \int \mathbf{dr}\, \mathrm{d}t\, (\bar{\Phi}_0, \bar{\Phi}^{\mathrm{q}}) \begin{pmatrix} 0 & \mathrm{i}\partial_t - H_{\mathrm{GP}} \\ \mathrm{i}\partial_t - H_{\mathrm{GP}} & 0 \end{pmatrix} \begin{pmatrix} \Phi_0 \\ \Phi^{\mathrm{q}} \end{pmatrix}, \tag{8.15}$$

where

$$H_{\mathrm{GP}} = -\frac{\hat{\boldsymbol{\partial}}_{\mathbf{r}}^2}{2m} + V^{\mathrm{cl}} + \frac{g}{2}\,|\Phi_0|^2 + 2g\rho_{\mathrm{qp}} - \partial_t \mathcal{K}, \tag{8.16}$$

and $\rho_{\mathrm{qp}}(x) = \langle \varphi(x)\bar{\varphi}(x)\rangle/2$ is the quasiparticle density at $x = \mathbf{r}, t$. The long derivative is defined as

$$\hat{\boldsymbol{\partial}}_{\mathbf{r}} = \nabla_{\mathbf{r}} - \mathrm{i}(\nabla_{\mathbf{r}}\mathcal{K}). \tag{8.17}$$

Variation with respect to $\bar{\Phi}^{\mathrm{q}}$ leads to the saddle-point (modified Gross–Pitaevskii) equation

$$(\mathrm{i}\partial_t - H_{\mathrm{GP}})\Phi_0 = 0. \tag{8.18}$$

If there is a nontrivial static *equilibrium* solution of this equation, the phase $\mathcal{K}$ (so far arbitrary) may be expressed through the slow densities of the condensate $\rho_0 = |\Phi_0|^2/2$ and the quasiparticles ρ_{qp} as

$$\partial_t \mathcal{K} = V^{\mathrm{cl}}(x) + g\rho_0(x) + 2g\rho_{\mathrm{qp}}(x) = \mu \tag{8.19}$$

(we neglected the space gradients of the equilibrium solution). The chemical potential μ may then be fixed using the condition $\int \mathbf{dr}(\rho_0 + \rho_{\mathrm{qp}}) = N$. In a spatially uniform case ($V^{\mathrm{cl}} = 0$) at zero temperature $\rho_{\mathrm{qp}} \ll \rho$ and $\rho_0 \approx \rho$, thus $\mu(0) = g\rho > 0$. On the other hand, at $T = T_{\mathrm{c}}$ one has $\rho_0 = 0$ and $\rho_{\mathrm{qp}} = \rho$, as a result $\mu(T_{\mathrm{c}}) = 2g\rho = 2\mu(0)$. This is a consequence of the exchange interactions between the quasiparticles, which are absent within the condensate. Once the temperature grows above T_{c}, the only solution of the saddle-point equation is $\Phi_0 = 0$, Eq. (8.19) is not applicable, while the chemical potential rapidly decreases and becomes negative, as it should for a classical gas.

To write down the quasiparticle action it is useful to define the *four*-component field vector $\vec{\varphi} = (\varphi, \bar{\varphi}, \varphi^{\mathrm{q}}, \bar{\varphi}^{\mathrm{q}})^{\mathrm{T}}$, along with its complex conjugate vector $\bar{\vec{\varphi}}$. With their help one may write

$$S_{\mathrm{qp}} = \frac{1}{2}\int \mathbf{dr}\, \mathrm{d}t\, \bar{\vec{\varphi}} \left(\begin{array}{c|c} 0 & \check{\tau}_3 \mathrm{i}\partial_t - \check{H}_{\mathrm{qp}} \\ \hline \check{\tau}_3 \mathrm{i}\partial_t - \check{H}_{\mathrm{qp}} & 0 \end{array}\right) \vec{\varphi}, \tag{8.20}$$

where the check symbol ($\vee$) denotes matrices in the Bogoliubov subspace and $\check{\tau}_i$ with $i = 0, 1, 2, 3$ are the corresponding Pauli matrices. The Hermitian operator $\check{H}_{\rm qp}$ acting in the Bogoliubov subspace is defined as

$$\check{H}_{\rm qp} = \begin{pmatrix} -\frac{\hat{\partial}_{\mathbf{r}}^2}{2m} + V^{\rm cl} + g|\Phi_0|^2 + 2g\rho_{\rm qp} - \partial_t \mathcal{K} & \frac{g}{2}\Phi_0^2 \\ \frac{g}{2}\bar{\Phi}_0^2 & -\frac{\hat{\partial}_{\mathbf{r}}^2}{2m} + V^{\rm cl} + g|\Phi_0|^2 + 2g\rho_{\rm qp} - \partial_t \mathcal{K} \end{pmatrix}. \tag{8.21}$$

Notice the absence of 1/2 in front of $|\Phi_0|^2$ in comparison with Eq. (8.16). It is due to the presence of the exchange interactions between the condensate and the quasiparticles. If in some region of space there is an *equilibrium* condensate and thus expression (8.19) for the chemical potential is valid, one may rewrite the quasiparticle Hamiltonian as

$$\check{H}_{\rm qp} = \begin{pmatrix} -\frac{\hat{\partial}_{\mathbf{r}}^2}{2m} + \frac{g}{2}|\Phi_0|^2 & \frac{g}{2}\Phi_0^2 \\ \frac{g}{2}\bar{\Phi}_0^2 & -\frac{\hat{\partial}_{\mathbf{r}}^2}{2m} + \frac{g}{2}|\Phi_0|^2 \end{pmatrix}. \tag{8.22}$$

The quasiparticles may exist outside of the condensate region. There they are described by the diagonal Hamiltonian with the effective potential given by $V^{\rm cl}(x) + 2g\rho(x) - \partial_t\mathcal{K}$. Here $\rho(x)$ is the total density outside of the condensate support area, interacting with the quasiparticles through the Hartree–Fock potential.

Finally the remaining part of the action, which is responsible for the collisions of the quasiparticles and exchange of particles with the condensate, has the form $S^{\rm coll} = S_2^{\rm coll} + S_3^{\rm coll} + S_4^{\rm coll}$, where

$$\begin{aligned}
S_2^{\rm coll} &= -\frac{g}{2}\int d\mathbf{r}\, dt \Big[\bar{\Phi}^{\rm q}\Phi_0\big(2\bar{\varphi}\varphi - 2\langle\bar{\varphi}\varphi\rangle + 2\bar{\varphi}^{\rm q}\varphi^{\rm q}\big) + \bar{\Phi}^{\rm q}\bar{\Phi}_0\big(\varphi\varphi + \varphi^{\rm q}\varphi^{\rm q}\big) + \text{c.c.}\Big];\\
S_3^{\rm coll} &= -\frac{g}{2}\int d\mathbf{r}\, dt \Big[\bar{\Phi}^{\rm q}\big(\bar{\varphi}\varphi\varphi + 2\bar{\varphi}^{\rm q}\varphi^{\rm q}\varphi + \bar{\varphi}\varphi^{\rm q}\varphi^{\rm q}\big) + \text{c.c.} \\
&\quad + \big(2\bar{\varphi}^{\rm q}\bar{\varphi}\varphi + \bar{\varphi}\bar{\varphi}\varphi^{\rm q} + \bar{\varphi}^{\rm q}\bar{\varphi}^{\rm q}\varphi^{\rm q}\big)\Phi_0 + \text{c.c.}\Big];\\
S_4^{\rm coll} &= -\frac{g}{2}\int d\mathbf{r}\, dt \Big[\bar{\varphi}^{\rm q}\bar{\varphi}\varphi\varphi - 2\bar{\varphi}^{\rm q}\varphi\langle\bar{\varphi}\varphi\rangle + \bar{\varphi}\bar{\varphi}^{\rm q}\varphi^{\rm q}\varphi^{\rm q} + \text{c.c.}\Big],
\end{aligned} \tag{8.23}$$

here the subscripts indicate the number of quasiparticles involved in the corresponding collision processes.

We turn now to the quasiparticle dynamics specified by Eqs. (8.20) and (8.21). Notice that the equation of motion $(\check{\tau}_3 i\partial_t - \check{H}_{\rm qp})(\varphi, \bar{\varphi})^{\rm T} = 0$, with $\check{H}_{\rm qp}$ given by Eq. (8.22), coincides with Eq. (8.8) for the linearized dynamics of the Gross–Pitaevskii system. Of course, in the present case the condensate density

$\rho_0(x) = |\Phi_0|^2/2$ is not the same as the total density and must be determined through the modified Gross–Pitaevskii equation, which in turn needs the information about the quasiparticle density $\rho_{\rm qp}(x)$. To diagonalize the quasiparticle Hamiltonian one needs to solve the Bogoliubov–deGennes equations,

$$\left(\check{\tau}_3\epsilon_k - \check{H}_{\rm qp}\right)\begin{pmatrix} u_k(\mathbf{r}) \\ v_k(\mathbf{r}) \end{pmatrix} = 0, \tag{8.24}$$

where k labels the full set of solutions, and $u_k(\mathbf{r})$ and $v_k(\mathbf{r})$ are two functions which may be normalized according to

$$\int \mathrm{d}\mathbf{r}\left[u_k^*(\mathbf{r})u_l(\mathbf{r}) - v_k^*(\mathbf{r})v_l(\mathbf{r})\right] = \delta_{kl}. \tag{8.25}$$

One can now define a new set of complex fields $\chi_k^{\rm cl}(t)$ and $\chi_k^{\rm q}(t)$ through the Bogoliubov transformation (cf. problem 2.10.2)

$$\varphi(\mathbf{r},t) = \sum_k \left[u_k(\mathbf{r})\chi_k^{\rm cl}(t) + v_k^*(\mathbf{r})\bar{\chi}_k^{\rm cl}(t)\right]; \tag{8.26}$$

$$\varphi^q(\mathbf{r},t) = \sum_k \left[u_k(\mathbf{r})\chi_k^{\rm q}(t) + v_k^*(\mathbf{r})\bar{\chi}_k^{\rm q}(t)\right].$$

In terms of these new fields the quasiparticle action (8.20) takes the standard Keldysh form

$$S_{\rm qp} = \int \mathrm{d}t \sum_k \left(\bar{\chi}_k^{\rm cl}, \bar{\chi}_k^{\rm q}\right)\begin{pmatrix} 0 & \mathrm{i}\partial_t - \epsilon_k \\ \mathrm{i}\partial_t - \epsilon_k & 0 \end{pmatrix}\begin{pmatrix} \chi_k^{\rm cl} \\ \chi_k^{\rm q} \end{pmatrix}. \tag{8.27}$$

Using the strategy of Chapter 6, one can proceed with writing the kinetic equation for the quasiparticles. To this end one needs an expression for the quasiparticle spectrum ϵ_k, which is so far implicit in the solutions of the Bogoliubov–deGennes equations (8.24). Focusing on the adiabatic effective potential and applying the Wigner transformation, one may label the spectrum with the set of local momenta $\mathbf{k}$, defined in a vicinity of a spatial point $\mathbf{r}$. As a result the spectrum is labeled as $\epsilon_{\mathbf{k}}(\mathbf{r})$, where a certain coarse graining of the coordinate space must be understood along with the fact that the momentum $\mathbf{k}$ cannot be smaller than the inverse size of the spatial coarse graining. The latter is in agreement with the choice of the quasiparticle field φ as a *fast* part of the full bosonic field ϕ. The small wavenumbers $\mathbf{k}$ are thus explicitly excluded from φ and therefore from χ. With this understanding the local spectrum $\epsilon_{\mathbf{k}}(\mathbf{r})$ may be found by diagonalization of the quasiparticle Hamiltonian (8.21), with $\nabla_{\mathbf{r}} \to \mathrm{i}\mathbf{k}$. This way, one finds for the spectrum:

$$\epsilon_{\mathbf{k}}(\mathbf{r}) = \sqrt{\left(\frac{(\mathbf{k} - \nabla_{\mathbf{r}}\mathcal{K})^2}{2m} + V^{\rm cl} + g|\Phi_0|^2 + 2g\rho_{\rm qp} - \partial_t\mathcal{K}\right)^2 - \left|\frac{g}{2}\,\Phi_0^2\right|^2}. \tag{8.28}$$

Notice that if the condensate is in equilibrium, so that the relation (8.19) for the chemical potential holds, the quasiparticle spectrum coincides with the Bogoliubov spectrum for the condensate excitations, Eq. (8.11). In such a case the quasiparticle spectrum is gapless and linear at small wavenumbers with the local sound velocity $c(\mathbf{r}) = \sqrt{g\rho_0(\mathbf{r})/m}$, determined by the local condensate density. In the same approximation the Bogoliubov coefficients are given by $u_k(\mathbf{r}) = u_{\mathbf{k}} e^{i\mathbf{kr}}$ and $v_k(\mathbf{r}) = v_{\mathbf{k}} e^{i\mathbf{kr}}$, where the $\mathbf{k}$ index represents the Fourier transform over the fast part of the $\mathbf{r}$-dependence and the slow functions $u_{\mathbf{k}}$ and $v_{\mathbf{k}}$ are given by

$$|u_{\mathbf{k}}|^2, |v_{\mathbf{k}}|^2 = \frac{1}{2\epsilon_{\mathbf{k}}}\left(\pm\epsilon_{\mathbf{k}} + \frac{(\mathbf{k}-\nabla_{\mathbf{r}}\mathcal{K})^2}{2m} + V^{\text{cl}} + g|\Phi_0|^2 + 2g\rho_{\text{qp}} - \partial_t\mathcal{K}\right). \quad (8.29)$$

One can now define the quasiparticle distribution functions $F(x, x')$ in the usual way through Eq. (6.22). Performing the Wigner transformation and restricting it to the mass-shell, one obtains the distribution function $\tilde{F}(\mathbf{r}, t, \mathbf{k}) = F(\mathbf{r}, t, \mathbf{k}, \epsilon_{\mathbf{k}}(\mathbf{r}))$, which obeys the kinetic equation

$$\left[\partial_t + \nabla_{\mathbf{k}}\epsilon_{\mathbf{k}}(\mathbf{r})\nabla_{\mathbf{r}} - \nabla_{\mathbf{r}}\epsilon_{\mathbf{k}}(\mathbf{r})\nabla_{\mathbf{k}}\right]\tilde{F}(\mathbf{r}, t, \mathbf{k}) = I^{\text{coll}}[\tilde{F}, \Phi_0]. \quad (8.30)$$

We put the quasiparticle weight $\tilde{Z} = 1$, which is appropriate for the weakly interacting Bose gas. The collision integral on the right-hand side is a functional of the quasiparticle distribution function as well as the space- and time-dependent condensate wavefunction $\Phi_0(x)$. Once the solution of the kinetic equation (8.30) is known one can determine the quasiparticle density $\rho_{\text{qp}}(x) = \langle\varphi(x)\bar{\varphi}(x)\rangle/2$ to feed it self-consistently into the local energy spectrum (8.28):

$$\langle\varphi(x)\bar{\varphi}(x)\rangle = \sum_k \left[|u_k(x)|^2 + |v_k(x)|^2\right]\langle\chi_k\bar{\chi}_k\rangle = \sum_{\mathbf{k}} \left[|u_{\mathbf{k}}|^2 + |v_{\mathbf{k}}|^2\right]\tilde{F}(x, \mathbf{k}). \quad (8.31)$$

The modified Gross–Pitaevskii equation (8.18) for the condensate wavefunction and the kinetic equation (8.30) for the quasiparticle distribution function are the basis of the quasi-classical description of Bose condensation.

8.3 Collisionless Relaxation of the Condensate Fluctuations

Hereafter we shall restrict ourselves to the intermediate range of temperatures (or, better, characteristic quasiparticle energies, if the temperature is not well defined) given by

$$\mu(0) < T; \epsilon_{\mathbf{k}} < T_{\text{c}}, \quad (8.32)$$

where $\mu(0) = g\rho = mc^2$ is the zero-temperature chemical potential and $T_{\text{c}} \propto \rho^{2/3}/m$ is the critical temperature of condensation. For higher temperatures the system is normal and may be described by the usual kinetic equation. On the other hand, at lower temperatures it is already very close to the zero-temperature condensate, described by the Gross–Pitaevskii equation (8.5). (There are interesting and important phenomena, such as, for example, Beliaev damping [129], which are present in the low-temperature regime and which we are going to miss, see Problem 8.9.4.) Under condition (8.32) the expression for the energy spectrum may be substantially simplified to yield

$$\epsilon_{\mathbf{k}}(\mathbf{r}) \approx \frac{(\mathbf{k}-\nabla_{\mathbf{r}}\mathcal{K})^2}{2m} + V^{\text{cl}} + g|\Phi_0|^2 + 2g\rho_{\text{qp}} - \partial_t\mathcal{K}, \tag{8.33}$$

whereas $u_k \approx 1$ and $v_k \approx 0$. This is the Popov approximation, already adopted previously, where one neglects anomalous averages of the φ fields as being small in factor $\sim \mu(0)/\epsilon_{\mathbf{k}} \ll 1$. In essence one avoids Bogoliubov rotation and disregards the difference between the initial quasiparticle field φ and the rotated one χ.

With the quasiparticle spectrum given by Eq. (8.33) (and neglecting $\nabla_{\mathbf{r}}^2\mathcal{K}$) the kinetic equation (8.30) acquires the form

$$\Big[\partial_t + \mathbf{v}_{\mathbf{k}}\nabla_{\mathbf{r}} - \nabla_{\mathbf{r}}\left(V^{\text{cl}} + g|\Phi_0|^2 + 2g\rho_{\text{qp}} - \partial_t\mathcal{K}\right)\nabla_{\mathbf{k}}\Big]\tilde{F}(\mathbf{r},t,\mathbf{k}) = 0, \tag{8.34}$$

where $\mathbf{v}_{\mathbf{k}} = (\mathbf{k}-\nabla_{\mathbf{r}}\mathcal{K})/m$ and we have omitted the collision integral for a while. The quasiparticle density is given by $\rho_{\text{qp}}(x) = \sum_{\mathbf{k}} n_{\text{qp}}(x,\mathbf{k})$, where $\tilde{F}(x,\mathbf{k}) = 2n_{\text{qp}}(x,\mathbf{k}) + 1$.

The system of the collisionless kinetic equation (8.34) and the modified Gross–Pitaevskii equation (8.18) is analogous to the system of Vlasov equations (7.8), (7.11) for plasma. In the latter case one has the Poisson equation instead of the Gross–Pitaevskii one. One can thus adopt the strategy of Section 7.2 to investigate collisionless Landau damping of small fluctuations of the condensate. To this end we shall assume that the quasiparticle distribution function is close to the equilibrium one, $\tilde{F}(x,\mathbf{k}) = \tilde{F}^{eq}(\epsilon_{\mathbf{k}}) + \tilde{F}^{(1)}(x,\mathbf{k})$, where $\tilde{F}^{eq}(\epsilon_{\mathbf{k}}) = \coth \epsilon_{\mathbf{k}}/2T$. The corresponding perturbation of the quasiparticle density is given by $\delta\rho_{\text{qp}}(x) = \frac{1}{2}\sum_{\mathbf{k}} \tilde{F}^{(1)}(x,\mathbf{k})$. We shall also assume that the system is translationally invariant ($V^{\text{cl}} = 0$) and the condensate wavefunction is close to a constant solution, $\Phi_0(x) = \Phi_0 + \phi(x)$, where Φ_0 is found from Eq. (8.19) and particle conservation, while the phase is fixed as $\mathcal{K} = \mu t$. The change in the condensate density is given by $\delta\rho_0 = \frac{1}{2}(\Phi_0\phi(x) + \bar{\phi}(x)\Phi_0)$.

The system of *linearized* kinetic and modified Gross–Pitaevskii equations takes the form

$$\big[\partial_t^{\text{R}} + \mathbf{v}_{\mathbf{k}}\nabla_{\mathbf{r}}\big]\tilde{F}^{(1)}(x,\mathbf{k}) = 2g\nabla_{\mathbf{r}}\big[\delta\rho_0(x) + \delta\rho_{\text{qp}}(x)\big]\nabla_{\mathbf{k}}\tilde{F}^{\text{eq}}(\epsilon_{\mathbf{k}});$$

$$\text{i}\partial_t\phi(x) = -\frac{\nabla_{\mathbf{r}}^2\phi(x)}{2m} + \frac{g}{2}\,|\Phi_0|^2\phi(x) + \frac{g}{2}\,\Phi_0^2\,\bar{\phi}(x) + 2g\delta\rho_{\text{qp}}(x)\Phi_0,$$

where in the second line we employed Eq. (8.19) for unperturbed densities. We have indicated the retarded nature of the time derivative in the kinetic operator, since it is going to play an important role in what follows. We perform now a Fourier transformation from $x = \mathbf{r}, t$ to $q = \mathbf{q}, \omega$, and solve the kinetic equation as $\tilde{F}^{(1)}(q, \mathbf{k}) = -2g[\delta\rho_0(q) + \delta\rho_{\rm qp}(q)]\mathbf{q}\nabla_{\mathbf{k}}\tilde{F}^{\rm eq}/(\omega + {\rm i}0 - \mathbf{v_k q})$. We then substitute this solution into the expression for the quasiparticle density,

$$\delta\rho_{\rm qp}(q) = \frac{1}{2}\sum_{\mathbf{k}} \tilde{F}^{(1)}(q, \mathbf{k}) = -2g\Pi^{\rm R}(q)[\delta\rho_0(q) + \delta\rho_{\rm qp}(q)], \tag{8.35}$$

where the retarded component of the quasiparticle polarization matrix is defined as (cf. Eq. (7.13))

$$\Pi^{\rm R}(\mathbf{q}, \omega) = \frac{1}{2}\sum_{\mathbf{k}} \frac{\mathbf{q}\nabla_{\mathbf{k}} F^{\rm eq}(\epsilon_{\mathbf{k}})}{\omega + {\rm i}0 - \mathbf{v_k q}} = \sum_{\mathbf{k}} \frac{\mathbf{q}\nabla_{\mathbf{k}} n_{\rm B}(\epsilon_{\mathbf{k}})}{\omega + {\rm i}0 - \mathbf{v_k q}}. \tag{8.36}$$

The quasiparticle density is thus $\delta\rho_{\rm qp}(q) = -\delta\rho_0(q)2g\Pi^{\rm R}(q)/[1 + 2g\Pi^{\rm R}(q)]$. In the limit of strong interactions this expression enforces the constant total density, that is, $\delta\rho_{\rm qp} \approx -\delta\rho_0$. On the contrary, we focus on weakly interacting gas, where $g\Pi^{\rm R} \ll 1$ and therefore

$$\delta\rho_{\rm qp}(q) = -2g\Pi^{\rm R}(q)\,\delta\rho_0(q). \tag{8.37}$$

The linearized Gross–Pitaevskii equation and its complex conjugate take the form

$$\begin{pmatrix} \omega - \mathbf{q}^2/(2m) - \frac{g}{2}|\Phi_0|^2 & -\frac{g}{2}\Phi_0^2 \\ -\frac{g}{2}\bar{\Phi}_0^2 & -\omega - \mathbf{q}^2/(2m) - \frac{g}{2}|\Phi_0|^2 \end{pmatrix} \begin{pmatrix} \phi \\ \bar{\phi} \end{pmatrix} = 2g\delta\rho_{\rm qp} \begin{pmatrix} \Phi_0 \\ \bar{\Phi}_0 \end{pmatrix}.$$

Finding from here $\phi(q)$ and $\bar{\phi}(q)$ and substituting them into the expression for the condensate density perturbation $\delta\rho_0(q) = \frac{1}{2}(\bar{\Phi}_0\phi(q) + \bar{\phi}(q)\Phi_0)$, one finds

$$\delta\rho_0(q) = \frac{g}{m}|\Phi_0|^2 \frac{\mathbf{q}^2}{\omega^2 - \omega_{\rm B}^2(\mathbf{q})}\,\delta\rho_{\rm qp}(q). \tag{8.38}$$

This expression is consistent with Eq. (8.37) only if

$$\omega^2 - \omega_{\rm B}^2(\mathbf{q}) = -c^2\mathbf{q}^2 4g\Pi^{\rm R}(\mathbf{q}, \omega), \tag{8.39}$$

where we have used that the sound velocity is $c^2 = g|\Phi_0|^2/(2m)$. Solving the last expression for $\omega = \tilde{\omega}_{\rm B}(\mathbf{q})$ gives the dispersion relation for the combined condensate and quasiparticle cloud oscillation mode. The real part of the right-hand side leads to a small renormalization of the sound velocity, which we disregard. More interestingly, the right-hand side contains the *imaginary* part, which describes the damping of the condensate Bogoliubov mode,

$$\tilde{\omega}_{\rm B}(\mathbf{q}) = \omega_{\rm B}(\mathbf{q}) - {\rm i}\Gamma_2(\mathbf{q}). \tag{8.40}$$

According to Eq. (8.39) the damping rate is given by

$$\Gamma_2(\mathbf{q}) = \frac{c^2\mathbf{q}^2}{\omega_B(\mathbf{q})}\, 2g\, \mathrm{Im}\Pi^R(\mathbf{q}, \omega_B(\mathbf{q})). \tag{8.41}$$

Restricting ourselves to the small wavenumbers $q < mc$, approximating the quasiparticle dispersion as $\epsilon_\mathbf{k} \approx \mathbf{k}^2/(2m)$ and employing Eqs. (8.36) and (7.15) we find [131]

$$\Gamma_2(\mathbf{q}) = -cq\, 2\pi g \int \frac{d\mathbf{k}}{(2\pi)^3}\, q\partial_{k_z} n_B(\epsilon_\mathbf{k})\, \delta\left(cq - \frac{k_z q}{m}\right) = \frac{gmTq}{\pi}, \tag{8.42}$$

where the z-axis is taken to be along the $\mathbf{q}$ vector. At small wavenumbers Γ_2 is linear in q, and thus the relative damping rate $\Gamma_2/\omega_B \propto T\sqrt{\mu(0)}/T_c^{3/2} \ll 1$ is a $\mathbf{q}$-independent constant, which is linear in temperature and small in the entire temperature range (8.32). (Since the integral is dominated by the $k \sim mc$ region, where the Popov approximation is not quantitatively accurate, the numerical constant in Eq. (8.42) should not be taken too seriously.)

The nature of the damping rate (8.41) is very similar to the Landau damping of plasma oscillations. Quanta of the condensate fluctuations are absorbed by the relatively *high*-energy resonant quasiparticles. Therefore, if a non-equilibrium condensate vibration is excited, its energy is eventually channeled into the above-the-condensate quasiparticle cloud. This damping mechanism is very different from the zero-temperature Beliaev damping [129]. The latter describes a condensate excitation decaying into two *smaller* energy condensate excitations. Such a process is energetically allowed due to a small positive curvature of the Bogoliubov dispersion (8.11) and exhibits a very sharp momentum dependence of the form $\Gamma_{\text{Beliaev}}(\mathbf{q}) \propto q^5/(m\rho_0)$ for $q < mc$, see Problem 8.9.4. Beliaev damping is also present at finite temperatures. However, for $T > mc^2$ the collisionless damping (8.41) is by far the dominant mechanism.

According to FDT any damping mechanism must be accompanied by fluctuations. We shall discuss the corresponding fluctuation mechanism in Section 8.5.

8.4 Condensate Growth and Collapse

Although the modified Gross–Pitaevskii equation (8.18) and the collisionless kinetic equation (8.34) describe energy exchange between the condensate and the quasiparticles, they do not describe the condensate growth or collapse. The reason is that Eq. (8.18) conserves the total number of particles in the condensate, while Eq. (8.34) conserves the total number of quasiparticles above the condensate. On the other hand, the condensate growth and collapse are clearly associated with the particle exchange between the two. To take into account this effect one has to resort to the three-particle part of the collision action, $S_3^{\rm coll}$, Eq. (8.23). To this end we need to average $\exp\{iS^{\rm coll}\}$ over the fluctuations of the quasiparticle fields φ. For a weakly interacting gas this may be achieved with the following approximation:

$$\left\langle e^{iS^{\rm coll}}\right\rangle \approx 1 + i\left\langle S^{\rm coll}\right\rangle - \frac{1}{2}\left\langle \left(S^{\rm coll}\right)^2\right\rangle \approx e^{-\frac{1}{2}\left\langle \left(S^{\rm coll}\right)^2\right\rangle}; \qquad \delta S = \frac{i}{2}\left\langle \left(S^{\rm coll}\right)^2\right\rangle, \quad (8.43)$$

where the angular brackets denote averaging over Gaussian φ-fluctuations, governed by the quasiparticle action (8.20), and we took into account that $\left\langle S^{\rm coll}\right\rangle = 0$.

Applying Eq. (8.43) to $S_3^{\rm coll}$, one finds the second-order correction to the Gross–Pitaevskii action, δS_3, which is a quadratic form of Φ_0 and $\Phi^{\rm q}$ as well as their complex conjugates. According to the causality structure, this quadratic form does not contain terms with only classical components Φ_0 and $\bar{\Phi}_0$, since all the terms in the action must have at least the first power of quantum fields $\Phi^{\rm q}$ or $\bar{\Phi}^{\rm q}$. In the Popov approximation, where one disregards the anomalous averages $\langle\varphi\varphi\rangle$ and $\langle\bar{\varphi}\bar{\varphi}\rangle$ and keeps only the normal ones $\langle\varphi\bar{\varphi}\rangle$, no anomalous terms, such as, for example, $\Phi^{\rm q}\Phi_0$, are generated in δS_3. As a result, the δS_3 action possesses the same structure as the Gross–Pitaevskii action (8.15) and contains only the terms proportional to $\bar{\Phi}^{\rm q}\Phi_0$, $\bar{\Phi}_0\Phi^{\rm q}$ and $\bar{\Phi}^{\rm q}\Phi^{\rm q}$. The first two describe particle exchange between the condensate and the quasiparticle cloud, while the last one is responsible for the fluctuations of the condensate.

We first focus on the particle exchange terms $\sim \bar{\Phi}^{\rm q}\Phi_0$. They are generated by the product of the first and second lines in $S_3^{\rm coll}$, Eq. (8.23). Due to the causality constraints only the first term from the first line $\sim \bar{\varphi}\varphi\varphi$, being multiplied by the three terms from the second line, provides nonzero contributions. The corresponding three diagrams are depicted in Fig. 8.1 and the corresponding contribution to the action δS_3 takes the following form (since in the Popov approximation we have disregarded anomalous averages, all the Green functions are the normal bosonic Green functions; see Chapter 6):

$$\begin{aligned}\delta S_{3a} = \frac{g^2}{4}\int dx\, dx'\, \bar{\Phi}^{\rm q}(x)\Phi_0(x')\Big[4G^{\rm K}(x',x)G^{\rm K}(x,x')G^{\rm R}(x,x')\\ + 2G^{\rm A}(x',x)G^{\rm K}(x,x')G^{\rm K}(x,x') + 2G^{\rm A}(x',x)G^{\rm R}(x,x')G^{\rm R}(x,x')\Big]. \quad (8.44)\end{aligned}$$

Now we (i) change both $G^{\rm R}(x,x')$ in the last term on the right-hand side of Eq. (8.44) to $G^{\rm R}(x,x') - G^{\rm A}(x,x')$, since $G^{\rm A}(x',x)G^{\rm A}(x,x') = 0$ due to causality; (ii) perform a Wigner transformation, using the fact that the condensate fields are slow, and write $G^{\rm K} \overset{\rm WT}{\to} -2\pi i F(x,k)\delta(\epsilon - \epsilon_{\mathbf{k}}(x))$, while $G^{\rm R} - G^{\rm A} \overset{\rm WT}{\to} -2\pi i\delta(\epsilon - \epsilon_{\mathbf{k}}(x))$, where the spectrum $\epsilon_{\mathbf{k}}(x)$ is given by Eq. (8.33); (iii) perform the energy integrals using delta-functions; (iv) symmetrize the first term on the right-hand side of Eq. (8.44) with respect to $\mathbf{k}$ and $\mathbf{p} - \mathbf{k}$ arguments (see Fig. 8.1(a)). Including also

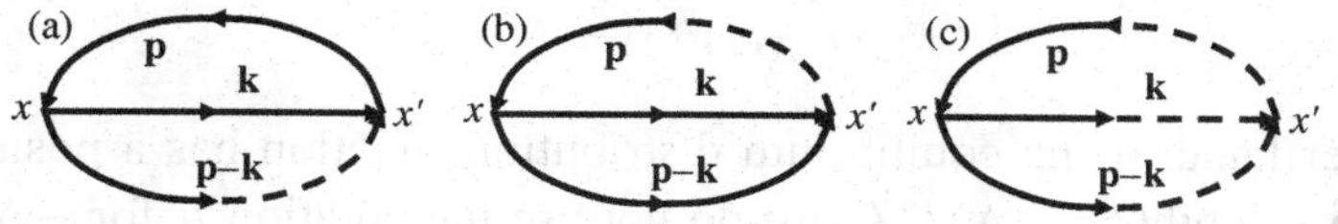

Figure 8.1 Three diagrams for the $\delta S_{3a} \sim \bar{\Phi}^{q}(x)\Phi_0(x')$ contribution to the effective Gross–Pitaevskii action. Diagram (a) carries a combinatorial factor of 4, while (b) and (c) carry a factor of 2. Full lines correspond to φ and dashed lines to φ^{q} fields.

the conjugated part $\sim \bar{\Phi}_0\Phi^{q}$ and keeping only the imaginary part of the square brackets in Eq. (8.44) (the real part constitutes an insignificant renormalization of the quasiparticle density in Eq. (8.16)), we find

$$\delta S_{3a} = \mathrm{i}\int dx\,\big(\bar{\Phi}^{q}\Phi_0 - \bar{\Phi}_0\Phi^{q}\big)\Gamma_3(x), \tag{8.45}$$

where

$$\Gamma_3(x;\tilde{F}) = \frac{\pi g^2}{2}\sum_{\mathbf{p},\mathbf{k}} \delta(\epsilon_{\mathbf{p}} - \epsilon_{\mathbf{k}} - \epsilon_{\mathbf{p}-\mathbf{k}}) \tag{8.46}$$
$$\times\Big[\tilde{F}(x,\mathbf{p})\big(\tilde{F}(x,\mathbf{k}) + \tilde{F}(x,\mathbf{p}-\mathbf{k})\big) - \tilde{F}(x,\mathbf{k})\tilde{F}(x,\mathbf{p}-\mathbf{k}) - 1\Big].$$

This is a particle-non-conserving term in the modified Gross–Pitaevskii equation. The latter is derived by variation of the action with respect to $\bar{\Phi}^{q}$ and reads as $[\mathrm{i}\partial_t - H_{\mathrm{GP}}]\Phi_0 = -\mathrm{i}\Gamma_3\Phi_0$. As a result, the total number of particles in the condensate evolves according to

$$\partial_t\int d\mathbf{r}\,|\Phi_0|^2 = -\int d\mathbf{r}\,2\Gamma_3(\mathbf{r},t)\,|\Phi_0|^2, \tag{8.47}$$

where we took into account that the superfluid current is absent at the boundaries. Notice that growth of the condensate corresponds to a negative Γ_3, while collapse corresponds to a positive one.

The physical process behind these expressions is that of a fast quasiparticle with momentum $\mathbf{p}$ hitting the condensate and knocking out *two* quasiparticles with momenta $\mathbf{k}$ and $\mathbf{p}-\mathbf{k}$. To see it most clearly one may substitute $\tilde{F}(\mathbf{k}) = 2n_{\mathrm{qp}}(\mathbf{k}) + 1$ to find for the expression in the square brackets on the right-hand side of Eq. (8.46), $4[n_{\mathrm{qp}}(\mathbf{p})(n_{\mathrm{qp}}(\mathbf{k}) + n_{\mathrm{qp}}(\mathbf{p}-\mathbf{k}) + 1) - n_{\mathrm{qp}}(\mathbf{k})n_{\mathrm{qp}}(\mathbf{p}-\mathbf{k})]$. The first part here includes induced and spontaneous processes, mentioned earlier. The last term is the opposite process where two quasiparticles $\mathbf{k}$ and $\mathbf{p}-\mathbf{k}$ "sink" into the condensate creating a single particle $\mathbf{p}$. If the quasiparticle occupation is the equilibrium one $\tilde{F}^{\mathrm{eq}}(\mathbf{k}) = \coth \epsilon_{\mathbf{k}}/2T$, the two processes exactly cancel each other and the condensate is neither growing nor shrinking. Indeed, in view of the "magic" identity (6.59) one finds

$$\Gamma_3(\mathbf{r};\tilde{F}^{\text{eq}}) = 0. \tag{8.48}$$

On the other hand, if the equilibrium distribution function has a positive energy offset $\tilde{F}(\mathbf{k}) = \coth(\epsilon_{\mathbf{k}} + \Lambda)/2T$ (we do not use the notation μ for $-\Lambda$, since the latter differs from the thermodynamic chemical potential in the interacting system), the condensate can't be stable and *collapses* with the rate

$$\begin{aligned}\Gamma_3(x) &= \frac{g^2}{2}\frac{1}{(2\pi)^3}\int_0^\infty p^2 \mathrm{d}p \int_0^\infty k^2 \mathrm{d}k \int_{-1}^{1} \mathrm{d}\cos\theta\; \delta\left(\frac{pk}{m}\cos\theta - \frac{k^2}{m} - V\right) \mathcal{I}(\epsilon_p, \epsilon_k) \\ &= \frac{g^2}{2}\frac{m^3}{(2\pi)^3}\int_{2V}^{\infty} \mathrm{d}\epsilon_p \int_V^{\epsilon_p - V} \mathrm{d}\epsilon_k\; \mathcal{I}(\epsilon_p, \epsilon_k) = \frac{g^2 m^3 T^2}{4\pi^3}\, C\left(\frac{\Lambda}{T}, \frac{V^{\text{cl}}(x)}{\Lambda}\right),\end{aligned} \tag{8.49}$$

where the quasiparticle spectrum is taken as $\epsilon_k(x) = k^2/2m + V^{\text{cl}}(x)$ and we used the notation $V = V^{\text{cl}}(x)$. The combination of the distribution functions under the integrals with the help of identity (6.59) may be written as

$$\mathcal{I}(\epsilon_p, \epsilon_k) = \left(\coth\frac{\epsilon_p + \Lambda}{2T} - \coth\frac{\epsilon_p + 2\Lambda}{2T}\right)\left(\coth\frac{\epsilon_k + \Lambda}{2T} + \coth\frac{\epsilon_p - \epsilon_k + \Lambda}{2T}\right).$$

Performing the integrals, one finds the dimensionless function $C(x,y)$. It has the following asymptotic behavior: $C = \pi^2/6$ for $x \ll 1$ and $y \ll 1$, while $C = 2\ln 2/y$ for $x \ll 1$ and $y \gg 1$ and $C = \mathrm{e}^{-x}$ for $x \gg 1$. The first of these asymptotic results tells us that in a wide range of parameters $V^{\text{cl}}(x) < \Lambda < T$ the condensate collapse rate $\Gamma_3(x)$ is practically a constant given by

$$\Gamma_3^{\max} = \frac{2\pi}{3}\, m a_s^2\, T^2, \tag{8.50}$$

where we took into account that $g = 4\pi a_s/m$. For smaller $\Lambda < V^{\text{cl}}(x), T$ the rate decreases to zero as $\Gamma_3(x) \propto \Lambda g^2 m^3 T^2 / V^{\text{cl}}(x)$ in agreement with Eq. (8.48). Notice that for the small central region of the trap $V^{\text{cl}}(x) < g\rho_0$ the Popov approximation is not valid and one should use $g\rho_0$ instead of $V^{\text{cl}}(x)$ in the last expression for Γ_3. To fix the numerical coefficient in such a regime one needs to go beyond the Popov approximation.

As an example of a growing condensate, consider a step of *evaporative cooling*. In this process, the high-energy tail of the Bose distribution is removed (evaporated) above some threshold energy ϵ_0, that is, $n_{\text{qp}}(\mathbf{k}) = 0$ for $\epsilon_{\mathbf{k}} > \epsilon_0$, while $n_{\text{qp}}(\mathbf{k}) = [\mathrm{e}^{\epsilon_{\mathbf{k}}/T} - 1]^{-1} = n_{\text{B}}(\epsilon_{\mathbf{k}})$ for $\epsilon_{\mathbf{k}} < \epsilon_0$, here $T \le T_{\text{c}}$. We focus on the central region of the trap, possibly already occupied by the condensate. The quasiparticle dispersion relation in this region is given by $\epsilon_{\mathbf{k}}(x) = \mathbf{k}^2/2m + g\rho_0(x) \approx \mathbf{k}^2/2m$. The quasiparticle cloud cools down below T and the condensate starts to grow with the rate $-\Gamma_3$ given by Eq. (8.46). The effect comes from the non-equilibrium region, where $\epsilon_{\mathbf{p}} > \epsilon_0$, while both $\epsilon_{\mathbf{k}} < \epsilon_0$ and $\epsilon_{\mathbf{p}-\mathbf{k}} = \epsilon_{\mathbf{p}} - \epsilon_{\mathbf{k}} < \epsilon_0$. As a result only the term with $n_{\rm qp}(\mathbf{k})n_{\rm qp}(\mathbf{p}-\mathbf{k})$ contributes:

$$-\Gamma_3 = \frac{2g^2}{(2\pi)^3}\int_0^\infty p^2 \mathrm{d}p \int_0^\infty k^2 \mathrm{d}k \int_{-1}^{1} \mathrm{d}\cos\theta\, \delta\left(\frac{pk}{m}\cos\theta - \frac{k^2}{m}\right) n_{\rm qp}(\mathbf{k})n_{\rm qp}(\mathbf{p}-\mathbf{k})$$

$$= \frac{2g^2m^3}{(2\pi)^3}\int_0^{\epsilon_0} \mathrm{d}\epsilon_k\, n_{\rm B}(\epsilon_k) \int_{\epsilon_0}^{\epsilon_0+\epsilon_k} \mathrm{d}\epsilon_p\, n_{\rm B}(\epsilon_p - \epsilon_k) = \frac{g^2m^3T^2}{4\pi^3}\, B\big(n_{\rm B}(\epsilon_0)\big), \quad (8.51)$$

where the dimensionless coefficient $B(n_0) = -n_0\int_0^1 \mathrm{d}s \ln(1-s)/[s(s+n_0)]$ depends on the Bose occupation number of the topmost unevaporated state $n_0 = [\mathrm{e}^{\epsilon_0/T} - 1]^{-1}$. For $n_0 \gg 1$ one finds $B(n_0) = \pi^2/6$, while for $n_0 \ll 1$, $B(n_0) = (1 - \ln n_0)n_0 + O(n_0^2) \ll 1$. Notice that the condensate growth rate saturates for $\epsilon_0 \lesssim T$ at the already familiar constant value $-\Gamma_3(x) = \Gamma_3^{\rm max}$, Eq. (8.50), and practically does not increase upon farther reducing the threshold energy ϵ_0 (as long as $\epsilon_0 > \mu(0) = g\rho_0$ so the Popov approximation is applicable). According to Eq. (8.50) the time needed to grow the condensate diverges as T^{-2} at small temperatures. Also notice that $\Gamma_3^{\rm max}/\mu(0) \propto T^2\mu(0)/T_{\rm c}^3 \ll 1$ in the entire temperature range (8.32) and therefore the amplitude growth is indeed a slow process. The initial exponential growth with the rate $-\Gamma_3$ saturates once the number of particles in the condensate approaches the value prescribed by the particle and energy conservation laws after equilibration. To describe this process quantitatively [132] one needs to follow the equilibration of the quasiparticle cloud; see Section 8.7.

Comparing the damping rate $\Gamma_2(\mathbf{q}) \sim a_{\rm s}Tq$, Eq. (8.42), and the characteristic growth rate $\Gamma_3^{\rm max} \sim ma_{\rm s}^2T^2$, Eq. (8.50), one finds the characteristic length scale

$$L_{\rm c} = q_{\rm c}^{-1} = (ma_{\rm s}T)^{-1}. \quad (8.52)$$

If the size of the condensate $L_{\rm trap}$ is less than $L_{\rm c}$, then for all relevant wavenumbers $\Gamma_2(\mathbf{q}) > \Gamma_3^{\rm max}$. This means that the internal vibrations of the condensate are damped faster than they have time to develop. As a result, the condensate growth may be viewed as a smooth and "peaceful" increase of its size without excitation of internal degrees of freedom. In the opposite limit, $L_{\rm trap} \gg L_{\rm c}$, the condensate growth may be a turbulent process, which creates long-lived local structures on the scale $L_{\rm c}$ or larger.

8.5 Fluctuations

Collisionless damping, Section 8.3, must be accompanied by fluctuations (noise). Such noise is produced by the cloud of quasiparticles and acts on the condensate wavefunction. In equilibrium the corresponding noise amplitude is given by FDT. To evaluate it in a generic case we notice that it originates from the $2(\bar{\Phi}^{\mathrm{q}}\Phi_0 + \bar{\Phi}_0\Phi^{\mathrm{q}})(\bar{\varphi}\varphi + \bar{\varphi}^{\mathrm{q}}\varphi^{\mathrm{q}})$ term in the two-particle collision action S_2^{coll}, Eq. (8.23). The corresponding diagrams for $\langle (S_2^{\mathrm{coll}})^2 \rangle$ are depicted in Fig. 8.2, and the part of action δS_2 (see Eq. (8.43)) reads as

$$\delta S_2 = g^2 \int \mathrm{d}x\, \mathrm{d}x' \left(\bar{\Phi}^{\mathrm{q}}\Phi_0 + \bar{\Phi}_0\Phi^{\mathrm{q}}\right)_x \Pi^{\mathrm{K}}(x,x') \left(\bar{\Phi}^{\mathrm{q}}\Phi_0 + \bar{\Phi}_0\Phi^{\mathrm{q}}\right)_{x'}; \tag{8.53}$$

$$\Pi^{\mathrm{K}}(x,x') = \frac{1}{2\mathrm{i}} \left[G^{\mathrm{K}}(x',x)G^{\mathrm{K}}(x,x') + G^{\mathrm{R}}(x',x)G^{\mathrm{A}}(x,x') + G^{\mathrm{A}}(x',x)G^{\mathrm{R}}(x,x') \right].$$

The expression for the Keldysh component of the polarization operator coincides with Eq. (7.24). Proceeding exactly as in Section 7.4, one finds for the Wigner transform of $\Pi^{\mathrm{K}}(x,x')$

$$\Pi^{\mathrm{K}}(x,\mathbf{q},\omega) = \mathrm{i}\pi \sum_{\mathbf{k}} \delta(\omega + \epsilon_{\mathbf{k}} - \epsilon_{\mathbf{k}+\mathbf{q}}) \left[\tilde{F}(x,\mathbf{k})\tilde{F}(x,\mathbf{k}+\mathbf{q}) - 1 \right]. \tag{8.54}$$

As discussed there, it satisfies the FDT relation (7.26) in equilibrium. Notice that the $\bar{\Phi}^{\mathrm{q}}\bar{\Phi}_0(\varphi\varphi + \varphi^{\mathrm{q}}\varphi^{\mathrm{q}})$ part of the S_2^{coll} action (8.23), being taken to second order, leads to the diagrams with the energy balance given by $\delta(\omega - \epsilon_{-\mathbf{k}} - \epsilon_{\mathbf{k}+\mathbf{q}})$. For our situation where $\mathbf{q}$ and ω are slow condensate variables, while $\mathbf{k}$ is the fast quasiparticle variable, this delta-function cannot be satisfied.

One can now perform the Hubbard–Stratonovich transformation with the help of an auxiliary *real* field $\xi(x)$:

$$e^{\mathrm{i}\delta S_2} = \int \mathbf{D}[\xi]\; e^{-\frac{\mathrm{i}}{4}\int \mathrm{d}x\mathrm{d}x'\, \xi(x)\Pi^{-1}(x,x')\xi(x') - \mathrm{i}g\int \mathrm{d}x\, \xi(x)\left(\bar{\Phi}^{\mathrm{q}}\Phi_0 + \bar{\Phi}_0\Phi^{\mathrm{q}}\right)_x}, \tag{8.55}$$

where $\int \mathrm{d}x''\, \Pi^{-1}(x,x'')\Pi^{\mathrm{K}}(x'',x') = \delta(x-x')$. The first term in the exponent on the right-hand side of Eq. (8.55) specifies the real Gaussian colored noise with the correlator

$$\langle \xi(x+y/2)\xi(x-y/2) \rangle = -2\mathrm{i} \sum_q{}' e^{\mathrm{i}qy}\, \Pi^{\mathrm{K}}(x,q), \tag{8.56}$$

where summation over $q = \mathbf{q},\omega$ is limited by the band designated for slow condensate degrees of freedom. Because of this limitation it is tempting to substitute

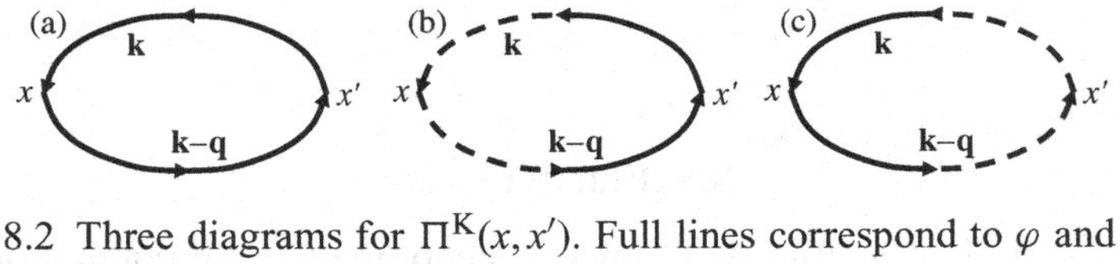

Figure 8.2 Three diagrams for $\Pi^{\mathrm{K}}(x,x')$. Full lines correspond to φ and dashed lines to φ^{q} fields.

$\Pi^K(x, q)$ by $\Pi^K(x, 0)$, which would lead to the local noise correlator $\langle \xi(x)\xi(x')\rangle = -2\mathrm{i}\Pi^K(x, 0)\delta(x - x')$. One should be aware, though, that the limit $\mathbf{q} \to 0$ and $\omega \to 0$ in Eq. (8.54) is not unique. The most sensible way is to put $\omega = \omega_B(\mathbf{q})$ (see Eq. (8.41)), but the limit $\mathbf{q} \to 0$ may still be a singular one. For example, employing Eq. (8.42) and FDT, one finds in equilibrium $\Pi^K(\mathbf{q}, cq) = 2\mathrm{i}mT^2/(\pi c^2 q)$. In the coordinate space this implies the nonlocal long-ranged correlator $\Pi^K(\mathbf{r}, \mathbf{r}') \sim 1/(\mathbf{r} - \mathbf{r}')^2$.

The last term in the exponent in Eq. (8.55) is a part of the Gross–Pitaevskii action (8.15), which adds the real noise $g\xi(x)\Phi_0$ to the condensate equation of motion [133]

$$\left[\mathrm{i}\partial_t - H_{\mathrm{GP}}\right]\Phi_0 = -\mathrm{i}\Gamma_3\Phi_0 + g\xi(x)\Phi_0. \tag{8.57}$$

Notice that one can gauge out the noise as $\Phi_0 \to \Phi_0 \mathrm{e}^{-\mathrm{i}g\int \mathrm{d}t\,\xi(x)}$. It then modifies the superfluid velocity as $\nabla_{\mathbf{r}}\mathcal{K} \to \nabla_{\mathbf{r}}\mathcal{K} + g\int \mathrm{d}t\nabla_{\mathbf{r}}\xi$. Therefore, it is the spatial gradient of the noise $\nabla_{\mathbf{r}}\xi$, rather than the noise ξ itself, which has a physical significance. In other words, one can always add a spatially independent function $\tilde{\xi}(t)$ to the noise without changing any physics. It is possible and convenient thus to normalize the noise as $\int \mathrm{d}\mathbf{r}\,\xi(\mathbf{r}, t) = 0$. It is also clear from Eq. (8.57) that the noise changes the phase, but not the amplitude of the condensate wavefunction Φ_0.

Because of the last observation and the homogeneous form of the modified Gross–Pitaevskii equation (8.57) one may worry that, if at some initial time $\Phi_0 = 0$, the condensate never forms. This is, of course, not the case. One way to understand it is to acknowledge that even in the absence of the macroscopic condensate $|\Phi_0|^2 = 2\rho_0 \neq 0$. Indeed, we have defined the condensate as a band of slow degrees of freedom and therefore $\rho_0 = \sum_{\mathbf{q}}' n_B(\mathbf{q}) > 0$, where the summation runs over such a slow band. The precise value of this initial ρ_0 is somewhat ambiguous. It is not really important, however, since once the condensate starts to form it grows exponentially with the rate Γ_3. The memory of the precise initial conditions is therefore rapidly lost.

We shall discuss now how these considerations may be generalized away from equilibrium. To find fluctuations of the condensate density one needs to consider the $\bar{\Phi}^{\mathrm{q}}\Phi^{\mathrm{q}}$ term in the effective action. With the help of Eq. (8.43), it is found as an expectation value of the product of the first line in S_3^{coll}, Eq. (8.23), and its complex conjugate. There are five nontrivial diagrams generated upon Gaussian averaging over the φ-fluctuations; see Fig. 8.3. The corresponding analytical expression reads as

$$\delta S_{3b} = \frac{g^2}{2}\int \mathrm{d}x\,\mathrm{d}x'\,\bar{\Phi}^{\mathrm{q}}(x)\Phi^{\mathrm{q}}(x')\Big[2G^{\mathrm{A}}_{x',x}G^{\mathrm{R}}_{x,x'}G^{\mathrm{K}}_{x,x'} + 2G^{\mathrm{R}}_{x',x}G^{\mathrm{A}}_{x,x'}G^{\mathrm{K}}_{x,x'}$$
$$+ G^{\mathrm{K}}_{x',x}G^{\mathrm{K}}_{x,x'}G^{\mathrm{K}}_{x,x'} + G^{\mathrm{K}}_{x',x}G^{\mathrm{R}}_{x,x'}G^{\mathrm{R}}_{x,x'} + G^{\mathrm{K}}_{x',x}G^{\mathrm{A}}_{x,x'}G^{\mathrm{A}}_{x,x'}\Big]$$

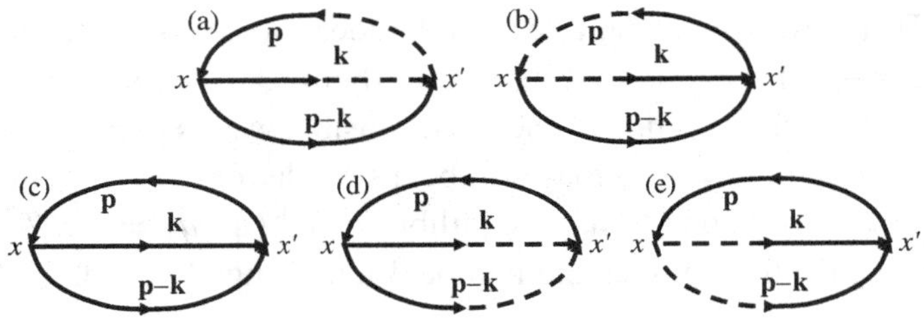

Figure 8.3 Five diagrams for the $\delta S_{3b} \sim \bar{\Phi}^{\mathrm{q}}(x)\Phi^{\mathrm{q}}(x')$ contribution to the effective Gross–Pitaevskii action. Full lines correspond to φ and dashed lines to φ^{q} fields. Diagrams (a), (b) carry a combinatorial factor 4, while (c)–(e) carry a factor 2.

$$= \frac{g^2}{2}\int \mathrm{d}x\,\mathrm{d}x'\,\bar{\Phi}^{\mathrm{q}}(x)\Phi^{\mathrm{q}}(x')\Big[-2(G^{\mathrm{R}}_{x',x}-G^{\mathrm{A}}_{x',x})(G^{\mathrm{R}}_{x,x'}-G^{\mathrm{A}}_{x,x'})G^{\mathrm{K}}_{x,x'}$$
$$+\,G^{\mathrm{K}}_{x',x}\big(G^{\mathrm{K}}_{x,x'}G^{\mathrm{K}}_{x,x'}+(G^{\mathrm{R}}_{x,x'}-G^{\mathrm{A}}_{x,x'})(G^{\mathrm{R}}_{x,x'}-G^{\mathrm{A}}_{x,x'})\big)\Big], \tag{8.58}$$

where we employed the fact that $G^{\mathrm{R}}(x',x)G^{\mathrm{R}}(x,x') = G^{\mathrm{R}}(x,x')G^{\mathrm{A}}(x,x') = 0$, due to causality. Now we (i) perform a Wigner transformation, using the slowness of the condensate fields, to write $G^{\mathrm{K}} \overset{\mathrm{WT}}{\to} -2\pi\mathrm{i}F(x,k)\delta(\epsilon-\epsilon_{\mathbf{k}}(x))$ and $G^{\mathrm{R}}-G^{\mathrm{A}} \overset{\mathrm{WT}}{\to} -2\pi\mathrm{i}\delta(\epsilon-\epsilon_{\mathbf{k}}(x))$; (ii) perform the energy integrals using delta-functions; (iii) symmetrize the first term on the right-hand side of Eq. (8.58) with respect to $\mathbf{k}$ and $\mathbf{p}-\mathbf{k}$ arguments (see Fig. 8.3(a),(b)). This way we find for the complete Gross–Pitaevskii action

$$S_{\mathrm{GP}} = \int \mathrm{d}x\,(\bar{\Phi}_0, \bar{\Phi}^{\mathrm{q}})\begin{pmatrix} 0 & \mathrm{i}\partial_t - H_{\mathrm{GP}} - \mathrm{i}\Gamma_3 \\ \mathrm{i}\partial_t - H_{\mathrm{GP}} + \mathrm{i}\Gamma_3 & \mathrm{i}\Gamma_3^{\mathrm{K}} \end{pmatrix}\begin{pmatrix}\Phi_0 \\ \Phi^{\mathrm{q}}\end{pmatrix}, \tag{8.59}$$

where δS_{3b} provides the Keldysh component of the matrix, given by

$$\Gamma_3^K(x;\tilde{F}) = \pi g^2 \sum_{\mathbf{p},\mathbf{k}} \delta(\epsilon_{\mathbf{p}} - \epsilon_{\mathbf{k}} - \epsilon_{\mathbf{p}-\mathbf{k}}) \tag{8.60}$$
$$\times\Big[-\tilde{F}(x,\mathbf{k}) - \tilde{F}(x,\mathbf{p}-\mathbf{k}) + \tilde{F}(x,\mathbf{p})\big(\tilde{F}(x,\mathbf{k})\tilde{F}(x,\mathbf{p}-\mathbf{k}) + 1\big)\Big].$$

If the distribution function is the equilibrium one with a positive energy offset, $\tilde{F}(\mathbf{k}) = \coth(\epsilon_{\mathbf{k}} + \Lambda)/2T$, one finds, employing identity (6.59),

$$\Gamma_3^{\mathrm{K}} = \coth\frac{\Lambda}{2T}\,2\Gamma_3,$$

where Γ_3 is given by Eq. (8.49). This is a manifestation of FDT. Indeed, the energy offset Λ may be gauged away at the expense of giving the condensate field the rapidly rotating phase $\Phi_0 \to \Phi_0 e^{\mathrm{i}\Lambda t}$. In such a gauge Λ is nothing but the frequency argument of the Keldysh matrix propagator $\hat{\Gamma}_3(\Lambda)$, where $\Gamma_3^{\mathrm{R}}(\Lambda) = -\Gamma_3^{\mathrm{A}}(\Lambda) = \Gamma_3$.

In the wide range of parameters $V^{\rm cl}(x) < \Lambda < T$ one finds a spatially independent result $\Gamma_3^{\rm K}(\Lambda) = 4T\Gamma_3^{\max}/\Lambda$ (cf. Eq. (8.49)). For $\Lambda < V^{\rm cl}(x)$ (including zero offset) $\Gamma_3^{\rm K}$ approaches a Λ-independent value $\Gamma_3^{\rm K}(0) \propto T\Gamma_3^{\max}/V^{\rm cl}(x)$. (Similarly to the discussion after Eq. (8.49), one should actually use $\max\{V^{\rm cl}(x), g\rho_0\}$ in the last expression.) Having a nonzero value of the Keldysh component $\Gamma_3^{\rm K}(0)$ seemingly contradicts the absence of the condensate growth rate for $\Lambda = 0$, Eq. (8.48). In fact there is no contradiction: the omitted next term of the Wigner transformation in Eq. (8.45), $\sim \partial_\epsilon \partial_t$, has the form $\sim (\bar{\Phi}^{\rm q}\partial_t\Phi_0 - \bar{\Phi}_0\partial_t\Phi^{\rm q})\Gamma_3^{\rm K}(0)/4T$, which is the exact FDT counterpart of the nonzero $\Gamma_3^{\rm K}(0)$. This term provides damping of the condensate fluctuations, which is due to the three-particle collisions. Such damping, however, is weaker than the collisionless damping Γ_2, Eqs. (8.41), (8.42), and therefore may be omitted. Nevertheless, in equilibrium $\Gamma_3^{\rm K}/2\Gamma_3 = 2T/\max\{\Lambda, V^{\rm cl}(x), g\rho_0\}$ in agreement with FDT.

Away from equilibrium one may evaluate $\langle|\Phi_0|^2\rangle$ in the situation where the macroscopic condensate is unstable, that is, $\Gamma_3 > 0$ and $\langle\Phi_0\rangle = 0$. This is exactly the case where fluctuations are especially important. One may thus neglect the nonlinear terms in $H_{\rm GP}$ and find by inversion of the matrix in Eq. (8.59)

$$\rho_0 = \frac{1}{2}\langle|\Phi_0|^2\rangle = \frac{1}{2}\sum_{\mathbf{q}}{}' \int \frac{d\omega}{2\pi}\frac{\Gamma_3^{\rm K}}{(\omega - \mathbf{q}^2/2m)^2 + \Gamma_3^2} = \sum_{\mathbf{q}}{}' \frac{\Gamma_3^{\rm K}}{4\Gamma_3}, \tag{8.61}$$

where it is important to understand that the momentum sum runs only over the band of slow degrees of freedom. In equilibrium one is back exactly to the value of ρ_0, discussed previously. Away from equilibrium one may evaluate it employing Eqs. (8.46) and (8.60). In a generic case one can apply the Hubbard–Stratonovich transformation to the $\bar{\Phi}^{\rm q}(x)\Phi^{\rm q}(x)\,\Gamma_3^{\rm K}$ term of the action (8.59) with the help of the auxiliary *complex* field $\zeta(x)$,

$$e^{i\delta S_{3b}} = \int \mathbf{D}[\zeta]\, e^{-\int dx[\Gamma_3^{\rm K}]^{-1}|\zeta(x)|^2 - i\int dx\left(\bar{\Phi}^{\rm q}(x)\zeta(x) + \bar{\zeta}(x)\Phi^{\rm q}(x)\right)}. \tag{8.62}$$

It defines the complex Gaussian noise with the correlator $\langle\bar{\zeta}(x)\zeta(x')\rangle = \Gamma_3^{\rm K}(x)\,\delta(x - x')$, where only slow components of the noise should be retained. Such noise acts as a source in the modified Gross–Pitaevskii equation $i\partial_t\Phi_0 - \cdots = \zeta(x)$. Its main effect is to generate a *deterministic* centrifugal potential in the complex plane of Φ_0 for the evolution of the radial variable $|\Phi_0|$. Indeed, the complex noise ζ induces diffusion on the complex plane $\Phi_0(\mathbf{q})$. The corresponding 2D diffusion equation for the probability distribution function is $\partial_t\mathcal{P} = (\Gamma_3^{\rm K}/2)\nabla_\Phi^2\mathcal{P}$, where $\nabla_\Phi^2 = \partial^2_{{\rm Re}\Phi_0} + \partial^2_{{\rm Im}\Phi_0}$. Transforming it to the radial coordinate $r = |\Phi_0|$ and properly normalized distribution $W = r\mathcal{P}$, one finds $\partial_t W = (\Gamma_3^{\rm K}/2)\partial_r(-W/r + \partial_r W)$, which in addition to the diffusion contains the centrifugal drift potential $\partial_r U = -\Gamma_3^{\rm K}/2r$. The corresponding Langevin equation is $\dot{r} = \Gamma_3^{\rm K}/2r+$ noise, or finally in terms of

$\rho_0 = r^2/2$ one has $\dot{\rho}_0 = \Gamma_3^{\mathrm{K}}/2 + \sqrt{2\rho_0} \times$ noise. While the stochastic noise part is less important than the real noise ξ, Eq. (8.56), the induced radial drift term $\Gamma_3^{\mathrm{K}}/2$ is crucial to initiate nucleation of the condensate.

8.6 Semiclassical Dynamics of the Condensate

We put together now bits and pieces of the condensate dynamics in the presence of the quasiparticles (treated in the Popov approximation). The condensate amplitude $\Phi_0(x)$ obeys the *stochastic* [133] modified Gross–Pitaevskii equation (8.57). In addition to the Hartree–Fock interaction terms, incorporated into the Hamiltonian H_{GP}, it also includes particle exchange with the quasiparticle cloud Γ_3, along with the Gaussian noise $\xi(x)$. We shall look for a solution of Eq. (8.57) in the form $\Phi_0(x) = \sqrt{2\rho_0(x)}\exp\{i\theta(x)\}$. The complex equation results in the two real ones: the radial one for the local condensate density $\rho_0(x)$ and the angular one for the superfluid velocity, defined as $\mathbf{v}_{\mathrm{sf}}(x) = \nabla_{\mathbf{r}}(\theta(x) - \mathcal{K}(x))/m$,

$$\partial_t \rho_0 = -\nabla_{\mathbf{r}}(\rho_0 \mathbf{v}_{\mathrm{sf}}) - 2\Gamma_3\rho_0 + {\sum_{\mathbf{q}}}' \frac{\Gamma_3^{\mathrm{K}}}{2}\,; \tag{8.63}$$

$$\partial_t \mathbf{v}_{\mathrm{sf}} = -\frac{1}{2}\nabla_{\mathbf{r}}\mathbf{v}_{\mathrm{sf}}^2 - \frac{1}{m}\nabla_{\mathbf{r}}\big(V^{\mathrm{cl}} + g\rho_0 + 2g\rho_{\mathrm{qp}}\big) - g\nabla_{\mathbf{r}}\xi\,. \tag{8.64}$$

In the last equation we have neglected $\nabla_{\mathbf{r}}^2\sqrt{\rho_0}$ as going beyond the accuracy with which we handled the Wigner transformations. Equation (8.63) is the modified continuity relation for the condensate density, which takes into account particle exchange with the quasiparticle cloud. The exchange rate Γ_3 and the noise-induced drift Γ_3^{K} are given by Eqs. (8.46) and (8.60), respectively. As discussed previously, one should understand summation in the last term as running only over the band of slow degrees of freedom. For $T > T_{\mathrm{c}}$, where $\Gamma_3 > 0$, it generates a small seed condensate density, which starts to grow exponentially when $\Gamma_3 < 0$. Equation (8.64) is the Euler equation with noise, associated with the collisionless damping of the condensate fluctuations by the quasiparticle cloud. Interaction with the latter occurs through the quasiparticle density ρ_{qp}, appearing on the right-hand side of Eq. (8.64). It is defined as $\rho_{\mathrm{qp}}(x) = \frac{1}{2}\sum_{\mathbf{k}}[\tilde{F}(x,\mathbf{k}) - 1]$. The corresponding Gaussian noise is specified by Eqs. (8.54) and (8.56).

To find the quasiparticle density $\rho_{\mathrm{qp}}(x)$, the exchange rate $\Gamma_3(x)$ and the noise amplitude, one needs to know the quasiparticle distribution function $\tilde{F}(x,\mathbf{k})$. The evolution of the latter is given by the kinetic equation (8.30):

$$\partial_t \tilde{F} = -\mathbf{v}_{\mathbf{k}}\nabla_{\mathbf{r}}\tilde{F} + \nabla_{\mathbf{r}}\epsilon_{\mathbf{k}}\nabla_{\mathbf{k}}\tilde{F} + I_3^{\mathrm{coll}}[\tilde{F}, \rho_0, \mathbf{v}_{\mathrm{sf}}] + I_4^{\mathrm{coll}}[\tilde{F}, \rho_0, \mathbf{v}_{\mathrm{sf}}], \tag{8.65}$$

where $\mathbf{v}_{\mathbf{k}} = \nabla_{\mathbf{k}}\epsilon_{\mathbf{k}}$. The three- and four-particle collision integrals $I_{3,4}^{\mathrm{coll}}$ are discussed in Section 8.7 and are given by Eqs. (8.73) and (8.79), respectively. Equations

(8.63)–(8.65) are the three evolution equations for the condensate density $\rho_0(x)$, the superfluid velocity $\mathbf{v}_{\rm sf}(x)$, and the quasiparticle distribution function $\tilde{F}(x,\mathbf{k})$. If some initial conditions are specified, for example, a non-equilibrium quasiparticle distribution and zero condensate density, one may iterate them forward in time to follow formation of the condensate and equilibration of the quasiparticle cloud [132].

To complete the scheme one needs to specify the choice of the local phase $\mathcal{K}(x)$, which in turn determines the quasiparticle energies $\epsilon_{\mathbf{k}}(x)$ entering the kinetic equation (8.65). The criterion for such a choice is to facilitate a maximal scale separation between the fast and slow degrees of freedom. If the condensate is already present in the entire volume, one wants to make its wavefunction $\Phi_0(x)$ the slowest variable. This is the case if Φ_0 is purely real. Indeed, this eliminates relatively fast phase rotations and leaves only slow amplitude variations. It is achieved by the choice of $\mathcal{K}$ that makes $\theta(x)$ as small as possible. Putting $\theta = 0$ in Eq. (8.64), one finds $\partial_t\mathcal{K}(x) = V^{\rm cl} + g\rho_0 + 2g\rho_{\rm qp}$, which upon substitution in the quasiparticle energy spectrum in the Popov approximation, Eq. (8.33), yields

$$\epsilon_{\mathbf{k}}(x) = \frac{\mathbf{k}^2}{2m} + \mathbf{k}\mathbf{v}_{\rm sf}(x) + g\rho_0(x). \tag{8.66}$$

Notice that the external potential $V^{\rm cl}(x)$ does *not* enter the local energy spectrum (8.66) nor does the drift part of the kinetic equation $\nabla_{\mathbf{r}}\epsilon_{\mathbf{k}}\nabla_{\mathbf{k}}\tilde{F}$. (It still appears, of course, in the Euler equation.) This effect is due to the Hartree–Fock screening of the potential by a nonuniform condensate.

The last feature makes the spectrum (8.66) inconvenient for the trap geometry. In a trap the condensate occupies only a relatively small central region. The quasiparticles, on the other hand, spread over a much wider volume. The choice of gauge, where $\theta = 0$, makes the Euler equation for the phase $\mathcal{K}$ formally applicable everywhere, even outside the condensate region. In the latter case it specifies the "vector potential" gauge for the quasiparticles. This is the reason why outside the condensate region (where there is no screening) the quasiparticle spectrum (8.66) does not include the scalar potential of the trap, $V^{\rm cl}(x)$. It is traded for the "vector potential" $\mathbf{v}_{\rm sf}(x)$. The disadvantage is that the latter exhibits fast time-dependence, even if the trap potential $V^{\rm cl}(\mathbf{r})$ is completely static. Therefore, this choice of $\mathcal{K}$, while making the condensate slow, exposes the peripheral quasiparticles to a fast vector potential.

To avoid this effect it is convenient to choose the phase $\mathcal{K}$ to be spatially uniform. It then drops out of the Euler equation, which is now solely limited to the condensate region, where $\rho_0 > 0$. A way to make the *complex* condensate wavefunction as slow as possible is to demand that its *average* phase does not evolve, that is, $\partial_t \int_{\rho_0>0} d\mathbf{r}\,\theta(\mathbf{r},t) = 0$, where the integral runs over the volume occupied by the condensate. Finding $\theta(\mathbf{r},t) - \mathcal{K}(t)$ from the real part of Eq. (8.57), one obtains

$$\partial_t \mathcal{K}(t) = \frac{1}{\mathcal{V}(t)} \int\limits_{\rho_0>0} d\mathbf{r} \left[\frac{\mathbf{v}_{\rm sf}^2}{2m} + V^{\rm cl}(\mathbf{r},t) + g\rho_0(\mathbf{r},t) + 2g\rho_{\rm qp}(\mathbf{r},t) \right], \qquad (8.67)$$

where $\mathcal{V}(t) = \int_{\rho_0>0} d\mathbf{r}$ is the volume of the condensate. Notice that in equilibrium this condition coincides with Eq. (8.19), as it should. The quasiparticle spectrum (8.33) in the Popov approximation is now given by

$$\epsilon_{\mathbf{k}}(\mathbf{r},t) \approx \frac{\mathbf{k}^2}{2m} + V^{\rm cl}(\mathbf{r},t) + 2g\rho_0(\mathbf{r},t) + 2g\rho_{\rm qp}(\mathbf{r},t) - \partial_t \mathcal{K}(t), \qquad (8.68)$$

where $\partial_t \mathcal{K}(t)$ is a spatially uniform, time-dependent chemical potential. One should not fix it with the particle number conservation, as was done in equilibrium; see Eq. (8.19). Instead, it is determined by the self-consistent evolution of the condensate density and current through Eq. (8.67). This gauge is convenient for treating the quasiparticles outside the condensate region. It fails to acknowledge, though, the quasiparticles' drag by the superfluid velocity $\mathbf{v}_{\rm sf}$ inside the condensate region. The corresponding effect is now delegated to the continuity equation for ρ_0, which in turn affects the quasiparticles through Eqs. (8.67) and (8.68). If superfluid currents are not expected to last for a long time, it seems to be worth the trade.

8.7 Quasiparticle Collision Integral

The three-quasiparticle self-energy in second order in the interaction constant is given by the square of the second line in $S_3^{\rm coll}$, Eq. (8.23). In the Popov approximation, where one disregards anomalous averages, there are six diagrams for $\Sigma_3^{\rm K}(x,x')$, Fig. 8.4(a)–(f), and three for $\Sigma_3^{\rm R}(x,x')$, Fig. 8.4(g)–(i). The corresponding analytical expressions are

$$\begin{aligned}\Sigma_3^{\rm K}(x,x') = \frac{{\rm i}g^2}{4}\,|\Phi_0|^2\Big[&4G^{\rm K}_{x',x}G^{\rm K}_{x,x'} + 4G^{\rm R}_{x',x}G^{\rm A}_{x,x'} + 4G^{\rm A}_{x',x}G^{\rm R}_{x,x'}\\ &+ 2G^{\rm K}_{x,x'}G^{\rm K}_{x,x'} + 2G^{\rm A}_{x,x'}G^{\rm A}_{x,x'} + 2G^{\rm R}_{x,x'}G^{\rm R}_{x,x'}\Big]; \end{aligned} \qquad (8.69)$$

$$\Sigma_3^{\rm R}(x,x') = \frac{{\rm i}g^2}{4}\,|\Phi_0|^2\Big[4G^{\rm A}_{x',x}G^{\rm K}_{x,x'} + 4G^{\rm K}_{x',x}G^{\rm R}_{x,x'} + 4G^{\rm R}_{x,x'}G^{\rm K}_{x,x'}\Big], \qquad (8.70)$$

where we employed slowness of Φ_0 fields to suppress their x-dependence. Due to causality $G^{\rm R}_{x',x}G^{\rm A}_{x,x'} + G^{\rm A}_{x',x}G^{\rm R}_{x,x'} = -(G^{\rm R}_{x',x} - G^{\rm A}_{x',x})(G^{\rm R}_{x,x'} - G^{\rm A}_{x,x'})$ and $(G^{\rm A}_{x,x'})^2 + (G^{\rm R}_{x,x'})^2 = (G^{\rm R}_{x,x'} - G^{\rm A}_{x,x'})^2$. We perform now a Wigner transformation with the energy momentum assignments specified in Fig. 8.4. We use $G^{\rm K} \overset{\rm WT}{\to} -2\pi {\rm i}\tilde{F}(\,\mathbf{p})\delta(\epsilon - \epsilon_{\mathbf{p}})$ and $G^{\rm R} - G^{\rm A} = 2{\rm i}{\rm Im}G^{\rm R} \overset{\rm WT}{\to} -2\pi{\rm i}\delta(\epsilon - \epsilon_{\mathbf{p}})$ and integrate over the intermediate energy using a delta-function to find

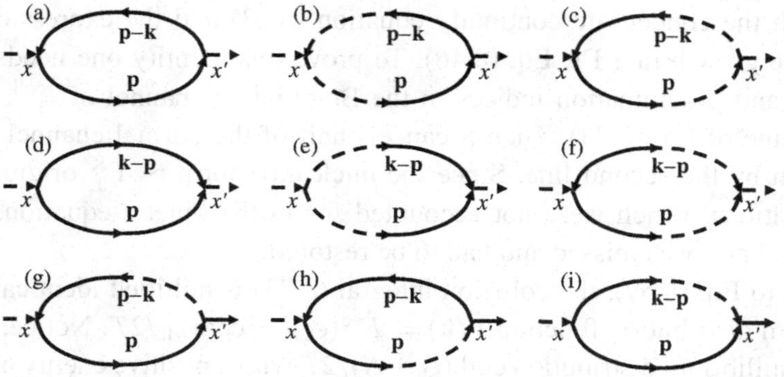

Figure 8.4 (a)–(f) Six diagrams for $\Sigma_3^K(x,x')$. The normal diagrams (a)–(c) carry a combinatorial factor of 4, while the Bogoliubov ones (d)–(f) carry a factor of 2. (g)–(i) Three diagrams for $\Sigma_3^R(x,x')$, all carry a factor of 4.

$$\Sigma_3^K(\mathbf{k}) = \frac{\pi g^2}{2\mathrm{i}}\,|\Phi_0|^2 \sum_{\mathbf{p}} \Big\{ 4\delta(\epsilon_{\mathbf{p}} - \epsilon_{\mathbf{k}} - \epsilon_{\mathbf{p-k}})\big[\tilde{F}(\mathbf{p}-\mathbf{k})\tilde{F}(\mathbf{p}) - 1\big] + 2\delta(\epsilon_{\mathbf{k}} - \epsilon_{\mathbf{p}} - \epsilon_{\mathbf{k-p}})\big[\tilde{F}(\mathbf{k}-\mathbf{p})\tilde{F}(\mathbf{p}) + 1\big]\Big\}; \tag{8.71}$$

$$\mathrm{Im}\big[\Sigma_3^R(\mathbf{k})\big] = \frac{\pi g^2}{4}\,|\Phi_0|^2 \sum_{\mathbf{p}} \Big\{ 4\delta(\epsilon_{\mathbf{p}} - \epsilon_{\mathbf{k}} - \epsilon_{\mathbf{p-k}})\big[\tilde{F}(\mathbf{p}) - \tilde{F}(\mathbf{p}-\mathbf{k})\big] - 2\delta(\epsilon_{\mathbf{k}} - \epsilon_{\mathbf{p}} - \epsilon_{\mathbf{k-p}})\big[\tilde{F}(\mathbf{p}) + \tilde{F}(\mathbf{k}-\mathbf{p})\big]\Big\}, \tag{8.72}$$

where we have suppressed the central coordinate x dependence of Σ as well as $|\Phi_0|^2$ and $\tilde{F}$ for brevity. We have symmetrized the last (Bogoliubov) term on the right-hand side of Eq. (8.70) (see Fig. 8.4(i)) with respect to **p** and **k** − **p** arguments. Employing Eq. (6.47), one obtains for the three-particle part of the collision integral $I_3^{\mathrm{coll}}[\tilde{F}(\mathbf{k})] = \mathrm{i}\Sigma_3^K(\mathbf{k}) + 2\tilde{F}(\mathbf{k})\,\mathrm{Im}\big[\Sigma_3^R(\mathbf{k})\big]$

$$I_3^{\mathrm{coll}}[\tilde{F}(x,\mathbf{k}),\rho_0(x),\mathbf{v}_{\mathrm{sf}}(x)] = 2\pi g^2\Big[\rho_0(x) - \sum_{\mathbf{q}}{}' \Gamma_3^K/4\Gamma_3\Big] \tag{8.73}$$

$$\sum_{\mathbf{p}}\Big\{2\,\delta(\epsilon_{\mathbf{p}} - \epsilon_{\mathbf{k}} - \epsilon_{\mathbf{p-k}})\big[\tilde{F}(x,\mathbf{p}-\mathbf{k})\tilde{F}(x,\mathbf{p}) - 1 + \tilde{F}(x,\,\mathbf{k})\big(\tilde{F}(x,\mathbf{p}) - \tilde{F}(x,\mathbf{p}-\mathbf{k})\big)\big]$$

$$+\,\delta(\epsilon_{\mathbf{k}} - \epsilon_{\mathbf{p}} - \epsilon_{\mathbf{k-p}})\big[\tilde{F}(x,\mathbf{k}-\mathbf{p})\tilde{F}(x,\mathbf{p}) + 1 - \tilde{F}(x,\,\mathbf{k})\big(\tilde{F}(x,\mathbf{p}) + \tilde{F}(x,\mathbf{k}-\mathbf{p})\big)\big]\Big\},$$

where we had to subtract the nucleation term $\sum_{\mathbf{q}}' \Gamma_3^K/4\Gamma_3$ "by hand" to secure the local particle conservation. The latter is now guaranteed by the identity

$$\frac{1}{2}\sum_{\mathbf{k}} I_3^{\mathrm{coll}}[\tilde{F}(x,\mathbf{k})] = 2\Gamma_3(x)\rho_0(x) - \sum_{\mathbf{q}}{}' \frac{\Gamma_3^K(x)}{2} \tag{8.74}$$

along with the condensate continuity equation (8.63) and the expression for the condensate growth rate Γ_3, Eq. (8.46). To prove this identity one needs to interchange $\mathbf{k}$ and $\mathbf{p}$ summation indices in the Bogoliubov channel of $I_3^{\rm coll}$, given by the third line of Eq. (8.73). Then it cancels half of the normal channel contribution, given by the second line. Since the nucleation term $\sim \Gamma_3^{\rm K}$ originates from the fluctuations, which were not accounted for in the kinetic equation, it is not surprising that it was missed and had to be restored.

Thanks to Eq. (6.59), the collision integral (8.73) is nullified identically by the equilibrium distribution function $\tilde{F}(\mathbf{k}) = \tilde{F}^{\rm eq}(\epsilon_{\mathbf{k}}) = \coth \epsilon_{\mathbf{k}}/2T$. Notice, however, that the equilibrium distribution $\coth(\epsilon_{\mathbf{k}}+\Lambda)/2T$ with a positive energy offset does *not* nullify the three-particle collision integral. Indeed, such an energy offset violates the energy balance specified by the delta-functions. The latter are fixed by the assumption of the slowness of the condensate field $\Phi_0(\mathbf{r},t)$. In other words, fixing the gauge where the condensate field is slow, fixes also the "floor" of energy. This is consistent with the fact that for such a quasiparticle distribution the condensate amplitude is not stable. It collapses with a rate given by Eq. (8.49). Therefore, the quasiparticle distribution function cannot be stable either.

The three-particle collisions involve a particle absorbed or emitted by the condensate. This is the origin of the factor of $\rho_0(x)$ on the right-hand side of Eq. (8.73). They are thus absent in the absence of the condensate and are relatively ineffective at early stages of the condensate formation. In these instances one has to resort to four-particle collisions to describe the evolution of the quasiparticle distribution. The four-quasiparticle self-energy in second order in the interaction constant is given by the square of $S_4^{\rm coll}$, Eq. (8.23). In the Popov approximation there are five diagrams for the Keldysh component $\Sigma_4^{\rm K}(x,x')$, Fig. 8.5(a)–(e), and four for the retarded component $\Sigma_4^{\rm R}(x,x')$, Fig. 8.5(f)–(i). Utilizing causality relations, one finds

$$\begin{aligned}\Sigma_4^{\rm K}(x,x') = -\frac{g^2}{8}\Big[4G^{\rm K}_{x',x}\Big(G^{\rm K}_{x,x'}G^{\rm K}_{x,x'} + (G^{\rm R}_{x,x'} - G^{\rm A}_{x,x'})(G^{\rm R}_{x,x'} - G^{\rm A}_{x,x'})\Big)\\ - 8\,(G^{\rm R}_{x',x} - G^{\rm A}_{x',x})(G^{\rm R}_{x,x'} - G^{\rm A}_{x,x'})G^{\rm K}_{x,x'}\Big];\end{aligned} \tag{8.75}$$

$$\begin{aligned}\Sigma_4^{\rm R}(x,x') = -\frac{g^2}{8}\Big[4G^{\rm A}_{x',x}\Big(G^{\rm K}_{x,x'}G^{\rm K}_{x,x'} + (G^{\rm R}_{x,x'} - G^{\rm A}_{x,x'})(G^{\rm R}_{x,x'} - G^{\rm A}_{x,x'})\Big)\\ + 8\,G^{\rm K}_{x',x}G^{\rm R}_{x,x'}G^{\rm K}_{x,x'}\Big].\end{aligned} \tag{8.76}$$

A Wigner transformation with the energy–momentum assignments indicated in Fig. 8.5, followed by energy integrations, leads to

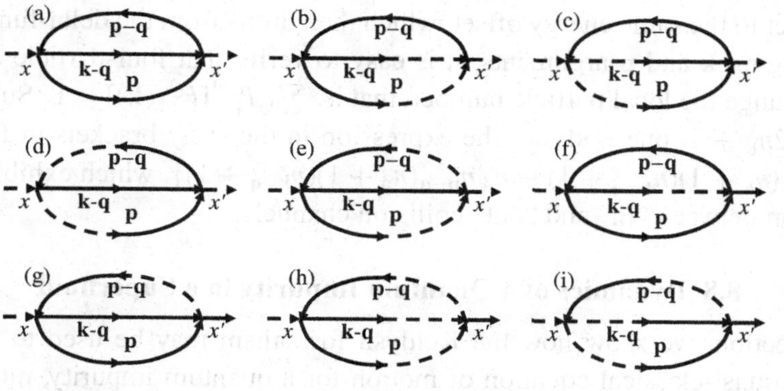

Figure 8.5 (a)–(e) Five diagrams for $\Sigma_4^{\mathrm{K}}(x,x')$. Diagrams (a)–(c) carry a combinatorial factor of 4, (d)–(e) carry a factor of 8. (f)–(i) Four diagrams for $\Sigma_4^{\mathrm{R}}(x,x')$: (f) carries a factor of 8, and (g)–(i) carry a factor 4.

$$\Sigma_4^{\mathrm{K}}(\mathbf{k}) = \frac{\pi g^2}{\mathrm{i}} \sum_{\mathbf{p}\,\mathbf{q}} \delta(\epsilon_{\mathbf{p}} + \epsilon_{\mathbf{k-q}} - \epsilon_{\mathbf{k}} - \epsilon_{\mathbf{p-q}})$$
$$\times \big[\tilde F(\mathbf{p-q})\big(\tilde F(\mathbf{p})\tilde F(\mathbf{k-q}) + 1\big) - \tilde F(\mathbf{p}) - \tilde F(\mathbf{k-q})\big]; \tag{8.77}$$

$$\Sigma_4^{\mathrm{R}}(\mathbf{k}) = \frac{\pi g^2}{2} \sum_{\mathbf{p}\,\mathbf{q}} \delta(\epsilon_{\mathbf{p}} + \epsilon_{\mathbf{k-q}} - \epsilon_{\mathbf{k}} - \epsilon_{\mathbf{p-q}})$$
$$\times \big[\tilde F(\mathbf{p})\tilde F(\mathbf{k-q}) + 1 - \tilde F(\mathbf{p-q})\big(\tilde F(\mathbf{p}) + \tilde F(\mathbf{k-q})\big)\big], \tag{8.78}$$

where we have suppressed the central coordinate x dependence of Σ and $\tilde F$. The last terms on the right-hand sides of these expressions are symmetrized with respect to $\mathbf{p}$ and $\mathbf{k-q}$ arguments; see Fig. 8.5(d)–(f). Employing Eq. (6.47), one finally obtains $I_4^{\mathrm{coll}}[\tilde F(\mathbf{k})] = \mathrm{i}\Sigma_4^{\mathrm{K}}(\mathbf{k}) + 2\tilde F(\mathbf{k})\,\mathrm{Im}\big[\Sigma_4^{\mathrm{R}}(\mathbf{k})\big]$ for the four-particle part of the collision integral

$$I_4^{\mathrm{coll}}[\tilde F(x,\mathbf{k}), \rho_0(x), \mathbf{v}_{\mathrm{sf}}(x)] = \pi g^2 \sum_{\mathbf{p}\,\mathbf{q}} \delta(\epsilon_{\mathbf{p}} + \epsilon_{\mathbf{k-q}} - \epsilon_{\mathbf{k}} - \epsilon_{\mathbf{p-q}})$$
$$\times \Big\{\big[\tilde F(x,\mathbf{p})\tilde F(x,\mathbf{k-q}) + 1\big]\big[\tilde F(x,\mathbf{p-q}) + \tilde F(x,\mathbf{k})\big]$$
$$- \big[\tilde F(x,\mathbf{k})\tilde F(x,\mathbf{p-q}) + 1\big]\big[\tilde F(x,\mathbf{k-q}) + \tilde F(x,\mathbf{p})\big]\Big\}, \tag{8.79}$$

where dependence on ρ_0 and $\mathbf{v}_{\mathrm{sf}}$ enters through the quasiparticle dispersion relation, Eq. (8.66) or (8.68). This collision integral is nullified by the equilibrium distribution function *with or without* the energy offset Λ, that is, by $\coth(\epsilon_{\mathbf{k}} + \Lambda)/2T$; see Eq. (6.59). Indeed, two incoming particles $\mathbf{k}$ and $\mathbf{p-q}$ as well as two outgoing ones $\mathbf{p}$ and $\mathbf{k-q}$ are all above the condensate and therefore are subject to the same energy offset, which then drops from the delta-function. By interchanging **k** and **p** arguments, it is easy to verify that four-particle collisions do not change the local particle number, that is, $\sum_{\mathbf{k}} I_4^{\text{coll}}[\tilde{F}(x,\mathbf{k})] = 0$. Substituting $\tilde{F}(\mathbf{k}) = 2n_{\mathbf{k}} + 1$, one finds for the expression in the curly brackets in Eq. (8.79) $8\{n_{\mathbf{p}}n_{\mathbf{k}-\mathbf{q}}(n_{\mathbf{k}}+1)(n_{\mathbf{p}-\mathbf{q}}+1) - n_{\mathbf{k}}n_{\mathbf{p}-\mathbf{q}}(n_{\mathbf{p}}+1)(n_{\mathbf{k}-\mathbf{q}}+1)\}$, which exhibits a clear distinction between "in" and "out" collision channels.

8.8 Dynamics of a Quantum Impurity in a Superfluid

In this section, we show how the Keldysh formalism may be used to derive an effective quasi-classical equation of motion for a quantum impurity, moving in a superfluid environment. Such an impurity may be a He^3 atom, or a slow neutron moving in He^4. Alternatively it may be a spin-flipped atom in a cold gas of spin-polarized Bose atoms. Having in mind the latter case, we restrict ourselves to a weakly interacting superfluid, while the He^4 case was considered in [134]. According to the Landau criterion, if the impurity is moving with a velocity which is less than the critical one (for a weakly interacting gas the critical velocity is close to the speed of sound c), one may expect that it does not radiate Bogoliubov phonons. However, as was first understood by Landau and Khalatnikov [135], the radiation is still possible via two-phonon Raman scattering. The idea is illustrated in Fig. 8.6(a), which shows that by absorbing and then re-emitting a phonon the final state of the impurity may satisfy the energy and momentum conservation laws. For such processes to occur one needs a finite concentration of phonons in the superfluid, namely a finite temperature T. Here we shall assume it to be small, $T \ll mc^2$.

We describe the quantum impurity with mass m_{d} and coordinate $\mathbf{R}(t)$ by a Feynman path integral with the action (3.4) on a closed time contour. The short-range interaction of the impurity with the superfluid density $\rho \sim |\phi|^2$ is described by the action $S_{\text{d-sf}} = g_{\text{d}} \int_C dt\, d\mathbf{r}\, \rho_{\text{d}}(\mathbf{r},t)|\phi(\mathbf{r},t)|^2$, where the impurity density is $\rho_{\text{d}}(\mathbf{r},t) = \delta(\mathbf{r} - \mathbf{R}(t))$.[1] It is convenient to parametrize the Bose field by its density and phase fluctuations, $\phi(\mathbf{r},t) = \sqrt{\rho_0 + \varrho(\mathbf{r},t)}\, e^{i\theta(\mathbf{r},t)}$; see Eqs. (8.12) and (8.13). The superfluid action, Eqs. (6.9), (6.13), and (8.1), acquires the form

$$S = \int_C dt\, d\mathbf{r} \left[\theta \partial_t \varrho - \frac{m\rho_0}{2}\, \mathbf{v}_{\text{sf}}^2 - \frac{g}{2}\, \varrho^2 - \frac{1}{8m\rho_0}\, (\nabla_{\mathbf{r}}\varrho)^2 - \frac{m}{2}\, \varrho \mathbf{v}_{\text{sf}}^2 \right], \qquad (8.80)$$

where $\mathbf{v}_{\text{sf}} = \nabla_{\mathbf{r}}\theta/m$. The first four terms on the right-hand side are the Lagrangian of the Bogoliubov phonons with the dispersion relation (8.11) and the speed of sound $c = \sqrt{g\rho_0/m}$. The last term is the nonlinear three-phonon interaction, and

[1] The impurity also interacts with the local superfluid velocity $\mathbf{v}_{\text{sf}}$ [134]. One can neglect this effect as long as the effective mass of the impurity is close to its bare mass. This is the case for weak-enough interactions, $g_d\rho_0 \ll T_c$.

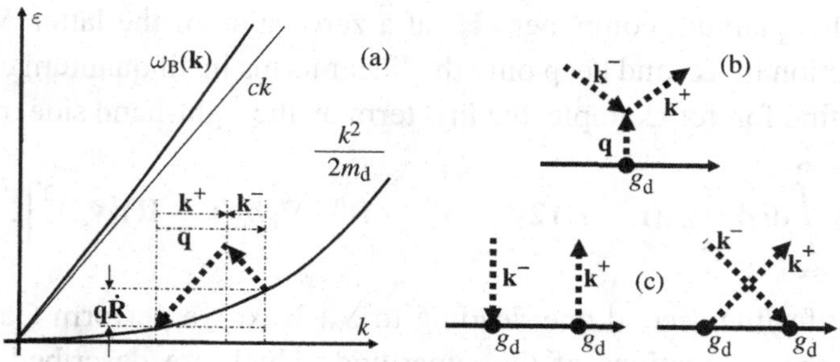

Figure 8.6 (a) Two-phonon Raman processes, which conserve energy and momentum of the slow impurity. Two-phonon amplitudes Γ_θ (b) and Γ_ϱ (c).

we have neglected other nonlinear terms, such as, for example, $\varrho(\nabla_{\mathbf{r}}\varrho)^2$, as being less important at $T \ll mc^2$.

The interaction with the impurity is now given by $S_{\text{d-sf}} = g_d \int_C dt\, d\mathbf{r}\, \rho_d \varrho$, which describes one phonon excitation by the impurity. As Fig. 8.6(a) shows, the real processes must involve two phonons and thus should be described by the composite two-phonon vertices depicted in Fig. 8.6(b), (c). In the limit of small temperature $T \ll mc^2$, the characteristic phonon wavelength c/T is much larger than the condensate *healing* (or correlation) length $(mc)^{-1}$ and therefore the two-phonon vertices may be considered as local. As a result, the effective impurity–superfluid interaction action takes the form

$$S_{\text{d-sf}} = \frac{1}{2}\int_C dt\, d\mathbf{r} \left[\Gamma_\theta\, \rho_d(\mathbf{r} - \mathbf{R}(t))\mathbf{v}_{\text{sf}}^2(\mathbf{r}, t) - \Gamma_\varrho\, \rho_d(\mathbf{r} - \mathbf{R}(t))\varrho^2(\mathbf{r}, t)\right], \quad (8.81)$$

where $\Gamma_\theta = g_d\rho_0/c^2$ and $\Gamma_\varrho = (g_d^2/m_d c^2)\cos\alpha$, where α is the angle between $\mathbf{k}^+$ and $\mathbf{k}^-$, that is, $\cos\alpha = \mathbf{k}^+ \cdot \mathbf{k}^-/k^+k^-$.[2]

To proceed we split the fields $\varrho(\mathbf{r}, t)$, $\theta(\mathbf{r}, t)$, and $\mathbf{R}(t)$ into those residing on forward and backward branches of the contour and perform a Keldysh rotation as, for example, $\mathbf{R}^\pm = \mathbf{R}^{\text{cl}} \pm \mathbf{R}^{\text{q}}$, and similarly for ϱ and θ. The semiclassical equation of motion for the impurity coordinate $\mathbf{R}^{\text{cl}}(t) = \mathbf{R}(t)$ is obtained by variation of the

[2] Since the intermediate virtual lines in the diagrams of Fig. 8.6(b), (c) are far from the mass-shell, they do not involve Keldysh Green functions and may be evaluated using the usual second-order perturbation theory. This way, one finds $\Gamma_\theta = g_d\langle\varrho\varrho\rangle m$; see Fig. 8.6(b), where according to Eq. (8.80) $\langle\varrho\varrho\rangle = (\rho_0/m)q^2/(\omega_B^2(\mathbf{q}) - \omega^2)$. As shown in what follows, $\omega \approx \mathbf{q}\cdot\dot{\mathbf{R}}$ and therefore $\omega_B(\mathbf{q}) \approx cq \gg \omega$, one finds that $\langle\varrho\varrho\rangle \approx \rho_0/(mc^2)$ and thus $\Gamma_\theta = g_d\rho_0/c^2$. One also finds that $\Gamma_\varrho = g_d^2\times$

$$\left[\frac{1}{\epsilon_{\mathbf{p}} + \epsilon^- - \epsilon_{\mathbf{p}+\mathbf{k}^-}} + \frac{1}{\epsilon_{\mathbf{p}} - \epsilon^+ - \epsilon_{\mathbf{p}-\mathbf{k}^+}}\right] \approx \frac{(\mathbf{k}^+ - \mathbf{k}^-)^2 - (k^+)^2 - (k^-)^2}{2m_d\epsilon^-(-\epsilon^+)} \approx \frac{\mathbf{k}^+\cdot\mathbf{k}^-}{m_d c^2 k^+ k^-},$$

where $\epsilon_{\mathbf{p}} = p^2/(2m_d)$ and one uses the mass-shell conditions $\epsilon_{\mathbf{p}} + \epsilon^- - \epsilon^+ = \epsilon_{\mathbf{p}+\mathbf{k}^- - \mathbf{k}^+}$ along with $\epsilon^\pm = \omega_B(\mathbf{k}^\pm) \gg |\epsilon_{\mathbf{p}} - \epsilon_{\mathbf{p}\mp\mathbf{k}^\pm}|$. In general, the amplitudes Γ_θ and Γ_ϱ may be expressed through the thermodynamic compressibility of the superfluid and effective mass of the impurity, without relying on the weakness of the interactions [134, 136].

action over the quantum component $\mathbf{R}^{\mathrm{q}}$ at a zero value of the latter. We can thus expand the action in $\mathbf{R}^{\mathrm{q}}$ and keep only the linear terms in all quantum components. This way we find for, for example, the first term on the right-hand side of Eq. (8.81),

$$S_\theta = \Gamma_\theta \int_{-\infty}^{\infty} dt\, d\mathbf{r} \left[\rho_{\mathrm{d}}(\mathbf{r}-\mathbf{R})\, 2\mathbf{v}_{\mathrm{sf}}^{\mathrm{cl}} \cdot \mathbf{v}_{\mathrm{sf}}^{\mathrm{q}} - \mathbf{R}^{\mathrm{q}} \cdot \nabla_{\mathbf{r}} \rho_{\mathrm{d}}(\mathbf{r}-\mathbf{R}) \left(\mathbf{v}_{\mathrm{sf}}^{\mathrm{cl}}\right)^2 \right], \tag{8.82}$$

and similarly for the second one, leading to S_ϱ. Next we perform Gaussian integration over the fluctuations of the superfluid, which are described by the first four terms in Eq. (8.80). The last nonlinear term was already used to derive the effective two-phonon amplitude Γ_θ and may be disregarded hereafter. Expanding in powers of $\mathrm{i}(S_\theta + S_\varrho)$ and performing Gaussian integrations, one notices that $\langle S_\theta \rangle = \langle S_\varrho \rangle = 0$ (because of causality and integration of the full derivatives), where the angular brackets stay for the integration over ϱ and θ. The first nonvanishing contribution to the equation of motion comes thus from $S_{\mathrm{eff}}[\mathbf{R}, \mathbf{R}^{\mathrm{q}}] = \frac{\mathrm{i}}{2}\langle (S_\theta + S_\varrho)^2 \rangle$. Keeping again only the term linear in $\mathbf{R}^{\mathrm{q}}$, one finds for, for example,

$$\begin{aligned}\left\langle (S_\theta)^2 \right\rangle = -2\Gamma_\theta^2 \int_{-\infty}^{\infty} dt_1 d\,\mathbf{r}_1 dt_2 d\mathbf{r}_2 \Big[&\mathbf{R}^{\mathrm{q}}(t_1) \nabla_{\mathbf{r}_1} \rho_{\mathrm{d}}(\mathbf{r}_1 - \mathbf{R}(t_1)) \rho_{\mathrm{d}}(\mathbf{r}_2 - \mathbf{R}(t_2)) \\ &\times \left\langle \left(\mathbf{v}_{\mathrm{sf}}^{\mathrm{cl}}(x_1)\right)^2 \mathbf{v}_{\mathrm{sf}}^{\mathrm{cl}}(x_2) \mathbf{v}_{\mathrm{sf}}^{\mathrm{q}}(x_2) \right\rangle + (\mathbf{r}_1 \leftrightarrow \mathbf{r}_2, t_1 \leftrightarrow t_2) \Big],\end{aligned} \tag{8.83}$$

where $x_i = (\mathbf{r}_i, t_i)$. We now introduce $t = (t_1 + t_2)/2$ and $\tau = t_1 - t_2$, that is, the central and relative times, and approximate the impurity trajectory as $\mathbf{R}(t_{1,2}) \approx \mathbf{R}(t) \pm \tau \dot{\mathbf{R}}(t)/2$. In the quantum component one may disregard the relative time, putting $\mathbf{R}^{\mathrm{q}}(t_{1,2}) \approx \mathbf{R}^{\mathrm{q}}(t)$. According to the Wick theorem $\langle (\mathbf{v}_{\mathrm{sf}}^{\mathrm{cl}}(x_1))^2 \mathbf{v}_{\mathrm{sf}}^{\mathrm{cl}}(x_2) \mathbf{v}_{\mathrm{sf}}^{\mathrm{q}}(x_2) \rangle = -(2/m^4) \nabla_{\mathbf{r}_1}^\alpha \nabla_{\mathbf{r}_2}^\beta D_{\theta\theta}^{\mathrm{R}}(x_1 - x_2) \nabla_{\mathbf{r}_2}^\beta \nabla_{\mathbf{r}_1}^\alpha D_{\theta\theta}^{\mathrm{K}}(x_2 - x_1)$, where $D_{\theta\theta}^{\mathrm{R}}(x_1 - x_2) = -\mathrm{i}\langle \theta^{\mathrm{cl}}(x_1) \theta^{\mathrm{q}}(x_2) \rangle$ and $D_{\theta\theta}^{\mathrm{K}}(x_1 - x_2) = -\mathrm{i}\langle \theta^{\mathrm{cl}}(x_1) \theta^{\mathrm{cl}}(x_2) \rangle$ in accordance with the standard definitions. Finally, we go to the momentum representation, using $\rho_{\mathrm{d}}(\mathbf{r} - \mathbf{R}(t)) = \sum_{\mathbf{q}} \mathrm{e}^{\mathrm{i}\mathbf{q}(\mathbf{r} - \mathbf{R}(t))}$, and obtain

$$\left\langle (S_\theta)^2 \right\rangle = 4\mathrm{i} \int_{-\infty}^{\infty} dt\, \mathbf{R}^{\mathrm{q}}(t) \cdot \sum_{\mathbf{q}} \mathbf{q}\, \Pi_{\theta\theta}\left(\mathbf{q}, \mathbf{q} \cdot \dot{\mathbf{R}}(t)\right),$$

where

$$\begin{aligned}\Pi_{\theta\theta}(\mathbf{q}, \omega) &= \frac{\Gamma_\theta^2}{m^4} \int_{-\infty}^{\infty} d\tau\, d\mathbf{r}\, \mathrm{e}^{\mathrm{i}(\mathbf{q}\cdot\mathbf{r} - \omega\tau)} \nabla_{\mathbf{r}}^\alpha \nabla_{\mathbf{r}}^\beta \left[D_{\theta\theta}^{\mathrm{R}}(\mathbf{r}, \tau) - D_{\theta\theta}^{\mathrm{A}}(\mathbf{r}, \tau) \right] \nabla_{\mathbf{r}}^\beta \nabla_{\mathbf{r}}^\alpha D_{\theta\theta}^{\mathrm{K}}(\mathbf{r}, \tau) \\ &= \frac{\Gamma_\theta^2}{m^4} \int \frac{d\epsilon\, d\mathbf{k}}{(2\pi)^{d+1}} (\mathbf{k}^+ \mathbf{k}^-) \left[D_{\theta\theta}^{\mathrm{R}}(\mathbf{k}^+, \epsilon^+) - D_{\theta\theta}^{\mathrm{A}}(\mathbf{k}^+, \epsilon^+) \right] \times (\mathbf{k}^+ \mathbf{k}^-) D_{\theta\theta}^{\mathrm{K}}(\mathbf{k}^-, \epsilon^-),\end{aligned}$$

where $\mathbf{k}^{\pm} = \mathbf{k} \pm \mathbf{q}/2$, $\epsilon^{\pm} = \epsilon \pm \omega/2$, and d is the dimension. The $D^{\mathrm{A}}_{\theta\theta}$ term originates from the last term on the right-hand side of Eq. (8.83) with interchanged arguments. Performing similar calculations for $\langle (S_{\varrho})^2 \rangle$ and $2\langle S_{\varrho} S_{\theta} \rangle$, one finds

$$S_{\mathrm{eff}}[\mathbf{R}, \mathbf{R}^{\mathrm{q}}] = \frac{\mathrm{i}}{2} \langle (S_{\theta} + S_{\varrho})^2 \rangle = -2 \int_{-\infty}^{\infty} dt\, \mathbf{R}^{\mathrm{q}}(t) \cdot \sum_{\mathbf{q}} \mathbf{q}\, \Pi\big(\mathbf{q}, \mathbf{q} \cdot \dot{\mathbf{R}}(t)\big), \tag{8.84}$$

where

$$\Pi(\mathbf{q}, \omega) = \int \frac{d\epsilon\, d\mathbf{k}\, (\mathbf{k}^{+}\mathbf{k}^{-})^2}{(2\pi)^{d+1}} \operatorname{Tr} \left\{ \hat{\Gamma} \big[\hat{D}^{\mathrm{R}}(\mathbf{k}^{+}, \epsilon^{+}) - \hat{D}^{\mathrm{A}}(\mathbf{k}^{+}, \epsilon^{+}) \big] \hat{\Gamma} \hat{D}^{\mathrm{K}}(\mathbf{k}^{-}, \epsilon^{-}) \right\}. \tag{8.85}$$

The matrix Green functions of the linearized phonon modes are found from Eq. (8.80) as

$$\hat{D}^{\mathrm{R(A)}}(\mathbf{k}, \epsilon) = \begin{pmatrix} D^{\mathrm{R(A)}}_{\theta,\theta} & D^{\mathrm{R(A)}}_{\theta,\varrho/k} \\ D^{\mathrm{R(A)}}_{\varrho/k,\theta} & D^{\mathrm{R(A)}}_{\varrho/k,\varrho/k} \end{pmatrix} = \frac{1}{2} \frac{1}{(\epsilon \pm \mathrm{i}0)^2 - (ck)^2} \begin{pmatrix} g & \mathrm{i}\epsilon/k \\ -\mathrm{i}\epsilon/k & \rho_0/m \end{pmatrix}, \tag{8.86}$$

where we have neglected the curvature of the Bogoliubov dispersion relation. The matrix Keldysh component of the Green function in thermal equilibrium is given by FDT $\hat{D}^{\mathrm{K}}(\mathbf{k}, \epsilon) = \coth(\epsilon/2T)\big[\hat{D}^{\mathrm{R}}(\mathbf{k}, \epsilon) - \hat{D}^{\mathrm{A}}(\mathbf{k}, \epsilon)\big]$. Finally, the vertex matrix is defined as $\hat{\Gamma} = \mathrm{diag}\{g_{\mathrm{d}}\rho_0/(m^2c^2), -g_{\mathrm{d}}^2/(m_{\mathrm{d}}c^2)\}$.

We notice that both involved phonons are real, that is, their energies and momenta are related through the dispersion relation, $\epsilon = \pm ck$. This is due to the fact that both $\hat{D}^{\mathrm{R}} - \hat{D}^{\mathrm{A}}$ and $\hat{D}^{\mathrm{K}}$ are proportional to $[\delta(\epsilon - ck) - \delta(\epsilon + ck)]$.[3] This fact was already anticipated in calculating the amplitudes Γ_{θ} and Γ_{ϱ} previously. Adding the free part (3.4) and Eq. (8.84), we find for the action of the slow, $\dot{R} \ll c$, impurity in the superfluid

[3] Taking the trace and performing the energy integration in Eq. (8.85) with the help of delta-functions, one finds

$$\Pi(\mathbf{q}, \omega) = \frac{\pi}{8} \left(\frac{g_{\mathrm{d}}}{m} - \frac{g_{\mathrm{d}}^2}{g m_{\mathrm{d}}} \right)^2 \int \frac{d\mathbf{k}}{(2\pi)^d} \frac{(\mathbf{k}^{+}\mathbf{k}^{-})^2}{c^2 k^{+} k^{-}} \coth \frac{ck^{-}}{2T} \big[\delta\left(ck^{+} - ck^{-} - \omega\right) - \delta\left(ck^{+} - ck^{-} + \omega\right) \big]$$

$$\approx \frac{1}{64\pi} \left(\frac{g_{\mathrm{d}}}{m} - \frac{g_{\mathrm{d}}^2}{g m_{\mathrm{d}}} \right)^2 \frac{\omega}{c^3 T} \int_0^{\infty} k^2 dk \frac{(k^2 - q^2/4)^2}{k^2 + q^2/4} \sinh^{-2} \left(\frac{c}{2T} \sqrt{k^2 + \frac{q^2}{4}} \right) \frac{\sqrt{k^2 + q^2/4}}{kq},$$

where we put $d = 3$, performed the angular integration using $k^{\pm} = \sqrt{k^2 + q^2/4 \pm kq \cos\vartheta}$ and expanded to first order in $\omega = \mathbf{q} \cdot \dot{\mathbf{R}} \ll cq$. Introducing dimensionless variables $x = ck/(2T)$ and $y = cq/(4T)$, one finds

$$\sum_{\mathbf{q}} \mathbf{q}\, \Pi(\mathbf{q}, \mathbf{q} \cdot \dot{\mathbf{R}}) = \dot{\mathbf{R}} \frac{64}{3\pi^3} \frac{T^8}{c^{12}} \left(\frac{g_{\mathrm{d}}}{m} - \frac{g_{\mathrm{d}}^2}{g m_{\mathrm{d}}} \right)^2 \iint_0^{\infty} dy dx\, y^3 x \frac{(x^2 - y^2)^2}{\sqrt{x^2 + y^2}} \sinh^{-2} \sqrt{x^2 + y^2}.$$

The double integral may be calculated in polar coordinates, bringing in a factor of $\pi^8/(12 \times 30)$.

$$S_{\mathrm{imp}}[\mathbf{R},\mathbf{R}^{\mathrm{q}}]=\int_{-\infty}^{\infty}\mathrm{d}t\left[-2\,\mathbf{R}^{\mathrm{q}}\cdot\left(m_{\mathrm{d}}\ddot{\mathbf{R}}+\gamma\dot{\mathbf{R}}+\nabla_{\mathbf{R}}V(\mathbf{R})\right)+4\mathrm{i}\gamma T\;\mathbf{R}^{\mathrm{q}}\cdot\mathbf{R}^{\mathrm{q}}\right],\quad(8.87)$$

where the friction coefficient is given by

$$\gamma=\frac{8\pi^5}{135}\frac{T^8}{c^{12}}\left(\frac{g_{\mathrm{d}}}{m}-\frac{g_{\mathrm{d}}^2}{gm_{\mathrm{d}}}\right)^2.\quad(8.88)$$

The fluctuation term $\sim \mathbf{R}^{\mathrm{q}}\cdot\mathbf{R}^{\mathrm{q}}$ on the right-hand side of Eq. (8.87) may be derived in exactly the same way as the viscous term $\sim \mathbf{R}^{\mathrm{q}}\cdot\dot{\mathbf{R}}$ was dealt with earlier. However, we can simply rely on FDT to write it down. The classical (i.e. local in time) form of this term follows if one approximates $\mathbf{R}^{\mathrm{q}}(t_1)\cdot\mathbf{R}^{\mathrm{q}}(t_2)\approx\mathbf{R}^{\mathrm{q}}(t)\cdot\mathbf{R}^{\mathrm{q}}(t)$, which is the case as long as the characteristic frequency of the impurity motion is much less than the temperature. Comparing the resulting action (8.87) with Eq. (4.2), we notice that the slow impurities are executing classical viscous Brownian motion. The latter may be described either by the Langevin equation (4.5) or the Fokker–Planck equation (4.32). In the long-time limit the latter is reduced to the diffusion equation (4.34) with the diffusion coefficient $D=T/\gamma\propto T^{-7}$ (Einstein relation). On the other hand, the diffusion coefficient in momentum space is $D_K=\gamma T\propto T^9$; see Eq. (4.32). Such high powers of the temperature dependence are due to the two-phonon nature of the process. In d dimensions it is given by $\gamma\propto T^{2d+2}$ and $D\propto T^{-2d-1}$. We found thus that the superfluid acts on the slow impurity as an effective *Ohmic* bath, introduced phenomenologically in Section 3.2. The advantage of the microscopic calculation is access to the value and the peculiar temperature dependence of the friction coefficient, Eq. (8.88).

The last factor in Eq. (8.88) is a result of the destructive interference between the two-phonon processes depicted in Figs. 8.6(b), (c). One may show [136] that in $d=1$ and in the symmetric case $g_{\mathrm{d}}=g$, $m_{\mathrm{d}}=m$, the cancelation is exact in all orders of the perturbation theory and thus $\gamma=0$. This fact is due to the exact integrability of the corresponding symmetric 1D model with local interactions [137]. The exact integrability prohibits the dissipative friction and thermalization of the impurity because of the presence of an infinite number of conservation laws. In $d>1$ one finds a nonzero γ even in the symmetric case in higher orders in g_{d} and g [134].

8.9 Problems

8.9.1 Condensate in a Harmonic Trap

Solve the static Gross–Pitaevskii equation (8.7) in an isotropic harmonic trap, $V^{\mathrm{cl}}(\mathbf{r})=m\omega_0^2r^2/2$, in the Thomas–Fermi approximation (i.e. neglecting $\nabla_{\mathbf{r}}^2/2m$

term). Show that the condensate density is given by $\rho_0(\mathbf{r}) = (m\omega_0^2/2g)(R^2 - r^2)$ (remember that in the present notations $|\Phi_0(\mathbf{r})|^2 = 2\rho_0(\mathbf{r})$), where the condensate radius, R, is related to the chemical potential as $\mu = m\omega_0^2 R^2/2$. Employ the normalization condition $\int \mathrm{d}^d\mathbf{r}\, \rho_0(\mathbf{r}) = N$ to find the chemical potential $\mu(N)$, the condensate radius $R(N)$, and the density in the center $\rho_0(0, N)$ as functions of N in $d = 1, 2, 3$.

Compare the condensate radius $R(N)$ with that of the Gaussian ground-state density of N noninteracting bosons in the same harmonic trap. What is a condition on N and the interaction parameter g for $R(N)$ to exceed the size of the Gaussian ground-state?

Evaluate the interaction energy as $E_{\text{int}} = \int \mathrm{d}^d\mathbf{r}\, [(V^{\text{cl}}(\mathbf{r}) - \mu)\rho_0(\mathbf{r}) + g\rho_0^2(\mathbf{r})/2]$ and verify that $\partial E_{\text{int}}(\mu)/\partial\mu = -N$.

8.9.2 Collective Modes of the Condensate in the Harmonic Trap

In a translationally invariant setting, the linearized dynamic fluctuations of the Gross–Pitaevskii equation are Bogoliubov quasiparticles with the continuous spectrum (8.11); see Eqs. (8.8)–(8.11). Find a spectrum of linearized excitations in the isotropic harmonic trap, considered in Problem 8.9.1 (Stringari [138]).

To this end, linearize the hydrodynamic equations (8.12), (8.13) for $\mathbf{v}_{\text{sf}}(\mathbf{r}, t)$ and $\varrho(\mathbf{r}, t) = \rho(\mathbf{r}, t) - \rho_0(\mathbf{r})$, neglecting the quantum pressure term. This leads to

$$\partial_t\varrho + \nabla_{\mathbf{r}}\big(\rho_0\mathbf{v}_{\text{sf}}\big) = 0; \qquad \partial_t\mathbf{v}_{\text{sf}} = -\frac{g}{m}\nabla_{\mathbf{r}}\varrho. \tag{8.89}$$

From here one finds $\partial_t^2\varrho = (g/m)\nabla_{\mathbf{r}}\big(\rho_0\nabla_{\mathbf{r}}\varrho\big)$. Finally, using results of Problem 8.9.1 for $\rho_0(\mathbf{r})$ and looking for harmonics in time oscillations $\varrho(\mathbf{r}, t) = \varrho(\mathbf{r})\,\mathrm{e}^{\mathrm{i}\omega t}$, one obtains

$$-\frac{\omega_0^2}{2}\nabla_{\mathbf{r}}\big((R^2 - r^2)\nabla_{\mathbf{r}}\varrho(\mathbf{r})\big) = \omega^2\varrho(\mathbf{r}). \tag{8.90}$$

One can look for solutions of this equation in the spherical coordinates in the form $\varrho(\mathbf{r}) = \varrho(r)Y_l^m(\theta, \varphi)$, where $Y_l^m(\theta, \varphi)$ are standard spherical harmonics. For example, the $l = 0$ s-wave part of the equation takes the form (here $r/R \to r$)

$$-(1 - r^2)\,\partial_r^2\varrho - \left(\frac{2}{r} - 4r\right)\partial_r\varrho - \frac{2\omega^2}{\omega_0^2}\varrho = 0. \tag{8.91}$$

Its solution, bounded at $r = 1$, is $\varrho(r) = r^{-1}P_\lambda(r)$ – the Legendre function of the first kind, where $\lambda = \left(\sqrt{9 + 8\omega^2/\omega_0^2} - 1\right)/2$. This solution is finite at $r = 0$, only if $\lambda = 2n_r + 1$, where $n_r = 0, 1, 2, \ldots$.[4] From here one finds quantization

[4] $n_r = l = 0$ results in $\varrho = \text{const}$, which is forbidden if particle number is conserved.

of the oscillation frequencies $\omega(n_r, 0) = \omega_0\sqrt{2n_r^2 + 3n_r}$. Solve the equation for $l = 1, 2, \ldots$ to find Stringari spectrum [138]

$$\omega(n_r, l) = \omega_0\sqrt{2n_r^2 + 3n_r + 2n_r l + l}. \tag{8.92}$$

8.9.3 Dark Solitons

Consider the 1d Gross–Pitaevskii equation (8.5):

$$i\partial_t\Phi_0 = -\frac{1}{2m}\,\partial_x^2\Phi_0 + \frac{g}{2}\left(|\Phi_0|^2 - 2\rho_0\right)\Phi_0. \tag{8.93}$$

Let us look for a localized solution traveling with a certain velocity v,

$$\Phi_0(x, t) = \Phi_s(x - vt) = \sqrt{2\rho_0}\,\chi(x - vt)\,e^{i\vartheta(x-vt)}. \tag{8.94}$$

Substituting it in Eq. (8.93), one finds two equations for the phase $\vartheta(\xi)$ and the normalized amplitude $\chi(\xi) = |\Phi_s(\xi)|/\sqrt{2\rho_0}$, which are functions of $\xi = x - vt$. Show that the first of these equations acquires the form of the continuity relation

$$\left[\chi^2\left(\vartheta' - mv\right)\right]' = 0, \tag{8.95}$$

where primes denote derivatives with respect to ξ. Using the fact that far from the soliton $\chi(\pm\infty) = 1$ and $\vartheta'(\pm\infty) = 0$, one finds $\vartheta' = mv(1 - 1/\chi^2)$. Employing this relation, show that the remaining equation for the amplitude may be written in the "Newtonian" form:

$$\chi'' = -\frac{\partial U(\chi)}{\partial \chi} \tag{8.96}$$

and find the effective potential $U(\chi)$. Think about "Newtonian" particle motion in such a potential as if χ is its "coordinate" and ξ is the "time." Notice that at $\xi \to \pm\infty$ one needs to enforce $\chi = 1$. Also notice that this implies a qualitative difference for the motion with $v > c = \sqrt{g\rho_0/m}$ (the sound velocity) and $v < c$.

In the former case, $v > c$, the only way to satisfy the boundary conditions is if the "Newtonian particle" undergoes small oscillations around the minimum of $U(\chi)$ at $\chi = 1$. Find the frequency of such small oscillations $\omega = \omega(v)$ and show that it is identical to the Bogoliubov $\omega_B(q)$ (8.11), where $v = \partial\omega_B(q)/\partial q$.

In case $v < c$, $\chi = 1$ corresponds to an unstable *maximum* of $U(\chi)$. Therefore a *bounce* solution is possible. This is the *dark soliton* solution. Show that

$$\Phi_s(x - vt) = \sqrt{2\rho_0}\left[\cos\frac{\theta}{2} - i\sin\frac{\theta}{2}\,\tanh\left(\frac{x - vt}{l}\right)\right], \tag{8.97}$$

where

$$\cos(\theta/2) = v/c < 1, \tag{8.98}$$

and $l^{-1} = mc\sin(\theta/2)$ is indeed such a bounce solution. Here θ is a change of the phase of the wave function across the soliton. Show that the number of particles "expelled" from the soliton core is

$$N_s = \int dx\, \left(\rho_0 - |\Phi_s|^2/2\right) = \frac{2K}{\pi}\sin\frac{\theta}{2}, \tag{8.99}$$

where the "quantum parameter" $K = \pi\rho_0/(mc)$. Notice that the particle number may be large, $N_s \gg 1$, in the limit of the weakly interacting gas, $K \gg 1$.

In a similar way, the energy of the soliton is given by

$$E_s = \frac{1}{2}\int dx\left[\frac{1}{2m}|\partial_x\Phi_s|^2 + g\left(\rho_0 - |\Phi_s|^2/2\right)^2\right] = \frac{4\rho_0 c}{3}\sin^3\frac{\theta}{2}, \tag{8.100}$$

while the soliton *core* momentum is

$$p_{sc} = \frac{1}{2}\,\mathrm{Im}\int dx\, \Phi_s^*\partial_x\Phi_s = -\rho_0\sin\theta. \tag{8.101}$$

However, one should take into account the periodic boundary conditions that ensure that dark soliton phase shift θ is uniformly spread over the length of the entire system L. Although this does not change the energy of the system in the thermodynamic limit (indeed the corresponding contribution to the energy scales as $\rho_0\theta^2/(mL)$), it produces a finite contribution $\rho_0\theta$ to the momentum. As a result the total (core plus the rest of the condensate) momentum of the dark soliton solution is

$$p_s = p_{sc} + \rho_0\theta = \rho_0\left(\theta - \sin\theta\right). \tag{8.102}$$

Equations (8.100) and (8.102) provide an implicit form of the dark soliton dispersion relation $E_s(p_s)$. Plot this dispersion relation along with the Bogoliubov quasiparticles spectrum $\omega_B(p) = \sqrt{c^2p^2 + (p^2/2m)^2}$. These two excitations appear naturally in the Bethe ansatz solution of the Lieb–Liniger model [125]. They are known as Lieb II and Lieb I modes correspondingly. One may show [139] that the soliton dispersion (Lieb II mode) provides a lower bound for the many-body excitation spectrum.

8.9.4 Beliaev Damping

It is shown in Section 8.8 that the friction coefficient of a slow subsonic impurity in 3d superfluid scales as T^8. It thus rapidly goes to zero as $T \to 0$. The situation is very different vis-à-vis Bogoliubov quasiparticles (phonons). Since their dispersion relation (8.11) exhibits positive curvature, the phonons are *supersonic* (cf. Fig. 8.6). This allows for a phonon decay even at $T = 0$. The dominant processes leading to such decay is splitting of a phonon with momentum $\mathbf{q}$ into two phonons with momenta $\mathbf{k}$ and $\mathbf{q}-\mathbf{k}$. Here we focus on small momenta $q \ll mc$ and evaluate the corresponding decay rate.

To this end, one may employ the hydrodynamic action (8.80) and calculate renormalizations of its Gaussian part in the second order in the three-phonon interaction vertex $m\varrho\mathbf{v}_{\rm sf}^2/2$. For example, let's focus on the kinetic term $\theta\partial_t\varrho$, which, as a result of such renormalization, acquires the form $\theta\big(\partial_t + \gamma_{\rm Bel}\big)\varrho$, where $\gamma_{\rm Bel}(q)$ is the Beliaev damping rate. Show that in the second order in $m\varrho\mathbf{v}_{\rm sf}^2/2$ at $T=0$,

$$\gamma_{\rm Bel}(q) = \frac{q}{m^2}\sum_{\mathbf{k}}\int\frac{d\epsilon}{2\pi}\,{\rm Im}\langle\varrho\nabla_{\mathbf{r}}\theta\rangle^{\rm R}_{\mathbf{k},\epsilon}\,{\rm Im}\langle\nabla_{\mathbf{r}}\theta\nabla_{\mathbf{r}}\theta\rangle^{\rm R}_{\mathbf{q}-\mathbf{k},\omega_{\rm B}(q)-\epsilon}. \tag{8.103}$$

The propagators can be read out from Eq. (8.86). The ϵ-integration leads to the energy conservation of the form $\delta(\omega_{\rm B}(q)-\omega_{\rm B}(k)-\omega_{\rm B}(q-k))$. In the limit $q \ll mc$ one may approximate $\omega_{\rm B}(q) \approx cq + cq^3/k_0^2$, where $k_0^2 = 8(mc)^2$. Moreover, in this limit all three momenta $\mathbf{q}$, $\mathbf{k}$, and $\mathbf{q}-\mathbf{k}$ are almost collinear. Show that this allows one to approximate

$$\omega_{\rm B}(q)-\omega_{\rm B}(k)-\omega_{\rm B}(q-k) \approx \frac{cqk}{|q-k|}\left(1-\cos\theta_k - \frac{3(q-k)^2}{k_0^2}\right),$$

where θ_k is the angle between $\mathbf{k}$ and $\mathbf{q}$. Perform now the integrations: first over $d\cos\theta_k$ and finally over $k^2 dk$ [5] to show that [129]

$$\gamma_{\rm Bel}(q) = \frac{1}{960\pi}\frac{q^5}{m\rho_0}, \tag{8.104}$$

where you need to employ the fact that $g/mc^2 = \rho_0^{-1}$. Find second-order corrections to the other terms in the quadratic part of the action (8.80). Show that they do not change this damping rate (other than renormalizing the numerical factor). Remarkably the Beliaev phonon decay rate (8.104) is independent of the interaction strength g and the speed of sound c. The latter only shows up in the applicability regime of Eq. (8.104), through $q \ll mc$. Show that in this entire range $\gamma_{\rm Bel}(q) \ll \omega_{\rm B}(q)$ and therefore the phonon mode is very much underdamped.

[5] The latter integral acquires the form $\int_0^q dk\,k^2(q-k)^2 = q^5/30$.

9
Dynamics of Phase Transitions

The formalism of Chapter 4 is applied to spatially extended systems. We focus on the slow dynamics near the first- and second-order phase transitions. In the latter case the systems exhibit scale-invariant behavior, which is tackled with the dynamical renormalization group. We discuss both equilibrium (following the Hohenberg–Halperin classification) and essentially non-equilibrium models. Examples of the latter include reaction-diffusion systems (directed percolation plus other universality classes) as well as surface growth dynamics (the Kardar–Parisi–Zhang universality class).

9.1 Dissipative Chains and Membranes

In Chapter 4 we considered dynamics of an overdamped classical particle connected to a thermal bath. We shall generalize it now to spatially extended models, representing, for example, an elastic chain or membrane connected to a bath and subject to certain forces. Let us denote displacement of such an object as $\varphi(\mathbf{r}, t)$, where $\mathbf{r}$ is a spatial index labeling lattice sites along the d-dimensional membrane (hereafter we shall employ the continuum notation). We use scalar real displacement φ for simplicity, since generalization to vector (or complex) fields is straightforward. Dynamical equations of motion for such a field acquire a form

$$\partial_t \varphi(\mathbf{r}, t) = -\frac{\delta \mathcal{F}[\varphi]}{\delta \varphi(\mathbf{r}, t)} + \xi(\mathbf{r}, t). \tag{9.1}$$

The generalized potential $\mathcal{F}[\varphi]$, known also as Landau free energy, may be written as

$$\mathcal{F}[\varphi] = \int d\mathbf{r} \left[\frac{D}{2} \left(\nabla_{\mathbf{r}} \varphi \right)^2 + V(\varphi) \right], \tag{9.2}$$

where the first term is elastic compression energy and the second one is an external potential. The simplest form of the random force is spatially uncorrelated Gaussian white noise:

$$\langle \xi(\mathbf{r}, t)\xi(\mathbf{r}', t')\rangle = 2\,T\,\delta(\mathbf{r} - \mathbf{r}')\,\delta(t - t'). \tag{9.3}$$

The fact that the deterministic force on the right-hand side of Eq. (9.1) is of potential form is by no means the most general case. We shall consider non-potential forces later on. However, there are certain important consequences of the potential nature of the force that deserve special attention.

We can now employ the MSR method, as described in Section 4.3, to transform the problem to a functional representation. To this end we introduce a functional delta-function of the Langevin equation (9.1) and employ exponential representation of it with the help of an auxiliary field $\varphi^{\mathrm{q}}(\mathbf{r}, t)$. In doing so it is important to keep the retarded (Ito) regularization of the time derivative in Eq. (9.1) to ensure that the corresponding Jacobian is one (see Section 4.3 for more detailed discussion). Performing Gaussian averaging over the white noise (9.2), one obtains the MSR action

$$S[\varphi, \varphi^{\mathrm{q}}] = \int_{-\infty}^{+\infty} \mathrm{d}t \int \mathrm{d}\mathbf{r} \left\{-2\varphi^{\mathrm{q}}\left[\partial_t^{\mathrm{R}}\varphi - D\nabla_{\mathbf{r}}^2\varphi + V'(\varphi)\right] + 4\mathrm{i}T\left(\varphi^{\mathrm{q}}\right)^2\right\}. \tag{9.4}$$

To emphasize connections with Keldysh formalism it is convenient to split the first term on the right-hand side into two equal parts and perform integration by parts on one of them. In the discretized time representation this amounts to the transposition of the ∂_t^{R} operator, which transforms its retarded (lower triangular) form into the advanced (upper triangular) one. This leads to

$$S[\varphi, \varphi^{\mathrm{q}}] = \int \mathrm{d}t\, \mathrm{d}\mathbf{r}\, (\varphi, \varphi^{\mathrm{q}}) \begin{pmatrix} 0 & \partial_t^{\mathrm{A}} + D\nabla_{\mathbf{r}}^2 - \delta - g\varphi^2 \\ -\partial_t^{\mathrm{R}} + D\nabla_{\mathbf{r}}^2 - \delta - g\varphi^2 & 4\mathrm{i}T \end{pmatrix} \begin{pmatrix} \varphi \\ \varphi^{\mathrm{q}} \end{pmatrix}, \tag{9.5}$$

where as a potential we took

$$V(\varphi) = \frac{\delta}{2}\,\varphi^2 + \frac{g}{4}\,\varphi^4 \tag{9.6}$$

for its important applications. This form of the action could be directly derived starting from the quantum formalism of Chapter 6, by introducing the Ohmic bath of oscillators and taking the classical limit. The derivation is completely equivalent to the one presented in Section 4.1. Comparing the classical MSR action (9.5) with its quantum counterparts, for example, Eqs. (6.9) and (6.13), one may notice several differences: (i) the time derivatives come without imaginary i, reflecting the overdamped dissipative dynamics; (ii) the term $4\mathrm{i}T(\varphi^{\mathrm{q}})^2$, which is due to the noise associated with the bath, is absent in the quantum version; (iii) on the other hand, the term with $(\varphi^{\mathrm{q}})^3\varphi$ is absent in the MSR action. This is due to the fact that in the classical limit one keeps terms up to second order in φ^{q}. The ratio of the coefficients in terms (i) and (ii) is fixed by FDT. The latter is valid due to the assumed equilibrium in the bath. Notice that the action (9.5) does *not* have the form $S[\varphi^+] - S[\varphi^-]$, where $\varphi^\pm = \varphi \pm \varphi^{\mathrm{q}}$ (unlike Eqs. (6.9) and (6.13)). This is due to the fact that the time-reversal symmetry (i.e. equivalence of the two branches of the contour) is explicitly broken by integrating out the thermal bath with the continuum spectrum.

9.2 Equilibrium Statistical Mechanics

Let us first investigate the MSR action (9.4) in the stationary path (or rather stationary field) approximation. To this end we take variational derivatives of the action with respect to $\varphi(\mathbf{r},t)$ and $\varphi^{\mathrm{q}}(\mathbf{r},t)$. As explained in Section 4.4, the stationary field $\varphi^{\mathrm{q}}(\mathbf{r},t)$ appears to be pure imaginary (or zero). This means that the path of integration over $\mathbf{D}[\varphi^{\mathrm{q}}]$ should be distorted in the complex planes of $\varphi^{\mathrm{q}}(\mathbf{r},t_j)$ to pass through stationary field configurations. It is thus convenient to rename the auxiliary field as $\varphi^{\mathrm{q}}(\mathbf{r},t) = \pi(\mathbf{r},t)/(2\mathrm{i})$. The corresponding equations of motion for the fields $\varphi(\mathbf{r},t)$ and $\pi(\mathbf{r},t)$ are real and acquire the Hamiltonian structure

$$\partial_t\varphi(\mathbf{r},t) = \frac{\delta\mathcal{H}[\pi,\varphi]}{\delta\pi(\mathbf{r},t)}, \qquad \partial_t\pi(\mathbf{r},t) = -\frac{\delta\mathcal{H}[\pi,\varphi]}{\delta\varphi(\mathbf{r},t)}; \tag{9.7}$$

$$\mathcal{H}[\pi,\varphi] = \int d\mathbf{r}\left[\pi\left(D\nabla_{\mathbf{r}}^2\varphi - V'(\varphi)\right) + T\pi^2\right] = \int d\mathbf{r}\,\pi\left[-\frac{\delta\mathcal{F}[\varphi]}{\delta\varphi(\mathbf{r},t)} + T\pi\right]. \tag{9.8}$$

There is one obvious way to solve these equations by putting $\pi(\mathbf{r},t) = 0$, while $\partial_t\varphi(\mathbf{r},t) = \delta\mathcal{H}[\pi,\varphi]/\delta\pi(\mathbf{r},t)\big|_{\pi=0} = D\nabla_{\mathbf{r}}^2\varphi - V'(\varphi)$. This corresponds, of course, to noiseless relaxation dynamics of the field φ. Such a solution evolves on the zero-energy manifold $\mathcal{H} = 0$. The long-time activation dynamics of the system is described by another zero-energy manifold. For the one degree of freedom example of Section 4.4 such a manifold (curve) is given by $P = V'(X)/T$ and the corresponding activation trajectories are time-reversed from the relaxation ones. Remarkably, this solution may be directly generalized for the spatially extended *potential* model (9.1) (see also the footnote in Section 4.13). To this end, let us try to impose that the Hamiltonian (9.8) is nullified locally in every spatial point, that is,

$$\pi(\mathbf{r},t) = \frac{-D\nabla_{\mathbf{r}}^2\varphi + V'(\varphi)}{T}; \qquad \partial_t\varphi(\mathbf{r},t) = -D\nabla_{\mathbf{r}}^2\varphi + V'(\varphi), \tag{9.9}$$

where the second equality is an immediate consequence of the first one and the equation of motion for the φ-field. It is easy to check that, if Eqs. (9.9) are true, the equation of motion for the π-field is also satisfied.[1] Notice that not only is the total energy $\mathcal{H} = 0$ conserved, but it is also zero at every point. This observation is related to the fact that in the stationary distribution the probability current is zero at every point (this is true only for the potential forces, of course). Also notice that on the zero-energy activation solution (9.9) the φ-field evolution is time-reversed from the noiseless relaxation.

We can address now the long-time probability of finding the extended system in a field configuration $\varphi_0(\mathbf{r})$. It is given by the exponentiated action, calculated along the trajectory that leads from $\varphi = 0$ to $\varphi_0(\mathbf{r})$. Such a trajectory is the time-reversal of the noiseless relaxation motion from an arbitrary initial profile $\varphi_0(\mathbf{r})$ down to the lowest energy state $\varphi = 0$. Employing that $\mathcal{H} = 0$ along such a trajectory, the corresponding action (9.4) $S = \int dt(\pi\, \partial_t\varphi - \mathcal{H})$ is[2]

$$\int \mathrm{d}\mathbf{r}\mathrm{d}t\, \pi\, \partial_t\varphi = \int \mathrm{d}\mathbf{r}\mathrm{d}t\, \frac{-D\nabla_{\mathbf{r}}^2\varphi + V'(\varphi)}{T}\, \partial_t\varphi = \frac{1}{T}\int \mathrm{d}\mathbf{r}\mathrm{d}t\, \partial_t \left[\frac{D}{2}(\nabla_{\mathbf{r}}\varphi)^2 + V(\varphi)\right].$$

One can now perform the integral over time explicitly. It is given by the expression in the square brackets calculated at the upper limit, which is $\varphi_0(\mathbf{r})$. As a result, we found that the corresponding action is $S = \mathcal{F}[\varphi_0]/T$. Finally, the long-time probability of finding the system in the field configuration $\varphi_0(\mathbf{r})$ is

$$\mathcal{P}[\varphi_0] = Z^{-1}\, \mathrm{e}^{-\mathcal{F}[\varphi_0]/T}; \qquad Z = \int \mathbf{D}[\varphi]\, \mathrm{e}^{-\mathcal{F}[\varphi]/T}, \tag{9.10}$$

provided the *partition function* Z exists. This is, of course, nothing but the equilibrium statistical mechanics of the static field $\varphi(\mathbf{r})$. We have found thus that the dynamical model (9.1) with the potential forces (9.2) evolves in the long-time limit toward the stationary Boltzmann distribution (9.10) (if the potential $V(\varphi)$ is bounded from below).

Strictly speaking, so far we have established Eq. (9.10) with exponential accuracy only, that is, we did not prove that the pre-exponential factor Z^{-1} is φ_0-independent. To do this one needs to go beyond the stationary field approximation and write down the Fokker–Planck equation for the probability $\mathcal{P}([\varphi], t)$, where the square brackets indicate that probability is a *functional* of $\varphi(\mathbf{r})$ field configuration at time t. In a complete analogy with Eq. (4.30) such a Fokker–Planck equation takes the form

[1] Indeed, we first notice that $\partial_t\varphi = T\pi$, then take the time derivative of the second equation in (9.9), $\partial_t(\partial_t\varphi) = -D\nabla_{\mathbf{r}}^2\partial_t\varphi + V''(\varphi)\partial_t\varphi$, which may be written as $\partial_t\pi = -D\nabla_{\mathbf{r}}^2\pi + V''(\varphi)\pi$, which is exactly the equation of motion for the π-field.

[2] For the rest of this chapter we suppress the imaginary unit i in the MSR action.

$$\partial_t \mathcal{P}([\varphi], t) = \int d\mathbf{r} \, \frac{\delta}{\delta\varphi(\mathbf{r})} \left[\frac{\delta \mathcal{F}[\varphi]}{\delta\varphi(\mathbf{r})} \mathcal{P}([\varphi], t) + T \frac{\delta \mathcal{P}([\varphi], t)}{\delta\varphi(\mathbf{r})} \right]. \tag{9.11}$$

The expression in the square brackets is the (minus) probability current $J([\varphi], \mathbf{r}, t)$. Demanding that each component, labeled by $\mathbf{r}$, of the latter vanishes, one immediately finds the stationary distribution (9.10). This proves that the only pre-exponential factor in Eq. (9.10) is the φ_0-independent normalization constant, that is, the inverse partition function.

9.3 Critical Nucleation in First-Order Transitions

As an example of a non-equilibrium situation we consider a "supercooled" phase, beyond the point of first-order phase transition. It is described by a field trapped in a metastable potential minimum depicted in Fig. 9.1. Since the corresponding potential is not bounded from below, one can't apply the equilibrium statistical mechanics specified by Eq. (9.10). Instead, one should consider nucleation of the more stable phase, which occurs due to rare large fluctuations of the field. If the latter activates a sufficiently large region of the system beyond the potential maximum, it becomes energetically favorable for the stable phase to grow. The long-time dynamics of such a rare fluctuation is described by $\mathcal{H} = 0$ stationary field equations (9.9). To overcome the potential barrier the fluctuation must reach a saddle-point field configuration $\varphi_s(\mathbf{r})$. The latter is a generalization of the unstable fixed point X_s in Fig. 4.1(b). As such, it must satisfy the stationary noiseless equation of motion

$$0 = -\left.\frac{\delta \mathcal{F}[\varphi]}{\delta\varphi(\mathbf{r}, t)}\right|_{\varphi_s(\mathbf{r})} = D\nabla_{\mathbf{r}}^2 \varphi_s - V'(\varphi_s) \tag{9.12}$$

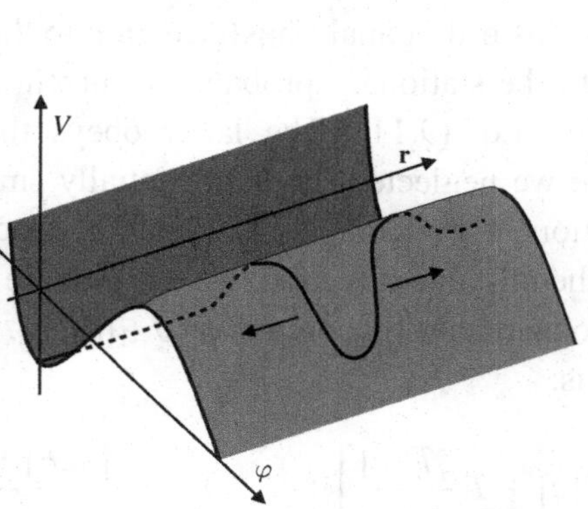

Figure 9.1 Metastable potential $V(\varphi)$. The arrows indicate the direction of the domain wall motion after the field fluctuation reaches the saddle configuration $\varphi_s(\mathbf{r})$ (bold line).

with the boundary condition $\varphi_s(\infty) = 0$, where zero denotes the metastable minimum. According to the calculations leading to Eq. (9.10), the long-time action of such a fluctuation is $\mathcal{F}[\varphi_s]/T$. As a result, the nucleation rate of critical domains is given by $R \sim e^{-\mathcal{F}[\varphi_s]/T}$. Once the critical domain is formed, it starts to grow according to the deterministic noiseless equation of motion $\partial_t \varphi = D\nabla^2_{\mathbf{r}}\varphi - V'(\varphi)$, analog of the $P = 0$ line in Fig. 4.1(b) for $X > X_s$. This latter process is associated with the domain walls motion, Fig. 9.1.

To establish the pre-exponential factor in the nucleation rate R we need to generalize treatment of the Kramers problem, Section 4.8, for the spatially extended case [140]. To this end consider the effective potential in a vicinity of the saddle configuration $\varphi_s(\mathbf{r})$. Expanding it to second order (the first order is absent, since φ_s is a stationary function of the $\mathcal{F}[\varphi]$ functional), one finds

$$\mathcal{F}[\varphi] = \mathcal{F}[\varphi_s] + \frac{1}{2}\int d\mathbf{r}d\mathbf{r'}\, \frac{\delta^2\mathcal{F}[\varphi]}{\delta\varphi(\mathbf{r})\delta\varphi(\mathbf{r'})}\bigg|_{\varphi_s} \delta\varphi(\mathbf{r})\delta\varphi(\mathbf{r'}) + \cdots,$$

where $\delta\varphi(\mathbf{r}) = \varphi(\mathbf{r}) - \varphi_s(\mathbf{r})$. It is convenient to diagonalize the symmetric form $\delta^2\mathcal{F}[\varphi]/\delta\varphi(\mathbf{r})\delta\varphi(\mathbf{r'})\big|_{\varphi_s}$ and find its eigenvalues λ_n and eigenfunctions $\chi_n(\mathbf{r})$, where $n = -1, 0, 1, \ldots, N-2$, and N is the total number of degrees of freedom. Since φ_s is a saddle rather than a minimum, there is one negative eigenvalue $\lambda_{-1} < 0$, while $\lambda_0, \ldots, \lambda_{N-2} \geq 0$. The eigenfunctions are orthogonal and can be normalized as $\int d\mathbf{r}\chi_n(\mathbf{r})\chi_m(\mathbf{r}) = \delta_{nm}$. One can now expand any variation $\delta\varphi(\mathbf{r})$ in this basis and write

$$\delta\varphi(\mathbf{r}) = \sum_{n=-1}^{N-2} c_n\chi_n(\mathbf{r}); \qquad \mathcal{F}[c] = \mathcal{F}[\varphi_s] + \frac{1}{2}\sum_{n=-1}^{N-2} \lambda_n c_n^2 + \cdots. \tag{9.13}$$

Similarly, one can apply this orthogonal transformation to the metastable distribution $\mathcal{P}[\varphi] \to \mathcal{P}[c]$ and to the stationary probability current out of the metastable state $J([\varphi], \mathbf{r}) \to J_n[c]$, Eq. (9.11). The latter obeys the continuity relation $\sum_n \partial_{c_n} J_n[c] = 0$, where we neglected the exponentially small time derivative of the metastable distribution in a vicinity of the saddle. Since we expect that the escape proceeds along the unstable direction, we can assume that $J_0, \ldots, J_{N-2} = 0$ and therefore the only nonzero current component $J_{-1}[c_0, \ldots, c_{N-2}]$ is independent of c_{-1}. We can write thus:

$$J_{-1}[c] = -\left[\frac{\partial\mathcal{F}[c]}{\partial c_{-1}}\mathcal{P}[c] + T\frac{\delta\mathcal{P}[c]}{\partial c_{-1}}\right]; \qquad 0 = -\left[\frac{\partial\mathcal{F}[c]}{\partial c_n}\mathcal{P}[c] + T\frac{\delta\mathcal{P}[c]}{\partial c_n}\right],$$

where $n = 0, 1, \ldots, N-2$. Solving the second equation here with the help of Eq. (9.13), one obtains $\mathcal{P}[c] = e^{-\mathcal{F}[c]/T}u(c_{-1})$, where $u(c_{-1})$ is a yet unknown function. Substituting it into the first equation and employing the continuity relation $\partial_{c_{-1}} J_{-1} = 0$, one finds

$$J_{-1}[c] = I e^{-\left(\mathcal{F}[\varphi_s] + \frac{1}{2}\sum_{n=0}^{N-2} \lambda_n c_n^2\right)/T}\,; \qquad \mathcal{P}[c] = \frac{I}{T}\, e^{-\mathcal{F}[c]/T} \int_{c_{-1}}^{\infty} d\tilde{c}_{-1}\, e^{-|\lambda_{-1}|\tilde{c}_{-1}^2/2T}$$

(compare with Eq. (4.48)), where I is a normalization constant. To determine the latter, one notices that the $d\tilde{c}_{-1}$ integral reaches practically a constant value $\sqrt{2\pi T/|\lambda_{-1}|}$ for $c_{-1} \lesssim -\sqrt{2T/|\lambda_{-1}|}$. For a low temperature this is a small distance, and the maximum of the metastable distribution is expected to be at much more negative c_{-1}. Therefore, around the maximum the distribution is $\mathcal{P}[\varphi] = I\sqrt{2\pi/T|\lambda_{-1}|}\, e^{-\mathcal{F}[\varphi]/T}$. To normalize it one needs to find eigenvalues of $\delta^2\mathcal{F}[\varphi]/\delta\varphi(\mathbf{r})\delta\varphi(\mathbf{r}')\big|_{\varphi=0}$ around the metastable configuration $\varphi = 0$. Due to the local stability of $\varphi = 0$ they are all positive, $\lambda_k^{(0)} > 0$, where $k = 1, \ldots, N$. Requiring $\int \mathbf{D}[\varphi]\mathcal{P}[\varphi] = 1$, one finds $I = \sqrt{T|\lambda_{-1}|/2\pi}\, e^{\mathcal{F}[0]/T} \prod_{k=1}^{N} \sqrt{\lambda_k^{(0)}/2\pi T}$. Finally, the critical nucleation rate R is given by the total normalized current across the surface, $c_{-1} = 0$, that is, $R = \int dc_0 \ldots dc_{N-2} J_{-1}[c]$. This way we find [140]

$$R = \sqrt{\frac{T|\lambda_{-1}|}{2\pi}} \prod_{k=1}^{N} \sqrt{\frac{\lambda_k^{(0)}}{2\pi T}}\; e^{-\left(\mathcal{F}[\varphi_s] - \mathcal{F}[0]\right)/T} \int dc_0 \ldots dc_{N-2}\, e^{-\frac{1}{2T}\sum_{n=0} \lambda_n c_n^2}. \tag{9.14}$$

Integrals over nonnegative modes $dc_0 \ldots dc_{N-2}$ could be performed in the Gaussian approximation, if not for the presence of the zero modes in the spectrum $\lambda_0, \lambda_1, \ldots$. The latter certainly exist due to the translational invariance of the saddle configuration $\varphi_s(\mathbf{r})$.

To illustrate how to deal with them, let us consider a one-dimensional example of Fig. 9.1. The saddle configuration $\varphi_s(x - x_0)$ may be centered around an arbitrary spatial point x_0. The zero eigenmode $\chi_0(x)$, where $\lambda_0 = 0$, associated with the translational invariance, may be thus identified as $\varphi(x) = \varphi_s(x) + x_0 \nabla_x \varphi_s(x) + \cdots$. Comparing it with Eq. (9.13), one finds $\chi_0(x) = \nabla_x\varphi_s(x)/[\int dx(\nabla_x\varphi_s)^2]^{1/2}$ and $c_0 = x_0[\int dx(\nabla_x\varphi_s)^2]^{1/2}$. Taking into account that for the saddle configuration the total elastic energy is equal to the potential one, we find $\int dx(\nabla_x\varphi_s)^2 = \mathcal{F}[\varphi_s]/D$, where we put $\mathcal{F}[0] = 0$. We thus obtain $dc_0 = dx_0\sqrt{\mathcal{F}[\varphi_s]/D}$. Since $\lambda_0 = 0$, the dc_0 integration results in the factor $L\sqrt{\mathcal{F}[\varphi_s]/D}$, where $L = \int dx_0$ is the total spatial length of the string. The integrals over remained modes with eigenvalues $\lambda_1, \ldots, \lambda_{N-2} > 0$ in Eq. (9.14) may be evaluated in the Gaussian approximation. One thus obtains for the *nucleation rate per unit length*

$$\frac{R}{L} = \left[\frac{\mathcal{F}[\varphi_s]}{T} \frac{|\lambda_{-1}|}{(2\pi)^3 D} \prod_{k=1}^{N} \lambda_k^{(0)} \prod_{n=1}^{N-2} \frac{1}{\lambda_n}\right]^{1/2} e^{-\mathcal{F}[\varphi_s]/T}. \tag{9.15}$$

The dimensionality is given by $[\lambda] = 1/\text{time}$ and $[D] = \text{length}^2/\text{time}$, the right-hand side is thus $[\text{time} \times \text{length}]^{-1}$, as it should be.

To proceed with Eq. (9.15) let us consider a cubic potential, cf. Eq. (3.34),

$$V(\varphi) = V_0 \left[\delta \left(\frac{\varphi}{\bar{\varphi}} \right)^2 - \left(\frac{\varphi}{\bar{\varphi}} \right)^3 \right].$$

The corresponding equation for $\varphi_s(x)$ takes the form $D\nabla_x^2 \varphi_s - V'(\varphi_s) = 0$ with the boundary conditions $\varphi_s(\pm\infty) = 0$. The equation may be integrated once to yield $\frac{D}{2}(\nabla_x \varphi_s)^2 = V(\varphi_s)$, which may now be easily solved, leading to $\varphi_s(x) = \bar{\varphi}\delta \cosh^{-2}(x/\bar{x})$, where $\bar{x} = \bar{\varphi}\sqrt{2D/V_0\delta}$. The activation energy is then $\mathcal{F}[\varphi_s] = (8/15)\sqrt{2V_0 D\bar{\varphi}^2}\, \delta^{5/2}$; compare Eq. (3.36). The corresponding equation for the eigenvalues λ_n acquires the form of the Schrödinger equation with a $\cosh^{-2}(x/\bar{x})$ potential. Its solutions are well known [36]. As expected, there is a single negative eigenvalue $\lambda_{-1} = -(5/2)V_0\delta/\bar{\varphi}^2$, the zero eigenvalue $\lambda_0 = 0$, a positive discrete eigenvalue $\lambda_1 = (3/2)V_0\delta/\bar{\varphi}^2$ and the continuum of positive eigenvalues $\lambda_p = (2 + p^2/2)V_0\delta/\bar{\varphi}^2$, where p is a real continuum label. The eigenvalues near the metastable minimum are also continuous, given by $\lambda_k^{(0)} = (2 + k^2/2)V_0\delta/\bar{\varphi}^2$. As a result, the pre-exponential factor in Eq. (9.15) is

$$\left[C \frac{\sqrt{V_0 D\bar{\varphi}^2}\, \delta^{5/2}}{T} \frac{V_0\delta/\bar{\varphi}^2}{D} (V_0\delta/\bar{\varphi}^2)^2 \right]^{1/2} = C^{1/2} \frac{V_0^{7/4}\, \delta^{11/4}}{T^{1/2}\, D^{1/4}\, \bar{\varphi}^{5/2}}, \tag{9.16}$$

where C is a numerical constant.[3] Notice the very strong $\delta^{11/4}$ dependence of the pre-exponential factor on the bifurcation parameter δ. It should be compared with the linear dependence in the corresponding $d = 0$ Kramers problem result, Eq. (4.53).

[3] $C = \frac{8\sqrt{2}}{15} \frac{5}{2} \frac{1}{(2\pi)^3} \prod_k \lambda_k^{(0)} \frac{2}{3} \prod_p \lambda_p^{-1}$. To evaluate the products over the continuous spectra, we notice that the quantization condition for k is $kL = 2\pi n$, where n is an integer. One thus finds $\prod_k \lambda_k^{(0)} = e^{\sum_k \ln \lambda_k^{(0)}} = e^{\int L/(2\pi) dk \ln(2+k^2/2)}$. The quantization condition at the saddle for p is $pL + \delta(p) = 2\pi n$, where $\delta(p)$ is the phase-shift in the *reflectionless* $\cosh^{-2}(x/\bar{x})$ potential, which is [36]

$$\delta(p) = \arg \left[\frac{(-1 + ip)(-2 + ip)(-3 + ip)}{(1 + ip)(2 + ip)(3 + ip)} \right].$$

As a result, $\prod_p \lambda_p^{-1} = e^{-\sum_p \ln \lambda_p} = e^{-\int (L+\delta'(p))/(2\pi) dp \ln(2+p^2/2)}$, where $\delta'(p)/(2\pi)$ is a (negative) correction to the density of the continuous states. Notice that $-\int \delta'(p)/(2\pi) dp = 3$ is a number of discrete states, which correctly takes care of the dimensional factor $V_0\delta/\bar{\varphi}^2$. Finally $\prod_k \lambda_k^{(0)} \prod_p \lambda_p^{-1} = e^{-\int \delta'(p)/(2\pi) dp \ln(2+p^2/2)} = 450$. Putting it all together, $C^{1/2} = 1.51$.

9.4 Dynamics Near a Classical Second-Order Transition

In this section we focus on the model with the potential (9.6), described by the action (9.5). Neglecting for a moment the nonlinear term $\propto g$ and assuming that $\delta > 0$, one finds Gaussian response and correlation functions; compare Eqs. (3.13),

$$D_0^{\mathrm{R}}(\mathbf{q},\omega) = \frac{1}{\mathrm{i}} \int \mathrm{d}\mathbf{r}\mathrm{d}t\, \mathrm{e}^{\mathrm{i}(\omega t - \mathbf{q}\mathbf{r})} \langle \varphi(\mathbf{r},t)\varphi^{\mathrm{q}}(\mathbf{0},0)\rangle = \frac{1}{2}\frac{1}{\mathrm{i}\omega - D\mathbf{q}^2 - \delta}; \tag{9.17a}$$

$$D_0^{\mathrm{A}}(\mathbf{q},\omega) = \frac{1}{\mathrm{i}} \int \mathrm{d}\mathbf{r}\mathrm{d}t\, \mathrm{e}^{\mathrm{i}(\omega t - \mathbf{q}\mathbf{r})} \langle \varphi^{\mathrm{q}}(\mathbf{r},t)\varphi(\mathbf{0},0)\rangle = \frac{1}{2}\frac{-1}{\mathrm{i}\omega + D\mathbf{q}^2 + \delta}; \tag{9.17b}$$

$$D_0^{\mathrm{K}}(\mathbf{q},\omega) = \frac{1}{\mathrm{i}} \int \mathrm{d}\mathbf{r}\mathrm{d}t\, \mathrm{e}^{\mathrm{i}(\omega t - \mathbf{q}\mathbf{r})} \langle \varphi(\mathbf{r},t)\varphi(\mathbf{0},0)\rangle = \frac{-\mathrm{i}\,2T}{\omega^2 + (D\mathbf{q}^2 + \delta)^2} \tag{9.17c}$$

and, as always, $\langle \varphi^{\mathrm{q}}\varphi^{\mathrm{q}}\rangle = 0$. Notice that, in agreement with the classical limit of FDT, $D^{\mathrm{K}} = (2T/\omega)(D^{\mathrm{R}} - D^{\mathrm{A}})$, which holds even for the full Green functions including nonlinear effects. From the last expression one finds for the equal time, $t=0$, the Gaussian correlation function

$$\langle \varphi(\mathbf{r},0)\varphi(\mathbf{0},0)\rangle = \int \frac{\mathrm{d}\mathbf{q}\,\mathrm{d}\omega}{(2\pi)^{d+1}} \frac{2T\mathrm{e}^{\mathrm{i}\mathbf{q}\mathbf{r}}}{\omega^2 + (D\mathbf{q}^2+\delta)^2} = T\int^{\Lambda} \frac{\mathrm{d}\mathbf{q}}{(2\pi)^d}\frac{\mathrm{e}^{\mathrm{i}\,\mathbf{q}\mathbf{r}}}{D\mathbf{q}^2+\delta}. \tag{9.18}$$

This is exactly what one expects from the static equilibrium statistical mechanics of a membrane in the Gaussian potential $V(\varphi) = \delta\varphi^2/2$, Eq. (9.10). The short distance divergence of the integral in $d > 1$ is regularized by a cutoff $\Lambda \approx 1/a$, where a is the lattice constant. On the other hand, divergence at long distances in the limit $\delta \to 0$ signals the importance of the nonlinear terms, so far omitted.

The $\delta \to 0$ limit describes the vicinity of the second-order phase transition. Indeed, for $\delta > 0$ the potential $V(\varphi)$, Eq. (9.6), has a single minimum at $\varphi = 0$, describing, say, a nonmagnetic state. If $\delta < 0$ there are two distinct minima at $\varphi = \pm(-\delta/g)^{1/2}$, describing a magnetized state; see Fig. 9.3. In the latter case the system spontaneously chooses (for $d > 1$) one of the two minima, breaking the symmetry $\varphi, \varphi^{\mathrm{q}} \to -\varphi, -\varphi^{\mathrm{q}}$ present in the action (9.5). It is thus said that $\delta \sim T - T_{\mathrm{c}}$ is a critical parameter, where T_{c} is a critical temperature of the second-order phase transition. It was shown experimentally, numerically, and eventually analytically that various observable and correlation functions exhibit power-law scaling dependence on $|\delta|$ in the limit $|\delta| \to 0$. In particular, the characteristic spatial scale of the static correlation function (9.18), known as the *correlation length* ξ, diverges as $\xi \propto |\delta|^{-\nu}$. The dynamic correlation functions, for example, (9.17c), possess a characteristic timescale τ, which diverges as $\tau \propto |\delta|^{-\nu z} \propto \xi^z$. Inspecting Eqs. (9.17) and (9.18), one concludes that in the Gaussian theory the corresponding *critical exponents* are $\nu = 1/2$ and $\nu z = 1$. We shall show in what follows that this is indeed the case for $d > d_c$, where in the present case the *critical dimension* is $d_c = 4$. Below the critical dimension, $d < 4$, the critical exponents deviate from the Gaussian predictions, maintaining nevertheless remarkable universality. The way to understand such a universality and evaluate critical exponents was suggested by Wilson [141], see also [142, 143], and is known as the *renormalization group* (RG). We shall not provide here a detailed account of RG ideology, referring the reader to a number of excellent presentations [5, 9, 144, 142, 143]. Instead, we shall focus on some peculiarities of the perturbative RG in a dynamical setting.

The idea is to gradually integrate out fast degrees of freedom, monitoring the evolution of the effective action for the remaining slow degrees of freedom. The corresponding evolution equations for the action parameters, being linearized around a proper fixed point, reveal the critical exponents. The differences between the static and dynamic theory are as follows: (i) in the latter case the degrees of freedom are labeled by momentum $\mathbf{q}$ and frequency ω, while in the former only by $\mathbf{q}$; (ii) due to the doubling of the number of fields φ and φ^{q}, the dynamic action has room for many more possible terms than the static one. One needs to discuss thus how to split the degrees of freedom into slow and fast ones and how to select the relevant terms in the action.

A convenient way to separate degrees of freedom is depicted in Fig. 9.2. We write, thus, $\varphi(\mathbf{q},\omega) = \varphi_{\mathrm{s}}(\mathbf{q}_{\mathrm{s}},\omega) + \varphi_{\mathrm{f}}(\mathbf{q}_{\mathrm{f}},\omega)$ and the same for $\varphi^{\mathrm{q}}(\mathbf{q},\omega)$, where subscripts s and f denote slow and fast parts, respectively. The fast part contains momenta in the shell $\Lambda/b < q < \Lambda$, where $b > 1$, and unrestricted frequencies, while the slow part has momenta $q < \Lambda/b$ and also unrestricted frequencies. The drawback of this approach is that the slow fields have all frequency components and thus their effective action technically acquires nonlocal (in time) form upon integration over the fast fields. It is only if one focuses on the low-frequency components that the time locality is restored. As a result, the high-frequency components of the slow fields are simply ignored rather than being integrated out. One could, in principle, introduce shells in both momentum and frequency directions

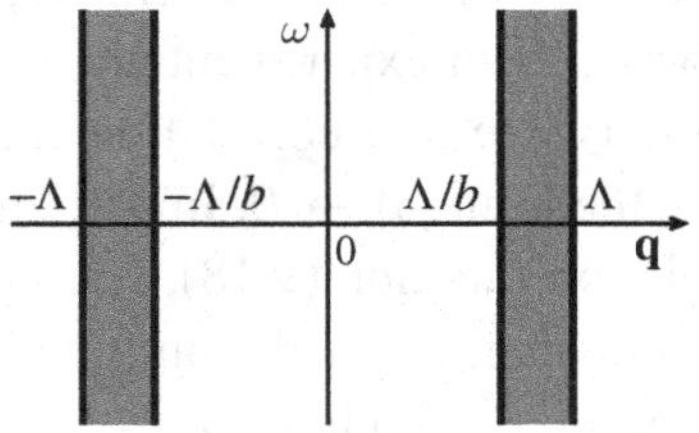

Figure 9.2 Separation of degrees of freedom into slow (central area) and fast (shaded areas).

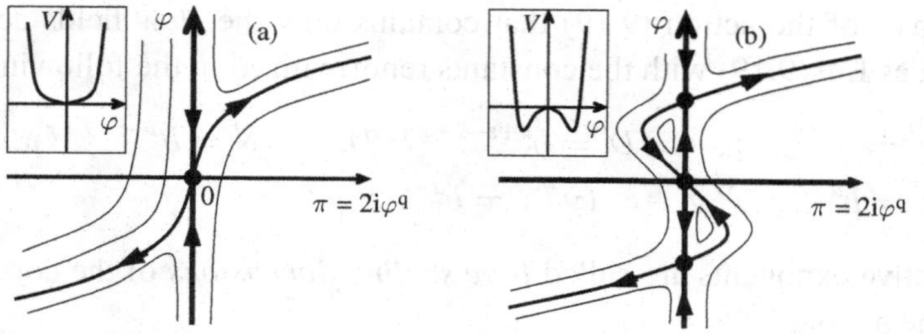

Figure 9.3 Phase portraits of the system undergoing a second-order phase transition: (a) nonmagnetic state $\delta > 0$; (b) magnetic state $\delta < 0$. The insets show the corresponding potentials $V(\varphi)$.

[9]. It complicates calculations, but seems not to change the results (technically due to the fast convergence of the frequency integrals, as, for example, in Eq. (9.18)).

To visualize what terms in the effective action are important to keep, it is convenient to involve the Hamiltonian language of Eq. (9.8), which is obtained by treating $\pi = 2i\varphi^{\mathrm{q}}$ as the canonical pair of φ. The phase portraits of the corresponding Hamiltonian (9.8) (without the gradient terms) for $\delta > 0$ and $\delta < 0$ are plotted in Fig. 9.3. As discussed in Section 4.4, one needs to focus on the two lines of zero energy $\pi = 0$ and $\pi = V'(\varphi)/T$. The second-order transition may be identified with the qualitative rearrangement of their geometry: three intersections versus one. Notice also that the reflection symmetry $(\pi, \varphi) \to (-\pi, -\varphi)$ of phase space is intact in both cases. One should thus keep the minimal number of terms, which (i) allow for the transition between three and one intersections and (ii) obey the reflection symmetry. This is achieved with the Hamiltonian $H = \pi(\gamma T\pi - \delta\varphi - g\varphi^3)$. Notice that other allowed terms, for example, $\pi^2\varphi^2$, do not change the qualitative shape of Fig. 9.3 for small δ and thus may be neglected. The minimal action to be considered is thus

$$S[\varphi, \pi] = \int d\mathbf{r}\, dt \left[\gamma\pi\partial_t\varphi - D\pi\nabla_{\mathbf{r}}^2\varphi + \delta\pi\varphi + g\pi\varphi^3 - \gamma T\pi^2\right], \tag{9.19}$$

where we reintroduced the friction coefficient γ, see Eq. (4.2). It contains five constants γ, D, δ, g, and γT. One then substitutes $\varphi = \varphi_{\mathrm{s}} + \varphi_{\mathrm{f}}$ and $\pi = \pi_{\mathrm{s}} + \pi_{\mathrm{f}}$ and proceeds in two steps.

In the first step the space of slow momenta $|\mathbf{q}_{\mathrm{s}}| < \Lambda/b$ is rescaled to the original size Λ, that is, $\mathbf{q}_{\mathrm{s}} \to \mathbf{q}/b$, where $|\mathbf{q}| < \Lambda$. In coordinate space it means $\mathbf{r} \to b\mathbf{r}$. We shall also rescale time and the slow fields in the following manner:

$$\mathbf{r} \to b\mathbf{r}, \quad t \to b^z t, \quad \varphi_{\mathrm{s}}(\mathbf{r}, t) \to b^{\chi}\varphi(b\mathbf{r}, b^z t), \quad \pi_{\mathrm{s}}(\mathbf{r}, t) \to b^{\tilde{\chi}}\pi(b\mathbf{r}, b^z t). \tag{9.20}$$

Then the part of the action (9.19) that contains only the *slow* fields acquires the same form as Eq. (9.19) with the constants renormalized in the following way:

$$\gamma' = b^{d+\tilde{\chi}+\chi}\gamma, \qquad D' = b^{d+z-2+\tilde{\chi}+\chi}D, \qquad \delta' = b^{d+z+\tilde{\chi}+\chi}\delta, \qquad (9.21)$$
$$g' = b^{d+z+\tilde{\chi}+3\chi}g, \qquad (\gamma T)' = b^{d+z+2\tilde{\chi}}\gamma T.$$

The respective exponents are called *bare scaling dimensions* of the corresponding terms in the action.

The second step is to expand e^{-S}, where S is given by Eq. (9.19), in a power series in terms containing interactions of the slow and fast fields. In particular, one is interested in terms (i) $3g\pi_s\varphi_s\varphi_f^2$ and (ii) $3g\pi_f\varphi_f\varphi_s^2$. One then performs Gaussian integration over the *fast* fields, employing the Green functions (9.17) (recall that $\pi_f = 2i\varphi_f^q$). After the rescaling (9.20) the fast momenta are running in $\Lambda < q_f < b\Lambda$. The one-loop diagram expressing correction to the terms in the action $\propto \pi\varphi$ employs vertex (i)[4] and is depicted in Fig. 9.4(a). The corresponding loop is given by

$$3g\frac{1}{i}\int_{\Lambda}^{b\Lambda}\frac{d\mathbf{q}_f}{(2\pi)^d}\int\frac{d\omega}{2\pi}D_0^K(\mathbf{q}_f,\omega) = -3g\int_{\Lambda}^{b\Lambda}\frac{d\mathbf{q}_f}{(2\pi)^d}\frac{T}{D\mathbf{q}_f^2+\delta}$$
$$\approx -3g\int_{\Lambda}^{b\Lambda}\frac{d\mathbf{q}_f}{(2\pi)^d}\frac{T}{D\mathbf{q}_f^2} + 3g\delta\int_{\Lambda}^{b\Lambda}\frac{d\mathbf{q}_f}{(2\pi)^d}\frac{T}{(D\mathbf{q}_f^2)^2}. \qquad (9.22)$$

The first δ-independent term constitutes additive correction to δ (or, in other words, it shifts the point of the transition, i.e. T_c). Since in our approach the latter is anyway a phenomenological parameter, the first term on the right-hand side of Eq. (9.22) may be omitted. The second term $\propto g\delta$ provides an additional (beyond the bare scaling (9.21)) *multiplicative* correction to δ, which must be kept. Since there is no dependence on external momentum and frequency in the diagram of Fig. 9.4(a), there are no one-loop corrections to γ and D. They appear, however, along with a

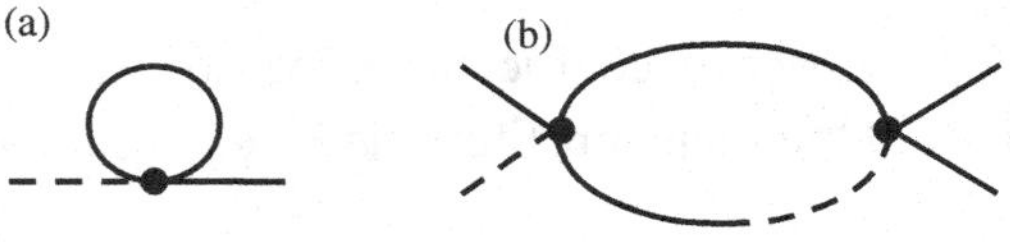

Figure 9.4 One-loop renormalization of running constants δ (a) and g (b). Full lines represent φ, dashed lines $\varphi^q = \pi/(2i)$, and each bold dot carries a factor of $3g$.

[4] Vertex (ii) would result in terms $\propto \varphi^2$, which violate the fundamental property $S[\varphi, 0] = 0$. Fortunately they are proportional to $D^R(0,0) + D^A(0,0) = 0$, due to coinciding time arguments.

correction to the γT vertex, in the two-loop order, coming from the vertices $g\pi_s\varphi_f^3$ and $3g\pi_f\varphi_s\varphi_f^2$.

The one-loop diagram, expressing corrections to the interaction vertex $\propto g\pi\varphi^3$, employs vertices (i) × (ii)[5] and is depicted in Fig. 9.4(b). It is given by

$$9g^2\frac{4}{\mathrm{i}}\int_{\Lambda}^{b\Lambda}\frac{d\mathbf{q}_f}{(2\pi)^d}\int\frac{d\omega}{2\pi}D_0^K(\mathbf{q}_f,\omega)D_0^R(\mathbf{q}_f,\omega)=9g^2\int_{\Lambda}^{b\Lambda}\frac{d\mathbf{q}_f}{(2\pi)^d}\frac{T}{(D\,\mathbf{q}_f^2)^2},\qquad(9.23)$$

where we have neglected the small critical parameter δ in comparison with $D\Lambda^2$. Exponentiating back the resulting expressions, one finds the one-loop corrections to the running coupling constants δ and g.

We now put $b = 1 + l$, where l is an infinitesimal increment, and perform the integrals in Eqs. (9.22) and (9.23). We then combine the loop corrections with the bare scaling (9.21) and write them both in the differential form

$$\begin{aligned}&\partial_l\gamma=(d+\tilde\chi+\chi)\gamma,\qquad \partial_l D=(d+z-2+\tilde\chi+\chi)D,\\&\partial_l(\gamma T)=(d+z+2\tilde\chi)\gamma T,\\&\partial_l\delta=(d+z+\tilde\chi+\chi)\delta-3g\delta K_d,\quad \partial_l g=(d+z+\tilde\chi+3\chi)g-9g^2K_d,\end{aligned}\qquad(9.24)$$

where $K_d = (T/D^2)\Lambda^{d-4}2^{1-d}\pi^{-d/2}\Gamma(d/2)$ is $\Lambda^{d-4}(2\pi)^{-d}T/D^2$ times the area of the unit sphere in d dimensions. We can now use the freedom of choosing $\tilde\chi$, χ along with the scale of time to fix three of these five parameters. It is convenient to fix γ, D, and γT, that is, the parameters of the Gaussian action at the critical point. This is achieved by demanding that $d+\tilde\chi+\chi=0$, $d+z-2+\tilde\chi+\chi=0$ as well as $d+z+2\tilde\chi=0$. This results in the following scaling dimensions: $z=2$, $\chi=1-d/2$, and $\tilde\chi=-1-d/2$. The remaining two RG equations read as

$$\partial_l\delta=\big(2-3gK_d\big)\delta,\qquad \partial_l g=\big(\epsilon-9gK_d\big)g,\qquad(9.25)$$

where $\epsilon = 4-d$.[6] For $\epsilon<0$, that is, $d>d_c=4$, the nonlinearity is irrelevant, meaning $g\to 0$ as l increases. The equation for the critical parameter δ may then be easily integrated, leading to $\delta=\delta_0 e^{2l}=\delta_0 b^2$. To carry out rescaling until $\delta\propto 1$, where the renormalized system is not critical, thus requires $b\propto\delta_0^{-1/2}$. Since b is the spatial scaling factor, see Eq. (9.20), it implies $\xi\propto\delta_0^{-1/2}$ and accordingly $\nu=1/2$, as expected for Gaussian theory.

If $d<4$ and $\epsilon>0$, the equation for the interaction constant g predicts that it evolves toward the Wilson–Fisher fixed point $g^*=\epsilon/(9K_4)$. Our perturbative scheme is justified as long as g^* is small, requiring $\epsilon\ll 1$ (this is the reason one

5 Combination (i) × (i) generates a new interaction vertex $\pi^2\varphi^2$, which is irrelevant (i.e. scales to zero); see what follows. On the other hand, (ii) × (ii) is zero, because $\langle\pi_f\pi_f\rangle=0$.

6 The bare scaling dimension of the term $\pi^2\varphi^2$ is $d+z+2\tilde\chi+2\chi=2-d$ and therefore it is indeed less relevant than both $\pi\varphi$ and $\pi\varphi^3$, as was already conjectured from the phase portrait, Fig. 9.3.

substitutes K_d by K_4). The corresponding perturbative scheme is thus known as the ϵ-expansion [141]. Then the RG equation for the critical parameter δ acquires the form $\partial_l \delta = (2 - \epsilon/3)\delta$. Integrating it, one finds for the correlation length critical exponent $\nu = (2 - \epsilon/3)^{-1} \approx 1/2 + \epsilon/12$, while the correlation time exponent is $\nu z = 1 + \epsilon/6$. We thus find that both spatial and temporal scales diverge on approaching the critical point with non-Gaussian exponents. In the model at hand the dynamical exponent $z = 2 + O(\epsilon^2)$ is not altered within the one-loop approximation; it is modified, however, at the two-loop level [145]. In what follows we shall consider examples where z is already nontrivial in the order ϵ.

The previous calculations may be straightforwardly extended to the case where φ (and φ^{q}) are N-component vectors and the equations of motion are $O(N)$ rotationally invariant. The Gaussian part of the corresponding action has the form $\sum_{a=1}^{N} \pi_a(\ldots)\varphi_a$, where dots stand for the linear kinetic operator. The interaction part acquires the form $\sum_{a,b=1}^{N} g\pi_a\varphi_a\varphi_b\varphi_b$. Splitting the fields into slow and fast components gives rise to four vertices: (i) $g\pi_{as}\varphi_{as}\varphi_{bf}\varphi_{bf}$; (ii) $2g\pi_{as}\varphi_{af}\varphi_{bs}\varphi_{bf}$; (iii) $g\pi_{af}\varphi_{af}\varphi_{bs}\varphi_{bs}$; and (iv) $2g\pi_{af}\varphi_{as}\varphi_{bs}\varphi_{bf}$. The bare scaling (9.21) is, of course, the same. The one-loop correction to δ, Fig. 9.4(a), originates from vertices (i) and (ii), which contribute to the coefficient as $N + 2$. The one-loop correction to g, Fig. 9.4(b), comes from the [(i) + (ii)] × [(iii) + (iv)] combination of vertices, which brings the coefficient $N + 2 + 2 + 4 = N + 8$. As a result the RG equations (9.25) acquire the form $\partial_l \delta = \big(2 - (N + 2)gK_4\big)\delta$ and $\partial_l g = \big(\epsilon - (N + 8)gK_4\big)g$, leading to, for example, $\nu^{-1} = 2 - \epsilon(N + 2)/(N + 8) + O(\epsilon^2)$.

9.5 Hohenberg–Halperin Classification

So far we have seen that the universality class, that is, the set of critical exponents, depends on the dimensionality d and the number of the field components N. This is similar to what is known in equilibrium statistical mechanics [144, 142, 143]. However, the dynamics of the transition, in particular the dynamic critical exponent z, also depends on the presence or absence of the conservation laws. This relation and emerging classification of the dynamic universality classes was elucidated by Hohenberg and Halperin [145]. The model considered in Section 9.4 (model A in the Hohenberg–Halperin classification) does not have any conserved quantities. It may be reformulated in a way to conserve $\int d\mathbf{r}\, \varphi(\mathbf{r}, t) = \varphi(\mathbf{q} = \mathbf{0}, t)$, describing the order parameter φ of, for example, a uniaxial magnet with conserved total spin. To ensure the conservation law, the equation of motion must have a form of the continuity relation

$$\partial_t \varphi(\mathbf{r}, t) = -\nabla_{\mathbf{r}} \mathbf{J} \qquad \mathbf{J} = -\nabla_{\mathbf{r}} \frac{\delta \mathcal{F}[\varphi]}{\delta \varphi(\mathbf{r}, t)} - \vec{\zeta}(\mathbf{r}, t), \tag{9.26}$$

where the Landau free energy $\mathcal{F}[\varphi]$ is given by Eqs. (9.2) and (9.6) and $\vec{\zeta}(\mathbf{r},t)$ is a random Gaussian vector noise with the isotropic correlation function

$$\langle \zeta^{\mu}(\mathbf{r},t)\zeta^{\nu}(\mathbf{r'},t')\rangle = 2T\delta^{\mu\nu}\delta(\mathbf{r}-\mathbf{r'})\delta(t-t'), \tag{9.27}$$

where $\mu,\nu = x,y,z$. This is model B. Its MSR action in the Hamiltonian notation takes the form $S[\varphi,\pi] = \int dt\left(\int d\mathbf{r}\, \pi\, \partial_t\varphi - \mathcal{H}[\pi,\varphi]\right)$, where the Hamiltonian (upon integration by parts) is

$$\mathcal{H}[\pi,\varphi] = \int d\mathbf{r}\left[-(\nabla_{\mathbf{r}}\pi)\cdot\nabla_{\mathbf{r}}\frac{\delta\mathcal{F}[\varphi]}{\delta\varphi(\mathbf{r},t)} + T(\nabla_{\mathbf{r}}\pi)^2\right]. \tag{9.28}$$

Again there are two $\mathcal{H} = 0$ solutions of the stationary field equations: noiseless relaxation $\pi = 0$, while $\partial_t\varphi = \nabla^2_{\mathbf{r}}(\delta\mathcal{F}/\delta\varphi)$, and the activation trajectory $\pi = (\delta\mathcal{F}/\delta\varphi)/T$, while $\partial_t\varphi = -\nabla^2_{\mathbf{r}}(\delta\mathcal{F}/\delta\varphi)$. The last equation, which is an immediate consequence of the equation of motion for φ, shows that activation trajectories are time-reversed copies of the relaxational ones. The action on the activation trajectory, leading to a configuration $\varphi_0(\mathbf{r})$, is

$$S = \int dt\, d\mathbf{r}\, \pi\, \partial_t\varphi = \frac{1}{T}\int dt\, d\mathbf{r}\, (\delta\mathcal{F}/\delta\varphi)\, \partial_t\varphi = \frac{1}{T}\int^{\varphi_0} dt\, \partial_t\mathcal{F}[\varphi] = \frac{\mathcal{F}[\varphi_0]}{T}.$$

This shows that in the long-time limit model B obeys the same equilibrium statistical mechanics, Eq. (9.10), as model A. It implies, in particular, that the static critical exponents, for example, ν, are the same as in model A. The dynamics, however, is different and so is the dynamic critical exponent z.

To develop an RG treatment of model B we employ the potential in the form (9.6), restore the friction coefficient γ, and write the MSR action as

$$S = \int d\mathbf{r}\, dt\left[\gamma\pi\,\partial_t\varphi + (\nabla^2_{\mathbf{r}}\pi)\big(D(\nabla^2_{\mathbf{r}}\varphi) - \delta\varphi - g\varphi^3\big) - \gamma T(\nabla_{\mathbf{r}}\pi)^2\right]. \tag{9.29}$$

Neglecting the nonlinear vertex $g(\nabla^2_{\mathbf{r}}\pi)\varphi^3$, one finds the Gaussian response and correlation functions (cf. Eqs. (9.17))

$$\int d\mathbf{r}dt\, e^{i(\omega t-\mathbf{q}\mathbf{r})}\langle\pi(\mathbf{r},t)\varphi(\mathbf{0},0)\rangle = \frac{1}{i\gamma\omega + D\mathbf{q}^4 + \delta\mathbf{q}^2}; \tag{9.30a}$$

$$\int d\mathbf{r}dt\, e^{i(\omega t-\mathbf{q}\mathbf{r})}\langle\varphi(\mathbf{r},t)\varphi(\mathbf{0},0)\rangle = \frac{2\gamma T\mathbf{q}^2}{(\gamma\omega)^2 + (D\mathbf{q}^4 + \delta\mathbf{q}^2)^2} \tag{9.30b}$$

and, of course, $\langle\pi\pi\rangle = 0$. Integrating Eq. (9.30b) over the frequency ω, one finds that the static correlation function $\langle\varphi(\mathbf{r},0)\varphi(\mathbf{0},0)\rangle$ coincides with that of model A, Eq. (9.18), and agrees with the equilibrium statistical mechanics, Eq. (9.10). We then split the fields φ and π into slow and fast components, rescale the action, and integrate out the fast fields in the one-loop approximation; see Section 9.4.

Two nonlinear vertices contribute to this calculation: (i) $-3g(\nabla_{\mathbf{r}}^2\pi_{\mathrm{s}})\varphi_{\mathrm{s}}\varphi_{\mathrm{f}}^2$ and (ii) $-3g(\nabla_{\mathbf{r}}^2\pi_{\mathrm{f}})\varphi_{\mathrm{f}}\varphi_{\mathrm{s}}^2$. The diagram of Fig. 9.4(a) includes only the static (equal time) correlation function (9.18) and thus is exactly the same in models A and B. The diagram of Fig. 9.4(b) acquires an additional factor $\mathbf{q}_{\mathrm{f}}^2$, coming from the vertex (ii). It compensates for the corresponding factor in the denominator, leading again to the same result as in model A, Eq. (9.23). The RG equations acquire the form (cf. Eq. (9.24))

$$\partial_l\gamma = (d+\tilde{\chi}+\chi)\gamma, \qquad \partial_l D = (d+z-4+\tilde{\chi}+\chi)D,$$
$$\partial_l(\gamma T) = (d+z-2+2\tilde{\chi})\gamma T, \qquad \partial_l\delta = (d+z-2+\tilde{\chi}+\chi-3gK_d)\delta,$$
$$\partial_l g = (d+z-2+\tilde{\chi}+3\chi-9gK_d)g.$$

Notice that the bare scaling dimensions of D, γT, δ, and g are two less than those in model A. This is due to the presence of additional gradients in the action (9.29). The rest of the calculation follows the same steps as after Eq. (9.24), and the only difference in the results is that $z = 4$ for $d \geq 4$ and $z = 4 + O(\epsilon^2)$ for $d < 4$. The former statement may be already recognized from the Gaussian correlators (9.30). Notice also that a small perturbation, violating the conservation law, for example, a term in the action $\propto \pi\varphi$, acquires the bare scaling dimension $d+z+\tilde{\chi}+\chi = 4$ and is strongly relevant. Therefore, such a perturbation grows fast upon renormalization, eventually bringing the system into the universality class of model A.

Other Hohenberg–Halperin universality classes are associated with the order parameter φ being coupled to one or more conserved fields, that is, hydrodynamic modes. For example, a non-conserved order parameter of model A may be coupled to a conserved scalar density $\varrho(\mathbf{r},t)$ of, say, annealed mobile impurities. The dynamics of the latter is governed by the conserved Langevin equation:

$$\partial_t\varphi(\mathbf{r},t) = -\frac{\delta\mathcal{F}[\varphi,\varrho]}{\delta\varphi(\mathbf{r},t)} + \xi(\mathbf{r},t), \qquad \partial_t\varrho(\mathbf{r},t) = \nabla_{\mathbf{r}}\left[\nabla_{\mathbf{r}}\frac{\delta\mathcal{F}[\varphi,\varrho]}{\delta\varrho(\mathbf{r},t)} + \vec{\zeta}(\mathbf{r},t)\right],$$
$$\mathcal{F}[\varphi,\varrho] = \int d\mathbf{r}\left[\frac{D}{2}(\nabla_{\mathbf{r}}\varphi)^2 + V(\varphi) + \frac{A}{2}\varphi^2\varrho + \frac{C}{2}\varrho^2\right], \tag{9.31}$$

where the noise correlators are given by Eqs. (9.3) and (9.27). This is model C. One can introduce now two auxiliary MSR fields φ^{q} and ϱ^{q}, or in the Hamiltonian notations $\pi = 2i\varphi^{\mathrm{q}}$ and $\rho = 2i\varrho^{\mathrm{q}}$, to write the action in the form $S = \int dt\left[\int d\mathbf{r}\left(\pi\,\partial_t\varphi + \rho\partial_t\varrho\right) - \mathcal{H}\right]$, where the Hamiltonian is

$$\mathcal{H}[\pi,\varphi;\rho,\varrho] = \int d\mathbf{r}\left[-\pi\frac{\delta\mathcal{F}[\varphi,\varrho]}{\delta\varphi(\mathbf{r},t)} - (\nabla_{\mathbf{r}}\rho)\cdot\nabla_{\mathbf{r}}\frac{\delta\mathcal{F}[\varphi,\varrho]}{\delta\varrho(\mathbf{r},t)} + T\pi^2 + T(\nabla_{\mathbf{r}}\rho)^2\right].$$

Investigating $\mathcal{H} = 0$ equations of motion, one may show that the long-time probability of finding a field configuration $\varphi_0(\mathbf{r})$, $\varrho_0(\mathbf{r})$ is $\mathcal{P} = Z^{-1}\mathrm{e}^{-\mathcal{F}[\varphi_0,\varrho_0]/T}$, where the equilibrium partition function is $Z = \int \mathbf{D}[\varphi,\varrho]\,\mathrm{e}^{-\mathcal{F}[\varphi,\varrho]/T}$. The fact that ϱ is conserved does not show up in this limit. One may perform Gaussian integration over $\varrho(\mathbf{r})$, which only leads to the renormalization of the quartic coupling constant $g \to g - A^2/C$ in the potential $V(\varphi)$. Therefore the static critical exponents of the field φ are the same as in model A. The Gaussian integration over the fields $\varrho(\mathbf{r},t)$ and $\rho(\mathbf{r},t)$ can be done, in principle, in the dynamic action as well. However, it generates an essentially nonlocal interaction vertex $(\pi\varphi)_{-\mathbf{q},-\omega}[A^2\mathbf{q}^2/(\mathrm{i}\omega - C\mathbf{q}^2)](\varphi^2)_{\mathbf{q},\omega}$, which reflects the conserved nature of the field ϱ. It is therefore better to work with the local dynamic action, which depends on four fields, and perform the perturbative RG procedure in it. We shall not go into its details here, but only mention that the dynamic critical exponent in $d < 4$ turns out to be $z = 2 + \epsilon/3$, [145], different from models A and B.

9.6 Quantum Phase Transitions

So far we have been discussing *classical* phase transitions, based on the action (9.5) and its generalizations. The latter is a limiting case of the quantum dissipative action (3.20), where all involved frequencies are much less than the critical temperature, $\epsilon \ll T_c$. If this is indeed the case, one may approximate $\coth(\epsilon/2T) \approx 2T/\epsilon$ in Eq. (3.18), leading to the classical time-local term $4\mathrm{i}\gamma T \int \mathrm{d}t \big[X^{\mathrm{q}}(t)\big]^2$ in Eq. (3.15). Being generalized to the spatially extended case, the latter leads to the $4\mathrm{i}\gamma T \int \mathrm{d}t\mathrm{d}\mathbf{r}\big[\varphi^{\mathrm{q}}(\mathbf{r},t)\big]^2$ fluctuation term in the action (9.5). As discussed in Section 9.4, the vicinity of the second-order phase transition is characterized by the divergent time scale $\tau \propto |\delta|^{-\nu z}$. The corresponding characteristic frequency goes to zero as $\epsilon \propto |\delta|^{\nu z}$ and is bound to become less than T_c sufficiently close to the transition, *if* T_c is finite.

Here we discuss a special case when, by tuning some external parameter (e.g. pressure or magnetic field), the critical temperature is tuned to be exactly zero, $T_c = 0$. Formally, this situation is equivalent to a *quantum* string (or membrane) placed in a potential, for example, as depicted in Fig. 9.1. Since $T = 0$, the string can't cross the barrier by thermal activation, but rather has to do it via quantum tunneling. The dynamics of such tunneling is assumed to be *overdamped* by a coupling to an external bosonic Ohmic bath.[7] It is indeed often the case close to the phase transition (i.e. bifurcation point of the potential), as discussed in Section 3.4. The corresponding damping kernel $2\mathrm{i}\gamma\epsilon\coth(\epsilon/2T) \to 2\mathrm{i}\gamma|\epsilon|$ is a nonlocal function

[7] In Section 10.8 we discuss the "bath" originating from fermions, which makes the problem considerably more complicated.

of time, given by the $T \to 0$ limit of the last term in Eq. (3.20). The corresponding quantum dissipative action takes the form (cf. Eq. (9.5))

$$\begin{aligned} S[\varphi, \varphi^{\rm q}] &= -2\int {\rm d}t\,{\rm d}\mathbf{r}\Big[\varphi^{\rm q}\left(\gamma\partial_t - D\nabla_{\mathbf{r}}^2 + \delta\right)\varphi + g\varphi^{\rm q}\varphi^3 + g(\varphi^{\rm q})^3\varphi\Big] \\ &+\frac{{\rm i}\gamma}{\pi}\iint {\rm d}\mathbf{r}\,\frac{{\rm d}t{\rm d}t'}{(t-t')^2}\Big[\varphi^{\rm q}(\mathbf{r},t) - \varphi^{\rm q}(\mathbf{r},t')\Big]^2. \end{aligned} \tag{9.32}$$

We again took $V(\varphi) = \delta\varphi^2/2 + g\varphi^4/4$ as a potential and, in accordance with Eq. (3.20), calculated $V(\varphi + \varphi^{\rm q}) - V(\varphi - \varphi^{\rm q})$. Since we are dealing with the quantum problem, there is no reason to omit $(\varphi^{\rm q})^3\varphi$, as was done in the classical setting following Section 4.1. The Gaussian propagators of this action are given by Eqs. (9.17) with the substitutions $\omega \to \gamma\omega$ and, in the numerator of Eq. (9.17c), $2T \to \gamma|\omega|$.

One can now split the fields into slow and fast components and rescale the space, time, and slow fields in accordance with Eq. (9.20). The coupling constants acquire bare scaling dimensions given by Eqs. (9.21), with the exception that there is no separate scaling for γT. Instead, the last term in Eq. (9.32) implies that $\gamma' = b^{d+2\tilde{\chi}}\gamma$. Comparing it with the first of Eqs. (9.21), one concludes that the quantum and classical components of the field scale in the same way, $\tilde{\chi} = \chi$. This is in accord with the expectation that both terms $\varphi^{\rm q}\varphi^3$ and $(\varphi^{\rm q})^3\varphi$ should be kept in the quantum problem. Demanding that the quadratic critical action is invariant under the scale transformation, that is, that γ and D are not renormalized, one finds $z = 2$ and $\chi = \tilde{\chi} = -d/2$. Then the bare scaling dimension of the critical parameter δ is $[\delta] = 2$, in agreement with the classical case. However, the scaling dimension of the nonlinear coupling g is very different: $[g] = 2 - d$, instead of $4 - d$ in the classical case. In other words, instead of $d_{\rm c} = 4$, we find $d_{\rm c} + z = 4$. Effectively quantum mechanics adds z additional spatial dimensions!

For $d > 2$ the nonlinearity is irrelevant and the Gaussian critical exponents, for example, $\nu = 1/2$ and $z = 2$, are exact. This is true not only at $T = 0$, but also for $T > 0$ in a certain range of the critical parameter δ, away from the immediate vicinity of the $T > 0$ classical transition, [146]; see Fig. 9.5. In the latter case the temperature provides a scale for the frequency and momentum arguments of the correlation functions, that is, the latter depend on ω/T and Dq^2/T. For $d \leq 2$ there are corrections to the Gaussian exponents, which one may access with the $\epsilon = 2 - d$ expansion. The corresponding diagrams for the one-loop corrections to δ and g[8] are plotted in Fig. 9.4. The calculations follow Eqs. (9.22) and (9.23)

[8] The diagram of Fig. 9.4(b) renormalizes the coefficient of $\varphi^{\rm q}\varphi^3$. The renormalization of the coefficient of $(\varphi^{\rm q})^3\varphi$ is provided by the same loop (with three dashed and one full legs). Therefore if initially these two constants are the same, the (one-loop) renormalization preserves their equality. Notice that a $(\varphi^{\rm q})^2\varphi^2$ term with the same scaling dimension as g is also generated.

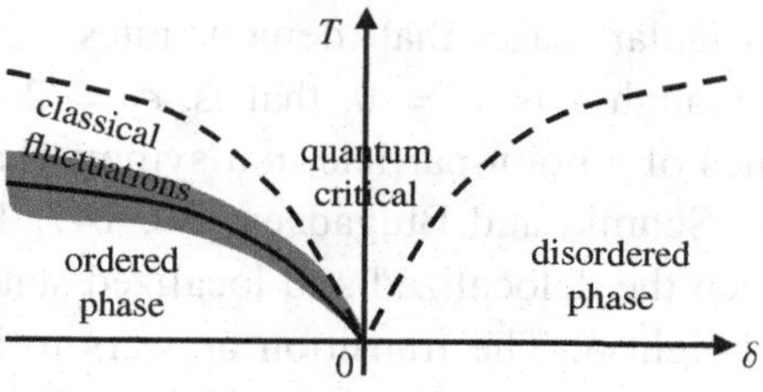

Figure 9.5 Possible phase diagram: critical temperature T_c versus a tuning parameter δ. The quantum phase transition takes place at the point where $T_c(\delta) = 0$. The shaded region is the Ginzburg regime where the classical scaling deviates from the mean field, if $d < d_c$. The left dashed line indicates a crossover between quantum critical and Gaussian regimes in the case $d + z < d_c$. The right dashed line $T \propto \delta^{\nu z}$ is a crossover between the quantum critical regime, where the correlation time is $\tau \sim T^{-1}$, and the disordered phase, where the correlation time is $\tau \propto |\delta|^{-\nu z}$.

with the substitution $2T \to \gamma|\omega|$ in the $D_0^K(\mathbf{q}, \omega)$ Green function. This shifts the logarithmic dimension of the loops from $d = 4$ to $d = 2$, but does not change the relative factor between the two diagrams of Fig. 9.4. As a result, the two nontrivial RG equations (9.25) preserve their form:

$$\partial_l \delta = \left(2 - 3g\tilde{K}_d\right)\delta, \qquad \partial_l g = \left(\epsilon - 9g\tilde{K}_d\right)g, \tag{9.33}$$

where, however, $\epsilon = 2 - d$ and $\tilde{K}_d = (\pi)^{-1}\partial_l \int_\Lambda^{\Lambda(1+l)} \mathrm{d}\,\mathbf{q}_\mathrm{f}/[(2\pi)^d D\mathbf{q}_\mathrm{f}^2]$. If $d < 2$ the second equation predicts the stable fixed point $g^* = \epsilon/(9\tilde{K}_2)$, where the critical parameter scales as $\partial_l\delta = (2 - \epsilon/3)\delta$. As a result, one finds in the ϵ-expansion for the correlation length critical exponent $\nu = (2 - \epsilon/3)^{-1} \approx 1/2 + \epsilon/12$, while the correlation time exponent is $\nu z = 1 + \epsilon/6$.

This non-Gaussian critical behavior is expected to persist for $T > 0$ and a certain range of the critical parameter δ around zero, known as a *quantum critical region* [146]. In this regime the (inverse) temperature serves as the only correlation time scale, $\tau \sim T^{-1}$. Due to the critical scaling the spatial correlation length must be $\xi \propto T^{-1/z}$. As a result, for example, the $\langle\varphi\varphi\rangle$ correlation function (cf. Eq. (9.17c)) is expected to have the form

$$D^K(\mathbf{q}, \omega, T) = \frac{1}{T}\,\Phi\left(\frac{cq}{T^{1/z}}, \frac{\omega}{T}\right), \tag{9.34}$$

where Φ is a universal scaling function of the two arguments and c is a cutoff-dependent constant. The factor $1/T$ in front simply takes care of the overall dimensionality of the correlation function; compare Eq. (9.17c). Of course, if $|\delta|$ is so large that the critical correlation length $|\delta|^{-\nu}$ is shorter than $T^{-1/z}$, the universal scaling (9.34) breaks down. This dictates the boundaries of the quantum critical region as $T \gtrsim |\delta|^{\nu z}$, Fig. 9.5.

An interesting particular case that demonstrates eventual failure of the ϵ-expansion at large enough ϵ is $d = 0$, that is, $\epsilon = 2$. It describes a dissipative quantum mechanics of a point particle in a symmetric double-well potential. As was first shown by Schmid and Bulgadaev [39, 147, 148], there is a $T = 0$ phase transition between the delocalized and localized states as a function of, for example, the barrier height δ. The transition appears to be of the Berezinskii–Kosterlitz–Thouless (BKT) universality class [9], known from the classical $d = 2$ XY model. This supports the assertion that a d-dimensional quantum model may be mapped onto a $d + z$-dimensional classical one. On the other hand, the ϵ-expansion fails to describe critical scaling near the BKT transition. For example, in the latter case the correlation length (i.e. correlation time of the dissipative quantum problem) is known to diverge exponentially $\propto e^{b/\sqrt{|\delta-\delta_c|}}$ and not as a power-law as expected from the ϵ-expansion.

9.7 Absorbing State Transitions

The Hohenberg–Halperin classification deals only with the models based on the Landau free energy $\mathcal{F}[\varphi, \ldots]$. In the long-time limit such models approach thermodynamic equilibrium, described by the equilibrium statistical mechanics, Eq. (9.10). On the other hand, one may consider essentially non-equilibrium models, which do *not* approach thermal equilibrium. A big class of such systems is based on reaction models without detailed balance, which may undergo transitions into the *absorbing state*. Their zero-dimensional versions are discussed in Section 4.11. We put them now onto a d-dimensional lattice and allow for the random walk of agents on such a lattice. A state of the system may be described by $\mathcal{P}(n_1, n_2, \ldots, t)$, which is a probability of finding an integer n_j number of agents on the lattice site j at time t. The part of the Master equation that describes the random walk takes the form, for example, in $d = 1$,

$$\begin{aligned}\partial_t \mathcal{P}([n_j], t) &= \sum_j D\Big[(n_{j-1}+1)\mathcal{P}(\ldots, n_{j-1}+1, n_j-1, \ldots, t)\\ &\quad - 2n_j\mathcal{P}(\ldots, n_{j-1}, n_j, n_{j+1}, \ldots, t)\\ &\quad + (n_{j+1}+1)\mathcal{P}(\ldots, n_j-1, n_{j+1}+1, \ldots, t)\Big]\\ &= \sum_j D\Big(e^{\hat{p}_j}\, e^{-\hat{p}_{j-1}} n_{j-1} - 2n_j + e^{\hat{p}_j}\, e^{-\hat{p}_{j+1}} n_{j+1}\Big)\, \mathcal{P}([n_j], t), \quad (9.35)\end{aligned}$$

where D is the rate of random hops between neighboring sites and, following Section 4.11, we have defined operators $\hat{p}_j = -\partial/\partial n_j$. As a result, the full Master equation may be written as $\partial_t \mathcal{P} = \hat{\mathcal{H}}[\hat{p}, n]\mathcal{P}$, where the Hamiltonian operator $\hat{\mathcal{H}}$ is

$$\hat{\mathcal{H}}[\hat{p},n] = \sum_j \Big[D\Big(e^{\hat{p}_j}\, e^{-\hat{p}_{j-1}} n_{j-1} - 2n_j + e^{\hat{p}_j}\, e^{-\hat{p}_{j+1}} n_{j+1} \Big) + H(\hat{p}_j, n_j) \Big], \qquad (9.36)$$

where $H(\hat{p}_j, n_j)$ is the on-site reaction Hamiltonian (4.78).

To avoid exponentials of the operators, it is convenient to perform the local Cole–Hopf canonical *operator* transformation

$$\hat{\pi} = e^{\hat{p}}, \qquad \hat{\varphi} = e^{-\hat{p}}\, n, \qquad [\hat{\varphi}, \hat{\pi}] = 1. \qquad (9.37)$$

The random walk part of the Hamiltonian (9.36) takes the much simpler form $\sum_j D\,\hat{\pi}_j\,(\hat{\varphi}_{j-1} - 2\hat{\varphi}_j + \hat{\varphi}_{j+1}) = \int d\mathbf{r} D\,\hat{\pi}(\mathbf{r}) \nabla_{\mathbf{r}}^2 \hat{\varphi}(\mathbf{r})$, where we have switched to continuous notation. To transform the local reaction Hamiltonian (4.78) consider, for example, the reaction $kA \stackrel{\lambda}{\rightarrow} (k+r)A$. The corresponding reaction rates are proportional to the number of ways to select k agents, which enter the reaction, out of n, that is, $W_{n\to n+r} = \lambda n(n-1)\dots(n-k+1)/k! = \lambda\, \hat{\pi}^k \hat{\varphi}^k / k!$.[9] As a result, the corresponding reaction Hamiltonian (4.78) takes the form

$$kA \stackrel{\lambda_{kr}}{\rightarrow} (k+r)A; \qquad \hat{H}_{kr}(\hat{\pi}, \hat{\varphi}) = \frac{\lambda_{kr}}{k!}\, (\hat{\pi}^{\,r+k} - \hat{\pi}^{\,k})\hat{\varphi}^{\,k}. \qquad (9.38)$$

The total reaction Hamiltonian is $H(\hat{\pi}, \hat{\varphi}) = \sum_{k=0}^{\infty} \sum_{r=-k}^{\infty} H_{kr}(\hat{\pi}, \hat{\varphi})$. For example, the reaction set (4.76) leads to the following Cole–Hopf transformed Hamiltonian (cf. Eq. (4.81)):

$$H(\hat{\pi}, \hat{\varphi}) = \frac{\lambda}{2}\,(1 - \hat{\pi}^2)\hat{\varphi}^2 + \mu(1 - \hat{\pi})\hat{\varphi} + \sigma(\hat{\pi}^2 - \hat{\pi})\hat{\varphi}. \qquad (9.39)$$

Notice that since $\hat{\pi} = e^{\hat{p}}$, the fundamental normalization $H(p,n)\big|_{p=0} = 0$, Eq. (4.24), now reads as $H(\pi, \varphi)\big|_{\pi=1} = 0$. Notice also that the corresponding classical equation of motion $\partial_t \varphi = \partial_\pi H(\pi, \varphi)\big|_{\pi=1}$ is the same as the mean-field rate equation (4.80), if $\varphi = n \gg k$.

A convenient representation of the Cole–Hopf operator formalism is achieved with the help of the *generating function*, defined as

$$\mathcal{G}(\pi, t) = \sum_{n=0}^{\infty} \pi^n\, \mathcal{P}(n,t); \qquad \partial_t \mathcal{G}(\pi, t) = \hat{H}(\pi, \hat{\varphi})\, \mathcal{G}(\pi, t). \qquad (9.40)$$

Here π is a real (or complex) number, while $\hat{\varphi} = \partial_\pi$, in agreement with the commutation relation (9.37). The equation of motion (9.40) for the generating function $\mathcal{G}(\pi, t)$ is an immediate consequence of its definition and the Master equation for the probability $\mathcal{P}(n,t)$.[10] Notice that due to the conservation of probability

[9] Employing the commutation relation (9.37), one finds, e.g. $\hat{\pi}^3 \hat{\varphi}^3 = e^{-3\partial_n} e^{\partial_n} n e^{\partial_n} n e^{\partial_n} n = e^{-2\partial_n} n e^{\partial_n} n(n+1)\, e^{\partial_n} = e^{-2\partial_n} n(n+1)(n+2) e^{2\partial_n} = (n-2)(n-1)n$.

[10] Consider, e.g. reaction $A \stackrel{\mu}{\rightarrow} \emptyset$. The Master equation for the probability distribution is given by $\partial_t \mathcal{P}(n,t) = \mu(n+1)\mathcal{P}(n+1,t) - \mu n \mathcal{P}(n,t)$. Multiplying both sides by π^n and summing over n, one finds

$\mathcal{G}(1,t) = 1$, and thus the condition $H(1,\varphi) = 0$ simply reflects the fact that $\mathcal{G}(1,t)$ is conserved.

Employing now the coherent state representation for the canonical pair π and $\hat{\varphi}$ along with the fact that the Hamiltonian is *normally ordered*, one may write the evolution operator for the generating function in the form of a functional integral over the set of fields $\pi_j(t)$ and $\varphi_j(t)$; compare Eq. (4.79), where j is the lattice index. The corresponding action in the continuous notations acquires the form[11]

$$S[\varphi,\pi] = \int dt\, d\mathbf{r}\big[\pi\,\partial_t\varphi - D\pi\,\nabla_{\mathbf{r}}^2\varphi - H(\pi,\varphi)\big], \tag{9.41}$$

where the local reaction Hamiltonian function $H(\pi(\mathbf{r},t),\varphi(\mathbf{r},t))$ is given by Eqs. (9.38) or, for example, (9.39). Despite close similarities with the action (9.19), the physical consequences of the action (9.41) are dramatically different. The formal reason for these differences is that the Hamiltonian $H(\pi,\varphi)$ does *not* have the structure $-\pi V'(\varphi) + T\pi^2$ and thus there is no Landau free energy $\mathcal{F}[\varphi]$ underlying the action (9.41). As a result, the system does not equilibrate to the static state described by the equilibrium statistical mechanics (9.10). The phase transitions are still possible, but their universality classes are different from those in the Hohenberg–Halperin classification.

If no agents are created from the empty state, that is, $k \neq 0$ in Eq. (9.38), the zero-dimensional models of Section 4.11 go to extinction in the long-time limit. If they are put on an infinite lattice, their behavior becomes more interesting. Depending on the parameters, the lattice sites may either go extinct as in $d = 0$, or maintain a fluctuating, but nonzero on average, population. In the latter case the extinct sites are repopulated by agents diffusing from neighboring non-empty sites. The transition between the two scenarios exhibits all the phenomenology of the dynamic second-order phase transitions. In particular, close to the transition there is a divergent correlation length $\xi \propto |\delta|^{-\nu}$ and divergent correlation time $\tau \propto |\delta|^{-\nu z}$, where δ is a critical parameter passing through zero at the transition point. In the active phase the average on-site population scales as $\langle n\rangle \propto \delta^{\beta}$. The set of critical exponents ν, z, β, and so on is remarkably universal and depends only on a few internal

$\partial_t\mathcal{G}(\pi,t) = \mu\partial_\pi\mathcal{G}(\pi,t) - \mu\pi\partial_\pi\mathcal{G}(\pi,t) = \mu(1-\pi)\partial_\pi\mathcal{G}(\pi,t) = \mu(1-\pi)\hat{\varphi}\,\mathcal{G}(\pi,t)$, in agreement with Eqs. (9.38) and (9.40).

[11] Another representation of the same operators is the Doi–Peliti formalism [149]. By analogy with quantum mechanics let us denote the state of the system with n agents as $|n\rangle$ and introduce creation/annihilation operators that act as $\hat{a}^\dagger|n\rangle = |n+1\rangle$ and $\hat{a}|n\rangle = n|n-1\rangle$ and thus obey the commutation relation $[\hat{a},\hat{a}^\dagger] = 1$. A generic state of the system, defined as $|\Psi(t)\rangle = \sum_{n=0}^{\infty}\mathcal{P}(n,t)|n\rangle$, obeys the imaginary time "Schrödinger" equation $\partial_t|\Psi(t)\rangle = \hat{H}(\hat{a}^\dagger,\hat{a})|\Psi(t)\rangle$, where the Hamiltonian may be derived from the Master equation. For example, for the simplest reaction $A \xrightarrow{\mu} \emptyset$, $\hat{H}(\hat{a}^\dagger,\hat{a}) = \mu\hat{a} - \mu\hat{a}^\dagger\hat{a} = \mu(1-\hat{a}^\dagger)\hat{a}$; indeed, $\hat{a}|\Psi(t)\rangle = \sum_n(n+1)\mathcal{P}(n+1,t)|n\rangle$, while $\hat{a}^\dagger\hat{a}|\Psi(t)\rangle = \sum_n n\mathcal{P}(n,t)|n\rangle$. Comparing it with Eq. (9.38), one concludes that $\hat{a}^\dagger = \hat{\pi}$ and $\hat{a} = \hat{\varphi}$, which may be verified for an arbitrary reaction. Therefore the resulting coherent state action is the same as Eq. (9.41).

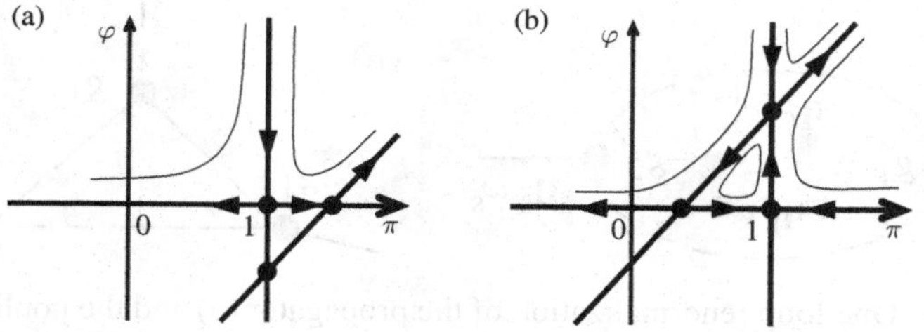

Figure 9.6 Phase portraits of the DP universality class: (a) empty state $\delta < 0$; (b) active state $\delta > 0$. At the transition all three zero-energy lines intersect at the same point $(1, 0)$. Notice the opposite sign convention of the critical parameter in comparison with Fig. 9.3. Indeed, here the "magnetic" state corresponds to $\delta > 0$.

symmetries of the model. In what follows we consider some of the universality classes and their geometric interpretation.

9.7.0.1 Directed Percolation

Consider a lattice version of the reaction model specified by Eq. (4.76). The corresponding action is given by Eqs. (9.41) and (9.39). Unlike the equilibrium models considered previously, here there is no potential function $V(\varphi)$ that bifurcates into two symmetric minima at the transition point. Nevertheless, the Hamiltonian function $H(\pi, \varphi)$ does exist and the phase transition may be associated with a qualitative rearrangement of its phase portrait; see Fig. 9.3 for the equilibrium case. As depicted in Fig. 4.4, the set of zero energy lines forms a triangle close to the bifurcation point.[12] One can thus identify the phase transition as the rearrangement of such a triangle, Fig. 9.6. The corresponding universality class is known as *directed percolation* (DP). Its geometric image is associated with Fig. 9.6, in the same way as the model A (i.e. dynamic Ising model) universality class is associated with Fig. 9.3.

To proceed with the RG description one needs to specify the simplest Hamiltonian $H(\pi, \varphi)$ that exhibits the rearrangement of its zero energy lines as a function of the critical parameter δ according to Fig. 9.6. As discussed in Section 4.11, such a universal Hamiltonian is given by Eq. (4.83). In (π, φ) variables it takes the form $H(\pi, \varphi) = (\pi - 1)(\delta - g_1\varphi + g_2(\pi - 1))\varphi$, where δ is the critical parameter and g_1, g_2 are running constants, whose bare values may be found from the comparison

[12] Figure 4.4 is plotted in the coordinates (p, n), which are related to (π, φ), through the area-preserving canonical transformation (9.37). Its qualitative structure is thus the same in (π, φ) coordinates.

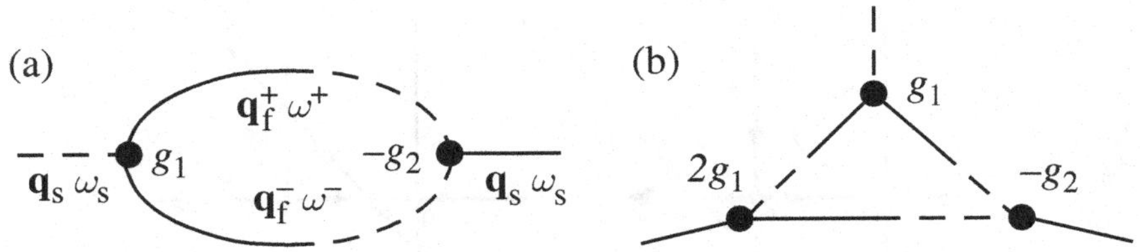

Figure 9.7 One-loop renormalization of the propagator (a) and the nonlinear vertex g_1 (b) in the DP model. Full lines represent φ, while dashed lines represent π. The type of vertex is indicated next to the bold dots and $\mathbf{q}_\mathrm{f}^{\pm} = \mathbf{q}_\mathrm{f} \pm \mathbf{q}_\mathrm{s}/2$; $\omega^{\pm} = \omega \pm \omega_\mathrm{s}/2$. There are combinatorial factors of 2 for diagram (a) and 4 for (b).

with Eq. (9.39). It is convenient to shift the momentum as $\pi - 1 \to \pi$ and write the resulting Hamiltonian as

$$H(\pi,\varphi) = \pi\,(\delta - g_1\varphi + g_2\pi)\,\varphi. \tag{9.42}$$

Notice that the kinetic and random walk parts of the action (9.41) are not affected by the shift. The corresponding field theory, known in high-energy physics as the Reggeon field theory [151], was identified as the proper representation of DP by Janssen, Grassberger [152], and Cardy [155]. Its Gaussian propagators $\langle\pi\varphi\rangle$ are given by Eqs. (9.17a,b), where $\varphi^\mathrm{q} = \pi/(2\mathrm{i})$. As in any non-equilibrium theory $\langle\pi\pi\rangle = 0$, which may be traced back to the conservation of probability, that is, $H(0,\varphi) = 0$ (after the shift). The peculiarity of reaction-diffusion models with the absorbing state is that $\langle\varphi\varphi\rangle = 0$ (there is no FDT!). This property originates from the fact that $H(\pi,0) = 0$, due to the absence of creation from the empty state (cf. Eq. (9.38)), and therefore no terms $\propto \pi^2\varphi^0$ are allowed. This latter property appears to be intact in the process of renormalization. There are two nonlinear three-leg vertices $g_1\pi\varphi^2$ and $-g_2\pi^2\varphi$, which renormalize both the propagators and the coupling constants g_1 and g_2.

The RG proceeds along the same lines as outlined in Section 9.4. The bare scaling of γ (i.e. constant in front of $\pi\,\partial_t\varphi$), D, and δ is the same as in Eq. (9.21), while the bare scaling of the nonlinear vertices is

$$g_1' = b^{d+z+\tilde{\chi}+2\chi}g_1, \qquad g_2' = b^{d+z+2\tilde{\chi}+\chi}g_2.$$

Splitting the fields into slow and fast components gives rise to the following interaction vertices: (i) $g_1\pi_\mathrm{s}\varphi_\mathrm{f}^2$; (ii) $2g_1\pi_\mathrm{f}\varphi_\mathrm{f}\varphi_\mathrm{s}$; (iii) $-g_2\pi_\mathrm{f}^2\varphi_\mathrm{s}$; and (iv) $-2g_2\pi_\mathrm{s}\pi_\mathrm{f}\varphi_\mathrm{f}$. Expanding e^{-S} in a power series in these vertices and performing Gaussian integrations over the fast fields with the help of Eqs. (9.17a,b), one finds the perturbative RG corrections. The one-loop renormalization of the propagator originates from

(i) × (iii) combination of vertices.[13] The corresponding diagram is plotted in Fig. 9.7(a) and is given by

$$
\begin{aligned}
&8g_1g_2\int\frac{d\mathbf{q}_f d\omega}{(2\pi)^{d+1}}D_0^R(\mathbf{q}_f^+,\omega^+)D_0^A(\mathbf{q}_f^-,\omega^-)=\frac{g_1g_2}{\gamma}\int_\Lambda^{b\Lambda}\frac{d\mathbf{q}_f}{(2\pi)^d}\frac{1}{D\mathbf{q}_f^2+\delta+\frac{D\,\mathbf{q}_s^2}{4}-\frac{i\gamma\omega_s}{2}}\\
&\approx\frac{g_1g_2}{\gamma}\int_\Lambda^{b\Lambda}\frac{d\,\mathbf{q}_f}{(2\pi)^d}\frac{1}{D\mathbf{q}_f^2}-\left(\delta+\frac{D\mathbf{q}_s^2}{4}-\frac{i\gamma\omega_s}{2}\right)\frac{g_1g_2}{\gamma}\int_\Lambda^{b\Lambda}\frac{d\mathbf{q}_f}{(2\pi)^d}\frac{1}{(D\mathbf{q}_f^2)^2},
\end{aligned}
\tag{9.43}
$$

where $\mathbf{q}_f^\pm=\mathbf{q}_f\pm\mathbf{q}_s/2$; $\omega^\pm=\omega\pm\omega_s/2$ and $\mathbf{q}_s$, ω_s are the external slow momentum and frequency. They should be understood as $\mathbf{q}_s^2\pi_s\varphi_s\ \to\ -\pi_s\nabla_\mathbf{r}^2\varphi_s$ and $-i\omega_s\pi_s\varphi_s\ \to\ \pi_s\partial_t\varphi_s$. As a result, the last term on the right-hand side provides the one-loop renormalization of the running constants δ, D, and γ, respectively. The first δ-independent term on the right-hand side results in a shift of the critical point away from $\delta=0$ and thus may be neglected.

The renormalization of the nonlinear coupling constant g_1 comes from the (i) × (ii) × (iii) combination of vertices.[14] The corresponding diagram is plotted in Fig. 9.7(b) and is given by

$$
-64g_1^2g_2\int\frac{d\mathbf{q}_f d\omega}{(2\pi)^{d+1}}\left[D_0^R(\mathbf{q}_f,\omega)\right]^2D_0^A(\mathbf{q}_f,\omega)=\frac{2g_1^2g_2}{\gamma}\int_\Lambda^{b\Lambda}\frac{d\mathbf{q}_f}{(2\pi)^d}\frac{1}{(D\mathbf{q}_f^2)^2}.
\tag{9.44}
$$

The renormalization of g_2 comes from the (i) × (iv) × (iii) combination and is given by Eq. (9.44) with g_2 and $-g_1$ interchanged. We now put $b=1+l$, evaluate the integrals in Eqs. (9.43), (9.44), exponentiate the result, and combine it with the bare scaling. As a result we find the set of RG equations

$$
\begin{aligned}
&\partial_l\gamma=(d+\tilde{\chi}+\chi-\frac{g_1g_2}{2}K_d)\gamma,\quad \partial_lD=(d+z-2+\tilde{\chi}+\chi-\frac{g_1g_2}{4}K_d)D,\\
&\partial_l\delta=(d+z+\tilde{\chi}+\chi-g_1g_2K_d)\delta,\\
&\partial_lg_1=(d+z+\tilde{\chi}+2\chi-2g_1g_2K_d)g_1,\quad \partial_lg_2=(d+z+2\tilde{\chi}+\chi-2g_1g_2K_d)g_2,
\end{aligned}
\tag{9.45}
$$

where $K_d=(1/\gamma D^2)\Lambda^{d-4}2^{1-d}\pi^{-d/2}\Gamma(d/2)$. We can now use the freedom of choosing $\tilde{\chi}$, χ along with the scale of time to fix three of these five parameters. It is convenient to fix $\gamma=1$ and D, that is, the parameters of the Gaussian action at the critical point, along with $\tilde{\chi}=\chi$ to maintain the symmetry of the action between π and φ. This is achieved by demanding that $d+2\chi-g_1g_2K_d/2=0$

[13] The (ii) × (iv) combination is proportional to $D_0^R(t)D_0^A(t)$, which is zero due to causality. All other combinations vanish due to $\langle\pi\pi\rangle=\langle\varphi\varphi\rangle=0$.

[14] The (ii) × (ii) × (iv) combination is zero due to causality. All other combinations vanish due to $\langle\pi\pi\rangle=\langle\varphi\varphi\rangle=0$.

and $d + z - 2 + 2\chi - g_1 g_2 K_d/4 = 0$, leading to the following scaling dimensions: $z = 2 - g_1 g_2 K_d/4$, and $\tilde{\chi} = \chi = -d/2 + g_1 g_2 K_d/4$. The remaining RG equations are

$$\partial_l \delta = \left(2 - \frac{3}{4} g_1 g_2 K_d\right) \delta, \qquad \partial_l g_{1,2} = \left(\frac{\epsilon}{2} - \frac{3}{2} g_1 g_2 K_d\right) g_{1,2}, \tag{9.46}$$

where $\epsilon = 4 - d$. For $d < d_c = 4$ one thus finds the nontrivial fixed point, where $g_1^* g_2^* = \epsilon/(3K_4) + O(\epsilon^2)$. In a vicinity of this fixed point the equation for the critical parameter is $\partial_l \delta = (2 - \epsilon/4)\delta$, resulting in the correlation length critical exponent $\nu^{-1} = 2 - \epsilon/4 + O(\epsilon^2)$. We also find the dynamic critical exponent $z = 2 - \epsilon/12 + O(\epsilon^2)$ as well as the exponent $\chi = -d/2 + \epsilon/12 = -2 + 7\epsilon/12 + O(\epsilon^2)$. The order parameter exponent β is defined through $\langle \varphi \rangle \propto |\delta|^\beta$. Because of the scaling $\varphi \to b^\chi \varphi$ and $b \propto \xi \propto |\delta|^{-\nu}$, one finds $\beta = -\chi\nu = 1 - \epsilon/6 + O(\epsilon^2)$. This set of critical exponents is different from all of the equilibrium Hohenberg–Halperin universality classes.

9.7.0.2 Other Universality Classes of Reaction–Diffusion Models

The triangular structure of the phase portrait, Fig. 9.6, is very robust. Indeed, the two zero-energy lines are fixed: $\pi = 1$ by the conservation of probability and $\varphi = 0$ by the presence of the absorbing empty state. Therefore, changing a single critical parameter, one can adjust the position of only one additional line. As a result, most of the reaction-diffusion models with an absorbing state possess the triangular phase portrait and fall into the DP universality class [152, 156, 157, 158]. Yet there are some exceptions, which are achieved by fixing some additional zero-energy lines, or by increased degeneracy of the $\varphi = 0$ line.

Parity-Conserving Models Consider a set of reactions that *all* conserve parity of the state, for example, $2A \stackrel{2g}{\to} \emptyset$ and $A \stackrel{\delta}{\to} 3A$. Notice that the $d = 0$ model with initially an odd number of agents never goes extinct. A random walk on the lattice violates on-site parity conservation, allowing for an extinction. If $g \gg \delta$ one expects in the long run to find only empty or singly occupied sites. The further decrease of the population is possible only due to diffusion and annihilation of two agents who accidentally meet. This is obviously very different from DP models, where each site may go extinct due to internal reactions only. The corresponding Hamiltonian (9.38) takes the form

$$H(\pi, \varphi) = g(1 - \pi^2)\varphi^2 + \delta(\pi^3 - \pi)\varphi = (1 - \pi^2)(g\varphi - \delta\pi)\varphi. \tag{9.47}$$

Its phase portrait is depicted in Fig. 9.8(a). Notice that the Hamiltonian and thus the action (9.41) possess the reflection symmetry $\varphi \to -\varphi$ and $\pi \to -\pi$, which is generic for any parity-conserving set of reactions. This symmetry is not violated

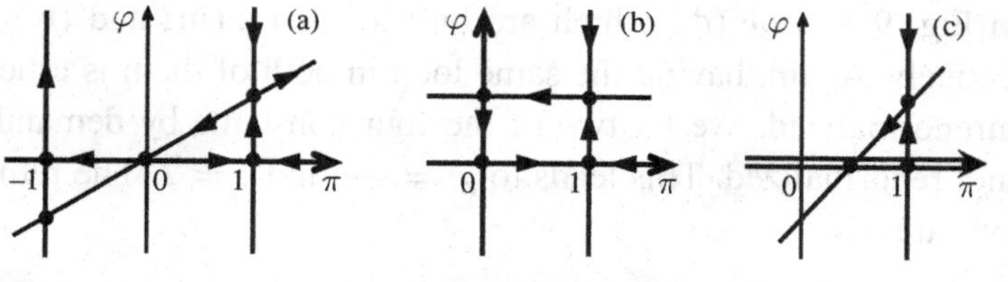

Figure 9.8 Near the transition phase portraits of (a) parity-conserving models; (b) models with non-accessible absorbing state; (c) multi-agent contact processes with $k = 2$. After [159].

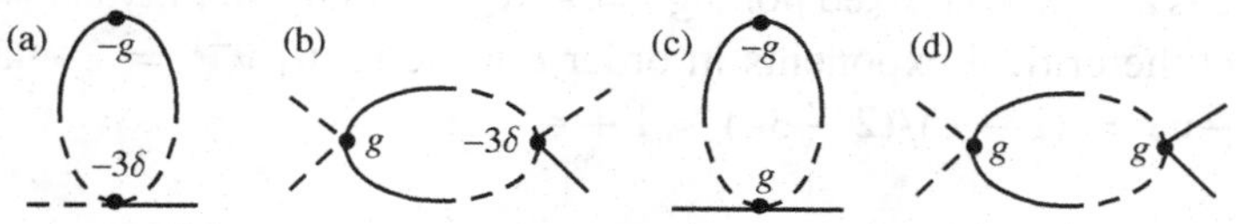

Figure 9.9 One-loop renormalization of the critical parameter δ, (a) or (b); and the nonlinear vertex g, (c) or (d) in the parity-conserving model. All diagrams have a combinatorial factor of 2.

by the renormalization process. As a result, *three* zero-energy lines $\pi = \pm 1$ and $\varphi = 0$ are protected, while the fourth line (in our example $\varphi = \delta\pi/g$) must retain its reflection symmetry. Therefore, the Hamiltonian (9.47) is indeed the simplest representative of the parity-conserving class. The phase transition takes place when the line $\varphi = \delta\pi/g$ is approaching the horizontal position, that is, the *renormalized* $\delta \to 0$.[15]

Turning to the fluctuations, one notices that it is not possible to perform the shift of momentum $\pi - 1 \to \pi$, as in the DP case, and focus on the immediate vicinity of the $(1, 0)$ point of the phase plane. Because of the reflection symmetry, one has to keep the entire interval $\pi \in [-1, 1]$ under consideration; see Fig. 9.8(a). Therefore, one must choose the scaling dimension of the π-field to be zero, $\tilde{\chi} = 0$. The bare scaling of the constants thus reads

$$\gamma' = b^{d+\chi}\gamma, \quad D' = b^{d+z-2+\chi}D, \quad \delta' = b^{d+z+\chi}\delta, \quad g' = b^{d+z+2\chi}g. \tag{9.48}$$

Splitting the fields into slow and fast components, one obtains a number of vertices, of which only four contribute to the one-loop renormalization: (i) $-3\delta\pi_{\rm f}^2\pi_{\rm s}\varphi_{\rm s}$; (ii) $-g\varphi_{\rm f}^2$; (iii) $g\pi_{\rm f}^2\varphi_{\rm s}^2$; and (iv) $g\pi_{\rm s}^2\varphi_{\rm f}^2$. There is no one-loop renormalization of γ and D. The critical parameter δ is renormalized by diagram Fig. 9.9(a) *or* (b), which are built on (ii) $\times$ (i) and (iv) $\times$ (i) vertices, respectively. The fact that both of them contain the same loop (up to a sign) guarantees that the expression $(1 - \pi^2)$ in the Hamiltonian (9.47) is not renormalized. The running constant g is renormalized

15 Since δ is renormalized the transition point may not coincide with the bare rate of the $A \to 3A$ reaction being zero [160].

by diagram Fig. 9.9(c) *or* (d), which are built on (ii) × (iii) and (iv) × (iii) vertices, respectively. Again, having the same loop in both of them is crucial to keep $(1-\pi^2)$ unrenormalized. We fix two of the four constants by demanding that γ and D are not renormalized. This leads to $\chi = -d$ and $z = 2$. The remaining two RG equations are

$$\partial_l \delta = \left(2 - 3g\tilde{K}_d\right)\delta, \qquad \partial_l g = \left(\epsilon - g\tilde{K}_d\right)g, \tag{9.49}$$

where $\epsilon = 2 - d$ and $\tilde{K}_d = D\Lambda^2 K_d$. The critical dimension is thus $d_c = 2$. For $d < d_c$ there is a nontrivial fixed point $g^* = \epsilon/\tilde{K}_2$ and in its vicinity $\partial_l \delta = (2-3\epsilon)\delta$. We thus find the critical exponents in order ϵ to be [160] $\nu^{-1} = 2 - 3\epsilon$, $z = 2$, while $\beta = -\nu\chi = (2-\epsilon)/(2-3\epsilon) \approx 1 + \epsilon$.

Models with Non-accessible Absorbing State In the previous example, sites with an odd number of agents were protected from extinction. One can extend this idea to all sites, forbidding any reaction that brings a site into the absorbing state Ø. (Of course, a site may still become empty due to diffusion out of it.) An example is $2A \xrightarrow{2g} A$ and $A \xrightarrow{\delta} 2A$. The corresponding Hamiltonian (9.38) takes the form

$$H(\pi, \varphi) = g(\pi - \pi^2)\varphi^2 + \delta(\pi^2 - \pi)\varphi = (\pi - \pi^2)\,(g\varphi - \delta)\,\varphi. \tag{9.50}$$

Its phase portrait is plotted in Fig. 9.8(b). Notice that $\pi = 0$ is a new line of zero energy. Inspecting the Hamiltonian (9.38), one notices that it is a generic feature of any reaction that has a non-empty final state $k + r \geq 1$. We shall see that this property is intact in the process of renormalization [159, 162]. To keep both the $\pi = 1$ and $\pi = 0$ lines, one must again choose the scaling dimension of π to be zero, $\tilde{\chi} = 0$. The bare scaling of the constants is thus given by Eq. (9.48). Splitting the fields into slow and fast components, one obtains four vertices that contribute to the renormalization: (i) $-\delta\pi_f^2\varphi_s$; (ii) $-g\pi_s\varphi_f^2$; (iii) $g\pi_f^2\varphi_s^2$; and (iv) $g\pi_s^2\varphi_f^2$. The critical parameter δ is renormalized by diagram Fig. 9.10(a) *or* (b), which are built on the (ii) × (i) and (iv) × (i) vertices, respectively. The fact that both of them contain the same loop (up to a sign) guarantees that the expression $(\pi - \pi^2)$ in the Hamiltonian (9.50) is not renormalized. The running constant g is renormalized by

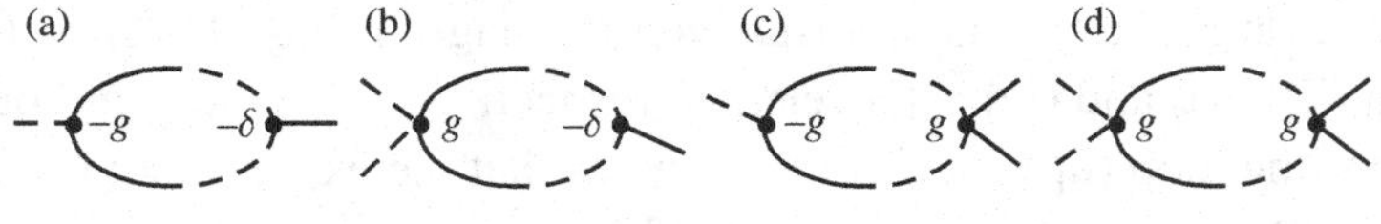

Figure 9.10 One-loop renormalization of the critical parameter δ, (a) or (b); and the nonlinear vertex g, (c) or (d) in the model with a non-accessible absorbing state. All diagrams have a combinatorial factor of 2.

diagram Fig. 9.10(c) *or* (d), which are built on the (ii) × (iii) and (iv) × (iii) vertices, respectively. Again, having the same loop in both of them is crucial to keep $(\pi - \pi^2)$ unrenormalized.[16] Fixing γ and D leads to $\chi = -d$ and $z = 2$, while the remaining RG equations are

$$\partial_l \delta = (2 - g\tilde{K}_d)\,\delta, \qquad \partial_l g = (\epsilon - g\tilde{K}_d)\,g, \tag{9.51}$$

where $\epsilon = 2 - d$. The critical dimension is thus $d_c = 2$. For $d < d_c$ there is a nontrivial fixed point $g^* = \epsilon/\tilde{K}_d$ and in its vicinity $\partial_l \delta = (2 - \epsilon)\delta = d\delta$. We thus find the critical exponents $\nu^{-1} = d$, $z = 2$, and $\beta = -\nu\chi = 1$. Using the fact that the only possible diagrams are those including the chains of "bubbles," one may show [162] that Eqs. (9.51) are actually exact to all orders in ϵ and thus we have found the critical exponents for any $d \leq 2$. For $d > 2$ the Gaussian exponents are $\nu = 1/2$, $z = 2$, and $\beta = 1$.

One may consider different reaction sets, for example, $2A \overset{2g}{\to} A$ and $A \overset{\delta}{\to} 3A$, with the Hamiltonian $H = (\pi - \pi^2)(g\varphi - \delta - \tilde{\delta}\pi)\varphi$, where $\tilde{\delta} = \delta$. The corresponding phase portrait has a tilted zero energy line $\varphi = (\delta + \tilde{\delta}\pi)/g$. Although the naive scaling of δ and $\tilde{\delta}$ is the same, one may argue [162] that the one-loop corrections make $\tilde{\delta}$ less relevant than δ and thus the phase portrait approaches that of the Hamiltonian (9.50), Fig. 9.8(b). As a result, these (and other similar) reactions belong to the same universality class.

Multi-agent Contact Processes One may consider reaction sets where *all* the reactions require *at least* k agents to be initiated. An example with $k = 2$, that is, a two-particle contact process with diffusion, is given by $2A \to \emptyset$, $2A \to 3A$, and $3A \to A$. It is easy to see that all terms in the corresponding Hamiltonian (9.38) are proportional to φ^k and therefore the $\varphi = 0$ zero-energy line is k times degenerate. One may show that this degeneracy is not removed in the process of renormalization. That is, no new vertices $\propto \varphi, \varphi^2, \ldots, \varphi^{k-1}$ are generated. The corresponding phase portrait in the vicinity of the transition is plotted in Fig. 9.8(c) for $k = 2$. After the shift $\pi - 1 \to \pi$ the minimal reaction Hamiltonian near the transition acquires the form

$$H(\pi, \varphi) = \pi\,(\delta - g_1\varphi + g_2\pi)\,\varphi^2, \tag{9.52}$$

where we again took $k = 2$. As in the DP case, to keep the triangular structure of the phase portrait, one has to impose the same scaling dimension for both fields, that is, $\tilde{\chi} = \chi$. Since both nonlinear vertexes include four fields, there is no one-loop renormalization of the propagator, similarly to Fig. 9.9. One thus finds in the

[16] In addition, the diagram Fig. 9.10(a) renormalizes the propagator parameters γ and D in a way similar to Eqs. (9.43) and (9.45). Since, however, the corresponding loop is proportional to the critical parameter δ, one may disregard such a renormalization in the vicinity of the transition.

one-loop order $z = 2$ and $\tilde{\chi} = \chi = -d/2$. The three remaining one-loop RG equations are [163]

$$\partial_l \delta = \left(1 + \epsilon/2 + g_2 \tilde{K}_d\right) \delta, \quad \partial_l g_1 = \left(\epsilon + 3 g_2 \tilde{K}_d\right) g_1, \quad \partial_l g_2 = \left(\epsilon + g_2 \tilde{K}_d\right) g_2, \tag{9.53}$$

where $\epsilon = 2 - d$. Notice the opposite sign of the one-loop terms in comparison with Eqs. (9.49) and (9.51). It may be traced back to the opposite sign of the $\pi^2 \varphi^2$ term in Eq. (9.52) versus Eqs. (9.47) and (9.50). As a result, for $d < d_c = 2$ the Gaussian fixed point, $g^*_{1,2} = 0$, is unstable and the ϵ-expansion does not predict any new fixed point in its ϵ-vicinity. For $d > 2$ the Gaussian fixed point is locally stable, but for $g_2 > |\epsilon|/\tilde{K}_2 \sim |\epsilon| D$ the RG flows away from it. There is numerical evidence [157] that the strong-coupling fixed point, unreachable in the ϵ-expansion, may belong to the DP universality class.

9.8 KPZ Dynamics of Growing Interfaces

Another example of an intrinsically non-equilibrium system is a surface growing by a random deposition of atoms. A model that describes a wide class of such processes was suggested by Kardar, Parisi, and Zhang (KPZ) [164]. It is written in terms of the evolution equation for the height $h(\mathbf{r}, t)$ of a growing d-dimensional surface in a $d + 1$-dimensional embedding space:

$$\partial_t h = D \nabla^2_{\mathbf{r}} h + \frac{\lambda}{2} (\nabla_{\mathbf{r}} h)^2 + \xi. \tag{9.54}$$

The first term on the right-hand side describes the diffusion of atoms along the surface. As explained by KPZ [164], the crucially important nonlinear term $\lambda(\nabla_{\mathbf{r}} h)^2/2$ has a purely geometric origin. Indeed, the growth occurs in the direction locally normal to the surface. Its angle θ with the direction of the h-axis is $\tan\theta = |\nabla_{\mathbf{r}} h|$. If an increment $\lambda \delta t$ is added along the normal, it is projected on the h-axis as $\delta h = \lambda \delta t / \cos\theta = \lambda \delta t (1 + (\nabla_{\mathbf{r}} h)^2)^{1/2}$. As a result, the rate of change of h is $\delta h / \delta t \approx \lambda + \lambda (\nabla_{\mathbf{r}} h)^2 / 2$. The average deposition rate λ may then be removed from the equation by going to the moving reference frame $h(\mathbf{r}, t) \to h(\mathbf{r}, t) + \lambda t$. The fluctuations in the deposition rate are represented by $\xi(\mathbf{r}, t)$, which may be taken to be a local Gaussian white noise

$$\langle \xi(\mathbf{r}, t) \xi(\mathbf{r}', t') \rangle = 2 \Delta \delta(\mathbf{r} - \mathbf{r}') \, \delta(t - t'). \tag{9.55}$$

There is an important relation between KPZ and the Burgers equation with conservative noise [165]. Introducing the slope field $\mathbf{v} = -\nabla_{\mathbf{r}} h$ and taking the gradient of the KPZ equation (9.54), one finds

$$\left(\partial_t + \lambda \mathbf{v} \nabla_{\mathbf{r}}\right) \mathbf{v} = D \nabla^2_{\mathbf{r}} \mathbf{v} + \vec{\zeta}, \tag{9.56}$$

where $\vec{\zeta}(\mathbf{r},t) = -\nabla_{\mathbf{r}}\xi(\mathbf{r},t)$ is the conservative (since $\int d\mathbf{r}\vec{\zeta}(\mathbf{r},t) = 0$) vector noise. For $\lambda = 1$ this is the Burgers equation for the velocity of an irrotational fluid, curl $\mathbf{v} = 0$. Its left-hand side is written in terms of the convective time derivative $d/dt = \partial_t + (\partial\mathbf{r}/\partial t)\nabla_{\mathbf{r}}$, where $\mathbf{r}$ is the coordinate of a particle and $\mathbf{v} = \partial\mathbf{r}/\partial t$ is the velocity of particles at a point $\mathbf{r}$. This interpretation suggests the invariance of Eq. (9.56) with respect to the Galilean transformation

$$t' = t, \qquad \mathbf{r}' = \mathbf{r} - \lambda\mathbf{v}_0 t, \qquad \mathbf{v}'(\mathbf{r}', t') = \mathbf{v}(\mathbf{r},t) - \mathbf{v}_0, \tag{9.57}$$

where $\mathbf{v}_0$ is an arbitrary constant vector and the invariance follows from $\partial_{t'}\mathbf{v}' = \partial_t\mathbf{v} + \lambda\mathbf{v}_0\nabla_{\mathbf{r}}\mathbf{v}$ and $\nabla_{\mathbf{r}'}\mathbf{v}' = \nabla_{\mathbf{r}}\mathbf{v}$. The Galilean invariance must also preserve the noise correlator, which is indeed the case for the local (or even nonlocal) white noise, that is, $\sim \delta(t-t')$. One may notice that λ is just a constant parameter of the invariance transformation (much the same way as the speed of light is a parameter of the Lorentz transformation). In the original KPZ variables the last equation of the Galilean transformation (9.57) is changed to $h'(\mathbf{r}', t') = h(\mathbf{r},t) + \mathbf{v}_0 \cdot \mathbf{r}$, which expresses the *tilt* invariance of the KPZ equation.

Since the KPZ equation (9.54) is scale-invariant, one expects (and indeed observes) its correlation functions to have a scaling form. Following the notation of Section 9.4, upon the scale transformation $\mathbf{r} \to b\mathbf{r}$ and $t \to b^z t$ the height field transforms as $h(\mathbf{r},t) \to b^\chi h(b\mathbf{r}, b^z t)$. As a result, for example, the pair correlation function is expected to have the form

$$\left\langle [h(\mathbf{r},t) - h(\mathbf{0},0)]^2 \right\rangle = r^{2\chi} f_{\text{KPZ}}(t/r^z), \tag{9.58}$$

where $f_{\text{KPZ}}(u)$ is a scaling function with asymptotic properties $f_{\text{KPZ}}(0) = \text{const}$ and $f_{\text{KPZ}}(u) \sim u^{2\chi/z}$ if $u \to \infty$. In the present context the exponent χ is known as the interface *roughness* exponent. Following [165] and [164], we shall attempt to find the exponents using the RG scheme.

Employing the MSR procedure of Section 4.3, one elevates the KPZ equation (9.54) to the exponent with the help of the auxiliary field $h^{\mathrm{q}}(\mathbf{r},t)$ and then performs averaging over the Gaussian white noise (9.55). As a result one obtains the action, which we write in the Hamiltonian notation, using the field $p(\mathbf{r},t) = 2ih^{\mathrm{q}}(\mathbf{r},t)$ canonically conjugated to $h(\mathbf{r},t)$,

$$S[h,p] = \int dt\, d\mathbf{r} \Big[p\partial_t h - Dp\nabla_{\mathbf{r}}^2 h - \frac{\lambda}{2} p(\nabla_{\mathbf{r}} h)^2 - \Delta p^2 \Big]. \tag{9.59}$$

We now split the fields into slow and fast components, as explained in Section 9.4, and integrate out the fast ones. The propagators of the fast components are, in fact, given by Eqs. (9.17) with $\varphi = h$, $\varphi^{\mathrm{q}} = p/(2i)$, $\delta = 0$, and $T = \Delta$. The only nonlinear term in the action, $-\lambda p(\nabla_{\mathbf{r}} h)^2/2$, generates two types of vertex: (i) $-\lambda p_{\mathrm{s}}(\nabla_{\mathbf{r}} h_{\mathrm{f}})^2/2$ and (ii) $-\lambda p_{\mathrm{f}}(\nabla_{\mathbf{r}} h_{\mathrm{f}})(\nabla_{\mathbf{r}} h_{\mathrm{s}})$. Notice that the slow field $h_{\mathrm{s}}(\mathbf{r},t)$ appears in the nonlinear vertices only through its gradient ($\nabla_\mathbf{r} h_\mathrm{s}$), in other words, *spatially* flat realizations $h_\mathrm{s}(t)$ do not enter the nonlinear vertices. This immediately implies that the coefficient in the term $p\partial_t h$ is *not* renormalized. Indeed, such a renormalization would require one to generate $p_\mathrm{s}\partial_t h_\mathrm{s}$ with the help of loops built on the (i) and (ii) vertices. Taking h_s to be time-dependent, but spatially flat, nullifies any such diagram, but not $p_\mathrm{s}\partial_t h_\mathrm{s}$. Therefore, no terms $\sim p_\mathrm{s}\partial_t h_\mathrm{s}$ can appear in any order of renormalization. We can thus choose the bare scaling dimension $\tilde{\chi}$ of the auxiliary field p in a way to rigidly fix the coefficient (one) in front of $p\partial_t h$ in any order. This requires (cf. Eq. (9.21)) that $d + \tilde{\chi} + \chi = 0$.

The non-renormalized coefficient of $\partial_t h$ along with the Galilean (i.e. tilt) invariance (9.57) of the KPZ equation dictates that λ is not renormalized either. Indeed, the Galilean invariance of the Burgers equation (9.56) preserves the convective nature of its left-hand side with the transformation parameter λ being a relative coefficient of the ∂_t and $\mathbf{v}\nabla_\mathbf{r}$ terms. Since the coefficient of $\partial_t \mathbf{v}$ is not renormalized, neither is the coefficient of $\mathbf{v}\nabla_\mathbf{r}\mathbf{v}$.[17] This statement manifests itself in, for example, cancelation of the two diagrams Fig. 9.11(a), (b) (in the limit of zero external momenta and frequencies), representing the one-loop correction to λ. The same must happen in all higher orders [166]. Employing thus $\partial_l \lambda = (d+z-2+\tilde{\chi}+2\chi)\lambda$ and demanding that λ is not renormalized, one arrives at the exact[18] identity between the two scaling exponents for any fixed point with $\lambda \neq 0$,

$$\chi + z = 2, \tag{9.60}$$

where we employed $d + \tilde{\chi} + \chi = 0$, established earlier. The only two constants that exhibit perturbative renormalization are thus D and Δ.

The one-loop renormalization of Δ is given by the diagram of Fig. 9.11(c), which employs (i) $\times$ (i) vertices and reads as

$$\left(\frac{\lambda}{2}\right)^2 \int_\Lambda^{b\Lambda} \frac{d\mathbf{q}_\mathrm{f}}{(2\pi)^d} \int \frac{d\omega}{2\pi}\, \mathbf{q}_\mathrm{f}^4 \left[D_0^\mathrm{K}(\mathbf{q}_\mathrm{f},\omega)\right]^2 = -\frac{\lambda^2\Delta^2}{4} \int_\Lambda^{b\Lambda} \frac{d\mathbf{q}_\mathrm{f}}{(2\pi)^d} \frac{\mathbf{q}_\mathrm{f}^4}{(D\mathbf{q}_\mathrm{f}^2)^3}, \tag{9.61}$$

where the factor $\mathbf{q}_\mathrm{f}^4$ originates from the fact that each vertex (i) includes two gradients of the fast fields. The one-loop renormalization of D utilizes the (i)$\times$(ii) combination of vertices, Fig. 9.11(d). One has to keep track of the slow momentum $\mathbf{q}_\mathrm{s}$ dependence of the external legs to get things right. The corresponding expression is

[17] On a more formal level, the Galilean invariance implies certain Ward identities for the exact generating function [166]. They connect the two-leg vertex $p\partial_t h$ with the three-leg vertex $p(\nabla_\mathbf{r} h)^2$. The non-renormalizability of the former thus implies the non-renormalizability of the latter. I am indebted to Uwe Täuber for clarification of this point.

[18] Strictly speaking, our proof is only applicable to any fixed point that may be reached in the perturbative expansion. There are numerical indications, though, that the statement may be more general.

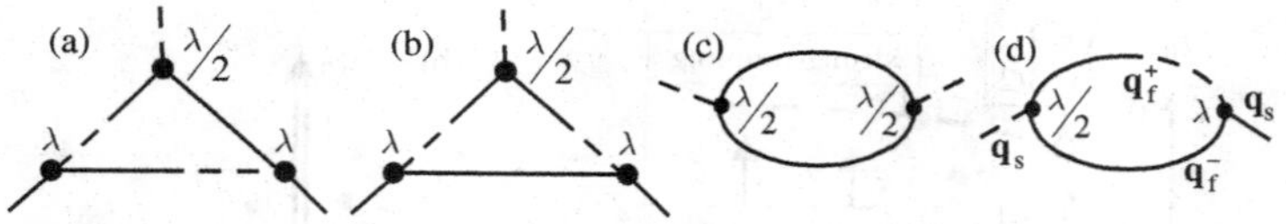

Figure 9.11 One-loop renormalization of the KPZ action (9.59). Diagrams (a) and (b), renormalizing λ, cancel each other (notice that (a) carries a combinatorial factor of 4, while (b) carries a factor of 2). Diagram (c) renormalizes Δ and (d) renormalizes D.

$$2\lambda^2 \int \frac{d\mathbf{q}_f d\omega}{(2\pi)^{d+1}} (\mathbf{q}_s \cdot \mathbf{q}_f^-)(\mathbf{q}_f^+ \cdot \mathbf{q}_f^-) D_0^R(\mathbf{q}_f^+, \omega) D_0^K(\mathbf{q}_f^-, \omega)$$

$$= \lambda^2 \Delta \int_\Lambda^{b\Lambda} \frac{d\mathbf{q}_f}{(2\pi)^d} \frac{(\mathbf{q}_s \cdot \mathbf{q}_f^-)(\mathbf{q}_f^+ \cdot \mathbf{q}_f^-)}{\left[D(\mathbf{q}_f^-)^2 + D(\mathbf{q}_f^+)^2\right] D(\mathbf{q}_f^-)^2} \approx \frac{\lambda^2 \Delta}{2D^2} \int_\Lambda^{b\Lambda} \frac{d\,\mathbf{q}_f}{(2\pi)^d} \frac{\mathbf{q}_s \cdot (\mathbf{q}_f - \mathbf{q}_s/2)}{\mathbf{q}_f^2 - (\mathbf{q}_s \cdot \mathbf{q}_f)}$$

$$\approx \frac{\lambda^2 \Delta}{2D^2} \int_\Lambda^{b\Lambda} \frac{d\,\mathbf{q}_f}{(2\pi)^d} \frac{1}{\mathbf{q}_f^2} \left(\frac{(\mathbf{q}_s \cdot \mathbf{q}_f)^2}{\mathbf{q}_f^2} - \frac{\mathbf{q}_s^2}{2} \right) = \frac{\lambda^2 \Delta}{2D^2} \left(\frac{1}{d} - \frac{1}{2} \right) \mathbf{q}_s^2 \tilde{K}_d, \tag{9.62}$$

where $\mathbf{q}_f^{\pm} = \mathbf{q}_f \pm \mathbf{q}_s/2$ and $\tilde{K}_d = \Lambda^{d-2} 2^{1-d} \pi^{-d/2} \Gamma(d/2)$. We kept here only the terms $\sim \mathbf{q}_s^2 = -\nabla_\mathbf{r}^2$. Combining these loop calculations with the bare scaling and using $d + \tilde{\chi} + \chi = 0$, one obtains the two nontrivial RG equations

$$\partial_l \Delta = \left(z - d - 2\chi + \bar{\lambda}^2 \tilde{K}_d\right) \Delta, \qquad \partial_l D = \left(z - 2 + \frac{2-d}{d} \bar{\lambda}^2 \tilde{K}_d\right) D, \tag{9.63}$$

where $\bar{\lambda}^2 = \lambda^2 \Delta / 4D^3$ is an effective running constant. Employing Eq. (9.60) along with $\partial_l \lambda = 0$, one may rewrite these two equations as a single closed RG equation for $\bar{\lambda}^2$:

$$\partial_l \bar{\lambda}^2 = \left(\epsilon + \frac{1 - 2\epsilon}{1 - \epsilon/2} \bar{\lambda}^2 \tilde{K}_d\right) \bar{\lambda}^2, \tag{9.64}$$

where $\epsilon = 2 - d$. We have found thus that the critical dimension is $d_c = 2$, but the RG flow is very different from the equilibrium transitions, or even DP. Indeed, according to Eq. (9.64) there is no stable fixed point of order ϵ for $d \le 2$, Fig. 9.12(a). Curiously, the RG predicts the order 1 fixed point for $d < 3/2$; we shall come back to this observation later. Most notably, for $d > 2$ the Gaussian fixed point $\bar{\lambda}^* = 0$ is stable, if the bare values of the parameters are such that $\bar{\lambda} < \bar{\lambda}_c = (|\epsilon|/\tilde{K}_2)^{1/2}$. In the Gaussian fixed point the correlation function (9.58) is given by Eq. (9.17c) with $\delta = 0$ and $T = \Delta$ and thus the critical exponents are $z = 2$ and $\chi = 1 - d/2$. The fact that $\chi < 0$ implies that the growing surface is essentially flat at large scales. For $\bar{\lambda} > \bar{\lambda}_c$ the RG runs away toward a strong coupling fixed point, describing a *rough* surface. We have found thus a phase transition

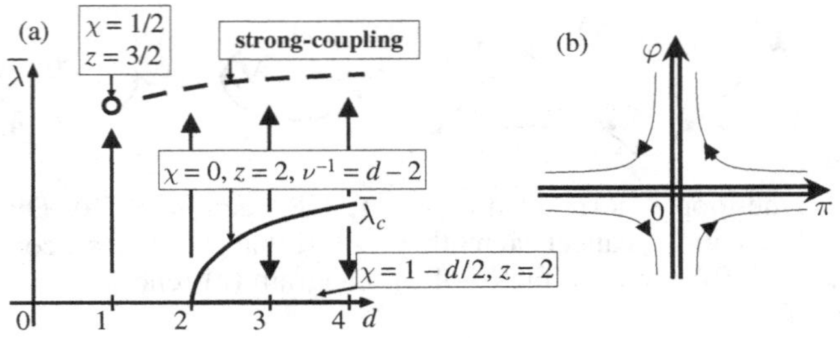

Figure 9.12 (a) KPZ critical exponents on the coupling constant $\bar{\lambda}$ versus surface dimensionality d plane. (b) Phase portrait of the critical KPZ Hamiltonian after a Cole–Hopf transformation $H(\pi, \varphi) = D\bar{\lambda}^2\pi^2\varphi^2$; see Eq. (9.66).

between the smooth and rough modes of the surface growth as a function of the effective coupling constant $\bar{\lambda}$ for $d > 2$. For $d \leq 2$ the smooth mode of the growth is absent and the surface is always rough. Properties of the strong coupling fixed point are inaccessible in any perturbative RG treatment.

One can discuss, however, scaling properties of the surface, which happens to be right at the phase transition between the smooth and rough modes, that is, having $\bar{\lambda} = \bar{\lambda}_c$. They are described by the *unstable* fixed point $\bar{\lambda}^* = \bar{\lambda}_c$. Demanding that both Δ and D are not renormalized at this fixed point, one finds, for example, $z = 2 + \epsilon^2/(2 - 4\epsilon)$. Since the calculations were done in one-loop only, one should not trust ϵ^2 terms and the only conclusion is that $z = 2 + O(\epsilon^2)$ and $\chi = O(\epsilon^2)$. In fact, the two-loop calculation of Frey and Täuber [166] showed that there are no ϵ^2 corrections to $z = 2$ and $\chi = 0$ either, suggesting that these exponents may be exact at the unstable fixed point [167].

To show that this is indeed the case one may apply the Cole–Hopf transformation [168]

$$\varphi = \exp\left\{\frac{\lambda}{2D}h - \frac{\lambda^2\Delta}{4D^2}t\right\}, \tag{9.65}$$

which transforms the KPZ equation (9.54) into a linear equation with multiplicative Stratonovich noise, $\partial_t\varphi = D\nabla_{\mathbf{r}}^2\varphi - (\Delta\lambda^2/4D^2)\varphi + (\lambda/2D)\varphi\xi$. With the help of the Stratonovich to Ito transformation (4.40) the latter is shown to be equivalent to the Ito–Langevin equation $\partial_t\varphi = D\nabla_{\mathbf{r}}^2\varphi + (\lambda/2D)\varphi\xi$. The corresponding MSR action is

$$S[\varphi, \pi] = \int \mathrm{d}t\, \mathrm{d}\mathbf{r}\left[\pi\left(\partial_t\varphi - D\nabla_{\mathbf{r}}^2\varphi\right) - D\bar{\lambda}^2\pi^2\varphi^2\right]. \tag{9.66}$$

From the fact that the only nonlinear vertex is $\pi^2\varphi^2$ one observes that there are no corrections to the propagator $\langle\pi\varphi\rangle$ in any order of the perturbative expansion [168]. As a result, $z = 2$ in any perturbatively reachable fixed point, while D is

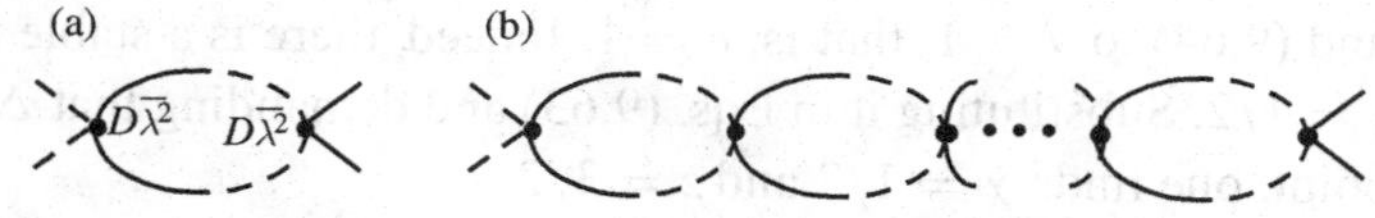

Figure 9.13 Renormalization of the Cole–Hopf transformed KPZ action (9.66). (a) One-loop diagram; (b) the only possible sequence of diagrams. It makes the one-loop RG equation perturbatively exact.

not renormalized. The only renormalization is that of $\bar\lambda^2$, given by the diagrams of Fig. 9.13. Since this sequence is nothing but successive applications of the one-loop renormalization, one concludes that the following one-loop RG equation

$$\partial_l \bar\lambda^2 = \left(\epsilon + \bar\lambda^2 \tilde{K}_d\right) \bar\lambda^2 \qquad (9.67)$$

is actually exact. It coincides with Eq. (9.64) in the leading order of ϵ-expansion (i.e. to the precision Eq. (9.64) was derived). Due to the identity (9.60) one finds that $\chi = 0$ at the unstable fixed point, which exists for $d > 2$. Linearizing Eq. (9.67) near the unstable fixed point $(\bar\lambda^2)^* = |\epsilon|/\tilde{K}_d$, one finds the spatial correlation length critical exponent $\nu^{-1} = |\epsilon| = d - 2$ [168]. The scaling properties of the KPZ problem are summarized in Fig. 9.12(a).

Finally, let us mention that for $d = 1$ (i.e. $\mathbf{r} \to x$) one knows the exact *stationary* distribution function of the surface slopes[19]

$$\mathcal{P}[h] = Z^{-1} \exp\left\{-\frac{D}{2\Delta}\int dx\, (\partial_x h)^2\right\}. \qquad (9.68)$$

This distribution function is exactly the same as the equilibrium distribution (9.10)[20] for the linear problem with $\lambda = 0$. In the latter case the KPZ equation (9.54) is a potential one (cf. Eq. (9.1)), with $\mathcal{F}[h] = \int dx\, D(\partial_x h)^2/2$. As a result, all *static* exponents in $d = 1$ coincide with their Gaussian, that is, $\lambda = 0$, counterparts. This way one finds [175] $\chi = 1 - d/2 = 1/2$. The dynamic critical exponent $z = 3/2$ follows then from the exact relation (9.60). Curiously, but completely unjustifiably, one can arrive at the same exact results, applying the one-loop

19 To prove this statement one needs to consider the Fokker–Planck equation, which corresponds to the KPZ-Langevin process, cf. Eqs. (9.11) and (4.30),

$$\partial_t \mathcal{P}([h],t) = \int d\mathbf{r}\, \frac{\delta}{\delta h(\mathbf{r})}\left[-\left(D\nabla_{\mathbf{r}}^2 h + \frac{\lambda}{2}(\nabla_{\mathbf{r}} h)^2\right)\mathcal{P}([h],t) + \Delta\, \frac{\delta \mathcal{P}([h],t)}{\delta h(\mathbf{r})}\right].$$

Let us show that the term $\sim \lambda$ with $\mathcal{P}$ given by Eq. (9.68) is the full derivative in $d = 1$. Indeed, it follows from: $\lambda[\partial_x^2 h - (D/2\Delta)(\partial_x h)^2 \partial_x^2 h]\mathcal{P} = \lambda \mathcal{P} \partial_x[\partial_x h - (D/6\Delta)(\partial_x h)^3]$. As a result it vanishes upon integration over dx and Eq. (9.68) is thus a solution of the stationary Fokker–Planck equation. Notice that the trick does not work in $d > 1$, indeed $(\nabla^\alpha h \nabla^\alpha h)(\nabla^\beta \nabla^\beta h)$ is not a full derivative. For the stationary path interpretation of the exact distribution function (9.68) see [169].

20 An even more spectacular example, where the intrinsically non-equilibrium problem admits the static equilibrium distribution function in $d = 1$, is offered by the symmetric exclusion process [171, 174].

Eqs. (9.63) and (9.64) to $d = 1$, that is, $\epsilon = 1$. Indeed, there is a stable fixed point with $(\bar{\lambda}^2)^* \tilde{K}_1 = 1/2$. Substituting it in Eqs. (9.63) and demanding that Δ and D are at the fixed point, one finds $\chi = 1/2$ and $z = 3/2$.

9.9 Large Deviations in Stochastic Systems

Here we discuss probability of certain atypical outcomes of the dynamical evolution of classical stochastic systems. Such rare events, or large deviations, although coming with an exponentially small probability, may nevertheless carry a large significance. An example is provided by a population extinction. The technique, discussed here is known as a *weak noise theory*, or a *macroscopic fluctuation theory* (MFT). It calls for a solution of semiclassical equations of motion obtained from a proper Martin–Siggia–Rose action with the boundary conditions encoding the rare event. We have already employed this strategy in the zero-dimensional setting in Section 4.11. Here we present its generalization to extended systems using the 1d KPZ model (see Section 9.8) as an example [169, 176, 179, 180, 181, 182]. Closely related methods appeared in the studies of turbulence and turbulent transport [183, 184, 185], diffusive lattice gases [171, 186, 187, 188] and stochastic reactions on lattices [69].

The question is as follows: find a probability distribution function $\mathcal{P}(H, t_0)$ for the interface height at $x = 0$ to advance by $H = h(0, t_0) - h(0, 0)$ during the time t_0. One should also specify an initial state of the interface at $t = 0$, for example, as being flat, $h(x, 0) = 0$, or as $h(x, 0)$ being randomly taken from the stationary distribution (9.68). In the long time limit, $t_0 \gg t_* = D^5/(\lambda^4\Delta^2)$, equation (9.58) dictates that $\langle H^2\rangle \propto t_0^{2\chi/z}$ and, since the 1d KPZ fixed point is characterized by $\chi = 1/2$ and $z = 3/2$ (cf. Fig. 9.12(a)), one observes that typically $(H/h_*) \propto (t_0/t_*)^{1/3}$, where $h_* = D/\lambda$. One thus expects that in the long time limit the distribution function depends on the single dimensionless scaling variable

$$\mathcal{P}(H, t_0) = \mathcal{P}(y); \qquad y = \left(\frac{H}{h_*}\right)\left(\frac{t_*}{t_0}\right)^{1/3} = \left(\frac{H^3 D^2}{t_0 \lambda \Delta^2}\right)^{1/3}. \tag{9.69}$$

Indeed, it was shown [189, 190, 192, 194] that $\mathcal{P}(y)$ is given by the orthogonal Tracy–Widom [195] distribution for the initially flat interface and by the Baik–Rains [196] distribution for the initial height taken from the stationary distribution. Both of them have non-Gaussian asymptotics: $\log \mathcal{P}(y) \propto -y^{3/2}$ for $y \gg 1$ and $\log \mathcal{P}(y) \propto -|y|^3$ for $y \ll -1$.

As we will see momentarily, MFT tackles the opposite limit of short times,[21] $t_0 \ll t_*$. To this end let us transform to dimensionless coordinates $t/t_0 \to t$,

[21] Since $t_* \sim \Delta^{-2}$, the short time limit $t_0 \ll t_*$ may be interpreted as the weak noise limit, $\Delta \ll D^{5/2}/(\lambda^2\sqrt{t_0})$.

$x/\sqrt{Dt_0} \to x$, and fields $h/h_* \to h$, $p/p_* \to p$, where $p_* = D/(t_0\lambda\Delta)$. Then the MSR action (9.59) acquires the form

$$S[h,p] = \sqrt{\frac{t_*}{t_0}} \int_0^1 \mathrm{d}t \int \mathrm{d}x \left[p\,\partial_t h - p\,\partial_x^2 h - \frac{1}{2} p(\partial_x h)^2 - p^2 \right]. \tag{9.70}$$

The corresponding functional integral should be equipped with the constraint, specifying the desired (rare) outcome of the evolution. In our example it takes the form

$$\delta\big(H/h_* - h(0,t_0) + h(0,0)\big) = \int \frac{\mathrm{d}\Lambda}{2\pi}\, \mathrm{e}^{\mathrm{i}\Lambda\left(H/h_* - h(0,t_0) + h(0,0)\right)}, \tag{9.71}$$

which adds the following boundary action

$$S_\mathrm{b}[h,\Lambda] = -\mathrm{i}\Lambda \frac{H}{h_*} + \mathrm{i}\Lambda \int_0^1 \mathrm{d}t \int \mathrm{d}x\, h(x,t)\,\delta(x)\big(\delta(t-1) - \delta(t)\big). \tag{9.72}$$

In the short time limit $t_0 \ll t_*$ there is a large parameter in front of the parameterless bulk action (9.70), which justifies calculating the functional integral in the saddle-point approximation. Taking variations of $S[h,p] + S_\mathrm{b}[h,\Lambda]$ with respect to $p(x,t)$ and $h(x,t)$, one finds

$$\partial_t h = \partial_x^2 h + \frac{1}{2}(\partial_x h)^2 + 2p; \tag{9.73a}$$

$$\partial_t p = -\partial_x^2 p + \partial_x(p\partial_x h) + \tilde{\Lambda}\delta(x)\delta(t-1), \tag{9.73b}$$

where $\tilde{\Lambda} = \mathrm{i}\Lambda\sqrt{t_0/t_*}$ and we assumed an initially flat interface, $h(x,0) = 0$, and thus no variation with respect to $h(x,0)$. Therefore in the time interval $0 < t < 1$ the two fields obey the Hamiltonian equations of motion

$$\partial_t h = \frac{\delta \mathcal{H}}{\delta p}; \qquad \partial_t p = -\frac{\delta \mathcal{H}}{\delta h}; \qquad \mathcal{H} = \int \mathrm{d}x \Big[p\,\partial_x^2 h + \frac{1}{2} p(\partial_x h)^2 + p^2 \Big], \tag{9.74}$$

supplemented with the boundary conditions

$$h(x,0) = 0; \qquad\qquad p(x,1) = \tilde{\Lambda}\delta(x). \tag{9.75}$$

The "future" boundary condition, imposed on the MSR "quantum" field $p(x,t)$ at the observation time $t = t_0$, follows from integrating Eq. (9.73b) over the infinitesimal area around the point $x = 0$, $t = 1$ and assuming that at any time $t < 1$ the fields are non-singular. Finally, the so-far-undetermined parameter $\tilde{\Lambda}$ should be fixed by the constraint $h(0,1) = H/h_*$. The fact that $\tilde{\Lambda}$ happens to be real means that the integration contour over Λ in Eq. (9.71) is distorted to pass through a purely imaginary saddle point. Once the equations of motion (9.74) with the boundary conditions (9.75) are solved, the probability of the rare event is calculated as

$$\log \mathcal{P}(H,t_0) = -S[h,p] = -\sqrt{\frac{t_*}{t_0}}\, s\left(\frac{H}{h_*}\right); \qquad s\left(\frac{H}{h_*}\right) = \int_0^1 \mathrm{d}t \int \mathrm{d}x\, p^2(x,t), \tag{9.76}$$

where the last equality follows from the fact that on the solutions of the equations of motion $\int \mathrm{d}x(p\,\partial_t h - p^2) = \mathcal{H}$; compare Eq. (9.73a).

In general, dynamic MFT equations (9.74) are hard to solve analytically. A fast and efficient numerical algorithm was suggested by Chernykh and Stepanov [197, 69]. It calls for an iterative alternation of forward-in-time propagation of $h(x,t)$, with $p(x,t)$ fixed by the previous iteration, followed by the backward-in-time propagation of $p(x,t)$, with $h(x,t)$ fixed by the previous iteration. This way, evolution of both fields is numerically stable due to the proper sign of the diffusion term.

An analytic progress is possible in limiting cases[22] where $|H| \gg h_*$. Let us first consider a large positive $H \gg h_*$. In this case it is convenient to perform the canonical Cole–Hopf transformation (9.65), which brings MFT equations (9.74) to a symmetric form

$$\varphi = \mathrm{e}^{h/2}, \qquad \pi = 2p\,\mathrm{e}^{-h/2}; \qquad h = 2\log\varphi, \qquad p = \frac{1}{2}\pi\varphi;$$
$$\partial_t\varphi = \partial_x^2\varphi + \frac{1}{2}\pi\varphi^2; \qquad \partial_t\pi = -\partial_x^2\pi - \frac{1}{2}\varphi\pi^2. \tag{9.77}$$

These are Ablowitz–Kaup–Newell–Segur (AKNS) equations [198], solvable with the inverse scattering technique. Recently the technique was extended to treat some problems with the mixed boundary conditions, like (9.75) [199, 201]. The large H (and thus large $\tilde{\Lambda}$) problem is progressively well approximated by soliton solutions of AKNS equations. Generic N-soliton solutions are outlined in problem 9.10.3. The boundary conditions (9.75) are satisfied (in the limit $H \gg h_*$) by the following particular solution of (9.77); see Fig. 9.14:

$$\varphi(x,t) = \frac{\mathrm{e}^{c^2t} + \cosh cx}{\mathrm{e}^{-c^2t} + \cosh cx}; \qquad \pi(x,t) = \frac{4c^2\mathrm{e}^{-c^2t}}{\mathrm{e}^{-c^2t} + \cosh cx}. \tag{9.78}$$

From $h(0,1) = H/h_*$ and $h(0,t) = 2\log\varphi(0,t) = 2c^2t$, one finds $c^2 = H/(2h_*)$. For $c \gg 1$, one may approximate $p = \pi\varphi/2 \approx 2c^2/\cosh^2 cx$, which leads to

[22] One can also solve the case of $|H| \ll h_*$, where nonlinearity may be disregarded and the problem reduces to the Edwards–Wilkinson model; see Problem 9.10.2, where one finds Gaussian law $s(H) \propto H^2$, [179, 182]).

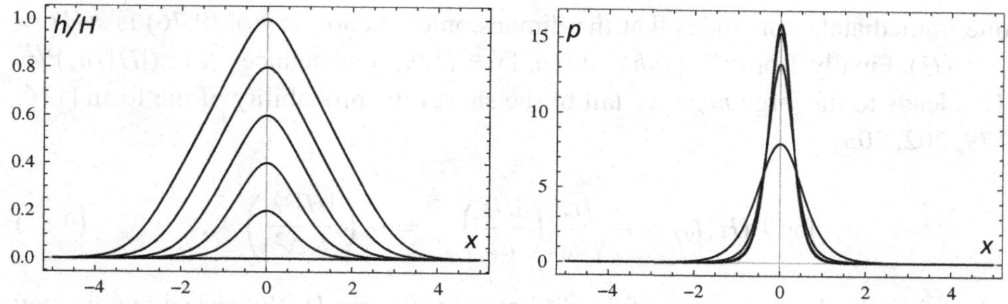

Figure 9.14 Optimal interface growth according to Eq. (9.78) with $H = 16h_*$. From bottom to top, $t/t_0 = 0, .2, .4, .6, .8, 1$. Notice that after a short initial transient time $\sim 2h_*/H \ll 1$, $p(x,t)$ (the right panel) saturates to a stationary soliton peak $p(x,t) \approx 2c^2/\cosh^2 cx$, which solely determines the optimal action (9.76). Numerically one sees an equally short transient near $t = 1$, when the sharp soliton evolves into the delta-function $p(x,1) = 4c\delta(x)$, due to the anti-diffusion $\partial_t p = -\partial_x^2 p + \dots$. Both transients do not affect the leading asymptotic (in $H/h_* \gg 1$) of the action (9.79).

the following expression for the probability of a large *positive* height; compare Eq. (9.76),[23] [176, 179, 181]:

$$\log \mathcal{P}(H, t_0) = -\sqrt{\frac{t_*}{t_0}}\,\frac{16}{3}\left(\frac{H}{2h_*}\right)^{3/2} = -\frac{8}{3\sqrt{2}}\left(\frac{H^3 D^2}{t_0 \lambda \Delta^2}\right)^{1/2}. \tag{9.79}$$

This coincides (including the right numerical factor!) with the positive (Airy) tail of the Tracy–Widom distribution [195], obtained in the *long time* limit, Eq. (9.69) [190]. This remarkable coincidence is responsible for the validity of Eq. (9.79) at any time (as long as its right-hand side is larger than one), though our saddle-point MFT derivation is only justified at $t_0 \ll t_*$.

The situation is very different for a large *negative* height, where the (cubic) tail of the long time Tracy–Widom distribution is qualitatively distinct from the short time asymptotic. To derive the latter one notices [176, 179] that for $H \ll -h_*$ one may disregard the diffusive terms in the MFT equations (9.74), which brings them to the hydrodynamic form

$$\partial_t V = V\partial_x V + 2\partial_x p; \qquad \partial_t p = \partial_x(pV), \tag{9.80}$$

where $V(x,t) = \partial_x h(x,t)$. These equations describe a nonstationary inviscid flow of an effective gas with density p, velocity V, and a negative pressure $P(p) = -p^2$. They are supplemented with the boundary conditions $V(x,0) = 0$ and $p(x,1) = \tilde{\Lambda}\delta(x)$. Remarkably, they may be brought to the completely parameter-free form by the following rescaling: $x/\tilde{\Lambda}^{1/3} \to x$, $V/\tilde{\Lambda}^{1/3} \to V$, and $p/\tilde{\Lambda}^{2/3} \to p$. From here

[23] We used the fact that $\int_{-\infty}^{\infty} \mathrm{d}x/\cosh^4 x = 4/3$.

one immediately concludes that the dimensionless action in Eq. (9.76) is $s(H) \propto \tilde{\Lambda}^{5/3}(H)$. Finally, from $V = \partial_x h$ and $h(0,1) = H/h_*$, one deduces $\tilde{\Lambda} \propto (|H|/h_*)^{3/2}$. This leads to the large *negative* tail of the short time probability of the form [176, 179, 202, 203]

$$\log \mathcal{P}(H, t_0) \propto -\sqrt{\frac{t_*}{t_0}} \left(\frac{|H|}{h_*}\right)^{5/2} = -\left(\frac{|H|^5 \lambda}{t_0 \Delta^2}\right)^{1/2}, \tag{9.81}$$

which is indeed independent of the diffusion coefficient D. Numerical factors may be found [179, 202] from solving the hydrodynamic equations (9.80). Notice that the negative tail does *not* conform to the large time scaling behavior (9.69) and its functional form, $|H|^{5/2}$, is different from the scaling of the negative Tracy–Widom tail, $|H|^3$ [195]. Recently the short time $|H|^{5/2}$ law was confirmed by mathematical tools [204].

We now briefly comment on the problem where the initial height profile $h(x,0) = h_0(x)$, instead of being flat, is randomly drawn from the 1d exact stationary distribution (9.68). In this case the action $S[h,p] + S_b[h,\Lambda]$, (9.70), (9.72), should be supplemented with the statistical weight of an initial height, $S_{ini}[h_0] = \frac{1}{2}\sqrt{\frac{t_*}{t_0}} \int dx(\partial_x h_0)^2$; compare Eq. (9.68). Variation over $h_0(x)$ leads then to the modified initial, $t = 0$, boundary condition of the form [182]

$$p(x,0) - \partial_x^2 h(x,0) = \tilde{\Lambda}\delta(x); \qquad p(x,1) = \tilde{\Lambda}\delta(x), \tag{9.82}$$

instead of Eq. (9.75). Notice that these boundary conditions are consistent with the exact conservation law $\int dx\, p(x,t) = \text{const}$, which follows from (9.73b).

One finds [182, 205, 202, 199] that at $H = H_c \approx 3.7063 h_*$ the second derivative of the large deviation function, $\partial_H^2 \log \mathcal{P}(H, t_0)$, exhibits discontinuity, in the limit $t_0 \ll t_*$. It is caused by the dynamical second-order phase transition, associated with the spontaneous breaking of a symmetry at $H = H_c$ [182]. The symmetry, which is broken, is the *reflection* symmetry, $x \to -x$, of the optimal initial profile $h_0(x)$, which leads to the rare event $h(0,t_0) - h(0,0) = H$. For $H < H_c$ the optimal initial profile, $h_0(x)$, is a symmetric (even) function of x. On the other hand, at $H > H_c$ the reflection symmetry is broken and there are two inequivalent asymmetric optimal initial profiles $h_0(x)$ and $h_0(-x)$. The corresponding optimal configuration is depicted in Fig. 9.15.

One may define an order parameter for this non-equilibrium phase transition as [202]

$$M = \int_0^\infty dx(\partial_x h_0)^2 - \int_{-\infty}^0 dx(\partial_x h_0)^2, \tag{9.83}$$

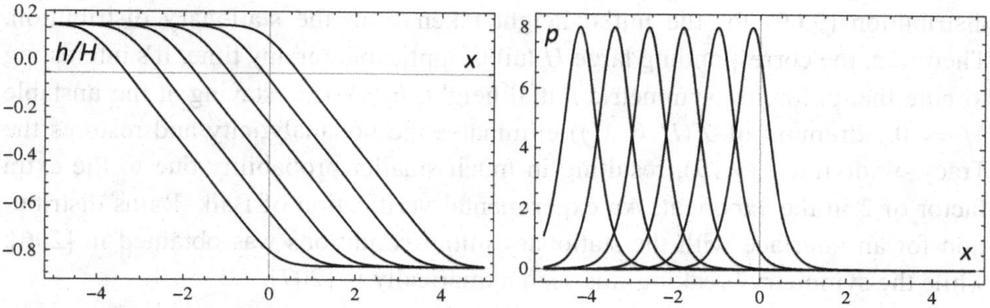

Figure 9.15 Optimal interface evolution for stationary initial conditions, calculated using Eq. (9.90) for $H = 32h_* > H_c$. From left to right, $t/t_0 = 0, .2, .4, .6, .8, 1$. There is another optimal solution obtained by $x \to -x$ transformation. Since the h-ramp has height H/h_* and length c, the weight of such initial configuration is $S_{\text{ini}}[h_0] = \frac{1}{2}\sqrt{\frac{t_*}{t_0}}\, c\left(\frac{H}{h_* c}\right)^2 = \sqrt{\frac{t_*}{t_0}}\, 2\left(\frac{H}{2h_*}\right)^{3/2}$, where the ramp velocity $c = \sqrt{H/2h_*}$; Problem 9.10.3. The dynamical action (9.76) adds $S[h,p] = \sqrt{\frac{t_*}{t_0}}\,\frac{2}{3}\left(\frac{H}{2h_*}\right)^{3/2}$. The resulting total action is half that of the flat initial interface (9.79).

where $h_0(x) = h(x,0)$ is the solution of Eqs. (9.74) with the boundary conditions (9.82). To explicitly break the reflection symmetry (analog of the magnetic field in a magnetic phase transition), one may study the probability of a "tilted" rare event $h(L,t_0) - h(0,0) = H$. The analog of the Landau free energy is defined as $\lim_{t_0 \to 0} \sqrt{t_0} \log \mathcal{P}(H,L,t_0;M) = -\mathcal{F}(H,L;M)$, where all the realizations are conditioned by both the rare event, H, L, and the order parameter, M, through Eq. (9.83). One finds that $\mathcal{F}(H,L;M)$ exhibits all the properties of the mean-field second-order transition (e.g. long-ranged Ising) with the identification of H as an inverse temperature, L as an external magnetic field, and M as magnetization. Close to the transition it takes the form [202]

$$\mathcal{F}(H,L;M) = \alpha_1(H_c - H)M^2 + \alpha_2 M^4 - \alpha_3 LM, \tag{9.84}$$

where $\alpha_{1,2,3} > 0$ are some effective parameters. The rare event probability is found by minimization of this function over the order parameter M. Notice that the phase transition takes place in the rare events probability function and not in the equilibrium free energy. It is a manifestation of the nonanalyticity in the rare events function, discussed in Section 4.13 in the context of few degrees of freedom.

In the symmetry-broken phase at $H \gg H_c$, the tail of the probability distribution function $\log \mathcal{P}(H,t_0)$ is given by Eq. (9.79) with the extra factor $1/2$ on its right-hand side, Fig. 9.15. This coincides with the proper tail of the Baik–Rains distribution [196]. The latter is known [192, 194] to describe the long time probability distribution (9.69) for the initial height taken from the stationary distribution. Therefore, the corresponding large H tail is applicable for any time. It's interesting to note that enforcing symmetric initial height, $h_0(x)$, (i.e. staying at the unstable $M = 0$ extremum of $\mathcal{F}(H, 0; M)$) eliminates the nonanalyticity and restores the Tracy–Widom tail (9.79), resulting in much smaller probability due to the extra factor of 2 in the exponent. An experimental verification of Baik–Rains distribution for an interface with the stationary initial conditions was obtained in [206], while the symmetry breaking was seen numerically in [207].

9.10 Problems

9.10.1 Edwards–Wilkinson Universality Class

Consider a linear stochastic system [208]:

$$\partial_t h = D\nabla_{\mathbf{r}}^2 h + \xi, \tag{9.85}$$

where $\langle \xi(\mathbf{r}, t)\xi(\mathbf{r}', t')\rangle = 2\Delta\delta(\mathbf{r} - \mathbf{r}')\,\delta(t - t')$. Derive a corresponding MSR action and the scaling relation:

$$\left\langle [h(\mathbf{r}, t) - h(\mathbf{0}, 0)]^2 \right\rangle = r^{2-d} f_{\mathrm{EW}}(t/r^2), \tag{9.86}$$

where $f_{\mathrm{EW}}(u)$ is a scaling function with asymptotic properties $f_{\mathrm{EW}}(0) = \mathrm{const}$ and $f_{\mathrm{EW}}(u) \sim u^{(2-d)/2}$ if $u \to \infty$; compare Eq. (9.58). That is, the roughness exponent is $\chi = 1 - d/2$ and $z = 2$ (cf. Fig. 9.12). Derive correlation function (9.86) in $d = 2$ case [209].

9.10.2 Height Distribution in 1d Edwards–Wilkinson Model

Show that the height $H = h(0, t_0) - h(0, 0)$ distribution function in the 1d Edwards–Wilkinson model of Problem 9.10.1 with the initially flat interface, $h(x, 0) = 0$, is Gaussian:

$$\mathcal{P}(H, t_0) \propto \exp\left\{ -\sqrt{\frac{\pi D}{t_0}}\,\frac{H^2}{2\Delta} \right\}. \tag{9.87}$$

From here it follows that $\langle H^2 \rangle \propto \sqrt{t_0}$ in agreement with the $d = 1$ case of (9.86). To this end solve explicitly the proper MFT equations $\partial_t h = \partial_x^2 h + 2p$ and $\partial_t p = -\partial_x^2 p$ with the boundary conditions (9.75) and evaluate the action (9.76).

Solve the same problem with the initial height $h(x, 0)$ taken from the stationary distribution (9.68). To this end repeat the same strategy with the modified boundary conditions (9.82) [179, 182]. Show that, provided $h(0, 0) = 0$, the optimal trajectory at all times satisfies [202]:

$$h(-x, t) = h(x, t); \qquad h(x, t) + h(x, 1 - t) = H. \tag{9.88}$$

9.10.3 Soliton Solutions of Large Deviation Equations

Using Hirota substitution [210]

$$\varphi = \frac{v}{u}; \qquad \pi = \frac{w}{u},$$

show that AKNS equations (9.77) are satisfied if the new functions u, v, w satisfy

$$(D_t - D_x^2)(v \cdot u) = 0; \qquad (D_t + D_x^2)(w \cdot u) = 0; \qquad D_x^2(u \cdot u) = \frac{1}{2} v w, \quad (9.89)$$

where Hirota derivatives [210] are defined as $D_t(A \cdot B) = (\partial_t A)B - A(\partial_t B)$ and $D_x^2(A \cdot B) = (\partial_x^2 A)B - 2(\partial_x A)(\partial_x B) + A(\partial_x^2 B)$. Show that equations (9.89) are solved by the following N-soliton ansatz [182]:

$$u = \sum_{i=1}^{N} \eta_i; \qquad v = \frac{1}{C} \sum_{i,j=1}^{N} (c_i - c_j)^2 \eta_i \eta_j; \qquad w = 2C, \qquad (9.90)$$

where $\eta_i(x,t) = \mathrm{e}^{c_i^2 t - c_i(x - X_i)}$. Here c_i, X_i, C are $2N+1$ arbitrary constants specifying soliton velocities and initial positions. (There is also a time-reversed family of solutions with $t \to -t$ and $v \leftrightarrow w$.) Equation (9.78) corresponds to $N = 3$ with $c_1 = 0$ and $c_2 = -c_3 = c$; $X_2 = -X_3 = -\log 2/c$; $C = 2c^2$. If one is not concerned about the short, $\sim c^{-2} \ll 1$ initial transient, Fig. 9.14, the $i = 1$ static soliton may be discarded, leading to a simpler $N = 2$ solution: $\varphi(x,t) = \mathrm{e}^{c^2 t}/\cosh cx$ and $\pi(x,t) = 4c^2 \mathrm{e}^{-c^2 t}/\cosh cx$. The optimal asymmetric solitons for stationary initial conditions are given by $N = 3$ with $c_1 = 0$, $c_2 = c$, and $c_3 = -c\,\mathrm{e}^{-c^2/2}$, while $X_i = -c_i$, where again $c^2 = H/(2h_*)$. It is depicted in Fig. 9.15.

Part III

Fermions

10
Fermions

In this chapter we reformulate the basic constructions of Chapters 2 and 6 for fermionic particles. To this end, we introduce Grassmann variables and integrals and formulate the fermionic evolution operator on the closed time contour as a coherent state functional integral. We then use it to introduce the fermion Green functions, perturbation theory, and kinetic equation.

10.1 Grassmann Variables and Integrals

Following the same route as in Section 2.1, we start by considering a single quantum level occupied by fermionic particles. Due to the Pauli principle one may have either zero or one fermion in such a state and therefore the entire *many-body* space is spanned by *two* orthonormal basis states, $|0\rangle$ and $|1\rangle$. It is convenient to introduce fermionic annihilation and creation operators, $\hat{c}$ and $\hat{c}^\dagger$, which operate in this many-body Hilbert space according to the following rules:

$$\hat{c}\,|0\rangle = 0; \qquad \hat{c}\,|1\rangle = |0\rangle; \qquad \hat{c}^\dagger|0\rangle = |1\rangle; \qquad \hat{c}^\dagger|1\rangle = 0. \tag{10.1}$$

One may notice the following properties of these operators

$$\hat{c}^\dagger\hat{c}|n\rangle = n|n\rangle; \qquad \{\hat{c},\hat{c}^\dagger\} = \hat{1}; \qquad \hat{c}^2 = \left(\hat{c}^\dagger\right)^2 = 0, \tag{10.2}$$

where the curly brackets denote the anti-commutator $\{\hat{c},\hat{c}^\dagger\} = \hat{c}\hat{c}^\dagger + \hat{c}^\dagger\hat{c}$ and $n = 0, 1$.

To introduce fermionic coherent states, one needs an algebra of anti-commuting Grassmann numbers. Those are the objects, denoted by ψ, ψ', ..., which formally satisfy the following multiplication rules:

$$\psi\psi' = -\psi'\psi; \qquad \psi^2 = 0, \qquad f(\psi) = f_0 + f_1\psi, \tag{10.3}$$

where $f(\psi)$ is an arbitrary function of the Grassmann ψ, which is *defined* by its first two Taylor expansion coefficients. Similarly, for a function of two variables one has $f(\psi,\psi')=f_{00}+f_{10}\psi+f_{01}\psi'+f_{11}\psi\psi'$, and so on. This fact allows us to introduce derivatives in the natural way, that is, $\partial\psi/\partial\psi = 1$ and therefore $\partial f(\psi)/\partial\psi = f_1$. Notice that the derivatives anti-commute, indeed

$$\frac{\partial}{\partial\psi}\frac{\partial}{\partial\psi'}f(\psi,\psi') = \frac{\partial}{\partial\psi}(f_{01}-f_{11}\psi) = -f_{11} = -\frac{\partial}{\partial\psi'}\frac{\partial}{\partial\psi}f(\psi,\psi').$$

We shall also need the concept of integration over the Grassmann variables [212], which is *defined* as

$$\int d\psi\, 1 = 0; \qquad \int d\psi\, \psi = 1. \tag{10.4}$$

We stress that this is a definition, which can't be derived from "first principles." It is also convenient to agree that Grassmann numbers anti-commute with the fermionic annihilation and creation operators, that is,

$$\{\psi,\hat{c}\} = \{\psi,\hat{c}^\dagger\} = 0.$$

By analogy with bosons, a fermionic coherent state is defined as an eigenstate of the annihilation operator $\hat{c}$. Since the Hilbert space has only two basis vectors, it must be a linear superposition of $|0\rangle$ and $|1\rangle$. It is easy to see that such a linear combination may not have ordinary numbers as coefficients, indeed $\hat{c}\,(x|0\rangle + y|1\rangle) = y|0\rangle \neq \lambda(x|0\rangle + y|1\rangle)$, unless $y = 0$ and then the eigenvalue is zero too. The difficulty may be avoided by using Grassmann numbers. Indeed, consider a state parametrized by the Grassmann number ψ,

$$|\psi\rangle = |0\rangle - \psi|1\rangle = (1-\psi\hat{c}^\dagger)|0\rangle = e^{-\psi\hat{c}^\dagger}|0\rangle. \tag{10.5}$$

Then $\hat{c}|\psi\rangle = -\hat{c}\psi|1\rangle = \psi|0\rangle = \psi|\psi\rangle$, and therefore the state $|\psi\rangle$ is indeed an eigenstate of the fermionic annihilation operator with the Grassmann eigenvalue ψ. In an analogous way we define the left coherent state $\langle\psi|$ as a left eigenstate of $\hat{c}^\dagger$ with the Grassmann eigenvalue $\bar{\psi}$,

$$\langle\psi| = \langle 0|e^{-\hat{c}\bar{\psi}} = \langle 0|(1-\hat{c}\,\bar{\psi}) = \langle 0| - \langle 1|\,\bar{\psi}. \tag{10.6}$$

Indeed $\langle\psi|c^\dagger = (\langle 0| - \langle 1|\,\bar{\psi})c^\dagger = \langle 0|\bar{\psi} = \langle\psi|\bar{\psi}$. Here $\bar{\psi}$ is just another Grassmann number, completely *unrelated* to ψ, which we use to parametrize left states. The set of coherent states is not orthonormal and the overlap of any two coherent states is

$$\langle\psi|\psi'\rangle = (\langle 0| - \langle 1|\,\bar{\psi})(|0\rangle - \psi'|1\rangle) = 1 + \bar{\psi}\psi' = e^{\bar{\psi}\psi'}. \tag{10.7}$$

One can write a resolution of unity in the fermionic coherent state representation as (compare with the bosonic expression (2.7))

$$\hat{1} = \int d\bar{\psi} \int d\psi \; e^{-\bar{\psi}\psi} \, |\psi\rangle\langle\psi|. \tag{10.8}$$

Indeed, employing $e^{-\bar{\psi}\psi} = 1 - \bar{\psi}\psi$ along with Eqs. (10.5) and (10.6) and keeping only the terms which contain both $\bar{\psi}$ and ψ (all others vanish upon integration, Eq. (10.4)), one finds $-\bar{\psi}\psi(|0\rangle\langle 0| + |1\rangle\langle 1|) = -\bar{\psi}\psi\,\hat{1}$ for the expression under the integral. Finally, performing the integrations according to Eq. (10.4), one proves Eq. (10.8). (Notice that one has to commute $d\psi$ and $\bar{\psi}$, which brings in a factor of minus one.)

Matrix elements of any *normally ordered* operator, that is, such that all fermionic creation operators are preceded by the annihilation ones, take the form

$$\langle\psi|\hat{H}(\hat{c}^\dagger, \hat{c})|\psi'\rangle = H(\bar{\psi}, \psi')\langle\psi|\psi'\rangle = H(\bar{\psi}, \psi')\, e^{\bar{\psi}\psi'}. \tag{10.9}$$

The trace of an operator $\hat{\mathcal{O}}$ is calculated as

$$\begin{aligned}\mathrm{Tr}\{\hat{\mathcal{O}}\} &= \sum_{n=0,1} \langle n|\hat{\mathcal{O}}|n\rangle = \sum_{n=0,1} \iint d\bar{\psi} d\psi \; e^{-\bar{\psi}\psi} \, \langle n|\psi\rangle\langle\psi|\hat{\mathcal{O}}|n\rangle \\ &= \iint d\bar{\psi} d\psi \; e^{-\bar{\psi}\psi} \sum_{n=0,1} \langle\psi|\hat{\mathcal{O}}|n\rangle\langle n| - \psi\rangle \\ &= \iint d\bar{\psi} d\psi \; e^{-\bar{\psi}\psi} \, \langle\psi|\hat{\mathcal{O}}| - \psi\rangle,\end{aligned} \tag{10.10}$$

where we have employed resolution of unity first in the coherent state basis and then in the number state basis. The minus sign in $| - \psi\rangle = |0\rangle + \psi|1\rangle$ comes about from commuting left and right coherent states.

We shall also need rules of Gaussian integration over two sets of *independent* Grassmann variables, $\bar{\psi}_j$ and ψ_j, where $j = 1, 2, \ldots, N$. The following identity,

$$Z[\bar{\chi}, \chi] = \int \prod_{j=1}^{N} [d\bar{\psi}_j d\psi_j] \, e^{-\sum_{ij}^{N} \bar{\psi}_i \hat{A}_{ij} \psi_j + \sum_{j}^{N} [\bar{\psi}_j \chi_j + \bar{\chi}_j \psi_j]} = \det\hat{A} \; e^{\sum_{ij}^{N} \bar{\chi}_i (\hat{A}^{-1})_{ij} \chi_j}, \tag{10.11}$$

is valid for *any invertible* complex matrix $\hat{A}_{ij}$.[1] Here $\bar{\chi}_j$ and χ_j are two additional mutually independent (and independent from $\bar{\psi}_j$ and ψ_j) sets of Grassmann numbers. We first notice that $\int \mathbf{D}[\bar{\psi}\psi]\, e^{-\sum \bar{\psi}\hat{A}\psi} = \det\hat{A}$ (and *not* $1/\det\hat{A}$ as is the case for bosons). This follows from the fact that the only nonvanishing term comes from the Nth order expansion of the exponent, that is, $\left(-\sum \bar{\psi}_i \hat{A}_{ij} \psi_j\right)^N / N!$, since

[1] Actually, for existence of the integral even this restriction is not needed. For example, for $N = 1$ and $A = 0$ one has $Z = \bar{\chi}\chi$. More generally, if matrix $\hat{A}$ has a zero eigenvalue, Z is proportional to the projection of the vector χ on the direction of the corresponding zero-mode eigenfunction.

the number of variables should be exactly equal to the number of integrals. Within this expression only terms with all ψ_j and all $\bar{\psi}_i$ distinct survive the integration. The surviving terms have signs reflecting the parity of permutations, leading directly to $\det\hat{A}$. Finally, one notices that $\bar{\psi}\hat{A}\,\psi - \bar{\psi}\chi - \bar{\chi}\psi = (\bar{\psi} - \bar{\chi}\hat{A}^{-1})\hat{A}\,(\psi - \hat{A}^{-1}\chi) - \bar{\chi}\hat{A}^{-1}\chi$ and the shift of integration variables $\psi \to \psi + \hat{A}^{-1}\chi$ (and similarly for $\bar{\psi}$) works exactly as in ordinary integrals.

The Wick theorem is formulated in a way similar to the complex bosonic case, Eq. (2.21), with the exception that every combination is multiplied by the parity of the corresponding permutation. This follows from the anti-commutativity of the Grassmann derivatives, mentioned previously. For example,

$$\langle\psi_a\bar{\psi}_b\rangle = \frac{1}{Z[0,0]}\frac{\delta^2 Z[\bar{\chi},\chi]}{\delta\chi_b\delta\bar{\chi}_a}\bigg|_{\chi=0} = \hat{A}^{-1}_{ab}, \tag{10.12}$$

$$\langle\psi_a\psi_b\bar{\psi}_c\bar{\psi}_d\rangle = \frac{1}{Z[0,0]}\frac{\delta^4 Z[\bar{\chi},\chi]}{\delta\chi_d\delta\chi_c\delta\bar{\chi}_b\delta\bar{\chi}_a}\bigg|_{\chi=0} = -\hat{A}^{-1}_{ac}\hat{A}^{-1}_{bd} + \hat{A}^{-1}_{ad}\hat{A}^{-1}_{bc}.$$

Notice that the first term on the right-hand side of the second expression comes with a sign opposite to that in Eq. (2.21).

10.2 Partition Function

Consider a single quantum state with energy ϵ_0. This state is populated by spinless fermions (i.e. particles obeying the Pauli exclusion principle). In fact, one may have either zero or one particle in this state. The secondary quantized Hamiltonian of such a system has the form

$$\hat{H} = \epsilon_0\,\hat{c}^\dagger\hat{c}, \tag{10.13}$$

where $\hat{c}^\dagger$ and $\hat{c}$ are creation and annihilation operators of fermions on the state ϵ_0.

One can consider the evolution operator along the closed time contour, C and the corresponding partition function, $Z = 1$, defined in exactly the same way as for bosonic systems, Eq. (2.13). The trace of the equilibrium density matrix is $\mathrm{Tr}\{\hat{\rho}_0\} = 1 + \rho(\epsilon_0)$, where the two terms stand for the empty and singly occupied states. One divides the contour into $(2N-2)$ time intervals, Fig. 2.1, of length $\delta_t \sim 1/N \to 0$, and introduces fermionic resolutions of unity in $2N$ points along the contour, C, in the form (10.8). The rest of the algebra goes through exactly as in the bosonic case; see Section 2.2. As a result, one arrives at

$$Z = \frac{1}{\mathrm{Tr}\{\hat{\rho}_0\}}\iint\prod_{j=1}^{2N}[\mathrm{d}\bar{\psi}_j\,\mathrm{d}\psi_j]\,\exp\left(\mathrm{i}\sum_{j,j'=1}^{2N}\bar{\psi}_j\,G^{-1}_{jj'}\,\psi_{j'}\right), \tag{10.14}$$

where the $2N \times 2N$ matrix $G^{-1}_{jj'}$ is (for $N=3$)

$$\mathrm{i}G^{-1}_{jj'} \equiv \left[\begin{array}{ccc|ccc} -1 & & & & & -\rho(\epsilon_0) \\ h_- & -1 & & & & \\ & h_- & -1 & & & \\ \hline & & 1 & -1 & & \\ & & & h_+ & -1 & \\ & & & & h_+ & -1 \end{array}\right], \tag{10.15}$$

where $h_\mp \equiv 1 \mp \mathrm{i}\epsilon_0\delta_t$. The main diagonal of this matrix originates from the resolutions of unity, Eq. (10.8), while the lower sub-diagonal comes from the matrix elements (10.9). Finally, the upper right element is $\langle\psi_1|\hat{\rho}_0|-\psi_{2N}\rangle$; compare Eq. (2.16). The only difference from the bosonic case is the negative sign in front of the $\rho(\epsilon_0)$ matrix element, originating from the minus sign in the $|-\psi_{2N}\rangle$ coherent state in the expression (10.10) for the fermionic trace. To check the normalization identity, let us evaluate the determinant of such a matrix:

$$\det\big[\mathrm{i}\hat{G}^{-1}\big] = 1+\rho(\epsilon_0)(1-h^2)^{N-1} \approx 1+\rho(\epsilon_0)\,\mathrm{e}^{(\epsilon_0\delta_t)^2(N-1)} \overset{N\to\infty}{\to} 1+\rho(\epsilon_0). \tag{10.16}$$

Employing the fact, Eq. (10.11), that the fermionic Gaussian integral is given by the determinant (unlike the inverse determinant for bosons) of the correlation matrix (10.15), one finds

$$Z = \frac{\det\big[\mathrm{i}\hat{G}^{-1}\big]}{\mathrm{Tr}\{\hat{\rho}_0\}} = 1, \tag{10.17}$$

as it should be. Once again, the upper right element of the discrete matrix (10.15) is crucial for maintaining the correct normalization. Taking the limit $N \to \infty$ and introducing the continuum notation, $\psi_j \to \psi(t)$, one obtains

$$Z = \int \mathbf{D}[\bar{\psi}\psi]\, \exp\big(\mathrm{i}S[\bar{\psi},\psi]\big) = \int \mathbf{D}[\bar{\psi}\psi]\, \exp\left(\mathrm{i}\int_C \mathrm{d}t\,\big[\bar{\psi}(t)\,\hat{G}^{-1}\psi(t)\big]\right), \tag{10.18}$$

where according to (10.14) and (10.15) the action is given by

$$S[\bar{\psi},\psi] = \sum_{j=2}^{2N}\left[\mathrm{i}\bar{\psi}_j\,\frac{\psi_j-\psi_{j-1}}{\delta t_j} - \epsilon_0\bar{\psi}_j\,\psi_{j-1}\right]\delta t_j + \mathrm{i}\,\bar{\psi}_1\Big[\psi_1+\rho(\epsilon_0)\psi_{2N}\Big], \tag{10.19}$$

with $\delta t_j \equiv t_j - t_{j-1} = \pm\delta_t$, where $\pm$ signs correspond to the forward and backward branches of the time contour. Thus, the continuum form of the operator $\hat{G}^{-1}$ is the same as for bosons, Eq. (2.29): $\hat{G}^{-1} = \mathrm{i}\partial_t - \epsilon_0$. Again, the upper right element of the discrete matrix (the last term in Eq. (10.19)), which contains information about the distribution function, is seemingly absent in the continuum notation.

Splitting the Grassmann field $\psi(t)$ into the two components $\psi^+(t)$ and $\psi^-(t)$ that reside on the forward and backward parts of the time contour, respectively, one may rewrite the action as

$$S[\bar{\psi},\psi] = \int_{-\infty}^{+\infty} \mathrm{d}t \left[\bar{\psi}^+(t)(\mathrm{i}\partial_t - \epsilon_0)\psi^+(t) - \bar{\psi}^-(t)(\mathrm{i}\partial_t - \epsilon_0)\psi^-(t)\right], \quad (10.20)$$

where the dynamics of ψ^+ and ψ^- are actually *not* independent from each other, owing to the presence of nonzero off-diagonal blocks in the discrete matrix (10.15).

10.3 Green Functions and Keldysh Rotation

The four fermionic Green functions: $G^{\mathbb{T}(\tilde{\mathbb{T}})}$ and $G^{<(>)}$ are defined in the same way as their bosonic counterparts; see Eq. (2.33). Inverting the discrete matrix (10.15), as required by the Gaussian identity (10.12), and going to the continuum limit, one finds (cf. Eq. (2.35)),

$$\langle\psi^+(t)\bar{\psi}^-(t')\rangle \equiv \mathrm{i}G^<(t,t') = -n_{\mathrm{F}}\exp\{-\mathrm{i}\epsilon_0(t-t')\}, \quad (10.21a)$$

$$\langle\psi^-(t)\bar{\psi}^+(t')\rangle \equiv \mathrm{i}G^>(t,t') = (1-n_{\mathrm{F}})\exp\{-\mathrm{i}\epsilon_0(t-t')\}, \quad (10.21b)$$

$$\langle\psi^+(t)\bar{\psi}^+(t')\rangle \equiv \mathrm{i}G^{\mathbb{T}}(t,t') = \theta(t-t')\mathrm{i}G^>(t,t') + \theta(t'-t)\mathrm{i}G^<(t,t'), \quad (10.21c)$$

$$\langle\psi^-(t)\bar{\psi}^-(t')\rangle \equiv \mathrm{i}G^{\tilde{\mathbb{T}}}(t,t') = \theta(t'-t)\mathrm{i}G^>(t,t') + \theta(t-t')\mathrm{i}G^<(t,t'). \quad (10.21d)$$

The difference from bosons is in the minus sign in the expression for $G^<$, due to the anti-commutation relations, and Bose occupation number is exchanged for the Fermi occupation number: $n_{\mathrm{B}} \to n_{\mathrm{F}} \equiv \rho(\epsilon_0)/(1+\rho(\epsilon_0))$. Equation (10.22)

$$G^{\mathbb{T}}(t,t') + G^{\tilde{\mathbb{T}}}(t,t') - G^>(t,t') - G^<(t,t') = 0 \quad (10.22)$$

holds for the fermion fermionic Green functions in the same sense as for the boson ones; see the discussion following Eq. (2.38). Thus, one would like to perform the Keldysh rotation to take advantage of this relation.

It is customary to perform the Keldysh rotation in the fermionic case in a different manner from the bosonic one. Define the new fields as

$$\psi_1(t) = \frac{1}{\sqrt{2}}\big(\psi^+(t) + \psi^-(t)\big), \qquad \psi_2(t) = \frac{1}{\sqrt{2}}\big(\psi^+(t) - \psi^-(t)\big). \quad (10.23)$$

Following Larkin and Ovchinnikov [28], it is agreed that the *bar* fields transform in a different way:

$$\bar{\psi}_1(t) = \frac{1}{\sqrt{2}}\big(\bar{\psi}^+(t) - \bar{\psi}^-(t)\big), \qquad \bar{\psi}_2(t) = \frac{1}{\sqrt{2}}\big(\bar{\psi}^+(t) + \bar{\psi}^-(t)\big). \quad (10.24)$$

Since the Grassmann fields $\bar{\psi}$ are *not* conjugated to ψ, but rather are completely independent fields, they may be transformed in an arbitrary manner (as long as the transformation matrix has a nonzero determinant). Notice that there is no issue regarding the convergence of the integrals, since the Grassmann integrals are always convergent. We also avoid the superscripts cl and q, because the Grassmann variables never have a classical meaning. Indeed, one can never write a stationary point or any other equation in terms of $\bar{\psi}, \psi$; rather, they must always be integrated out at some stage of the calculations.

Employing Eqs. (10.23) and (10.24) along with Eqs. (10.21), one finds

$$-\mathrm{i}\langle\psi_a(t)\bar{\psi}_b(t')\rangle = \hat{G}_{ab}(t,t') = \begin{pmatrix} G^{\mathrm{R}}(t,t') & G^{\mathrm{K}}(t,t') \\ 0 & G^{\mathrm{A}}(t,t') \end{pmatrix}, \tag{10.25}$$

where hereafter $a, b = 1, 2$. The fact that the (2, 1) element of this matrix is zero is a manifestation of the identity (10.22). The *retarded*, *advanced*, and *Keldysh* components of the Green function (10.25) are expressed in terms of $G^{\mathbb{T}(\tilde{\mathbb{T}})}$ and $G^{<(>)}$ in exactly the same way as their bosonic analogs, Eqs. (2.41):

$$G^{\mathrm{R}}(t,t') = G^{1,1}(t,t') = \frac{1}{2}\left(G^{\mathbb{T}} - G^{\tilde{\mathbb{T}}} + G^{>} - G^{<}\right) = \theta(t-t')\left(G^{>} - G^{<}\right), \tag{10.26a}$$

$$G^{\mathrm{A}}(t,t') = G^{2,2}(t,t') = \frac{1}{2}\left(G^{\mathbb{T}} - G^{\tilde{\mathbb{T}}} - G^{>} + G^{<}\right) = \theta(t'-t)\left(G^{<} - G^{>}\right), \tag{10.26b}$$

$$G^{\mathrm{K}}(t,t') = G^{1,2}(t,t') = \frac{1}{2}\left(G^{\mathbb{T}} + G^{\tilde{\mathbb{T}}} + G^{>} + G^{<}\right) = G^{>} + G^{<}. \tag{10.26c}$$

They therefore possess the same symmetry properties, Eqs. (2.42)–(2.45). In particular,

$$G^{\mathrm{A}} = \left[G^{\mathrm{R}}\right]^{\dagger}, \qquad G^{\mathrm{K}} = -\left[G^{\mathrm{K}}\right]^{\dagger}, \tag{10.27}$$

where the Green functions are understood as matrices in the time domain and the Hermitian conjugation includes complex conjugation along with interchanging of the two time arguments. Both retarded and advanced matrices have nonzero main diagonals, that is, $t = t'$, which are opposite to each other:

$$G^{\mathrm{R}}(t,t) + G^{\mathrm{A}}(t,t) = 0; \qquad \int \frac{\mathrm{d}\epsilon}{2\pi}\left[G^{\mathrm{R}}(\epsilon) + G^{\mathrm{A}}(\epsilon)\right] = 0. \tag{10.28}$$

This equation is not restricted to our toy model, but is completely general. An important consequence of Eq. (10.28) is

$$\mathrm{Tr}\left\{\hat{G}^{(1)} \circ \hat{G}^{(2)} \circ \cdots \circ \hat{G}^{(l)}\right\}(t,t) = 0, \tag{10.29}$$

where the circular multiplication sign involves convolution in the time domain along with the 2×2 matrix multiplication. The argument (t, t) states that the first time argument of $\hat{G}^{(1)}$ and the last time argument of $\hat{G}^{(l)}$ are the same.

Note that the fermionic Green function has a different structure compared to its bosonic counterpart, Eq. (2.40), that is, positions of the R, A, and K components in the matrix are exchanged. The reason, of course, is the different convention for transformation of the *bar* fields. One could choose the fermionic convention to be the same as the bosonic (but *not* the other way around), thus having the same structure (2.40) for the fermions as for the bosons. The rationale for the Larkin–Ovchinnikov choice (10.25) is that the inverse Green function, $\hat{G}^{-1}$ and fermionic self-energy $\hat{\Sigma}_{\mathrm{F}}$ have the same appearance as $\hat{G}$, namely

$$\hat{G}^{-1} = \begin{pmatrix} \left[G^{\mathrm{R}}\right]^{-1} & \left[G^{-1}\right]^{\mathrm{K}} \\ 0 & \left[G^{\mathrm{A}}\right]^{-1} \end{pmatrix}, \qquad \hat{\Sigma} = \begin{pmatrix} \Sigma^{\mathrm{R}} & \Sigma^{\mathrm{K}} \\ 0 & \Sigma^{\mathrm{A}} \end{pmatrix}, \tag{10.30}$$

whereas in the case of bosons $\hat{G}^{-1}$ (see Eq. (2.51)) and $\hat{\Sigma}$ (see Eq. (6.17)) look different from $\hat{G}$ (see Eq. (2.40)). This fact gives the form (10.25) and (10.30) a certain technical advantage.

For the single fermionic state, after the Keldysh rotation, the Green functions (10.21) allow us to find components of the matrix (10.25):

$$G^{\mathrm{R}}(t,t') = -\mathrm{i}\theta(t-t')\mathrm{e}^{-\mathrm{i}\epsilon_0(t-t')} \overset{\mathrm{FT}}{\to} (\epsilon - \epsilon_0 + \mathrm{i}0)^{-1}, \tag{10.31a}$$

$$G^{\mathrm{A}}(t,t') = \mathrm{i}\theta(t'-t)\mathrm{e}^{-\mathrm{i}\epsilon_0(t-t')} \overset{\mathrm{FT}}{\to} (\epsilon - \epsilon_0 - \mathrm{i}0)^{-1}, \tag{10.31b}$$

$$G^{\mathrm{K}}(t,t') = -\mathrm{i}(1-2n_{\mathrm{F}})\mathrm{e}^{-\mathrm{i}\epsilon_0(t-t')} \overset{\mathrm{FT}}{\to} -2\pi\mathrm{i}(1-2n_{\mathrm{F}})\delta(\epsilon-\epsilon_0), \tag{10.31c}$$

where the right-hand sides provide also the Fourier transforms with respect to $t - t'$. In thermal equilibrium $n_{\mathrm{F}}(\epsilon) = (\mathrm{e}^{(\epsilon-\mu)/T} + 1)^{-1}$ and one obtains

$$G^{\mathrm{K}}(\epsilon) = \tanh\frac{\epsilon-\mu}{2T}\left[G^{\mathrm{R}}(\epsilon) - G^{\mathrm{A}}(\epsilon)\right]. \tag{10.32}$$

This is the FDT for fermions. As in the case of bosons, the FDT is a generic feature of an equilibrium system, not restricted to the toy model. In general, it is convenient to parametrize the anti-Hermitian Keldysh Green function by a Hermitian matrix $F = F^{\dagger}$ as

$$G^{\mathrm{K}} = G^{\mathrm{R}} \circ F - F \circ G^{\mathrm{A}}. \tag{10.33}$$

The Wigner transform of $F = 1 - 2n_{\mathrm{F}}$ is the fermionic distribution function. Equation (10.33) allows one to rewrite the fermionic Green function (10.25) in the following convenient form:

$$\hat{G} = \begin{pmatrix} G^{\mathrm{R}} & G^{\mathrm{K}} \\ 0 & G^{\mathrm{A}} \end{pmatrix} = \hat{\mathcal{U}}^{-1} \circ \begin{pmatrix} G^{\mathrm{R}} & 0 \\ 0 & G^{\mathrm{A}} \end{pmatrix} \circ \hat{\mathcal{U}}, \qquad \hat{\mathcal{U}} = \begin{pmatrix} 1 & F \\ 0 & 1 \end{pmatrix} = \mathrm{e}^{\hat{\sigma}_+ F}, \tag{10.34}$$

where $\hat{\sigma}_+ = (\hat{\sigma}_1 + \mathrm{i}\hat{\sigma}_2)/2$ is acting in the Keldysh space.

10.4 Free Fermionic Fields and Their Action

One may proceed now to a system with many degrees of freedom. We shall assume that they are labeled by a momentum index $\mathbf{k}$ and spin index $\sigma = \uparrow, \downarrow = \pm 1$. To this end, one changes $\epsilon_0 \to \epsilon_{\mathbf{k},\sigma} = \mathbf{k}^2/(2m) + H_Z\sigma$, where H_Z is a Zeeman magnetic field,[2] and performs summations over $\mathbf{k}$ and σ. It is instructive to transform to the coordinate space representation $\psi_\sigma(\mathbf{r},t) = \sum_{\mathbf{k}} \psi_\sigma(\mathbf{k},t)\mathrm{e}^{\mathrm{i}\mathbf{k}\mathbf{r}}$, while $\bar{\psi}_\sigma(\mathbf{r},t) = \sum_{\mathbf{k}} \bar{\psi}_\sigma(\mathbf{k},t)\mathrm{e}^{-\mathrm{i}\mathbf{k}\mathbf{r}}$. Although $\bar{\psi}$ is not a complex conjugate of ψ, but is rather an independent Grassmann number, we choose the opposite Fourier transform convention, as if they are mutual complex conjugates. Then $\mathbf{k}^2/(2m) \to -\nabla_{\mathbf{r}}^2/(2m)$ and the Keldysh action for a noninteracting gas of fermions takes the form

$$S_0[\bar{\psi},\psi] = \sum_{\sigma,\sigma'} \iint \mathrm{d}x\, \mathrm{d}x' \sum_{a,b=1}^{2} \bar{\psi}_{a,\sigma}(x) \big[\hat{G}^{-1}_{\sigma\sigma'}(x,x')\big]^{ab}\, \psi_{b,\sigma'}(x'), \tag{10.35}$$

where $x = (\mathbf{r}, t)$ and the matrix correlator $[\hat{G}^{-1}]^{ab}$ has the structure (10.30) with

$$\big[G^{\mathrm{R(A)}}_{\sigma\sigma'}(x,x')\big]^{-1} = \delta(x-x')\left(\mathrm{i}\partial_{t'} + \frac{1}{2m}\nabla^2_{\mathbf{r}'} - \mathbf{H}_Z \cdot \vec{s}_{\sigma\sigma'} \pm \mathrm{i}0\right), \tag{10.36}$$

where $\vec{s}_{\sigma'\sigma}$ is the usual three-component vector of Pauli matrices in spin space. Although in continuum notation the R and A components seem to be the same, one has to remember that in the discrete time representation they are matrices with structure below and above the main diagonal, respectively. The Keldysh component is a pure regularization, in the sense that it does not have a continuum limit. (The Keldysh component of the self-energy does have a nonzero continuum representation.) All this information is already properly taken into account in the structure of the Green function (10.25). The explicit form of the bare fermionic Green functions is

$$G^{\mathrm{R(A)}}_{\sigma\sigma'}(\mathbf{k},\epsilon) = \delta_{\sigma\sigma'}(\epsilon - \epsilon_{\mathbf{k},\sigma} \pm \mathrm{i}0)^{-1}, \tag{10.37a}$$

$$G^{\mathrm{K}}_{\sigma\sigma'}(\mathbf{k},\epsilon) = -2\pi\mathrm{i}\,\delta_{\sigma\sigma'}F(\epsilon)\,\delta(\epsilon - \epsilon_{\mathbf{k},\sigma}), \tag{10.37b}$$

where $F(\epsilon) = 1 - 2n_{\mathrm{F}}(\epsilon)$. In general, the Keldysh component of the Green function may be parametrized by the Hermitian matrix $F_{\sigma\sigma'}(x,x')$,

[2] We have absorbed the factor $\hbar e/mc$ into the magnetic field units.

$$G^{\mathrm{K}}_{\sigma\sigma'}(x,x') = \sum_{\sigma''}\int \mathrm{d}x''\left[G^{\mathrm{R}}_{\sigma\sigma''}(x,x'')F_{\sigma''\sigma'}(x'',x') - F_{\sigma\sigma''}(x,x'')G^{\mathrm{A}}_{\sigma''\sigma'}(x'',x')\right]. \tag{10.38}$$

Projecting its Wigner transform onto the four Pauli matrices, one may define $F(x,p) = \mathrm{Tr}\{F_{\sigma\sigma}(x,p)\}$ and $\mathbf{F}^{\mu}(x,p) = \mathrm{Tr}\{F_{\sigma\sigma'}(x,p)\hat{s}^{\mu}_{\sigma'\sigma}\}$ (where $\mu = x,y,z$), as the distribution functions of the fermionic density and the spin-density, respectively. One should not be alarmed that all components of the spin are determined simultaneously. This has the same semiclassical interpretation as the simultaneous specification of $x = (\mathbf{r},t)$ and $p = (\mathbf{k},\epsilon)$.

10.5 External Fields and Sources

According to the basic idea of the Keldysh technique, the partition function $Z = 1$ is normalized by construction; see Eq. (10.17). To make the entire theory meaningful one should introduce auxiliary source fields, which enable one to compute various observable quantities: density of particles, currents, and so on. For example, one may introduce an external time-dependent scalar potential $V(\mathbf{r},t)$ defined along the contour $\mathcal{C}$. Its interaction with the fermion density leads to the action $S_V = -\int \mathrm{d}\mathbf{r}\int_{\mathcal{C}}\mathrm{d}t\, V(\mathbf{r},t)\bar{\psi}(\mathbf{r},t)\psi(\mathbf{r},t)$, where we have suppressed the spin index for brevity. Expressing it via the field components residing on the forward and backward contour branches, one finds

$$\begin{aligned} S_V &= -\int \mathrm{d}\mathbf{r}\int_{-\infty}^{+\infty}\mathrm{d}t\left[V^{+}\bar{\psi}^{+}\psi^{+} - V^{-}\bar{\psi}^{-}\psi^{-}\right] = -2\int \mathrm{d}\mathbf{r}\mathrm{d}t\left[V^{\mathrm{cl}}\rho^{\mathrm{q}} + V^{\mathrm{q}}\rho^{\mathrm{cl}}\right] \\ &= -\int \mathrm{d}\mathbf{r}\mathrm{d}t\left[V^{\mathrm{cl}}(\bar{\psi}_1\psi_1 + \bar{\psi}_2\psi_2) + V^{\mathrm{q}}(\bar{\psi}_1\psi_2 + \bar{\psi}_2\psi_1)\right], \end{aligned} \tag{10.39}$$

where the $V^{\mathrm{cl(q)}}(\mathbf{r},t)$ components are defined in the standard way for real boson fields, $V^{\mathrm{cl(q)}} = (V^{+} \pm V^{-})/2$. The physical density of fermions (symmetrized over the two branches of the contour), given by $\rho^{\mathrm{cl}} = \frac{1}{2}(\bar{\psi}^{+}\psi^{+} + \bar{\psi}^{-}\psi^{-})$, is coupled to the *quantum* component of the source field, V^{q}. On the other hand, the quantum component of density, defined as $\rho^{\mathrm{q}} = \frac{1}{2}(\bar{\psi}^{+}\psi^{+} - \bar{\psi}^{-}\psi^{-})$, is coupled to the *classical* source component, V^{cl}, which is nothing but an external physical scalar potential, the same on the two branches. In the last line of Eq. (10.39) we performed rotation from $\psi^{\pm}$ to $\psi_{1(2)}$ according to Eqs. (10.23) and (10.24).

The notation may be substantially compactified by introducing two vertex fermionic $\hat{\gamma}$-matrices:

$$\hat{\gamma}^{\mathrm{cl}} \equiv \begin{pmatrix} 1 & 0 \\ 0 & 1 \end{pmatrix}, \qquad \hat{\gamma}^{\mathrm{q}} \equiv \begin{pmatrix} 0 & 1 \\ 1 & 0 \end{pmatrix}. \tag{10.40}$$

With the help of these definitions, the source action (10.39) may be written as

$$S_V = -\int d\mathbf{r} \int_{-\infty}^{+\infty} dt \sum_{a,b=1}^{2} \left[V^{\rm cl} \bar{\psi}_a \gamma^{\rm cl}_{ab} \psi_b + V^{\rm q} \bar{\psi}_a \gamma^{\rm q}_{ab} \psi_b \right] = -{\rm Tr}\{\vec{\bar{\Psi}} \hat{V} \vec{\Psi}\}, \quad (10.41)$$

where we have introduced the Keldysh doublet $\vec{\Psi}$ and matrix $\hat{V}$, defined as

$$\vec{\Psi} = \begin{pmatrix} \psi_1 \\ \psi_2 \end{pmatrix}, \qquad \hat{V} = V^\alpha \hat{\gamma}^\alpha = \begin{pmatrix} V^{\rm cl} & V^{\rm q} \\ V^{\rm q} & V^{\rm cl} \end{pmatrix}, \quad (10.42)$$

where $\alpha = ({\rm cl}, {\rm q})$.

In a similar way one may introduce an external vector potential into the formalism. The corresponding part of the action[3] $S_A = -\int d\mathbf{r} \int_{\mathcal{C}} dt\, \mathbf{A}(\mathbf{r},t)\mathbf{j}(\mathbf{r},t)$ represents the coupling between $\mathbf{A}(\mathbf{r},t)$ and the fermion current density defined as $\mathbf{j}(\mathbf{r},t) = \frac{1}{2mi}[\bar{\psi}(\mathbf{r},t)\nabla_{\mathbf{r}}\psi(\mathbf{r},t) - \nabla_{\mathbf{r}}\bar{\psi}(\mathbf{r},t)\psi(\mathbf{r},t)]$. By splitting $\int_{\mathcal{C}} dt$ into forward and backward parts and performing Keldysh rotation, one finds by analogy with the scalar potential case (10.39) that

$$S_A = -{\rm Tr}\{\vec{\bar{\Psi}} \hat{\mathbf{A}} \mathbf{v}_{\rm F} \vec{\Psi}\}, \qquad \hat{\mathbf{A}} = \mathbf{A}^\alpha \hat{\gamma}^\alpha = \begin{pmatrix} \mathbf{A}^{\rm cl} & \mathbf{A}^{\rm q} \\ \mathbf{A}^{\rm q} & \mathbf{A}^{\rm cl} \end{pmatrix}. \quad (10.43)$$

We have linearized the fermionic dispersion relation near the Fermi energy and employed the fact that $-i\nabla_{\mathbf{r}} \approx \mathbf{k}_{\rm F}$ and $\mathbf{v}_{\rm F} = \mathbf{k}_{\rm F}/m$.

Let us now define the generating function as

$$Z[V^{\rm cl}, V^{\rm q}] \equiv \langle \exp(iS_V) \rangle, \quad (10.44)$$

where the angular brackets denote the functional integration over the Grassmann fields $\bar{\psi}$ and ψ with weight $\exp(iS_0)$, specified by the fermionic action (10.35). In the absence of the quantum component, $V^{\rm q} = 0$, the source field is the same on both branches of the time contour. Therefore, the evolution along the contour brings the system back to its exact original state. Thus, one expects that the classical component alone does not change the fundamental normalization identity, $Z = 1$. As a result,

$$Z[V^{\rm cl}, 0] \equiv 1, \quad (10.45)$$

as we already discussed in Chapter 2; see Eq. (2.63). Indeed, one may verify this statement explicitly by expanding the generating function (10.44) in powers of $V^{\rm cl}$ and employing the Wick theorem. For example, to first order in $V^{\rm cl}$ one finds $Z[V^{\rm cl}, 0] = 1 - \int dt\, V^{\rm cl}(t){\rm Tr}[\hat{\gamma}^{\rm cl}\hat{G}(t,t)] + \cdots = 1$, where one uses that $\hat{\gamma}^{\rm cl} = \hat{1}$ along with Eq. (10.29). It is straightforward to see that for exactly the same reasons all higher-order terms in $V^{\rm cl}$ vanish as well.

[3] The vector source $\mathbf{A}(\mathbf{r},t)$ that we are using here differs from the actual vector potential by a factor of e/c. However, we refer to it as the vector potential and restore electron charge only in the final expressions.

A lesson from Eq. (10.45) is that one necessarily has to introduce *quantum* sources (which change sign between the forward and backward branches of the contour). The presence of such source fields explicitly violates causality, thus making the generating function nontrivial. On the other hand, these fields usually do not have a physical meaning and play only an auxiliary role. In most cases one uses them to generate observables by an appropriate differentiation. Indeed, as was mentioned, the physical density is coupled to the quantum component of the source. In the end, one takes the quantum sources to be zero, restoring the causality of the action. Note that the classical component, V^{cl}, does *not* have to be taken to zero.

Let us see how it works. Suppose one is interested in an average fermion density $\rho^{\text{cl}}(x)$ at a space-time point x in the presence of a certain physical scalar potential V^{cl}. According to Eqs. (10.39) and (10.44) it is given by

$$\rho^{\text{cl}}(x; V^{\text{cl}}) = \frac{\mathrm{i}}{2} \frac{\delta}{\delta V^{\text{q}}(x)} Z[V^{\text{cl}}, V^{\text{q}}]\bigg|_{V^{\text{q}}=0}, \tag{10.46}$$

where $x = (\mathbf{r}, t)$. The problem is simplified if the external field, V^{cl}, is weak in some sense. One may then restrict oneself to the linear response, by defining the polarization (or susceptibility)

$$\Pi^{\text{R}}(x, x') \equiv -\frac{\delta \rho^{\text{cl}}(x; V^{\text{cl}})}{\delta V^{\text{cl}}(x')}\bigg|_{V^{\text{cl}}=0} = -\frac{\mathrm{i}}{2} \frac{\delta^2 Z[V^{\text{cl}}, V^{\text{q}}]}{\delta V^{\text{cl}}(x')\delta V^{\text{q}}(x)}\bigg|_{V^{\text{q}}=V^{\text{cl}}=0}. \tag{10.47}$$

We add the superscript R, anticipating on physical grounds that the response function must be *retarded* (causality). We shall demonstrate it shortly. According to Eqs. (10.39) and (10.44), the response function (10.47) is given by

$$\Pi^{\text{R}}(x, x') = 2\mathrm{i}\left\langle \rho^{\text{cl}}(x)\rho^{\text{q}}(x')\right\rangle, \qquad \delta\rho^{\text{cl}}(x; V^{\text{cl}}) = -\int \mathrm{d}x'\, \Pi^{\text{R}}(x, x')V^{\text{cl}}(x'), \tag{10.48}$$

where the angular brackets again denote averaging with the fermionic action without the sources. This is the *Kubo* formula for the linear response.

In a similar way let us introduce the *polarization* matrix as

$$\hat{\Pi}^{\alpha\beta}(x, x') \equiv -\frac{\mathrm{i}}{2} \frac{\delta^2 \ln Z[\hat{V}]}{\delta V^{\beta}(x')\delta V^{\alpha}(x)}\bigg|_{\hat{V}=0} = \begin{pmatrix} 0 & \Pi^{\text{A}}(x, x') \\ \Pi^{\text{R}}(x, x') & \Pi^{\text{K}}(x, x') \end{pmatrix}. \tag{10.49}$$

Owing to the normalization identity, Eq. (10.45), the logarithm is redundant for the R and A components and therefore the two definitions (10.47) and (10.49) are in agreement. The fact that $\Pi^{\text{cl,cl}} = 0$ is obvious from Eq. (10.45). To evaluate the polarization matrix, $\hat{\Pi}$, consider the Gaussian action (10.35). Adding the source term (10.41), one finds $S_0 + S_V = \int \mathrm{d}x\, \vec{\bar{\Psi}}[\hat{G}^{-1} - V^{\alpha}\hat{\gamma}^{\alpha}]\vec{\Psi}$. Integrating over the

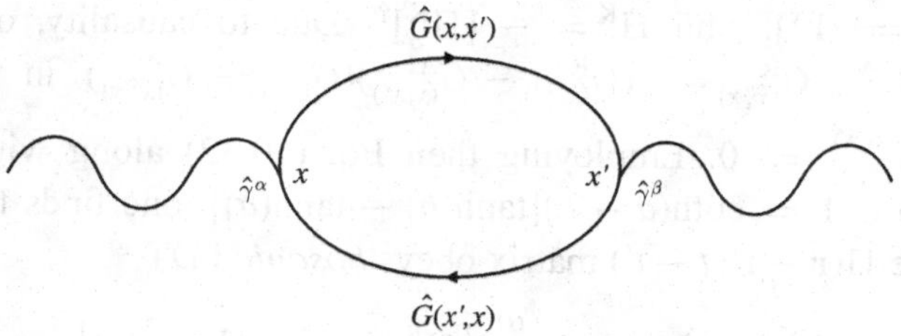

Figure 10.1 Polarization matrix $\hat{\Pi}_0^{\alpha\beta}(x,x')$: each solid line stands for the fermion matrix Green function (10.25), wavy lines represent external classical or quantum potentials $V^{\mathrm{cl(q)}}$, and $x = (\mathbf{r}, t)$. The loop diagram is a graphic representation of the trace in Eq. (10.51).

Grassmann fields $\bar{\psi}$, ψ, one obtains according to the Gaussian identity (10.11) (cf. Eq. (2.68))

$$Z[V] = \frac{\det\left[\mathrm{i}\hat{G}^{-1} - \mathrm{i}V^\alpha\hat{\gamma}^\alpha\right]}{\mathrm{Tr}\{\hat{\rho}_0\}} = \det\left[\hat{1} - \hat{G}\,V^\alpha\hat{\gamma}^\alpha\right] = \mathrm{e}^{\mathrm{Tr}\ln[\hat{1}-\hat{G}\,V^\alpha\hat{\gamma}^\alpha]}, \tag{10.50}$$

where Eq. (10.17) has been used. Since $Z[0] = 1$, the normalization is exactly right. One may now differentiate $\ln Z = \mathrm{Tr}\ln[\hat{1} - \hat{G}\,V^\alpha\hat{\gamma}^\alpha]$ over V^α, according to Eq. (10.49). As a result, one finds for the polarization matrix of the Gaussian (i.e. noninteracting) fermions

$$\hat{\Pi}_0^{\alpha\beta}(x,x') = \frac{\mathrm{i}}{2}\,\mathrm{Tr}\left\{\hat{\gamma}^\alpha\hat{G}(x,x')\hat{\gamma}^\beta\hat{G}(x',x)\right\}, \tag{10.51}$$

which has a transparent diagrammatic representation, Fig. 10.1.

Substituting the explicit form of the gamma matrices (10.40) and the Green functions (10.25), one obtains the *response* and *correlation* components of the polarization matrix[4]

$$\Pi_0^{\mathrm{R(A)}}(x,x') = \frac{\mathrm{i}}{2}\left(G^{\mathrm{R(A)}}(x,x')G^{\mathrm{K}}(x',x) + G^{\mathrm{K}}(x,x')G^{\mathrm{A(R)}}(x',x)\right), \tag{10.52a}$$

$$\Pi_0^{\mathrm{K}}(x,x') = \frac{\mathrm{i}}{2}\left(G^{\mathrm{K}}(x,x')G^{\mathrm{K}}(x',x) + G^{\mathrm{R}}(x,x')G^{\mathrm{A}}(x',x) + G^{\mathrm{A}}(x,x')G^{\mathrm{R}}(x',x)\right). \tag{10.52b}$$

Notice the opposite sign of the fermionic polarization loop with respect to its bosonic analog, Eqs. (7.22)–(7.24) and Eq. (8.53). From the first line here and the properties of the Green functions it is obvious that $\Pi^{\mathrm{R(A)}}(x,x')$ is indeed a lower (upper) triangular matrix in the time domain, justifying their superscripts. Moreover, from the symmetry properties of the fermionic Green functions

[4] The fact that $\Pi^{\mathrm{cl,cl}} = 0$ is an immediate consequence of $\hat{\gamma}^{\mathrm{cl}} = \hat{1}$ and Eq. (10.29).

one finds $\Pi^{\rm R}=[\Pi^{\rm A}]^\dagger$ and $\Pi^{\rm K}= -[\Pi^{\rm K}]^\dagger$. Due to causality, one may write $G^{\rm R}_{(x,x')}G^{\rm A}_{(x',x)}+G^{\rm A}_{(x,x')}G^{\rm R}_{(x',x)}=-(G^{\rm R}_{(x,x')}-G^{\rm A}_{(x,x')})(G^{\rm R}_{(x',x)}-G^{\rm A}_{(x',x)})$ in Eq. (10.52b); indeed $G^{\rm R(A)}_{(t,t')}G^{\rm R(A)}_{(t',t)}=0$. Employing then Eq. (10.32) along with the identity $\tanh(a)\tanh(b)-1=\coth(a-b)[\tanh(b)-\tanh(a)]$, one finds that in thermal equilibrium the $\hat{\Pi}(\mathbf{r}-\mathbf{r}',t-t')$ matrix obeys *bosonic* FDT

$$\Pi^{\rm K}(\mathbf{q},\omega)=\coth\frac{\omega}{2T}\big[\Pi^{\rm R}(\mathbf{q},\omega)-\Pi^{\rm A}(\mathbf{q},\omega)\big]. \tag{10.53}$$

As a result, the polarization matrix, $\hat{\Pi}$, possesses all the properties of the *bosonic* self-energy $\hat{\Sigma}$; see Eq. (6.17). Employing the explicit form (10.37) of the bare Green functions and integrating over energy with the help of the delta-function, one finds for the components of the fermionic polarization matrix

$$\Pi_0^{\rm R(A)}(\mathbf{q},\omega)=-\frac{1}{2}\sum_{\mathbf{k},\sigma}\frac{F(\epsilon_{\mathbf{k}+\mathbf{q},\sigma})-F(\epsilon_{\mathbf{k},\sigma})}{\omega\pm {\rm i}0+\epsilon_{\mathbf{k},\sigma}-\epsilon_{\mathbf{k}+\mathbf{q},\sigma}}; \tag{10.54a}$$

$$\Pi_0^{\rm K}(\mathbf{q},\omega)=-{\rm i}\pi\sum_{\mathbf{k},\sigma}\delta(\omega+\epsilon_{\mathbf{k},\sigma}-\epsilon_{\mathbf{k}+\mathbf{q},\sigma})\big[F(\epsilon_{\mathbf{k},\sigma})F(\epsilon_{\mathbf{k}+\mathbf{q},\sigma})-1\big]. \tag{10.54b}$$

Again, notice the opposite sign in comparison with the corresponding bosonic expressions, Eqs. (7.25). However, in the present case $F=1-2n_{\rm F}$, while for bosons $F=2n_{\rm B}+1$. As a result, in terms of the occupation numbers $n_{\rm F(B)}$, Eqs. (7.25a) and (10.54a) agree, while Eqs. (7.25b) and (10.54b) differ only by a term $\propto n^2_{\rm F(B)}$. Therefore in the classical limit, where $n_{\rm F}\approx n_{\rm B}\approx {\rm e}^{-(\epsilon_{\mathbf{k}}-\mu)/T}\ll 1$, the polarization matrix does not depend on the underlying quantum statistics.

10.6 Free Degenerate Fermi Gas

In view of the last remark, we already know the high-temperature limit of the fermionic polarization matrix. Indeed, it coincides with that of the bosonic one and was evaluated in Section 7.2. Here we focus on the opposite limit of the degenerate electron gas and restrict ourselves to $q\ll k_{\rm F}$. Then we find for the retarded component of the polarization matrix

$$\Pi_0^{\rm R}(\mathbf{q},\omega)=\sum_{\mathbf{k}}\frac{n_{\rm F}(\mathbf{k}_+)-n_{\rm F}(\mathbf{k}_-)}{\omega+{\rm i}0-\mathbf{v}_{\mathbf{k}}\mathbf{q}}=\int_{-1}^{1}{\rm d}z\int_0^\infty\frac{k^2{\rm d}k}{(2\pi)^2}\,\frac{n_{\rm F}(\mathbf{k}_+)-n_{\rm F}(\mathbf{k}_-)}{\omega+{\rm i}0-v_kqz},$$

where we suppressed the spin index, put $\mathbf{k}_\pm=\mathbf{k}\pm\mathbf{q}/2$, and introduced notation $z=\cos\theta$ and $\mathbf{v}_{\mathbf{k}}=\mathbf{k}/m$. We also notice that the Fermi occupation numbers are given by $n_{\rm F}(\mathbf{k}_\pm)\approx n_{\rm F}((k^2\pm kqz)/2m)$, restricting the integration to the narrow energy strip of width $v_{\rm F}qz$ around the Fermi energy. We can now go from the momentum integral to the energy one, introducing the density of states (DOS) $\nu(\epsilon)=m\sqrt{2m\epsilon}/2\pi^2$. Since $v_F q \ll \epsilon_F$, we can take the latter as a constant $\nu \equiv \nu(\epsilon_F)$. This way we find for the retarded polarization component

$$\Pi_0^R(\mathbf{q},\omega) \approx \frac{\nu}{2}\int_{-1}^{1} dz\,\frac{-v_F q z}{\omega + i0 - v_F q z}$$
$$= \nu\left[1 + \frac{\omega}{2v_F q}\ln\left|\frac{v_F q - \omega}{v_F q + \omega}\right|\right] + i\pi\nu\,\frac{\omega}{2v_F q}\,\theta(v_F q - |\omega|). \qquad (10.55)$$

In this approximation there is only one scale of energy dependence $v_F q$. It misses the effects associated with the smaller scale $q^2/2m$ as well as the temperature T. Those bring corrections to Eq. (10.55), which sometimes are important to keep in mind. First, according to Eq. (10.55), $\Pi_0^R(\mathbf{q},0) = \nu$. More careful consideration, which keeps corrections $\sim q/k_F$, shows

$$\Pi_0^R(\mathbf{q},0) \approx \nu\left[1 - q^2/(12k_F^2)\right]. \qquad (10.56)$$

Moreover, according to Eq. (10.55), at $\omega = v_F q$ there is a logarithmic singularity in $\mathrm{Re}\Pi_0^R(\mathbf{q},\omega)$, while $\mathrm{Im}\Pi_0^R(\mathbf{q},\omega)$ abruptly drops to zero at the same point. The singularities in the real and imaginary parts are related through the Kramers–Kronig relation. They both are smeared by temperature. In particular, the imaginary part at $\omega \gg v_F q$ is larger than zero, although it is exponentially small. Approximating the Fermi distribution by the Maxwell one, that is, $n_F(\epsilon_\mathbf{k}) \approx e^{-(\epsilon_\mathbf{k}-\epsilon_F)/T}$, one finds

$$\mathrm{Im}\,\Pi_0^R(\mathbf{q},\omega) \approx \frac{\pi\nu}{2}\,\frac{\omega}{v_F q}\, e^{-\frac{\epsilon_F}{T}\left[\left(\frac{\omega}{v_F q}\right)^2 - 1\right]}. \qquad (10.57)$$

This is essentially the classical Landau damping result, Eq. (7.17). The real and imaginary parts of $\Pi_0^R(\mathbf{q},\omega)$ are plotted in Fig. 10.2 as functions of ω for a fixed $q \ll k_F$. The Keldysh component of the polarization matrix in equilibrium is given by $\Pi_0^K(\mathbf{q},\omega) = 2i\coth(\omega/2T)\mathrm{Im}\Pi_0^R(\mathbf{q},\omega)$; see Eq. (10.53).

10.7 Interactions

The *normally ordered* Hamiltonian of the two-particle fermion–fermion interaction has the form, cf. Eq. (6.11),

$$\hat{H}_{\rm int} = \frac{1}{2}\sum_{\mathbf{q},\mathbf{k},\mathbf{k}'}\sum_{\sigma,\sigma'} U(\mathbf{q})\,\hat{c}^\dagger_{\mathbf{k}\sigma}\hat{c}^\dagger_{\mathbf{k}'\sigma'}\hat{c}_{\mathbf{k}'+\mathbf{q}\sigma'}\hat{c}_{\mathbf{k}-\mathbf{q}\sigma}, \qquad (10.58)$$

where $U(\mathbf{q})$ is the Fourier transform of the interaction potential, which we assume to be spin independent. The corresponding Grassmann action on the closed time contour is

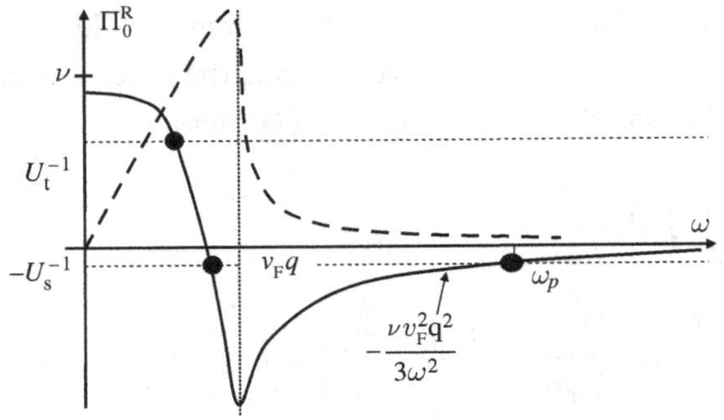

Figure 10.2 Real (full line) and imaginary (dashed line) parts of the retarded polarization component $\Pi_0^R(\mathbf{q}, \omega)$ of a degenerate Fermi gas, as functions of ω at a fixed $q \ll k_F$. Here U_s and U_t are singlet and triplet interaction constants; see Eqs. (10.64) and (10.65).

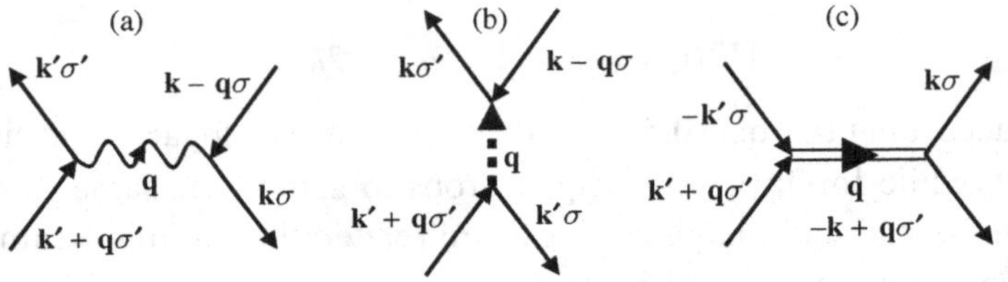

Figure 10.3 Three channels of fermion–fermion interactions, Eq. (10.60): (a) singlet channel, the wavy line denotes the interaction potential $U(\mathbf{q})$; (b) spin-triplet channel, the dashed line denotes U_t; (c) Cooper channel, the full double line denotes λ/ν.

$$S_{\text{int}} = -\frac{1}{2}\int_C \mathrm{d}t \sum_{\mathbf{q},\mathbf{k},\,\mathbf{k}'} \sum_{\sigma,\sigma'} U(\mathbf{q})\, \bar{\psi}_{\mathbf{k}\sigma} \bar{\psi}_{\mathbf{k}'\sigma'} \psi_{\mathbf{k}'+\mathbf{q}\sigma'} \psi_{\mathbf{k}-\mathbf{q}\sigma}, \tag{10.59}$$

where we have suppressed the time index t, running along the contour. Although Eq. (10.59) is an exact expression for the two-particle interaction, it is often convenient to rewrite it in such a way that the wavenumber $\mathbf{q}$ summation includes only momenta that are small with respect to the Fermi momentum. To this end we split the $\sum_{\mathbf{q}}$ into $q \ll k_F$ and $q \sim k_F$ parts, keep the former as is, and in the latter relabel momenta indices in two complementary ways, Fig. 10.3. As a result the four-fermion product in Eq. (10.59) takes the form

$$\sum_{\mathbf{k},\mathbf{k}',\sigma,\sigma'} \Big[U(\mathbf{q})\, \bar{\psi}_{\mathbf{k}\sigma} \bar{\psi}_{\mathbf{k}'\sigma'} \psi_{\mathbf{k}'+\,\mathbf{q}\sigma'} \psi_{\mathbf{k}-\mathbf{q}\sigma} + U(\mathbf{k}' + \mathbf{q} - \mathbf{k}) \bar{\psi}_{\mathbf{k}'\sigma} \bar{\psi}_{\mathbf{k}\sigma'} \psi_{\,\mathbf{k}'+\mathbf{q}\sigma'} \psi_{\mathbf{k}-\mathbf{q}\sigma}$$

$$+ U(\mathbf{k}' + \mathbf{k}) \bar{\psi}_{\mathbf{k}\sigma} \bar{\psi}_{-\mathbf{k}+\mathbf{q}\sigma'} \psi_{\,\mathbf{k}'+\mathbf{q}\sigma'} \psi_{-\mathbf{k}'\sigma} \Big], \tag{10.60}$$

where $k, k' \sim k_F$, and $q \ll k_F$. This expression may be simplified if one neglects the momentum dependence of the interaction potential at large momenta. Then the second four-fermion product may be rewritten as

$$\sum_{\mathbf{k},\mathbf{k}',\sigma,\sigma'} \bar\psi_{\mathbf{k}'\sigma}\bar\psi_{\mathbf{k}\sigma'}\psi_{\mathbf{k}'+\mathbf{q}\sigma'}\psi_{\mathbf{k}-\mathbf{q}\sigma} = -\sum_{\mathbf{k},\mathbf{k}',\sigma,\sigma'} \bar\psi_{\mathbf{k}\sigma'}\psi_{\mathbf{k}-\mathbf{q}\sigma}\bar\psi_{\mathbf{k}'\sigma}\psi_{\mathbf{k}'+\mathbf{q}\sigma'}$$
$$= -\frac{1}{2}\sum_{\mathbf{k},\sigma',\sigma} \bar\psi_{\mathbf{k}\sigma'}\vec{\hat{s}}_{\sigma'\sigma}\psi_{\mathbf{k}-\mathbf{q}\sigma}\sum_{\mathbf{k}',\sigma,\sigma'} \bar\psi_{\mathbf{k}'\sigma}\vec{\hat{s}}_{\sigma\sigma'}\psi_{\mathbf{k}'+\mathbf{q}\sigma'}$$
$$-\frac{1}{2}\sum_{\mathbf{k},\sigma} \bar\psi_{\mathbf{k}\sigma}\psi_{\mathbf{k}-\mathbf{q}\sigma}\sum_{\mathbf{k}',\sigma} \bar\psi_{\mathbf{k}'\sigma}\psi_{\mathbf{k}'+\mathbf{q}\sigma}. \tag{10.61}$$

As for the last term in Eq. (10.60), one may use that

$$\sum_{\mathbf{k},\mathbf{k}',\sigma,\sigma'} \bar\psi_{\mathbf{k}\sigma}\bar\psi_{-\mathbf{k}+\mathbf{q}\sigma'}\psi_{\mathbf{k}'+\mathbf{q}\sigma'}\psi_{-\mathbf{k}'\sigma} = 2\sum_{\mathbf{k}} \bar\psi_{\mathbf{k}\uparrow}\bar\psi_{-\mathbf{k}+\mathbf{q}\downarrow}\sum_{\mathbf{k}'} \psi_{\mathbf{k}'+\mathbf{q}\downarrow}\psi_{-\mathbf{k}'\uparrow}. \tag{10.62}$$

Indeed, the diagonal terms with $\sigma' = \sigma$ vanish due to the anti-commutativity of Grassmann variables. One can now introduce the fermion density, vector spin-density, and Cooper pair fields as

$$\rho(\mathbf{q},t) = \sum_{\mathbf{k},\sigma} \bar\psi_{\mathbf{k}\sigma}(t)\psi_{\mathbf{k}+\mathbf{q}\sigma}(t); \qquad \mathbf{s}(\mathbf{q},t) = \frac{1}{2}\sum_{\mathbf{k},\sigma,\sigma'} \bar\psi_{\mathbf{k}\sigma}(t)\vec{\hat{s}}_{\sigma\sigma'}\psi_{\mathbf{k}+\mathbf{q}\sigma'}(t);$$
$$\Phi(\mathbf{q},t) = \sum_{\mathbf{k}} \psi_{\mathbf{k}+\mathbf{q}\downarrow}(t)\psi_{-\mathbf{k}\uparrow}(t); \qquad \bar\Phi(\mathbf{q},t) = \sum_{\mathbf{k}} \bar\psi_{-\mathbf{k}\uparrow}(t)\bar\psi_{\mathbf{k}+\mathbf{q}\downarrow}(t).$$

In coordinate space one has the following slowly varying functions on the scale of the Fermi wavelength (indeed, by construction $q \ll k_F$):

$$\rho(\mathbf{r},t) = \sum_{\sigma} \bar\psi_\sigma(\mathbf{r},t)\psi_\sigma(\mathbf{r},t); \qquad \mathbf{s}(\mathbf{r},t) = \frac{1}{2}\sum_{\sigma,\sigma'} \bar\psi_\sigma(\mathbf{r},t)\vec{\hat{s}}_{\sigma\sigma'}\psi_{\sigma'}(\mathbf{r},t);$$
$$\Phi(\mathbf{r},t) = \psi_\downarrow(\mathbf{r},t)\psi_\uparrow(\mathbf{r},t); \qquad \bar\Phi(\mathbf{r},t) = \bar\psi_\uparrow(\mathbf{r},t)\bar\psi_\downarrow(\mathbf{r},t). \tag{10.63}$$

In terms of these slow quantities the interaction action (10.60) takes the form

$$S_{\text{int}} = -\int_C dt \left\{ \frac{1}{2}\iint d\mathbf{r}d\mathbf{r}'\, \rho(\mathbf{r},t)U_s(\mathbf{r}-\mathbf{r}')\rho(\mathbf{r}',t) \right. \tag{10.64}$$
$$\left. -\,2U_t\int d\mathbf{r}\,\mathbf{s}(\mathbf{r},t)\cdot\mathbf{s}(\mathbf{r},t) + \frac{\lambda}{\nu}\int d\mathbf{r}\,\bar\Phi(\mathbf{r},t)\Phi(\mathbf{r},t) \right\},$$

where U_t and $(\lambda/\nu) \sim U(k_F)$ are *triplet and Cooper channel* interaction constants ($\nu = \nu(\epsilon_F)$ is DOS at the Fermi energy), and $U_s(\mathbf{q}) = U(\mathbf{q}) - U_t$ is the *singlet channel* interaction potential.

We shall postpone discussion of the triplet channel until Section 10.8 and the Cooper channel until Chapter 17 and focus here on the singlet, that is, density–density, interactions. To proceed we perform a Hubbard–Stratonovich transformation with the help of the real scalar boson field $\varphi(\mathbf{r},t)$ (cf. Eq. (7.3)),

$$e^{-\frac{i}{2}\int_C dt \iint d\mathbf{r}d\mathbf{r}' \rho U_s \rho} = \int \mathbf{D}[\varphi]\, e^{i\int_C dt \left\{\frac{1}{2}\iint d\mathbf{r}d\mathbf{r}' \varphi(\mathbf{r},t) U_s^{-1}(\mathbf{r}-\mathbf{r}')\varphi(\mathbf{r}',t) - \int d\mathbf{r}\, \varphi(\mathbf{r},t)\rho(\mathbf{r},t)\right\}}, \tag{10.65}$$

where U_s^{-1} is an inverse singlet interaction potential, that is, $U_s^{-1} \circ U_s = 1$. One notices that the auxiliary bosonic field, $\varphi(\mathbf{r},t)$, enters the fermionic action in exactly the same way as a scalar source field, V, see Section 10.5. Following the same approach, one introduces $\varphi^{\mathrm{cl(q)}} = (\varphi^+ \pm \varphi^-)/2$ and rewrites the fermion–boson interaction term as $\varphi^\alpha \sum_\sigma \bar\psi_{a\sigma}\gamma^\alpha_{ab}\psi_{b\sigma}$, where the summations over $a,b = (1,2)$ and $\alpha = (\mathrm{cl},\mathrm{q})$ are assumed. The free bosonic term takes the form $\varphi^\alpha U_s^{-1}\hat\sigma_1^{\alpha\beta}\varphi^\beta$, where $\hat\sigma_1$ is the first Pauli matrix acting in the bosonic Keldysh space.

At this stage the Fermionic action is quadratic in Grassmann fields and one may perform the Gaussian integration according to Eq. (10.11). The result is the effective *bosonic* action written in terms of the two-component fluctuating scalar potential φ^α, as well as the source fields V^α:

$$S[\varphi,V] = \int dt\, d\mathbf{r}d\mathbf{r}'\, \varphi(\mathbf{r},t)\, U_s^{-1}(\mathbf{r}-\mathbf{r}')\hat\sigma_1\varphi(\mathbf{r}',t) - i\,\mathrm{Tr}\ln\left\{1 - \hat G[V^\alpha + \varphi^\alpha]\hat\gamma^\alpha\right\}; \tag{10.66}$$

compare Eq. (7.4). We have reduced an interacting fermionic problem to a theory of an effective nonlinear bosonic field. The latter is nothing but the longitudinal component of the photon field. Its dynamics and self-interactions originate from the polarization of the Fermi gas, through the expansion of the logarithm.

To proceed we shall restrict ourselves to the random phase approximation (RPA), already discussed in Section 7.4. It neglects all terms in the expansion of the logarithm beyond the second order (the first order in V_{cl} vanishes due to Eq. (10.29), while the first order in V_{q} is absent due to the charge neutrality of the unperturbed system). The second-order term in the expansion is conveniently expressed through the polarization matrix $\Pi_0^{\alpha\beta}$, Eqs. (10.49)–(10.51), of the *noninteracting* fermions. The resulting effective bosonic theory is Gaussian with the action

$$S_{\mathrm{RPA}}[\varphi,V] = \iint dx dx' \left\{\varphi\left[\hat U_{\mathrm{RPA}}\right]^{-1}\varphi + \varphi\hat\Pi_0 V + V\hat\Pi_0\varphi + V\hat\Pi_0 V\right\}, \tag{10.67}$$

where the inverse RPA screened interaction in the singlet channel is

$$\left[\hat U_{\mathrm{RPA}}(x,x')\right]^{-1} = U_s^{-1}(\mathbf{r}-\mathbf{r}')\hat\sigma_1 + \hat\Pi_0(x,x'); \tag{10.68}$$

compare Eq. (7.20). The fermionic polarization matrix $\hat\Pi_0$ plays the role of the (minus) self-energy, $\hat\Sigma$ (cf. Eqs. (6.16) and (6.17)), in the effective bosonic theory. As was mentioned at the end of Section 10.5, it indeed possesses all the proper symmetries and the causality properties. In particular, $\mathrm{i}\Pi_0^{\mathrm{K}} = -2F\,\mathrm{Im}\Pi_0^{\mathrm{R}}$ is negative, see Fig. 10.2 and Eqs. (10.55) and (10.57), providing convergence of the functional integrals over φ^α. It describes thermal and quantum fluctuations of the photon field.

Since the RPA action (10.67) is quadratic in the photon fields φ^α, one may integrate them out and evaluate the generating function $Z[V]$, Eq. (10.44). Performing the corresponding Gaussian integral according to Eq. (2.22), one finds $\ln Z_{\mathrm{RPA}}[V] = \mathrm{i}V\left(\hat{\Pi}_0 - \hat{\Pi}_0 \circ \hat{U}_{\mathrm{RPA}} \circ \hat{\Pi}_0\right)V = \mathrm{i}V\hat{\Pi}_{\mathrm{RPA}}V$. Since, according to Eq. (10.68), $U_{\mathrm{RPA}} = (1 + U_{\mathrm{s}}\hat{\sigma}_1\hat{\Pi}_0)^{-1}U_{\mathrm{s}}\hat{\sigma}_1$, simple matrix algebra leads to $\hat{\Pi}_{\mathrm{RPA}} = \hat{\Pi}_0 \circ (1 + U_{\mathrm{s}}\hat{\sigma}_1\hat{\Pi}_0)^{-1}$. The response component of this RPA screened polarization matrix acquires the form

$$\Pi_{\mathrm{RPA}}^{\mathrm{R(A)}}(\mathbf{q},\omega) = \frac{\Pi_0^{\mathrm{R(A)}}(\mathbf{q},\omega)}{1 + U_{\mathrm{s}}(\mathbf{q})\Pi_0^{\mathrm{R(A)}}(\mathbf{q},\omega)}. \tag{10.69}$$

The Keldysh component is $\Pi_{\mathrm{RPA}}^{\mathrm{K}} = (1 + U_{\mathrm{s}}\Pi_0^{\mathrm{R}})^{-1}\Pi_0^{\mathrm{K}}(1 + U_{\mathrm{s}}\Pi_0^{\mathrm{A}})^{-1}$. If one writes that $\Pi_0^{\mathrm{K}} = B(\Pi_0^{\mathrm{R}} - \Pi_0^{\mathrm{A}})$, where B is the boson distribution function of the scalar potential φ, one finds $\Pi_{\mathrm{RPA}}^{\mathrm{K}} = B(\Pi_{\mathrm{RPA}}^{\mathrm{R}} - \Pi_{\mathrm{RPA}}^{\mathrm{A}})$. In equilibrium, where $B = \coth(\omega/2T)$, this relation is a manifestation of FDT. It is valid, however, even out of equilibrium and states that the RPA screened correlation component $\Pi_{\mathrm{RPA}}^{\mathrm{K}}$ is expressed through the same bosonic distribution function B as the bare polarization Π_0^{K}. With the help of Eqs. (10.54) the bosonic distribution function B may be found in terms of the fermionic distribution F. The latter is a solution of the fermionic kinetic equation, discussed in the next chapter.

The response is resonant if the equation $1 + U_{\mathrm{s}}(\mathbf{q})\Pi_0^{\mathrm{R}}(\mathbf{q},\omega) = 0$ has a complex solution $\omega = \omega(\mathbf{q})$ with $\mathrm{Im}\,\omega \ll \mathrm{Re}\,\omega$. For the long-range Coulomb interactions $U_{\mathrm{s}}(\mathbf{q}) \approx 4\pi e^2/\mathbf{q}^2$, this condition determines the plasma mode with the frequency $\omega_{\mathrm{p}}^2 = 4\pi e^2\rho_0/m$ (indeed, $\nu m v_{\mathrm{F}}^2/3 = \rho_0$ is the fermion density); see Fig. 10.2.[5] For $q \ll \kappa_{\mathrm{TF}}^{-1}$, where $\kappa_{\mathrm{TF}}^{-1} = \sqrt{4\pi e^2\nu} \sim \omega_{\mathrm{p}}/v_{\mathrm{F}}$ is the (inverse) Thomas–Fermi screening radius, the plasmon mode is underdamped. The small Landau damping, Eq. (10.57), originates from the excitations of high-energy resonant electrons with velocity $v = \omega_{\mathrm{p}}/q \gg v_{\mathrm{F}}$. Their occupation number is well approximated by the classical Maxwell distribution $n_{\mathrm{F}} \approx \mathrm{e}^{-(mv^2/2-\epsilon_{\mathrm{F}})/T} \ll 1$. Therefore, if the long-wavelength plasmon excitations are excited (e.g. by a strong optical pulse), their kinetics is described by the semiclassical theory developed in Sections 7.4 and 7.5. Indeed, the real part of the retarded polarization (10.55) is provided by fermions in the narrow strip of energies $\sim v_{\mathrm{F}}q$ around the Fermi surface. On the other

5 As shown in Fig. 10.2, the equation $-U_{\mathrm{s}}^{-1} = \mathrm{Re}\Pi_0^{\mathrm{R}}(\mathbf{q},\omega)$ has another solution with $\omega_{\mathrm{ea}} \approx 0.83v_{\mathrm{F}}q$. Such an electron-acoustic mode is heavily overdamped (indeed, $\mathrm{Im}\Pi_0^{\mathrm{R}}(\mathbf{q},\omega_{\mathrm{ea}}) \approx 1.3\nu$) and therefore is not resonant; see also Section 7.3.

hand, fermions responsible for Landau damping live far above the Fermi surface, suggesting a description in terms of coupled plasmon and high-energy electron distribution functions; see Section 7.5.

For short-range interactions, $U(\mathbf{q}) \stackrel{q\to 0}{\to} U_\mathrm{s}$, the resonant excitation is the acoustic *zero sound* mode $\omega = v_\mathrm{zs} q$. According to Fig. 10.2 and Eq. (10.55) the corresponding speed of sound is $v_\mathrm{zs} = v_\mathrm{F}\sqrt{U_\mathrm{s}\nu/3}$ in the limit $U_\mathrm{s}\nu \gg 1$ and $v_\mathrm{zs} \approx v_\mathrm{F}\big(1 + 2\mathrm{e}^{-2/(U_\mathrm{s}\nu)-2}\big)$ in the opposite case $U_\mathrm{s}\nu < 1$. The zero sound mode should be distinguished from the ordinary sound, which is also present in Fermi liquids and is discussed in Section 11.2. The zero sound is a collisionless phenomenon originating due to the presence of the dynamic collective field φ. On the other hand, the ordinary sound is a result of hydrodynamic oscillations of the fermionic density and pressure, established by the fast local equilibration due to the frequent collisions. Therefore, their domains of existence are separated by the condition $\omega\tau_\mathrm{sc} \gg 1$ for the zero sound versus $\omega\tau_\mathrm{sc} \ll 1$ for the ordinary one. Here $\tau_\mathrm{sc} \sim \epsilon_F/T^2$ is the characteristic fermion collision time, Sections 11.1 and 11.2. Notice that for weak repulsive interactions the zero sound velocity exceeds the Fermi velocity, $v_\mathrm{zs} \gtrsim v_\mathrm{F}$, while at the same time the ordinary sound velocity approaches $v_\mathrm{s} \approx v_\mathrm{F}/\sqrt{3}$.

10.8 Stoner Instability

We focus now on the triplet interaction channel in Eq. (10.64). The crucial observation is that it comes with the opposite sign (i.e. effective attraction) in comparison with the repulsive singlet channel interaction. The reason for this peculiarity is the minus sign obtained due to commutation of the Grassmann variables in Eq. (10.61), which is a manifestation of the Fermi statistics. One can split the triplet term in Eq. (10.59) with the help of an auxiliary vector field $\mathbf{h}(\mathbf{r}, t)$ defined along the closed contour:

$$\mathrm{e}^{2\mathrm{i}\int_C dt d\mathbf{r}\, \mathbf{s} U_t \mathbf{s}} = \int \mathbf{D}[\mathbf{h}]\; \mathrm{e}^{\mathrm{i}\int_C dt d\mathbf{r} \left\{-\frac{1}{2}\mathbf{h} U_t^{-1}\mathbf{h} - 2\mathbf{h}(\mathbf{r},t)\cdot\mathbf{s}(\mathbf{r},t)\right\}}. \tag{10.70}$$

The fluctuating *exchange* magnetic field $\mathbf{h}(\mathbf{r}, t)$ enters the fermionic action in exactly the same way as the Zeeman field $\mathbf{H}_\mathrm{Z}$; see Eq. (10.36). One can now separate it into classical and quantum components $\mathbf{h}^\alpha = (\mathbf{h}^+ \pm \mathbf{h}^-)/2$ in exactly the same way as was done previously for the fluctuating scalar potential. At this stage one may perform Gaussian integration over the fermionic fields. The result is the effective bosonic action written in terms of the exchange magnetic field.[6]

[6] Strictly speaking, here one has to include all three interaction channels to have the scalar potential φ^α, exchange magnetic field $\mathbf{h}^\alpha$, and the complex pairing potential Δ^α inside the logarithm simultaneously. However, in the expansion of the logarithm to the second order, discussed here, the different channels do not mix. They start interacting with each other in higher orders, which describe nonlinear interactions of the bosonic fields.

$$S[\mathbf{h}, \mathbf{H}_Z] = -\int_{-\infty}^{\infty} \mathrm{d}t\, \mathrm{d}\mathbf{r}\, \mathbf{h}(\mathbf{r}, t) U_t^{-1} \hat{\sigma}_1\, \mathbf{h}(\mathbf{r}, t) - \mathrm{i}\, \mathrm{Tr} \ln \left\{ 1 - \hat{G}\big[(\mathbf{H}_Z^{\alpha} + \mathbf{h}^{\alpha})\vec{\hat{s}}\,\big]\hat{\gamma}^{\alpha} \right\}. \tag{10.71}$$

Notice again the opposite sign of the free boson term in comparison with Eq. (10.66); $\hat{\sigma}$ and $\hat{s}$ are Pauli matrices in the Keldysh and spin spaces correspondingly.

One can now expand the logarithm in powers of the effective magnetic field $\mathbf{H}_Z^{\alpha} + \mathbf{h}^{\alpha}$. We focus first on the quadratic term. Since the bare fermionic Green functions are diagonal in the spin indices, Eqs. (10.37), the resulting term acquires the form of the scalar product $(\mathbf{H}_Z^{\alpha} + \mathbf{h}^{\alpha}) \cdot (\mathbf{H}_Z^{\beta} + \mathbf{h}^{\beta})$. The coefficient is nothing but the bare polarization matrix $\hat{\Pi}_0^{\alpha\beta}$. As a result one finds for the quadratic RPA action in the triplet channel (we omit the Zeeman field for brevity)

$$S_{\mathrm{RPA}}[\mathbf{h}] = \iint \mathrm{d}x \mathrm{d}x'\, \mathbf{h}^{\alpha}(x) \big[-U_t^{-1} \hat{\sigma}_1^{\alpha\beta} \delta(x - x') + \hat{\Pi}_0^{\alpha\beta}(x, x') \big] \mathbf{h}^{\beta}(x'). \tag{10.72}$$

We employ now Eqs. (10.55) and (10.56) for the free fermion polarization operator at $\omega < v_F q$ along with the $T = 0$ FDT to obtain for the zero-temperature action[7]

$$S_{\mathrm{RPA}}[\mathbf{h}] = \sum_{\mathbf{q}} \Bigg\{ -2 \int \mathrm{d}t\, \mathbf{h}^{q*}(\mathbf{q}, t) \left[\gamma(\mathbf{q})\, \partial_t + D\mathbf{q}^2 + \delta \right] \mathbf{h}^{\mathrm{cl}}(\mathbf{q}, t) + \frac{\mathrm{i}\gamma(\mathbf{q})}{\pi} \iint \frac{\mathrm{d}t \mathrm{d}t'}{(t - t')^2} \left| \mathbf{h}^{\mathrm{q}}(\mathbf{q}, t) - \mathbf{h}^{\mathrm{q}}(\mathbf{q}, t') \right|^2 \Bigg\}, \tag{10.73}$$

where $\gamma(\mathbf{q}) = \pi\nu/(2v_F q)$, while $D = \nu/(12k_F^2)$ and $\delta = U_t^{-1} - \nu$. This action ought to be compared with the model action (9.32), used to discuss quantum phase transitions in Section 9.6. The main difference between the two is that the damping coefficient $\gamma(\mathbf{q})$ appears to be q-dependent (we shall discuss nonlinear terms in what follows). Remarkably, the parameter δ may become negative if $\nu U_t > 1$, see Fig. 10.2. As we already know, this indicates the tendency of the system to develop an ordered state with a nonzero expectation value of $\mathbf{h}^{\mathrm{cl}}$. Such transition is known as the Stoner ferromagnetic transition.

To find the order parameter in the magnetic phase, one needs to add nonlinear terms to the action (10.73). They may be found by expanding the logarithm in Eq. (10.71) in powers of $\mathbf{h}$. This is a cumbersome procedure. One may try to use the fact that the Pauli magnetization of the free electron gas is given by $M(H_Z) = -\frac{\mathrm{i}}{2}\, \partial Z[H_Z]/\partial H_Z^{\mathrm{q}}|_{H_Z^{\mathrm{q}}=0}$. For a *static* field it may be easily calculated by counting the number of up and down spins below a common H_Z-dependent

[7] We have omitted higher orders of the expansion of $\Pi_0^R(\mathbf{q}, \omega)$, Eq. (10.55), in powers of $(\omega/v_F q)^2$, since they bring only terms irrelevant in the RG sense, i.e. at large distances and times.

chemical potential, resulting in $M(H_Z) = \nu H_Z - g H_Z^3 + \cdots$, where $g = 3\nu/(16\epsilon_F^2)$ in $d = 3$. Since $\mathbf{H}_Z$ and $\mathbf{h}$ enter the action (10.71) in the same way, the last term in the Pauli magnetization translates into the following nonlinear action for *quasi-static* exchange magnetic field $\mathbf{h}$:

$$S_{\text{int}}[\mathbf{h}] = -2\int dt d\mathbf{r}\; g\big(\mathbf{h}^{\text{q}} \cdot \mathbf{h}^{\text{cl}}\big)\big(\mathbf{h}^{\text{cl}} \cdot \mathbf{h}^{\text{cl}}\big). \tag{10.74}$$

This term provides saturation of the magnetization in the ferromagnetic phase $\delta < 0$. Other possible nonlinear terms, such as, for example, $\big(\mathbf{h}^{\text{q}} \cdot \mathbf{h}^{\text{q}}\big)\big(\mathbf{h}^{\text{q}} \cdot \mathbf{h}^{\text{cl}}\big)$, have the same scaling dimensions as (10.74), as discussed in Section 9.6.

The resulting so-called Hertz–Millis [213] model, Eqs. (10.73) and (10.74), of the Stoner transition may be analyzed using the RG machinery. The only difference with the model of Section 9.6 is the damping $\gamma(\mathbf{q})$ diverging at small q. This is a manifestation of the conserved nature of the order parameter $\mathbf{h}$. Indeed, the corresponding classical equation of motion takes the form $\partial_t \mathbf{h}(\mathbf{q}, t) = [\gamma(\mathbf{q})]^{-1}(\ldots)$. Since $[\gamma(0)]^{-1} = 0$, the $\mathbf{q} = 0$ component of the magnetization does not evolve. The bare scaling dimensions of the coupling constants, read out of the action (10.73) and (10.74), are $[\gamma] = d+1+2\chi$, $[D] = d+z-2+2\chi$, $[\delta] = d+z+2\chi$, and $[g] = d+z+4\chi$. Demanding that the Gaussian part of the critical action is scale invariant, that is, $[\gamma] = [D] = 0$, one finds $z = 3$ and $\chi = -(d+1)/2$, while $[\delta] = 2$. One thus finds that $[g] = 1 - d$ and therefore the critical dimension is $d_c = 1$. For $d > 1$, the Gaussian exponents are exact and we conclude that $\nu = 1/2$, while $z = 3$. Since the characteristic momentum scale is the inverse correlation length, that is, $q \propto |\delta|^{\nu}$, while the characteristic frequency is $\omega \propto |\delta|^{\nu z}$, we conclude that $\omega \ll v_F q \ll \epsilon_F$, justifying approximations made while writing the quadratic part of the action (10.73). Gaussian scaling is valid throughout the quantum critical region, Fig. 9.5, and crosses over to the classical one in the vicinity of a finite T transition. For $d < 4$ the latter is $\nu = 1/2 + \epsilon/12$ and $z = 3 + O(\epsilon^2)$, where $\epsilon = 4 - d$.

It was recently understood [215, 216], however, that the Hertz–Millis model actually fails to describe the ferromagnetic transition in the fermionic system. The reason is that the effective bosonic theory (10.71) is obtained by integrating out *gapless* fermionic degrees of freedom. As a result the effective vertices of the bosonic theory are nonlocal functions of time (and space). Following Hertz–Millis [213], we have incorporated this fact into the Gaussian part of the action (10.73), but disregarded it in the nonlinear part (10.74). This happens to be badly inconsistent[8] (as was probably first noticed in [217]). In contrast to Eq. (10.74), the actual quartic vertex includes a long-ranged time-nonlocal part [216] (there is also a nonlocal cubic term of the form $\mathbf{h} \cdot [\mathbf{h} \times \mathbf{h}]$). The nonlocality is effectively cut off by

[8] I am indebted to A. Chubukov and J. Schmalian for discussing this issue.

the field $\mathbf{h}$ itself, leading to a nonanalytic effective bosonic potential of the form $\mathcal{F}[\mathbf{h}] = \delta h^2 + g_1 h^4 \ln h + g h^4$ [215].[9] The nonanalytic $h^4 \ln h$ term transforms the continuous transition into the weak first-order transition.

10.9 Problems

10.9.1 Fermionic Bogoliubov Transformation

Consider a quadratic Hamiltonian for a spin-1/2 fermionic state:

$$\hat{H} = \sum_{\sigma=\uparrow,\downarrow} \epsilon_0 \hat{c}_\sigma^\dagger \hat{c}_\sigma + \Delta(\hat{c}_\uparrow^\dagger \hat{c}_\downarrow^\dagger + \hat{c}_\downarrow \hat{c}_\uparrow) = \left(\hat{c}_\uparrow^\dagger, \hat{c}_\downarrow\right) \begin{pmatrix} \epsilon_0 & \Delta \\ \Delta & -\epsilon_0 \end{pmatrix} \begin{pmatrix} \hat{c}_\uparrow \\ \hat{c}_\downarrow^\dagger \end{pmatrix} + \epsilon_0. \tag{10.75}$$

Show that with the help of the *unitary* transformation

$$\begin{pmatrix} \hat{c}_\uparrow \\ \hat{c}_\downarrow^\dagger \end{pmatrix} = \begin{pmatrix} \cos(\alpha) & \sin(\alpha) \\ -\sin(\alpha) & \cos(\alpha) \end{pmatrix} \begin{pmatrix} \hat{\mathfrak{c}}_1 \\ \hat{\mathfrak{c}}_2^\dagger \end{pmatrix}, \tag{10.76}$$

where $\{\hat{\mathfrak{c}}_a, \hat{\mathfrak{c}}_b^\dagger\} = \delta_{a,b}$, $a, b = 1, 2$, and $\tan(2\alpha) = -\Delta/\epsilon_0$, the Hamiltonian (10.75) is diagonalized: $\hat{H} = \sum_{a=1,2} \sqrt{\epsilon_0^2 + \Delta^2}\, \hat{\mathfrak{c}}_a^\dagger \hat{\mathfrak{c}}_a + \epsilon_0 - \sqrt{\epsilon_0^2 + \Delta^2}$. Show that its normalized ground state, $|\mathrm{o}\rangle$, found from the condition $\hat{\mathfrak{c}}_a|\mathrm{o}\rangle = 0$, is given by the coherent state of the fermionic pair

$$|\mathrm{o}\rangle = \cos(\alpha)\, e^{\tanh(\alpha)\hat{c}_\uparrow^\dagger \hat{c}_\downarrow^\dagger}|0\rangle, \tag{10.77}$$

[9] The nonanalyticity may be noticed already in the $\mathrm{Re}\Sigma^{\mathrm{R}}(\epsilon)$ at $\xi_k = \epsilon_{\mathbf{k}} - \epsilon_{\mathrm{F}} = 0$ calculated at the quantum transition point, $\delta = 0$ and $T = 0$. To this end let us calculate $\Sigma^{\mathrm{K}}(\epsilon)$, Eq. (11.6),

$$\Sigma^{\mathrm{K}}(\epsilon) = \frac{\mathrm{i}}{2} \sum_{\mathbf{q},\omega} (\mathrm{sign}(\epsilon+\omega)\mathrm{sign}(\omega) - 1)(-2\pi \mathrm{i})\delta(\epsilon + \omega - v_{\mathrm{F}}\mathbf{q}) \left[\frac{3}{\mathrm{i}\gamma(\mathbf{q})\omega - D\mathbf{q}^2} - \frac{-3}{\mathrm{i}\gamma(\mathbf{q})\omega + D\mathbf{q}^2} \right],$$

where we used the fact that $G^{\mathrm{R}} - G^{\mathrm{A}} = -2\pi \mathrm{i}\delta(\epsilon + \omega - v_{\mathrm{F}} q \cos\theta)$ along with the retarded/advanced RPA propagators of the three-component vector $\mathbf{h}$-field, Eq. (10.73), and $\coth(\omega/2T) \stackrel{T\to 0}{\to} \mathrm{sign}(\omega)$. Integration over the angle $z = \cos\theta$ is performed with the help of the energy delta-function, bringing the factor $(v_{\mathrm{F}} q)^{-1}$. The remaining radial momentum integration takes the form

$$\int_0^\infty \frac{q^2 dq}{(2\pi)^2} \frac{1}{v_{\mathrm{F}} q} \frac{-6\mathrm{i}\gamma(q)\omega}{(Dq^2)^2 + \gamma^2(q)\omega^2} = -\frac{6\mathrm{i}}{(2\pi)^2} \frac{\gamma\omega}{v_{\mathrm{F}} D} \frac{1}{3} \int_0^\infty \frac{d(Dq^3)}{(Dq^3)^2 + \gamma^2\omega^2} = -\frac{\mathrm{i}}{(2\pi)^2} \frac{\pi\, \mathrm{sign}(\omega)}{v_{\mathrm{F}} D},$$

where $\gamma = q\gamma(q)$. Performing ω-integration, one finally finds $\Sigma^{\mathrm{K}}(\epsilon) = -\mathrm{i}\epsilon/(4\pi v_{\mathrm{F}} D) = -3\mathrm{i}\epsilon k_{\mathrm{F}}^2/(\pi v_{\mathrm{F}} \nu) = -3\pi \mathrm{i}\epsilon$. According to $T = 0$ FDT, $\Sigma^{\mathrm{K}}(\epsilon) = \mathrm{sign}(\epsilon) 2\mathrm{i}\mathrm{Im}\Sigma^{\mathrm{R}}(\epsilon)$ and thus $\mathrm{Im}\Sigma^{\mathrm{R}}(\epsilon) = -3\pi|\epsilon|/2$. Finally, using the Kramers–Kronig relation, the absolute value nonanalyticity in the imaginary part translates into the nonanalytic term $\mathrm{Re}\Sigma^{\mathrm{R}} = (3/2)\epsilon \ln|\epsilon|$ in the real part of the self-energy. As a result, the inverse dressed Green function $[G_0^{\mathrm{R}}]^{-1} - \Sigma^{\mathrm{R}} \sim \epsilon \ln|\epsilon|$ indicates nonlocality in time of the renormalized fermionic action. In $d = 2$ the corresponding singularity is much stronger, $\mathrm{Re}\Sigma^{\mathrm{R}} \propto \epsilon^{2/3}$ [218].

where $|0\rangle$ is the fermionic vacuum state, $\hat{c}_\sigma|0\rangle = 0$. Find three other states in the new Hilbert state basis formed by $\hat{c}_1^\dagger|o\rangle$, $\hat{c}_2^\dagger|o\rangle$, and $\hat{c}_1^\dagger\hat{c}_2^\dagger|o\rangle$. Verify that the new basis is orthonormal.

Write the Hamiltonian (10.75) as a 4×4 matrix in the fermionic two-body space, using $|0\rangle, \hat{c}_\uparrow^\dagger|0\rangle, \hat{c}_\downarrow^\dagger|0\rangle, \hat{c}_\uparrow^\dagger\hat{c}_\downarrow^\dagger|0\rangle$ as the basis. Diagonalize it and verify that the new basis is formed by the eigenstates of the Hamiltonian with the proper eigenenergies. Notice that, since the Hamiltonian (10.75) conserves the fermion *parity* (but not the fermion number), the even and odd particle number subspaces block-diagonalize the two-body matrix Hamiltonian. The advantage of the Bogoliubov unitary transformation over this brute-force calculation is that the former may be immediately generalized to the case of N spin-1/2 fermions, where both ϵ_0 and Δ are $N\times N$ matrices. The brute force approach requires handling $2^{2N}\times 2^{2N}$ Hamiltonian, which is block-diagonalized into two $2^{2N-1}\times 2^{2N-1}$ matrices. This is not an easy exercise even for $N=2$.

10.9.2 Random Matrices via Fermionic Replica

Consider a Hermitian $N\times N$ matrix $\hat{H}$. Its real eigenvalues are denoted as ϵ_j, where $j = 1, 2, \ldots, N$. Verify that its density of states may be written as

$$\nu(\epsilon) = \sum_{j=1}^{N}\delta(\epsilon-\epsilon_j) = \frac{1}{\pi}\mathrm{Im}\mathrm{Tr}\big(\hat{E}-\hat{H}\big)^{-1} = \frac{1}{\pi}\frac{\partial}{\partial\epsilon}\mathrm{Im}\log\det(\hat{E}-\hat{H}), \qquad (10.78)$$

where $\hat{E} = (\epsilon - \mathrm{i}0)\hat{\mathbb{1}}$ is proportional to the unit matrix. The determinant may be written as the fermionic multiple integral (10.11), but the logarithm presents a problem. To overcome it one may use the replica trick [279], stating

$$\log Z = \lim_{n\to 0}\frac{Z^n-1}{n}, \qquad \nu(\epsilon) = \lim_{n\to 0}\frac{1}{\pi n}\,\mathrm{Im}\,\partial_\epsilon Z^n \qquad (10.79)$$

where $Z = \det(\hat{E}-\hat{H})$. For an integer $n>0$ one therefore obtains

$$Z^n = \int\prod_{a=1}^{n}\prod_{j=1}^{N}[\mathrm{d}\bar{\psi}_j^{(a)}\mathrm{d}\psi_j^{(a)}]\;\mathrm{e}^{-\sum\limits_{a=1}^{n}\sum\limits_{i,j=1}^{N}\bar{\psi}_i^{(a)}(\hat{E}-\hat{H})_{ij}\psi_j^{(a)}}. \qquad (10.80)$$

Let us assume now that the matrix $\hat{H}$ has random entrees (obeying Hermiticity conditions) taken from independent Gaussian distributions. The corresponding probability distribution is

$$P(\hat{H}) = \mathcal{N}\,\mathrm{e}^{-\frac{N}{2}\mathrm{Tr}\hat{H}^2} = \mathcal{N}\,\mathrm{e}^{-\frac{N}{2}\sum\limits_{i,j=1}^{N}|H_{ij}|^2}, \qquad (10.81)$$

where the factor $N/2$ in the exponent is chosen to maintain a finite support for the density of states in the $N\to\infty$ limit (see the following discussion).

Find the normalization factor, $\mathcal{N}$, from $\int \mathbf{D}[\hat{H}]P(\hat{H}) = 1$, where $\mathbf{D}[\hat{H}] = \prod_{i<j}^{N} \mathrm{d}(\mathrm{Re}H_{ij})\mathrm{d}(\mathrm{Im}H_{ij}) \prod_{i}^{N} \mathrm{d}H_{ii}$. Average the replicated partition function over the Gaussian ensemble of Hermitian matrices and show that

$$\langle Z^n\rangle_{\hat{H}} = \int \prod_{a=1}^{n}\prod_{j=1}^{N}[\mathrm{d}\bar{\psi}_j^{(a)}\mathrm{d}\psi_j^{(a)}]\, \mathrm{e}^{-\sum\limits_{a=1}^{n}\sum\limits_{i,j=1}^{N}\bar{\psi}_i^{(a)}\check{E}_{ij}\psi_j^{(a)} - \frac{1}{2N}\sum\limits_{a,b=1}^{n}\check{T}_{ab}\check{T}_{ba}}, \tag{10.82}$$

where $\langle Z^n\rangle_{\hat{H}} = \int \mathbf{D}[\hat{H}]P(\hat{H})\, Z^n$ and $\check{T}_{ab} = \sum_{j=1}^{N} \bar{\psi}_j^{(a)}\psi_j^{(b)}$ is an $n\times n$ matrix. Pay special attention on rearranging the product of four Grassmann variables in the exponent, keeping a proper sign. Decouple the $\sum_{a,b=1}^{n} \check{T}_{ab}\check{T}_{ba}$ term in the exponent with the help of an auxiliary Hubbard–Stratonovich Hermitian $n\times n$ matrix $\check{Q}_{ab}$ and perform the Gaussian integral over the Grassmann variables $[\mathrm{d}\bar{\psi}_j^{(a)}\mathrm{d}\psi_j^{(a)}]$. Show that

$$\langle Z^n\rangle_{\hat{H}} = \int \mathbf{D}[\hat{H}]\, \mathrm{e}^{-\frac{N}{2}\mathrm{Tr}\hat{H}^2 + n\mathrm{Tr}\log(\hat{E}-\hat{H})} = \int \mathbf{D}[\check{Q}]\, \mathrm{e}^{N\left[-\frac{1}{2}\mathrm{Tr}\check{Q}^2 + \mathrm{Tr}\log(\check{E}-\mathrm{i}\check{Q})\right]}, \tag{10.83}$$

where $\check{E}_{ab} = (\epsilon - \mathrm{i}0)\delta_{ab}$. This is a remarkable duality between the integrals over $N\times N$ and $n\times n$ Hermitian matrices, proven with the help of the Grassmann integrals. Its advantage is that the $\check{Q}$-integral contains $N \gg 1$ as a factor in the exponent. This allows one to perform the $\check{Q}$-integration in the stationary point approximation. First let's shift the integration variable $\check{Q} \to \check{Q} - \mathrm{i}\check{E}$ to find

$$\partial_\epsilon \langle Z^n\rangle_{\hat{H}} = \int \mathbf{D}[\check{Q}]\; \mathrm{i}N\,\mathrm{Tr}(\check{Q} - \mathrm{i}\check{E})\; \mathrm{e}^{N\left[-\frac{1}{2}\mathrm{Tr}(\check{Q}-\mathrm{i}\check{E})^2 + \mathrm{Tr}\log(-\mathrm{i}\check{Q})\right]}. \tag{10.84}$$

Perform variation of the exponent over the $\check{Q}$-matrix and show that the saddle-point condition is given by the matrix equation

$$(\check{Q} - \mathrm{i}\check{E}) = \check{Q}^{-1}. \tag{10.85}$$

Since $\check{E} \propto \check{\mathbb{1}}$, the stationary $\check{Q}$-matrix may be diagonalized by a unitary transformation and the stationary point condition becomes an algebraic equation for the set of its eigenvalues λ_a, found from $(\lambda - \mathrm{i}\epsilon) = 1/\lambda$. This admits two solutions: $\lambda_\pm = \pm\sqrt{1-\epsilon^2/4} + \mathrm{i}\epsilon/2$. Consider first a *replica symmetric* stationary point $\lambda_a = \lambda_+$, for $a = 1,\ldots,n$. Then

$$\partial_\epsilon \langle Z^n\rangle_{\hat{H}} \propto \mathrm{i}nN(\lambda_+ - \mathrm{i}\epsilon)\, \mathrm{e}^{nN\left[-\frac{1}{2}(\lambda_+ - \mathrm{i}\epsilon)^2 + \log(-\mathrm{i}\lambda_+)\right]}. \tag{10.86}$$

The proportionality coefficient, which includes Gaussian fluctuations around the stationary point, may be shown to have a form of $[\ldots]^n$, and therefore it goes to one, if $n \to 0$. Substitute this into (10.79) and show that

$$\nu_0(\epsilon) = \frac{N}{\pi}\sqrt{1 - \frac{\epsilon^2}{4}}; \qquad |\epsilon| < 2, \tag{10.87}$$

and $\nu_0(\epsilon) = 0$ for $|\epsilon| > 2$. This result is known as the Wigner semicircle law for the average density of states of random matrices.

Since each λ_a may take two stationary values, $\lambda_\pm$, there are 2^n stationary configurations in total. One may show [285] that in the limit $n \to 0$, besides one replica symmetric configuration, considered previously, only n other survive. Those are given by $n-1$ eigenvalues set to λ_+ and one eigenvalue set to λ_-. They lead to a $1/N$ oscillatory correction to the average density of states of the form $\delta\nu(\epsilon) \propto \cos[2\pi \int_{-2}^{\epsilon} d\epsilon' \nu_0(\epsilon') + \pi]$. This is a function with exactly N maxima.[10] It reflects the discrete nature of the underlying spectra before the ensemble averaging. The same technique may be applied to the correlation functions of the density of states, for example, $\langle \nu(\epsilon)\nu(\epsilon')\rangle_{\hat{H}}$; see [285] and Section 15.2 for more information about this object.

10.9.3 Fermionic Lindbladian Action

Consider a fermionic analog of the bosonic parametrically driven oscillator, discussed in Section 5.5. The model consists of two fermionic states, labelled as $\sigma = \uparrow, \downarrow$ and thus having *four*-dimensional many-body Hilbert space. These two states are coupled to a Gaussian fermionic bath, through the fermion exchange. The corresponding closed time contour action is $S = S_0 + S_{\text{bath}}$, where

$$S_0 = \int_C dt \left[\sum_\sigma \bar{\psi}_\sigma (i\partial_t - \varepsilon_\sigma)\psi_\sigma - \Delta \left(e^{2i\omega_p t} \bar{\psi}_\uparrow \bar{\psi}_\downarrow + e^{-2i\omega_p t} \psi_\downarrow \psi_\uparrow \right) \right], \quad (10.88a)$$

$$S_{\text{bath}} = \int_C dt \sum_{s,\sigma} \left[\bar{\Psi}_{\sigma s} (i\partial_t - \varepsilon_s)\Psi_{\sigma s} + g_s (\bar{\Psi}_{\sigma s}\psi_\sigma + \bar{\psi}_\sigma \Psi_{\sigma s}) \right], \quad (10.88b)$$

where ε_s and g_s are spectra of the bath and the fermion exchange amplitudes; both are assumed to be spin-independent. Perform the gauge transformation $e^{-i\omega_p t}\psi_\sigma \to \psi_\sigma$ and the fermionic Keldysh rotation. Integrate out the Gaussian bath (assume it to be at equilibrium with temperature T) to arrive at the effective dissipative action for the system:

$$S_{\text{diss}} = -\int d\epsilon \sum_{a,b;\sigma} \bar{\psi}_\sigma^a(\epsilon)\Sigma^{ab}(\epsilon)\psi_\sigma^b(\epsilon), \quad (10.89)$$

where $a, b = 1, 2$ are the Keldysh indexes and

$$\Sigma^{R,A}(\epsilon) = \mp \frac{i}{2} J(\epsilon + \omega_p); \qquad \Sigma^K(\epsilon) = -iJ(\epsilon + \omega_p) \tanh\left(\frac{\epsilon + \omega_p}{2T}\right), \quad (10.90)$$

10 The oscillatory function is a consequence of the exponent in (10.84), which takes the form $e^{(n-1)A(\lambda_+)+A(\lambda_-)} \overset{n\to 0}{\longrightarrow} e^{A(\lambda_-)-A(\lambda_+)}$, where $A(\lambda) = N\left[-(\lambda - i\epsilon)^2/2 + \log(-i\lambda)\right]$. It is easy to check that $i(A(\lambda_-) - A(\lambda_+))$ is given by the expression under the cosine.

where $J(\epsilon) = \pi \sum_s g_s^2 \delta(\epsilon - \varepsilon_s)$ is the bath spectral function. For the nearly resonant drive $|\varepsilon_\sigma - \omega_p| \ll \omega_p$, one may define $\kappa_\sigma = J(\varepsilon_\sigma)/2$ to approximate the dissipative action with the local in time form. As a result, the effective action of the two-state system $S_0 + S_{\rm diss}$ is

$$S = \int \mathrm{d}t \Bigg[\sum_\sigma (\bar\psi_\sigma^1, \bar\psi_\sigma^2) \begin{pmatrix} \mathrm{i}\partial_t - \delta_\sigma + \mathrm{i}\kappa_\sigma & 2\mathrm{i}\kappa_\sigma \tanh(\varepsilon_\sigma/2T) \\ 0 & \mathrm{i}\partial_t - \delta_\sigma - \mathrm{i}\kappa_\sigma \end{pmatrix} \begin{pmatrix} \psi_\sigma^1 \\ \psi_\sigma^2 \end{pmatrix}$$
$$- \Delta(\bar\psi_\uparrow^1 \bar\psi_\downarrow^2 + \bar\psi_\uparrow^2 \bar\psi_\downarrow^1 - \psi_\uparrow^1 \psi_\downarrow^2 - \psi_\uparrow^2 \psi_\downarrow^1) \Bigg], \quad (10.91)$$

where $\delta_\sigma = \varepsilon_\sigma - \omega_p$.

Consider now the Lindblad equation for the reduced density matrix, $\hat\rho$, of the two fermionic states system. It is a 4×4 matrix, obeying the following Lindblad equation:

$$\partial_t \hat\rho(t) = -\mathrm{i}[\hat H_{\rm p}, \hat\rho(t)] + \sum_a \gamma_a \left(\hat L_a \hat\rho(t) \hat L_a^\dagger - \frac{1}{2} \left\{ \hat L_a^\dagger \hat L_a, \hat\rho(t) \right\} \right), \quad (10.92)$$

where for our model $\hat H_{\rm p}[\hat c_\sigma^\dagger, \hat c_\sigma] = \sum_\sigma \delta_\sigma \hat c_\sigma^\dagger \hat c_\sigma + \Delta(\hat c_\uparrow^\dagger \hat c_\downarrow^\dagger + \hat c_\downarrow \hat c_\uparrow)$, $a = 1, \ldots, 4$ with $\hat L_{1,2} = \hat c_\sigma$ and $\hat L_{3,4} = \hat c_\sigma^\dagger$. Following Section 5.3 develop the functional integral representation for the evolution operator corresponding to the fermionic Lindblad equation. To this end, discretize the time axis and introduce two sets of coherent states, $|\psi_{j,\sigma}^+\rangle$ and $|\psi_{j,\sigma}^-\rangle$, such that $\hat c_\sigma |\psi_{j,\sigma}^\pm\rangle = \psi_{j,\sigma}^\pm |\psi_{j,\sigma}^\pm\rangle$, at each discretization point. With their help write the instantaneous density matrix operator as

$$\hat\rho_j = \iint \mathrm{d}[\bar\psi_j^+, \psi_j^+]\, \mathrm{d}[\bar\psi_j^-, \psi_j^-]\, e^{-\bar\psi_j^+ \psi_j^+ - \bar\psi_j^- \psi_j^-}\, |\psi_j^+\rangle \langle \psi_j^+ | \hat\rho_j | \psi_j^- \rangle \langle \psi_j^- |. \quad (10.93)$$

Proceeding as in Section 5.3, show that the partition function $Z = \mathrm{Tr}\{\hat\rho(t)\} = 1$ may be represented as a functional integral $Z = \int \mathbf{D}[\psi_\sigma, \psi_\sigma]\, \mathrm{e}^{\mathrm{i}S}$ with the action

$$S(\bar\psi_\sigma^+, \psi_\sigma^+, \bar\psi_\sigma^-, \psi_\sigma^-) = \sum_\sigma \int \mathrm{d}t \left[\bar\psi_\sigma^+ \mathrm{i}\partial_t \psi_\sigma^+ - \bar\psi_\sigma^- \mathrm{i}\partial_t \psi_\sigma^- - K(\bar\psi_\sigma^+, \psi_\sigma^+, \bar\psi_\sigma^-, \psi_\sigma^-) \right]; \quad (10.94)$$

$$K(\bar\psi_\sigma^+, \psi_\sigma^+, \bar\psi_\sigma^-, \psi_\sigma^-) = H_{\rm p}(\bar\psi_\sigma^+, \psi_\sigma^+) - H_{\rm p}(\bar\psi_\sigma^-, \psi_\sigma^-)$$
$$+ \mathrm{i} \sum_a \gamma_a \left(\bar L_a(\bar\psi_\sigma^-, \psi_\sigma^-) L_a(\bar\psi_\sigma^+, \psi_\sigma^+) - \frac{1}{2} \bar L_a L_a(\bar\psi_\sigma^+, \psi_\sigma^+) - \frac{1}{2} \bar L_a L_a(\bar\psi_\sigma^-, \psi_\sigma^-) \right), \quad (10.95)$$

where the *three* types of *normally ordered* operators $\hat L_a^\dagger[\hat c_\sigma^\dagger, \hat c_\sigma]$, $\hat L_a[\hat c_\sigma^\dagger, \hat c_\sigma]$, and $\hat L_a^\dagger \hat L_a[\hat c_\sigma^\dagger, \hat c_\sigma]$ correspondingly yield the three functions $\bar L_a(\bar\psi_\sigma, \psi_\sigma)$, $L_a(\bar\psi_\sigma, \psi_\sigma)$, and $\bar L_a L_a(\bar\psi_\sigma, \psi_\sigma)$ upon the substitution $\hat c_\sigma^\dagger \to \bar\psi_\sigma$ and $\hat c_\sigma \to \psi_\sigma$.

In particular, for $\hat{L}_{1,2} = \hat{c}_{\uparrow,\downarrow}$ and $\hat{L}_{3,4} = \hat{c}^\dagger_{\uparrow,\downarrow}$ the dissipative part of (10.95) takes the form

$$\mathrm{i}\sum_\sigma \gamma_\sigma\left(\bar{\psi}^-_\sigma\psi^+_\sigma - \frac{1}{2}\bar{\psi}^+_\sigma\psi^+_\sigma - \frac{1}{2}\bar{\psi}^-_\sigma\psi^-_\sigma\right) - \mathrm{i}\sum_\sigma \tilde{\gamma}_\sigma\left(\bar{\psi}^+_\sigma\psi^-_\sigma - \frac{1}{2}\bar{\psi}^+_\sigma\psi^+_\sigma - \frac{1}{2}\bar{\psi}^-_\sigma\psi^-_\sigma\right)$$
$$= -\frac{\mathrm{i}}{2}\sum_\sigma (\bar{\psi}^1_\sigma,\ \bar{\psi}^2_\sigma)\begin{pmatrix}\gamma_\sigma + \tilde{\gamma}_\sigma & 2(\gamma_\sigma - \tilde{\gamma}_\sigma)\\ 0 & -\gamma_\sigma - \tilde{\gamma}_\sigma\end{pmatrix}\begin{pmatrix}\psi^1_\sigma\\ \psi^2_\sigma\end{pmatrix},$$

where $\gamma_{\uparrow,\downarrow} = \gamma_{1,2}$ and $\tilde{\gamma}_{\uparrow,\downarrow} = \gamma_{3,4}$. Notice the minus sign in front of $\hat{L}_{3,4} = \hat{c}^\dagger_{\uparrow,\downarrow}$ part, which is due to the normal ordering and Grassmann anti-commutation. Show that the two actions (10.91) and (10.94) are equivalent, provided

$$\gamma_{1,2} = \gamma_\sigma = \kappa_\sigma\left[1 + \tanh\left(\frac{\varepsilon_\sigma}{2T}\right)\right]; \quad \gamma_{3,4} = \tilde{\gamma}_\sigma = \kappa_\sigma\left[1 - \tanh\left(\frac{\varepsilon_\sigma}{2T}\right)\right]. \tag{10.96}$$

This establishes the fact that the Lindblad equation (10.92), with $\hat{L}_{1,2} = \hat{c}_\sigma$ and $\hat{L}_{3,4} = \hat{c}^\dagger_\sigma$ and the coupling constants (10.96), describes the evolution of the reduced density matrix of the driven-dissipative fermionic system, described earlier. The fact that the bath is in equilibrium translates into the Lindbladian analog of the FDT: the ratio of up, $\hat{c}^\dagger_\sigma$, and down, $\hat{c}_\sigma$, transition rates is $\tilde{\gamma}_\sigma/\gamma_\sigma = \mathrm{e}^{-\varepsilon_\sigma/T}$.

A generic Keldysh Lindbladian *quadratic* action for a system with N fermion states may be written as[11]

$$S = \int dt\, \overline{\Psi}\begin{pmatrix}\mathrm{i}\partial_t - \check{H}_\mathrm{p} + \mathrm{i}\check{Q} & \mathrm{i}\check{D}\\ 0 & \mathrm{i}\partial_t - \check{H}_\mathrm{p} - \mathrm{i}\check{Q}\end{pmatrix}\Psi, \tag{10.97}$$

where the $4N$ component Keldysh–Nambu spinors are

$$\overline{\Psi} = (\bar{\psi}^1_j,\ \psi^2_j,\ \bar{\psi}^2_j,\ \psi^1_j); \qquad \Psi = (\psi^1_j,\ \bar{\psi}^2_j,\ \psi^2_j,\ \bar{\psi}^1_j)^\mathrm{T}, \tag{10.98}$$

$j = 1,\ldots, N$, and $\check{H}_\mathrm{p}$, $\check{Q}$, and $\check{D}$ are $2N \times 2N$ Hermitian matrices

$$\check{Q} = \frac{1}{2}\begin{pmatrix}\gamma + \tilde{\gamma} & \eta^\dagger_\mathrm{s}\\ \eta_\mathrm{s} & \gamma^\mathrm{T} + \tilde{\gamma}^\mathrm{T}\end{pmatrix}; \qquad \check{D} = \begin{pmatrix}\gamma - \tilde{\gamma} & \eta^\dagger_\mathrm{a}\\ \eta_\mathrm{a} & \tilde{\gamma}^\mathrm{T} - \gamma^\mathrm{T}\end{pmatrix}, \tag{10.99}$$

where $N \times N$ matrices γ and $\tilde{\gamma}$ are Hermitian, while η_s is complex symmetric and η_a is complex antisymmetric. This action is equivalent to the Lindblad equation (10.92) with the Hamiltonian $\hat{H}_\mathrm{p} = (\hat{c}^\dagger, \hat{c})\check{H}_\mathrm{p}(\hat{c}, \hat{c}^\dagger)^\mathrm{T}$ and the set of jump operators $\hat{L}_a = \sum_{j=1}^N(\mu_{aj}\hat{c}_j + \nu_{aj}\hat{c}^\dagger_j)$, where

$$\gamma_{ij} = \sum_{a=1}^{2N}\mu^*_{ai}\mu_{aj}; \qquad \tilde{\gamma}_{ij} = \sum_{a=1}^{2N}\nu_{ai}\nu^*_{aj}; \qquad \eta_{ij} = \sum_{a=1}^{2N}\nu^*_{ai}\mu_{aj}, \tag{10.100}$$

and $\eta_{\mathrm{s,a}} = \eta \pm \eta^\mathrm{T}$.

[11] I am indebted to Foster Thompson for developing this representation and explaining it to me.

Green functions corresponding to the action (10.97) are given by

$$\check{G}^{\mathrm{R,A}}(\epsilon) = \left(\epsilon - \check{H}_{\mathrm{p}} \pm \mathrm{i}\check{Q}\right)^{-1}; \qquad \check{G}^{\mathrm{K}}(\epsilon) = -\check{G}^{\mathrm{R}}(\epsilon)\,\mathrm{i}\check{D}\,\check{G}^{\mathrm{A}}(\epsilon). \tag{10.101}$$

There are $2N$ complex poles of $\check{G}^{\mathrm{R}}(\epsilon)$ at $\epsilon = \epsilon_a$, $a = 1, \ldots, 2N$. They form pairs with opposite real parts and same nonpositive imaginary part. These poles determine $\mathcal{N}^2 = (2^N)^2$ eigenvalues of the Lindbladian super-operator, given by $\lambda_{\{n_a\}} = -\mathrm{i}\sum_{a=1}^{2N} n_a\epsilon_a$, where $n_a = 0, 1$. There is at least one zero eigenvalue $\lambda_{\{0\}} = 0$. To find the corresponding stationary density matrix, one needs to diagonalize the Hermitian matrix $\mathrm{i}\check{G}^{\mathrm{K}}(t = 0) = \int \frac{d\epsilon}{2\pi}\,\mathrm{i}\check{G}^{\mathrm{K}}(\epsilon)$ with a unitary matrix $\check{U}$ and real eigenvalues, denoted as $\pm\tanh(\beta_j/2)$. The stationary state of the Lindblad equation is given by $\hat{\rho}_{\mathrm{ss}} \propto \mathrm{e}^{-\sum_j^N \beta_j \hat{d}_j^\dagger \hat{d}_j}$, where $(\hat{d}^\dagger, \hat{d}) = (\hat{c}^\dagger, \hat{c})\check{U}^\dagger$.

11

Kinetic Theory and Hydrodynamics

In this chapter we derive the kinetic equation and collision integral for electron–electron interactions. We then show how the hydrodynamic description of the Fermi liquid emerges from such a kinetic equation. Finally collective hydrodynamic modes are discussed.

11.1 Kinetic Equation

We now consider the kinetic equation for the fermionic distribution function $F_{\sigma\sigma'}(x,x')$, defined in Eq. (10.38). In this section we restrict ourselves to the singlet component $F = F^0 = \mathrm{Tr}F_{\sigma\sigma}$, while the spin-density components defined as $\mathbf{F}^\mu = \mathrm{Tr}F_{\sigma\sigma'}\hat{s}^\mu_{\sigma'\sigma}$ with $\mu = x,y,z$ are discussed in Section 11.4. According to Eqs. (10.54), knowledge of the fermionic distribution function is crucial to evaluate, for example, the polarization matrix away from equilibrium.

Following Section 6.3, one starts from the Dyson equation for the dressed fermionic Green function:

$$\left(\hat{G}_0^{-1} - \hat{\Sigma}\right) \circ \hat{G} = \hat{1}. \tag{11.1}$$

The subscript "0" indicates the bare inverse Green function. It has only diagonal components $[G_0^{\mathrm{R(A)}}]^{-1}$, given by Eq. (10.36), while its Keldysh component is a pure regularization. The fermionic self-energy matrix $\hat{\Sigma}$ has the same structure as $\hat{G}^{-1}$, Eq. (10.30), with a finite Keldysh component. Thus the R and A components of the Dyson equation take the simple form

$$\left(\mathrm{i}\partial_t + \frac{1}{2m}\nabla_{\mathrm{r}}^2 - V^{\mathrm{cl}} - \Sigma^{\mathrm{R(A)}}\right) G^{\mathrm{R(A)}} = \delta(t-t')\delta(r-r'). \tag{11.2}$$

Employing the parametrizations (10.38) $G^{\mathrm{K}} = G^{\mathrm{R}} \circ F - F \circ G^{\mathrm{A}}$, where F is a Hermitian matrix, one may rewrite the Keldysh component of the Dyson equation as (cf. Eq. (6.24))

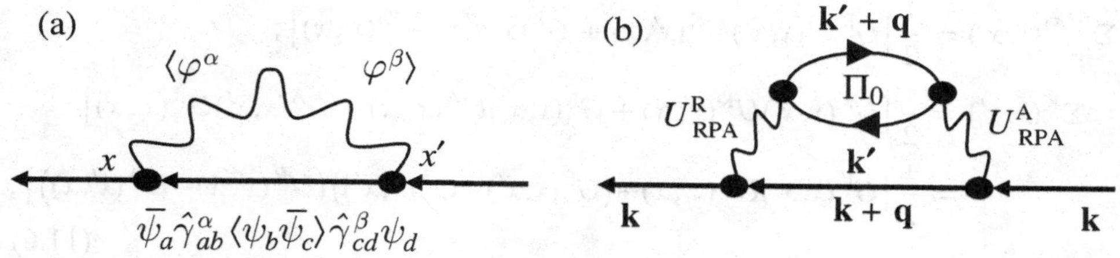

Figure 11.1 (a) Fermionic self-energy. (b) States entering the collision integral (11.11); incoming states are **k** and $\mathbf{k}'+\mathbf{q}$, while outgoing states are $\mathbf{k}'$ and $\mathbf{k}+\mathbf{q}$. Only states that conserved total energy participate in the collision integral.

$$-\left[\left(\mathrm{i}\partial_t+\frac{1}{2m}\nabla_{\mathrm{r}}^2-V^{\mathrm{cl}}\right)\overset{\circ}{,}F\right]=\Sigma^{\mathrm{K}}-\left(\Sigma^{\mathrm{R}}\circ F-F\circ\Sigma^{\mathrm{A}}\right). \tag{11.3}$$

This equation is the quantum kinetic equation for the fermionic distribution F. Its left-hand side is the kinetic term, while the right-hand side contains the collision integral. Upon Wigner transformation and restriction of the distribution function to the mass-shell, the kinetic term acquires exactly the same form (6.51) as in the case of a complex boson field. In what follows we focus on the structure of the fermionic collision integral.

The simplest diagram, Fig. 11.1(a), for the fermionic self-energy matrix, $\hat{\Sigma}_{ad}$, where $a,d=(1,2)$ are the Keldysh indices, is obtained by expanding the Hubbard–Stratonovich vertex $\mathrm{e}^{-\mathrm{i}(\bar{\psi}_a\hat{\gamma}^\alpha_{ab}\psi_b)\varphi^\alpha}$, Eq. (10.65), to second order and averaging with the help of the Wick theorem over both fermion and boson fields. As a result, one finds

$$\begin{aligned}\mathrm{i}\,\hat{\Sigma}_{ad}(x,x')&=\hat{\gamma}^\alpha_{ab}\left\langle\psi_b(x)\bar{\psi}_c(x')\right\rangle\hat{\gamma}^\beta_{cd}\times\langle\varphi^\beta(x')\varphi^\alpha(x)\rangle\\&=\left(\hat{\gamma}^\alpha\,\mathrm{i}\,\hat{G}(x,x')\,\hat{\gamma}^\beta\right)_{ad}\times\frac{\mathrm{i}}{2}\,\hat{U}^{\beta\alpha}(x',x),\end{aligned} \tag{11.4}$$

where we used $\langle\psi_b(x)\bar{\psi}_c(x')\rangle=\mathrm{i}\hat{G}_{bc}(x,x')$ for the fermion fields while the real boson fields' averages are $\langle\varphi^\beta(x')\varphi^\alpha(x)\rangle=\frac{\mathrm{i}}{2}\hat{U}^{\beta\alpha}(x',x)$, see, for example, Eq. (10.65). Here summations over all repeated indices are understood. This leads to

$$\hat{\Sigma}_{ad}=\frac{\mathrm{i}}{2}\left[\left(\hat{\gamma}^{\mathrm{cl}}\hat{G}\hat{\gamma}^{\mathrm{cl}}\right)_{ad}U^{\mathrm{K}}+\left(\hat{\gamma}^{\mathrm{cl}}\hat{G}\hat{\gamma}^{\mathrm{q}}\right)_{ad}U^{\mathrm{A}}+\left(\hat{\gamma}^{\mathrm{q}}\hat{G}\hat{\gamma}^{\mathrm{cl}}\right)_{ad}U^{\mathrm{R}}\right]. \tag{11.5}$$

Finally, one finds for the $\Sigma^{\mathrm{R}}=\Sigma_{11}$, $\Sigma^{\mathrm{A}}=\Sigma_{22}$, and $\Sigma^{\mathrm{K}}=\Sigma_{12}$ components of the fermionic self-energy

$$\begin{aligned}\Sigma^{R(A)}(x,x') &= \frac{i}{2}\left[G^{R(A)}(x,x')U^{K}(x',x) + G^{K}(x,x')U^{A(R)}(x',x)\right];\\ \Sigma^{K}(x,x') &= \frac{i}{2}\left[G^{K}(x,x')U^{K}(x',x)+G^{R}(x,x')U^{A}(x',x)+G^{A}(x,x')U^{R}(x',x)\right]\\ &= \frac{i}{2}\left[G^{K}(x,x')U^{K}(x',x)-\left(G^{R}(x,x')-G^{A}(x,x')\right)\left(U^{R}(x',x)-U^{A}(x',x)\right)\right],\end{aligned} \tag{11.6}$$

where in the last equality one has used that $G^{R(A)}(x,x')U^{R(A)}(x',x) = 0$, due to causality. For the same reason, $\Sigma_{21} = \frac{i}{2}\left(G^{A}U^{A} + G^{R}U^{R}\right) = 0$. As expected, the retarded and advanced components are lower and upper triangular matrices in time space, correspondingly, with $\Sigma^{R} = [\Sigma^{A}]^{\dagger}$, while $\Sigma^{K} = -[\Sigma^{K}]^{\dagger}$. Notice the close resemblance of expressions (11.6) to their bosonic counterparts, Eqs. (6.36)–(6.38).

We now perform a Wigner transform and introduce the mass-shell fermionic distribution function $\tilde{F}(\mathbf{r}, t, \mathbf{k})$, as discussed in Section 6.7. This way, Eq. (11.3) takes the form of the kinetic equation

$$\left[\partial_t + \mathbf{v_k}\nabla_{\mathbf{r}} - (\nabla_{\mathbf{r}}V^{cl})\nabla_{\mathbf{k}}\right]\tilde{F}(\mathbf{r},t,\mathbf{k}) = I^{coll}[\tilde{F}], \tag{11.7}$$

where we neglected renormalizations coming from $\mathrm{Re}\,\Sigma$ for simplicity. The collision integral is given by the right-hand side of Eq. (11.3), where Eqs. (11.6) are substituted. The latter are algebraic in the real space-time, but they take the form of the convolutions in the momentum-energy space. One may simplify it by noticing that $G^{R} - G^{A} = -2\pi i\delta(\epsilon + \omega - \epsilon_{\mathbf{k+q}})$, to perform the energy ω integration. As a result, one finds for the collision integral

$$I^{coll}[\tilde{F}] = \frac{i}{2}\sum_{\mathbf{q}}\left\{[\tilde{F}(\mathbf{k+q}) - \tilde{F}(\mathbf{k})]U^{K} + [\tilde{F}(\mathbf{k+q})\tilde{F}(\mathbf{k}) - 1][U^{R} - U^{A}]\right\}, \tag{11.8}$$

where we have suppressed the space and time dependence of $\tilde{F}$ and arguments of the interaction propagators are $\hat{U}(\mathbf{q},\omega) = \hat{U}(\mathbf{q},\epsilon_{\mathbf{k+q}} - \epsilon_{\mathbf{k}})$. If one understands such an interaction as the bare *instantaneous* interaction potential, Eq. (10.65) (i.e. $U^{R} = U^{A} = U_{s}(\mathbf{q})$, while $U^{K} = 0$), one finds $I^{coll} = 0$ and therefore there is no collisional relaxation. Thus, one has to employ an approximation for $\hat{U}$ that contains some retardation. The simplest and most convenient one is RPA, where $\hat{U}_{RPA} = (U_s^{-1}\hat{\sigma}_1 + \hat{\Pi}_0)^{-1}$, see Eqs. (10.67) and (10.68), with the frequency-dependent matrix $\hat{\Pi}_0$, which is thus nonlocal in time. Employing Eq. (10.49), it is convenient to write the components of this matrix relation in the following way:

$$U^{R(A)}_{RPA} = U^{R}_{RPA}\left(U_s^{-1} + \Pi_0^{A(R)}\right)U^{A}_{RPA}; \qquad U^{K}_{RPA} = -U^{R}_{RPA}\Pi_0^{K}U^{A}_{RPA}. \tag{11.9}$$

This brings Eq. (11.8) to the form

$$I^{\rm coll}[\tilde F] = -\frac{i}{2}\sum_{\mathbf q}\left|U^{\rm R}_{\rm RPA}(\mathbf q,\omega)\right|^2 \qquad (11.10)$$
$$\times\left\{[\tilde F(\mathbf k+\mathbf q)-\tilde F(\mathbf k)]\,\Pi^{\rm K}_0(\mathbf q,\omega)-[\tilde F(\mathbf k+\mathbf q)\tilde F(\mathbf k)-1][\Pi^{\rm R}_0(\mathbf q,\omega)-\Pi^{\rm A}_0(\mathbf q,\omega)]\right\},$$

where $\omega = \epsilon_{\mathbf k+\mathbf q}-\epsilon_{\mathbf k}$. We employ now Eqs. (10.54), which express components of the polarization matrix $\hat\Pi_0(\mathbf q,\omega)$ through the fermionic distribution function, and substitute them in Eq. (11.10). This way, we find the collision integral depicted in Fig. 11.1(b):

$$I^{\rm coll}[\tilde F] = -\frac{\pi}{2}\sum_{\mathbf q,\mathbf k'}\left|U^{\rm R}_{\rm RPA}(\mathbf q,\epsilon_{\mathbf k+\mathbf q}-\epsilon_{\mathbf k})\right|^2\delta(\epsilon_{\mathbf k+\mathbf q}-\epsilon_{\mathbf k}-\epsilon_{\mathbf k'+\mathbf q}+\epsilon_{\mathbf k'}) \qquad (11.11)$$
$$\times\left\{[\tilde F(\mathbf k+\mathbf q)-\tilde F(\mathbf k)][\tilde F(\mathbf k')\tilde F(\mathbf k'+\mathbf q)-1]-[\tilde F(\mathbf k+\mathbf q)\tilde F(\mathbf k)-1][\tilde F(\mathbf k'+\mathbf q)-\tilde F(\mathbf k')]\right\};$$

compare it with the corresponding bosonic result (8.79). Notice that the total energy and momentum in the inelastic cross section of the diagram in Fig. 11.1(b) are equal to the external energy and momentum. The right-hand side acquires a more familiar form if expressed in terms of the fermionic occupation numbers $\tilde F(\mathbf k) = 1-2n_{\mathbf k}$. Then the expression in the curly brackets on the right-hand side is $8\{n_{\mathbf k+\mathbf q}n_{\mathbf k'}(1-n_{\mathbf k})(1-n_{\mathbf k'+\mathbf q})-n_{\mathbf k'+\mathbf q}n_{\mathbf k}(1-n_{\mathbf k'})(1-n_{\mathbf k+\mathbf q})\}$; recall also that the left-hand side of the kinetic equation is $-2\partial_t n_{\mathbf k}+\cdots$. This shows that for the "in" process states $\mathbf k$ and $\mathbf k'+\mathbf q$ should be empty, while states $\mathbf k'$ and $\mathbf k+\mathbf q$ are occupied, and vice versa for the "out" process. Because of the energy conservation $\epsilon_{\mathbf k+\mathbf q}-\epsilon_{\mathbf k} = \epsilon_{\mathbf k'+\mathbf q}-\epsilon_{\mathbf k'}$, one may show that the collision integral is nullified if $\tilde F(\mathbf k) = F^{\rm eq}(\epsilon_{\mathbf k}) = \tanh(\epsilon_{\mathbf k}-\mu)/2T$; this is the equilibrium fermionic distribution function. As a result, the right-hand side of the kinetic equation (11.3) is zero in equilibrium, thus $\Sigma^{\rm K} = (\Sigma^{\rm R}-\Sigma^{\rm A})\tanh(\epsilon-\mu)/2T$. Since the bare Green functions obey the same relation, it justifies the FDT (10.32) for the full *dressed* equilibrium Green function.

The square of the matrix element in the collision integral is given by the RPA screened interaction potential $\left|U^{\rm R}_{\rm RPA}(\mathbf q,\epsilon_{\mathbf k+\mathbf q}-\epsilon_{\mathbf k})\right|^2$. The latter is itself a function of the fermionic distribution function $\tilde F$. It is often the case, however, that $\omega = \epsilon_{\mathbf k+\mathbf q}-\epsilon_{\mathbf k}\approx v_{\rm F}q \ll \epsilon_{\rm F}$. In this limit $\Pi^{\rm R}_0(\mathbf q,\omega)\approx\nu$ is independent of the distribution; see Eq. (10.55). If also $q\ll\kappa_{\rm TF}^{-1} = \sqrt{4\pi e^2\nu}$, one finds the universal result $U^{\rm R}_{\rm RPA}\approx 1/\nu$.

Let us look at the relaxation of a particle excited to a state $\mathbf k$ with the energy $\epsilon_{\mathbf k}$ above the Fermi energy. If the rest of the system is in equilibrium, say at zero temperature, then one may focus on the "out" term of the collision integral. This leads to $\partial_t n_{\mathbf k} = -n_{\mathbf k}/\tau_{\rm sc}$, where the "out" relaxation rate is

$$\frac{1}{\tau_{sc}} = \frac{2\pi}{\nu^2}\sum_{\mathbf{q},\mathbf{k}'} n_{\mathbf{k}'+\mathbf{q}}(1-n_{\mathbf{k}'})(1-n_{\mathbf{k}+\mathbf{q}})\delta(\epsilon_{\mathbf{k}+\mathbf{q}}-\epsilon_{\mathbf{k}}-\epsilon_{\mathbf{k}'+\mathbf{q}}+\epsilon_{\mathbf{k}'}) \propto \frac{\xi_{\mathbf{k}}^2}{\epsilon_F}, \quad (11.12)$$

where $\xi_{\mathbf{k}} \equiv \epsilon_{\mathbf{k}} - \epsilon_F$.[1] The fact that $1/\tau_{sc} \ll |\xi_{\mathbf{k}}| \ll \epsilon_F$ is crucial for the consistency of the semiclassical quasiparticle representation employed here.

11.2 Hydrodynamics of the Fermi Liquid

The hydrodynamics considers non-equilibrium excitations, which are characterized by spatial and temporal scales that are much larger than the mean free path and the mean free time between two-particle collisions. From the point of view of such excitations the collisions are very frequent and thus they tend to locally equilibrate the liquid. The local equilibrium distribution is constrained by the local values of conserved quantities (aka conserved charges). The conserved charges are special because they cannot equilibrate locally on the fast timescale of the mean free time. Indeed, being conserved, they require transport across large spatial scales dictated by long wavelength perturbations. Hydrodynamics provides equations of motion for the conserved charges.

In our example of spinless electrons there are three conserved charges: number of particles (i.e. electric charge), momentum, and energy. Indeed the collision integral (11.11) describes two incoming and two ongoing particles with the incoming momentum $\mathbf{k}+(\mathbf{k}'+\mathbf{q})$ and the equal outgoing momentum $(\mathbf{k}+\mathbf{q})+\mathbf{k}'$ and energies $\epsilon_{\mathbf{k}}+\epsilon_{\mathbf{k}'+\mathbf{q}} = \epsilon_{\mathbf{k}+\mathbf{q}}+\epsilon_{\mathbf{k}'}$ (cf. delta-function in Eq. (11.11)). These conservation laws imply

$$\sum_{\mathbf{k}} I^{\text{coll}}[n] = \sum_{\mathbf{k}} \mathbf{k}\, I^{\text{coll}}[n] = \sum_{\mathbf{k}} \epsilon_{\mathbf{k}} I^{\text{coll}}[n] = 0, \quad (11.13)$$

valid locally at every point $(\mathbf{r}, t)$. These equalities may be, of course, checked directly using Eq. (11.11) and its symmetry properties upon exchange of $\mathbf{k}$ and

[1] To see this one may use that at $T = 0$

$$1/\tau_{sc} = -\sum_{\mathbf{q}} \int d\omega |U^R_{\text{RPA}}(\mathbf{q},\omega)|^2\, 2\text{Im}\Pi^R(\mathbf{q},\omega)\,(1-n_{\mathbf{k}+\mathbf{q}})\,\delta(\omega+\epsilon_{\mathbf{k}}-\epsilon_{\mathbf{k}+\mathbf{q}}).$$

According to Eq. (10.55), $\text{Im}\Pi^R = \pi\nu\omega/(2v_F q)$ for $|\omega| < v_F q$, where $\omega = \epsilon_{\mathbf{k}+\mathbf{q}} - \epsilon_{\mathbf{k}} \approx \mathbf{k}\cdot\mathbf{q}/m = kqz/m \approx v_F q z$ with $z = \cos\theta$. Because of the factor $(1-n_{\mathbf{k}+\mathbf{q}})$ the state $\mathbf{k}+\mathbf{q}$ is empty, leading to $\epsilon_F < \epsilon_{\mathbf{k}+\mathbf{q}} < \epsilon_{\mathbf{k}}$. This leads to $-\xi_{\mathbf{k}} = \epsilon_F - \epsilon_{\mathbf{k}} < kqz/m < 0$ and therefore the lower limit of the angular z-integration is $-\xi_{\mathbf{k}}/(v_F q)$. As a result

$$\frac{1}{\tau_{sc}} = -\pi\nu \int \frac{q^2 dq}{4\pi^2} |U^R_{\text{RPA}}(q)|^2 \int_{-\xi_{\mathbf{k}}/(v_F q)}^{0} dz\, z = \frac{\nu\xi_{\mathbf{k}}^2}{8\pi v_F^2}\int dq\, |U^R_{\text{RPA}}(q)|^2 \propto \frac{\nu\xi_{\mathbf{k}}^2}{v_F^2}\frac{k_F}{\nu^2},$$

where we used that $U^R_{\text{RPA}}(q) \approx 1/\nu$, while the q-integration is to be cut off at $q \approx k_F$. Recalling that $\nu \sim mk_F$, one obtains Eq. (11.12).

$\mathbf{k}'$. Therefore the kinetic equation (11.7), which we write here in terms of the occupation numbers $n_\mathbf{k} = (1 - \tilde{F}(\mathbf{k}))/2$, takes the form

$$\Big[\partial_t + \mathbf{v_k} \cdot \nabla_\mathbf{r} - \nabla_\mathbf{r} V \cdot \nabla_\mathbf{k}\Big] n_\mathbf{k}(\mathbf{r}, t) = -\frac{1}{2} I^{\mathrm{coll}}[n], \tag{11.14}$$

where $V(\mathbf{r}, t)$ is an effective potential. It leads to the three continuity relations:

$$\partial_t \rho + \mathrm{div}\, \mathbf{J} = 0; \qquad \partial_t p^\mu + \partial_\nu \Pi^{\mu\nu} = -\partial_\mu V \rho; \qquad \partial_t E_e + \mathrm{div}\, \mathbf{J}_h = -\frac{1}{m} \nabla_\mathbf{r} V \cdot \mathbf{p}, \tag{11.15}$$

for the charge, momentum, and energy densities, defined as

$$\rho(\mathbf{r}, t) = \sum_\mathbf{k} n_\mathbf{k}(\mathbf{r}, t); \qquad \mathbf{p}(\mathbf{r}, t) = \sum_\mathbf{k} \mathbf{k}\, n_\mathbf{k}(\mathbf{r}, t); \qquad E_e(\mathbf{r}, t) = \sum_\mathbf{k} \xi_\mathbf{k} n_\mathbf{k}(\mathbf{r}, t). \tag{11.16}$$

Here $\mu = x, y, z$ and $\xi_\mathbf{k} = \epsilon_\mathbf{k} - \mu$. The corresponding fluxes, according to Eq. (11.14), are given by

$$\begin{aligned} \mathbf{J}(\mathbf{r}, t) &= \sum_\mathbf{k} \mathbf{v_k}\, n_\mathbf{k}(\mathbf{r}, t); \\ \Pi^{\mu\nu}(\mathbf{r}, t) &= \sum_\mathbf{k} k^\mu v^\nu_\mathbf{k}\, n_\mathbf{k}(\mathbf{r}, t); \\ \mathbf{J}_h(\mathbf{r}, t) &= \sum_\mathbf{k} \xi_\mathbf{k} \mathbf{v_k}\, n_\mathbf{k}(\mathbf{r}, t). \end{aligned} \tag{11.17}$$

In deriving the right-hand sides of Eqs. (11.15) we have used integration by parts, for example, $\sum_\mathbf{k} k^\mu \nabla^\nu_\mathbf{k} n_\mathbf{k} = -\delta^{\mu\nu} \sum_\mathbf{k} n_\mathbf{k} = -\delta^{\mu\nu}\rho$, and assumed absence of particles at $k \to \infty$. The right-hand side of the momentum equation reflects change of the total momentum by an external force, $-\partial_\mu V$, while that of the energy equation reflects the work done on a moving liquid by this force. Though formally exact, these equations are of no practical use yet, since they require knowledge of the distribution function $n_\mathbf{k}(\mathbf{r}, t)$. The latter is to be determined from solving the kinetic equation (11.14).

To make a progress we use the assumed slowness of charges and fluxes on the scale of the mean free path/time. This means that in the zeroth approximation the sought distribution function nullifies the collision integral. To find a generic form of such distribution, let us rewrite the "in minus out" part of the collision integral as

$$\begin{aligned} & n_\mathbf{k} n_{\mathbf{k}_1} (1 - n_{\mathbf{k}'})(1 - n_{\mathbf{k}'_1}) - n_{\mathbf{k}'} n_{\mathbf{k}'_1} (1 - n_\mathbf{k})(1 - n_{\mathbf{k}_1}) \\ &= n_\mathbf{k} n_{\mathbf{k}_1} n_{\mathbf{k}'} n_{\mathbf{k}'_1} \left[\frac{1 - n_{\mathbf{k}'}}{n_{\mathbf{k}'}} \frac{1 - n_{\mathbf{k}'_1}}{n_{\mathbf{k}'_1}} - \frac{1 - n_\mathbf{k}}{n_\mathbf{k}} \frac{1 - n_{\mathbf{k}_1}}{n_{\mathbf{k}_1}} \right], \end{aligned} \tag{11.18}$$

where $\mathbf{k}$ and $\mathbf{k}_1 = \mathbf{k}' + \mathbf{q}$ are the incoming momenta, while $\mathbf{k}'$ and $\mathbf{k}_1' = \mathbf{k} + \mathbf{q}$ – the outgoing ones. To nullify the square bracket on the right-hand side of this expression it is necessary and sufficient that $\ln[(1 - n_{\mathbf{k}})/n_{\mathbf{k}}]$ is a linear function of the three conserved charges. This way, one finds

$$\frac{1 - n_{\mathbf{k}}}{n_{\mathbf{k}}} = \exp\left\{\frac{\epsilon_{\mathbf{k}} - \mathbf{k} \cdot \mathbf{u} - \mu}{T}\right\}, \tag{11.19}$$

where μ, $\mathbf{u}$, and T^{-1} are three arbitrary multiplayers for the particle number, momentum, and energy, correspondingly. If these three quantities are space- and time-independent constants, the kinetic equation is identically satisfied (in the absence of an external force). The hydrodynamic approximation promotes them to being slow functions: $\mu \to \mu + \delta\mu(\mathbf{r}, t)$, $\mathbf{u} \to \mathbf{u}(\mathbf{r}, t)$, and $T \to T + \delta T(\mathbf{r}, t)$. This way, the collision integral is still nullified identically, while the left-hand side of the kinetic equation (11.14), although nonzero, is small as long as all the variations are slow. The corresponding distribution function is

$$n_{\mathbf{k}}^{(0)} = \frac{1}{e^{\frac{\epsilon_{\mathbf{k}} - \mathbf{k} \cdot \mathbf{u} - \mu}{T}} + 1} \approx n_F(\xi_{\mathbf{k}}) - \frac{\partial n_F(\xi_{\mathbf{k}})}{\partial \xi_{\mathbf{k}}} \left(\mathbf{k} \cdot \mathbf{u} + \delta\mu + \frac{\xi_{\mathbf{k}} \delta T}{T}\right), \tag{11.20}$$

where n_F is the Fermi distribution and we expanded to the first order in slowly varying components.

We now use this quasi-equilibrium distribution to evaluate the conserved charges (11.16) and their fluxes (11.17). We use $\sum_{\mathbf{k}} = \int d\xi_{\mathbf{k}} \nu(\xi_{\mathbf{k}}) \int d\Omega_{\mathbf{k}}$, where $\nu(\xi_{\mathbf{k}})$ is the density of states and $\int d\Omega_{\mathbf{k}}$ is the angular integral in the momentum space. Furthermore we assume sufficiently low temperature, which allows us to put $\partial n_F/\partial \xi_{\mathbf{k}} \approx -\delta(\xi_{\mathbf{k}})$, though $\int d\xi\, \xi^2 \partial n_F/\partial \xi = -T^2\pi^2/3$. Finally, to simplify notations, we use an isotropic dispersion relation $\epsilon_{\mathbf{k}} = \epsilon_k$. This way, one finds

$$\delta\rho(\mathbf{r}, t) = \nu\, \delta\mu(\mathbf{r}, t); \qquad \mathbf{p}(\mathbf{r}, t) = m\rho\, \mathbf{u}(\mathbf{r}, t); \qquad \delta E_e(\mathbf{r}, t) = c_V \delta T(\mathbf{r}, t), \tag{11.21}$$

where $\nu = \nu(\epsilon_F) = \delta\rho/\delta\mu$ is the compressibility, $\rho_0 = \nu k_F v_F/3$ is the average density and $c_V(T) = \nu T\pi^2/3$ is the specific heat. In the first and third expressions here we have subtracted space-time-independent background density, $\rho_0(\mu)$, and the energy density, $E_0(\mu, T)$, that is, the Fermi pressure. Turning to the fluxes (11.17), one obtains

$$\mathbf{J}(\mathbf{r}, t) = \rho\, \mathbf{u}(\mathbf{r}, t); \qquad \Pi^{\mu\nu}(\mathbf{r}, t) = E_0(\mu + \delta\mu(\mathbf{r}, t), T)\, \delta^{\mu\nu}; \qquad \mathbf{J}_h(\mathbf{r}, t) = 0. \tag{11.22}$$

Substituting these charges and fluxes into the conservation laws (11.15), one obtains dissipationless hydrodynamic equations of the Fermi liquid. They are the continuity and the Newton–Euler equations:

$$\partial_t \rho + \mathrm{div}(\rho\mathbf{u}) = 0; \qquad m\, \partial_t(\rho\mathbf{u}) = -\nabla_{\mathbf{r}} E_0(\rho, T) - \rho \nabla_{\mathbf{r}} V. \tag{11.23}$$

Due to the absence of the heat current, $\mathbf{J}_h = 0$, in this approximation, the temperature should be time independent $T = \text{const}$.

To include the heat flux as well as the *viscosity* of the Fermi liquid one needs to go beyond the quasi-equilibrium approximation (11.19) for the distribution function. To this end, we look for a solution of the kinetic equation (11.14) in the form of $n_\mathbf{k} = n_\mathbf{k}^{(0)} + \delta n_\mathbf{k}$, where $n_\mathbf{k}^{(0)}(\mathbf{r}, t)$ is the quasi-equilibrium solution (11.19) and $\delta n_\mathbf{k}(\mathbf{r}, t)$ is a small deviation. As we will see momentarily, the smallness of $\delta n_\mathbf{k}$ is controlled by the ratio of the mean scattering time(length), τ_{sc} ($v_F \tau_{\mathrm{sc}}$) and characteristic time(space) scale of the perturbations. On the left-hand side of the kinetic equation (11.14) one may keep only $n_\mathbf{k}^{(0)}$ and disregard $\delta n_\mathbf{k}$. On the right-hand side, $n_\mathbf{k}^{(0)}$ nullifies the collision integral and therefore $\delta n_\mathbf{k}$ ought to be kept. One may linearize the collision integral in the small deviation, reducing it to a linear *functional* of $\delta n_\mathbf{k}(\mathbf{r}, t)$, which is in general nonlocal in **k**-space. To simplify presentation we treat this functional in the relaxation (scattering) time approximation, that is, substitute it by a local linear relation:

$$-\frac{1}{2} I^{\mathrm{coll}}[n] \approx -\frac{1}{\tau_{\mathrm{sc}}} \delta n_\mathbf{k}(\mathbf{r}, t); \qquad \frac{1}{\tau_{\mathrm{sc}}} \propto \frac{T^2}{\epsilon_F}, \tag{11.24}$$

where we used Eq. (11.12) and put $\xi_\mathbf{k} = \epsilon_\mathbf{k} - \mu \approx T$. The more accurate treatment (see Refs. [245, 246] and Problem 12.6.1) leads to qualitatively the same results. With the help of Eqs. (11.14) and (11.24) one finds

$$\delta n_\mathbf{k} = -\tau_{\mathrm{sc}} \left[\partial_t + \mathbf{v}_\mathbf{k} \nabla_\mathbf{r}\right] n_\mathbf{k}^{(0)} \approx \tau_{\mathrm{sc}} \frac{\partial n_F}{\partial \xi_\mathbf{k}} \left[\partial_t + \mathbf{v}_\mathbf{k} \nabla_\mathbf{r}\right] \left(\mathbf{k} \cdot \mathbf{u} + \delta\mu + \frac{\xi_\mathbf{k} \delta T}{T} \right). \tag{11.25}$$

Let us first use this expression to find a heat current $\mathbf{J}_h$, Eq. (11.17). Since $\partial n_F / \partial \xi$ is an even function of ξ, only the last term on the RHS of Eq. (11.25), which is odd in ξ, contributes to $\mathbf{J}_h$. Integrating over ξ and angles, one finds $\mathbf{J}_h = -\kappa \nabla_\mathbf{r} \delta T$, where the thermal conductivity is given by $\kappa = \tau_{\mathrm{sc}} c_V v_F^2 / 3$. Since $\tau_{\mathrm{sc}} \sim T^{-2}$ and $c_V \sim T$, the thermal conductivity scales as $\kappa \sim T^{-1}$ and grows upon decreasing the temperature. One should keep in mind, though, that the hydrodynamic treatment runs out of its validity at very small temperatures, where $l_{\mathrm{sc}} = v_F \tau_{\mathrm{sc}} > (\nabla_\mathbf{r})^{-1}$ – the characteristic length of the thermal gradient. Employing the energy continuity relation (11.15) along with Eq. (11.21), one finds the Fick's diffusion equation for the temperature variations

$$\partial_t \delta T(\mathbf{r}, t) = D_h \nabla_\mathbf{r}^2 \delta T(\mathbf{r}, t) \qquad D_h = \frac{\kappa}{c_V} = \frac{v_F^2 \tau_{\mathrm{sc}}}{3}. \tag{11.26}$$

Finally, we look for the $\sim \tau_{\mathrm{sc}}$ contribution to the momentum flux tensor $\Pi^{\mu\nu}$. Only even in ξ terms on the right-hand side of Eq. (11.25) contribute. Due to angular integration, only $\mathbf{v_k}\nabla_\mathbf{r}(\mathbf{k}\cdot\mathbf{u})$ and $\partial_t\delta\mu$ remain. Straightforward calculation[2] yields

$$\delta\Pi^{\mu\nu} = -\tau_{\rm sc}\frac{\nu v_F^2 k_F^2}{15}\left[\delta^{\mu\nu}\left(\partial_\lambda u^\lambda + \frac{5}{v_F k_F}\,\partial_t\delta\mu\right) + \partial_\mu u^\nu + \partial_\nu u^\mu\right], \qquad (11.27)$$

according to the continuity relation (11.23), $\partial_t\delta\mu = -\frac{\rho_0}{\nu}\,\mathrm{div}\,\mathbf{u} = -\frac{v_F k_F}{3}\,\partial_\lambda u^\lambda$. As a result one finds for the momentum flux tensor

$$\Pi^{\mu\nu} = E_0\delta^{\mu\nu} - \eta\left[\partial_\mu u^\nu + \partial_\nu u^\mu - \frac{2}{3}\delta^{\mu\nu}\partial_\lambda u^\lambda\right]; \qquad \eta = \tau_{\rm sc}\frac{\nu v_F^2 k_F^2}{15}. \qquad (11.28)$$

The coefficient η in front of the traceless symmetric part of the momentum flux tensor is called *viscosity*. We found that viscosity of the Fermi liquid increases at low temperatures as $\sim T^{-2}$, [219]. The viscosity term results in a dissipation added to the Euler equation (11.23), which acquires the form

$$m\,\partial_t(\rho\mathbf{u}) = -\nabla_\mathbf{r}\left(E_0 - \frac{\eta}{3}\,\mathrm{div}\,\mathbf{u}\right) + \eta\,\nabla_\mathbf{r}^2\mathbf{u} - \rho\nabla_\mathbf{r}V. \qquad (11.29)$$

This equation is employed in the next section to discuss damping of the hydrodynamic modes.

11.3 Hydrodynamic Modes

Here we look at small harmonic deviations from a stationary uniform Fermi liquid. To this end we use $\nabla_\mathbf{r}E_0 = (\partial E_0/\partial\mu)(\partial\mu/\partial\rho)\nabla_\mathbf{r}\rho$ and $\partial E_0/\partial\mu = \rho_0$ along with $\partial\mu/\partial\rho = 1/\nu$. Neglecting for a moment viscosity, we take divergence of both sides of the Euler equation (11.23) and use the continuity equation to find

$$m\,\partial_t^2\rho = \frac{\rho_0}{\nu}\,\nabla_\mathbf{r}^2\rho + \rho_0\nabla_\mathbf{r}^2 V. \qquad (11.30)$$

Here V is an effective potential, which we use to incorporate the collisionless part of the inter-particle interactions.

For a short-range interacting Fermi liquid like ^{3}He or atomic gases, one may employ the local density approximation to write $V(\mathbf{r},t) = g(\rho(\mathbf{r},t) - \rho_0)$, where g is an interaction constant. This brings Eq. (11.30) to the form of the acoustic wave equation $\partial_t^2\rho - v_s^2\,\nabla_\mathbf{r}^2\,\rho = 0$, where the sound velocity is

2 Here we used $\int d\Omega_\mathbf{k}\, n^\mu n^\nu n^\lambda n^\eta = \frac{1}{d(d+2)}\left(\delta^{\mu\nu}\delta^{\eta\lambda} + \delta^{\mu\eta}\delta^{\nu\lambda} + \delta^{\mu\lambda}\delta^{\nu\eta}\right)$, where $\mathbf{n}$ is a unit vector in momentum space, along with $\int d\Omega_\mathbf{k}\, n^\mu n^\nu = \frac{1}{d}\,\delta^{\mu\nu}$.

given by $v_s^2 = \rho_0(\nu^{-1} + g)/m$. (In other words, $\rho_0 V$ should have been considered as a part of the pressure E_0, which serves to renormalize the thermodynamic compressibility $\partial\mu/\partial\rho = \nu^{-1} + g$.) This is the hydrodynamic (or ordinary) sound mode, mentioned at the end of Section 10.7. Using compressibility of the noninteracting Fermi gas $\nu = 3\rho_0/v_F k_F$, one finds $v_s^2 = v_F^2/3$ (valid if $g \to 0$). The hydrodynamic sound mode is different from the collisionless zero-sound mode, Section 10.7. For not-too-strong interactions its velocity, v_s, is less than the Fermi velocity, and therefore it overlaps with the particle-hole continuum where $\mathrm{Im}\,\Pi_0^R(\mathbf{q}, \omega) \neq 0$, Fig. 10.2. Yet, as is discussed in what follows, it is underdamped for $\omega\tau_{sc} \ll 1$. This should be contrasted with the zero sound-mode with the velocity above the particle-hole continuum, $v_{zs} \gtrsim v_F$, yet overdamped by the Landau damping mechanism in the same regime $\omega\tau_{sc} \ll 1$.

For charged particles, like electrons, the effective potential satisfies the Poisson equation $\nabla_{\mathbf{r}}^2 V = 4\pi e^2(\rho(\mathbf{r}, t) - \rho_0)$. Substituting it in Eq. (11.30) and passing to the Fourier representation, one finds the dispersion relation of the collective hydrodynamic mode

$$\omega_{\mathbf{q}}^2 = \omega_{\mathrm{p}}^2 + v_F^2 q^2/3, \tag{11.31}$$

where $\omega_p = \sqrt{4\pi e^2 \rho_0/m}$ is the plasma frequency. Therefore, the dispersion of the hydrodynamic mode is basically the same as that of the collisionless plasmon mode, Eq. (7.14). However, its nature is quite different and, in particular, its damping is due to the viscosity rather than to the Landau damping.

To discuss dissipation of the hydrodynamic modes we include the viscous terms through Eq. (11.29). Along with the continuity equation it leads to the friction term in the equation for the Fourier components of the density:

$$\partial_t^2 \rho_{\mathbf{q}} = -\omega_{\mathbf{q}}^2 \rho_{\mathbf{q}} - 2\gamma_{\mathbf{q}} \partial_t \rho_{\mathbf{q}}; \qquad \gamma_{\mathbf{q}} = \frac{2}{3}\frac{\eta q^2}{\rho_0 m}, \tag{11.32}$$

where $\omega_{\mathbf{q}}$ is either the plasmon dispersion (11.31) for electrons, or $\omega_{\mathbf{q}} = v_s q$ for neutral gases, while $\gamma_{\mathbf{q}}$ is the friction coefficient. This viscous damping of hydrodynamic modes should be compared with the Landau damping, given by $\mathrm{Im}\,\Pi_0^R(\mathbf{q}, \omega_{\mathbf{q}})$, Eq. (10.57). The latter is applicable for the collective modes in the collisionless regime $\omega\tau_{sc} \gg 1$. Notice that the viscous friction $\gamma_{\mathbf{q}} \sim T^{-2}$ is decreasing upon increasing the temperature. On the other hand, the Landau damping is rapidly increasing with the temperature. Focusing on the short-range acoustic case, one observes that the hydrodynamic sound mode is underdamped as long as $\gamma_q \ll v_s q$, dictating $q \ll 3v_s\rho_0 m/(2\eta) \propto l_{sc}^{-1}$, where $l_{sc} = v_F\tau_{sc}$ and we put

$v_s \sim v_F$. This coincides with the region of validity of the hydrodynamic approximation, $ql_{sc} \ll 1$. Therefore, throughout the hydrodynamic regime the sound mode is underdamped. It is propagating with the inverse decay length, $\xi_q^{-1} = \gamma_q / v_s \sim q^2 l_{sc} \ll q$.

11.4 Magnetization Dynamics

We now discuss the role of the spin degrees of freedom in the kinetics of the Fermi liquid. In general, the spin part of the kinetic equation is written for the *matrix* distribution function $\hat{F}_{\sigma\sigma'}(x,x')$ defined in Eq. (10.38). Keeping the Zeeman magnetic field in the Hamiltonian, one writes the kinetic equation as

$$-\left[\left(\mathrm{i}\partial_t + \frac{1}{2m}\nabla_r^2 - V^{\mathrm{cl}} - \mathbf{H}_Z \cdot \vec{\hat{s}}\right) \circ, \hat{F}\right] = -\mathrm{i}\hat{I}^{\mathrm{coll}}[\hat{F}], \tag{11.33}$$

where both the distribution function and the collision integral are matrices in spin space as well as in (x,x') space. One can now introduce the scalar and spin-vector components of the distribution function, as $F(x,p) = \mathrm{Tr}\{\hat{F}(x,p)\}$ and $\mathbf{F}^\mu(x,p) = \mathrm{Tr}\{\hat{F}(x,p)\hat{s}^\mu\}$ (where $\mu = x,y,z$, and $\hat{s}^\mu$ are Pauli matrices in the spin space). We now perform the Wigner transformations and restrict the distribution function to the mass-shell, as discussed in Section 6.7. This way we find the scalar and vector on-shell distribution functions $\tilde{F}(\mathbf{r},t,\mathbf{k})$ and $\tilde{\mathbf{F}}(\mathbf{r},t,\mathbf{k})$. The former one was the subject of the previous sections, while here we focus on the latter.

There is a subtlety involved in restricting the distribution *matrix* onto the mass-shell. Indeed, in the presence of, for example, a static Zeeman field, $\mathbf{H}_Z$, the mass-shells of spin up and down electrons are different by the respective shifts $\pm H_Z$ in the reference frame where the magnetic field is in the z-direction. This calls for two different mass-shell restrictions for the diagonal elements of $F_{\sigma\sigma'}$, but creates ambiguity in treatment of the off-diagonal ones. We avoid it by treating magnetic fields (including the one induced by the ferromagnetic Stoner instability) as small perturbations, which do not affect the mass-shell definition. There is a price to pay for this approach, which is a presence of a vector component $\tilde{\mathbf{F}}(\mathbf{k})$ in equilibrium. Notice that the FDT, $F_{\sigma\sigma'}(\epsilon) = \delta_{\sigma\sigma'} \tanh \epsilon/(2T)$, is still intact and it is only the substitution $\epsilon \to \epsilon_{\mathbf{k}\sigma}$ that is the reason why the equilibrium $\hat{\tilde{F}}(\mathbf{k})$ deviates from the unit matrix.

We now project the matrix kinetic equation (11.33) onto the spin directions by taking traces with $\hat{s}^\mu$. The resulting kinetic equation for the vector spin distribution function, $\tilde{\mathbf{F}}(\mathbf{r},t,\mathbf{k})$, takes the form

$$\hat{Z}^{-1}\,\partial_t\tilde{\mathbf{F}} + \mathbf{v_k}\nabla_\mathbf{r}\tilde{\mathbf{F}} - (\nabla_\mathbf{r}V)\nabla_\mathbf{k}\tilde{\mathbf{F}} - 2\big[\mathbf{H}_Z \times \tilde{\mathbf{F}}\big] = \mathbf{I}^{\mathrm{coll}}[\tilde{F}, \tilde{\mathbf{F}}], \tag{11.34}$$

where we used that $\hat{\tilde{F}} = (\tilde{F} + \tilde{\mathbf{F}} \cdot \vec{\hat{s}})/2$ and $\mathrm{Tr}\{[\hat{s}^\mu, \hat{s}^\nu]\hat{s}^\lambda\} = 4i\epsilon^{\mu\nu\lambda}$ as well as $[\mathbf{H}_Z \times \tilde{\mathbf{F}}]^\mu = \epsilon^{\mu\nu\lambda}\mathbf{H}_Z^\nu \tilde{\mathbf{F}}^\lambda$. The quasiparticle *tensor* weight is defined (cf. Eq. (6.51)) as $[\hat{Z}^{-1}]^{\mu\nu} = \delta^{\mu\nu} - \partial_\epsilon \mathrm{Tr}\{(\hat{\Sigma}^{\mathrm{R}}\hat{s}^\mu + \hat{s}^\mu \hat{\Sigma}^{\mathrm{A}})\hat{s}^\nu\}/4$. In what follows we will approximate it as $\hat{Z} = 1$. The vector collision integral is found as $\mathbf{I}^{\mathrm{coll}} = \mathrm{Tr}\{\hat{I}^{\mathrm{coll}}\hat{s}^\mu\}$. The factor of two in front of the Zeeman field is an electron g-factor.

Since the interaction Hamiltonian (10.58) does not flip spins of two interacting particles, the corresponding collisions are spin conserving. This leads to (cf. Eq. (11.13))

$$\sum_{\mathbf{k}} \mathrm{Tr}\{\hat{s}^\mu \hat{I}^{\mathrm{coll}}\} = \sum_{\mathbf{k}} \mathbf{I}^{\mathrm{coll}} = 0, \tag{11.35}$$

Therefore the local *magnetization density*, defined as

$$\mathbf{S}(\mathbf{r}, t) = -\frac{1}{2}\sum_{\mathbf{k}} \tilde{\mathbf{F}}(\mathbf{r}, t, \mathbf{k}), \tag{11.36}$$

is an extra conserved charge, in addition to the particle number, momentum, and energy considered in Section 11.2. Taking the sum over all states $\mathbf{k}$ in Eq. (11.34), one finds the corresponding continuity relation

$$\partial_t S^\mu + \partial_\nu J_{\mathrm{S}}^{\mu\nu} = 2\big[\mathbf{H}_Z \times \mathbf{S}\big]^\mu, \tag{11.37}$$

where we put $\hat{Z} = 1$ for brevity. The magnetization flux tensor is given by $J_{\mathrm{S}}^{\mu\nu}(\mathbf{r}, t) = -\frac{1}{2}\sum_{\mathbf{k}} v_{\mathbf{k}}^\nu \tilde{F}^\mu(\mathbf{r}, t, \mathbf{k})$.

Here we apply this equation to discuss magnetization dynamics in an isotropic itinerant ferromagnet. Neglecting for a moment the magnetization flux tensor part, we focus on the dissipationless part, $\partial_t \mathbf{S} = 2\big[\mathbf{H}_Z \times \mathbf{S}\big]$, known as the *Landau–Lifshitz* equation. The effective magnetic field, $\mathbf{H}_Z$, should be considered as a way to incorporate collisionless interactions between spins. It is given $\mathbf{H}_Z = -\delta\mathcal{F}[\mathbf{S}]/\delta\mathbf{S}$, where $\mathcal{F}[\mathbf{S}]$ is the Landau free energy functional of the magnetization $\mathbf{S}(\mathbf{r}, t)$. In a vicinity of the second-order Stoner transition in the isotropic ferromagnet it may be approximated as

$$\mathcal{F}[\mathbf{S}] \propto \int d\mathbf{r} \left[-S^2 + \frac{1}{2S_0^2}S^4 + \frac{\xi_0^2}{2}(\nabla_{\mathbf{r}}\mathbf{S})^2\right], \tag{11.38}$$

where S_0 is a saturation magnetization and ξ_0 a correlation length. The Landau–Lifshitz equation forces magnetization to a state with $S^2 = S_0^2$, which we can assign to be in the z-direction, $S_z = S_0$. The remaining dynamics of small deviations from this state takes the form

$$\partial_t S_x = 2S_0\xi_0^2 \nabla_{\mathbf{r}}^2 S_y; \qquad \partial_t S_y = -2S_0\xi_0^2 \nabla_{\mathbf{r}}^2 S_x. \tag{11.39}$$

Eliminating one of the polarizations, one finds the harmonic motion for the Fourier components: $\partial_t^2 S_{x,y}(\mathbf{q},t) = -(2S_0\xi_0^2)^2 q^4 S_{x,y}(\mathbf{q},t)$. This leads to the hydrodynamic ferromagnetic spin waves with the dispersion relation $\omega_{\mathbf{q}} \propto q^2$. There are two orthogonal polarizations of them.

The flux tensor part, $\partial_\nu J_S^{\mu\nu}$, leads to the dumping of the magnetization precession. In the context of the Landau–Lifshitz equation it is known as the Gilbert damping. In a far-from-equilibrium case it may also lead to an opposite effect of transferring the angular momentum from a spin-polarized electron flow to a local magnetization. This effect is known as the Slonczewski–Berger [220, 221] spin-torque phenomenon. We will derive both Gilbert damping and spin-torque terms of the magnetization equation in Section 13.6 for a specific model of a metallic ferromagnetic grain in a weak contact with ferromagnetic leads. The grain magnetization can rotate relative to the leads, governed by Landaui–Lifshitzi–Gilberti–Slonczewskii–Berger equation (13.68).

11.5 Problems

11.5.1 H-Theorem

The entropy of the Fermi gas is given by[3]

$$S = -\sum_{\mathbf{k}} \big[n_{\mathbf{k}} \log n_{\mathbf{k}} + (1-n_{\mathbf{k}}) \log(1-n_{\mathbf{k}}) \big]. \tag{11.40}$$

Its time derivative is thus (cf. Eq. (11.14))

$$\frac{dS}{dt} = -\sum_{\mathbf{k}} (\partial_t n_{\mathbf{k}}) \big[\log n_{\mathbf{k}} - \log(1-n_{\mathbf{k}}) \big] = \frac{1}{2} \sum_{\mathbf{k}} I^{\mathrm{coll}}[n_{\mathbf{k}}] \log \frac{n_{\mathbf{k}}}{1-n_{\mathbf{k}}},$$

where we have omitted the spatial and momentum drift terms of the kinetic equation for brevity. Using the explicit form of the collision integral (11.18), prove that $dS/dt \geq 0$. Moreover, the equality is only reached when $I^{\mathrm{coll}}[n_{\mathbf{k}}] = 0$, that is, in the equilibrium state. This shows that the entropy monotonically increases due to collisions and reaches its maximum value when the system equilibrates.

[3] Indeed, consider $N \gg 1$ states, which may be either empty or occupied with the average occupation $n \leq 1$. The total number of accessible many-body states is

$$e^{NS} = \prod_{i=1}^{N} \sum_{s_i=0,1} \delta\left(\sum_{i=1}^{N} s_i - nN \right) = \int_{-i\infty}^{i\infty} \frac{d\lambda}{2\pi} \prod_{i=1}^{N} \sum_{s_i=0,1} e^{\lambda(\sum_{i=1}^{N} s_i - nN)} = \int_{-i\infty}^{i\infty} \frac{d\lambda}{2\pi} e^{N[\log(1+e^{\lambda}) - \lambda n]}.$$

In the large N limit the integral over λ may be calculated in the saddle-point approximation. The latter is determined from $e^{\lambda}/(1+e^{\lambda}) = n$. This leads to $e^{\lambda} = n/(1-n)$. Substituting it back into the exponent, one finds expression (11.40) for the entropy per degree of freedom.

Hint: Consider first single-particle elastic collisions with the collision integral $I^{\rm coll}[n_{\bf k}] = \sum_{{\bf k}'} \mathcal{M}_{{\bf k},{\bf k}'}\big[n_{\bf k}(1-n_{{\bf k}'})-(1-n_{\bf k})n_{{\bf k}'}\big]$, Section 14.1, where the scattering rate $\mathcal{M}_{{\bf k},{\bf k}'} \propto \delta(\epsilon_{\bf k} - \epsilon_{{\bf k}'})$ is positive and symmetric with respect to interchange of the "in and out" states $\mathbf{k}$ and $\mathbf{k}'$. Substituting this into the expression for $\mathrm{d}S/\mathrm{d}t$ and symmetrizing with respect to $\mathbf{k}$ and $\mathbf{k}'$, one finds

$$\frac{\mathrm{d}S}{\mathrm{d}t} = \frac{1}{4}\sum_{{\bf k},{\bf k}'} \mathcal{M}_{{\bf k},{\bf k}'}\big[n_{\bf k}(1-n_{{\bf k}'}) - (1-n_{\bf k})n_{{\bf k}'}\big] \log \frac{n_{\bf k}(1-n_{{\bf k}'})}{(1-n_{\bf k})n_{{\bf k}'}}\,.$$

Nonnegativity of $\mathrm{d}S/\mathrm{d}t$ follows from that of $(A - B)(\log A - \log B) \geq 0$, for any positive A and B. This quantity is zero if and only if $A = B$. Extend this reasoning to electron–electron collisions, employing the corresponding collision integral given by (11.18).

11.5.2 Hydrodynamics of Electron Flow

In modern ultraclean 2d materials there may be a window of temperatures, where electron–electron collision rate greatly exceeds both electron-impurity and electron–phonon relaxation rates. Under these conditions electron flow in an external electric field is governed by the Fermi liquid hydrodynamics. Consider a quasi-1d strip of width $2w$ of a 2d material with an electric field E_x applied along it. The electron velocity, $u_x(y)$, is determined by Eq. (11.29), where $E_0 = \text{const}$ and $\operatorname{div}\mathbf{u} = 0$ due to the incompressible nature of the electronic liquid (dictated by the strong Coulomb repulsion). It reads as

$$\eta\, \partial_y^2 u_x = -e\rho E_x, \tag{11.41}$$

where η is viscosity. Adopt no-slip boundary conditions $u_x(-w) = u_x(w) = 0$ to solve the equation and find the total current $I_x = e\rho\int_{-w}^{w} \mathrm{d}y\, u_x(y)$. Notice that $I_x \propto w^3 E_x$, in contrast to the Ohm's law expectation $I_x \propto wE_x$. This is the electronic analog of the Poiseuille law. One can formally compare it with the Drude formula $I_x = e^2\rho\tau_w 2wE_x/m$, where τ_w is an effective scattering time. Estimate τ_w using (11.28) (with 2d density of states) for the viscosity. Compare it with the ballistic result for a characteristic time due to the boundary scattering $\tau \approx 2w/v_{\rm F}$. Show that for $wk_{\rm F} \gg 1$ and $T \approx E_{\rm F}$ the former may greatly exceed the latter. This remarkable observation was first made by Gurzhi [222]. See reference [223] for a discussion of nonlocal effects in electronic hydrodynamic flow.

12

Aspects of Kinetic Theory

In this chapter we discuss some (disconnected) aspects of the kinetic theory. We start from kinetics in presence of a band structure-induced Berry curvature, followed by kinetics of periodically driven systems in Floquet basis. We then show how the kinetic equation may be viewed as the saddle-point condition for some effective nonlocal field theory. We use this representation to derive semiclassical corrections to the kinetic description, which are known as Boltzmann–Langevin theory. Finally, we delve into the butterfly effect and chaos propagation, as described by out-of-time-order correlation (OTOC) functions.

12.1 Kinetic Equation in Presence of the Berry Curvature

Band structure effects not only modify electron dispersion relation, but also result in additional modifications of the equations of motion due to induced Berry curvature. The latter takes its origin in the Abelian Berry connection $\mathcal{A}_n(\mathbf{k}) = -\mathrm{i}\langle u_{n\mathbf{k}}|\nabla_{\mathbf{k}}|u_{n\mathbf{k}}\rangle$, where $|u_{n\mathbf{k}}\rangle = u_{\mathbf{k}}(\mathbf{r})$ is a spinor in the space of orbits and $\psi_{n\mathbf{k}}(\mathbf{r}) = \mathrm{e}^{\mathrm{i}\mathbf{kr}}u_{n\mathbf{k}}(\mathbf{r})$ is a Bloch wave function of a band n with the quasi-momentum $\mathbf{k}$ in the Brillouin zone. The Berry curvature is given by $\mathbf{\Omega}(\mathbf{k}) = \nabla_{\mathbf{k}} \times \mathcal{A}(\mathbf{k})$ (hereafter we focus on a single band and thus suppress the band index n) and it enters semiclassical equations of motions as [224]

$$\dot{\mathbf{r}} = \nabla_{\mathbf{k}}h - \dot{\mathbf{k}} \times \mathbf{\Omega}; \tag{12.1a}$$

$$\dot{\mathbf{k}} = -\nabla_{\mathbf{r}}h + \dot{\mathbf{r}} \times \mathbf{H}, \tag{12.1b}$$

where $h(\mathbf{r},\mathbf{k}) = \epsilon_{\mathbf{k}} + V(\mathbf{r})$ is a Hamiltonian and we also allowed for an orbital magnetic field $\mathbf{H} = \nabla_{\mathbf{r}} \times \mathbf{A}(\mathbf{r})$. The origin of the Berry curvature is in a projection of an initial multiband system onto an effective carrier's dynamics in a *single* (e.g.

conductance) band.[1] If there is more than one conductance band (in addition to some number of projected-out valence bands), the Berry curvature is non-Abelian. The corresponding non-Abelian kinetic theory is developed in Ref. [225]. Here we focus on the case of the single conduction band and therefore the Abelian Berry curvature.

Equations (12.1) are still Hamiltonian, though written in the *noncanonical* phase-space coordinates $\xi = (\mathbf{r}, \mathbf{k})$. To show this we need a brief reminder of the analytical mechanics [226, 227]. A generic first-order Lagrangian may be written as $L = a_\alpha(\xi)\dot{\xi}^\alpha - h(\xi)$, where $\alpha = 1, \ldots, 2d$ labels phase-space coordinates. Its extrema conditions are

$$\frac{\delta L}{\delta \xi^\beta} = \frac{\partial L}{\partial \xi^\beta} - \frac{\mathrm{d}}{\mathrm{d}t}\frac{\partial L}{\partial \dot{\xi}^\beta} = \dot{\xi}^\alpha \omega_{\alpha\beta} - \partial_\beta h = 0, \tag{12.2}$$

where $\omega_{\alpha\beta}(\xi) = \partial_\beta a_\alpha - \partial_\alpha a_\beta$ is the antisymmetric matrix, called the symplectic form. In canonical coordinates $(\mathbf{q}, \mathbf{p})$, $a_\mathbf{q} = \mathbf{p}$ and $a_\mathbf{p} = 0$ with the Lagrangian $L = \mathbf{p}\dot{\mathbf{q}} - h(\mathbf{q}, \mathbf{p})$ and $\hat{\omega} = \begin{pmatrix} 0 & 1 \\ -1 & 0 \end{pmatrix} = \hat{J}$, which leads to the familiar Hamilton pair $\dot{\mathbf{q}} = \partial_\mathbf{p} h$ and $\dot{\mathbf{p}} = -\partial_\mathbf{q} h$. In generic noncanonical coordinates the Hamilton equations acquire the form

$$\dot{\xi}^\alpha \omega_{\alpha\beta} = \partial_\beta h. \tag{12.3}$$

They may be formulated in the Poisson brackets notations $\dot{\xi}^\alpha = \{h, \xi^\alpha\}$, if the Poisson brackets are defined as

$$\{f, g\} = \partial_\beta f \, \omega^{\beta\alpha} \, \partial_\alpha g, \tag{12.4}$$

and $\omega^{\beta\alpha}\omega_{\alpha\gamma} = \delta^\beta_\gamma$, that is, $\omega^{\beta\alpha}$ is the inverse of the symplectic form $\omega_{\alpha\beta}$.

Comparing (12.3) with (12.1), where all time derivatives are moved to the left-hand side, one finds that the corresponding symplectic form is

$$\omega_{\alpha\beta} = \begin{pmatrix} -\varepsilon_{\mu\nu\sigma} H^\sigma & \delta_{\mu\nu} \\ -\delta_{\mu\nu} & -\varepsilon_{\mu\nu\sigma}\Omega_\sigma \end{pmatrix}, \tag{12.5}$$

[1] I do not attempt a rigorous derivation of this fact, but rather provide some motivation. Consider the coordinate operator $\hat{\mathbf{r}} = -\mathrm{i}\nabla_\mathbf{k}$; its projection on the conductance band is $\mathbf{R} = \langle \psi_{n\mathbf{k}} | \hat{\mathbf{r}} | \psi_{n\mathbf{k}} \rangle = \mathbf{r} + A_n(\mathbf{k})$, where $\mathbf{r}$ comes from the differentiation of $\mathrm{e}^{\mathrm{i}\mathbf{kr}}$. If $\mathbf{r}$ and $\mathbf{k}$ obey the usual commutation relations $[r^\mu, r^\nu] = [k_\mu, k_\nu] = 0$ and $[k_\mu, r^\nu] = \mathrm{i}\delta^\nu_\mu$, then components of $\mathbf{R}$ do not commute with each other: $[R^\mu, R^\nu] = -\mathrm{i}(\partial_{k_\mu} A^\nu - \partial_{k_\nu} A^\mu) = -\mathrm{i}\mathcal{F}^{\mu\nu}$. This should be compared with the momentum operator in the presence of the usual vector potential $\mathbf{K} = \mathrm{i}\nabla_\mathbf{r} + \mathbf{A}(\mathbf{r})$, with the commutation relations $[K_\mu, K_\nu] = \mathrm{i}(\partial_{r^\mu} A_\nu - \partial_{r^\nu} A_\mu) = \mathrm{i}F_{\mu\nu}$. While the latter results in the extra component of the force (Lorentz), $\dot{\mathbf{r}} \times \mathbf{H}$, the former leads to the extra component of velocity, $-\dot{\mathbf{k}} \times \boldsymbol{\Omega}$, where $H^\sigma = \varepsilon^{\sigma\mu\nu} F_{\mu\nu}$ and $\Omega_\sigma = \varepsilon_{\sigma\mu\nu}\mathcal{F}^{\mu\nu}$.

where $\mu, \nu, \sigma = 1, 2, 3$ and $\varepsilon_{\mu\nu\sigma}$ is the Levi–Civita symbol. Its determinant is found to be $\det(\omega_{\beta\alpha}) = (1 - \mathbf{\Omega} \cdot \mathbf{H})^2$. And the inverse matrix is

$$\omega^{\alpha\beta} = \frac{1}{1 - \mathbf{\Omega} \cdot \mathbf{H}} \begin{pmatrix} -\varepsilon^{\mu\nu\sigma}\Omega_\sigma & -\delta^{\mu\nu} + H^\mu \Omega^\nu \\ \delta^{\mu\nu} - \Omega^\mu H^\nu & -\varepsilon^{\mu\nu\sigma} H^\sigma \end{pmatrix}. \tag{12.6}$$

Consider now a probability distribution function over the phase space volume[2]

$$n(\mathbf{r}, \mathbf{k}, t)\, \mathrm{d}V = n(\mathbf{r}, \mathbf{k}, t)\sqrt{\det \omega_{\alpha\beta}}\, \mathrm{d}^d\mathbf{r}\, \mathrm{d}^d\mathbf{k}, \tag{12.7}$$

expressed through the noncanonical coordinates $\xi = (\mathbf{r}, \mathbf{k})$. Upon a Hamiltonian flow (no collisions) it evolves according to

$$\frac{\mathrm{d}n}{\mathrm{d}t} = \frac{\partial n}{\partial t} + \{h, n\} = 0, \tag{12.8}$$

where

$$\{h, n\} = \partial_\beta h\, \omega^{\beta\alpha}\, \partial_\alpha n = \partial_\alpha\big(\partial_\beta h\, \omega^{\beta\alpha}\, n\big) - \partial_\alpha\big(\partial_\beta h\, \omega^{\beta\alpha}\big)n = \partial_\alpha\big(\dot{\xi}^\alpha\, n\big) - n\, \partial_\alpha \dot{\xi}^\alpha.$$

The last term $\partial_\alpha \dot{\xi}^\alpha = \partial_\alpha(\partial_\beta h\, \omega^{\beta\alpha}) = \partial_\beta h\, \partial_\alpha \omega^{\beta\alpha}$, where we used the antisymmetry of $\omega^{\beta\alpha}$. Using (12.6) along with the fact that $\nabla_{\mathbf{r}} \cdot \mathbf{H} = \nabla_{\mathbf{k}} \cdot \mathbf{\Omega} = 0$, one notices that only derivatives of the $(1 - \mathbf{\Omega} \cdot \mathbf{H})^{-1} = 1/\sqrt{\det \omega_{\alpha\beta}}$ contribute to $\partial_\alpha \omega^{\beta\alpha}$, while derivatives of all other factors vanish. As a result,

$$\partial_\alpha \dot{\xi}^\alpha = \partial_\beta h\, \partial_\alpha \left(\frac{1}{\sqrt{\det \omega_{\alpha\beta}}}\right) \sqrt{\det \omega_{\alpha\beta}}\; \omega^{\beta\alpha} = -\dot{\xi}^\alpha \partial_\alpha \log\big(\sqrt{\det \omega_{\alpha\beta}}\,\big), \tag{12.9}$$

and therefore

$$\partial_\alpha \dot{\xi}^\alpha = \frac{\mathrm{d}_t(\mathbf{\Omega} \cdot \mathbf{H})}{1 - \mathbf{\Omega} \cdot \mathbf{H}}, \tag{12.10}$$

where $\mathrm{d}_t\big(\mathbf{\Omega} \cdot \mathbf{H}\big) = \nabla_{\mathbf{r}}\big(\mathbf{\Omega} \cdot \mathbf{H}\big) \cdot \dot{\mathbf{r}} + \nabla_{\mathbf{k}}\big(\mathbf{\Omega} \cdot \mathbf{H}\big) \cdot \dot{\mathbf{k}}$. Putting everything together, one finds

$$\frac{\partial n}{\partial t} + \nabla_{\mathbf{r}}\big(n\, \dot{\mathbf{r}}\big) + \nabla_{\mathbf{k}}\big(n\, \dot{\mathbf{k}}\big) = n\, \frac{\mathrm{d}_t(\mathbf{\Omega} \cdot \mathbf{H})}{1 - \mathbf{\Omega} \cdot \mathbf{H}}. \tag{12.11}$$

This equation does *not* have a form of the continuity relation, due to the presence of the right-hand side. This reflects the fact that $\iint \mathrm{d}^d\mathbf{r}\, \mathrm{d}^d\mathbf{k}\, n(\mathbf{r}, \mathbf{k}, t)$ is not conserved. However, $\iint \mathrm{d}^d\mathbf{r}\, \mathrm{d}^d\mathbf{k}\, \rho(\mathbf{r}, \mathbf{k}, t)$, where $\rho(\mathbf{r}, \mathbf{k}, t) = n(\mathbf{r}, \mathbf{k}, t)\sqrt{\det \omega_{\alpha\beta}}$, is conserved.

[2] To see this, consider the canonical coordinates $x = (\mathbf{q}, \mathbf{p})$, where the phase-space volume element is $\mathrm{d}V = \mathrm{d}^d\mathbf{q}\, \mathrm{d}^d\mathbf{p}$. Changing variables to noncanonical $x^\alpha \to \xi^\alpha$, the symplectic form changes as

$$J_{\alpha\beta} \to \omega_{\alpha\beta} = \frac{\partial x^\gamma}{\partial \xi^\alpha} \frac{\partial x^\delta}{\partial \xi^\beta} J_{\gamma\delta}.$$

The volume element transforms as $\mathrm{d}V = \prod_\alpha \mathrm{d}x^\alpha = \det\big(\frac{\partial x}{\partial \xi}\big) \prod_\alpha \mathrm{d}\xi^\alpha = \sqrt{\det \omega_{\alpha\beta}} \prod_\alpha \mathrm{d}\xi^\alpha$. I am grateful to Zach Raines for clarifying these issues.

Repeating similar calculations, one finds that ρ indeed satisfies the continuity equation $\partial_t\rho + \nabla_{\mathbf{r}}(\rho\,\dot{\mathbf{r}}) + \nabla_{\mathbf{k}}(\rho\,\dot{\mathbf{k}}) = 0$. On the other hand, $\partial_t\rho + \{h,\rho\} \neq 0$.

One may work with either of these two functions. It is important to remember, however, that observables $\mathcal{O}(t) = \langle\mathcal{O}(\mathbf{r},\mathbf{k},t)\rangle_{\mathbf{r},\mathbf{k}}$ are expressed either as $\mathcal{O} = \iint \mathrm{d}^d\mathbf{r}\,\mathrm{d}^d\mathbf{k}\,\sqrt{\det\omega_{\alpha\beta}}\,n(\mathbf{r},\mathbf{k},t)\mathcal{O}(\mathbf{r},\mathbf{k},t)$ or $\mathcal{O} = \iint \mathrm{d}^d\mathbf{r}\,\mathrm{d}^d\mathbf{k}\,\rho(\mathbf{r},\mathbf{k},t)\mathcal{O}(\mathbf{r},\mathbf{k},t)$. In my opinion, $n(\mathbf{r},\mathbf{k},t)$ is preferred, since in equilibrium $n = n_{\mathrm{F}}(\epsilon_{\mathbf{k}}/T)$, while ρ still includes the factor $\sqrt{\det\omega_{\alpha\beta}}$. Since any collision integral is nullified in equilibrium, it takes the familiar "in minus out" form in terms of the n-function, but not the ρ-function.

An equivalent alternative way to write the kinetic term (12.11) is to substitute $\rho = n\sqrt{\det\omega_{\alpha\beta}}$ in the continuity equation for ρ. This leads to the n-equation in the form of the divergence in the curved phase space

$$\frac{\partial n}{\partial t} + \frac{1}{\sqrt{\det\omega_{\alpha\beta}}}\nabla_{\mathbf{r}}\big(n\sqrt{\det\omega_{\alpha\beta}}\,\dot{\mathbf{r}}\big) + \frac{1}{\sqrt{\det\omega_{\alpha\beta}}}\nabla_{\mathbf{k}}\big(n\sqrt{\det\omega_{\alpha\beta}}\,\dot{\mathbf{k}}\big) = 0. \tag{12.12}$$

Notice that velocity $\dot{\mathbf{r}} \neq \nabla_{\mathbf{k}}h$ and the force $\dot{\mathbf{k}} \neq -\nabla_{\mathbf{r}}h$, because both of them include anomalous terms. To account for them explicitly we use $\dot{\xi}^\alpha = \{h,\xi^\alpha\}$, or in components: $\dot{\mathbf{r}} = -\mathbf{E}\,\omega^{\mathbf{r},\mathbf{r}} + \mathbf{v}_{\mathbf{k}}\,\omega^{\mathbf{k},\mathbf{r}}$ and $\dot{\mathbf{k}} = -\mathbf{E}\,\omega^{\mathbf{r},\mathbf{k}} + \mathbf{v}_{\mathbf{k}}\,\omega^{\mathbf{k},\mathbf{k}}$, where $\mathbf{v}_{\mathbf{k}} = \nabla_{\mathbf{k}}h = \nabla_{\mathbf{k}}\epsilon_{\mathbf{k}}$ is the normal velocity and $\mathbf{E}(\mathbf{r}) = -\nabla_{\mathbf{r}}h = -\nabla_{\mathbf{r}}V(\mathbf{r})$ is the electric field. Employing (12.6), one finds

$$\sqrt{\det\omega_{\alpha\beta}}\,\dot{\mathbf{r}} = \mathbf{v}_{\mathbf{k}} - \mathbf{E}\times\mathbf{\Omega} - (\mathbf{v}_{\mathbf{k}}\cdot\mathbf{\Omega})\mathbf{H};$$
$$\sqrt{\det\omega_{\alpha\beta}}\,\dot{\mathbf{k}} = \mathbf{E} + \mathbf{v}_{\mathbf{k}}\times\mathbf{H} - (\mathbf{E}\cdot\mathbf{H})\,\mathbf{\Omega}.$$

With this Eq. (12.12) takes the form:

$$\frac{\partial n}{\partial t} + \frac{\nabla_{\mathbf{r}}\Big[\big(\mathbf{v}_{\mathbf{k}} - \mathbf{E}\times\mathbf{\Omega} - (\mathbf{v}_{\mathbf{k}}\cdot\mathbf{\Omega})\,\mathbf{H}\big)n\Big] + \nabla_{\mathbf{k}}\Big[\big(\mathbf{E} + \mathbf{v}_{\mathbf{k}}\times\mathbf{H} - (\mathbf{E}\cdot\mathbf{H})\,\mathbf{\Omega}\big)n\Big]}{1 - \mathbf{\Omega}\cdot\mathbf{H}} = 0. \tag{12.13}$$

In the absence of the magnetic field, $\mathbf{H} = 0$, the equation simplifies considerably,

$$\frac{\partial n}{\partial t} + \nabla_{\mathbf{r}}\Big[\big(\mathbf{v}_{\mathbf{k}} - \mathbf{E}\times\mathbf{\Omega}\big)n\Big] + \mathbf{E}\cdot\nabla_{\mathbf{k}}n = 0, \tag{12.14}$$

and it boils down to adding the anomalous Berry contribution to the velocity: $\mathbf{v}_{\mathbf{k}} \to \mathbf{v}_{\mathbf{k}} + \mathbf{v}^{\mathrm{a}}_{\mathbf{k}}$, where $\mathbf{v}^{\mathrm{a}}_{\mathbf{k}}(\mathbf{r}) = -\mathbf{E}(\mathbf{r})\times\mathbf{\Omega}(\mathbf{k})$. Notice that the anomalous velocity may depend on location through an inhomogeneous electric field. In this case it needs to be kept under the spatial gradient, as in Eq. (12.14). Also notice that, since $\det\omega_{\alpha\beta} = 1$, in this case $n(\mathbf{r},\mathbf{k},t)$ is conserved and the kinetic term acquires the form of the continuity relation.

In presence of collisions the right-hand side of the previous equations contains a collision integral, $-I^{\mathrm{coll}}[n]/2$, cf. Eq. (11.14). In zeroth approximation, the latter may be taken from the theory with no Berry curvature and magnetic field. However, a more careful consideration shows that the collisions are also affected by the geometric effects. In the context of elastic scattering, they lead to the two distinct effects known as the *skew-scattering* and the *side-jump* scattering mechanisms. The skew-scattering is a result of the asymmetry of the scattering rates $\mathcal{M}_{\mathbf{k},\mathbf{k}'} \neq \mathcal{M}_{\mathbf{k}',\mathbf{k}}$, where $\mathbf{k}$ and $\mathbf{k}'$ are initial and final scattering states. It follows from the solution of the Lipman–Schwinger equation for the scattering amplitude, where the Green function, describing free propagation between the scattering events, includes Berry curvature through its spinor structure (in the orbits space) [228]. The side-jump mechanism originates from the spatial nonlocality of the distribution function and the peculiarity of the spatial Wigner transform in the Bloch wave basis [229]. Finally, the anomalous component of the velocity, $\mathbf{v}^{\mathrm{a}}_{\mathbf{k}}$, may show up in observables. For example, the current, $\mathbf{j} = \sum_{\mathbf{k}}(\mathbf{v}_{\mathbf{k}} + \mathbf{v}^{\mathrm{a}}_{\mathbf{k}})n(\mathbf{r}, \mathbf{k}, t)$, acquires the anomalous velocity contribution. Notice that, since the anomalous velocity is proportional to the electric field, one may use here the equilibrium distribution to calculate the current in the linear response. By this reason the corresponding contribution is known as *intrinsic*.

12.2 Floquet Kinetic Theory

Here we address a non-perturbative treatment of periodic time-dependent Hamiltonians. Consider first a *single-particle* quantum mechanics with a matrix time-dependent Hermitian Hamiltonian $H_{ij}(t)$, where $i,j = 1, 2, \ldots N$, and $H_{ij}(t) = H_{ij}(t + t_0)$ for all i,j. Here t_0 is the period of a drive and $\Omega = 2\pi/t_0$ is the corresponding fundamental frequency. The indexes i,j may represent, for example, space of orbitals and/or lattice sites. Alternatively, one may imagine diagonalizing such a matrix in momentum space and treating i,j as orbitals and quasi-momenta (H_{ij} is often assumed to be diagonal in the latter subspace).

According to the Floquet theorem, one may find N solutions of the time-dependent Schrödinger equation $\mathrm{i}\partial_t\Psi_i(t) = \sum_j^N H_{ij}(t)\Psi_j(t)$ of the form

$$\Psi^\alpha_j(t) = \mathrm{e}^{-\mathrm{i}\epsilon_\alpha t}\phi^\alpha_j(t); \qquad \phi^\alpha_j(t + t_0) = \phi^\alpha_j(t), \qquad \alpha = 1, \ldots, N. \tag{12.15}$$

Since multiplication $\phi^\alpha_j(t) \to \mathrm{e}^{-in\Omega t}\phi^\alpha_j(t)$, with an integer n, does not affect the periodicity of $\phi^\alpha_j(t)$, one needs to limit all the energies to the Floquet zone $0 \leq \epsilon_\alpha < \Omega$, for Eq. (12.15) to be well defined. The periodic functions $\phi^\alpha_j(t)$ obey $\mathrm{i}\partial_t\phi^\alpha_i(t) = \sum_j^N (H_{ij}(t) - \epsilon_\alpha\delta_{ij})\phi^\alpha_j(t)$ and constitute an instantaneous full orthonormal basis

$$\sum_{j=1}^{N} \bar{\phi}^\alpha_j(t)\phi^\beta_j(t) = \delta^{\alpha\beta}; \qquad \sum_{\alpha=1}^{N} \phi^\alpha_i(t)\bar{\phi}^\alpha_j(t) = \delta_{ij}. \tag{12.16}$$

One may expand the periodic functions $\phi_j^\alpha(t)$ into Fourier series as

$$\phi_j^\alpha(t) = \sum_m \mathrm{e}^{-im\Omega t}\, \phi_{j,m}^\alpha, \tag{12.17}$$

where the $2MN$ component vectors $\phi_{j,m}^\alpha$ (here M is a reasonably chosen cutoff in the frequency space) may be found from the matrix eigenvalue equation

$$\sum_{m=-M}^{M-1}\sum_{j=1}^{N} H_{ij}^{n-m}\phi_{j,m}^\alpha = (\epsilon_\alpha + n\Omega)\phi_{i,n}^\alpha, \tag{12.18}$$

where $H_{ij}^{n-m} = \int_0^{t_0}(\mathrm{d}t/t_0)\,\mathrm{e}^{in\Omega t}H_{ij}(t)\,\mathrm{e}^{-im\Omega t}$, and $2MN$ corresponding eigenvalues are $\lambda_{\alpha,n} = \epsilon_\alpha + n\Omega$, with $\alpha = 1,\ldots,N$ and $n = -M,\ldots,M-1$.

We turn now to *many-body* problems. First consider a noninteracting fermion system with the Hamiltonian

$$\hat{H}_0(t) = \sum_{ij}^{N} \hat{c}_i^\dagger H_{ij}(t)\hat{c}_j.$$

The corresponding Keldysh action is

$$S = \int \mathrm{d}t \sum_{ij}^{N}\left[\bar{\psi}_{1,i}(\mathrm{i}\partial_t^{\mathrm{R}}\delta_{ij} - H_{ij}(t))\psi_{1,j} + \bar{\psi}_{2,i}(\mathrm{i}\partial_t^{\mathrm{A}}\delta_{ij} - H_{ij}(t))\psi_{2,j}\right], \tag{12.19}$$

where $\psi_{a,j}(t)$ are Grassmann fields and superscripts R(A) specify retarded (advanced) regularizations of the corresponding discrete time derivatives, $a = 1,2$. The retarded/advanced Green functions are found as solutions of the following equations: $\sum_j^N[\mathrm{i}\partial_t^{\mathrm{R,A}}\delta_{ij} - H_{ij}(t)]G_{jk}^{\mathrm{R,A}}(t,t') = \delta_{ik}\delta(t-t')$ and are given by

$$G_{jk}^{\mathrm{R,A}}(t,t') = \mp\mathrm{i}\theta(\pm(t-t'))\sum_{\alpha=1}^{N}\phi_j^\alpha(t)\bar{\phi}_k^\alpha(t')\,\mathrm{e}^{-\mathrm{i}\epsilon_\alpha(t-t')}. \tag{12.20}$$

This may be directly checked using (12.15), (12.16). The Keldysh component is less obvious. To understand it, notice that the action (12.19) may be identically written in the Floquet-diagonal form

$$S = \int \mathrm{d}t \sum_{\alpha=1}^{N}\left[\bar{\chi}_1^\alpha(\mathrm{i}\partial_t^{\mathrm{R}} - \epsilon_\alpha)\chi_1^\alpha + \bar{\chi}_2^\alpha(\mathrm{i}\partial_t^{\mathrm{A}} - \epsilon_\alpha)\chi_2^\alpha\right], \tag{12.21}$$

where the new Grassmann fields are $\chi_a^\alpha(t) = \sum_j \bar{\phi}_j^\alpha(t)\psi_{a,j}(t)$, or conversely $\psi_{a,j}(t) = \sum_\alpha \phi_j^\alpha(t)\chi_a^\alpha(t)$; correspondingly $\bar{\chi}_a^\alpha(t) = \sum_j \bar{\psi}_{a,j}(t)\phi_j^\alpha(t)$ and also $\bar{\psi}_{a,j}(t) = \sum_\alpha \bar{\chi}_a^\alpha(t)\bar{\phi}_j^\alpha(t)$. This action dictates, for example,

$$G_{\alpha\beta}^{\mathrm{R}}(t,t') = -\mathrm{i}\langle\chi_1^\alpha(t)\bar{\chi}_1^\beta(t')\rangle = -\mathrm{i}\theta(t-t')\,\mathrm{e}^{-\mathrm{i}\epsilon_\alpha(t-t')}\delta_{\alpha\beta} = G_\alpha^{\mathrm{R}}(t-t'), \tag{12.22}$$

and similarly for the advanced function. Transforming to the original basis, this leads back to (12.20), as it should be, of course. It is convenient to define the Keldysh component of the Green function directly in the Floquet diagonal basis as $G^{\rm K}_{\alpha\beta}(t,t') = -{\rm i}\langle\chi_1^\alpha(t)\bar\chi_2^\beta(t')\rangle$. Its form is not determined by the continuum action (12.21), but rather by the implicit regularization. (Notice that the latter does not need to be Floquet-diagonal and thus the Keldysh component is, in general, a matrix in Floquet indexes α, β, despite the diagonal form of the continuum action (12.21).) The strategy is to use the implicit Keldysh component to define the distribution function $F_{\alpha\beta}(t,t')$ as (cf. Eq. (10.38))

$$G^{\rm K}_{\alpha\beta}(t,t') = \int {\rm d}t''\Big[G^{\rm R}_\alpha(t-t'')F_{\alpha\beta}(t'',t') - F_{\alpha\beta}(t,t'')G^{\rm A}_\beta(t''-t')\Big], \tag{12.23}$$

and to derive a kinetic equation governing the evolution of $F_{\alpha\beta}(t,t')$. Assuming one can find a proper solution of such a kinetic equation, $G^{\rm K}_{\alpha\beta}(t,t')$ is determined by Eqs. (12.22) and (12.23).

The left-hand side of the kinetic equation is given by the commutator, Eq. (11.3), $-[({\rm i}\partial_t - \epsilon_\alpha) \,\overset{\circ}{,}\, F]$. One thus arrives at the Floquet kinetic equation of the form

$$\Big[\partial_t + \partial_{t'} - {\rm i}(\epsilon_\alpha - \epsilon_\beta)\Big]F_{\alpha\beta}(t,t') = {\rm i}\Sigma^{\rm K} - {\rm i}\big(\Sigma^{\rm R}\circ F - F\circ\Sigma^{\rm A}\big) = I^{\rm coll}[F]. \tag{12.24}$$

In the absence of collisions, $I^{\rm coll}[F] = 0$, the time dependence of the distribution function is

$$F_{\alpha\beta}(t,t') = F_{\alpha\beta}(t-t')\,{\rm e}^{{\rm i}(\epsilon_\alpha-\epsilon_\beta)(t+t')/2}. \tag{12.25}$$

This shows that the diagonal elements, $F_{\alpha\alpha} = F_\alpha(t-t')$, are independent of the central time $(t+t')/2$, while the off-diagonal elements, $F_{\alpha\beta}$, oscillate with the frequency $\epsilon_\alpha - \epsilon_\beta$ as functions of the central time. Consequently, one may argue that upon a coarse-grained time-averaging procedure the latter disappear. We thus focus primarily on the diagonal elements F_α. Passing to the Wigner transform variables $t - t' \to \epsilon$ and $(t+t')/2 \to t$, one expects that $F_\alpha = F_\alpha(t,\epsilon)$ are slow functions of the central time t. Their slow (on the scale of t_0) dependence on the central time is induced by collisions, represented by the right-hand side of (12.24).

To proceed we need to introduce specific collision mechanisms, which determine components of the self-energy matrix $\hat\Sigma_{\alpha\beta}(t,t')$. Let's consider tunneling of fermions to reservoir states, labelled as k, described by a quadratic Hamiltonian $\hat H = \hat H_0(t) + \hat H_1$, where the tunneling and the reservoir part, $\hat H_1$, is time-independent and given by

$$\hat H_1 = \sum_{jk}\big[\hat c_j^\dagger W_{jk}\,\hat d_k + \hat d_k^\dagger W^*_{jk}\,\hat c_j\big] + \sum_k \varepsilon_k \hat d_k^\dagger \hat d_k, \tag{12.26}$$

where W_{jk} are time-independent tunneling amplitudes in the original basis. The tunneling part of the Keldysh action is

$$S_W = -\int dt \sum_{jk;ab} \left[W_{jk}\, \bar{\psi}_{a,j} \hat{\gamma}^{\rm cl}_{ab}\, d_{b,k} + W^*_{jk}\, \bar{d}_{a,k} \hat{\gamma}^{\rm cl}_{ab}\, \psi_{b,j} \right]$$

$$= -\int dt \sum_{\alpha\beta;ab} \left[W_{\alpha,k}(t)\, \bar{\chi}^{\alpha}_{a} \hat{\gamma}^{\rm cl}_{ab}\, d_{b,k} + W^*_{\alpha,k}(t)\, \bar{d}_{a,k} \hat{\gamma}^{\rm cl}_{ab}\, \chi^{\alpha}_{b} \right], \qquad (12.27)$$

where $d_{a,k}(t)$ (without hat) are Grassmann fields, $\hat{\gamma}^{\rm cl}_{ab} = \delta_{ab}$, and

$$W_{\alpha,k}(t) = \sum_{j=1}^{N} \bar{\phi}^{\alpha}_{j}(t) W_{jk} = \sum_{m} W^{m}_{\alpha,k} e^{im\Omega t}; \qquad W^{m}_{\alpha,k} = \sum_{j=1}^{N} \bar{\phi}^{\alpha}_{j,m} W_{jk}. \qquad (12.28)$$

Notice that, while the bare Floquet action (12.21) is time independent, the tunneling action acquires a periodic time dependence, $W_{\alpha,k}(t + t_0) = W_{\alpha,k}(t)$, through the Floquet functions, $\phi^{\alpha}(t)$.

Expanding e^{iS_W} to the second order and performing Gaussian Grassmann integration over $d_{a,k}$ fields, with the action $S_d = \int dt \sum_{a,k} \bar{d}_{a,k}(i\partial_t - \varepsilon_k) d_{a,k}$, one finds for the self-energy

$$\hat{\Sigma}_{\alpha\beta}(t,t') = \sum_{k} W_{\alpha,k}(t) \hat{G}_k(t-t') W^*_{\beta,k}(t'), \qquad (12.29)$$

where $\hat{G}_k(t-t')$ is the matrix Keldysh Green function of the reservoir. In the diagonal approximation, where $F_{\alpha\beta} \propto \delta_{\alpha\beta}$ and $\hat{\Sigma}_{\alpha\beta} \propto \delta_{\alpha\beta}$, the self-energy contains

$$W_{\alpha,k}(t) W^*_{\alpha,k}(t') = \sum_{n,m} W^{n}_{\alpha,k} W^{*m}_{\alpha,k}\, e^{i\Omega(nt - mt')} \approx \sum_{n} |W^{n}_{\alpha,k}|^2\, e^{in\Omega(t-t')},$$

where we have omitted all terms $\propto e^{i(n-m)\Omega(t+t')/2 + \frac{i}{2}(n+m)\Omega(t-t')}$, with $n \neq m$, since they are eliminated upon time-averaging over the central time $(t+t')/2$. Therefore in the diagonal approximation the self-energy is

$$\hat{\Sigma}_{\alpha\alpha}(\epsilon) \approx \sum_{k,n} |W^{n}_{\alpha,k}|^2 \hat{G}_k(\epsilon + n\Omega), \qquad (12.30)$$

and

$$\hat{G}_k(\epsilon) = \begin{pmatrix} (\epsilon - \varepsilon_k + i0)^{-1} & -2\pi i F_d(\varepsilon_k) \delta(\epsilon - \varepsilon_k) \\ 0 & (\epsilon - \varepsilon_k - i0)^{-1} \end{pmatrix}, \qquad (12.31)$$

where $F_d(\varepsilon_k)$ is fermion distribution function of the reservoir. Substituting this into (12.24), one finds for the Wigner function $F_\alpha(t, \epsilon_\alpha) = F_\alpha(t)$:

$$\partial_t F_\alpha(t) = 2\pi \sum_{k,n} |W^{n}_{\alpha,k}|^2 \left[F_d(\varepsilon_k) - F_\alpha(t) \right] \delta(\epsilon_\alpha + n\Omega - \varepsilon_k). \qquad (12.32)$$

This equation governs the evolution of $F_\alpha = 1 - 2n_\alpha$, where n_α are Floquet occupation numbers. It may be dubbed the *Floquet golden rule*. Its main difference from the conventional one is that, despite tunneling being time-independent, the energy conservation holds up to an integer of the driving frequency Ω. This is a consequence of the dynamic nature of the Floquet states. The corresponding matrix elements are proportional to Fourier harmonics of the Floquet functions, Eq. (12.28). Equation (12.32) signals the absence of the detailed balance, since a stationary state can't be achieved with $F_\alpha = F_d(\varepsilon_k)$. Indeed, there may be multiple reservoir states with energies $\varepsilon_k = \epsilon_\alpha + n\Omega$, which satisfy the energy conservation for a fixed α.

One may consider other collision mechanisms, such as electron–phonon, or electron–electron interactions [230]. The corresponding kinetic equations resemble the conventional ones with the energy conservation delta-functions allowing for $n\Omega$ mismatch. The corresponding matrix elements are decorated with the Fourier components of the Floquet functions $\phi^\alpha_{j,n}$. Various aspects of Floquet kinetics are reflected in [230, 232, 233, 234].

One may wonder about the validity of the diagonal approximation and the role of the off-diagonal elements $F_{\alpha\beta}(t,\epsilon)$. This is especially relevant in the presence of degeneracies in the Floquet spectrum, that is, states with $\epsilon_\alpha = \epsilon_\beta$. In this case the argument to neglect the off-diagonal elements due to their oscillatory central time, t, dependence $\propto e^{i(\epsilon_\alpha - \epsilon_\beta)t}$ is not applicable. In my opinion, the role and consequences of the off-diagonal coherences $F_{\alpha\beta}$ in Floquet kinetics deserve further investigation.[3]

12.3 Kinetic Equation as a Stationary Point of the $\hat{G}$-$\hat{\Sigma}$ Action

Here we provide an alternative view on the kinetic equation. The latter may be thought of as a stationary point condition for an effective action formulated in terms of bilocal matrix fields $\hat{G}(x,x')$ and $\hat{\Sigma}(x,x')$. This perspective will allow us to discuss fluctuations phenomena in the framework of the so-called Boltzmann–Langevin equation, Section 12.4, and will prove to be useful for derivation of kinetic equations in disordered systems.

To illustrate the approach we consider spinless fermions interacting with a scalar Gaussian degree of freedom, $\varphi(x)$. The latter may represent, for example, phonons, or electron–electron interactions treated within the RPA framework. The corresponding action is given by

$$S = \int dx \left[\bar{\psi}_a(x) \left(\hat{G}_0^{-1}\right)^{ab} \psi_b(x) - \bar{\psi}_a(x)\hat{\gamma}^\alpha_{ab}\psi_b(x)\,\varphi^\alpha(x)\right], \tag{12.33}$$

[3] I am grateful to Mark Rudner for useful discussions of this issue.

where $a, b = 1, 2$ and $\alpha = \mathrm{cl}, \mathrm{q}$. The Gaussian correlator of the scalar field is $\langle \varphi^\alpha(x)\varphi^\beta(x')\rangle = \frac{\mathrm{i}}{2} U^{\alpha\beta}(x, x')$. Performing Gaussian integration over $\varphi(x)$, one finds the effective nonlocal interaction action

$$S_{\mathrm{int}} = -\frac{1}{4}\iint \mathrm{d}x\mathrm{d}x'\, \bar{\psi}_a(x)\hat{\gamma}^\alpha_{ab}\psi_b(x)\, U^{\alpha\beta}(x, x')\, \bar{\psi}_c(x')\hat{\gamma}^\beta_{cd}\psi_d(x'). \tag{12.34}$$

Our goal now is to rewrite this action in terms of the bilocal fluctuating field $\hat{G}_{bc}(x, x') = -\mathrm{i}\,\psi_b(x)\bar{\psi}_c(x')$ (notice the absence of the expectation value symbol, $\langle \ldots \rangle$). To this end we introduce an identity using functional delta-function

$$\begin{aligned} 1 &= \int \mathbf{D}[\hat{G}]\,\delta\Big(\hat{G}_{bc}(x, x') + \mathrm{i}\,\psi_b(x)\bar{\psi}_c(x')\Big) \\ &= \int \mathbf{D}[\hat{G}, \hat{\Sigma}]\,\mathrm{e}^{\iint \mathrm{d}x\mathrm{d}x'\, \hat{\Sigma}_{cb}(x', x)\left(\hat{G}_{bc}(x, x') + \mathrm{i}\psi_b(x)\bar{\psi}_c(x')\right)}, \end{aligned} \tag{12.35}$$

where another independent fluctuating field $\hat{\Sigma}_{cb}(x', x)$ is introduced as a Lagrange multiplier to enforce the delta-function. The action then takes the form (here we employ short-hand matrix notations)

$$S = \bar{\psi}\left(\hat{G}_0^{-1} - \hat{\Sigma}\right)\psi - \mathrm{i}\,\hat{\Sigma} \circ \hat{G} - \frac{1}{4}\left\{\hat{\gamma}^\alpha \hat{G}\hat{\gamma}^\beta \hat{G}\right\} U^{\alpha\beta}, \tag{12.36}$$

where the last term is the interaction action (12.34), written in terms of the bilocal field $\hat{G}$. Finally, one can perform the Gaussian integration over the Grassmann fields, ψ and $\bar{\psi}$, to arrive at the so-called $\hat{G}$-$\hat{\Sigma}$ action:

$$\mathrm{i}S[\hat{G}, \hat{\Sigma}] = \mathrm{Tr}\ln\left(\hat{G}_0^{-1} - \hat{\Sigma}\right) + \hat{\Sigma} \circ \hat{G} - \frac{\mathrm{i}}{4}\left\{\hat{G}\hat{\gamma}^\beta \hat{G}\hat{\gamma}^\alpha\right\} U^{\alpha\beta}. \tag{12.37}$$

So far we have not done any approximations. Thus, in principle, this action contains the full exact information about all correlation functions of the interacting model.

We now look for $\hat{G}$ and $\hat{\Sigma}$ field configurations, which stabilize the $\hat{G}$-$\hat{\Sigma}$ action. The corresponding variational derivatives yield

$$\frac{\delta \mathrm{i}S}{\delta \hat{\Sigma}} = -\left(\hat{G}_0^{-1} - \hat{\Sigma}\right)^{-1} + \hat{G}; \qquad \frac{\delta \mathrm{i}S}{\delta \hat{G}} = \hat{\Sigma} - \frac{\mathrm{i}}{2}\,\hat{\gamma}^\beta \hat{G}\hat{\gamma}^\alpha\, U^{\alpha\beta}. \tag{12.38}$$

Requiring them to be zero and formally multiplying the first by $(\hat{G}_0^{-1} - \hat{\Sigma})$, one obtains two stationary point equations for the bilocal matrix fields:

$$\left(\hat{G}_0^{-1} - \hat{\Sigma}\right) \circ \hat{G} = \hat{1}; \qquad \hat{\Sigma} = \frac{\mathrm{i}}{2}\,\hat{\gamma}^\beta \hat{G}\hat{\gamma}^\alpha\, U^{\alpha\beta}. \tag{12.39}$$

One immediately recognizes the Dyson equation (11.1) along with the expression (11.4) for the fermionic self-energy; see also Fig. 11.1(a). Therefore, at the stationary point the fluctuating bilocal fields $\hat{\Sigma}$ and $\hat{G}$ are given by the (one loop) self-energy and the corresponding dressed Green function. As a result, both the kinetic equation (11.7) and the collision integral (11.8) are contained in the stationarity condition for the $\hat{G}$-$\hat{\Sigma}$ action.

To make this connection even more apparent we notice that the solutions of the stationary point equations, satisfying the causality structure, may be written as (cf. Eq. (10.34))

$$\hat{G} = \begin{pmatrix} G^{\mathrm{R}} & G^{\mathrm{K}} \\ 0 & G^{\mathrm{A}} \end{pmatrix} = \hat{\mathcal{U}}^{-1} \circ \begin{pmatrix} G^{\mathrm{R}} & 0 \\ 0 & G^{\mathrm{A}} \end{pmatrix} \circ \hat{\mathcal{U}}, \qquad \hat{\mathcal{U}} = \begin{pmatrix} 1 & F \\ 0 & 1 \end{pmatrix}, \tag{12.40}$$

where F, defined through $G^{\mathrm{K}} = G^{\mathrm{R}} \circ F - F \circ G^{\mathrm{A}}$, is a solution of the stationary point kinetic equation. Similarly

$$\hat{\Sigma} = \begin{pmatrix} \Sigma^{\mathrm{R}} & \Sigma^{\mathrm{K}} \\ 0 & \Sigma^{\mathrm{A}} \end{pmatrix} = \hat{\mathcal{U}}^{-1} \circ \begin{pmatrix} \Sigma^{\mathrm{R}} & \Sigma^{\mathrm{K}} - \Sigma^{\mathrm{R}} \circ F + F \circ \Sigma^{\mathrm{A}} \\ 0 & \Sigma^{\mathrm{A}} \end{pmatrix} \circ \hat{\mathcal{U}}. \tag{12.41}$$

The off-diagonal $(1,2)$ element constitutes the collision integral (cf. Eq. (11.3)). It vanishes in equilibrium.

Motivated by this form of the saddle-point solutions, one may introduce rotated *fluctuating* fields as

$$\hat{\hat{G}} = \hat{\mathcal{U}} \circ \hat{G} \circ \hat{\mathcal{U}}^{-1}; \qquad \hat{\hat{\Sigma}} = \hat{\mathcal{U}} \circ \hat{\Sigma} \circ \hat{\mathcal{U}}^{-1}, \tag{12.42}$$

so the new fields $\hat{\hat{G}}$ and $\hat{\hat{\Sigma}}$ are diagonal at the equilibrium stationary solution. In terms of these new rotated fields the $\hat{G}$-$\hat{\Sigma}$ action acquires the form

$$\mathrm{i}S[\hat{\hat{G}}, \hat{\hat{\Sigma}}] = \mathrm{Tr} \ln \left(\hat{G}_0^{-1} + \hat{\mathcal{U}}[\hat{G}_0^{-1}, \hat{\mathcal{U}}^{-1}] - \hat{\hat{\Sigma}} \right) + \hat{\hat{\Sigma}} \circ \hat{\hat{G}} - \frac{\mathrm{i}}{4} \left\{ \hat{\hat{G}} \hat{\Gamma}^{\beta}_{12} \hat{\hat{G}} \hat{\Gamma}^{\alpha}_{21} \right\} U^{\alpha\beta}, \tag{12.43}$$

where

$$\hat{\Gamma}^{\beta}_{12} = \hat{\mathcal{U}}_1 \circ \hat{\gamma}^{\beta} \circ \hat{\mathcal{U}}_2^{-1}; \qquad \hat{\Gamma}^{\mathrm{cl}}_{12} = \begin{pmatrix} 1 & F_1 - F_2 \\ 0 & 1 \end{pmatrix}; \qquad \hat{\Gamma}^{\mathrm{q}}_{12} = \begin{pmatrix} F_1 & 1 - F_2 F_1 \\ -1 & -F_2 \end{pmatrix}. \tag{12.44}$$

Here $F_1 = \tilde{F}(\mathbf{k}_1)$ and $F_2 = \tilde{F}(\mathbf{k}_2)$ and in the collision integral (11.8) $\mathbf{k}_1 = \mathbf{k}$, $\mathbf{k}_2 = \mathbf{k} + \mathbf{q}$. Notice that matrix $\hat{\mathcal{U}}[\hat{G}_0^{-1}, \hat{\mathcal{U}}^{-1}]$ has only the $(1,2)$ nonzero matrix element given by $[G_0^{-1}, F]$. Thus the $(1,2)$ matrix element of $\hat{\mathcal{U}}[\hat{G}_0^{-1}, \hat{\mathcal{U}}^{-1}] - \hat{\hat{\Sigma}}$ is given by the kinetic equation (11.3). This observation will be utilized in the next section.

12.4 Boltzmann–Langevin Kinetic Equation

As explained in the previous section, the Boltzmann equation may be viewed as a part of the stationary point conditions for the formally exact $\hat{G}$-$\hat{\Sigma}$ action. These conditions, (12.39), include Dyson equation and the expression for the fermionic self-energy. While the former may be considered as exact, the latter is certainly an approximation that treats the self-energy as the simplest self-consistent diagram, Fig. 11.1(a). Other diagrams are missed in the saddle-point treatment. Our goal thus is to go beyond the saddle-point approximation and to include *fluctuations* of $\hat{G}$ and $\hat{\Sigma}$ fields around their stationary values. We will restrict ourselves to Gaussian fluctuations only. This results in the so-called Boltzmann–Langevin kinetic equation [235, 236, 237].

To proceed it is convenient to employ the rotated form (12.43) of the $\hat{G}$-$\hat{\Sigma}$ action. The reason is that in the rotated form the kinetic equation appears solely as the $(1,2)$ element of the matrix saddle-point equations (12.39). The corresponding equation originates from the variation over the conjugated degree of freedom, which is the $(2,1)$ element of the fluctuating $\hat{\tilde{G}}$ and $\hat{\tilde{\Sigma}}$ matrices. We thus adopt the following parametrization (cf. Eqs. (12.40), (12.41)):

$$\hat{\tilde{G}} = \begin{pmatrix} G^{\mathrm{R}} & 0 \\ \delta_{\tilde{G}} & G^{\mathrm{A}} \end{pmatrix}; \qquad \hat{\tilde{\Sigma}} = \begin{pmatrix} \Sigma^{\mathrm{R}} & \Sigma^{\mathrm{K}} - \Sigma^{\mathrm{R}} \circ F + F \circ \Sigma^{\mathrm{A}} \\ \delta_{\tilde{\Sigma}} & \Sigma^{\mathrm{A}} \end{pmatrix}. \tag{12.45}$$

Here $\delta_{\tilde{G}}$ and $\delta_{\tilde{\Sigma}}$ are quantum fluctuating components of the bilocal fields. We now substitute this parametrization into the action (12.43) and expand it to the second order in $\delta_{\tilde{G}}$, $\delta_{\tilde{\Sigma}}$ quantum fluctuations (and also in the kinetic term $[\hat{G}_0^{-1}, F] + \Sigma^{\mathrm{K}} - (\Sigma^{\mathrm{R}} \circ F - F \circ \Sigma^{\mathrm{A}})$). This leads to

$$\begin{aligned} iS[\delta_{\tilde{G}}, \delta_{\tilde{\Sigma}}] = {} & \delta_{\tilde{\Sigma}} \circ G^{\mathrm{R}} \circ \Big\{ [\hat{G}_0^{-1}, F] + \Sigma^{\mathrm{K}} - (\Sigma^{\mathrm{R}} \circ F - F \circ \Sigma^{\mathrm{A}}) \Big\} \circ G^{\mathrm{A}} \\ & + \delta_{\tilde{G}} \circ \Big\{ \Sigma^{\mathrm{K}} - (\Sigma^{\mathrm{R}} \circ F - F \circ \Sigma^{\mathrm{A}}) + \frac{i}{2} \Big(\hat{\Gamma}^{\beta} \hat{\tilde{G}} \hat{\Gamma}^{\alpha} \Big)_{12} U^{\alpha\beta} \Big\} \\ & - \frac{i}{4}\, \delta_{\tilde{G}}(1)\, [F_1 - F_2] \Big\{ [F_2 - F_1]\, U^{\mathrm{K}}_{2-1} + [F_2 F_1 - 1] \big[U^{\mathrm{R}}_{2-1} - U^{\mathrm{A}}_{2-1} \big] \Big\} \delta_{\tilde{G}}(2), \end{aligned} \tag{12.46}$$

where $G^{\mathrm{R,A}} = (G_0^{-1} - \Sigma^{\mathrm{R,A}})^{-1}$ and $\hat{\tilde{G}} = \mathrm{diag}\{G^{\mathrm{R}}, G^{\mathrm{A}}\}$. Notations here are somewhat inconsistent: while the first two lines are written in the bilocal matrix form, the last line assumes the Wigner transform with $\delta_{\tilde{G}}(l) = \delta_{\tilde{G}}(\mathbf{r}, t, \mathbf{k}_l, \epsilon_l)$, where $l = 1, 2$ and implicit integrations over $\mathbf{r}, t, \mathbf{k}_1, \mathbf{k}_2, \epsilon_1, \epsilon_2$.

The first two lines in Eq. (12.46) are already familiar linear variations. Demanding that the expressions in the curly brackets vanish, one recovers the kinetic equation (11.3) along with the specific form of the collision integral, Eq. (11.8). We now focus on the last line, quadratic in $\delta_{\tilde{G}}$ fluctuations. The way to handle it is to split it with the Hubbard–Stratonovich auxiliary field $\xi_F(\mathbf{r}, t, \mathbf{k}, \epsilon)$ as $e^{i\delta_{\tilde{G}}(1)\{\ldots\}\delta_{\tilde{G}}(2)} = \int \mathbf{D}[\xi_F]\, \mathcal{P}(\xi_F)\, e^{-\delta_{\tilde{G}}(1)\xi_F(1)}$, where $\mathcal{P}(\xi_F)$ is a Gaussian stochastic measure such that $\langle \xi_F(1) \rangle = 0$ and $\langle \xi_F(1)\xi_F(2) \rangle \propto \{\ldots\}$. The term $\delta_{\tilde{G}}(1)\xi_F(1)$ in the exponent may now be incorporated into the second line in Eq. (12.46). This leads to a stochastic contribution to the self-energy. Notice that there is no stochastic correction to the Dyson equation, since terms like $\delta_{\tilde{G}}\delta_{\tilde{\Sigma}}$ and $\delta^2_{\tilde{\Sigma}}$ are absent in Eq. (12.46). This reflects the formal exactness of the Dyson equation, as opposed to an approximate nature of the specific form (11.4) of the self-energy.

The previous considerations lead to a stochastic contribution to the collision integral, $\xi_F(\mathbf{r}, t, \mathbf{k}, \epsilon)$. One should be mindful, however, that this object, originating from the bilocal field $\delta_{\tilde{G}}$, depends on four arguments. On the other hand, the distribution function is the on-shell object, which depends only on the three variables, $\tilde{F}(\mathbf{r}, t, \mathbf{k})$. Following the procedure of Section 6.7, we thus introduce $\tilde{\xi}_F(\mathbf{r}, t, \mathbf{k}) = \int d\epsilon\, \xi_F(\mathbf{r}, t, \mathbf{k}, \epsilon)\, \delta(\epsilon - \epsilon_{\mathbf{k}})$ – the on-shell part of the stochastic source. Then the kinetic equation acquires the Boltzmann–Langevin form:

$$\left[\partial_t + \mathbf{v}_{\mathbf{k}}\nabla_{\mathbf{r}} - (\nabla_{\mathbf{r}}V^{\mathrm{cl}})\nabla_{\mathbf{k}}\right]\tilde{F}(\mathbf{r}, t, \mathbf{k}) = I^{\mathrm{coll}}[\tilde{F}] + \tilde{\xi}_F(\mathbf{r}, t, \mathbf{k}). \tag{12.47}$$

The Gaussian correlator of the stochastic source may be read out of the last line in Eq. (12.46) and is given by

$$\langle \tilde{\xi}_F(\mathbf{r}_1, t_1, \mathbf{k}_1)\tilde{\xi}_F(\mathbf{r}_2, t_2, \mathbf{k}_2)\rangle = \delta(\mathbf{r}_1 - \mathbf{r}_2)\delta(t_1 - t_2)\left[\tilde{F}(\mathbf{k}_2) - \tilde{F}(\mathbf{k}_1)\right] I(\mathbf{k}_1, \mathbf{k}_2), \tag{12.48}$$

where

$$\begin{aligned} I(\mathbf{k}_1, \mathbf{k}_2) &= \tfrac{\mathrm{i}}{2}\{[\tilde{F}(\mathbf{k}_2) - \tilde{F}(\mathbf{k}_1)]U^{\mathrm{K}}_{2-1} + [\tilde{F}(\mathbf{k}_2)\tilde{F}(\mathbf{k}_1) - 1][U^{\mathrm{R}}_{2-1} - U^{\mathrm{A}}_{2-1}]\} \\ &= \tfrac{\mathrm{i}}{2}\left\{[\tilde{F}(\mathbf{k}_2) - \tilde{F}(\mathbf{k}_1)]\tilde{B}(\mathbf{k}_2 - \mathbf{k}_1) + [\tilde{F}(\mathbf{k}_2)\tilde{F}(\mathbf{k}_1) - 1]\right\}[U^{\mathrm{R}}_{2-1} - U^{\mathrm{A}}_{2-1}], \end{aligned} \tag{12.49}$$

and $\hat{U}_{2-1} = \hat{U}(\mathbf{k}_2 - \mathbf{k}_1, \epsilon_{\mathbf{k}_2} - \epsilon_{\mathbf{k}_1})$ is the bosonic propagator restricted to be on shell. Here $U^{\mathrm{K}} = U^{\mathrm{R}} \circ B - B \circ U^{\mathrm{A}}$ and $\tilde{B}(\mathbf{q})$ is the on-shell bosonic distribution. The factor $U^{\mathrm{R}}_{2-1} - U^{\mathrm{A}}_{2-1}$ enforces the energy conservation: $\omega_{\mathbf{k}_2 - \mathbf{k}_1} = \epsilon_{\mathbf{k}_2} - \epsilon_{\mathbf{k}_1}$ and thus in equilibrium $I(\mathbf{k}_1, \mathbf{k}_2) = 0$. Also notice that $I(\mathbf{k}_1, \mathbf{k}_2)$ is odd upon interchange of the arguments, $I(\mathbf{k}_1, \mathbf{k}_2) = -I(\mathbf{k}_2, \mathbf{k}_1)$. This makes $\langle \xi_F(\mathbf{k}_1)\xi_F(\mathbf{k}_2)\rangle$ even, as it should be. The expression (12.49) should be compared with the collision integral (11.8), which in this notation acquires the form

$$I^{\mathrm{coll}}[\tilde{F}] = \sum_{\mathbf{k}_2} I(\mathbf{k}, \mathbf{k}_2). \tag{12.50}$$

Notice that in equilibrium both the collision integral and the noise correlator vanish. Therefore the equilibrium distribution, $\tilde{F}(\mathbf{k}) = \tanh \epsilon_{\mathbf{k}}/2T$, is not affected by the Langevin stochasticity.

The physics behind the Langevin term is traced back to the random nature of the inelastic collisions. Since the noise correlator (12.48), (12.49) itself depends on the instantaneous distribution function, $\tilde{F}(\mathbf{r}, t, \mathbf{k})$, the Boltzmann–Langevin equation (12.47) constitutes an example of a *multiplicative* stochastic process. It therefore must be associated with a certain regularization procedure. Being derived from the Keldysh-type action with the causality structure, it inherits the retarded Ito regularization, see Sections 4.3 and 4.6.

To illustrate how the stochastic part of the distribution function manifests itself, one needs to take a closer look at the source fields, which have to be added to the $\hat{G}$-$\hat{\Sigma}$ action in order to generate observables. As an example we consider the quantum component of the vector potential, $\hat{\mathbf{A}}^{\mathrm{q}}(x) = \hat{\gamma}^{\mathrm{q}}\mathbf{A}^{\mathrm{q}}(x)$. It enters the bare inverse Green function as $\hat{G}_0^{-1} \to \hat{G}_0^{-1} - \mathbf{v}_{\mathbf{k}}\hat{\mathbf{A}}^{\mathrm{q}}(x)$, where $\mathbf{v}_{\mathbf{k}} = \partial_{\mathbf{k}}\epsilon_{\mathbf{k}}$ is velocity. The trace logarithm part of the action acquires the form $\mathrm{Tr}\ln\left(\hat{G}^{-1} - \mathbf{v}_{\mathbf{k}}\hat{\mathbf{A}}^{\mathrm{q}}\right) = \mathrm{Tr}\ln\left(1 - \hat{G}\,\mathbf{v}_{\mathbf{k}}\hat{\mathbf{A}}^{\mathrm{q}}\right)$, where $\hat{G}$ is the dressed Green function. One can now expand this expression to the second order in the source field $\mathbf{A}^{\mathrm{q}}(x)$ to find

$$\mathrm{i}S[\hat{G},\hat{\Sigma},\mathbf{A}^{\mathrm{q}}] = \mathrm{i}S[\hat{G},\hat{\Sigma}] - 2\mathrm{i}\mathbf{A}^{\mathrm{q}}(x)\mathbf{J}(x) - 2\mathbf{A}^{\mathrm{q}}(x)\mathcal{K}(x,x')\mathbf{A}^{\mathrm{q}}(x'), \tag{12.51}$$

where $\mathbf{J}(x) = -\frac{\mathrm{i}}{2}\mathrm{Tr}\{\hat{G}_{x,x'}\hat{\gamma}^{\mathrm{q}}\mathbf{v}_{\mathbf{k}}\}_{x'=x}$ and $\mathcal{K}(x,x') = \frac{1}{4}\mathrm{Tr}\{\hat{G}_{x,x'}\hat{\gamma}^{\mathrm{q}}\mathbf{v}_{\mathbf{k}}\hat{G}_{x',x}\hat{\gamma}^{\mathrm{q}}\mathbf{v}_{\mathbf{k}}\}$ is the rank-two tensor. At the stationary point, they acquire the form

$$\mathbf{J}(x) = -\frac{\mathrm{i}}{2}G^{\mathrm{K}}_{x,x}\mathbf{v}_{\mathbf{k}} = \frac{1}{2}\sum_{\mathbf{k}}\tilde{F}(x,\mathbf{k})\mathbf{v}_{\mathbf{k}}; \tag{12.52}$$

$$\mathcal{K}(x,x') \approx \delta(x-x')\sum_{\mathbf{k}}\left(1 - \tilde{F}^2(x,\mathbf{k})\right)\mathbf{v}_{\mathbf{k}}\mathbf{v}_{\mathbf{k}}\tau_{\mathrm{sc}}, \tag{12.53}$$

where $\tau_{\mathrm{sc}}(\mathbf{k}) = (2\mathrm{Im}\,\Sigma^{\mathrm{R}}_{\mathbf{k}})^{-1} = -\frac{1}{4}\int\frac{d\epsilon}{2\pi}(G^{\mathrm{R}}_{\epsilon\mathbf{k}} - G^{\mathrm{A}}_{\epsilon\mathbf{k}})^2$ is the mean free scattering time. The approximate locality of the tensor $\mathcal{K}(x,x')$ is based on the assumption that $\mathbf{A}^{\mathrm{q}}(x)$ changes slowly in time on the scale of τ_{sc} and in space on the scale of the mean free path $v_F\tau_{\mathrm{sc}}$.

One can evaluate the average current $\langle\mathbf{J}\rangle = \frac{\mathrm{i}}{2}\delta Z[\mathbf{A}]/\delta\mathbf{A}^{\mathrm{q}}(x)|_{\mathbf{A}^{\mathrm{q}}=0} = \mathbf{J}(x)$, Eq. (12.52), where the distribution function is found as a solution of the kinetic equation. To have a nonzero average current, the latter must include an external electric field. The stochastic part of the kinetic equation plays no role here, since it averages out to zero. Let's now look at the current *noise*, defined as the second cumulant,

$$\langle S\rangle = -\frac{1}{4}\left.\frac{\delta^2 Z[\mathbf{A}]}{\delta\mathbf{A}^{\mathrm{q}}(x)\delta\mathbf{A}^{\mathrm{q}}(x')}\right|_{\mathbf{A}^{\mathrm{q}}=0} - \langle\mathbf{J}\rangle^2 = \langle\mathcal{K}(x,x')\rangle + \langle\mathbf{J}(x)\mathbf{J}(x')\rangle - \langle\mathbf{J}\rangle^2. \tag{12.54}$$

The angular brackets here are understood as averaging over the Langevin noise with the correlator (12.48). Since both $\mathcal{K}(x,x')$ and $\mathbf{J}(x)\mathbf{J}(x')$ are quadratic in the distribution function, $\tilde{F}$, their averages are affected by the Langevin part of the Boltzmann equation (12.47).

In equilibrium the stochasticity is absent and the current noise is given by the Johnson–Nyquist expression, $\langle S\rangle = \mathcal{K} = 4Tg$, where $g = \nu\int d\Omega_{\mathbf{k}}\,\mathbf{v}_{\mathbf{k}}\mathbf{v}_{\mathbf{k}}\tau_{\mathrm{sc}}$ is the conductivity tensor and ν is the density of states (we have also used that

$\int d\epsilon_{\mathbf{k}}(1-\tanh^2\epsilon_{\mathbf{k}}/2T) = 4T$). This is, of course, a manifestation of the FDT, which guarantees a relation between $\mathcal{K}(x,x')$ and $\delta\mathbf{J}(x)/\delta\mathbf{A}^{\text{cl}}(x)$. The non-equilibrium part of the current noise is affected by the Langevin stochasticity. The latter leads to, for example, $\langle\mathbf{J}(x)\mathbf{J}(x')\rangle - \langle\mathbf{J}\rangle^2 \neq 0$ and also affects $\langle\mathcal{K}(x,x')\rangle$.

The field-theoretical treatment of the Langevin noise is somewhat different from the phenomenological one presented in the literature [235, 236, 237]. The latter treats the noise as $\langle\mathcal{S}\rangle = \langle\mathbf{J}(x)\mathbf{J}(x')\rangle - \langle\mathbf{J}\rangle^2$ (without the first term on the right-hand side of Eq. (12.54)). In this case, even the equilibrium Johnson–Nyquist noise comes as a result of the stochastic fluctuations of the distribution function. Thus, the corresponding noise correlator does not vanish in equilibrium (nevertheless, there is still a close relation between the collision integral and the noise correlator [237]).

12.5 The Butterfly Effect and OTOC

The butterfly effect refers to spreading of perturbations in chaotic many-body systems. In a classical context it is typically associated with the divergence of trajectories, characterized by a positive Lyapunov exponent. Quantum mechanics lacks the notion of infinitesimally closed trajectories and thus calls for another measure of chaoticity. The convenient characteristic is provided by the *out-of-time-order correlator* (OTOC). It was first introduced by Larkin and Ovchinnikov [238] and later revived by Kitaev [239] and Maldacena, Shenker, and Stanford [240]. For Heisenberg picture observables $\hat{\mathcal{O}}$ and $\hat{\hat{\mathcal{O}}}$ the OTOC is defined as

$$\mathcal{C}(t) = \text{Tr}\Big\{\hat{\mathcal{O}}(t)\hat{\hat{\mathcal{O}}}(0)\hat{\mathcal{O}}(t)\hat{\hat{\mathcal{O}}}(0)\hat{\rho}(-\infty)\Big\}. \tag{12.55}$$

For, for example, a zero-temperature ground state it is given by the overlap of the state $\hat{\mathcal{O}}(t)\hat{\hat{\mathcal{O}}}(0)|0\rangle$ with the state $\langle 0|\hat{\mathcal{O}}(t)\hat{\hat{\mathcal{O}}}(0)$. The former is obtained by first perturbing the ground state with $\hat{\hat{\mathcal{O}}}$ at time $t = 0$, then propagating the system forward in time until $t > 0$. The latter one requires a time machine, since first the ground state is hit by $\hat{\mathcal{O}}(t)$ and then the resulting state propagates back in time to $t = 0$ when $\hat{\hat{\mathcal{O}}}$ acts on it. Thus, OTOC measures how much the state is scrambled by the perturbation and the subsequent evolution. One expects that in a chaotic system the overlap and thus $\mathcal{C}(t)$ decays rapidly with time. Moreover, if $\hat{\mathcal{O}}$ and $\hat{\hat{\mathcal{O}}}$ are local operators at two spatially separated locations, $\mathcal{C}(t)$ is peaked at a specific time when the perturbation propagates between the two points.

To calculate OTOC we follow the approach of Aleiner, Faoro, and Ioffe [241]. They introduced two copies of the Keldysh contour, Fig. 12.1, which allow one to evaluate OTOC by introducing proper sources. Colloquially the contour represents

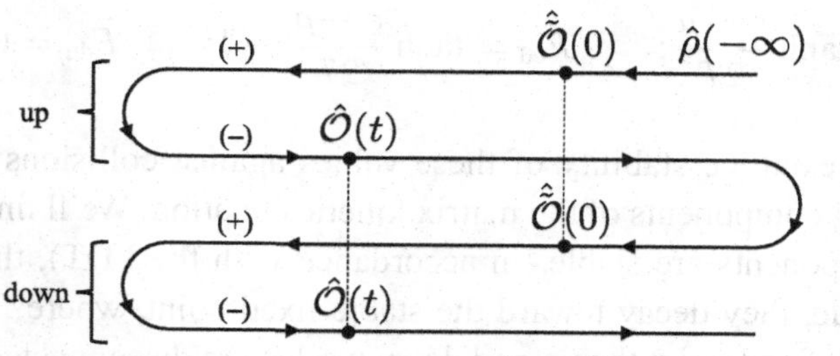

Figure 12.1 Extended time contour for calculation of OTOC. It includes two copies of Keldysh contour dubbed "up" and "down" worlds.

the two "worlds" called "up" and "down" with time opposite "butterflies." Development of the chaos and vanishing of the overlap between the two worlds leads to an effective decoupling of the up and down contours. On the other hand, in non-chaotic systems the two contours remain coupled.

To proceed we discretize time along the contour of Fig. 12.1 and introduce fermionic coherent states, which form a $4N$-component vector $(\psi_u^+, \psi_u^-, \psi_d^+, \psi_d^-)^T$, where "u" and "d" stand for up and down correspondingly and we suppressed the discrete time index, $j = 1, \ldots, N$, for brevity. Inverting the $4N \times 4N$ matrix, which straightforwardly generalizes the one in Eq. (10.15), one finds 4×4 bare Green function

$$\check{G}(t,t') = \begin{pmatrix} \hat{G}_{uu}(t,t') & \hat{G}_{ud}(t,t') \\ \hat{G}_{du}(t,t') & \hat{G}_{dd}(t,t') \end{pmatrix}, \tag{12.56}$$

where

$$\hat{G}_{uu} = \hat{G}_{dd} = \begin{pmatrix} G^{\mathbb{T}} & G^{<} \\ G^{>} & G^{\tilde{\mathbb{T}}} \end{pmatrix}; \quad \hat{G}_{ud} = \begin{pmatrix} G^{<} & G^{<} \\ G^{<} & G^{<} \end{pmatrix}; \quad \hat{G}_{du} = \begin{pmatrix} G^{>} & G^{>} \\ G^{>} & G^{>} \end{pmatrix}, \tag{12.57}$$

and $G^{\mathbb{T}}$, $G^{<}$, $G^{>}$, $G^{\tilde{\mathbb{T}}}$ are already familiar Green functions, defined in Eq. (10.21). We now perform Keldysh–Larkin–Ovchinnikov rotation separately in the up and down worlds, exactly as in Eqs. (10.23), (10.24). This brings the 4×4 Green function, $\check{G}$ into the retarded-advanced-Keldysh triangular form, Eq. (10.25), where the three nontrivial components are now 2×2 matrices in the up-down subspace:

$$\check{G}^{R} = \begin{pmatrix} G^{R} & 0 \\ 0 & G^{R} \end{pmatrix}; \quad \check{G}^{A} = \begin{pmatrix} G^{A} & 0 \\ 0 & G^{A} \end{pmatrix}; \quad \check{G}^{K} = \begin{pmatrix} G^{K}_{uu} & G^{K}_{ud} \\ G^{K}_{du} & G^{K}_{dd} \end{pmatrix}, \tag{12.58}$$

where the check symbol stands for matrices in the up-down two-world space. Here $G^{K}_{uu} = G^{K}_{dd} = G^{K} = G^{>} + G^{<}$ and $G^{K}_{ud} = 2G^{<}$, $G^{K}_{du} = 2G^{>}$.

The distribution *matrix*, $\check{F}$, is defined in the usual way (cf. Eq. (10.33)), through $\check{G}^{K} = \check{G}^{R} \circ \check{F} - \check{F} \circ \check{G}^{A}$. In an equilibrium noninteracting system its components are given by

$$F_{\mathrm{uu}} = F_{\mathrm{dd}} = \tanh\frac{\epsilon-\mu}{2T}; \qquad F_{\mathrm{ud}} = \tanh\frac{\epsilon-\mu}{2T} - 1; \qquad F_{\mathrm{du}} = \tanh\frac{\epsilon-\mu}{2T} + 1. \tag{12.59}$$

We will next examine stability of these values against collisions by considering corresponding components of the matrix kinetic equation. We'll find that while the diagonal components are stable (in accordance with the FDT), the off-diagonals are not. Instead, they decay toward the stable fixed point, where $F_{\mathrm{ud}} = F_{\mathrm{du}} = 0$. At this chaotic fixed point the up and down worlds are disconnected, leading to the decay of OTOC, $\mathcal{C}(t)$.

We'll take electron–electron collisions as a mechanism of chaotization (other interactions, such as electron–phonon interaction, were considered in Ref. [241] and shown to lead to the same conclusions). To derive the collision integral, one needs components of the self-energy. Specifically we'll be interested in the off-diagonal up-down and down-up components (since the diagonal components of the kinetic equation turn out to be unaffected by the presence of the other world). We first notice that, since $\check{G}^{\mathrm{R}}$ and $\check{G}^{\mathrm{A}}$ are diagonal, the non-diagonal components of the polarization operator $\check{\Pi}(x,x')$ contains only the Keldysh components $\Pi^{\mathrm{K}}_{\mathrm{ud}} = \Pi^{\mathrm{K}}_{\mathrm{du}} = (\mathrm{i}/2)G^{\mathrm{K}}_{\mathrm{ud}}G^{\mathrm{K}}_{\mathrm{du}}$ (cf. Eq. (10.52)) while $\Pi^{\mathrm{R}}_{\mathrm{ud}} = \Pi^{\mathrm{R}}_{\mathrm{du}} = 0$ and the same for the advanced components. This implies that the retarded and advanced components of the RPA screened interaction $\check{U}^{\mathrm{R}}_{\mathrm{RPA}}$ and $\check{U}^{\mathrm{A}}_{\mathrm{RPA}}$ are also diagonal in the up-down subspace, while its Keldysh component has off-diagonal components $\check{U}^{\mathrm{K}}_{\mathrm{RPA}} = -\check{U}^{\mathrm{R}}_{\mathrm{RPA}}\check{\Pi}^{\mathrm{K}}\check{U}^{\mathrm{A}}_{\mathrm{RPA}}$ (cf. Eq. (11.9)). According to Eq. (11.5) this leads to (recall that $\hat{\gamma}^{\mathrm{cl}} = 1$)

$$\hat{\Sigma}_{\mathrm{ud}} = \frac{\mathrm{i}}{2}\,\hat{G}_{\mathrm{ud}}U^{\mathrm{K}}_{\mathrm{ud}} = -\frac{\mathrm{i}}{2}\,\hat{G}_{\mathrm{ud}}U^{\mathrm{R}}_{\mathrm{RPA}}\Pi^{\mathrm{K}}_{\mathrm{ud}}U^{\mathrm{A}}_{\mathrm{RPA}} \tag{12.60}$$

and similarly for the down-up matrix self-energy. Notice that, since $\hat{G}_{\mathrm{ud}}$ contains only a Keldysh component, but no retarded and advanced one, the only nonzero component of the off-diagonal self-energy is $\Sigma^{\mathrm{K}}_{\mathrm{ud}} = G^{\mathrm{K}}_{\mathrm{ud}}|U^{\mathrm{R}}_{\mathrm{RPA}}|^2G^{\mathrm{K}}_{\mathrm{ud}}G^{\mathrm{K}}_{\mathrm{du}}/4$, while $\Sigma^{\mathrm{R}}_{\mathrm{ud}} = \Sigma^{\mathrm{A}}_{\mathrm{ud}} = 0$. Notice that $\Sigma^{\mathrm{K}}_{\mathrm{ud}} \propto F_{\mathrm{ud}}F_{\mathrm{ud}}F_{\mathrm{du}}$ but is independent of the diagonal component, F, of the distribution.

Neglecting for a moment spatial derivatives, one can write the off-diagonal components of the kinetic equation as

$$\partial_t\tilde{F}_{\mathrm{ud}}(t,\mathbf{k}) = I^{\mathrm{coll}}_{\mathrm{ud}}[\hat{F}] = \mathrm{i}\left[\Sigma^{\mathrm{K}}_{\mathrm{ud}} - \left(\Sigma^{\mathrm{R}}_{\mathrm{uu}}\circ F_{\mathrm{ud}} - F_{\mathrm{ud}}\circ\Sigma^{\mathrm{A}}_{\mathrm{dd}}\right)\right], \tag{12.61}$$

and similarly for the down-up component. Assuming $F_{\mathrm{uu}} = F_{\mathrm{dd}} = F$ for simplicity, and following the steps leading to the collision integral (11.11), one finds

$$I^{\mathrm{coll}}_{\mathrm{ud}}[\hat{F}] = -\frac{\pi}{2}\sum_{\mathbf{q},\mathbf{k}'}\left|U^{\mathrm{R}}_{\mathrm{RPA}}(\mathbf{q},\epsilon_{\mathbf{k}+\mathbf{q}}-\epsilon_{\mathbf{k}})\right|^2\delta(\epsilon_{\mathbf{k}+\mathbf{q}}-\epsilon_{\mathbf{k}}-\epsilon_{\mathbf{k}'+\mathbf{q}}+\epsilon_{\mathbf{k}'})$$
$$\times\left\{\tilde{F}_{\mathrm{ud}}(\mathbf{k}+\mathbf{q})\tilde{F}_{\mathrm{ud}}(\mathbf{k}')\tilde{F}_{\mathrm{du}}(\mathbf{k}'+\mathbf{q}) - \tilde{F}_{\mathrm{ud}}(\mathbf{k})\mathcal{L}[\tilde{F}]\right\}, \tag{12.62}$$

where the "out" part of the collision integral contains $\mathcal{L}[\tilde{F}] \sim \Sigma^{\rm R}_{\rm uu} - \Sigma^{\rm A}_{\rm dd}$, given by

$$\mathcal{L}[\tilde{F}] = \tilde{F}(\mathbf{k}' + \mathbf{q})\tilde{F}(\mathbf{k}') - 1 + \tilde{F}(\mathbf{k} + \mathbf{q})\left(\tilde{F}(\mathbf{k}' + \mathbf{q}) - \tilde{F}(\mathbf{k}')\right).$$

The "in" part, coming from $\Sigma^{\rm K}_{\rm ud}$, depends only on $\tilde{F}_{\rm ud}$ and $\tilde{F}_{\rm du}$, but not on the diagonal component $\tilde{F}$, as explained earlier. This makes the structure of the off-diagonal collision integral qualitatively different from the diagonal one. The latter may be shown to be unaffected by the presence of another world and is given by Eq. (11.11).

The off-diagonal collision integral (12.62) is exactly nullified by the equilibrium noninteracting distributions (12.59). To investigate stability of this solution one may assume that the diagonal distributions are quickly equilibrated to $\tilde{F}(\mathbf{k}) = \tanh(\epsilon_{\mathbf{k}} - \mu)/2T$. Then the off-diagonal ones may be looked for in the following form[4]

$$\tilde{F}_{\rm ud}(t, \mathbf{k}) = \phi(t)\left(\tanh\frac{\epsilon_{\mathbf{k}} - \mu}{2T} - 1\right); \qquad \tilde{F}_{\rm du}(t, \mathbf{k}) = \phi(t)\left(\tanh\frac{\epsilon_{\mathbf{k}} - \mu}{2T} + 1\right). \tag{12.63}$$

Substituting this into Eqs. (12.61), (12.62), one finds [241]

$$\frac{{\rm d}\phi}{{\rm d}t} = \frac{1}{\tau_{\rm sc}}\left(\phi^3 - \phi\right), \tag{12.64}$$

where $\tau_{\rm sc}$ is the "out" Fermi liquid electron–electron scattering time, given by Eq. (11.12). Equation (12.64) describes a viscous dynamics in $V(\phi) = \frac{1}{2}\phi^2 - \frac{1}{4}\phi^4$ potential. It shows that $\phi = 1$ noninteracting fixed point, represented by Eq. (12.59), is unstable against collisions. The stable fixed point, $\phi = 0$, corresponds to $\tilde{F}_{\rm ud} = \tilde{F}_{\rm du} = 0$, that is, completely decoupled up and down worlds, where $\mathcal{C} = 0$. Equation (12.64) implies that OTOC $\mathcal{C}(t) - \mathcal{C}(0) \propto -e^{+2t/\tau_{\rm sc}}$, where the butterfly instability rate $2/\tau_{\rm sc} \propto T^2/\epsilon_F \ll T$. This finding is in agreement with Ref. [240], which states that the maximum instability rate can't exceed $2\pi T$.

In spatially extended systems the left-hand side of the kinetic equation is often diffusive (examples are provided in Chapter 14 and subsequent chapters). In case of OTOC it is translated into a diffusion equation for $\phi(\mathbf{r}, t)$ of the form

$$\partial_t\phi - D\nabla^2_{\mathbf{r}}\phi = \frac{1}{\tau_{\rm sc}}\left(\phi^3 - \phi\right) + \xi_\phi(\mathbf{r}, t), \tag{12.65}$$

where D is a diffusion coefficient. The scattering time here may not be the same as the Fermi liquid one, Eq. (11.12), due to its renormalization by a mechanism responsible for diffusion. An example of this phenomenon is discussed in Section 16.5. The last term on the right-hand side of Eq. (12.65) is the Boltzmann–Langevin noise. To derive the corresponding noise correlation function, one should

[4] One can, in principle, consider distinct time-dependent functions $\phi_{\rm ud}(t)$ and $\phi_{\rm du}(t)$. It is easy to show, though, that $\phi_{\rm ud} - \phi_{\rm du}$ decays to zero fast.

generalize the approach of Sections 12.3 and 12.4 to the 4×4 matrix formalism. Somewhat lengthy but straightforward calculation yields

$$\langle \xi_\phi(\mathbf{r}_1, t_1)\xi_\phi(\mathbf{r}_2, t_2)\rangle \propto \nu^{-1}\delta(\mathbf{r}_1 - \mathbf{r}_2)\delta(t_1 - t_2)(\phi^2 - \phi^4), \tag{12.66}$$

where $\nu \approx U_{\mathrm{RPA}}^{-1}$ is the density of states at the Fermi energy.

Equation (12.65) is (a variant of) the Fisher–Kolmogoroff–Petrovsky–Piscounoff (F-KPP) equation [242, 243], which first appeared in population dynamics as well as a description of combustion fronts. This equation supports ballistically propagating 1D fronts of the form $\phi(x,t) = \phi(x - v_{\mathrm{b}}t)$. Such fronts describe an "invasion" of the stable state, $\phi = 0$, into a territory, previously occupied by the unstable state, $\phi = 1$. The front (aka *butterfly*) velocity, v_{b}, is not fixed by the F-KPP equation, which only provides its lower bound,[5] $v_{\mathrm{b}} \geq v_{\mathrm{b}}^* = \sqrt{8D/\tau_{\mathrm{sc}}}$. A linear stability analysis against noise, ξ_ϕ, suggests [244] that fronts with the smallest velocity, v_{b}^*, are the stable ones.

We thus arrived at a spectacular conclusion [241]: in systems with diffusive propagation of conserved quantities (such as heat, charge, spin, etc.), the chaotization, as described by OTOC, propagates *ballistically*. The chaos "invades" the ordered region in a form of the Fisher front, which propagates with a finite speed, v_{b}^*, dubbed the *butterfly velocity*. The latter depends on the conserved quantities diffusion constant, D, and the chaotic instability rate, $1/\tau_{\mathrm{sc}} \leq 2\pi T/\hbar$ [240], as $v_{\mathrm{b}}^* \propto \sqrt{D/\tau_{\mathrm{sc}}}$. The Boltzmann–Langevin–Ito stochastic force on the right-hand side of Eq. (12.65) generates fluctuations of the front speed and its broadening [244]. It is important to notice that the noise amplitude (12.66) vanishes at both stable, $\phi = 0$, and unstable, $\phi = 1$, equilibria. The latter fact ensures that the unstable equilibrium does not develop a spontaneous instability before the front (initiated by the perturbation $\hat{\hat{\mathcal{O}}}(0)$) arrives to the observation point (where the observable $\hat{\mathcal{O}}(t)$ is measured).

12.6 Problems

12.6.1 Linearized Kinetic Equation

Consider a spatially uniform and isotropic in momentum space situation, where the occupation numbers are functions of energy and time only. This allows us to introduce $n_\xi(t) = \sum_{\mathbf{k}} n_{\mathbf{k}}(t)\delta(\xi - \xi_{\mathbf{k}}) = \nu \int d\xi\, n_{\xi_{\mathbf{k}}}(t)\delta(\xi - \xi_{\mathbf{k}})$. The kinetic equation acquires the form (cf. Eq. (11.18))

[5] This is easy to see by substituting $\phi(z)$, where $z = x - v_{\mathrm{b}}t$, into Eq. (12.65). It acquires the form $D\ddot{\phi} = -\partial_\phi(-V(\phi)) - v_{\mathrm{b}}\dot{\phi}$, where the dots denote derivatives with respect to z. This is a Newtonian equation for a particle with mass D in the inverted potential $-V(\phi) = \phi^4/4 - \phi^2/2$ with the friction coefficient v_{b}. Propagating front solutions correspond to *overdamped* non-oscillatory sliding from $\phi = 0$ to $\phi = 1$. This dictates the lower bound for the friction coefficient v_{b}.

$$\partial_t n_\xi = \iiint d\xi_1 d\xi' d\xi_1' \, \mathcal{M}(\xi,\xi_1;\xi',\xi_1')\,\delta(\xi+\xi_1-\xi'-\xi_1') \qquad (12.67)$$
$$\times\Big[-n_\xi n_{\xi_1}(1-n_{\xi'})(1-n_{\xi_1'}) + n_{\xi'}n_{\xi_1'}(1-n_\xi)(1-n_{\xi_1})\Big],$$

where $\mathcal{M} = \frac{4\pi}{\nu}\sum_{\mathbf{k},\mathbf{q},\mathbf{k}'}\left|U^{\mathrm{R}}_{\mathrm{RPA}}(\mathbf{q},\xi_1'-\xi)\right|^2 \delta(\xi-\xi_{\mathbf{k}})\delta(\xi_1-\xi_{\mathbf{k}'+\mathbf{q}})\delta(\xi'-\xi_{\mathbf{k}'})\delta(\xi_1'-\xi_{\mathbf{k}+\mathbf{q}})$ is the scattering rate, which we will approximate with a positive constant, having in mind short-range screened interactions. Let us assume furthermore that deviations from equilibrium are small and therefore the occupation numbers may be written as

$$n_\xi(t) = n_{\mathrm{F}}(\xi) + \chi(\xi,t), \qquad (12.68)$$

where the equilibrium distribution is $n_{\mathrm{F}}(\xi) = (\mathrm{e}^\xi+1)^{-1}$ and from now on we have passed to dimensionless energy $\xi/T \to \xi$.

Substitute Eq. (12.68) in (12.67) and keep only the terms linear in $\chi(\xi,t)$. Assuming $\mathcal{M}(\xi,\xi_1;\xi',\xi_1') = \mathcal{M} = \text{const}$, perform[6] explicitly the energy integrals and show that $\chi(\xi,t)$ satisfy the following homogeneous linear equation [245, 246]:

$$\tau_{\mathrm{sc}}\,\partial_t\chi(\xi) = -\frac{\xi^2+\pi^2}{\pi^2}\chi(\xi)$$
$$+\int\frac{d\xi'}{\pi^2}\left[\frac{2(\xi-\xi')}{\sinh\left(\frac{\xi-\xi'}{2}\right)} - \frac{\xi+\xi'}{\sinh\left(\frac{\xi+\xi'}{2}\right)}\right]\frac{\cosh(\xi'/2)}{\cosh(\xi/2)}\chi(\xi'),$$

where $\tau_{\mathrm{sc}}^{-1} = \mathcal{M}(\pi T)^2/2$. To proceed one should distinguish between perturbations with different symmetries: *odd* (or temperature) $\chi(\xi,t) = -\chi(-\xi,t)$, which preserve the electron-hole symmetry, versus *even* (or chemical potential) $\chi(\xi,t) = \chi(-\xi,t)$, which change the total number of particles. For such functions the preceding equation acquires the form

$$\tau_{\mathrm{sc}}\,\partial_t\chi(\xi) = -\frac{\xi^2+\pi^2}{\pi^2}\chi(\xi) + \frac{\lambda(\lambda+1)}{2}\int\frac{d\xi'}{\pi^2}\frac{\xi-\xi'}{\sinh\left(\frac{\xi-\xi'}{2}\right)}\frac{\cosh(\xi'/2)}{\cosh(\xi/2)}\chi(\xi'), \qquad (12.69)$$

where for the even perturbations $\lambda = 1$ and for the odd ones $\lambda = 2$.

Define

$$\psi(x,t) = \int d\xi\,\mathrm{e}^{\mathrm{i}x\xi/\pi}\cosh(\xi/2)\,\chi(\xi,t), \qquad (12.70)$$

[6] Do not entrust the integrations to Mathematica, which may not simplify the result sufficiently. Employ $\int_0^\infty da \log a/((a-1)(ab+1)) = (\pi^2+\log^2 b)/(2+2b)$.

and show that the integral equation (12.69) acquires a form of the imaginary-time Schrödinger equation with the Pöschl–Teller potential

$$\tau_{\rm sc}\,\partial_t\psi = -\hat{H}\psi; \qquad \hat{H} = -\partial_x^2 + 1 - \frac{\lambda(\lambda+1)}{\cosh^2 x}. \tag{12.71}$$

Any initial perturbation $\chi(\xi,0)$ should be first decomposed into the even and odd components as $(\chi(\xi,0) \pm \chi(-\xi,0))/2$, which are then Fourier transformed according to (12.70) and then expanded into the basis of even eigenfunctions of $\lambda = 1$ Pöschl–Teller potential, $\psi_k^{(\lambda=1)}(x)$, and odd eigenfunctions of $\lambda = 2$ potential $\psi_k^{(\lambda=2)}(x)$, correspondingly. The Fourier image of the perturbation evolves then according to $\psi(x,t) = \sum_{k,\lambda} c_k^{(\lambda)}\psi_k^{(\lambda)}(x)\,\mathrm{e}^{-E_k^{(\lambda)}t/\tau_{\rm sc}}$, where $E_k^{(\lambda)}$ are eigenenergies of the Hamiltonian (12.71). Therefore, the spectrum of (12.71) determines the relaxation rates.

Clearly there is a continuous spectrum with $E_k^{(\lambda)} \geq 1$, leading to the relaxation rate $1/\tau_{\rm sc}$. There is, however, also a discrete spectrum, which for integer λ is given by $E_{\lambda-m}^{(\lambda)} = 1 - m^2$, where $m = \lambda, \lambda-1, \ldots, 1$. For the even case, $\lambda = 1$, this leads to a single (even) zero mode $E_0^{(1)} = 0$. For the odd case, $\lambda = 2$, this leads to $E_0^{(2)} = -3$ with an even eigenfunction (which is therefore irrelevant) and $E_1^{(2)} = 0$ with an odd eigenfunction. Find[7] two zero-mode eigenfunctions: even $\chi_0^{(\lambda=1)}(\xi)$ and odd $\chi_1^{(\lambda=2)}(\xi)$, which correspond to $E_0^{(1)} = E_1^{(2)} = 0$. The corresponding perturbations do *not* decay. Explain the physics behind these two non-decaying perturbations. **Hint**: think about the conservation laws (11.13).

12.6.2 Near-Field Heat Transfer

A standard mechanism of the heat transfer involves radiation of (on mass-shell, i.e. real) photons. It leads to the celebrated Stefan–Boltzmann law, stating that the heat flux (per unit area) is $J_h^{\rm SB} \propto c^{-2}T^4$. It was realized some time ago [247, 248, 249, 250] that, if two bodies are in a close proximity, there is the heat transfer mediated by virtual evanescent modes of the electromagnetic field. This mechanism is particularly clear in case of electrons in metals interacting via Coulomb interactions (longitudinal component of the electromagnetic field). In this case electron–hole pairs in the hotter metal can induce pairs in the colder one via the Coulomb-mediated scattering. These processes result in the net energy transfer.

[7] Notice that the Pöschl–Teller Hamiltonian $\hat{H} = -\partial_x^2 + \lambda^2 - \frac{\lambda(\lambda+1)}{\cosh^2 x} = \hat{B}^\dagger\hat{B}$ is supersymmetric (cf. footnote 2 in chapter 4) with $\hat{B} = \mathrm{i}\partial_x + \mathrm{i}\lambda\tanh x$. It therefore admits a zero-energy ground state, which is found as $\hat{B}\psi_0 = 0$.

Consider two parallel metallic layers kept at a distance d and having different temperatures $T_1 > T_2$. Show that to the second order in the screened *inter*layer interaction[8] $U_{12}(\mathbf{q},\omega)$, the heat flux between them is given by [251, 252]

$$J_h = \iint \frac{d\omega d^2\mathbf{q}}{(2\pi)^3} |U^R_{12}(\mathbf{q},\omega)|^2 \,\mathrm{Im}\,\Pi^R_1(\mathbf{q},\omega)\mathrm{Im}\,\Pi^R_2(\mathbf{q},\omega)\,\omega\big[n_B(\omega/T_1) - n_B(\omega/T_2)\big], \tag{12.72}$$

where $\mathbf{q}$ is the 2d in-plane momentum, $\Pi_j(\mathbf{q},\omega)$ is the *intra*layer polarization operator, $j = 1, 2$, and $n_B(\omega/T_j) = (e^{\omega/T_j} - 1)^{-1}$ are boson occupation numbers.

To this end define the heat flux as $J_h = \frac{d}{dt}\sum_{\mathbf{k}} \xi_{\mathbf{k}} n_{\mathbf{k}}(t) = -\frac{1}{2}\sum_{\mathbf{k}} \xi_{\mathbf{k}} I^{\mathrm{coll}}[n_{\mathbf{k}}]$, where the occupation numbers refer to, for example, the first layer. The collision integral is convenient to take in the form (11.10), where the $\tilde{F}$-functions refer to $j = 1$, while polarization operators to $j = 2$ and $U^R_{\mathrm{RPA}} = U^R_{12}$. Indeed, this reflects collisions mediated by interlayer interactions, which lead to heat exchange. The interlayer interactions lead to the internal equilibration, which is assumed to be much faster. As a result, the interlayer collision integrals are null and also, $\Pi^K_2 = 2\coth(\omega/T_2)\mathrm{Im}\,\Pi^R_2$ and $\tilde{F}_1(\mathbf{k}+\mathbf{q})\tilde{F}_1(\mathbf{k}) - 1 = \coth(\omega/T_1)[\tilde{F}_1(\mathbf{k}+\mathbf{q}) - \tilde{F}_1(\mathbf{k})]$. Also notice that $\sum_{\mathbf{k}} \xi_{\mathbf{k}}[\tilde{F}_1(\mathbf{k}+\mathbf{q}) - \tilde{F}_1(\mathbf{k})]\delta(\omega - \xi_{\mathbf{k}+\mathbf{q}} + \xi_{\mathbf{k}}) = (\omega/\pi)\mathrm{Im}\,\Pi^R_1$ (upon anti-symmetrization with respect to $\xi_{\mathbf{k}+\mathbf{q}}$ and $\xi_{\mathbf{k}}$); compare Eq. (10.54).

The integrals in (12.72) are effectively limited by $|\omega| \lesssim T_j$ and $q \lesssim d^{-1}$. Therefore, for sufficiently small temperatures and $k_F d > 1$ one may use small frequency and wave number approximation for $\Pi_j(\mathbf{q},\omega)$; compare Eq. (10.55). Put $\mathrm{Re}\,\Pi^R_j(\mathbf{q},\omega) \approx \nu$, while $\mathrm{Im}\,\Pi^R_j(\mathbf{q},\omega) \approx \pi\nu\omega/(2v_F q)\theta(v_F q - |\omega|)$ and show that $J_h = J^{\mathrm{NF}}_h(T_1) - J^{\mathrm{NF}}_h(T_2)$, where the near-field heat flux $J_h \propto v_F^{-2} T^4 \log T$. Compare this result with the radiative Stefan–Boltzmann prediction.

[8] In the case of the two layers the RPA screening (10.68) acquires the matrix structure in the $j = 1, 2$ space

$$[\check{U}^R(\mathbf{q},\omega)]^{-1} = [\check{U}_0(\mathbf{q})]^{-1} + \check{\Pi}^R_0(\mathbf{q},\omega),$$

where the polarization matrix is diagonal $\check{\Pi}^R_{0;ij} = \delta_{ij}\Pi^R_j$, and the bare 2d Coulomb interactions are $\check{U}_{0;11} = \check{U}_{0;22} = \int \frac{dq_z}{2\pi}\frac{4\pi}{q^2+q_z^2} = \frac{2\pi}{q}$, while $\check{U}_{0;12} = \check{U}_{0;21} = \int \frac{dq_z}{2\pi}\frac{4\pi}{q^2+q_z^2}e^{iq_z d} = \frac{2\pi}{q}e^{-qd}$.

13
Quantum Transport

This chapter is devoted to electron transport through quantum coherent circuits. We start from the multichannel Landauer formula for the conductance. We then discuss noise power, as well as higher-order current cumulants. The latter provide the knowledge of the so-called full counting statistics of the transmitted charge, given by the Levitov formula. We also discuss adiabatic pumping: the Brouwer formula for the mean pumped charge and its full counting statistics. Finally we address the spin-torque effect and spin-torque shot noise in nano-magnetic structures.

13.1 Landauer Formula

Consider a quasi-one-dimensional adiabatic constriction connected to two reservoirs, which we refer to as left (L) and right (R), Fig. 13.1. Such a constriction is called a quantum point contact (QPC). The electron motion within QPC is separable into transverse and longitudinal components. Owing to the confinement, the transverse motion is quantized and we assign a quantum number n to label transverse conduction channels, with $\phi_n(\mathbf{r}_\perp)$ being the corresponding transversal wave functions. The longitudinal motion is described in terms of the extended scattering states, that is, normalized electron plane waves incident from the left,

$$u_n^{\mathrm{L}}(k,\mathbf{r}) = \phi_n(\mathbf{r}_\perp)\begin{cases} \mathrm{e}^{ikx} + \mathbf{r}_n(k)\,\mathrm{e}^{-ikx}, & x \to -\infty \\ \mathbf{t}_n(k)\,\mathrm{e}^{ikx}, & x \to +\infty \end{cases}, \tag{13.1a}$$

and the right,

$$u_n^{\mathrm{R}}(k,\mathbf{r}) = \phi_n(\mathbf{r}_\perp)\begin{cases} \mathbf{t}_n(k)\,\mathrm{e}^{-ikx}, & x \to -\infty \\ \mathrm{e}^{-ikx} + \mathbf{r}_n(k)\,\mathrm{e}^{ikx}, & x \to +\infty \end{cases}, \tag{13.1b}$$

onto the scattering region, Fig. 13.1. Here k is the wave vector in the longitudinal directions and $\mathbf{t}_n(k)$ and $\mathbf{r}_n(k)$ are channel-specific transmission and reflection amplitudes. We have disregarded a possible phase difference between left and right

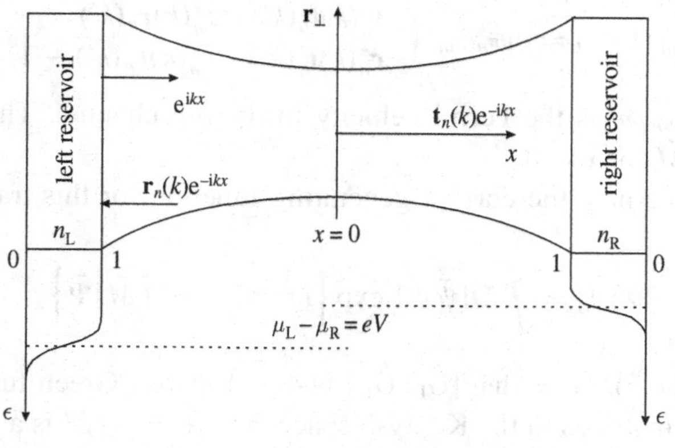

Figure 13.1 Two-terminal scattering problem for the quantum point contact.

reflection coefficients for brevity. The occupation numbers of the L(R) states in the L(R) reservoir are assumed to be the equilibrium Fermi–Dirac distributions $n_{L(R)}(\epsilon_k) = \left[\exp[(\epsilon_k - \mu_{L(R)})/T] + 1\right]^{-1}$, with electrochemical potentials shifted by an external voltage $\mu_L - \mu_R = eV$.

It is convenient to choose scattering states (Eq. (13.1)) as a basis and write the secondary quantized electron field operator in the coordinate space as

$$\hat{\Psi}(\mathbf{r}) = \sum_{nk} \left[\hat{c}^{L}_{nk} u^{L}_{n}(k, \mathbf{r}) + \hat{c}^{R}_{nk} u^{R}_{n}(k, \mathbf{r})\right], \tag{13.2}$$

where $\hat{c}^{L(R)}_{nk}$ are fermion annihilation operators in the corresponding states. For future use we also define the current operator in the longitudinal direction as $\hat{\mathcal{J}} = (e/m)\mathrm{Im} \int d\mathbf{r}_\perp \hat{\Psi}^\dagger \partial_x \hat{\Psi}$, leading to

$$\hat{\mathcal{J}}(x) = \sum_{nk,n'k'} \hat{c}^{\dagger a}_{nk} M^{ab}_{nn'}(x; k, k') \hat{c}^{b}_{n'k'}, \tag{13.3}$$

with the matrix elements

$$M^{ab}_{nn'}(x; k, k') = \frac{e}{2im} \int d\mathbf{r}_\perp \left[u^{*a}_{n}(k, \mathbf{r}) \partial_x u^{b}_{n'}(k', \mathbf{r}) - [\partial_x u^{*a}_{n}(k, \mathbf{r})] u^{b}_{n'}(k', \mathbf{r})\right], \tag{13.4}$$

where $a, b = \mathrm{L, R}$. Employing the orthogonality condition in the transverse direction, $\int d\mathbf{r}_\perp \phi_n(\mathbf{r}_\perp) \phi^*_{n'}(\mathbf{r}_\perp) = \delta_{nn'}$, direct calculation of $\hat{M}_{nn'}(x; k, k')$ for, for example, $x > 0$ gives[1]

[1] Equation (13.5) is obtained as a result of certain approximations. The exact expression for the current matrix explicitly depends on the coordinate x. There are two types of term. The first depends on x as $\exp(\pm i(k + k')x) \approx \exp(\pm 2ik_F x)$, where k_F is the Fermi momentum; it represents Friedel oscillations. Their contribution to the current is small as $(k - k')/k_F \ll 1$, and thus neglected. The second type of term contains $\exp(\pm i(k - k')x) \approx 1$, since $|k - k'| \sim L_T^{-1} \ll x^{-1}$, where $L_T = v_F/T$ is ballistic thermal length, and the coordinate x is confined by the sample size $L \ll L_T$. See the corresponding discussions in [253].

$$\hat{M}_{nn'}(k,k') = ev_{\mathrm{F}n}\delta_{nn'} \begin{pmatrix} \mathbf{t}_n^*(k)\mathbf{t}_n(k') & \mathbf{t}_n^*(k)\mathbf{r}_n(k') \\ \mathbf{r}_n^*(k)\mathbf{t}_n(k') & \mathbf{r}_n^*(k)\mathbf{r}_n(k') - 1 \end{pmatrix}, \tag{13.5}$$

where $v_{\mathrm{F}n} = k_{\mathrm{F}n}/m$ is the Fermi velocity in the nth channel. There is a similar expression for $\hat{M}$ for $x < 0$.

One can define now the current-generating function for this transport problem as

$$Z[A] = \int \mathbf{D}[\bar{\psi}\psi] \exp\left\{ \mathrm{i}\,\vec{\bar{\Psi}}[\check{\mathbf{G}}^{-1} - \hat{A}\hat{M}]\vec{\Psi} \right\}, \tag{13.6}$$

here $\vec{\bar{\Psi}} = (\bar{\psi}^{\mathrm{L}}, \bar{\psi}^{\mathrm{R}})$, $\check{\mathbf{G}} = \mathrm{diag}\{\hat{G}_{\mathrm{L}}, \hat{G}_{\mathrm{R}}\}$ is 4×4 matrix Green function, whereas $\hat{G}_{\mathrm{L/R}}$ are 2×2 matrices in the Keldysh space, and $\hat{A} = A^{\mathrm{q}}\hat{\gamma}^{\mathrm{q}}$ is a purely quantum vector potential (cf. Eq. (10.43)), used to generate the observable (i.e. current). Since the functional integral over fermionic fields in Eq. (13.6) is quadratic, while the generating function is normalized as $Z[0] = 1$, one finds upon Gaussian integration

$$\ln Z[A] = \mathrm{Tr}\ln\left[\hat{1} - \check{\mathbf{G}}\hat{A}\hat{M}\right]. \tag{13.7}$$

By analogy with Eq. (10.46) the average current is generated from $Z[A]$ via functional differentiation with respect to the quantum component of the vector potential, $\langle J\rangle = (\mathrm{i}/2)\delta \ln Z[A]/\delta A^{\mathrm{q}}(t)|_{A^{\mathrm{q}}=0}$. Expanding the trace of the logarithm to linear order in $\hat{A}$, as $\mathrm{Tr}\ln[\hat{1} - \check{\mathbf{G}}\hat{A}\hat{M}] \approx -\mathrm{Tr}[\check{\mathbf{G}}A^{\mathrm{q}}\hat{\gamma}^{\mathrm{q}}\hat{M}]$, one finds for the average current

$$\begin{aligned} \langle J\rangle &= -\frac{\mathrm{i}e}{2}\,\mathrm{Tr}\left\{ v_{\mathrm{F}n} \begin{pmatrix} \hat{G}_{\mathrm{L}}\hat{\gamma}^{\mathrm{q}} & 0 \\ 0 & \hat{G}_{\mathrm{R}}\hat{\gamma}^{\mathrm{q}} \end{pmatrix} \begin{pmatrix} \mathbf{t}_n^*(k)\mathbf{t}_n(k) & \mathbf{t}_n^*(k)\mathbf{r}_n(k) \\ \mathbf{r}_n^*(k)\mathbf{t}_n(k) & \mathbf{r}_n^*(k)\mathbf{r}_n(k) - 1 \end{pmatrix} \right\} \\ &= -\frac{\mathrm{i}e}{2}\sum_{nk} v_{\mathrm{F}n}\,|\mathbf{t}_n(k)|^2 \int \frac{d\epsilon}{2\pi}\left[G_{\mathrm{L}}^{\mathrm{K}}(\epsilon;n,k) - G_{\mathrm{R}}^{\mathrm{K}}(\epsilon;n,k)\right], \end{aligned} \tag{13.8}$$

where we used $|\mathbf{t}_n|^2 = 1 - |\mathbf{r}_n|^2$ and the Keldysh trace

$$\mathrm{Tr}\{\hat{G}_a\hat{\gamma}^{\mathrm{q}}\} = G_a^{\mathrm{K}}(t,t;n,k) = \int \frac{d\epsilon}{2\pi}\,G_a^{\mathrm{K}}(\epsilon;n,k).$$

The Green functions are $G_{\mathrm{L(R)}}^{\mathrm{K}}(\epsilon;n,k) = -2\pi\mathrm{i}\delta(\epsilon - \epsilon_{nk})[1 - 2n_{\mathrm{L(R)}}(\epsilon)]$, with $\epsilon_{nk} = v_{\mathrm{F}n}k$ (see Eq. (10.31)). Performing momentum k summation with the help of the delta-function, one finds

$$\langle J\rangle = \frac{e}{2\pi}\sum_n \int d\epsilon\,|\mathbf{t}_n(\epsilon)|^2\,[n_{\mathrm{L}}(\epsilon) - n_{\mathrm{R}}(\epsilon)]. \tag{13.9}$$

If the energy dependence of the transmission probability may be disregarded, one obtains the linear dependence between the average current and the applied voltage $\langle J\rangle = gV$, where the linear conductance *per one spin direction* is given by

$$g = \frac{e^2}{2\pi\hbar}\sum_n |\mathrm{t}_n|^2, \tag{13.10}$$

and all transmissions are taken at the Fermi energy $|\mathrm{t}_n|^2 = |\mathrm{t}_n(\epsilon_\mathrm{F})|^2$. (Notice that we restored the Planck constant $\hbar$ in the final expression for the conductance.) Equation (13.10) is known as the multichannel Landauer formula [254, 256] (see [257, 259] for detailed reviews on this subject).

13.2 Shot Noise

Based on the previous example we can make a step forward and calculate the second cumulant of the current fluctuations, the so-called noise power, defined as a symmetrized current–current correlation function:

$$\mathcal{S}(\omega, V) = \int \mathrm{d}t\, \mathrm{e}^{\mathrm{i}\omega t}\langle \delta\hat{J}(t)\delta\hat{J}(0) + \delta\hat{J}(0)\delta\hat{J}(t)\rangle, \quad \delta\hat{J}(t) = \hat{J}(t) - \langle J\rangle. \tag{13.11}$$

Within Keldysh technique this correlator may be deduced from $Z[A]$, Eq. (13.7). Indeed, one needs to expand the trace of the logarithm in Eq. (13.7) to second order in the auxiliary vector potential $\ln Z[A] \propto -\frac{1}{2}\mathrm{Tr}\big[\check{\mathbf{G}}\hat{A}\,\hat{M}\check{\mathbf{G}}\hat{A}\,\hat{M}\big]$ and differentiate twice over the quantum component A^q (cf. Eq. (10.49)),

$$\mathcal{S}(\omega, V) = -\frac{\mathrm{i}}{2}\,\frac{\delta^2 \ln Z[A]}{\delta A^\mathrm{q}(\omega)\delta A^\mathrm{q}(-\omega)}\bigg|_{A^\mathrm{q}=0}. \tag{13.12}$$

This expression automatically gives the properly symmetrized noise power (13.11). As a result of the differentiation one finds

$$\begin{aligned}\mathcal{S}(\omega, V) &= \frac{\mathrm{i}}{2}\,\mathrm{Tr}\left\{\check{\mathbf{G}}(\epsilon_+)\hat{\gamma}^\mathrm{q}\hat{M}\check{\mathbf{G}}(\epsilon_-)\hat{\gamma}^\mathrm{q}\hat{M}\right\}\\ &= \frac{\mathrm{i}e^2}{2}\sum_{nkk'} v_{Fn}^2 \int \frac{\mathrm{d}\epsilon}{2\pi}\Big[\,|\mathrm{t}_n|^4\mathrm{Tr}\{\hat{G}_\mathrm{L}(\epsilon_+)\hat{\gamma}^\mathrm{q}\hat{G}_\mathrm{L}(\epsilon_-)\hat{\gamma}^\mathrm{q}\}\\ &\quad + |\mathrm{t}_n|^2|\mathrm{r}_n|^2\mathrm{Tr}\{\hat{G}_\mathrm{L}(\epsilon_+)\hat{\gamma}^\mathrm{q}\hat{G}_\mathrm{R}(\epsilon_-)\hat{\gamma}^\mathrm{q}\} + |\mathrm{t}_n|^2|\mathrm{r}_n|^2\mathrm{Tr}\{\hat{G}_\mathrm{R}(\epsilon_+)\hat{\gamma}^\mathrm{q}\hat{G}_\mathrm{L}(\epsilon_-)\hat{\gamma}^\mathrm{q}\}\\ &\quad + |\mathrm{t}_n|^4\mathrm{Tr}\{\hat{G}_\mathrm{R}(\epsilon_+)\hat{\gamma}^\mathrm{q}\hat{G}_\mathrm{R}(\epsilon_-)\hat{\gamma}^\mathrm{q}\}\Big],\end{aligned} \tag{13.13}$$

where we already calculated the partial trace over the left/right subspace, assuming that transmissions are energy independent, and introduced the notation $\epsilon_\pm = \epsilon \pm \omega/2$. Traces over the Keldysh space give

$$\mathrm{Tr}\{\hat{G}_a\hat{\gamma}^\mathrm{q}\hat{G}_b\hat{\gamma}^\mathrm{q}\} = G_a^\mathrm{K}G_b^\mathrm{K} + G_a^\mathrm{R}G_b^\mathrm{A} + G_a^\mathrm{A}G_b^\mathrm{R} = G_a^\mathrm{K}G_b^\mathrm{K} - (G_a^\mathrm{R} - G_a^\mathrm{A})(G_b^\mathrm{R} - G_b^\mathrm{A}), \tag{13.14}$$

due to causality constraints. To perform the momentum k and k' summations one employs the fact that $G_a^\mathrm{K}(\epsilon; n, k) = -2\pi\mathrm{i}\delta(\epsilon - \mu_a - v_{\mathrm{F}n}k)F_a(\epsilon)$, as well as that

$G^{\rm R}_a - G^{\rm A}_a = -2\pi {\rm i}\delta(\epsilon - \mu_a - v_{{\rm F}n}k)$. Performing the summations with the help of delta-functions, one finds $\sum_{kk'} {\rm Tr}\{\hat{G}_a \hat{\gamma}^{\rm q} \hat{G}_b \hat{\gamma}^{\rm q}\} = v_{{\rm F}n}^{-2}(1 - F_a F_b)$. Finally, the expression for the noise power $\mathcal{S}(\omega, V)$ reads as [260]

$$\mathcal{S} = \frac{e^2}{2\pi\hbar}\sum_n \int {\rm d}\epsilon \Big[|\mathbf{t}_n|^4 \big(B_{\rm LL}(\epsilon) + B_{\rm RR}(\epsilon)\big) + |\mathbf{t}_n|^2 |\mathbf{r}_n|^2 \big(B_{\rm LR}(\epsilon) + B_{\rm RL}(\epsilon)\big)\Big], \tag{13.15}$$

where the statistical factors are $B_{ab}(\epsilon) = [1 - F_a(\epsilon_+)F_b(\epsilon_-)]/2$ and we again restored $\hbar$ at the end. Despite the complicated appearance, the energy integration in Eq. (13.15) can be performed in the closed form[2]

$$\mathcal{S}(\omega, V) = \frac{e^2}{2\pi\hbar}\sum_n \Big[2|\mathbf{t}_n|^4 \omega \coth\left(\frac{\omega}{2T}\right) + |\mathbf{t}_n|^2|\mathbf{r}_n|^2 (eV + \omega)\coth\left(\frac{eV + \omega}{2T}\right)$$
$$+ |\mathbf{t}_n|^2|\mathbf{r}_n|^2 (eV - \omega)\coth\left(\frac{eV - \omega}{2T}\right)\Big]. \tag{13.16}$$

There are two limiting cases, which can be deduced from Eq. (13.16). The first one corresponds to equilibrium current fluctuations, $V \to 0$. In this case

$$\mathcal{S}(\omega, 0) = \frac{e^2}{2\pi\hbar}\sum_n \left[2|\mathbf{t}_n|^4 + 2|\mathbf{t}_n|^2|\mathbf{r}_n|^2\right] \omega\coth\left(\frac{\omega}{2T}\right) = 2g\,\omega\coth\left(\frac{\omega}{2T}\right), \tag{13.17}$$

where we used Eq. (13.10) for the linear conductance g and $|\mathbf{t}_n|^2 + |\mathbf{r}_n|^2 = 1$. This result is nothing but the familiar Johnson–Nyquist fluctuation–dissipation relation for the current fluctuations. Note that, despite the complicated dependence on the transmission probabilities in Eq. (13.15), the equilibrium noise power (13.17) is written only in terms of the linear conductance (13.10).

The other limiting case is fully non-equilibrium noise at zero temperature $T \to 0$ and a finite bias V. For such a case one finds from Eq. (13.16) the Lesovik [260] formula for the excess part of the noise:

$$\mathcal{S}(\omega, V) - \mathcal{S}(\omega, 0) = \frac{e^2}{2\pi\hbar}\Big(|eV + \omega| + |eV - \omega| - 2|\omega|\Big)\sum_n |\mathbf{t}_n|^2|\mathbf{r}_n|^2, \tag{13.18}$$

which is called the *shot* noise. In particular, the low-frequency (compared to the applied voltage) part of the shot noise acquires the form

$$\mathcal{S}(0, V) - \mathcal{S}(0, 0) = \frac{e^2}{2\pi\hbar}|2eV|\sum_n |\mathbf{t}_n|^2(1 - |\mathbf{t}_n|^2). \tag{13.19}$$

[2] One employs that the two leads are in equilibrium, i.e. $F_a(\epsilon) = \tanh[(\epsilon - \mu_a)/2T]$ and uses the integral $\int_{-\infty}^{+\infty} {\rm d}x\,[1 - \tanh(x + y)\tanh(x - y)] = 4y\coth(2y)$.

An important observation is that, in contrast to the equilibrium noise (13.17), the shot noise cannot be written solely in terms of the conductance g. It can, only for the case of a tunnel junction, where all transmissions are small, $|\mathbf{t}_n|^2 \ll 1$. Equation (13.19) reduces to $\mathcal{S}(0,V) - \mathcal{S}(0,0) = 2eVg = 2e\langle I\rangle$, which is known as the Schottky formula (for reviews of shot noise in various systems see [261, 262, 263]).

13.3 Counting Statistics

We now develop an alternative treatment of quantum transport, which will allow us to access not only the first and second moments of the current, but the full statistics of its fluctuations. Most of the results discussed here were first derived by Levitov and colleagues [264, 267], while the method we adopt was pioneered by Nazarov and colleagues [268]; see also [269]. We restrict ourselves to the single channel, since generalization to the (two-lead) multichannel case is rather straightforward (the multi-lead setup requires somewhat more work [268]).

Instead of using the wavefunctions (13.1), we shall model a quantum scatterer setup with two *incoming* and two *outgoing* fermionic modes. The corresponding annihilation operators are denoted by $\hat{\psi}_{\mathrm{L}}(x)$ and $\hat{\psi}_{\mathrm{R}}(x)$ for incoming modes and $\hat{\chi}_{\mathrm{L}}(x)$ and $\hat{\chi}_{\mathrm{R}}(x)$ for outgoing ones (there are also four corresponding creation operators $\hat{\psi}^{\dagger}_{\mathrm{L(R)}}(x)$ and $\hat{\chi}^{\dagger}_{\mathrm{L(R)}}(x)$). The left (right) modes $\hat{\psi}_{\mathrm{L(R)}}(x)$ and $\hat{\chi}_{\mathrm{L(R)}}(x)$ occupy the left (right) lead, that is, $-\infty < x < 0$ $(0 < x < \infty)$. The quantum scatterer is represented by its $\mathcal{S}$-scattering matrix, which serves as the boundary condition at $x = 0$, connecting the incoming and outgoing modes (Fig. 13.2),

$$\hat{\chi}(0) = \mathcal{S}\hat{\psi}(0); \qquad \hat{\chi}^{\dagger}(0) = \hat{\psi}^{\dagger}(0)\mathcal{S}^{\dagger}; \qquad \mathcal{S} = \begin{pmatrix} \mathbf{r} & \mathbf{t} \\ \mathbf{t} & \mathbf{r} \end{pmatrix}, \tag{13.20}$$

where we introduced two-component vectors of incoming $\hat{\psi}$ and outgoing $\hat{\chi}$ modes. The corresponding Keldysh action acquires the form

$$S = \mathrm{i}\int_{\mathcal{C}} \mathrm{d}t \left\{ \int_{-\infty}^{0} \mathrm{d}x \left[\bar{\psi}_{\mathrm{L}}(\partial_t - v_{\mathrm{F}}\partial_x)\psi_{\mathrm{L}} + \bar{\chi}_{\mathrm{L}}(\partial_t + v_{\mathrm{F}}\partial_x)\chi_{\mathrm{L}}\right] \right.$$
$$\left. + \int_{0}^{\infty} \mathrm{d}x \left[\bar{\psi}_{\mathrm{R}}(\partial_t + v_{\mathrm{F}}\partial_x)\psi_{\mathrm{R}} + \bar{\chi}_{\mathrm{R}}(\partial_t - v_{\mathrm{F}}\partial_x)\chi_{\mathrm{R}}\right] \right\}, \tag{13.21}$$

where we have linearized the fermionic dispersion relation near the Fermi energy as $\xi_k = \epsilon_k - \epsilon_{\mathrm{F}} \approx \pm v_{\mathrm{F}} k \to \mp \mathrm{i} v_{\mathrm{F}}\partial_x$. The boundary conditions (13.20) are imposed by introducing the delta-function $\delta(\chi(0,t) - \mathcal{S}\psi(0,t))$ and another one for the bar fields into the functional integral over the $\psi(x,t)$ and $\chi(x,t)$ Grassmann fields.

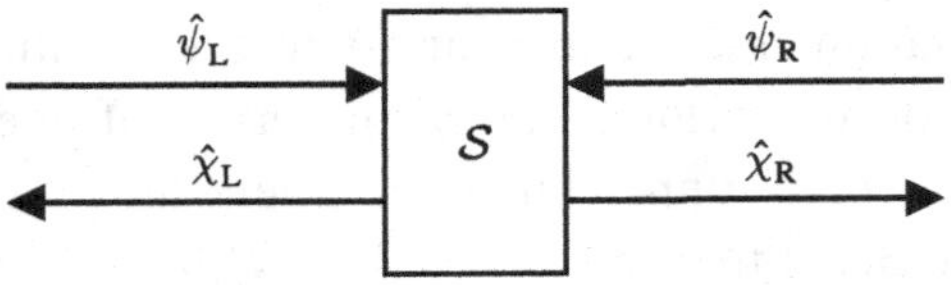

Figure 13.2 Scattering setup: two incoming modes $\hat{\psi}_{L(R)}(x)$ and two outgoing modes $\hat{\chi}_{L(R)}(x)$ are coupled through the scattering matrix $\hat{\chi}(0) = \mathcal{S}\hat{\psi}(0)$.

The quantity of interest is the transmitted charge, which is defined as the time integral of the current. The latter is convenient to represent as the half sum of the currents immediately to the left and right of the scatterer:

$$J(t) = \frac{1}{2}\left[\bar{\psi}_L v_F \psi_L - \bar{\chi}_L v_F \chi_L - \bar{\psi}_R v_F \psi_R + \bar{\chi}_R v_F \chi_R\right], \tag{13.22}$$

where all the fields are taken at $x = 0$. To generate the statistics of the current, one couples it to the auxiliary vector potential $\eta(t)/2$, defined along the contour, and evaluates the generating function

$$Z[\eta] = \left\langle e^{\frac{i}{2}\int_C \eta(t)J(t)} \right\rangle, \tag{13.23}$$

where the angular brackets denote averaging with the weight e^{iS}, Eq. (13.21), and imposed boundary conditions (13.20). To generate the observable, the auxiliary field should be chosen purely *quantum*, that is, changing sign between forward and backward branches (cf. Eq. (10.43)). Moreover, since we are interested in the statistics of charge transmitted during a long time interval $0 < t < t_0$, one may put it to be a constant $\eta(t) = \pm\eta\,\theta(t)\theta(t_0 - t)$, where the plus (minus) sign refers to the forward (backward) branch. One then notices that the exponent in Eq. (13.23) contains $iS \pm (i\eta/2)\int_0^{t_0} dt\, J(t)$ and therefore η may be eliminated from the exponent by the following gauge transformation of the Grassmann fields:

$$e^{\mp i\frac{\eta}{4}\theta(-x)}\psi_L^{\pm} \to \psi_L^{\pm}; \qquad e^{\pm i\frac{\eta}{4}\theta(x)}\psi_R^{\pm} \to \psi_R^{\pm}$$

and the same for χ, where plus/minus signs, as always, refer to the two branches of the time contour. Then the source (counting) field η modifies the boundary conditions (13.20) as

$$\chi^{\pm}(0,t) = \mathbb{S}_{\pm\eta}\psi^{\pm}(0,t); \qquad \mathbb{S}_{\pm\eta} = e^{\mp i\frac{\eta}{4}\hat{\varsigma}_3}\,\mathbb{S}\,e^{\pm i\frac{\eta}{4}\hat{\varsigma}_3} = \begin{pmatrix} \mathbf{r} & \mathbf{t}e^{\mp i\frac{\eta}{2}} \\ \mathbf{t}e^{\pm i\frac{\eta}{2}} & \mathbf{r} \end{pmatrix}, \tag{13.24}$$

and similarly for the bar fields. Here $\hat{\varsigma}_3$ is the third Pauli matrix in L, R-space. As a result the transmission coefficients $\mathbf{t}$ acquire a counting phase η (see Section 2.9), which depends on the direction of the charge transfer.

We are now in the position to integrate out all the fields $\psi(x,t)$ except those residing next to the scatterer, namely $\psi_0(t) = \psi(0,t)$. This is achieved with the help of the following identity:[3]

$$\int \mathbf{D}[\bar{\psi},\psi]\, e^{i\int dt dx\, \bar{\psi}\, \check{G}^{-1}\psi} = e^{-\frac{1}{\pi\nu}\int dt dt'\, \bar{\psi}_0(t)\check{\Lambda}_\psi^{-1}(t-t')\psi_0(t')}, \tag{13.25}$$

where $\check{G}^{-1} = \mathrm{diag}\{\hat{G}_L^{-1}, \hat{G}_R^{-1}\}$ is a 4×4 matrix with left (right) inverse Green functions $\hat{G}_{L(R)}^{-1} = \mathrm{diag}\{[G_{L(R)}^R]^{-1}, [G_{L(R)}^A]^{-1}\} = i\partial_t \mp iv_F\partial_x$ being 2×2 matrices in Keldysh space, Eq. (10.36). The Gaussian correlator of the *local* fields $\psi_0(t)$ is given by[4]

$$\pi\nu\check{\Lambda}_\psi(t-t') = i\check{G}(0,t-t') = i\sum_k \check{G}(k,t-t') = \pi\nu\begin{pmatrix}\hat{\delta}(t-t') & 2\hat{F}_\psi(t-t')\\ 0 & -\hat{\delta}(t-t')\end{pmatrix}, \tag{13.26}$$

where $\nu = (2\pi v_F)^{-1}$ is the density of states of incoming fermions and the Fourier transform of the distribution matrix is $\hat{F}_\psi(\epsilon) = \mathrm{diag}\{F_L(\epsilon), F_R(\epsilon)\}$ and $F_{L(R)} = 1 - 2n_{L(R)}$. Here $n_{L(R)}(\epsilon)$ is the occupation number in the left (right) lead, sending the electrons toward the scatterer. Notice that $\check{\Lambda}^2 = 1$, that is,

$$\check{\Lambda}\circ\check{\Lambda} = \int dt''\,\check{\Lambda}(t-t'')\check{\Lambda}(t''-t') = \check{\delta}(t-t')$$

for an arbitrary distribution matrix $\hat{F}_\psi$, and therefore $\check{\Lambda}^{-1} = \check{\Lambda}$. The integration over the outgoing fields, except for the local ones $\chi_0(t) = \chi(0,t)$, is performed in a way completely analogous to Eq. (13.25) with the only difference that $\hat{F}_\psi \to \hat{F}_\chi$. The outgoing modes are assumed to be perfectly coupled to massive leads, which ideally absorb all particles transmitted or reflected by the scatterer. To model this situation, one assumes that the outgoing modes are *empty* of the particles (i.e. do not Pauli block any of the scattered particles) and thus $\hat{F}_\chi(\epsilon) = \hat{1} - 2\hat{n}_\chi(\epsilon) = \hat{1}$.

[3] An "operational" way to derive this relation is to introduce resolution of unity in the form

$$1 = \int \mathbf{D}[\psi_0]\,\delta(\psi_0(t) - \psi(0,t)) = \int \mathbf{D}[\xi,\psi_0]\, e^{-\int dt\,\bar{\xi}(t)(\psi_0(t)-\psi(0,t))}$$

and similarly for $\bar{\psi}_0$. The last term in the exponent may be written as $\int dt dx\,\bar{\xi}(t)\delta(x)\psi(x,t)$. Integration over the fields $\psi(x,t)$ may then be carried out according to Eq. (10.11). This leads to $i\int dt dt'\bar{\xi}(t)\check{G}(0,t-t')\xi(t')$, which serves as the Gaussian weight for the ξ-integral. Taking into account $\int dt\,\bar{\xi}\psi_0$ along with its conjugate and performing Gaussian integration, Eq. (10.11), over $\bar{\xi}$ and ξ, one obtains Eq. (13.25).

[4] We have used the standard definition of the $\hat{F}$-function as $\hat{G}^K = \hat{G}^R\circ\hat{F}_\psi - \hat{F}_\psi\circ\hat{G}^A$ and

$$\sum_k G^{R(A)}(k,\epsilon) = \int \frac{dk}{2\pi}\frac{1}{\epsilon - v_F k \pm i0} = \mp\frac{i\pi}{2\pi v_F} \overset{FT}{\to} \mp i\pi\nu\,\delta(t-t').$$

As a result one finds for the generating function (13.23)

$$Z[\eta] = \int \mathbf{D}[\bar{\psi}_0, \psi_0, \bar{\chi}_0, \chi_0] e^{\frac{-1}{\pi\nu}\int dt dt' \left[\bar{\psi}_0 \check{\Lambda}_\psi^{-1} \psi_0 + \bar{\chi}_0 \check{\Lambda}_\chi^{-1} \chi_0\right]} \delta(\chi_0 - \check{\mathbb{S}}\psi_0)\delta(\bar{\chi}_0 - \bar{\psi}_0 \check{\mathbb{S}}^\dagger), \tag{13.27}$$

where the integration measure is normalized in a way to keep $Z[0] = 1$. Here $\psi_0(t)$ and $\chi_0(t)$ are four-component vectors on the left (right) along with the Keldysh spaces. The corresponding 4×4 scattering matrix $\check{\mathbb{S}}$ is built from 2×2 blocks (13.24), and after Keldysh rotation (10.23), (10.24) it acquires the following structure in Keldysh space (K):

$$\check{\mathbb{S}}(\eta) = \frac{1}{2}\begin{pmatrix} \mathbb{S}_\eta + \mathbb{S}_{-\eta} & \mathbb{S}_\eta - \mathbb{S}_{-\eta} \\ \mathbb{S}_\eta - \mathbb{S}_{-\eta} & \mathbb{S}_\eta + \mathbb{S}_{-\eta} \end{pmatrix}_K . \tag{13.28}$$

One performs now the integration over $\bar{\chi}_0, \chi_0$ in Eq. (13.27) with the help of delta-functions and remaining Grassmann Gaussian integration over $\bar{\psi}_0, \psi_0$ with the help of Eq. (10.11) and finds

$$Z[\eta] = \det\left[\check{\Lambda}_\psi + \check{\mathbb{S}}^\dagger(\eta)\check{\Lambda}_\chi \check{\mathbb{S}}(\eta)\right] / \det\left[\check{\Lambda}_\psi + \check{\mathbb{S}}^\dagger(0)\check{\Lambda}_\chi \check{\mathbb{S}}(0)\right], \tag{13.29}$$

where we used $\check{\Lambda}^{-1} = \check{\Lambda}$ and the explicitly enforced normalization identity $Z[0] = 1$. The determinants here are to be understood over the 4×4 (Keldysh $\times$ left-right) space along with the energy (or time) spaces. Employing $\check{\Lambda}_\psi = \begin{pmatrix} 1 & 2\hat{F}_\psi \\ 0 & -1 \end{pmatrix}$ and $\check{\Lambda}_\chi = \begin{pmatrix} 1 & 2 \\ 0 & -1 \end{pmatrix}$, unitarity of the scattering matrices $\mathbb{S}_\eta^\dagger \mathbb{S}_\eta = 1$ along with Eq. (13.28), one obtains[5] the Levitov [264] formula for the full counting statistics:

$$Z[\eta] = \det\left[1 + \begin{pmatrix} \hat{F}_\psi & \hat{F}_\psi \\ -1 & -1 \end{pmatrix} \frac{1}{2}(1 - \mathbb{S}_{-\eta}^\dagger \mathbb{S}_\eta)\right] = \det\left[1 - \hat{n}_{\mathrm{F}}(1 - \mathbb{S}_{-\eta}^\dagger \mathbb{S}_\eta)\right], \tag{13.30}$$

where $\hat{n}_{\mathrm{F}} = (1 - \hat{F}_\psi)/2 = \mathrm{diag}\{n_{\mathrm{L}}, n_{\mathrm{R}}\}$ and the last determinant is done over 2×2 left-right $\times$ energy spaces. We employ now the explicit form of the scattering matrix (13.24) along with the unitarity condition $\mathbf{r}^\dagger \mathbf{t} + \mathbf{t}^\dagger \mathbf{r} = 0$ to evaluate the 2×2 determinant. The remaining expression is diagonal in energy space and therefore its determinant is given by the product over all energies:

$$Z[\eta] = \prod_\epsilon \left[1 + n_{\mathrm{L}}(1 - n_{\mathrm{R}})(e^{i\eta} - 1)|\mathbf{t}|^2 + n_{\mathrm{R}}(1 - n_{\mathrm{L}})(e^{-i\eta} - 1)|\mathbf{t}|^2\right]. \tag{13.31}$$

[5] Denoting $\hat{C} = (1 - \mathbb{S}_{-\eta}^\dagger \mathbb{S}_\eta)/2$, one finds by first subtracting columns and then adding lines

$$\det\begin{pmatrix} 1 + \hat{F}_\psi \hat{C} & \hat{F}_\psi \hat{C} \\ -\hat{C} & 1 - \hat{C} \end{pmatrix} = \det\begin{pmatrix} 1 + \hat{F}_\psi \hat{C} & 1 \\ -\hat{C} & -1 \end{pmatrix} = \det\begin{pmatrix} 1 + \hat{F}_\psi \hat{C} - \hat{C} & 0 \\ -\hat{C} & -1 \end{pmatrix} = \det(1 - (1 - \hat{F}_\psi)\hat{C}).$$

The charge $q = \int_0^{t_0} \mathrm{d}t J(t)$ is accumulated over the long time t_0, which dictates the extent of the time contour $\mathcal{C}$. As a result, the energy is quantized in units of $2\pi/t_0$, making the preceding product well-defined.

Consider the case of zero temperature, where the expression in the square brackets on the right-hand side of Eq. (13.31) is different from unity only for $\mu_R < \epsilon < \mu_L$, where $\mu_L - \mu_R = eV$ is the voltage applied between the left and right leads. There are thus $N_0 = eVt_0/(2\pi\hbar)$ discrete energy values, which bring nontrivial factors to the generating function. The generating function is thus (cf. Eq. (2.71)),

$$Z[\eta] = \left[1 + (\mathrm{e}^{\mathrm{i}\eta} - 1)|\mathbf{t}|^2\right]^{N_0} = \left[|\mathbf{r}|^2 + \mathrm{e}^{\mathrm{i}\eta}|\mathbf{t}|^2\right]^{N_0} = \sum_{k=0}^{N_0} \binom{N_0}{k} |\mathbf{r}|^{2(N_0-k)}|\mathbf{t}|^{2k}\mathrm{e}^{ik\eta}. \tag{13.32}$$

According to the definition (13.23), $\partial_k Z[\eta]/\partial\eta^k\big|_{\eta=0} = (\mathrm{i}/e)^k\langle q^k\rangle$ provides cumulants of the charge, transmitted during the time interval t_0. Therefore

$$Z[\eta] = \int \mathrm{d}q\, \mathrm{e}^{\,\mathrm{i}q\eta/e}\,\mathcal{P}(q), \tag{13.33}$$

where $\mathcal{P}(q)$ is the transmitted charge probability distribution function. Performing the inverse Fourier transform of Eq. (13.32), one finds for the probability of transmitting charge q during the time interval t_0, upon an applied voltage V [264, 267],

$$\mathcal{P}(q) = \sum_{k=0}^{N_0} \delta(q - ek)\binom{N_0}{k}|\mathbf{r}|^{2(N_0-k)}|\mathbf{t}|^{2k}. \tag{13.34}$$

The charge is integer valued, and its distribution is *binomial*. The latter describes $N_0 = eVt_0/(2\pi\hbar)$ attempts, each one "succeeds" with the probability $|\mathbf{t}|^2$ and "fails" with the probability $1 - |\mathbf{t}|^2 = |\mathbf{r}|^2$. The average transmitted charge is given by the success rate times the number of attempts:

$$\langle q\rangle = e|\mathbf{t}|^2 N_0 = \frac{e^2}{2\pi\hbar}|\mathbf{t}|^2 Vt_0 = (gV)t_0,$$

where the linear conductance g is given by the Landauer formula (13.10). As expected, the average charge is given by the average current $\langle J\rangle = gV$ times the observation time t_0. The second cumulant of the binomial distribution is given by the product of the success and failure rates times the number of attempts:

$$\langle\langle q^2\rangle\rangle = e^2|\mathbf{t}|^2|\mathbf{r}|^2 N_0 = \frac{e^2}{2\pi\hbar}|\mathbf{t}|^2(1 - |\mathbf{t}|^2)eVt_0,$$

in agreement with the Lesovik formula (13.19).[6] The higher cumulants of the transmitted charge at $T = 0$ are given by the coefficients of the Taylor expansion of the logarithm of the generating function:

$$\ln Z\,[\eta, |\mathbf{t}|^2] = \frac{eVt_0}{2\pi}\ln\left[1 + (\mathrm{e}^{\mathrm{i}\eta} - 1)|\mathbf{t}|^2\right] \tag{13.35}$$

in powers of η. An alternative derivation of these results, valid for a multichannel system with disordered metallic leads, is given in Section 15.5.

13.4 Fluctuation Relation

If both leads are in thermal equilibrium at the same temperature T, one may notice a simple algebraic relation between their occupation numbers:

$$n_{\mathrm{L}}(\epsilon)(1 - n_{\mathrm{R}}(\epsilon)) = \mathrm{e}^{eV/T}\, n_{\mathrm{R}}(\epsilon)(1 - n_{\mathrm{L}}(\epsilon)), \tag{13.36}$$

valid for each energy ϵ. With its help one observes that Eq. (13.31) admits the following symmetry:

$$Z[-\eta + \frac{\mathrm{i}eV}{T}, V] = Z[\eta, V]. \tag{13.37}$$

Tobiska and Nazarov [270] argued that this relation is not restricted to the simple model, but holds as long as the underlying Hamiltonian possesses time-reversal symmetry. In particular they checked that it remains valid in the presence of interactions. (In the latter case the relation holds only after the product of all energies is taken, not for each energy separately, as is the case in the noninteracting case considered here.)

Putting $\eta = 0$ and employing the normalization identity $Z[0, V] = 1$, one obtains $Z[\mathrm{i}eV/T, V] = 1$. Employing Eq. (13.33), one finds

$$\int \mathrm{d}q\, \mathrm{e}^{-qV/T}\, \mathcal{P}(q) = \left\langle \mathrm{e}^{-W/T} \right\rangle = 1, \tag{13.38}$$

where $W = qV = \int \mathrm{d}t\, IV$ is the work done by a battery to transfer a charge q across the scatterer. This relation is the quantum transport analog of the Jarzynski relation (4.72). The only way it may be compatible with the normalization $\int \mathrm{d}q\, \mathcal{P}(q) = 1$ is if negative charges (i.e. transferred against the applied voltage) contribute to the integral. This may seem to be in contradiction to Levitov's binomial distribution (13.34), where only positive (and zero) charges have nonzero probability. The latter statement is strictly valid, though, only at $T = 0$, while at $T > 0$ there is a finite,

[6] The factor of two differences is due to the fact that the noise power (13.11) is defined as the symmetric combination of the current variances and therefore $\langle\langle q^2 \rangle\rangle = \frac{1}{2}\mathcal{S}(0, V)t_0$.

but exponentially small, probability of transferring the opposite charge. Indeed, combining Eqs. (13.37) and (13.33), one finds

$$\mathcal{P}(-q) = \mathcal{P}(q)\,\mathrm{e}^{-Vq/T}. \tag{13.39}$$

One notices thus that the dominant contribution to the integral in Eq. (13.38) comes from $q \approx -\langle q\rangle$, that is, rare events with the relative probability $\propto \mathrm{e}^{-\langle q\rangle V/T}$. This indicates that the fluctuation relation (13.37) is most useful in the limit $V \to 0$, where it deals with the typical, rather than the exponentially rare events.

To take advantage of it [270], we take the full derivative with respect to voltage of the logarithm of Eq. (13.37) and put $V = 0$. This way, one finds

$$\frac{\partial}{\partial V}\Big(\ln Z[\eta] - \ln Z[-\eta]\Big)\Big|_{V=0} = -\frac{ie}{T}\frac{\partial}{\partial \eta}\ln Z[\eta], \tag{13.40}$$

where we took into account that $Z[-\eta, 0] = Z[\eta, 0]$. By expanding the left- and right-hand sides of this relation in power series in η and employing the fact that $\ln Z[\eta] = \sum_l \langle\langle q^l\rangle\rangle (i\eta/e)^l/l!$ is the generating function of the charge cumulants, one finds

$$\langle\langle q^{2l}\rangle\rangle = 2T\frac{\partial}{\partial V}\langle\langle q^{2l-1}\rangle\rangle\Big|_{V=0}, \tag{13.41}$$

$l = 1, 2, \ldots$. The $l = 1$ relation of this series, $\langle\langle q^2\rangle\rangle = 2T\partial\langle q\rangle/\partial V$, is nothing but the FDT Johnson–Nyquist noise (i.e. $\omega \ll T$ limit of Eq. (13.17)). Taking the second derivative of the logarithm of Eq. (13.37) with respect to voltage, putting $V = 0$, and employing Eq. (13.40), one finds

$$\frac{\partial^2}{\partial V^2}\Big(\ln Z[\eta] - \ln Z[-\eta]\Big)\Big|_{V=0} = \frac{ie}{T}\frac{\partial^2}{\partial\eta\partial V}\Big(\ln Z[\eta] + \ln Z[-\eta]\Big)\Big|_{V=0}. \tag{13.42}$$

Expanding over η brings another series of exact relations:

$$\frac{\partial}{\partial V}\langle\langle q^{2l}\rangle\rangle\Big|_{V=0} = -T\frac{\partial^2}{\partial V^2}\langle\langle q^{2l-1}\rangle\rangle\Big|_{V=0}. \tag{13.43}$$

The $l = 1$ member of this family, $-T^{-1}\partial\langle\langle q^2\rangle\rangle/\partial V = \partial^2\langle q\rangle/\partial V^2$, describes the *photovoltaic*, that is, *rectification*, effect. Its right-hand side is the second-order dc current response to a low-frequency, $\omega \ll T$, ac voltage. The left-hand side relates it to the linear sensitivity of the low-frequency noise to the small dc voltage, $eV \ll T$. In terms of the rectified dc current and the noise power, it takes the form

$$J_{\mathrm{dc}} = -\frac{1}{2T}\frac{\partial S(\omega, V)}{\partial V}\Big|_{V=0}|\delta V_\omega|^2, \tag{13.44}$$

where δV_ω is a small ac voltage applied across the quantum scatterer.

13.5 Adiabatic Pumping

We turn now to the situation where the current is generated by a periodic modulation of the quantum scatterer, rather than by the applied bias voltage. In particular we focus on a slow *adiabatic* modulation, when the rate of change is smaller than any other relevant energy scale. Such a process is called *adiabatic pumping*. To proceed we need to acknowledge that the scattering matrix $\mathcal{S}(\epsilon)$, used earlier, may actually be a function of the energy ϵ of the incoming particles. In the time representation it translates into $\mathcal{S}(t-t')$, allowing for a certain time delay between incoming and outgoing modes, $\chi(t) = \int^t dt'\, \mathcal{S}(t-t')\psi(t')$. If the quantum scatterer is externally modulated, its scattering matrix loses time translational invariance and acquires the form $\mathcal{S}(t,t')$. Its Wigner transform $\mathcal{S}(t,\epsilon)$ is referred to as the time- and energy-dependent $\mathcal{S}$-matrix. The unitarity condition holds, but takes the form of the convolution $1 = \mathcal{S}^\dagger \circ \mathcal{S} = \int dt''\, \mathcal{S}^\dagger(t,t'')\mathcal{S}(t'',t') = \delta(t-t')$.

Multiplying the matrix under the determinant in Eq. (13.30) by $\mathcal{S}_\eta$ from the left and by $\mathcal{S}^\dagger_\eta$ from the right, one finds for the generating function

$$\ln Z[\eta] = \operatorname{Tr}\ln\left[1 - \mathcal{S}_\eta \hat{n}_{\mathrm{F}} \mathcal{S}^\dagger_\eta + \mathcal{S}_\eta \hat{n}_{\mathrm{F}} \mathcal{S}^\dagger_{-\eta}\right], \tag{13.45}$$

where $\hat{n}_{\mathrm{F}} = \hat{n}_{\mathrm{F}}(t-t')$ is the Fourier transform of $\hat{n}_{\mathrm{F}}(\epsilon) = \operatorname{diag}\{n_{\mathrm{L}}(\epsilon), n_{\mathrm{R}}(\epsilon)\}$. The trace is to be understood in the 2×2 left/right space as well as in the time-space. The average pumped charge is $\langle q\rangle = -\mathrm{i}e\partial \ln Z[\eta]/\partial\eta\big|_{\eta=0}$. We thus expand the expression on the right-hand side of Eq. (13.45) to linear order in η. To this end we notice that $\mathcal{S}_\eta \approx \mathcal{S} + \mathrm{i}\eta[\mathcal{S}, \hat{\varsigma}_3]/4$ and thus

$$\langle q\rangle = -\frac{e}{2}\operatorname{Tr}\left\{\mathcal{S}\hat{n}_{\mathrm{F}}[\mathcal{S}^\dagger, \hat{\varsigma}_3]\right\}. \tag{13.46}$$

If $\mathcal{S}(t,\epsilon) = \mathcal{S}(\epsilon)$ is static, but $n_{\mathrm{L}}(\epsilon) \neq n_{\mathrm{R}}(\epsilon)$, we return to the Landauer formula (13.9): $\langle q\rangle = e\int (dt d\epsilon)/(2\pi)\, |\mathrm{t}(\epsilon)|^2\, [n_{\mathrm{L}}(\epsilon) - n_{\mathrm{R}}(\epsilon)]$. We now examine the opposite case of the time-dependent scattering matrix $\mathcal{S}(t,\epsilon)$ with no external bias, that is, $n_{\mathrm{L}}(\epsilon) = n_{\mathrm{R}}(\epsilon) = n_{\mathrm{F}}(\epsilon)$. To evaluate the trace in Eq. (13.46) we employ the Wigner transform, which is justified because of the adiabatic nature of the time dependence. This way we find for the Wigner transform of the product (cf. Eq. (6.42))

$$\mathcal{S}\hat{n}_{\mathrm{F}}\mathcal{S}^\dagger \overset{\mathrm{WT}}{\rightarrow} \mathcal{S}\hat{n}_{\mathrm{F}}\mathcal{S}^\dagger + \frac{\mathrm{i}}{2}\left(\mathcal{S}\partial_\epsilon \hat{n}_{\mathrm{F}}\partial_t \mathcal{S}^\dagger - \partial_t \mathcal{S}\partial_\epsilon \hat{n}_{\mathrm{F}}\mathcal{S}^\dagger\right),$$

where we took into account that $\partial_\epsilon \mathcal{S}\hat{n}_{\mathrm{F}}\partial_t \mathcal{S}^\dagger - \partial_t \mathcal{S}\hat{n}_{\mathrm{F}}\partial_\epsilon \mathcal{S}^\dagger = 0$, due to the unitarity of the $\mathcal{S}$-matrix and $\hat{n}_{\mathrm{F}} = \hat{1}n_{\mathrm{F}}$. If the ϵ-dependence of the $\mathcal{S}(t,\epsilon)$-matrix is weak at the scale of temperature, one may use $\partial_\epsilon n_{\mathrm{F}} \approx -\delta(\epsilon-\epsilon_{\mathrm{F}})$ and evaluate the energy integral in the trace using the delta-function. As a result one finds for the average pumped charge

$$\langle q\rangle = \frac{e}{4\pi \mathrm{i}}\int \mathrm{d}t\,\mathrm{Tr}\left\{\frac{\partial \mathcal{S}}{\partial t}\,[\mathcal{S}^\dagger,\hat{\varsigma}_3]\right\}, \tag{13.47}$$

where $\mathcal{S} = \mathcal{S}(t,\epsilon_{\mathrm{F}})$. This is the Brouwer formula [271] for the average adiabatically pumped charge.

In principle Eq. (13.45) provides access not only to the average charge pumped per period, but also to its full counting statistics. In practice, however, calculations of the corresponding determinant require some care associated with the presence of the chiral anomaly [264, 269]. We restrict ourselves to the model case, where these subtleties do not play a role. In this example the tunneling barrier harmonically oscillates between the full and zero transparency. Its unitary time-dependent scattering matrix is specified by

$$\mathbf{t}(t) = \cos(\Omega t); \qquad \mathbf{r}(t) = \mathrm{i}\sin(\Omega t).$$

One can easily check with the help of the Brouwer formula that the average pumped charge is zero (which is obvious, given the presence of reflection symmetry at all times). Our aim, though, is to discuss fluctuations around the zero pumped charge. To this end it is convenient to evaluate

$$\begin{aligned}\left|Z[\eta]\right|^2 &= \det\left[\left(1-\hat{n}_{\mathrm{F}}+\hat{n}_{\mathrm{F}}\mathcal{S}^\dagger_{-\eta}\mathcal{S}_\eta\right)\left(1-\hat{n}_{\mathrm{F}}+\mathcal{S}^\dagger_\eta\mathcal{S}_{-\eta}\hat{n}_{\mathrm{F}}\right)\right]\\ &= \det\left[1+\hat{n}_{\mathrm{F}}\mathcal{S}^\dagger_{-\eta}\mathcal{S}_\eta(1-\hat{n}_{\mathrm{F}})+(1-\hat{n}_{\mathrm{F}})\mathcal{S}^\dagger_\eta\mathcal{S}_{-\eta}\hat{n}_{\mathrm{F}}\right],\end{aligned} \tag{13.48}$$

where we took into account that in the energy representation at zero temperature $n_{\mathrm{F}}^2(\epsilon) = n_{\mathrm{F}}(\epsilon) = \theta(-\epsilon+\mu)$. The product of the gauged transformed $\mathcal{S}$-matrices (13.24) is given by

$$\mathcal{S}^\dagger_\eta\mathcal{S}_{-\eta} = \begin{pmatrix}\frac{e^{-i\eta}+1}{2} & 0\\ 0 & \frac{e^{i\eta}+1}{2}\end{pmatrix} + \begin{pmatrix}\frac{e^{-i\eta}-1}{4} & \frac{\sin\eta/2}{2i}\\ \frac{i\sin\eta/2}{2} & \frac{e^{i\eta}-1}{4}\end{pmatrix}e^{2i\Omega t} + \begin{pmatrix}\frac{e^{-i\eta}-1}{4} & \frac{i\sin\eta/2}{2}\\ \frac{\sin\eta/2}{2i} & \frac{e^{i\eta}-1}{4}\end{pmatrix}e^{-2i\Omega t}.$$

To go to the energy representation we choose the duration of the time contour to be exactly one period of the quantum scatterer oscillation, that is, $2\pi/\Omega$, and therefore the fermionic energy is quantized as $\epsilon_m - \mu = \Omega(m+1/2)$, where m is an integer. As a result, we find a matrix in the energy representation $\left(\mathcal{S}^\dagger_\eta\mathcal{S}_{-\eta}\right)_{m,l} = \hat{A}(\eta)\delta_{m,l} + \hat{B}(\eta)\delta_{m,l+2} + \hat{C}(\eta)\delta_{m,l-2}$, where the 2×2 matrices $\hat{A}, \hat{B}, \hat{C}$ are given by the previous equation. One thus finds for, for example, the last term on the right-hand side of Eq. (13.48), $\theta(\epsilon_m)\left(\mathcal{S}^\dagger_\eta\mathcal{S}_{-\eta}\right)_{m,l}\theta(-\epsilon_l) = \hat{B}(\eta)(\delta_{m,1}\delta_{l,-1}+\delta_{m,0}\delta_{l,-2})$. Similarly, the middle term in Eq. (13.48) is given by $\theta(-\epsilon_m)\left(\mathcal{S}^\dagger_{-\eta}\mathcal{S}_\eta\right)_{m,l}\theta(\epsilon_l) = \hat{C}(-\eta)(\delta_{m,-2}\delta_{l,0}+\delta_{m,-1}\delta_{l,1})$. Therefore, the energy matrix under the determinant in Eq. (13.48) has only four nontrivial components in the energy space $m,l = -2,1,0,1$. Evaluation of the determinant of such a 4×4 matrix is straightforward, leading to

$$|Z[\eta]|^2 = \det\left[1 - \hat{B}(\eta)\hat{C}(-\eta)\right]^2,$$

where the remaining determinant is over the 2×2 left-right space. One thus finds for the single period pumped charge generating function

$$Z[\eta] = \det\left[1 - \hat{B}(\eta)\hat{C}(-\eta)\right] = \frac{1}{2}(1 + \cos\eta) = \frac{1}{2} + \frac{e^{i\eta}}{4} + \frac{e^{-i\eta}}{4}. \tag{13.49}$$

Its Fourier transform (13.33) is the probability distribution function of the charge, which is thus $\mathcal{P}(q) = \frac{1}{2}\,\delta(q) + \frac{1}{4}\left[\delta(q-e) + \delta(q+e)\right]$. We found that with a probability of one-half no charge is transferred during one period of oscillation, and with an equal probability of one-quarter a single electron is moved to the right or to the left. Such an adiabatic pump is thus a "half ideal" coin tosser. Choosing the duration of the time contour to be $2\pi N_0/\Omega$, where N_0 is an integer, changes the quantization condition for the energy. Repeating the calculations, one finds that the generating function for N_0 pumping periods is $Z_{N_0}[\eta] = \left(Z[\eta]\right)^{N_0}$. This means that each period is an independent try of the "coin toss" with the three possible outcomes. One finds thus for the mean square pumped charge after N_0 cycles: $\langle q_{N_0}^2 \rangle = eN_0/2$.

13.6 Spin Torque

Consider a small mono-domain metallic ferromagnet with total spin $\mathbf{S}(t)$, which may precess in an external magnetic field $\mathbf{H}$, Eq. (11.37). We'll refer to it as a "free" layer. The free layer is assumed to be coupled through a tunnel junction to a big ferromagnet with a fixed magnetization; see Fig. 13.3. Furthermore, we assume that a voltage V is applied across the tunnel junction. It drives the current of electrons preferentially polarized in the direction of the fixed layer polarization into the free layer. Upon entering the free layer the polarized electrons find themselves under the influence of the exchange field $J\mathbf{S}$, where J is a ferromagnetic exchange constant in the free layer. It leads to spin-flip processes, which tend to equilibrate the incoming electrons' spin with the magnetization direction of the free layer. This process has

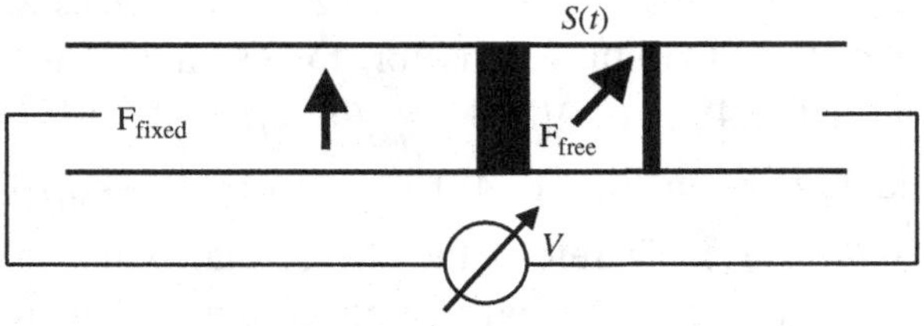

Figure 13.3 A free ferromagnet with spin $S(t)$ is coupled to a fixed ferromagnet with spin along the z-direction through a tunneling barrier (shaded region).

a back-reaction: the angular momentum of the tunneled electrons is transferred to the free layer and rotates its spin **S**. Therefore, the spin-polarized current exerts torque on the magnetic moment (which is opposite to the spin) of the free layer. The effect was discussed first by Slonczewski and Berger [220, 221] and is known as the spin torque.

Our goal is to modify the Landau–Lifshitz equation (11.37) to account for this effect. To this end we describe the electrons in the fixed and free layers by the creation and annihilation operators $\hat{c}^\dagger_{k\sigma}, \hat{c}_{k\sigma}$ and $\hat{d}^\dagger_{l\sigma'}, \hat{d}_{l\sigma'}$, respectively. It is convenient to choose *different* spin quantization axes in the fixed and free layers, taking them along the fixed magnetization and the *instantaneous* vector $\mathbf{S}(t)$, respectively. The Hamiltonian of the electronic subsystem takes the form

$$\hat{H} = \sum_{k,\sigma} \epsilon_{k\sigma} \hat{c}^\dagger_{k\sigma} \hat{c}_{k\sigma} + \sum_{l,\sigma'} \epsilon_{l\sigma'} \hat{d}^\dagger_{l\sigma'} \hat{d}_{l\sigma'} + \sum_{kl,\sigma\sigma'} \left[W^{\sigma\sigma'}_{kl} \hat{c}^\dagger_{k\sigma} \hat{d}_{l\sigma'} + \text{h.c.} \right] - 2J\hat{\mathbf{s}} \cdot \hat{\mathbf{S}}, \tag{13.50}$$

where h.c. stands for Hermitian conjugate and $\hat{\mathbf{s}} = \frac{1}{2} \sum_{ll',\sigma,\sigma'} \hat{d}^\dagger_{l\sigma} \vec{s}_{\sigma\sigma'} \hat{d}_{l'\sigma'}$ is the spin of itinerant electrons in the free layer. The tunneling matrix elements are denoted by $W^{\sigma\sigma'}_{kl}$. We assume spin-conserved tunneling and thus the spin-dependence of the matrix elements originate solely from the different quantization conventions in the two layers. Indeed, an "up" electron from the fixed layer, after tunneling into the free layer, has a certain amplitude to be counted as "up" or "down." The latter depends on the mutual orientation of the two quantization axes. As a result,

$$W^{\sigma\sigma'}_{kl} = \mathrm{e}^{-\mathrm{i}\frac{\theta}{2}\hat{s}_y} \mathrm{e}^{-\mathrm{i}\frac{\phi}{2}\hat{s}_z} W_{kl} = \begin{pmatrix} \mathrm{e}^{-\mathrm{i}\frac{\phi}{2}} \cos(\theta/2) & -\mathrm{e}^{\mathrm{i}\frac{\phi}{2}} \sin(\theta/2) \\ \mathrm{e}^{-\mathrm{i}\frac{\phi}{2}} \sin(\theta/2) & \mathrm{e}^{\mathrm{i}\frac{\phi}{2}} \cos(\theta/2) \end{pmatrix} W_{kl}, \tag{13.51}$$

where (θ, ϕ) are polar angles of the free layer quantization axis **S** in the "laboratory" frame, where the fixed layer is polarized along the z-direction. The operator $\mathrm{e}^{-\mathrm{i}\theta\hat{s}_y/2}\mathrm{e}^{-\mathrm{i}\phi\hat{s}_z/2}$ rotates the d-spinor to the instantaneous reference frame of the free layer.

We also need a way to describe the dynamics of the large spin **S**. A convenient way of doing it is to use Holstein–Primakoff parametrization [272] for the three components $\hat{S}_z; \hat{S}_\pm = \hat{S}_x \pm \mathrm{i}\hat{S}_y$ of the spin S operator

$$\hat{S}_z = S - \hat{b}^\dagger \hat{b}; \qquad \hat{S}_- = \hat{b}^\dagger \sqrt{2S - \hat{b}^\dagger \hat{b}}; \qquad \hat{S}_+ = \sqrt{2S - \hat{b}^\dagger \hat{b}}\, \hat{b}, \tag{13.52}$$

where $\hat{b}^\dagger, \hat{b}$ are the usual bosonic creation and annihilation operators. One can check that, given bosonic commutation relation $[\hat{b}, \hat{b}^\dagger] = 1$, the three components of $\hat{\mathbf{S}}$ satisfy the angular momentum algebra $[\hat{S}_\mu, \hat{S}_\nu] = \mathrm{i}\epsilon_{\mu\nu\lambda}\hat{S}_\lambda$. Since we are interested in the semiclassical dynamics of the vector $\mathbf{S}(t)$, we only need to follow small deviations of the large spin from its *instantaneous* direction. The latter is then adiabatically adjusted to accommodate such a deviation. This strategy allows us to restrict to second order in the Holstein–Primakoff bosonic operators, that is, $\hat{S}_- \approx \sqrt{2S}\,\hat{b}^\dagger$ and $\hat{S}_+ \approx \sqrt{2S}\,\hat{b}$. Employing further for the scalar product $\hat{\mathbf{s}}\cdot\hat{\mathbf{S}} = (\hat{s}_-\hat{S}_+ + \hat{s}_+\hat{S}_-)/2 + \hat{s}_z\hat{S}_z$, one writes the Hamiltonian (13.50) in the following form:

$$\hat{H} = \sum_{k,\sigma} \epsilon_{k\sigma}\hat{c}^\dagger_{k\sigma}\hat{c}_{k\sigma} + \sum_{l,\sigma'}(\epsilon_l - \sigma' JS)\hat{d}^\dagger_{l\sigma'}\hat{d}_{l\sigma'} + \sum_{kl,\sigma\sigma'}\left[W^{\sigma\sigma'}_{kl}\hat{c}^\dagger_{k\sigma}\hat{d}_{l\sigma'} + \text{h.c.}\right]$$
$$- \left[\sqrt{2S}\,\hat{b}\left(J\sum_{l,\sigma}\hat{d}^\dagger_{l\downarrow}\hat{d}_{l\uparrow} - H_-\right) + \text{h.c.}\right] + \hat{b}^\dagger\hat{b}\left(J\sum_{l,\sigma}\sigma\hat{d}^\dagger_{l\sigma}\hat{d}_{l\sigma} - 2H_z\right), \tag{13.53}$$

where H_z and $H_\pm = H_x \pm \mathrm{i}H_y$ are components of an external Zeeman magnetic field.

We now write the action along the closed time contour in terms of the complex bosonic field $b(t)$ (along with the fermionic fields for $\hat{c}$ and $\hat{d}$ operators), go to their classical and quantum components, and perform the Gaussian integration over the fermionic fields. This way we obtain an effective action for the bosonic fields

$$S = \int \mathrm{d}t \left\{\left[\bar{b}^{\mathrm{q}}(\mathrm{i}\partial_t + 2H_z)\,b^{\mathrm{cl}} - 2H_+\sqrt{S}\,\bar{b}^{\mathrm{q}} + \text{c.c.}\right] - \mathrm{i}\,\mathrm{Tr}\ln\begin{pmatrix}\hat{G}^{-1}_{(d)}[b^{\mathrm{cl}}, b^{\mathrm{q}}] & \hat{W}^\dagger \\ \hat{W} & \hat{G}^{-1}_{(c)}\end{pmatrix}\right\}, \tag{13.54}$$

where all elements of the matrix in the last term are matrices in the Keldysh and spin spaces, with $\hat{W}^{\sigma\sigma'}$ diagonal in the Keldysh space. The inverse Green function of the free layer is given by

$$\hat{G}^{-1}_{(d)} = \hat{G}^{-1}_{l\sigma'} + J\sqrt{S}\left[\hat{s}_- b^\alpha + \hat{s}_+\bar{b}^\alpha\right]\hat{\gamma}^\alpha$$
$$- \frac{J\hat{s}_3}{2}\left[(\bar{b}^{\mathrm{cl}}b^{\mathrm{cl}} + \bar{b}^{\mathrm{q}}b^{\mathrm{q}})\hat{\gamma}^{\mathrm{cl}} + (\bar{b}^{\mathrm{cl}}b^{\mathrm{q}} + \bar{b}^{\mathrm{q}}b^{\mathrm{cl}})\hat{\gamma}^{\mathrm{q}}\right], \tag{13.55}$$

where $\hat{G}^{-1}_{l\sigma'} = \hat{\gamma}^{\mathrm{q}}(\mathrm{i}\partial_t - \epsilon_l + \sigma' JS)$. As mentioned, we restrict ourselves to first and second orders in the bosonic operators. For simplicity we shall assume that the tunneling is weak and keep only second order in the tunneling matrix elements. We thus expand $\mathrm{Tr}\ln[1 + \check{G}(\check{W} + \check{G}^{-1}_{(d)} - \check{G}^{-1}_{l\sigma'})]$ in Eq. (13.54) to first and second order in b^{cl} and b^{q}, keeping only terms of second order in $\hat{W}$. According to the fundamental normalization $S[b^{\mathrm{cl}}, 0] = 0$, there can be no terms linear in b^{cl}. Terms linear in b^{q} are given by the diagram of Fig. 13.4(a), which leads to

$$S_1 = \frac{J\sqrt{S}}{\mathrm{i}}\mathrm{Tr}\Big\{b^{\mathrm{q}}\sum_{kl,\sigma}\bar{W}_{kl}^{\sigma\downarrow}\hat{G}_{k\sigma}^{(c)}W_{kl}^{\sigma\uparrow}\hat{G}_{l\uparrow}\hat{\gamma}^{\mathrm{q}}\hat{G}_{l\downarrow} + \bar{b}^{\mathrm{q}}\sum_{kl,\sigma}\bar{W}_{kl}^{\sigma\uparrow}\hat{G}_{k\sigma}^{(c)}W_{kl}^{\sigma\downarrow}\hat{G}_{l\downarrow}\hat{\gamma}^{\mathrm{q}}\hat{G}_{l\uparrow}\Big\}$$

$$= \frac{J\sqrt{S}}{2\mathrm{i}}\int \mathrm{d}t\left[b^{\mathrm{q}}\sin\theta\,\mathrm{e}^{-\mathrm{i}\phi}\sum_{kl}|W_{kl}|^2\mathrm{Tr}\{\hat{G}_{k\downarrow}^{(c)}\hat{G}_{l\uparrow}\hat{\gamma}^{\mathrm{q}}\hat{G}_{l\downarrow} - \hat{G}_{k\uparrow}^{(c)}\hat{G}_{l\uparrow}\hat{\gamma}^{\mathrm{q}}\hat{G}_{l\downarrow}\} + \mathrm{c.c.}\right],$$

where we employed Eq. (13.51) for the tunneling matrix elements $W_{kl}^{\sigma\sigma'}$. The corresponding traces are given by

$$\mathrm{Tr}\{\hat{G}_{k\sigma}^{(c)}\hat{G}_{l\uparrow}\hat{\gamma}^{\mathrm{q}}\hat{G}_{l\downarrow}\} = \int\frac{d\epsilon}{2\pi}\left[G_{k\sigma}^{(c)\mathrm{R}}G_{l\uparrow}^{\mathrm{K}}G_{l\downarrow}^{\mathrm{R}} + G_{k\sigma}^{(c)\mathrm{A}}G_{l\uparrow}^{\mathrm{A}}G_{l\downarrow}^{\mathrm{K}} + G_{k\sigma}^{(c)\mathrm{K}}G_{l\uparrow}^{\mathrm{A}}G_{l\downarrow}^{\mathrm{R}}\right]$$

$$= \int\frac{d\epsilon}{2\pi}\left[\sum_{\pm}\frac{-2\pi\mathrm{i}F_d(\epsilon)\delta(\epsilon-\epsilon_l\pm JS)}{(\epsilon^{\pm}-\epsilon_{k\sigma})(\epsilon^{\pm}-\epsilon_l\mp JS)} + \frac{-2\pi\mathrm{i}F_c(\epsilon)\delta(\epsilon-\epsilon_{k\sigma})}{(\epsilon^{-}-\epsilon_l+JS)(\epsilon^{+}-\epsilon_l-JS)}\right]$$

$$= \sum_{\pm}\frac{\pm\mathrm{i}}{2JS}\frac{F_d(\epsilon_l\mp JS)}{(\epsilon_l^{\pm}\mp JS-\epsilon_{k\sigma})} + \frac{\mathrm{i}F_c(\epsilon_{k\sigma})}{2JS}\left[\frac{1}{\epsilon_{k\sigma}^{-}-\epsilon_l+JS} - \frac{1}{\epsilon_{k\sigma}^{+}-\epsilon_l-JS}\right]$$

$$= \frac{\pi}{2JS}\sum_{\pm}\left[F_d(\epsilon_l\mp JS) - F_c(\epsilon_{k\sigma})\right]\delta(\epsilon_{k\sigma}-\epsilon_l\pm JS),$$

where $\epsilon^{\pm} = \epsilon \pm \mathrm{i}0$ and we neglected the principal parts of the Green functions, since they disappear upon summation over momenta k and l. We can now introduce spin components of the tunneling current between the fixed and free layers as

$$I_{\sigma\sigma'} = 4\pi s\sum_{kl}|W_{kl}|^2\left[n_c(\epsilon_{k\sigma}) - n_d(\epsilon_{l\sigma'})\right]\delta(\epsilon_{k\sigma}-\epsilon_{l\sigma'}), \tag{13.56}$$

where $n_{c(d)} = (1-F_{c(d)})/2$ are occupation numbers of the fixed (free) layers, $\epsilon_{l\sigma'} = \epsilon_l - \sigma' JS$, and $s = 1/2$ is the spin of an electron. The total spin current is given by

$$I_{\mathrm{s}} = \sum_{\sigma\sigma'}\sigma I_{\sigma\sigma'} = I_{\uparrow\uparrow} + I_{\uparrow\downarrow} - I_{\downarrow\uparrow} - I_{\downarrow\downarrow}. \tag{13.57}$$

With this notation one finds for the part of the action linear in $b(t)$ fields

$$S_1 = \frac{\mathrm{i}I_{\mathrm{s}}}{\sqrt{S}}\int \mathrm{d}t\left[b^{\mathrm{q}}(t)\sin\theta\,\mathrm{e}^{-\mathrm{i}\phi} - \bar{b}^{\mathrm{q}}(t)\sin\theta\,\mathrm{e}^{\mathrm{i}\phi}\right]. \tag{13.58}$$

Before tackling terms quadratic in the b-fields, let us discuss what we have found so far. Taking the first term on the right-hand side of Eq. (13.54) along with the first order induced action (13.58), one finds

$$S_0 + S_1 = \int \mathrm{d}t\Big[\bar{b}^{\mathrm{q}}(\mathrm{i}\partial_t + 2H_z)\,b^{\mathrm{cl}} - 2H_+\sqrt{S}\,\bar{b}^{\mathrm{q}} - \frac{\mathrm{i}I_{\mathrm{s}}}{\sqrt{S}}\,\bar{b}^{\mathrm{q}}\sin\theta\,\mathrm{e}^{\mathrm{i}\phi} + \mathrm{c.c.}\Big]. \tag{13.59}$$

Since this action is linear in $\bar{b}^{\mathrm{q}}$ and b^{q}, integration over them results in the functional delta-function, imposing classical equations of motion on b^{cl} and $\bar{b}^{\mathrm{cl}}$, respectively. Employing the fact that $b^{\mathrm{cl}} = S_+/\sqrt{S}$, one finds

$$\mathrm{i}\partial_t S_+ = -2(H_z S_+ - H_+ S) + \mathrm{i}I_{\mathrm{s}} \sin\theta\,(\cos\phi + \mathrm{i}\sin\phi), \tag{13.60}$$

along with the complex conjugate equation for S_-. The imaginary and real parts provide equations for S_x and S_y. These equations are written in the *instantaneous* reference frame, where the spin $\mathbf{S}$ points in the z-direction (i.e. $S_\pm = 0$ and $S_z = S$). The latter has relative polar angles θ and ϕ with the reference frame of the fixed layer. One can define a vector $\mathbf{I}_{\mathrm{s}}$ with length I_{s} and the direction of the fixed layer polarization and rewrite Eq. (13.60) in the vector form:

$$\partial_t \mathbf{S} = 2[\mathbf{H} \times \mathbf{S}] + \frac{1}{S^2}\big[\mathbf{S} \times [\,\mathbf{I}_{\mathrm{s}} \times \mathbf{S}]\big]. \tag{13.61}$$

The first term here is the Landau–Lifshitz precession, Eq. (11.37), while the last one is the Slonczewski–Berger [220, 221] spin-torque contribution. It is easy to check that its angular dependence is exactly one, which we derived for the tunneling spin-current, the last term in Eq. (13.60). Notice also that it is independent of the exchange coupling constant J. For the positive spin-current (i.e. flowing from the fixed to the free layer), the spin torque forces the spin of the free layer to align with the direction of the fixed layer polarization.

We turn now to the terms quadratic in the b-fields. There are two kinds of them, $\bar{b}^{\mathrm{q}} b^{\mathrm{cl}} + \mathrm{c.c.}$ and $|b^{\mathrm{q}}|^2$ (indeed, $|b^{\mathrm{cl}}|^2$ is forbidden by the normalization). As we shall see, they describe dissipation and noise, respectively. In equilibrium, that is, in the absence of the spin-current, the two are related through FDT. In the presence of the spin-current, however, the noise has an additional spin shot-noise component, which is a subject of the subsequent discussion. We first focus on second order in the $[\hat{\sigma}_- b^\alpha + \hat{\sigma}_+ \bar{b}^\alpha]\hat{\gamma}^\alpha$ terms, which leads to the diagrams depicted in Fig. 13.4(b),(c). These diagrams lead to

$$S_2 = \frac{\mathrm{i}}{S}\iint \mathrm{d}t\mathrm{d}t'\,\big(\bar{b}^{\mathrm{cl}}(t), \bar{b}^{\mathrm{q}}(t)\big)\begin{pmatrix} 0 & D^{\mathrm{A}}(t-t') \\ D^{\mathrm{R}}(t-t') & D^{\mathrm{K}}(t-t') \end{pmatrix}\begin{pmatrix} b^{\mathrm{cl}}(t') \\ b^{\mathrm{q}}(t') \end{pmatrix}, \tag{13.62}$$

where

$$\begin{aligned} D^{\mathrm{R}}(\omega) \;=&\; -\,(JS)^2 \sum_{kl,\sigma}\Big[\big|W_{kl}^{\sigma\uparrow}\big|^2\, \mathrm{Tr}\big\{\hat{G}^{(c)}_{k\sigma}(\epsilon)\hat{G}_{l\uparrow}(\epsilon)\hat{\gamma}^{\mathrm{cl}}\hat{G}_{l\downarrow}(\epsilon+\omega)\hat{\gamma}^{\mathrm{q}}\hat{G}_{l\uparrow}(\epsilon)\big\} \\ &+ \big|W_{kl}^{\sigma\downarrow}\big|^2\, \mathrm{Tr}\big\{\hat{G}^{(c)}_{k\sigma}(\epsilon)\hat{G}_{l\downarrow}(\epsilon)\hat{\gamma}^{\mathrm{q}}\hat{G}_{l\uparrow}(\epsilon-\omega)\hat{\gamma}^{\mathrm{cl}}\hat{G}_{l\downarrow}(\epsilon)\big\}\Big], \end{aligned}$$

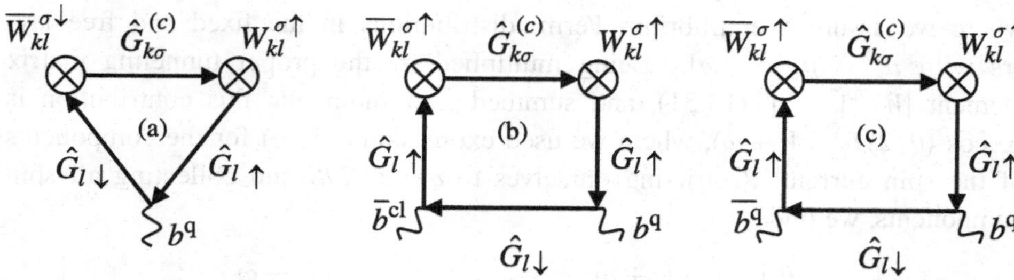

Figure 13.4 Contributions to the action in second order in tunneling amplitudes, $W_{kl}^{\sigma,\sigma'}$, denoted by crossed circles. (a) The S_1 action, linear in $b^{\mathrm{q}}(t)$, leads to the spin-torque term; (b) part of the S_2 action, proportional to $\bar{b}^{\mathrm{cl}}(t)b^{\mathrm{q}}(t')$, leads to the damping term; (c) part of the S_2 action, proportional to $\bar{b}^{\mathrm{q}}(t)b^{\mathrm{q}}(t')$, leads to the noise cumulant. Fields b^{α} flip down the spin of electrons in the free layer; $\bar{b}^{\alpha}$ flip it up.

while

$$D^{\mathrm{A}}(\omega) = -(JS)^2\sum_{kl,\sigma}\Big[\big|W_{kl}^{\sigma\uparrow}\big|^2\,\mathrm{Tr}\big\{\hat{G}_{k\sigma}^{(c)}(\epsilon)\hat{G}_{l\uparrow}(\epsilon)\hat{\gamma}^{\mathrm{q}}\hat{G}_{l\downarrow}(\epsilon+\omega)\hat{\gamma}^{\mathrm{cl}}\hat{G}_{l\uparrow}(\epsilon)\big\}$$
$$+\big|W_{kl}^{\sigma\downarrow}\big|^2\,\mathrm{Tr}\big\{\hat{G}_{k\sigma}^{(c)}(\epsilon)\hat{G}_{l\downarrow}(\epsilon)\hat{\gamma}^{\mathrm{cl}}\hat{G}_{l\uparrow}(\epsilon-\omega)\hat{\gamma}^{\mathrm{q}}\hat{G}_{l\downarrow}(\epsilon)\big\}\Big].$$

Notice that transformation from the retarded to the advanced component requires not only changing $\omega \to -\omega$, but also flipping all the spin indices, in particular $\sigma' \to -\sigma'$. This is a manifestation of the time-reversal symmetry in quantum mechanics with spin [36]. Finally,

$$D^{\mathrm{K}}(\omega) = -(JS)^2\sum_{kl,\sigma\sigma'}\big|W_{kl}^{\sigma\sigma'}\big|^2\,\mathrm{Tr}\big\{\hat{G}_{k\sigma}^{(c)}(\epsilon)\hat{G}_{l\sigma'}(\epsilon)\hat{\gamma}^{\mathrm{q}}\hat{G}_{l-\sigma'}(\epsilon+\sigma'\omega)\hat{\gamma}^{\mathrm{q}}\hat{G}_{l\sigma'}(\epsilon)\big\}.$$

Calculation of the traces of the four matrix Green functions is lengthy, but straightforward. For example, terms contributing to D^{K} for a particular spin configuration read as[7]

$$\hat{G}_{k\uparrow}^{(c)\mathrm{R}}\hat{G}_{l\uparrow}^{\mathrm{R}}\hat{G}_{l\downarrow}^{\mathrm{A}}(\epsilon+\omega)\hat{G}_{l\uparrow}^{\mathrm{R}} + \hat{G}_{k\uparrow}^{(c)\mathrm{A}}\hat{G}_{l\uparrow}^{\mathrm{A}}\hat{G}_{l\downarrow}^{\mathrm{R}}(\epsilon+\omega)\hat{G}_{l\uparrow}^{\mathrm{A}} + \hat{G}_{k\uparrow}^{(c)\mathrm{K}}\hat{G}_{l\uparrow}^{\mathrm{A}}\hat{G}_{l\downarrow}^{\mathrm{K}}(\epsilon+\omega)\hat{G}_{l\uparrow}^{\mathrm{R}},$$

if the energy argument is not written explicitly it is understood as ϵ. Integrating over the energy and employing $\epsilon_{l\sigma'} = \epsilon_l - \sigma' JS$, one finds for this specific contribution

$$\frac{2\pi}{(2JS-\omega)^2}\big[1 - F_c(\epsilon_{k\uparrow})F_d(\epsilon_{l\downarrow})\big]\delta(\epsilon_{k\uparrow} - \epsilon_{l\downarrow} + \omega)$$
$$= \frac{2\pi}{(2JS-\omega)^2}\,\coth\frac{eV+\omega}{2T}\,\big[F_c(\epsilon_{k\uparrow}) - F_d(\epsilon_{l\downarrow})\big]\delta(\epsilon_{k\uparrow} - \epsilon_{l\downarrow} + \omega),$$

[7] We have omitted terms $\propto G_{l\downarrow}^{\mathrm{K}}(\epsilon+\omega)G_{l\uparrow}^{\mathrm{K}}(\epsilon) \sim \delta(\epsilon+\omega-\epsilon_l-JS)\delta(\epsilon-\epsilon_l+JS)$, since the two delta-functions are incompatible unless $\omega = 2JS$.

where we assumed equilibrium Fermi distributions in the fixed and free layers with $\mu_c - \mu_d = eV$. Being multiplied by the proper tunneling matrix element $|W_{kl}^{\uparrow\uparrow}|^2$, Eq. (13.51), and summed over momenta, this contribution is $\propto \cos^2(\theta/2) I_{\uparrow\downarrow}(eV+\omega)$, where we used expression (13.56) for the components of the spin current. Restricting ourselves to $\omega \ll 2JS$ and collecting all spin components, we find

$$D^{\mathrm{K}}(\omega) = \frac{1}{2}\cos^2\frac{\theta}{2}\left[\coth\frac{eV+\omega}{2T} I_{\uparrow\downarrow}(eV+\omega) + \coth\frac{eV-\omega}{2T} I_{\downarrow\uparrow}(eV-\omega)\right] + \frac{1}{2}\sin^2\frac{\theta}{2}\left[\coth\frac{eV+\omega}{2T} I_{\downarrow\downarrow}(eV+\omega) + \coth\frac{eV-\omega}{2T} I_{\uparrow\uparrow}(eV-\omega)\right]. \tag{13.63}$$

The very similar calculations for $D^{\mathrm{R(A)}}$ yield

$$\begin{aligned} D^{\mathrm{R}}(\omega) &= \frac{1}{4}\cos^2\frac{\theta}{2}\Big[I_{\uparrow\downarrow}(eV+\omega) - I_{\downarrow\uparrow}(eV-\omega)\Big] + \frac{1}{4}\sin^2\frac{\theta}{2}\Big[I_{\downarrow\downarrow}(eV+\omega) - I_{\uparrow\uparrow}(eV-\omega)\Big], \\ D^{\mathrm{A}}(\omega) &= \frac{1}{4}\cos^2\frac{\theta}{2}\Big[I_{\downarrow\uparrow}(eV-\omega) - I_{\uparrow\downarrow}(eV+\omega)\Big] + \frac{1}{4}\sin^2\frac{\theta}{2}\Big[I_{\uparrow\uparrow}(eV-\omega) - I_{\downarrow\downarrow}(eV+\omega)\Big]. \end{aligned} \tag{13.64}$$

In equilibrium, $eV = 0$, FDT is satisfied, that is, $D^{\mathrm{K}} = \coth(\omega/2T)[D^{\mathrm{R}} - D^{\mathrm{A}}]$ (notice that having $\omega \leftrightarrow -\omega$ and $\uparrow\leftrightarrow\downarrow$ between the retarded and advanced components is crucial for FDT validity). We focus here on the opposite limit of strong non-equilibrium and small frequencies, $\omega \ll eV$. In the Keldysh component one may put $\omega = 0$, while the retarded/advanced components need to be expanded to first order in ω. The reason for the latter is that $\mathrm{i}D^{\mathrm{R}}(0) = -\mathrm{i}D^{\mathrm{A}}(0)$ enter the action (13.62) as an effective z-component of the magnetic field H_z; see Eq. (13.59). In the instantaneous reference frame, where $S_z = S$ and $S_x = S_y = 0$, the H_z component drops from the equation of motion (13.60) and does not influence the dynamics. We thus drop $D^{\mathrm{R(A)}}(0)$,[8] but keep the first order in ω, since $\mathrm{i}\bar{b}^{\mathrm{q}}\omega b^{\mathrm{cl}} \to -\bar{b}^{\mathrm{q}}\partial_t b^{\mathrm{cl}}$. As a result, the low-frequency second-order action (13.62) acquires a time-local dissipative form (cf. Eq. (4.2)),

$$S_2 = \frac{1}{S}\int dt\,\left(\bar{b}^{\mathrm{cl}}(t), \bar{b}^{\mathrm{q}}(t)\right)\begin{pmatrix} 0 & \alpha\partial_t \\ -\alpha\partial_t & 2\mathrm{i}D \end{pmatrix}\begin{pmatrix} b^{\mathrm{cl}}(t) \\ b^{\mathrm{q}}(t) \end{pmatrix}, \tag{13.65}$$

[8] For the same reason we drop terms originating from the expansion of $\mathrm{Tr}\ln(1+\cdots)$ to first order in the very last term on the right-hand side of Eq. (13.55). Indeed, all such terms are frequency-independent and thus inconsequential.

where the dissipation and noise amplitude are given by [273]

$$\alpha(\theta,V)=\frac{\partial I_{\rm sf}(\theta,V)}{\partial eV}\,;\qquad \mathcal{D}(\theta,V)=\coth\left(\frac{eV}{2T}\right)I_{\rm sf}(\theta,V). \tag{13.66}$$

We have introduced here the spin-flip current, defined as

$$I_{\rm sf}(\theta,V)=\frac{1}{4}\cos^2\frac{\theta}{2}\Big[I_{\uparrow\downarrow}(eV)+I_{\downarrow\uparrow}(eV)\Big]+\frac{1}{4}\sin^2\frac{\theta}{2}\Big[I_{\downarrow\downarrow}(eV)+I_{\uparrow\uparrow}(eV)\Big]. \tag{13.67}$$

To proceed it is convenient to split the $e^{-2\mathcal{D}|b^{q}|^2/S}$ part of e^{iS_2} using the Hubbard–Stratonovich transformation with the auxiliary complex field $j(t)$,

$$e^{-\frac{2\mathcal{D}}{S}\int dt\,|b^{q}(t)|^2}=\int \mathbf{D}[\bar{j},j]\,e^{-\int dt\left[\frac{1}{2\mathcal{D}}|j(t)|^2+\frac{i}{\sqrt{S}}\bar{b}^{q}(t)j(t)+\frac{i}{\sqrt{S}}\bar{j}(t)\,b^{q}(t)\right]}.$$

Then the action takes the form $S_2=-\int dt\,\bar{b}^{q}\big((\alpha/S)\partial_t b^{cl}+j(t)/\sqrt{S}\big)+\text{c.c.}$, which should be added to the S_0+S_1 part (13.59). Being still linear in $\bar{b}^{q}$, the combined action imposes the classical equation of motion for $b^{cl}(t)=S_+(t)/\sqrt{S}$. In the instantaneous rotating reference frame, where $S_\pm=0$ and $S_z=S$, it takes the form $i\partial_t S_+=\alpha\partial_t S_+/S+j(t)+\cdots$, while $i\partial_t S_z=0$. It is convenient to denote real and imaginary parts of the complex function j as $-j_y$ and j_x, respectively. Transforming then to the laboratory frame, one finally finds

$$\partial_t\mathbf{S}=2\big[\mathbf{H}\times\mathbf{S}\big]-\frac{\alpha}{S}\big[\mathbf{S}\times\partial_t\mathbf{S}\big]+\frac{1}{S^2}\big[\mathbf{S}\times[(\mathbf{I}_s+\mathbf{j}(t))\times\mathbf{S}]\big]. \tag{13.68}$$

The α-term here is Gilbert damping, which forces the spin $\mathbf{S}$ to align against the external magnetic field $\mathbf{H}$. In our model the damping is associated with the tunneling of the spin-polarized electrons from the fixed layer and equilibrating them with the instantaneous spin polarization direction of the free layer. The Gilbert damping constant α is thus given by the spin-flip tunneling conductance, as defined by Eqs. (13.66) and (13.67). In general, one should also add another contribution, say α_0, which accounts for all other mechanisms of dissipation.

The dissipation also induces fluctuations, which enter the equation as the fluctuating part of the spin-current $\mathbf{j}(t)$.[9] Since the random complex field j has Gaussian statistics with $\langle j(t)\bar{j}(t')\rangle=\langle j_y j_y\rangle+\langle j_x j_x\rangle=2\mathcal{D}\delta(t-t')$, one notices that the vector noise $\mathbf{j}(t)$ has an isotropic Gaussian correlator

$$\langle j_\mu(t)j_\nu(t')\rangle=\mathcal{D}(\theta,V)\,\delta(t-t')\,\delta_{\mu\nu}, \tag{13.69}$$

given by Eqs. (13.66) and (13.67). Notice that for small voltage, $eV\ll T$, the spin-flip current $I_{\rm sf}$ is linear in voltage, and therefore $\mathcal{D}=2T\alpha$. In this limit thus

[9] One could equally well rewrite the fluctuating term as the random part of the magnetic field $\mathbf{H}\to\mathbf{H}+\mathbf{h}(t)$, where $\mathbf{h}(t)$ is a Gaussian isotropic random vector with the correlator $\langle h_\mu(t)h_\nu(t')\rangle=\delta_{\mu\nu}\delta(t-t')\mathcal{D}/4S^2$.

$\langle j_\mu j_\nu \rangle = 2T\alpha\,\delta_{\mu\nu}$ in agreement with the FDT equilibrium magnetic noise, first discussed by Brown [274]. In the opposite limit $eV \gg T$, Eqs. (13.66)–(13.69) describe spin shot noise [273] with the correlator $\langle jj \rangle \approx 2s|I_{\rm sf}|$, where $s = 1/2$. This result should be compared with the Schottky (tunneling limit) charge shot noise $\langle \delta J \delta J \rangle = 2e|J|$, Eq. (13.19). The difference, which stems from the vector nature of the spin degree of freedom, is that the spin noise is not proportional to the spin-current $I_{\rm s}$ itself, but rather to a somewhat different object – the spin-flip current $I_{\rm sf}$.

13.7 Problems

13.7.1 Meir–Wingreen Formula

Consider electron transport through an interacting quantum dot, sandwiched between two noninteracting metallic leads, labeled as $s = \mathrm{L}, \mathrm{R}$. The Hamiltonian is given by

$$\hat{H} = \sum_{k,s} \epsilon_{k,s} \hat{c}^\dagger_{k,s} \hat{c}_{k,s} + \hat{H}_{\rm dot}[\hat{d}^\dagger_n, \hat{d}_n] + \sum_{n,k,s} \left[W^{(s)}_{kn} \hat{c}^\dagger_{k,s} \hat{d}_n + \text{h.c.} \right], \tag{13.70}$$

where n denotes some set of quantum numbers characterizing states of the dot and operators $\hat{d}^\dagger_n (\hat{d}_n)$ create (annihilate) electrons in these states; $\hat{H}_{\rm dot}$ is a Hamiltonian of the isolated dot, which may include arbitrary interactions. Finally, $W^{(\mathrm{L,R})}_{kn}$ are tunneling matrix elements between the leads and the dot.

Define current operators as $\hat{J}_s = \partial_t \hat{N}_s = \mathrm{i}[\hat{H}, \hat{N}_s]$, where the particle number operators in the leads are $\hat{N}_s = \sum_k \hat{c}^\dagger_{k,s} \hat{c}_{k,s}$. Perform commutations and show that $\hat{J}_s = \mathrm{i} \sum_{n,k} \left[W^{(s)}_{kn} \hat{c}^\dagger_{k,s} \hat{d}_n - \text{h.c.} \right]$.

Write down the Keldysh action, corresponding to the Hamiltonian (13.70), where two-component Grassmann vectors $\bar{\psi}_n, \psi_n$ encode $\hat{d}^\dagger_n, \hat{d}_n$ operators and $\bar{\chi}_{k,s}, \chi_{k,s}$ encode $\hat{c}^\dagger_{k,s}, \hat{c}_{k,s}$ operators. One includes quantum sources, which generate the two currents, as $\mathrm{i} \sum_s A^{\rm q}_s \sum_{n,k} \left[W^{(s)}_{kn} \, \bar{\chi}_{k,s} \hat{\gamma}^{\rm q} \, \psi_n - W^{*(s)}_{kn} \, \bar{\psi}_n \hat{\gamma}^{\rm q} \, \chi_{k,s} \right]$ term in the action. This way, the expectation values of the current operators are given by $J_s = \frac{\mathrm{i}}{2} \delta Z[A^{\rm q}_{\rm L}, A^{\rm q}_{\rm R}] / \delta A^{\rm q}_s |_{A^{\rm q}_s = 0}$, cf. (10.46). Use now the fact that the leads are noninteracting to perform the Gaussian integrations over $\bar{\chi}_{k,s}, \chi_{k,s}$ variables to find $Z[A^{\rm q}_{\rm L}, A^{\rm q}_{\rm R}] = \int \mathbf{D}[\bar{\psi}, \psi] \, \mathrm{e}^{\mathrm{i} S_{\rm eff}[\bar{\psi}, \psi; A^{\rm q}_{\rm L}, A^{\rm q}_{\rm R}]}$, with the effective action

$$\begin{aligned} S_{\rm eff}[\bar{\psi}, \psi; A^{\rm q}_{\rm L}, A^{\rm q}_{\rm R}] &= \int \mathrm{d}t \left[\sum_n \bar{\psi}_n \mathrm{i} \partial_t \psi_n - H_{\rm dot}[\bar{\psi}_n, \psi_n] \right] \\ &\quad - \iint \mathrm{d}t \mathrm{d}t' \sum_{n,n',s} \bar{\psi}_n(t) [1 - \mathrm{i} A^{\rm q}_s \hat{\gamma}^{\rm q}] \, \hat{\Gamma}^{(s)}_{nn'}(t - t') [1 + \mathrm{i} A^{\rm q}_s \hat{\gamma}^{\rm q}] \psi_{n'}(t'), \end{aligned} \tag{13.71}$$

where $\hat{\Gamma}^{(s)}_{nn'}(t-t') = \sum_k W^{*(s)}_{kn}\hat{G}^{(s)}_k(t-t')W^{(s)}_{kn'}$ and $\hat{G}^{(s)}_k(t-t')$ is the Keldysh matrix Green function of the lead s. Since the leads are assumed to be in local equilibrium, the FDT dictates that $\Gamma^{(s),\mathrm{K}}_{nn'}(\epsilon) = F_s(\epsilon)\big(\Gamma^{(s),\mathrm{R}}_{nn'}(\epsilon) - \Gamma^{(s),\mathrm{A}}_{nn'}(\epsilon)\big)$, where $F_s(\epsilon) = \tanh(\epsilon - \mu_s)/2T_s$.

Differentiate the effective action over the quantum sources to show that

$$J_s = \frac{1}{2}\int \frac{d\epsilon}{2\pi}\,\mathrm{Tr}\{[\hat{\Gamma}^{(s)}(\epsilon), \hat{\gamma}^q]\hat{G}(\epsilon)\} = \frac{\mathrm{i}}{2}\int d\epsilon\, \mathbf{\Gamma}^s_{nn'}\big[G^{\mathrm{K}}_{n'n} - F_s(G^{\mathrm{R}}_{n'n} - G^{\mathrm{A}}_{n'n})\big]. \quad (13.72)$$

Here $\hat{G}_{n'n} = -\mathrm{i}\langle\psi_{n'}\bar{\psi}_n\rangle$ are components of the *exact* Green function of the dot, where $\langle\ldots\rangle$ is averaging with the action $S = \bar{\psi}\mathrm{i}\partial_t\psi - H_{\mathrm{dot}}[\bar{\psi},\psi] - \bar{\psi}(\hat{\Gamma}^{(R)} + \hat{\Gamma}^{(L)})\psi$. We have also introduced the notation $\mathbf{\Gamma}^s_{nn'}(\epsilon) = \sum_k W^{*(s)}_{kn}\delta(\epsilon - \epsilon_{k,s})W^{(s)}_{kn'}$, where $\hat{G}^{(s),\mathrm{R}}_k(\epsilon) - \hat{G}^{(s),\mathrm{A}}_k(\epsilon) = -2\pi\mathrm{i}\delta(\epsilon - \epsilon_{k,s})$. Verify that, if the dot and the two leads are in equilibrium, then $J_{\mathrm{L}} = J_{\mathrm{R}} = 0$.

In a non-equilibrium stationary state $J_{\mathrm{L}} = -J_{\mathrm{R}}$. This leads to the expression for the dc current through the dot, $J = (J_{\mathrm{L}} - J_{\mathrm{R}})/2$:

$$J = \frac{\mathrm{i}}{4}\int d\epsilon\,\mathrm{Tr}\Big\{(\mathbf{\Gamma}^{\mathrm{L}} - \mathbf{\Gamma}^{\mathrm{R}})G^{\mathrm{K}} - (\mathbf{\Gamma}^{\mathrm{L}}F_{\mathrm{L}} - \mathbf{\Gamma}^{\mathrm{R}}F_{\mathrm{R}})(G^{\mathrm{R}} - G^{\mathrm{A}})\Big\}, \quad (13.73)$$

where the trace is understood over the orbital n, n' indexes. This result is known as Meir–Wingreen formula [275]. It expresses the current through the $\Gamma^{\mathrm{L,R}}(\epsilon)$ coupling matrices and the exact Green functions of the interacting dot. The latter refer *not* to the isolated dot, but rather to the dot coupled to the leads. This coupling is included through the $\bar{\psi}(\hat{\Gamma}^{(R)} + \hat{\Gamma}^{(L)})\psi$ term in the dot action.

Consider a particular case $\mathbf{\Gamma}^R(\epsilon) = \lambda\,\mathbf{\Gamma}^L(\epsilon)$, [275]. Let's also assume that $G^{\mathrm{K}}_{n'n}(\epsilon) = F_{\mathrm{dot}}(\epsilon)(G^{\mathrm{R}}_{n'n}(\epsilon) - G^{\mathrm{A}}_{n'n}(\epsilon))$ with the distribution function $F_{\mathrm{dot}}(\epsilon)$, which is independent on the orbital indexes. Show that the current conservation, $J_{\mathrm{L}} = -J_{\mathrm{R}}$, implies $F_{\mathrm{dot}} = (\lambda F_R + F_L)/(\lambda + 1)$ and therefore

$$J = \int d\epsilon\,(F_{\mathrm{R}}(\epsilon) - F_{\mathrm{L}}(\epsilon))\,\mathrm{Im}\,\mathrm{Tr}\left\{\frac{\mathbf{\Gamma}^R\mathbf{\Gamma}^L}{\mathbf{\Gamma}^R + \mathbf{\Gamma}^L}G^{\mathrm{R}}\right\}. \quad (13.74)$$

Although this expression looks like its noninteracting analog, it may still fully incorporate interaction effects in the dot through $G^{\mathrm{R}}(\epsilon)$.

13.7.2 Levitons: Optimal Voltage Pulses

The counting statistics of transmitted charge, given by Eq. (13.30), is derived for a constant voltage, V. One may generalize it for a time-dependent voltage, given by a periodic sequence of pulses. In particular, it is instructive to consider pulses with a period T_0, such that $\int_0^{T_0} dt\, eV(t) = 2\pi m\hbar$, where m is an integer. Hereafter, we put

$m = 1$, but this does not affect the conclusions. To this end, notice that the diagonal matrix $\hat{n}_F = \text{diag}\{n(\epsilon - eV), n(\epsilon)\}$ may be written in the time representation as

$$\hat{n}_F(t,t') = \begin{pmatrix} e^{-i\Phi(t)} & 0 \\ 0 & 1 \end{pmatrix} n(t-t')\hat{\mathbb{1}} \begin{pmatrix} e^{i\Phi(t')} & 0 \\ 0 & 1 \end{pmatrix}, \tag{13.75}$$

where $n(t-t')$ is the Fourier transform of the equilibrium distribution $n(\epsilon)$. Since we operate in the basis of T_0-periodic functions, the energies are quantized in units $\epsilon_n = 2\pi n/T_0$. One thus finds for the zero-temperature distribution function $n(\epsilon_n) = \theta(-\epsilon_n)$:

$$n(t-t') = \frac{1}{T_0}\sum_n n(\epsilon_n) e^{-i\epsilon_n(t-t')} = \frac{i}{2T_0}\frac{e^{i\pi(t-t')/T_0}}{\sin(\pi(t-t')/T_0)}. \tag{13.76}$$

For a constant voltage, $\Phi(t) = eVt$, which may be immediately generalized as $\Phi(t) = \int_0^t eV(t)dt$, or $\partial_t\Phi(t) = eV(t)$. For the normalization of voltage pulses, discussed earlier, $\Phi(T_0) - \Phi(0) = 2\pi$, and therefore $e^{\pm i\Phi(t)}$ is also a T_0-periodic function. Substitute these equations into the generation function (13.30), employ the fact that $\log Z[\eta] = \text{Tr}\log[1 - \hat{n}_F(t,t')(1 - \mathbb{S}^\dagger_{-\eta}\mathbb{S}_\eta)]$, where the trace is understood both in the time space and in the left-right 2×2 space, and calculate the first two irreducible moments of the transmitted charge as [276]

$$\langle q\rangle = e\left.\frac{d\log Z[\eta]}{di\eta}\right|_{\eta=0} = e|\mathbf{t}|^2 N_0 \int_0^{T_0}\frac{dt}{2\pi i}\, e^{-i\Phi(t)}\partial_t e^{i\Phi(t)}; \tag{13.77}$$

$$\langle\langle q^2\rangle\rangle = e^2\left.\frac{d^2\log Z[\eta]}{d\eta^2}\right|_{\eta=0} = e^2|\mathbf{t}|^2|\mathbf{r}|^2 N_0 \iint_0^{T_0}\frac{dt dt'}{T_0^2}\frac{\sin^2\left[\frac{\Phi(t)-\Phi(t')}{2}\right]}{\sin^2[\pi(t-t')/T_0]},$$

where $N_0 = t_0/T_0$ is the number of voltage pulse periods. From here one notices that the first cumulant, that is, the average current, is given by the Landauer formula independently of the shape of the voltage pulses, $\Phi(T_0) - \Phi(0) = 2\pi$. On the other hand, the second cumulant, that is, the noise, depends explicitly on the shape of the pulse. One can therefore ask what pulse shape (or shapes) minimize the noise (with a fixed average current $\langle q\rangle$).

Notice that the expression for the noise coincides with the Ambegoakar–Eckern–Schön action [40] (3.45), used in the theory of the Coulomb blockade; see Problem 3.6.2. Also notice that, while AES action is written in the *imaginary* time, the expression for the noise is in the *real* time. Yet they formally coincide with the periodicity given by $\beta = 1/T$ in the AES case and by the voltage pulse period T_0 in the present case. Therefore, minimization of noise is formally equivalent to looking for stationary configurations of AES action. The latter task is discussed in Problem 3.6.2. Calculate the variation of the noise with respect to $\Phi(t)$ and show

[276] that the resulting extremal condition is satisfied by the Korshunov instantons [41]

$$e^{i\Phi(t)} = \frac{z - z_1}{1 - z\bar{z}_1}, \tag{13.78}$$

where $z = e^{2\pi i t/T_0}$ and z_1 is an arbitrary complex number inside the unit circle, $|z_1| < 1$. Plot $\Phi(t)$ and $V(t)$ for various z_1. Calculate the noise for such pulses and show that it is given by the Lesovik formula $\langle\langle q^2 \rangle\rangle = e^2 |\mathbf{t}|^2 |\mathbf{r}|^2 N_0$, independent of z_1. This shows that the constant voltage, that is, $z_1 = 0$ and thus $\Phi(t) = 2\pi t/T_0$, results in a minimal possible noise. Amazingly this is not the only option: one may choose $|z_1| \to 1$, resulting in short and sharp voltage pulses. Such pulses, called *Levitons* after L. Levitov, carry the same average current and produce the same noise as the time-independent voltage. This observation is used extensively in quantum devices [277].

Part IV

Disordered Metals and Superconductors

14
Disordered Fermionic Systems

This chapter deals with a Fermi gas in the presence of a static random potential. It is first treated with the kinetic equation approach, which leads to the classical diffusion of density perturbations. We then derive an effective low-energy field-theory, known as the Keldysh nonlinear sigma-model. It contains classical diffusion as its stationary point, while fluctuations around such a stationary point provide quantum weak localization corrections as well as a scaling theory of localization transition in $2+\epsilon$ dimensions.

14.1 Kinetic Equation Approach

One is often interested in calculating various observables, such as current or density, in the presence of a static (quenched) space-dependent disorder potential $V_{\text{dis}}(\mathbf{r})$. As any other *single-particle* operator, the disorder potential modifies the quadratic part of the fermionic action, bringing the term

$$
\begin{aligned}
S_{\text{dis}}[V_{\text{dis}}] &= -\int_C \mathrm{d}t \int \mathbf{dr}\; V_{\text{dis}}(\mathbf{r})\, \bar{\psi}(\mathbf{r},t)\psi(\mathbf{r},t) \\
&= -\int \mathbf{dr}\; V_{\text{dis}}(\mathbf{r}) \int_{-\infty}^{\infty} \mathrm{d}t\, \bar{\psi}^a(\mathbf{r},t)\hat{\gamma}^{\text{cl}}_{ab}\psi^b(\mathbf{r},t),
\end{aligned}
\tag{14.1}
$$

where in the second line we used the fact that the disorder potential is purely classical, that is, the same on both branches of the time contour. Although we limit ourselves to a static potential, we still assume that it is adiabatically switched on and off sometime after and before $t=-\infty$. As a result, the initial density matrix $\hat{\rho}_0$ and therefore the factor $\mathrm{Tr}\{\hat{\rho}_0\}$ in Eq. (10.14) are disorder independent. We thus absorb it into the measure of the fermionic functional integration and do not show it explicitly.

Since the exact form of the disorder potential is usually unknown, one is typically interested in average values of observables over an ensemble of realizations of $V_{\rm dis}(\mathbf{r})$. To perform such an averaging one needs to specify certain statistical properties of the disorder potential. It is usually assumed that the latter is Gaussian, that is, that the third and higher irreducible cumulants of $V_{\rm dis}$ are absent. Although in practice this is probably not really the case, the higher cumulants turn out to be irrelevant in the RG sense. As a result they do not affect most of the large-scale observables. Under this assumption, one needs to specify only the second cumulant of the disorder potential:[1]

$$\left\langle V_{\rm dis}(\mathbf{r})V_{\rm dis}(\mathbf{r}')\right\rangle_{\rm dis} = g(\mathbf{r}-\mathbf{r}'). \tag{14.2}$$

We shall see that the Fourier transform $g(\mathbf{q}) = g(-\mathbf{q})$ of the correlation function $g(\mathbf{r})$ specifies the relative probability of scattering with the change of electronic wavenumber by $\mathbf{q}$.

We start by deriving the kinetic equation for the mass-shell electronic distribution function $\tilde{F}(\mathbf{k},\mathbf{r},t)$. In doing so, we shall treat the disorder potential $V_{\rm dis}(\mathbf{r})$ not as a part of the kinetic term (cf. Eq. (11.3)), but rather as a source for the self-energy and thus for the collision integral. Calculating the self-energy, we restrict ourselves to the second-order expansion in $V_{\rm dis}(\mathbf{r})$, which is then averaged over the disorder. This is certainly an approximation, which misses a number of important phenomena such as localization or mesoscopic fluctuations. The systematic way to improve on this approximation is a subject of the subsequent sections. Meanwhile, we proceed with the outlined program. Expanding $e^{iS_{\rm dis}[V_{\rm dis}]}$, Eq. (14.1), to second order and applying the fermionic Wick theorem (10.12), one obtains for the self-energy matrix in the Keldysh space, Fig. 14.1(a),

$$\hat{\Sigma}(x,x') = V_{\rm dis}(\mathbf{r})V_{\rm dis}(\mathbf{r}')\hat{G}(x,x'), \tag{14.3}$$

where $x = (\mathbf{r},t)$ and we used $\hat{\gamma}^{\rm cl}_{ab} = \delta_{ab}$. Performing averaging over disorder according to Eq. (14.2) and going to the Fourier representation, one finds $\hat{\Sigma}(\mathbf{k},\epsilon) = \sum_{\mathbf{k}'} g(\mathbf{k}-\mathbf{k}')\hat{G}(\mathbf{k}',\epsilon)$, Fig. 14.1(b). Employing $G^{\rm K} = F(G^{\rm R}-G^{\rm A})$ and $G^{\rm R}-G^{\rm A} = -2\pi i\delta(\epsilon-\xi_{\mathbf{k}'})$, one finds for the collision integral (11.3)

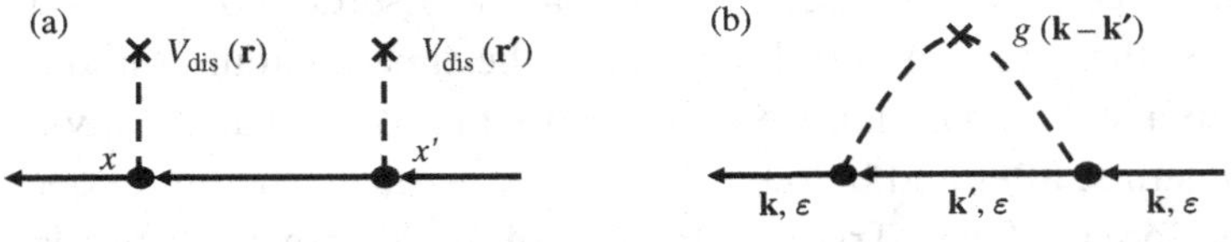

Figure 14.1 Fermionic self-energy to second order in $V_{\rm dis}(\mathbf{r})$: (a) before averaging over the disorder; (b) after the averaging.

[1] The first cumulant $\langle V_{\rm dis}(\mathbf{r})\rangle_{\rm dis}$ may be always absorbed into a redefinition of the chemical potential and thus thought of as being zero.

$$I^{\rm coll} = i\Sigma^{\rm K} - iF(\Sigma^{\rm R} - \Sigma^{\rm A}) = 2\pi \sum_{\mathbf{k}'} g(\mathbf{k}-\mathbf{k}')[\tilde{F}(\mathbf{k}') - \tilde{F}(\mathbf{k})]\delta(\xi_{\mathbf{k}} - \xi_{\mathbf{k}'}), \quad (14.4)$$

where we have restricted the distribution function to the mass-shell by putting $\epsilon = \xi_{\mathbf{k}}$. The collision integral is nothing but the Golden Rule with the scattering probability between $\mathbf{k} \leftrightarrow \mathbf{k}'$ being $g(\mathbf{k}-\mathbf{k}') = g(\mathbf{k}'-\mathbf{k})$, "in" and "out" occupation factors and energy conservation.

To proceed we notice that elastic scattering tends to make the distribution function uniform in momentum directions, that is, dependent mostly on the energy $\xi_{\mathbf{k}} = \epsilon_{\mathbf{k}} - \epsilon_{\rm F}$ and only weakly on the direction, characterized by the unit vector $\mathbf{n_k} = \mathbf{k}/k$. The residual weak dependence on the direction may be represented by the spherical harmonic expansion

$$\tilde{F}(\mathbf{k}, x) = F_0(\epsilon, x) + \mathbf{n_k} \cdot \mathbf{F}_1(\epsilon, x) + \cdots, \quad (14.5)$$

where we used that $\epsilon = \xi_{\mathbf{k}}$ and the first omitted term is $n_k^\mu n_k^\nu F_2^{\mu\nu}(\epsilon)$ with $F_2^{\mu\nu}$ a rank-two traceless tensor. We now employ essentially the same method, which leads from the Fokker–Planck equation (4.32) to the diffusion equation (4.34). We substitute spherical harmonic expansion (14.5) into the kinetic equation (11.3),

$$[\partial_t + \mathbf{v_k}\nabla_{\mathbf{r}}]\tilde{F}(\mathbf{k}, x) = I^{\rm coll}[\tilde{F}], \quad (14.6)$$

with the collision integral (14.4). We now (i) integrate it over the angular directions $\int d\Omega_{\mathbf{k}}$ of the unit vector $\mathbf{n_k}$; (ii) multiply by $\mathbf{n_k}$ and then integrate over the angle. This way one obtains the scalar and vector equations

$$\partial_t F_0 + \frac{v_{\rm F}}{d}\, {\rm div}\mathbf{F}_1 = 0, \quad (14.7)$$

$$\partial_t \mathbf{F}_1 + v_{\rm F}\nabla_{\mathbf{r}} F_0 = -\frac{1}{\tau_{\rm tr}}\,\mathbf{F}_1, \quad (14.8)$$

where we put $\nabla_{\mathbf{k}}\xi_{\mathbf{k}} = v_{\mathbf{k}} \approx v_{\rm F}\mathbf{n_k}$, employed the fact that $\int d\Omega_{\mathbf{k}}\, n_{\mathbf{k}}^\mu n_{\mathbf{k}}^\nu = \delta^{\mu\nu}/d$, where d is the dimensionality of space, and introduced the *transport* scattering time

$$\frac{1}{\tau_{\rm tr}} = 2\pi \sum_{\mathbf{k}'} g(\mathbf{k}-\mathbf{k}')[1 - \mathbf{n_k}\cdot\mathbf{n_{k'}}]\delta(\xi_{\mathbf{k}} - \xi_{\mathbf{k}'}) = 2\pi\nu \int_0^\pi \frac{d\theta}{\pi}\, g(\theta)(1-\cos\theta), \quad (14.9)$$

where ν is the density of states, defined as $\sum_{\mathbf{k}} \ldots = \nu \int d\xi_{\mathbf{k}} \ldots$. In the last expression θ is an angle between $\mathbf{n_k}$ and $\mathbf{n_{k'}}$ and $k = k' \approx k_{\rm F}$. If $g(\theta) = g$ is a constant, that is, scattering with any momentum transfer is equally probable, the cosine term ("in" term in the collision integral (14.4)) averages out to zero. In this case $1/\tau_{\rm tr} = 1/\tau_{\rm el} = 2\pi\nu g$ is an elastic or "out" relaxation rate and $\tau_{\rm el}$ is the elastic scattering time. Assuming that the characteristic frequency of external fields is

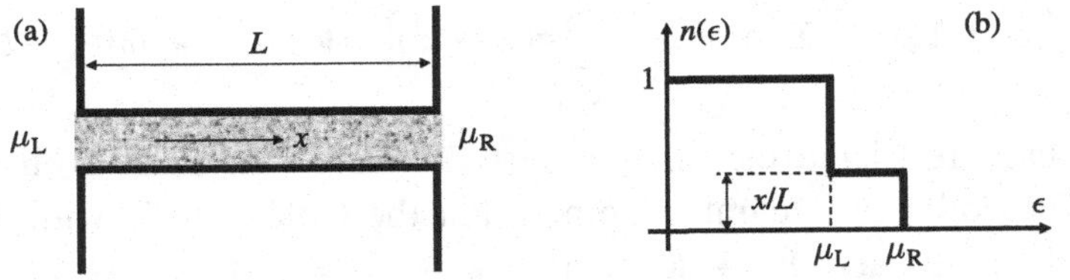

Figure 14.2 (a) Quasi-one-dimensional wire of length L attached to two leads with the chemical potentials $\mu_R - \mu_L = eV$. (b) The fermionic occupation number $n(\epsilon, x) = (1 - F_\epsilon(x))/2$.

much less than the transport relaxation rate, one may neglect the $\partial_t \mathbf{F}_1$ term on the left-hand side of Eq. (14.8). One then finds $\mathbf{F}_1 = -v_F \tau_{tr} \nabla_{\mathbf{r}} F_0$ and accordingly

$$\partial_t F_\epsilon(\mathbf{r}, t) = D \nabla_{\mathbf{r}}^2 F_\epsilon(\mathbf{r}, t)\,; \qquad D = v_F^2 \tau_{tr}/d, \tag{14.10}$$

where we used the shorthand notation $F_\epsilon(\mathbf{r}, t) \equiv F_0(\epsilon, \mathbf{r}, t) = \int d\Omega_{\mathbf{k}}\, \tilde{F}(\mathbf{k}, \mathbf{r}, t)$. We found thus that the average over the angle part of the distribution function obeys the diffusion equation with the diffusion constant proportional to the transport relaxation time. Equation (14.7) is the continuity relation with the current density given by $\mathbf{j}(\epsilon) = -v_F \mathbf{F}_1/(2d) = D \nabla_{\mathbf{r}} F_\epsilon / 2$.[2] The total electric current is therefore given by $\mathbf{J} = e \sum_{\mathbf{k}} \mathbf{j}(\epsilon) = (e\nu D/2) \int d\epsilon\, \nabla_{\mathbf{r}} F_\epsilon$.

As an example, let us consider a disordered quasi-one-dimensional wire of length L, attached to two leads, kept at different voltages [278], Fig. 14.2(a). We look for a space-dependent, *stationary* function $F_\epsilon(x)$, where x is the coordinate along the wire. It ought to satisfy the diffusion equation (14.10), $D\, \partial_x^2 F_\epsilon(x) = 0$, supplemented by the boundary conditions $F_\epsilon(0) = F_L(\epsilon)$ and $F_\epsilon(L) = F_R(\epsilon)$, where $F_{L(R)}(\epsilon)$ are the distribution functions of the left and right leads. The proper solution is

$$F_\epsilon(x) = F_L(\epsilon) + [F_R(\epsilon) - F_L(\epsilon)]\, \frac{x}{L}. \tag{14.11}$$

At low temperature, $T \ll eV$, the distribution function looks like a two-step function, Fig. 14.2(b), where the energy separation between the steps is the applied voltage, eV, while the relative height linearly depends on the position x. The total current along the wire is

$$J = \frac{e\nu D}{2} \int d\epsilon\, \partial_x F_\epsilon = \frac{e\nu D}{2L} \int d\epsilon\, [F_R(\epsilon) - F_L(\epsilon)] = \frac{e\nu D}{L} eV = \sigma_D \frac{V}{L}, \tag{14.12}$$

where the Drude conductivity of the diffusive wire is given by $\sigma_D = e^2 \nu D$.

We now discuss the Boltzmann–Langevin stochastic term in the kinetic equation; see Section 12.4:

$$[\partial_t + \mathbf{v}_{\mathbf{k}} \nabla_{\mathbf{r}}]\, \tilde{F}(\mathbf{k}, x) = I^{\text{coll}}[\tilde{F}] + \tilde{\xi}_F(\mathbf{k}, x), \tag{14.13}$$

[2] The factor of minus one-half is due to the fact that $\tilde{F}(\mathbf{k}) = 1 - 2n(\mathbf{k})$, while $j(\epsilon) = \int d\Omega_{\mathbf{k}} v_{\mathbf{k}} n(\mathbf{k})$.

where the noise correlator $\langle\tilde{\xi}_F(\mathbf{k}_1,x_1)\tilde{\xi}_F(\mathbf{k}_2,x_2)\rangle$ is given by Eqs. (12.48) and (12.50) with

$$I(\mathbf{k}_1,\mathbf{k}_2) = 2\pi g(\mathbf{k}_1-\mathbf{k}_2)\big[\tilde{F}(\mathbf{k}_2)-\tilde{F}(\mathbf{k}_1)\big]\delta(\xi_{\mathbf{k}_1}-\xi_{\mathbf{k}_2}). \tag{14.14}$$

The leading contribution comes from the zeroth spherical harmonic of the noise $\xi_\epsilon(x) = \int d\Omega_{\mathbf{k}}\, \tilde{\xi}_F(\mathbf{k},x)$. Using Eq. (14.5) and integrating over $\mathbf{k}_1$ and $\mathbf{k}_2$ angles, one finds its correlation function to be

$$\langle\xi_\epsilon(x)\xi_{\epsilon'}(x')\rangle = \delta(\epsilon-\epsilon')\delta(x-x')\frac{4}{\pi\nu d\tau_{tr}}\,\mathbf{F}_1^2 = \delta(\epsilon-\epsilon')\delta(x-x')\frac{4D}{\pi\nu}(\nabla_{\mathbf{r}}F_\epsilon)^2. \tag{14.15}$$

Clearly the noise is only present in the out-of-equilibrium current-carrying state. Following Eqs. (14.7) and (14.8), one finds that the distribution function satisfies the stochastic diffusion equation

$$\partial_t F_\epsilon(x) - D\nabla_{\mathbf{r}}^2 F_\epsilon(x) = \xi_\epsilon(x), \tag{14.16}$$

where $x = (\mathbf{r},t)$. This equation belongs to the Edwards–Wilkinson class of stochastic models, and appeared in Problems 9.10.1–9.10.2. We shall return to the discussion of the stochastic term in the end of Section 14.4.

14.2 Averaging over the Quenched Disorder

As was mentioned, the diffusion equation (14.10) is only an approximation, which in some cases (most notably in the localized state) completely fails to describe the system. To develop a more systematic approach it is desirable to perform the disorder averaging at an early stage and deduce an effective *deterministic*, that is, without randomness, field theory. The latter may be then either treated by perturbation theory in disorder strength, RG, or possibly solved exactly (e.g. in 1d, or on a graph). Notice that so far we have been doing things in the opposite order: first expanding and only then averaging. The idea that the averaging may be performed first is facilitated by the observation that it is seemingly easy to average $e^{iS_{dis}}$ over the Gaussian distribution of V_{dis}. Indeed, it brings the Gaussian integral over $\mathbf{D}[V_{dis}]$, which may be immediately evaluated.

This appealing observation can *not* be straightforwardly applied in the framework of the equilibrium formalism. The reason is that one has to average the *logarithm* of the partition function and not the partition function itself.[3] Observables are obtained then as derivatives of the (averaged) logarithm of the partition function with respect to various sources. Indeed, averaging of the partition function itself $\langle Z[V_{dis}]\rangle_{dis} = \big\langle \mathrm{Tr}\{e^{-H[V_{dis}]/T}\}\big\rangle_{dis}$, although technically straightforward, grossly overemphasizes rare disorder realizations with the smallest energy. On the

[3] In other words, there is a disorder-dependent denominator in Eq. (1.5).

other hand, averaging the logarithm $\big\langle \ln Z[V_{\text{dis}}]\big\rangle_{\text{dis}}$, gives the right weight to various disorder realizations, but technically is rather demanding.

Two techniques were invented to perform the averaging of the logarithm. Historically the first was the replica trick [279, 280, 281, 282, 283]. It is based on the following observation: $\ln Z = \lim_{n\to 0}(Z^n - 1)/n$. It calls for the introduction of (integer) n identical replicas of the same disordered system, calculating the disorder average of Z^n and finally taking the limit $n \to 0$. While evaluating $\langle Z^n\rangle_{\text{dis}}$ is relatively straightforward, it is the analytical continuation from integer to real n that produces most of the problems. The procedure certainly works within perturbation theory. Extending it to non-perturbative calculations, while successful in some cases, for example, [284, 285, 287], has not yet been achieved in full generality.

Another method, pioneered by Efetov [289, 290], is the supersymmetric technique. It is based on the fact that the Z^{-1} of a *noninteracting* fermionic system is equal to the Z of a bosonic system with the same disorder potential, see Eqs. (10.11) and (2.20). Therefore, the factor Z^{-1} in $\delta \ln Z/\delta\mathbf{A} = Z^{-1}\delta Z/\delta\mathbf{A}$, where $\mathbf{A}$ is a source field such as a vector or scalar potential, may be substituted by the bosonic partition function Z. One thus introduces a fictitious bosonic replica of the fermionic system at hand and performs the disorder averaging of such a fermion–boson pair. This approach has proven to be extremely powerful in deriving various non-perturbative results for disordered systems [290, 291]. It is strictly limited, however, to noninteracting fermions and is incapable of incorporating the effects of electron–electron interactions. Indeed, the basic relation between the fermionic and bosonic partition functions is based on the properties of the corresponding *Gaussian* integrals, Eqs. (10.11) and (2.20), and can't be generalized to non-Gaussian ones.

In the Keldysh formalism, observables are given by derivatives of the generating function (and *not the logarithm* of the generating function!) over appropriate quantum source fields, which are then put to zero. More precisely, the two definitions, with and without the logarithm, coincide owing to the normalization identity $Z = 1$ in the absence of quantum sources. The Keldysh formalism thus provides an alternative to replica and supersymmetry, ensuring that $Z = 1$ by construction [292, 294, 295, 296, 297, 298]. In the diagrammatic language any proper disorder averaging technique must exclude parasitic closed loop diagrams. The replica trick does it by assigning a factor of n to such closed loops and putting $n \to 0$ at the end. The supersymmetric technique is based on the mutual cancelation of bosonic and fermionic loops. In the Keldysh technique the corresponding closed loops are nullified in view of the identity (10.29). Another fundamental advantage of the Keldysh technique is that it allows one to calculate observables under non-equilibrium conditions, while replica and supersymmetry are usually limited to the linear response at equilibrium. The purpose of this chapter is to show how the effective field theory of a disordered electron liquid, known as the nonlinear sigma-model (NLSM), may be constructed within the Keldysh formalism.

The disorder action (14.1) brings the factor $e^{iS_{dis}[V_{dis}]}$ into the definition of the generating function (10.44). As was argued previously, it is the generating function and thus the factor $e^{iS_{dis}[V_{dis}]}$ that needs to be averaged over realizations of the disorder potential. To simplify the subsequent calculations we shall assume that the disorder correlation function $g(\mathbf{r})$, Eq. (14.2), is short-ranged and therefore may be written as $g(\mathbf{r}) = g\delta(\mathbf{r}) = \delta(\mathbf{r})/(2\pi\nu\tau_{el})$, where τ_{el} is the elastic, or "out," scattering time defined after Eq. (14.9). With the locality assumption the disorder averaging takes the form of the Gaussian integral over $V_{dis}(\mathbf{r})$ with the local weight

$$\langle \dots \rangle_{dis} = \int \mathbf{D}[V_{dis}] \exp\left\{-\pi\nu\tau_{el}\int d\mathbf{r}\, V_{dis}^2(\mathbf{r})\right\} \dots . \tag{14.17}$$

One can now perform the disorder averaging of the appropriate term in the generating function:

$$\begin{aligned}\left\langle e^{iS_{dis}}\right\rangle_{dis} &= \int \mathbf{D}[V_{dis}]\, e^{-\int d\mathbf{r}\left[\pi\nu\tau_{el}V_{dis}^2(\mathbf{r})+iV_{dis}(\mathbf{r})\int_{-\infty}^{+\infty}dt\,\bar\psi^a(\mathbf{r},t)\hat\gamma^{cl}_{ab}\psi^b(\mathbf{r},t)\right]} \\ &= \exp\left\{-\frac{1}{4\pi\nu\tau_{el}}\int d\mathbf{r}\iint_{-\infty}^{+\infty} dt dt' \left[\bar\psi^a(\mathbf{r},t)\psi^a(\mathbf{r},t)\right]\left[\bar\psi^b(\mathbf{r},t')\psi^b(\mathbf{r},t')\right]\right\},\end{aligned} \tag{14.18}$$

where $a, b = 1, 2$ and summations over all repeated indices are assumed. Grassmann fields in the exponent may be rearranged as (notice the sign change) $[\bar\psi^a(\mathbf{r},t)\psi^a(\mathbf{r},t)][\bar\psi^b(\mathbf{r},t')\psi^b(\mathbf{r},t')] = -[\bar\psi^a(\mathbf{r},t)\psi^b(\mathbf{r},t')][\bar\psi^b(\mathbf{r},t')\psi^a(\mathbf{r},t)]$. One can employ now the Hubbard–Stratonovich transformation with the matrix-valued field $\hat{Q} = Q^{ab}_{tt'}(\mathbf{r})$ to decouple the (time nonlocal) four-fermion term as[4]

$$\begin{aligned}&\exp\left\{\frac{1}{4\pi\nu\tau_{el}}\int d\mathbf{r}\iint_{-\infty}^{+\infty} dt dt' \left[\bar\psi^a(\mathbf{r},t)\psi^b(\mathbf{r},t')\right]\left[\bar\psi^b(\mathbf{r},t')\psi^a(\mathbf{r},t)\right]\right\} \\ &= \int \mathbf{D}[\hat Q]\, e^{-\frac{\pi\nu}{4\tau_{el}}\mathrm{Tr}\{\hat Q^2\} - \frac{1}{2\tau_{el}}\int d\mathbf{r}\iint_{-\infty}^{+\infty} dt dt'\, Q^{ab}_{tt'}(\mathbf{r})\bar\psi^a(\mathbf{r},t)\psi^b(\mathbf{r},t')},\end{aligned} \tag{14.19}$$

where $\hat Q$ is a Hermitian matrix in the Keldysh as well as in the time space, that is, $Q^{ab}_{tt'} = \left[Q^{ba}_{t't}\right]^*$. Here the trace of $\hat Q^2$ implies summation over the matrix indices as well as time and spatial integrations,

[4] Here we do not keep track of the time-reversal symmetry, i.e. the fact that the Hamiltonian is a real symmetric operator. As a result, the following considerations are restricted to the case where the time-reversal invariance is broken by, e.g. an external magnetic field (complex Hermitian Hamiltonian). This is the so-called *unitary* symmetry class. The *orthogonal* class, i.e. the one where the time-reversal symmetry is present, is considered in Section 14.7.

$$\mathrm{Tr}\{\hat{Q}^2\} = \int d\mathbf{r} \iint_{-\infty}^{+\infty} dt dt' \sum_{a,b=1}^{2} Q_{tt'}^{ab}(\mathbf{r}) Q_{t't}^{ba}(\mathbf{r}). \tag{14.20}$$

As a result, one has traded the initial functional integral over the static field $V_{\rm dis}(\mathbf{r})$ for the functional integral over the dynamic Hermitian matrix field $\hat{Q}_{tt'}^{ab}(\mathbf{r})$. At first glance, it does not strike as a terribly bright idea. Nevertheless, there is a great simplification hidden in this procedure. The advantage is that the disorder potential, being δ-correlated, is a rapidly oscillating function. On the other hand, as shown in what follows, the $\hat{Q}$-matrix field is slow (both in space and time). It thus represents the true macroscopic (or hydrodynamic) degrees of freedom of the system, which are diffusively propagating modes.

The Grassmann fields now enter the action only as the following quadratic form: $\mathrm{Tr}\{\bar{\vec{\Psi}}[\hat{G}^{-1} + \frac{i}{2\tau_{\rm el}}\hat{Q}]\vec{\Psi}\}$, where $\hat{G}^{-1}$ is the free fermion inverse Green function (10.36) and the $\hat{Q}$-matrix originates from the disorder averaging, Eq. (14.19). Since the action is quadratic in fermions, they may be explicitly integrated out using the rules of fermionic Gaussian integration (Eq. (10.11); cf. also Eq. (10.50)), leading to the determinant of the corresponding quadratic form, $\hat{G}^{-1} + \frac{\mathrm{i}}{2\tau_{\rm el}}\hat{Q}$. All of the matrices here should be understood as having 2×2 Keldysh structure along with the $N \times N$ structure in discrete time. One thus finds for the disorder-averaged generating function

$$\langle Z \rangle_{\rm dis} = \int \mathbf{D}[\hat{Q}] \exp\left\{\mathrm{i}S[\hat{Q}]\right\},$$
$$\mathrm{i}S[\hat{Q}] = -\frac{\pi\nu}{4\tau_{\rm el}} \mathrm{Tr}\{\hat{Q}^2\} + \mathrm{Tr}\ln\left[\hat{G}^{-1} + \frac{\mathrm{i}}{2\tau_{\rm el}}\hat{Q}\right], \tag{14.21}$$

where the source fields are hidden in the inverse bare Green function $\hat{G}^{-1}$; see Section 14.5.

14.3 Nonlinear Sigma-Model

To proceed, we need to understand what configurations of the matrix-valued field $\hat{Q}$ contribute most to the functional integral (14.21). To this end we look for stationary configurations of the action $S[\hat{Q}]$ in (14.21). Taking the variation over $\hat{Q}_{tt'}(\mathbf{r})$, one obtains the stationary point equation

$$\hat{\underline{Q}}_{tt'}(\mathbf{r}) = \frac{\mathrm{i}}{\pi\nu}\left(\hat{G}^{-1} + \frac{\mathrm{i}}{2\tau_{\rm el}}\hat{\underline{Q}}\right)^{-1}_{tt',\mathbf{rr}}, \tag{14.22}$$

where $\hat{\underline{Q}}_{tt'}(\mathbf{r})$ denotes a stationary configuration of the fluctuating field $\hat{Q}_{tt'}(\mathbf{r})$. We first look for a spatially uniform and time-translationally invariant solution $\hat{\underline{Q}}_{t-t'}$

of (14.22) and then consider space- and time-dependent deviations from such a solution. This strategy is adopted from the theory of magnetic systems, where one first finds a uniform static magnetized configuration and then treats spin-waves as smooth perturbations on top of such a static uniform solution. From the structure of (14.22) one expects that the stationary configuration $\underline{\hat{Q}}$ possesses the same form as the fermionic self-energy (10.30) (more accurately, one expects that among possible stationary configurations there is a *classical* configuration that admits the causality structure (10.30)). One looks, therefore, for a solution of (14.22) in the form of the matrix

$$\underline{\hat{Q}}_{t-t'} = \hat{\Lambda}_{t-t'} = \begin{pmatrix} \Lambda^{\mathrm{R}}_{t-t'} & \Lambda^{\mathrm{K}}_{t-t'} \\ 0 & \Lambda^{\mathrm{A}}_{t-t'} \end{pmatrix}. \tag{14.23}$$

The fact that the stationary $\hat{Q}$-matrix is not Hermitian implies that the integration path in the space of complex matrices is distorted from staying in the subspace of Hermitian matrices to pass through the stationary matrix (14.23).

Substituting this expression into (14.22), which in the energy–momentum representation reads as $\hat{\Lambda}_\epsilon = \frac{\mathrm{i}}{\pi\nu}\sum_{\mathbf{k}}\left(\epsilon - \xi_{\mathbf{k}} + \frac{\mathrm{i}}{2\tau_{\mathrm{el}}}\hat{\Lambda}_\epsilon\right)^{-1}$, with $\xi_{\mathbf{k}} \equiv k^2/2m - \epsilon_{\mathrm{F}}$, one finds

$$\Lambda^{\mathrm{R(A)}}_\epsilon = \frac{\mathrm{i}}{\pi\nu}\sum_{\mathbf{k}} \frac{1}{\epsilon - \xi_{\mathbf{k}} + \frac{\mathrm{i}}{2\tau_{\mathrm{el}}}\Lambda^{\mathrm{R(A)}}_\epsilon} = \pm 1, \tag{14.24}$$

where the way one performs momentum summation is $\sum_{\mathbf{k}} \ldots = \nu \int d\xi_{\mathbf{k}} \ldots$. The signs on the right-hand side are chosen so as to respect causality: the retarded (advanced) Green function is analytic in the entire upper (lower) half-plane of the complex energy ϵ. One has also assumed that $1/\tau_{\mathrm{el}} \ll \epsilon_{\mathrm{F}}$ to extend the energy integration to minus/plus infinity, while using constant density of states ν. The Keldysh component, as always, may be parametrized by a Hermitian distribution function: $\Lambda^{\mathrm{K}}_\epsilon = \Lambda^{\mathrm{R}} \circ F_\epsilon - F_\epsilon \circ \Lambda^{\mathrm{A}} = 2F_\epsilon$, where the distribution function F_ϵ is not fixed by the stationary point equation (14.22) and must be determined through the boundary conditions. In equilibrium, however, F_ϵ is nothing but the thermal fermionic distribution function $F^{\mathrm{eq}}_\epsilon = \tanh \epsilon/2T$. Finally, we have for the stationary $\underline{\hat{Q}}$-matrix configuration

$$\hat{\Lambda}_\epsilon = \begin{pmatrix} 1^{\mathrm{R}}_\epsilon & 2F_\epsilon \\ 0 & -1^{\mathrm{A}}_\epsilon \end{pmatrix}, \tag{14.25}$$

where we have introduced the retarded and advanced unit matrices to remind us about the causality structure. Transforming back to the time representation, one finds $\Lambda^{\mathrm{R(A)}}_{t-t'} = \pm\delta(t - t' \mp 0)$, where ∓ 0 indicates that the delta-function is shifted below (above) the main diagonal, $t = t'$. As a result, $\mathrm{Tr}\{\hat{\Lambda}\} = 0$ and $S[\hat{\Lambda}] = 0$, as it should be, of course, for any purely classical field configuration (14.23). One should note, however, that this particular form of the saddle-point solution (14.25) is a result of the approximation that the single-particle density of states ν is independent of energy. In general, it does depend on ϵ and thus the retarded (advanced) component of $\hat{\Lambda}_\epsilon$ is an analytic function of energy in the upper (lower) half-plane, which does depend on energy on the scale of the order of the Fermi energy ϵ_F. Therefore, the infinitesimally shifted delta-functions in $\Lambda^{\mathrm{R(A)}}_{t-t'} = \pm\delta(t-t'\mp 0)$ should be understood as $\delta_{t\mp 0} = f_\pm(t)\theta(\pm t)$, where $\theta(\pm t)$ is the Heaviside step function, and $f_\pm(t)$ are functions that are highly peaked for $|t| \lesssim \epsilon_\mathrm{F}^{-1}$ and satisfy the normalization $\int_0^{\pm\infty} \mathrm{d}t f_\pm(t) = \pm 1$. This high-energy regularization is important to remember in calculations to avoid spurious divergences. In particular, for this reason $1^\mathrm{R}_{t-t'}M^\mathrm{R}_{t',t} = 0$ and $1^\mathrm{A}_{t-t'}M^\mathrm{A}_{t',t} = 0$, where $M^{\mathrm{R(A)}}_{t',t}$ is an arbitrary retarded (advanced) two-point function in time space.

Now we are in a position to examine fluctuations around the stationary configuration (14.25). The fluctuations of $\hat{Q}$ fall into two general classes: (i) massive, with mass $\propto \nu/\tau_\mathrm{el}$ and (ii) massless or Goldstone modes, that is, such that the action depends only on gradients or time derivatives of these degrees of freedom. The fluctuations along the massive directions can be integrated out in the Gaussian approximation and lead to insignificant renormalization of the parameters in the action. The massless, or Goldstone, modes describe diffusive motion of the electrons. Fluctuations of the $\hat{Q}$-matrix along the massless directions may be not small and should be taken into account beyond small deviations. To this end one needs a way to parametrize a manifold of massless modes as a certain (nonlinear) constraint imposed on the allowed (i.e. massless) $\hat{Q}$-matrices. To identify the relevant Goldstone modes, consider the first term in the action $S[\hat{Q}]$ of (14.21). The stationary configuration given by (14.25) satisfies

$$\hat{Q}^2 = \begin{pmatrix} 1^\mathrm{R}_\epsilon & 0 \\ 0 & 1^\mathrm{A}_\epsilon \end{pmatrix} = \hat{1}. \tag{14.26}$$

Note that $\mathrm{Tr}\{\hat{Q}^2\} = \mathrm{Tr}\{\hat{1}^\mathrm{R}\} + \mathrm{Tr}\{\hat{1}^\mathrm{A}\} = 0$, owing to the definition of the retarded and advanced unit matrices; see Eq. (2.44). Fluctuations of $\hat{Q}$ that do not satisfy the constraint (14.26) are massive. The manifold of $\hat{Q}$-matrix configurations that obey the nonlinear constraint (14.26) is generated by rotations of the stationary matrix $\hat{\Lambda}_\epsilon$ and may be parametrized as follows:

$$\hat{Q} = \hat{\mathcal{R}}^{-1} \circ \hat{\Lambda} \circ \hat{\mathcal{R}}. \tag{14.27}$$

Indeed, if $\mathcal{R}_{tt'}(\mathbf{r}) = \mathcal{R}_{t-t'}$ is stationary and uniform, it commutes with $\hat{G}^{-1}$ and thus the rotated stationary matrix $\hat{\mathcal{R}}^{-1} \circ \hat{\Lambda} \circ \hat{\mathcal{R}}$ still satisfies the stationary point equation (14.22). The specific form of $\hat{\mathcal{R}}$ is not important at the moment and will be chosen later. The massless modes, or spin waves if one adopts the magnetic analogy, that are associated with $\hat{\mathcal{R}}_{tt'}(\mathbf{r})$ are slow functions of $t+t'$ and $\mathbf{r}$ and their gradients are small. Our goal now is to derive an effective action for the soft-mode $\hat{Q}$-field configurations given by Eqs. (14.26) and (14.27).

To this end one substitutes Eq. (14.27) into Eq. (14.21) and cyclically permutes $\hat{\mathcal{R}}$ matrices under the trace operation. This way, one obtains $\hat{\mathcal{R}}\circ\hat{G}^{-1}\circ\hat{\mathcal{R}}^{-1}\simeq\hat{G}^{-1}+\hat{\mathcal{R}}\circ[\hat{G}^{-1}\,;\,\hat{\mathcal{R}}^{-1}]=\hat{G}^{-1}+\mathrm{i}\hat{\mathcal{R}}\partial_t\hat{\mathcal{R}}^{-1}+\mathrm{i}\hat{\mathcal{R}}\mathbf{v}_\mathrm{F}\nabla_\mathbf{r}\hat{\mathcal{R}}^{-1}$, where one uses the explicit form of the bare inverse Green function (10.36) and linearizes the dispersion relation near the Fermi energy $\xi_\mathbf{k}=k^2/2m-\epsilon_\mathrm{F}\approx\mathbf{v}_\mathrm{F}\mathbf{k}\to-\mathrm{i}\mathbf{v}_\mathrm{F}\nabla_\mathbf{r}$. As a result, the desired action acquires the form

$$\mathrm{i}S[\hat{Q}]=\mathrm{Tr}\ln\left[\hat{G}^{-1}+\frac{\mathrm{i}}{2\tau_\mathrm{el}}\hat{\Lambda}+\mathrm{i}\hat{\mathcal{R}}\partial_t\hat{\mathcal{R}}^{-1}+\mathrm{i}\hat{\mathcal{R}}\mathbf{v}_\mathrm{F}\nabla_\mathbf{r}\hat{\mathcal{R}}^{-1}\right], \tag{14.28}$$

where we omit the circular multiplication sign for brevity. Let us define now the *impurity dressed* Green function matrix $\hat{\mathcal{G}}$ as the solution of the following Dyson equation:

$$(\hat{G}^{-1}+\frac{\mathrm{i}}{2\tau_\mathrm{el}}\hat{\Lambda})\hat{\mathcal{G}}=\hat{1}. \tag{14.29}$$

It is indeed straightforward to see that $\hat{\Sigma}=-\mathrm{i}\hat{\Lambda}/2\tau_\mathrm{el}$ is the self-energy matrix (cf. Eq. (14.3)). For subsequent calculations it is convenient to write $\mathcal{G}$, reflecting its causality structure, as

$$\hat{\mathcal{G}}=\begin{pmatrix}\mathcal{G}^\mathrm{R} & \mathcal{G}^\mathrm{K}\\ 0 & \mathcal{G}^\mathrm{A}\end{pmatrix}=\frac{1}{2}\,\mathcal{G}^\mathrm{R}[\hat{1}+\hat{\Lambda}]+\frac{1}{2}\,\mathcal{G}^\mathrm{A}[\hat{1}-\hat{\Lambda}], \tag{14.30}$$

with retarded, advanced, and Keldysh components given by

$$\mathcal{G}^\mathrm{R(A)}(\mathbf{k},\epsilon)=\left[\epsilon-\xi_\mathbf{k}\pm\mathrm{i}/2\tau_\mathrm{el}\right]^{-1},\qquad \mathcal{G}^\mathrm{K}(\mathbf{k},\epsilon)=\mathcal{G}^\mathrm{R}(\mathbf{k},\epsilon)F_\epsilon-F_\epsilon\mathcal{G}^\mathrm{A}(\mathbf{k},\epsilon). \tag{14.31}$$

Employing now the fact that $\mathrm{Tr}\ln\hat{\mathcal{G}}=0$ (cf. Eq. (10.29)), one may rewrite the action (14.28) as

$$\begin{aligned}\mathrm{i}S[\hat{Q}]&=\mathrm{Tr}\ln\left[\hat{1}+\mathrm{i}\hat{\mathcal{G}}\hat{\mathcal{R}}\partial_t\hat{\mathcal{R}}^{-1}+\mathrm{i}\hat{\mathcal{G}}\hat{\mathcal{R}}\mathbf{v}_\mathrm{F}\nabla_\mathbf{r}\hat{\mathcal{R}}^{-1}\right]\\ &\approx\mathrm{i}\,\mathrm{Tr}\{\hat{\mathcal{G}}\hat{\mathcal{R}}\partial_t\hat{\mathcal{R}}^{-1}\}+\frac{1}{2}\,\mathrm{Tr}\{\hat{\mathcal{G}}(\hat{\mathcal{R}}\mathbf{v}_\mathrm{F}\nabla_\mathbf{r}\hat{\mathcal{R}}^{-1})\hat{\mathcal{G}}(\hat{\mathcal{R}}\mathbf{v}_\mathrm{F}\nabla_\mathbf{r}\hat{\mathcal{R}}^{-1})\},\end{aligned} \tag{14.32}$$

where in the second line we have expanded the logarithm in gradients of the rotation matrices $\hat{\mathcal{R}}$ to linear order in $\partial_t\hat{\mathcal{R}}^{-1}$ and to quadratic order in $\nabla_\mathbf{r}\hat{\mathcal{R}}^{-1}$. (The term linear in the spatial gradient vanishes due to the angular integration.) Since $\sum_\mathbf{k}\hat{\mathcal{G}}(\mathbf{k},\epsilon)=-\mathrm{i}\pi\nu\hat{\Lambda}_\epsilon$, which directly follows from the stationary point equation (14.22), one finds for the ∂_t term in the effective action (14.28) $\mathrm{i}\mathrm{Tr}\{\hat{\mathcal{G}}\hat{\mathcal{R}}\partial_t\hat{\mathcal{R}}^{-1}\}=\pi\nu\mathrm{Tr}\{\hat{\Lambda}\hat{\mathcal{R}}\partial_t\hat{\mathcal{R}}^{-1}\}=\pi\nu\mathrm{Tr}\{\partial_t\hat{\mathcal{R}}^{-1}\hat{\Lambda}\hat{\mathcal{R}}\}=\pi\nu\mathrm{Tr}\{\partial_t\hat{Q}\}$, where

$\mathrm{Tr}\{\partial_t \hat{Q}\} = \int \mathrm{d}t \partial_t \mathrm{Tr}\{\hat{Q}_{tt'}\}|_{t'=t}$. For the $\nabla_{\mathbf{r}}$ term, one finds $-\frac{1}{4}\pi \nu D \mathrm{Tr}\{(\nabla_{\mathbf{r}} Q)^2\}$, where $D = v_F^2 \tau_{el}/d$ is the diffusion constant and d is the spatial dimensionality.[5]

Finally, one finds for the action of the soft-mode configurations [296, 297, 298]

$$\mathrm{i}S[\hat{Q}] = -\frac{\pi\nu}{4}\mathrm{Tr}\Big\{D(\nabla_{\mathbf{r}}\hat{Q})^2 - 4\partial_t \hat{Q}\Big\}. \tag{14.33}$$

Notice that, since we deal with the short-ranged potential, $\tau_{el} = \tau_{tr}$ and therefore the diffusion coefficient D is in agreement with Eq. (14.10). Should we work with a generic disorder (14.2), we would arrive at the action (14.33) with D given by Eq. (14.10). Despite its simple appearance, the action (14.33) is highly nonlinear owing to the constraint $\hat{Q}^2 = \hat{1}$. The theory specified by Eqs. (14.26) and (14.33) is called the *matrix nonlinear sigma-model*. The name came from the theory of magnetism, where the unit-length vector $\vec{\sigma}(\mathbf{r})$ represents a local classical spin rotating over the sphere $\vec{\sigma}^2 = 1$.

14.4 Stationary Point and Fluctuations

As a first step, one needs to determine the most probable (stationary) configuration, $\underline{\hat{Q}}_{t_1 t_2}(\mathbf{r})$, on the soft-mode manifold defined by the nonlinear constraint (14.26). In a time-independent and spatially uniform case we already know the answer: it is $\underline{\hat{Q}} = \hat{\Lambda}$, Eq. (14.25). Here we look for a stationary configuration in the presence of weak spatial and temporal gradients. To this end, one may parametrize deviations from $\underline{\hat{Q}}_{t_1 t_2}(\mathbf{r})$ as $\hat{Q} = \hat{\mathcal{R}}^{-1} \circ \underline{\hat{Q}} \circ \hat{\mathcal{R}}$ and choose $\hat{\mathcal{R}} = \exp(\hat{\mathcal{W}}/2)$, where $\hat{\mathcal{W}}_{t_1 t_2}(\mathbf{r})$ is a generator of rotations. Expanding to first order in $\hat{\mathcal{W}}$, one finds $\hat{Q} = \underline{\hat{Q}} - [\hat{\mathcal{W}} \,\overset{\circ}{,}\, \underline{\hat{Q}}]/2$. One may now substitute such a $\hat{Q}$-matrix into the action (14.33) and require that the terms linear in the generators $\hat{\mathcal{W}}$ vanish. This leads to the stationary point equation for $\underline{\hat{Q}}$. For the spatial gradient term in (14.33) one obtains $\frac{1}{2}\mathrm{Tr}\{\hat{\mathcal{W}}\nabla_{\mathbf{r}} D\left[(\nabla_{\mathbf{r}}\underline{\hat{Q}})\underline{\hat{Q}} - \underline{\hat{Q}}\nabla_{\mathbf{r}}\underline{\hat{Q}}\right]\} = -\mathrm{Tr}\{\hat{\mathcal{W}}\nabla_{\mathbf{r}} D\left(\underline{\hat{Q}}\nabla_{\mathbf{r}}\underline{\hat{Q}}\right)\}$, where one has employed $\nabla_{\mathbf{r}}\underline{\hat{Q}} \circ \underline{\hat{Q}} + \underline{\hat{Q}} \circ \nabla_{\mathbf{r}}\underline{\hat{Q}} = 0$, since $\underline{\hat{Q}}^2 = \hat{1}$. For the second term one finds $\mathrm{Tr}\{\hat{\mathcal{W}}_{t_1 t_2}(\partial_{t_1} + \partial_{t_2})\underline{\hat{Q}}_{t_1 t_2}\} = \mathrm{Tr}(\hat{\mathcal{W}}\{\partial_t, \underline{\hat{Q}}\})$, where the curly brackets denote the anti-commutator. Demanding that the linear term in $\hat{\mathcal{W}}$ vanishes, one obtains

[5] Indeed, for the product of the Green functions one uses

$$\sum_{\mathbf{k}} \mathcal{G}^{R}(\mathbf{k},\epsilon)\mathbf{v}_F^{\mu}\mathcal{G}^{A}(\mathbf{k},\epsilon)\mathbf{v}_F^{\nu} = \nu v_F^2 \int \frac{\mathrm{d}\xi_{\mathbf{k}}}{(\epsilon - \xi_{\mathbf{k}} + \mathrm{i}/2\tau_{el})(\epsilon - \xi_{\mathbf{k}} - \mathrm{i}/2\tau_{el})} \int \mathrm{d}\Omega_{\mathbf{k}} \mathbf{n}_{\mathbf{k}}^{\mu}\mathbf{n}_{\mathbf{k}}^{\nu} = \nu v_F^2 2\pi\tau_{el}\frac{\delta_{\mu\nu}}{d},$$

where $\mathbf{n}_{\mathbf{k}}$ is the unit vector and $\int \mathrm{d}\Omega_{\mathbf{k}}$ is the normalized integral over the unit sphere in d dimensions. The corresponding RR and AA terms vanish upon $\xi_{\mathbf{k}}$ integration. Employing Eq. (14.30) along with $\hat{\Lambda}^2 = \hat{1}$ and $\hat{\mathcal{R}}\nabla_{\mathbf{r}}\hat{\mathcal{R}}^{-1} = -\nabla_{\mathbf{r}}\hat{\mathcal{R}}\hat{\mathcal{R}}^{-1}$, one then notices that

$$\mathrm{Tr}\{[\hat{1} + \hat{\Lambda}](\hat{\mathcal{R}}\nabla_{\mathbf{r}}\hat{\mathcal{R}}^{-1})[\hat{1} - \hat{\Lambda}](\hat{\mathcal{R}}\nabla_{\mathbf{r}}\hat{\mathcal{R}}^{-1})\} = -\frac{1}{2}\mathrm{Tr}\{[\nabla_{\mathbf{r}}(\hat{\mathcal{R}}^{-1}\hat{\Lambda}\hat{\mathcal{R}})]^2\} = -\frac{1}{2}\mathrm{Tr}\{(\nabla_{\mathbf{r}}\hat{Q})^2\}.$$

$$\nabla_{\mathbf{r}}\big(D\,\underline{\hat{Q}}\circ\nabla_{\mathbf{r}}\underline{\hat{Q}}\big)-\{\partial_t,\underline{\hat{Q}}\}=0. \tag{14.34}$$

In the context of superconductivity this equation is known as the Usadel equation [299]. We'll keep this name even for the normal metal considered here.

Let us look for a solution of the Usadel equation in the subspace of "classical," that is, having causality structure, configurations. We then put $\underline{\hat{Q}}=\hat{\Lambda}$, Eq. (14.25). Since the lower-left component is zero by causality, the condition $\underline{\hat{Q}}^2=1$ dictates ± 1 on the diagonal. As a result, the only yet unspecified component is the upper-right distribution function $F_{t_1t_2}(\mathbf{r})$. Therefore, in this case the Usadel equation is reduced to the single equation for the distribution function $F_{t_1t_2}(\mathbf{r})$. Substituting $\hat{\Lambda}$ (14.25) into Eq. (14.34) and performing a Wigner transformation,

$$F_{t_1t_2}(\mathbf{r})=\int\frac{d\epsilon}{2\pi}F_\epsilon\,(\mathbf{r},t)\,e^{-i\epsilon(t_1-t_2)},\qquad t=\frac{t_1+t_2}{2}, \tag{14.35}$$

one obtains

$$\nabla_{\mathbf{r}}\big[D(\mathbf{r})\nabla_{\mathbf{r}}F_\epsilon(\mathbf{r},t)\big]-\partial_tF_\epsilon(\mathbf{r},t)=0, \tag{14.36}$$

where we allowed for a (smooth) spatial dependence of the diffusion constant. This is the already familiar diffusion kinetic equation (14.10) for the isotropic component of the fermionic distribution function. Note that it is the same equation for any energy ϵ and different energies do not "talk" to each other, which is natural for a noninteracting system. In the presence of interactions, the equation acquires the collision integral on the right-hand side that mixes different energies; see Section 16.5. It is worth mentioning that the elastic scattering does not show up in the collision integral. It was already fully taken into account in the derivation of the Usadel equation and went into the diffusion term.

We see now that the kinetic equation (14.10) describes only the stationary configuration $\underline{\hat{Q}}_{tt'}(\mathbf{r})=\hat{\Lambda}_{tt'}$ of the nonlinear sigma-model and does not account for the fluctuations around it. Discussing the fluctuations, we restrict ourselves to the Goldstone (soft) modes fluctuations, which satisfy $\hat{Q}^2=\hat{1}$ and neglect all massive modes, which deviate away from this manifold. To parametrize such soft modes we first notice that (cf. Eq. (10.34))

$$\hat{\Lambda}=\hat{\mathcal{U}}^{-1}\circ\hat{\sigma}_3\circ\hat{\mathcal{U}},\qquad \hat{\mathcal{U}}=\begin{pmatrix}1&F\\0&1\end{pmatrix}=e^{\hat{\sigma}_+F}, \tag{14.37}$$

where $\hat{\sigma}_3$ is the third Pauli matrix acting in the Keldysh subspace, which has 1^{R}_ϵ and -1^{A}_ϵ on its main diagonal; see Eq. (14.25). The distribution function $F_{tt'}(\mathbf{r})$ is a solution of the saddle-point equation (14.36). The soft-mode fluctuations, obeying $\hat{Q}^2=\hat{1}$, may be obtained by $\hat{\sigma}_3\to\hat{\mathcal{R}}^{-1}\hat{\sigma}_3\hat{\mathcal{R}}$, where $\hat{\mathcal{R}}_{t_1t_2}(\mathbf{r})$ is a rotation matrix, which is convenient to write through the rotation generators as $\hat{\mathcal{R}}=e^{\hat{\mathcal{W}}/2}$. As a result, all proper $\hat{Q}$-matrices may be parametrized as

$$\hat{Q} = \hat{\mathcal{U}}^{-1} \circ e^{-\hat{\mathcal{W}}/2} \circ \hat{\sigma}_3 \circ e^{\hat{\mathcal{W}}/2} \circ \hat{\mathcal{U}}. \tag{14.38}$$

To provide a nontrivial rotation of the $\hat{\sigma}_3$ matrix, the rotation generator $\hat{\mathcal{W}}$ must be a purely *off-diagonal* matrix in the Keldysh space. As a result, the rotation generator may be written as

$$\hat{\mathcal{W}}_{tt'}(\mathbf{r}) = \begin{pmatrix} 0 & d^{\mathrm{cl}}_{tt'}(\mathbf{r}) \\ d^{\mathrm{q}}_{tt'}(\mathbf{r}) & 0 \end{pmatrix}, \tag{14.39}$$

where $d^{\mathrm{cl}}_{tt'}(\mathbf{r})$ and $d^{\mathrm{q}}_{tt'}(\mathbf{r})$ are two independent two-point functions in time space. As we shall see, these two functions serve as the classical and quantum components of the $\hat{Q}$-matrix field, respectively. The classical component $d^{\mathrm{cl}}_{tt'}(\mathbf{r})$ provides deviations of the distribution function F from its stationary value, while the quantum component $d^{\mathrm{q}}_{tt'}(\mathbf{r})$ provides a source to differentiate over. In particular, one may check that $S[d^{\mathrm{cl}}, d^{\mathrm{q}} = 0] = 0$. From the fact that $d^{\mathrm{cl}}_{tt'}$ describes fluctuations of the Hermitian distribution matrix $F_{tt'}$, while $d^{\mathrm{q}}_{tt'}$ is the corresponding source field, it follows that both $d^{\mathrm{cl}}_{tt'}(\mathbf{r})$ and $d^{\mathrm{q}}_{tt'}(\mathbf{r})$ are Hermitian matrices in time space. One thus understands the functional integration over $\hat{Q}_{tt'}(\mathbf{r})$ in Eq. (14.21) as an integration over two mutually independent Hermitian matrices in the time domain, $d^{\mathrm{cl}}_{tt'}(\mathbf{r})$ and $d^{\mathrm{q}}_{tt'}(\mathbf{r})$.

One may expand now the action (14.33) in powers of $d^{\mathrm{cl}}_{tt'}(\mathbf{r})$ and $d^{\mathrm{q}}_{tt'}(\mathbf{r})$. Since $\underline{\hat{Q}}_{tt'}$ was chosen to be a stationary configuration, the expansion starts from second order. If the stationary distribution F is spatially uniform, so $\nabla_{\mathbf{r}}\hat{\mathcal{U}} = 0$, one obtains in the second order

$$\mathrm{i}S[d^{\mathrm{cl}}, d^{\mathrm{q}}] = -\frac{\pi\nu}{2} \int d\mathbf{r} \iint dt dt'\, d^{\mathrm{q}}_{tt'}(\mathbf{r}) \left[-D\nabla^2_{\mathbf{r}} + \partial_t + \partial_{t'}\right] d^{\mathrm{cl}}_{t't}(\mathbf{r}). \tag{14.40}$$

The quadratic form may be diagonalized by going to the energy–momentum representation according to

$$\hat{\mathcal{W}}_{\epsilon\epsilon'}(\mathbf{q}) = \int d\mathbf{r} \iint dt dt'\, \hat{\mathcal{W}}_{tt'}(\mathbf{r})\, e^{\mathrm{i}\epsilon t - \mathrm{i}\epsilon' t'}\, e^{-\mathrm{i}\mathbf{q}\mathbf{r}}, \tag{14.41}$$

which brings the quadratic action (14.40) into the diagonal form

$$\mathrm{i}S[d^{\mathrm{cl}}, d^{\mathrm{q}}] = -\frac{\pi\nu}{2} \sum_{\mathbf{q},\epsilon,\epsilon'} d^{\mathrm{q}}_{\epsilon\epsilon'}(-\mathbf{q}) \left[D\mathbf{q}^2 + \mathrm{i}\epsilon - \mathrm{i}\epsilon'\right] d^{\mathrm{cl}}_{\epsilon'\epsilon}(\mathbf{q}). \tag{14.42}$$

As a result, the Gaussian propagators of the $\hat{Q}$-matrix fluctuations are

$$\begin{aligned}
\langle d^{\mathrm{cl}}_{\epsilon_1\epsilon_2}(\mathbf{q}) d^{\mathrm{q}}_{\epsilon_3\epsilon_4}(-\mathbf{q})\rangle_{\mathcal{W}} &= -\frac{2}{\pi\nu} \frac{\delta_{\epsilon_2\epsilon_3}\delta_{\epsilon_1\epsilon_4}}{Dq^2 - \mathrm{i}\omega} \equiv -\frac{2}{\pi\nu} \delta_{\epsilon_2\epsilon_3}\delta_{\epsilon_1\epsilon_4}\, \mathcal{D}^{\mathrm{R}}(\mathbf{q},\omega), \\
\langle d^{\mathrm{q}}_{\epsilon_1\epsilon_2}(\mathbf{q}) d^{\mathrm{cl}}_{\epsilon_3\epsilon_4}(-\mathbf{q})\rangle_{\mathcal{W}} &= -\frac{2}{\pi\nu} \frac{\delta_{\epsilon_2\epsilon_3}\delta_{\epsilon_1\epsilon_4}}{Dq^2 + \mathrm{i}\omega} \equiv -\frac{2}{\pi\nu} \delta_{\epsilon_2\epsilon_3}\delta_{\epsilon_1\epsilon_4}\, \mathcal{D}^{\mathrm{A}}(\mathbf{q},\omega),
\end{aligned} \tag{14.43}$$

where $\omega \equiv \epsilon_1 - \epsilon_2 = \epsilon_4 - \epsilon_3$ and the object

$$\mathcal{D}^{\mathrm{R(A)}}(\mathbf{q}, \omega) = \frac{1}{Dq^2 \mp \mathrm{i}\omega} \tag{14.44}$$

is called the *diffuson*. As we shall see, it describes the diffusive motion of the electron density fluctuations at small frequency $\omega \ll 1/\tau_{\mathrm{el}}$ and small wavenumbers $Dq^2 \ll 1/\tau_{\mathrm{el}}$, that is, $q \ll 1/l_{\mathrm{el}}$, where the elastic mean free path is $l_{\mathrm{el}} = \sqrt{D\tau_{\mathrm{el}}} \sim v_{\mathrm{F}}\tau_{\mathrm{el}}$.

If the stationary-point distribution function $F_\epsilon(\mathbf{r})$ is spatially nonuniform, there is an additional quantum–quantum term in the quadratic action of the form

$$\mathrm{i}\tilde{S}[d^{\mathrm{q}}] = -\frac{\pi\nu D}{2}\int d\mathbf{r} \sum_{\epsilon,\epsilon'} d^{\mathrm{q}}_{\epsilon\epsilon'}(\mathbf{r})(\nabla_{\mathbf{r}}F_{\epsilon'})d^{\mathrm{q}}_{\epsilon'\epsilon}(\mathbf{r})(\nabla_{\mathbf{r}}F_\epsilon). \tag{14.45}$$

This term generates a nonzero Keldysh correlation function, which is given by

$$\langle d^{\mathrm{cl}}_{\epsilon_1\epsilon_2}(\mathbf{r})d^{\mathrm{cl}}_{\epsilon_2\epsilon_1}(\mathbf{r}')\rangle_{\mathcal{W}} = -\frac{4D}{\pi\nu}\int d\mathbf{r}''\mathcal{D}^{\mathrm{R}}(\mathbf{r}, \mathbf{r}'', \omega)(\nabla_{\mathbf{r}''}F_{\epsilon_1})(\nabla_{\mathbf{r}''}F_{\epsilon_2})\mathcal{D}^{\mathrm{A}}(\mathbf{r}'', \mathbf{r}', \omega), \tag{14.46}$$

where $\omega \equiv \epsilon_1 - \epsilon_2$. Unlike the usual bosons, the matrix field correlator $\langle d^{\mathrm{cl}}d^{\mathrm{cl}}\rangle_{\mathcal{W}}$ exists only out of equilibrium. Indeed, being proportional to $(\nabla_{\mathbf{r}}F_\epsilon)^2$, it scales as a square of the applied voltage. Nevertheless, it may show up even in linear response problems, such as, for example, mesoscopic conductance fluctuations, Section 15.3. For the use of this term for interaction effects in the shot-noise power, see [300]. Notice that, in accordance with the causality structure, there can *not* be a classical–classical, that is, $\sim d^{\mathrm{cl}}d^{\mathrm{cl}}$, term in the action, and therefore $\langle d^{\mathrm{q}}d^{\mathrm{q}}\rangle_{\mathcal{W}} = 0$.

An alternative interpretation of the quantum-quantum part of the action (14.45) is found by doing Wigner, $d^{\mathrm{q}}_{\epsilon\epsilon'}(\mathbf{r}) \to d^{\mathrm{q}}_\epsilon(\mathbf{r}, t)$, and Hubbard–Stratonovich transformations, $\mathrm{i}\tilde{S} \to -(\pi\nu/2)\int d\mathbf{r}dt \sum_\epsilon d^{\mathrm{q}}_\epsilon(\mathbf{r}, t)\xi_\epsilon(\mathbf{r}, t)$, where $\xi_\epsilon(\mathbf{r}, t)$ is a Gaussian stochastic field with the correlation function (14.15). Together with the diffuson action (14.40) this leads (upon integration over $d^{\mathrm{q}}_\epsilon(\mathbf{r}, t)$) to the Boltzmann–Langevin stochastic kinetic equation (14.16) for the Wigner transform of the classical component of the distribution function, $d^{\mathrm{cl}}_{\epsilon\epsilon'}(\mathbf{r}) \to F_\epsilon(\mathbf{r}, t)$.

The higher-order terms in the expansion of the action (14.33) in powers of $d^{\mathrm{cl}}_{tt'}(\mathbf{r})$ and $d^{\mathrm{q}}_{tt'}(\mathbf{r})$ describe mutual interactions of the diffuson modes. Such nonlinear corrections to the classical linear diffusion are consequences of the quantum nature of the electron motion. They may eventually lead to the Anderson localization [301] of the density fluctuations, instead of the diffusive spreading. The localization phenomenon is due to generation of the effective mass for the d-fluctuations, which causes exponential decay of the correlations. This effect is essentially nonperturbative. On the level of the perturbation theory the nonlinear interactions of the diffusive modes lead to singular *weak localization* corrections to the diffusion constant D [302, 303, 304, 305]. In the unitary symmetry class we have been working with so far, such corrections start from the two-loop level [302, 306]. We shall discuss them in Section 14.8 in the framework of the orthogonal symmetry class, where the weak-localization corrections are present already in one loop.

14.5 Sources and External Fields

One may now incorporate the source terms S_V and S_A, Eqs. (10.39) and (10.43), into the fermionic action: $\mathrm{Tr}\left\{\bar{\vec{\Psi}}\left[\hat{G}^{-1}+\frac{i}{2\tau_{\mathrm{el}}}\hat{Q}-\hat{V}-\mathbf{v}_{\mathrm{F}}\hat{\mathbf{A}}\right]\vec{\Psi}\right\}$, where, for example, $\hat{V}=V^{\alpha}\hat{\gamma}^{\alpha}$ and $\alpha=\mathrm{cl},\mathrm{q}$. After fermionic Gaussian integration over $\bar{\vec{\Psi}}$ and $\vec{\Psi}$ one finds for the generating function (cf. Eq. (14.21)),

$$\begin{aligned}\langle Z[\mathbf{A},V]\rangle_{\mathrm{dis}} &= \int \mathbf{D}[\hat{Q}]\exp\left\{\mathrm{i}S[\hat{Q},\mathbf{A},V]\right\},\\ \mathrm{i}S[\hat{Q},\mathbf{A},V] &= -\frac{\pi\nu}{4\tau_{\mathrm{el}}}\mathrm{Tr}\{\hat{Q}^2\}+\mathrm{Tr}\ln\left[\hat{G}^{-1}+\frac{\mathrm{i}}{2\tau_{\mathrm{el}}}\hat{Q}-\hat{V}-\mathbf{v}_{\mathrm{F}}\hat{\mathbf{A}}\right].\end{aligned} \tag{14.47}$$

We now expand the trace of the logarithm in *both* gradients of $\hat{Q}$ and in the source fields $\hat{V}$ and $\hat{\mathbf{A}}$. The latter assumes that the source fields are small on the scale of the elastic scattering rate and do not strongly disturb the stationary configuration (14.25) (see Chapter 16 for more discussions of this point). Then, similarly to Eq. (14.33), one finds from Eq. (14.47)

$$\mathrm{i}S[\hat{Q},\mathbf{A},V]=\frac{\mathrm{i}\nu}{2}\,\mathrm{Tr}\left\{\hat{V}\hat{\sigma}_1\hat{V}\right\}-\frac{\pi\nu}{4}\,\mathrm{Tr}\left\{D(\hat{\partial}_{\mathbf{r}}\hat{Q})^2-4\partial_t\hat{Q}-4\mathrm{i}\hat{V}\hat{Q}\right\}, \tag{14.48}$$

where $\hat{\sigma}_1$ is the first Pauli matrix acting in Keldysh space, and we have introduced the gauge-invariant space derivative

$$\hat{\partial}_{\mathbf{r}}\hat{Q}=\nabla_{\mathbf{r}}\hat{Q}+\mathrm{i}[\hat{\mathbf{A}},\hat{Q}]. \tag{14.49}$$

A few comments are in order regarding Eq. (14.48). First, it is still restricted to the manifold of $\hat{Q}$-matrices satisfying $\hat{Q}^2=\hat{1}$. The second trace on the right-hand side of Eq. (14.48), containing $\hat{Q}$, originates from the $\sum_{\mathbf{k}}\mathbf{v}_{\mathrm{F}}\mathcal{G}^{\mathrm{R}}\mathbf{v}_{\mathrm{F}}\mathcal{G}^{\mathrm{A}}$ and $\sum_{\mathbf{k}}\mathcal{G}^{\mathrm{R(A)}}$ combinations in the expansion of the logarithm. On the other hand, the first term on the right-hand side of Eq. (14.48) originates from $\sum_{\mathbf{k}}\mathcal{G}^{\mathrm{R}}\mathcal{G}^{\mathrm{R}}$ and $\sum_{\mathbf{k}}\mathcal{G}^{\mathrm{A}}\mathcal{G}^{\mathrm{A}}$ combinations. These terms should be retained since the matrix $V^{\alpha}(\epsilon-\epsilon')\hat{\gamma}^{\alpha}$ is not restricted to the $1/\tau_{\mathrm{el}}$ shell near the Fermi energy. This is so because the scalar potential shifts the entire electronic band and not only the energy shell $|\epsilon|,|\epsilon'|<1/\tau_{\mathrm{el}}$. Thus, it is essential to follow the variations of the electron spectrum all the way down to the bottom of the band to respect the charge neutrality.

Equations (14.48) and (14.49) generalize the sigma-model action (14.33) for the presence of the scalar and vector potentials.[6]

[6] Here we provide some technical details needed to derive the effective action (14.48) from Eq. (14.47). The gradient expansion of the logarithm in Eq. (14.47) uses a $\hat{\mathcal{Q}}$-matrix in the form of (14.27) and leads, in analogy with Eq. (14.28), to

$$iS = \mathrm{Tr}\ln\left[\hat{1} + i\hat{\mathcal{G}}\hat{\mathcal{R}}\partial_t\hat{\mathcal{R}}^{-1} + i\hat{\mathcal{G}}\hat{\mathcal{R}}\mathbf{v}_F\nabla_{\mathbf{r}}\hat{\mathcal{R}}^{-1} - \hat{\mathcal{G}}\hat{\mathcal{R}}\hat{V}\hat{\mathcal{R}}^{-1} - \hat{\mathcal{G}}\hat{\mathcal{R}}\mathbf{v}_F\hat{\mathbf{A}}\hat{\mathcal{R}}^{-1}\right]. \tag{14.50}$$

Expanding this expression to linear order in $\hat{\mathcal{G}}\hat{\mathcal{R}}\partial_t\hat{\mathcal{R}}^{-1}$ and quadratic order in $\hat{\mathcal{G}}\hat{\mathcal{R}}\mathbf{v}_F\nabla_{\mathbf{r}}\hat{\mathcal{R}}^{-1}$, one reproduces Eq. (14.32), which leads eventually to the action (14.33). To linear order in $\hat{V}$ and $\hat{\mathbf{A}}$ one finds

$$iS_1[\hat{\mathcal{Q}}, \mathbf{A}, V] = -\mathrm{Tr}\{\hat{\mathcal{G}}\hat{\mathcal{R}}\hat{V}\hat{\mathcal{R}}^{-1}\} + i\mathrm{Tr}\{\hat{\mathcal{G}}(\hat{\mathcal{R}}\mathbf{v}_F\nabla_{\mathbf{r}}\hat{\mathcal{R}}^{-1})\hat{\mathcal{G}}(\hat{\mathcal{R}}\mathbf{v}_F\hat{\mathbf{A}}\hat{\mathcal{R}}^{-1})\}. \tag{14.51}$$

Proceeding with the first term on the right-hand side in the same way as with the ∂_t term, one finds $-\mathrm{Tr}\{\hat{\mathcal{G}}\hat{\mathcal{R}}\hat{V}\hat{\mathcal{R}}^{-1}\} = i\pi\nu\mathrm{Tr}\{\hat{\Lambda}\hat{\mathcal{R}}\hat{V}\hat{\mathcal{R}}^{-1}\} = i\pi\nu\mathrm{Tr}\{\hat{V}\hat{\mathcal{Q}}\}$. For the second term, employing Eq. (14.30) for the disorder dressed Green function and retaining retarded–advanced products of the Green functions $\sum_{\mathbf{k}} \mathcal{G}^{\mathrm{R}}(\mathbf{k}, \epsilon)\mathbf{v}_F\mathcal{G}^{\mathrm{A}}(\mathbf{k}, \epsilon)\mathbf{v}_F = 2\pi\nu D$, one finds

$$-\pi\nu D\,\mathrm{Tr}\{(\hat{\mathcal{R}}^{-1}\nabla_{\mathbf{r}}\hat{\mathcal{R}} + \hat{\mathcal{R}}^{-1}\hat{\Lambda}\hat{\mathcal{R}}\nabla_{\mathbf{r}}\hat{\mathcal{R}}^{-1}\hat{\Lambda}\hat{\mathcal{R}})\hat{\mathbf{A}}\} = -\pi\nu D\,\mathrm{Tr}\{\hat{\mathcal{Q}}\nabla_{\mathbf{r}}\hat{\mathcal{Q}}\hat{\mathbf{A}}\},$$

where $\hat{\mathcal{R}}\nabla_{\mathbf{r}}\hat{\mathcal{R}}^{-1} = -\nabla_{\mathbf{r}}\hat{\mathcal{R}}\hat{\mathcal{R}}^{-1}$ was used. All together it gives for Eq. (14.51)

$$iS_1[\hat{\mathcal{Q}}, \mathbf{A}, V] = i\pi\nu\,\mathrm{Tr}\{\hat{V}\hat{\mathcal{Q}}\} - i\pi\nu D\,\mathrm{Tr}\{\hat{\mathcal{Q}}\nabla_{\mathbf{r}}\hat{\mathcal{Q}}\hat{\mathbf{A}}\}. \tag{14.52}$$

In the second order in $\hat{V}$ and $\hat{\mathbf{A}}$ one finds

$$iS_2[\hat{\mathcal{Q}}, \mathbf{A}, V] = -\frac{1}{2}\,\mathrm{Tr}\{\hat{\mathcal{G}}\hat{V}\hat{\mathcal{G}}\hat{V}\} - \frac{1}{2}\,\mathrm{Tr}\{\hat{\mathcal{G}}(\hat{\mathcal{R}}\mathbf{v}_F\hat{\mathbf{A}}\hat{\mathcal{R}}^{-1})\hat{\mathcal{G}}(\hat{\mathcal{R}}\mathbf{v}_F\hat{\mathbf{A}}\hat{\mathcal{R}}^{-1})\}. \tag{14.53}$$

Note that in the term $\sim \hat{V}^2$ we took $\hat{\mathcal{R}} = \hat{\mathcal{R}}^{-1} = \hat{1}$. This is because the $\hat{V}^2$ contribution represents essentially static compressibility of the electron gas, which is determined by the entire energy band, while the $\hat{\mathcal{R}}$ and $\hat{\mathcal{R}}^{-1}$ matrices deviate from the unit matrix only in the narrow energy shell around the Fermi energy. Thus, for the first term on the right-hand side of Eq. (14.53) one can write $\mathrm{Tr}\{\hat{\mathcal{G}}\hat{V}\hat{\mathcal{G}}\hat{V}\} = V^\alpha \Upsilon^{\alpha\beta} V^\beta$, where

$$\Upsilon^{\alpha\beta} = \sum_{\mathbf{k}}\int \frac{d\epsilon}{2\pi}\,\mathrm{Tr}\{\hat{\mathcal{G}}(\mathbf{k}, \epsilon)\,\hat{\gamma}^\alpha\,\hat{\mathcal{G}}(\mathbf{k}, \epsilon)\,\hat{\gamma}^\beta\},$$

where the last trace spans only over the Keldysh matrix structure. Using Eq. (14.30) for the matrix Green function, and retaining only retarded–retarded and advanced–advanced products, one finds

$$\Upsilon^{\alpha\beta} = \frac{1}{4}\sum_{\mathbf{k}}\int \frac{d\epsilon}{2\pi}\mathrm{Tr}\left\{(\mathcal{G}^{\mathrm{R}})^2[\hat{1} + \hat{\Lambda}]\hat{\gamma}^\alpha[\hat{1} + \hat{\Lambda}]\hat{\gamma}^\beta + (\mathcal{G}^{\mathrm{A}})^2[\hat{1} - \hat{\Lambda}]\hat{\gamma}^\alpha[\hat{1} - \hat{\Lambda}]\hat{\gamma}^\beta\right\}$$

$$= \sigma_1^{\alpha\beta}\int \frac{d\epsilon}{2\pi}F_\epsilon\sum_{\mathbf{k}}\left[[\mathcal{G}^{\mathrm{R}}]^2 - [\mathcal{G}^{\mathrm{A}}]^2\right] = \sigma_1^{\alpha\beta}\int \frac{d\epsilon}{2\pi}\frac{\partial F_\epsilon}{\partial\epsilon}\sum_{\mathbf{k}}\left[\mathcal{G}^{\mathrm{R}} - \mathcal{G}^{\mathrm{A}}\right] = -2i\nu\sigma_1^{\alpha\beta},$$

where we used the fact that $[\mathcal{G}^{\mathrm{R(A)}}(\mathbf{k}, \epsilon)]^2 = -\partial_\epsilon\mathcal{G}^{\mathrm{R(A)}}(\mathbf{k}, \epsilon)$, integrated by parts; employed the fact that $\sum_{\mathbf{k}}(\mathcal{G}^{\mathrm{R}}(\mathbf{k}, \epsilon) - \mathcal{G}^{\mathrm{A}}(\mathbf{k}, \epsilon)) = -2\pi i\nu$, and assumed that for any reasonable fermionic distribution $F_{\epsilon\to\pm\infty} \to \pm 1$. Because of the latter assumption, $\int d\epsilon\,\partial F_\epsilon/\partial\epsilon = 2$.

The second term on the right-hand side of Eq. (14.53) is dealt with in the same way as we did after Eq. (14.32), $\mathrm{Tr}\{\hat{\mathcal{G}}(\hat{\mathcal{R}}\mathbf{v}_F\hat{\mathbf{A}}\hat{\mathcal{R}}^{-1})\hat{\mathcal{G}}(\hat{\mathcal{R}}\mathbf{v}_F\hat{\mathbf{A}}\hat{\mathcal{R}}^{-1})\} = \pi\nu D\mathrm{Tr}\{[\hat{1} + \hat{\Lambda}]\hat{\mathcal{R}}\hat{\mathbf{A}}\hat{\mathcal{R}}^{-1}[\hat{1} - \hat{\Lambda}]\hat{\mathcal{R}}\hat{\mathbf{A}}\hat{\mathcal{R}}^{-1}\} = \pi\nu D\mathrm{Tr}\{\hat{\mathbf{A}}^2 - \hat{\mathbf{A}}\hat{\mathcal{Q}}\hat{\mathbf{A}}\hat{\mathcal{Q}}\}$, which finally gives for the $S_2[\hat{\mathcal{Q}}, \mathbf{A}, V]$ part of the action

$$iS_2[\hat{\mathcal{Q}}, \mathbf{A}, V] = \frac{i\nu}{2}\,\mathrm{Tr}\{\hat{V}\hat{\sigma}_1\hat{V}\} + \frac{\pi\nu D}{2}\,\mathrm{Tr}\{\hat{\mathbf{A}}\hat{\mathcal{Q}}\hat{\mathbf{A}}\hat{\mathcal{Q}} - \hat{\mathbf{A}}^2\}. \tag{14.54}$$

Combining now Eq. (14.33) with Eqs. (14.52) and (14.54), and taking into account that $\mathrm{Tr}\{(\nabla_{\mathbf{r}}\hat{\mathcal{Q}})^2 + 4i\hat{\mathbf{A}}\hat{\mathcal{Q}}\nabla_{\mathbf{r}}\hat{\mathcal{Q}} - 2(\hat{\mathbf{A}}\hat{\mathcal{Q}}\hat{\mathbf{A}}\hat{\mathcal{Q}} - \hat{\mathbf{A}}^2)\} = \mathrm{Tr}\{(\hat{\partial}_{\mathbf{r}}\hat{\mathcal{Q}})^2\}$, where the covariant derivative is defined by Eq. (14.49), one finds the full action in the form of Eq. (14.48).

The electric current is defined as a variational derivative of the generating function with respect to the quantum component of the vector potential $\mathbf{J}(\mathbf{r},t) = (ei/2)\delta\langle Z[\mathbf{A}]\rangle_{\text{dis}}/\delta\,\mathbf{A}^{\text{q}}(\mathbf{r},t)$; compare Eq. (10.46). According to Eq. (14.48) it gives

$$\mathbf{J}(\mathbf{r},t) = \frac{e\pi\nu D}{2}\Big\langle \text{Tr}\big\{\hat{\gamma}^{\text{q}}\hat{Q}(\mathbf{r})\,\hat{\boldsymbol{\partial}}_{\mathbf{r}}\hat{Q}(\mathbf{r})\big\}_{tt}\Big\rangle_{Q}, \tag{14.55}$$

where the $\hat{\boldsymbol{\partial}}_{\mathbf{r}}$ operator, Eq. (14.49), includes the vector potential. Neglecting the $\hat{Q}$-matrix fluctuations, that is, putting $\hat{Q} = \hat{\Lambda}$, and also $\hat{\mathbf{A}} = 0$, one finds that $\mathbf{J}(\mathbf{r},t) = (e\pi\nu D)\int(d\epsilon/2\pi)\nabla_{\mathbf{r}}F_{\epsilon}(\mathbf{r},t)$, which was already used at the end of Section 14.1. The latter expression, however, is only an approximation that neglects the fluctuations. As a result it misses, for example, mesoscopic fluctuations, discussed in Chapter 15, and interaction corrections, Chapter 16.

14.6 Kubo Formula and Linear Response

The linear response theory in the framework of the Keldysh technique was formulated in Section 10.5. Let us see now how the disorder-averaged response functions may be obtained from the sigma-model action. To this end one employs the general definition of the density response function $\Pi^{\text{R}}(x,x')$, Eq. (10.49), along with the disorder-averaged generating function (14.47), (14.48), which give

$$\begin{aligned}\Pi^{\text{R}}(x-x') &= -\frac{i}{2}\left.\frac{\delta^2\langle Z[V^{\text{cl}},V^{\text{q}}]\rangle_{\text{dis}}}{\delta V^{\text{cl}}(x')\delta V^{\text{q}}(x)}\right|_{\hat{V}=0} \\ &= \nu\delta(\mathbf{r}-\mathbf{r}')\delta(t-t') \\ &\quad+\frac{i}{2}(\pi\nu)^2\Big\langle \text{Tr}\big\{\hat{\gamma}^{\text{q}}\hat{Q}_{tt}(\mathbf{r})\big\}\text{Tr}\big\{\hat{\gamma}^{\text{cl}}\hat{Q}_{t't'}(\mathbf{r}')\big\}\Big\rangle_{Q},\end{aligned} \tag{14.56}$$

where $x = (\mathbf{r},t)$ and angular brackets stand for the averaging over the action (14.33). The first term on the right-hand side of Eq. (14.56) originates from the differentiation of the $\text{Tr}\{\hat{V}\hat{\sigma}_1\hat{V}\}$ part of the action (14.48), while the second term comes from double differentiation of the exponentiated $i\pi\nu\text{Tr}\{\hat{V}\hat{Q}\}$. Equation (14.56) represents the sigma-model equivalent of the Kubo formula for the linear density response. In the Fourier representation it takes the form

$$\Pi^{\text{R}}(\mathbf{q},\omega) = \nu + \frac{i}{2}(\pi\nu)^2\iint\frac{d\epsilon d\epsilon'}{(2\pi)^2}\Big\langle \text{Tr}\{\hat{\gamma}^{\text{q}}\hat{Q}_{\epsilon+\omega,\epsilon}(\mathbf{q})\}\text{Tr}\{\hat{\gamma}^{\text{cl}}\hat{Q}_{\epsilon',\epsilon'+\omega}(-\mathbf{q})\}\Big\rangle_{Q}. \tag{14.57}$$

Employing Eqs. (14.38) and (14.39), one finds to linear order in the diffuson modes (since $\text{Tr}\{\hat{\gamma}^{\text{cl}}\hat{\Lambda}\} = 0$, the only contribution in the zeroth order is ν):

$$\text{Tr}\big\{\hat{\gamma}^{\text{cl}}\hat{Q}_{\epsilon',\epsilon'+\omega}(-\mathbf{q})\big\} = d^{\text{q}}_{\epsilon',\epsilon'+\omega}(-\mathbf{q})(F_{\epsilon'+\omega} - F_{\epsilon'}), \tag{14.58a}$$

$$\text{Tr}\big\{\hat{\gamma}^{\text{q}}\hat{Q}_{\epsilon+\omega,\epsilon}(\mathbf{q})\big\} = d^{\text{q}}_{\epsilon+\omega,\epsilon}(\mathbf{q})(1 - F_{\epsilon}F_{\epsilon+\omega}) - d^{\text{cl}}_{\epsilon+\omega,\epsilon}(\mathbf{q}). \tag{14.58b}$$

Since $\langle d^{\mathrm{q}} d^{\mathrm{q}} \rangle_W = 0$, only the last term on the right-hand side of Eq. (14.58b), being paired with Eq. (14.58a), contributes to the average value in Eq. (14.57). The result is

$$\Pi^{\mathrm{R}} = \nu + \frac{\mathrm{i}\pi\nu^2}{4}\int \mathrm{d}\epsilon\,(F_\epsilon - F_{\epsilon+\omega})\,\langle d^{\mathrm{cl}}_{\epsilon+\omega,\epsilon}(\mathbf{q}) d^{\mathrm{q}}_{\epsilon,\epsilon+\omega}(-\mathbf{q})\rangle_W = \nu\left[1 + \frac{\mathrm{i}\omega}{Dq^2 - \mathrm{i}\omega}\right], \tag{14.59}$$

where we have used propagator (14.43) and the integral $\int \mathrm{d}\epsilon\,(F_\epsilon - F_{\epsilon+\omega}) = -2\omega$, which is valid for any distribution function satisfying $F_{\epsilon\to\mp\infty} = \mp 1$. One has obtained thus the diffusive form of the density–density response function (i.e. the retarded component of the polarization matrix)

$$\Pi^{\mathrm{R}}(\mathbf{q},\omega) = \nu\,\frac{Dq^2}{Dq^2 - \mathrm{i}\omega}. \tag{14.60}$$

The fact that $\Pi^{\mathrm{R}}(0,\omega) = 0$ is a consequence of particle number conservation. Also notice that this function is indeed retarded (i.e. analytic in the entire upper half-plane of the complex frequency ω), as it should be. The weak-localization corrections to the classical diffusion are obtained by expanding, for example, $\mathrm{Tr}\{(\nabla_{\mathbf{r}}\hat{Q})^2\}$ in the action to fourth and sixth orders in the diffusion generators d^α and contracting them with Eqs. (14.58). Equation (14.60) may be compared with Eq. (10.55) for the response function of a clean electron gas. The latter exhibits singularities at $\omega = \pm v_{\mathrm{F}} q$, which are characteristic for the ballistic motion. The diffusive expression (14.60) is valid for $\omega \lesssim 1/\tau_{\mathrm{el}}$ *and* $q \lesssim 1/l_{\mathrm{el}}$, while the ballistic one (10.55) is applicable outside this region.

The current–current response function, $K^{\mathrm{R}}(\mathbf{q},\omega)$, may be obtained in a similar way, by taking variations of the disorder-averaged generating function with respect to the vector potential. It may be also found, however, using the continuity relation $\mathbf{q}\cdot\mathbf{j} + \omega\rho = 0$, which implies the following relation between density and current responses: $K^{\mathrm{R}}(\mathbf{q},\omega) = e^2\omega^2\Pi^{\mathrm{R}}(\mathbf{q},\omega)/q^2$. As a result, the conductivity is given by[7]

$$\sigma(\mathbf{q},\omega) = \frac{e^2}{\mathrm{i}\omega}K^{\mathrm{R}}(\mathbf{q},\omega) = e^2\,\frac{-\mathrm{i}\omega}{q^2}\,\Pi^{\mathrm{R}}(\mathbf{q},\omega) = e^2\nu D\frac{-\mathrm{i}\omega}{Dq^2 - \mathrm{i}\omega}. \tag{14.61}$$

In the limit of the uniform external field, $\mathbf{q}\to 0$, this leads to the already familiar Drude result $\sigma(0,\omega) = e^2\nu D = \sigma_{\mathrm{D}}$; see Eq. (14.12).

14.7 Cooper Channel

If the system possesses time-reversal symmetry (TRS) (for the spinless case it means that there is a basis in which the Hamiltonian is real, $\hat{H}^{\mathrm{T}} = \hat{H}$), there

[7] The factor $1/(\mathrm{i}\omega)$ is due to the fact that K^{R} is a response function on the vector potential $\mathbf{A}$, while the conductivity is a response on the *electric field*, $\mathbf{E}$. The latter is related to the vector potential as $\mathbf{E} = -\partial_t\,\mathbf{A} = \mathrm{i}\omega\mathbf{A}$.

is an additional set of soft degrees of freedom, which has been neglected so far. It originates from another way of decoupling the four-fermion term (14.18), obtained upon averaging over the quenched disorder. Instead of rearranging it as $-[\bar{\psi}^a\psi'^b][\bar{\psi}'^b\psi^a]$, where the prime indicates the different time argument, one could also rewrite it as $[\bar{\psi}^a\bar{\psi}'^b][\psi'^b\psi^a]$. By decoupling it with a Hubbard–Stratonovich field, one obtains another set of soft modes. Due to apparent similarities with the theory of superconductivity, see Chapter 17, these soft modes are known as the Cooper channel. One should be aware, though, that superconductivity typically involves spin in a crucial way, while our treatment here is given for the spinless case.

To implement this idea in a systematic way, one notices that for a quadratic Hamiltonian,

$$\hat{H}(\bar{\psi},\psi) = \bar{\psi}\hat{H}\psi = \left[\bar{\psi}\hat{H}\psi\right]^{\mathrm{T}} = -\psi^{\mathrm{T}}\hat{H}^{\mathrm{T}}\bar{\psi}^{\mathrm{T}} = -\psi^{\mathrm{T}}\hat{H}\bar{\psi}^{\mathrm{T}},$$

where in the third equality we permuted Grassmann numbers and in the last one employed the time-reversal symmetry $\hat{H}^{\mathrm{T}} = \hat{H}$. This observation suggests that one may double the fermionic basis by introducing the time-reversal space. To this end we define two-component "spinors" as

$$\Psi = \frac{1}{\sqrt{2}}\begin{pmatrix} \psi \\ -\bar{\psi} \end{pmatrix}; \qquad \bar{\Psi} = \frac{1}{\sqrt{2}}\left(\bar{\psi},\psi\right); \qquad \bar{\Psi} = \Psi^{\mathrm{T}}(\mathrm{i}\hat{\tau}_2), \tag{14.62}$$

where $\hat{\tau}_\mu$ are Pauli matrices acting in the time-reversal space. With this notation the quadratic fermionic action on the closed time contour may be written as

$$S = \int_C \mathrm{d}t\, \bar{\Psi}\left(\mathrm{i}\hat{\tau}_3\partial_t - \hat{H}\right)\Psi. \tag{14.63}$$

The fact that the Hamiltonian enters with the unit matrix, that is, $\hat{\tau}_0$, is a manifestation of TRS. For our problem the Hamiltonian is $\hat{H} = -\nabla_{\mathbf{r}}^2/(2m) - \epsilon_F + V_{\mathrm{dis}}(\mathbf{r})$.

We now proceed exactly as before: (i) split the fields Ψ and $\bar{\Psi}$ into forward and backward components and perform Keldysh–Larkin–Ovchinnikov rotation (10.23), (10.24); (ii) average over the disorder according to Eq. (14.18); (iii) perform a Hubbard–Stratonovich transformation with the help of the Hermitian matrix $\check{Q}_{tt'}(\mathbf{r})$, which is a 4×4 matrix in the direct product of the Keldysh and time-reversal spaces, as well as an $N \times N$ matrix in the discretized time space. Since the $\check{Q}$-matrix has the same symmetry as $\Psi(t')\bar{\Psi}(t)$ it obeys the symplectic relation

$$\check{Q}^T = \hat{\tau}_2\check{Q}\hat{\tau}_2, \tag{14.64}$$

where the transposition operation involves matrix transposition as well as the interchange of the two time arguments. The presence of the additional symmetry (14.64) dictates that the Hubbard–Stratonovich part of the action takes the form

$(\pi\nu/8\tau_{\rm el})\mathrm{Tr}\{\check{Q}^2\}$; see Eq. (14.19). Finally, one performs Gaussian integration over the Ψ and $\bar{\Psi}$ fields. Since only half of the fermionic fields are independent (due to the symmetry $\bar{\Psi} = \Psi^{\rm T}(\mathrm{i}\hat{\tau}_2)$), it brings in the factor $\det^{1/2}[\check{G}^{-1} + \mathrm{i}\check{Q}/(2\tau_{\rm el})]$ (similarly to the bosonic Gaussian integral over the real variables, Eq. (2.22)), where $\check{G}^{-1} = \mathrm{i}\hat{\tau}_3\partial_t - \xi_{\mathbf{k}}$. Repeating the calculations that brought in the sigma-model action (14.33), we obtain its generalization for the TRS case:

$$\mathrm{i}S[\check{Q}] = -\frac{\pi\nu}{8}\,\mathrm{Tr}\Big\{D(\check{\boldsymbol{\partial}}_{\mathbf{r}}\check{Q})^2 - 4\hat{\tau}_3\partial_t\check{Q}\Big\}, \tag{14.65}$$

where the covariant derivative is given by $\check{\boldsymbol{\partial}}_{\mathbf{r}}\check{Q} = \nabla_{\mathbf{r}}\check{Q} + \mathrm{i}[\hat{\mathbf{A}}\hat{\tau}_3, \check{Q}]$. Since the magnetic field breaks TRS, the source vector potential enters the action with the symmetry-breaking factor $\hat{\tau}_3$. If TRS is completely broken by, for example, a strong enough magnetic field, the soft-mode manifold acquires the block-diagonal form $\check{Q} = \mathrm{diag}\{\hat{Q}, \hat{Q}^{\rm T}\}$ in time-reversal space, where the $\hat{Q}$-matrix is the one we used before in the unitary symmetry class. The action (14.65) reduces then back to (14.33). In the presence of TRS $\check{Q}$ also has the off-diagonal blocks in time-reversal space, which are the additional soft modes or *Cooperons*.

To parametrize this soft-mode manifold, one notices that the stationary point $\check{Q}$-matrix takes the form $\check{\Lambda} = \mathrm{diag}\{\hat{\Lambda}, \hat{\Lambda}^{\rm T}\}$, which may be written, using Eq. (14.37), as $\check{\Lambda} = \check{\mathcal{U}}^{-1}(\hat{\sigma}_3\hat{\tau}_0)\check{\mathcal{U}}$, where $\check{\mathcal{U}} = \mathrm{diag}\{\hat{\mathcal{U}}, \hat{\mathcal{U}}^{\rm T}\}$. The $\check{Q}$-matrix may be then parametrized as $\check{Q} = \check{\mathcal{U}}^{-1}\mathrm{e}^{-\check{\mathcal{W}}/2}(\hat{\sigma}_3\hat{\tau}_0)\mathrm{e}^{\check{\mathcal{W}}/2}\check{\mathcal{U}}$ (cf. Eq. (14.38)), where $\check{\mathcal{W}}$ matrices acquire structure in time-reversal space. The symplectic symmetry (14.64) dictates that $\check{\mathcal{W}}^{\rm T} = -\hat{\tau}_2\check{\mathcal{W}}\hat{\tau}_2$, which leads to

$$\check{\mathcal{W}} = \begin{pmatrix} \hat{\mathcal{W}} & \hat{\mathcal{B}}_1 \\ \hat{\mathcal{B}}_2 & -\hat{\mathcal{W}}^{\rm T} \end{pmatrix}_{\rm TR}; \qquad \hat{\mathcal{B}}_1^{\rm T} = \hat{\mathcal{B}}_1, \qquad \hat{\mathcal{B}}_2^{\rm T} = \hat{\mathcal{B}}_2, \tag{14.66}$$

where the subscript TR indicates that the displayed structure is in time-reversal space. To induce a rotation of the $(\hat{\sigma}_3\hat{\tau}_0)$ matrix, the generator $\check{\mathcal{W}}$ (and therefore all three matrices $\hat{\mathcal{W}}, \hat{\mathcal{B}}_1$ and $\hat{\mathcal{B}}_2$) must be off-diagonal in the Keldysh space, that is, $\check{\mathcal{W}}(\hat{\sigma}_3\hat{\tau}_0) + (\hat{\sigma}_3\hat{\tau}_0)\check{\mathcal{W}} = 0$. This leads to

$$\hat{\mathcal{W}} = \begin{pmatrix} 0 & d^{\rm cl} \\ d^{\rm q} & 0 \end{pmatrix}_{\rm K}; \qquad \hat{\mathcal{B}}_1 = \begin{pmatrix} 0 & c_1 \\ c_1^{\rm T} & 0 \end{pmatrix}_{\rm K}; \qquad \hat{\mathcal{B}}_2 = \begin{pmatrix} 0 & c_2^{\rm T} \\ c_2 & 0 \end{pmatrix}_{\rm K}, \tag{14.67}$$

where the displayed structure is in Keldysh space (K). The two-point functions $c_{1;tt'}(\mathbf{r})$ and $c_{2;tt'}(\mathbf{r})$ represent Cooperon modes, while $d^{\rm cl}_{tt'}(\mathbf{r})$ and $d^{\rm q}_{tt'}(\mathbf{r})$ are the already familiar diffuson modes. Unlike diffusons, which may be identified as classical and quantum, the Cooperons (similarly to the fermion fields) lack such a distinction and are labeled by the Keldysh index $a = 1, 2$. It is often convenient to deal with the parametrization (14.66), (14.67) in the energy representation. In doing so it is important to remember that, due to the chosen Fourier convention (14.41) and $(c_a^{\mathrm{T}})_{tt'} = c_{a;t't}$, the transposition operation in energy space takes the form

$$(c_a^{\mathrm{T}})_{\epsilon\epsilon'} = c_{a;-\epsilon'-\epsilon}. \tag{14.68}$$

Substituting the parametrization (14.66), (14.67) into the action (14.65) and expanding to second order in the d^α and c_a fields, one obtains the Gaussian action of the diffuson and Cooperon modes. The diffuson part is exactly that of Eq. (14.42), while the Cooperon part takes the form

$$\mathrm{i}S\,[c_1, c_2] = -\frac{\pi\nu}{2} \sum_{\mathbf{q},\epsilon,\epsilon'} c_{1;\epsilon\epsilon'}(-\mathbf{q}) \left[D\mathbf{q}^2 - \mathrm{i}\epsilon - \mathrm{i}\epsilon'\right] c_{2;\epsilon'\epsilon}(\mathbf{q}). \tag{14.69}$$

As a result, the Gaussian propagator of the Cooperon modes is

$$\langle c_{2;\epsilon_1\epsilon_2}(\mathbf{q}) c_{1;\epsilon_3\epsilon_4}(-\mathbf{q})\rangle = -\frac{2}{\pi\nu}\frac{\delta_{\epsilon_2\epsilon_3}\delta_{\epsilon_1\epsilon_4}}{Dq^2 - \mathrm{i}(\epsilon_1 + \epsilon_2)} \equiv \frac{-2}{\pi\nu}\delta_{\epsilon_2\epsilon_3}\delta_{\epsilon_1\epsilon_4}\,\mathcal{C}(\mathbf{q}, \epsilon_1 + \epsilon_2), \tag{14.70}$$

where the object $\mathcal{C}(\mathbf{q}, \epsilon) = (Dq^2 - \mathrm{i}\epsilon)^{-1}$ is called the *Cooperon*. In the absence of the magnetic field it is very similar to the diffuson (14.44) and, as was mentioned previously, provides an additional slow diffusive mode. Perturbations that break TRS, such as an external magnetic field or magnetic impurities, induce mass to the Cooperon (but not to the diffuson) and therefore gradually suppress the corresponding fluctuations.

14.8 Weak Localization and Scaling

We are now in a position to investigate the first quantum correction to the Drude conductivity $\sigma_{\mathrm{D}} = e^2\nu D$, Eqs. (14.12) and (14.61). To this end we focus on the $(\pi\nu D/8)\mathrm{Tr}\{[\hat{\mathbf{A}}\hat{\tau}_3, \check{Q}]^2\}$ term of the action (14.65), where $\hat{\mathbf{A}} = \hat{\gamma}^\alpha\, \mathbf{A}^\alpha(t)$. Other terms, such as $\sim \mathrm{Tr}\{[\hat{\mathbf{A}}\hat{\tau}_3, \check{Q}]\nabla_{\mathbf{r}}\check{Q}\}$, do not contribute to the one-loop correction considered here. According to linear response theory (Kubo formula), developed in Section 10.5 (see also the footnote at the end of Section 14.6), the linear conductivity is given by (cf. Eq. (10.47)),

$$\sigma^{\mathrm{R}}(\omega) = \frac{e^2}{\mathrm{i}\omega}\frac{1}{2\mathrm{i}}\left.\frac{\delta^2 Z[A^{\mathrm{cl}}, A^{\mathrm{q}}]}{\delta A^{\mathrm{cl}}_\omega \delta A^{\mathrm{q}}_{-\omega}}\right|_{A=0},$$

where both the classical vector potential and the current are directed along, say, the x-axis. For our purposes $Z[A^{\mathrm{cl}}, A^{\mathrm{q}}] = (\pi\nu D/8)\langle \mathrm{Tr}\{[\hat{\gamma}^\alpha A^\alpha\hat{\tau}_3, \check{Q}]^2\}\rangle$, where the angular brackets denote averaging over $\check{Q}$-matrix fluctuations. First, neglecting all the fluctuations, one puts the $\check{Q}$-matrix at its stationary point $\check{\Lambda}_\epsilon = \mathrm{diag}\{\hat{\Lambda}_\epsilon, \hat{\Lambda}^{\mathrm{T}}_{-\epsilon}\}$,

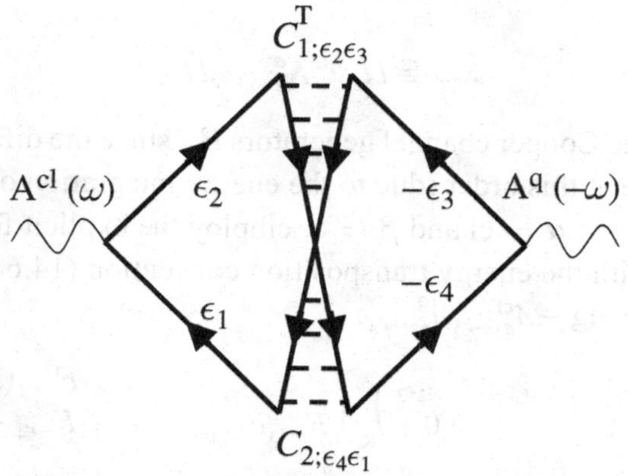

Figure 14.3 One loop weak-localization correction. The bold square is the local vertex $\mathrm{Tr}\{\hat{\mathcal{U}}_{\epsilon_1}\hat{\gamma}^{\mathrm{cl}}\hat{\mathcal{U}}_{\epsilon_2}\hat{\sigma}_-(\hat{\mathcal{U}}^{\mathrm{T}})_{\epsilon_3}\hat{\gamma}^{\mathrm{q}}(\hat{\mathcal{U}}^{\mathrm{T}})_{\epsilon_4}\hat{\sigma}_-\}$, known also as the Hikami box. The vertical ladder is the Cooperon propagator $\delta_{-\epsilon_3\epsilon_1}\delta_{-\epsilon_2\epsilon_4}\sum_{\mathbf{q}}\mathcal{C}(\mathbf{q},\omega)$, Eq. (14.70), between the coinciding spatial points. The fact that its legs cross indicates that the relevant contraction involves transposition, that is, it is between c_1^{T} and c_2.

where $\hat{\Lambda}_\epsilon$ is given by Eq. (14.25) and we employed Eq. (14.68) for the transposition operation in energy space. The generating function is then given by

$$Z = \frac{\pi\nu D}{8}\sum_{\epsilon,\epsilon'}\mathrm{Tr}\Big\{\big(\hat{\gamma}^\alpha A^\alpha_{\epsilon-\epsilon'}\hat{\tau}_3\check{\Lambda}_{\epsilon'} - \check{\Lambda}_\epsilon\hat{\gamma}^\alpha A^\alpha_{\epsilon-\epsilon'}\hat{\tau}_3\big)\big(\hat{\gamma}^\beta A^\beta_{\epsilon'-\epsilon}\hat{\tau}_3\check{\Lambda}_\epsilon - \check{\Lambda}_{\epsilon'}\hat{\gamma}^\beta A^\beta_{\epsilon'-\epsilon}\hat{\tau}_3\big)\Big\}.$$

Evaluating the trace and derivatives over the vector potential, one obtains

$$\sigma^{\mathrm{R}}(\omega) = -\frac{e^2\pi\nu D}{\omega}\int\frac{d\epsilon}{2\pi}(F_\epsilon - F_{\epsilon+\omega}) = e^2\nu D = \sigma_{\mathrm{D}},$$

as expected.

We now expand the two $\check{Q}$-matrices to first order in the fluctuations, that is, $\check{Q}\to\check{\mathcal{U}}^{-1}(\hat{\sigma}_3\hat{\tau}_0)\check{\mathcal{W}}\check{\mathcal{U}}$.[8] The typical term, resulting from such an expansion, acquires the form (Fig. 14.3)

$$\mathrm{Tr}\Big\{\check{\mathbb{A}}_{\varepsilon_1\varepsilon_2}(\hat{\sigma}_3\hat{\tau}_0)\check{\mathcal{W}}_{\epsilon_2\epsilon_3}\check{\mathbb{A}}_{\varepsilon_3\varepsilon_4}(\hat{\sigma}_3\hat{\tau}_0)\check{\mathcal{W}}_{\epsilon_4\epsilon_1}\Big\} = A^\alpha_{\epsilon_1-\epsilon_2}A^\beta_{\epsilon_3-\epsilon_4}$$

$$\times\mathrm{Tr}\begin{pmatrix}0 & \hat{\mathcal{U}}_{\epsilon_1}\hat{\gamma}^\alpha\hat{\mathcal{U}}_{\epsilon_2}\hat{\sigma}_3(\hat{\mathcal{B}}_1)_{\epsilon_2\epsilon_3}\\ -\hat{\mathcal{U}}^{\mathrm{T}}_{-\epsilon_1}\hat{\gamma}^\alpha\hat{\mathcal{U}}^{\mathrm{T}}_{-\epsilon_2}\hat{\sigma}_3(\hat{\mathcal{B}}_2)_{\epsilon_2\epsilon_3} & 0\end{pmatrix}\begin{pmatrix}0 & \hat{\mathcal{U}}_{\epsilon_3}\hat{\gamma}^\beta\hat{\mathcal{U}}_{\epsilon_4}\hat{\sigma}_3(\hat{\mathcal{B}}_1)_{\epsilon_4\epsilon_1}\\ -\hat{\mathcal{U}}^{\mathrm{T}}_{-\epsilon_3}\hat{\gamma}^\beta\hat{\mathcal{U}}^{\mathrm{T}}_{-\epsilon_4}\hat{\sigma}_3(\hat{\mathcal{B}}_2)_{\epsilon_4\epsilon_1} & 0\end{pmatrix},$$

[8] Expanding one $\check{Q}$-matrix to second order in $\check{\mathcal{W}}$, while putting the other one to be $\check{\Lambda}$, produces terms that all vanish due to the energy integration of purely retarded/advanced functions.

where

$$\check{\mathbb{A}}_{\varepsilon\varepsilon'} \equiv \check{\mathcal{U}}_{\varepsilon}\hat{\gamma}^{\alpha}\mathbf{A}^{\alpha}_{\varepsilon-\varepsilon'}\hat{\tau}_3\check{\mathcal{U}}^{-1}_{\varepsilon'} \tag{14.71}$$

and we kept only the Cooper channel generators $\hat{\mathcal{B}}_a$, since the diffuson generator $\hat{\mathcal{W}}$ does not contribute at this order (due to the energy integration of a purely retarded function). We now put $\alpha = \text{cl}$ and $\beta = \text{q}$, employ the explicit form (14.37) of the $\hat{\mathcal{U}}_{\epsilon}$-matrix along with the energy transposition convention (14.68), and find for the coefficient multiplying $-A^{\text{cl}}_{\epsilon_1-\epsilon_2}A^{\text{q}}_{\epsilon_3-\epsilon_4}$:

$$\text{Tr}\begin{pmatrix}(F_{\epsilon_1}-F_{\epsilon_2})(c_1^{\text{T}})_{\epsilon_2\epsilon_3} & c_{1;\epsilon_2\epsilon_3}\\ -(c_1^{\text{T}})_{\epsilon_2\epsilon_3} & 0\end{pmatrix}_{\text{K}}\begin{pmatrix}c_{2;\epsilon_4\epsilon_1} & F_{-\epsilon_4}(c_2^{\text{T}})_{\epsilon_4\epsilon_1}\\ F_{-\epsilon_3}c_{2;\epsilon_4\epsilon_1} & (F_{-\epsilon_3}F_{-\epsilon_4}-1)(c_2^{\text{T}})_{\epsilon_4\epsilon_1}\end{pmatrix}_{\text{K}}$$
$$= (F_{\epsilon_1}-F_{\epsilon_2})\langle c_{1;-\epsilon_3-\epsilon_2}c_{2;\epsilon_4\epsilon_1}\rangle + F_{-\epsilon_3}\langle c_{1;\epsilon_2\epsilon_3}c_{2;\epsilon_4\epsilon_1}\rangle - F_{-\epsilon_4}\langle c_{1;-\epsilon_3-\epsilon_2}c_{2;-\epsilon_1-\epsilon_2}\rangle,$$

where in the last line the angular brackets denote Gaussian averaging over $\check{Q}$-fluctuations according to Eq. (14.70). The last two terms here are proportional to $\delta_{\epsilon_2\epsilon_1}\delta_{\epsilon_3\epsilon_4}$ and vanish during the energy integrations. The first term, however, has a very different structure. It is proportional to $\delta_{-\epsilon_3\epsilon_1}\delta_{-\epsilon_2\epsilon_4}$ and thus to $A^{\text{cl}}_{\epsilon_1-\epsilon_2}A^{\text{q}}_{-\epsilon_1+\epsilon_2}$, while the Cooperon $\mathcal{C} \propto \langle c_{1;\epsilon_1\epsilon_4}c_{2;\epsilon_4\epsilon_1}\rangle$ depends on $\epsilon_4+\epsilon_1=\epsilon_1-\epsilon_2=\omega$. Collecting all such terms, differentiating over the vector potentials and integrating over $(\epsilon_1+\epsilon_2)/2$, one finally obtains for the one-loop correction to the frequency-dependent conductivity

$$\delta\sigma^{\text{R}}(\omega) = -\frac{2e^2D}{\pi}\sum_{\mathbf{q}}\frac{1}{Dq^2-i\omega}. \tag{14.72}$$

Summation over the wavenumber $\mathbf{q}$ is due to the fact that the c_1^{T} and c_2 fields are taken at the same spatial point and thus the Cooperons $\mathcal{C}(\mathbf{q},\omega)$ of all wavenumbers contribute equally.

One notices that for $d \le 2$ the correction to the ac conductivity diverges at small ω as

$$\delta\sigma(\omega) \propto -e^2\left(\frac{D}{\omega}\right)^{1-d/2}. \tag{14.73}$$

In particular, for $d = 2$ in the dc limit one finds

$$\delta\sigma = -\frac{e^2}{2\pi^2}\ln\frac{\tau_\phi}{\tau_{\text{el}}}, \tag{14.74}$$

where the upper limit of momentum integration in (14.72) is set by $Dq^2 \lesssim 1/\tau_{\text{el}}$. On the other hand, the lower limit $Dq^2 \gtrsim 1/\tau_\phi$ is given either by a phase-breaking mechanism (e.g. due to electron–electron interactions) or by a TRS-breaking perturbation, such as a magnetic field or magnetic impurities. The latter observation implies a weak field *positive* magneto-conductivity; see Problem 14.10.3, which

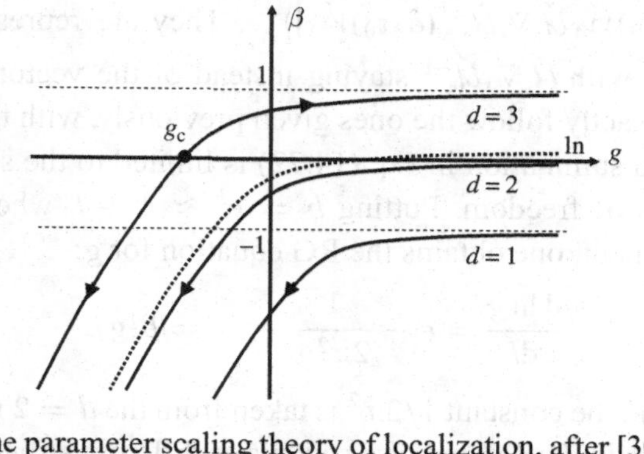

Figure 14.4 One parameter scaling theory of localization, after [308]. The scaling function $\beta = \mathrm{d}\ln g/\mathrm{d}l$ versus $\ln g$ for $d = 1, 2, 3$ (full lines) and $d = 2+\epsilon$ (dotted line). The arrows show the directions of RG flow.

is indeed observed in a wide class of disordered materials [307]. The divergent nature of the correction indicates that the metallic fixed point, characterized by small fluctuations around the stationary point $\check{\Lambda}$, is unstable. That is, the symmetry breaking does *not* occur and the $\check{Q}$-matrix field freely rotates around the entire manifold $\check{Q}^2 = \check{1}$. In the language of the magnetic analogy: the metal corresponds to a ferromagnetic (i.e. spontaneously broken symmetry) state, while the insulator corresponds to a paramagnetic (i.e. symmetric) state.

To put these observations at a more quantitative level one may develop the renormalization group (RG) treatment of the sigma-model [308, 280, 281, 306]; see Section 9.4. To this end, one splits the $\check{Q}$-matrix degrees of freedom into slow and fast ones, integrates out the fast ones, and then rescales the phase space and the fields to recast the action into the initial form, albeit with renormalized constants. A way to implement it is to write the $\check{Q}$-matrix as (cf. Eq. (14.38)),

$$\check{Q} = \check{\mathcal{U}}_{\mathrm{s}}^{-1} \circ \mathrm{e}^{-\check{\mathcal{W}}_{\mathrm{f}}/2} \circ (\hat{\sigma}_3 \hat{\tau}_0) \circ \mathrm{e}^{\check{\mathcal{W}}_{\mathrm{f}}/2} \circ \check{\mathcal{U}}_{\mathrm{s}}, \tag{14.75}$$

where the slow degrees of freedom $\check{\mathcal{U}}_{\mathrm{s}}$ are not restricted to the specific form (14.37), but rather encode an arbitrary slow $\check{Q}_{\mathrm{s}} = \check{\mathcal{U}}_{\mathrm{s}}^{-1}(\hat{\sigma}_3\hat{\tau}_0)\check{\mathcal{U}}_{\mathrm{s}}$ matrix. The rotation generators $\check{\mathcal{W}}_{\mathrm{f}}$, having the structure of Eqs. (14.66) and (14.67), represent now the fast degrees of freedom.

One now rescales $\mathbf{r} \to b\mathbf{r}$, $t \to b^z t$, where b is the contraction factor of the momentum space. The scaling dimension of $\check{\mathcal{W}}_{\mathrm{f}}$, $\check{\mathcal{U}}_{\mathrm{s}}$, and $\check{Q}$ is z, which is consistent with the fact that $\int \mathrm{d}t'' \check{Q}_{tt''}\check{Q}_{t''t'} = \check{\delta}_{t,t'}$. From here one deduces the bare scaling dimensions of the two constants in the action (14.65), $g \equiv \nu D$ and ν, to be $[g] = d-2$ and $[\nu] = z$. Next, one expands the action in powers of $\check{\mathcal{W}}_{\mathrm{f}}$ and integrates over the fast degrees of freedom. To second order one encounters terms like

$g\langle \mathrm{Tr}\{\check{\mathcal{U}}_s \nabla_{\mathbf{r}} \check{\mathcal{U}}_s^{-1}(\hat{\sigma}_3\hat{\tau}_0)\check{\mathcal{W}}_f \check{\mathcal{U}}_s \nabla_{\mathbf{r}} \check{\mathcal{U}}_s^{-1}(\hat{\sigma}_3\hat{\tau}_0)\check{\mathcal{W}}_f\}\rangle_{\mathcal{W}_f}$. They are represented by the diagram of Fig. 14.3, with $\check{\mathcal{U}}_s \nabla_{\mathbf{r}} \check{\mathcal{U}}_s^{-1}$ staying instead of the vector potential $\hat{\mathbf{A}}$. The calculations thus exactly follow the ones given previously, with the only difference that the momentum summation in Eq. (14.72) is limited to the shell $\Lambda < q < b\Lambda$ of the fast degrees of freedom. Putting $b = \mathrm{e}^l \approx 1 + l$, where $l = \ln b$ is an infinitesimal increment, one obtains the RG equation for g:

$$\frac{d\ln g}{dl} = \epsilon - \frac{1}{2\pi^2 g} + \cdots = \beta(g), \tag{14.76}$$

where $\epsilon = d-2$ and the constant $1/2\pi^2$ is taken from the $d = 2$ result (14.74). The time derivative term in the action has a somewhat different structure. Its expansion brings terms like $\nu\langle \mathrm{Tr}\{\check{\mathcal{U}}_s \partial_t \check{\mathcal{U}}_s^{-1}(\hat{\sigma}_3\hat{\tau}_0)\check{\mathcal{W}}_f\check{\mathcal{W}}_f\}\rangle_{\mathcal{W}_f}$, which may be shown to vanish due to the energy integration of the purely retarded function. The same is true in all higher orders, leading to the absence of renormalization for ν beyond the bare scaling. This is natural, since the disorder is not expected to alter the density of states (unlike the diffusion constant).

One is therefore left with the single parameter scaling description [308], given by Eq. (14.76). The RG flow suggested (since our calculations are valid only for $g \gg 1$, in units of $e^2/\hbar$) by Eq. (14.76) is schematically depicted in Fig. 14.4. One notices that for $d \leq 2$ the conductance g decreases with RG rescaling, indicating that a large enough system is always localized. On the other hand, for $d > 2$ there is a critical conductance g_c above which the system tends to be metallic. The latter means that the conductance $g \to L^{d-2}$, when the system size $L \to \infty$, which is exactly what is expected from Ohm's law. In the ϵ-expansion, that is, for $\epsilon \ll 1$, the critical conductance is large, $g_c \sim 1/\epsilon$, validating the perturbative RG calculations. In particular, the correlation (i.e. localization) length diverges as $\xi \propto (g_c - g)^{-\nu}$, where the critical exponent ν (not related to DOS) is $\nu^{-1} = d\beta/d\ln g\big|_{g_c} \approx \epsilon$; see Eq. (14.76).

14.9 Time-Dependent Perturbations

Here we briefly touch on an issue of a chaotic system subject to a fast time-dependent drive. We'll show that in the saddle-point approximation it leads to the diffusion in the energy direction. The fluctuation corrections may lead to dynamical localization, which curtails the system's ability to absorb the extra energy.

To be specific, let's assume that a disordered electronic system (e.g. a metallic ring) is subject to a uniform-in-space, time-dependent vector potential $\mathbf{A}(t)$. The latter may be provided by a time-dependent Aharonov–Bohm flux threaded through the ring. It is clear that such time-dependent flux generates the electric field

$e\mathbf{E}(t) = \partial_t \mathbf{A}(t)$, which leads to Joule heating of the metal. Neglecting spatial variations, the action (14.48) acquires the following form:[9]

$$i S[\hat{Q}, \mathbf{A}] = \frac{\pi}{4\delta} \mathrm{Tr}\Big\{ D\big([\mathbf{A}\hat{\gamma}^{\mathrm{cl}}, \hat{Q}]\big)^2 + 4\partial_t \hat{Q} \Big\}, \tag{14.77}$$

where δ is the mean level spacing at the chemical potential, $\delta^{-1} = \int \mathrm{d}^d\mathbf{r}\, \nu$. Taking the variation with respect to the $\hat{Q}$-field, under the condition $\hat{Q}^2 = \hat{1}$, as described in the opening paragraph of Section 14.4, one finds the stationary point condition (recall that $\hat{\gamma}^{\mathrm{cl}} = \hat{1}$):

$$D\big(\hat{Q} \circ \mathbf{A} \circ \hat{Q} \circ \mathbf{A} - \mathbf{A} \circ \hat{Q} \circ \mathbf{A} \circ \hat{Q}\big) - \{\partial_t, \hat{Q}\} = 0. \tag{14.78}$$

For a classical stationary point of the form (14.25) with a generic nonlocal in time distribution function, $F_{t_1 t_2}$, the nontrivial $(1,2)$ matrix component of this equation becomes

$$-D\big(\mathbf{A}(t_1) - \mathbf{A}(t_2)\big)^2 F_{t_1 t_2} - (\partial_{t_1} + \partial_{t_2}) F_{t_1 t_2} = 0. \tag{14.79}$$

Finally one performs the Wigner transform according to Eqs. (14.35) and (6.41). The first term, $\mathbf{A}_{t_1}^2 F_{t_1 t_2} - 2\mathbf{A}_{t_1} F_{t_1 t_2} \mathbf{A}_{t_2} + F_{t_1 t_2} \mathbf{A}_{t_2}^2 \overset{\mathrm{WT}}{\to} -(\partial_t \mathbf{A})^2 \partial_\epsilon^2 F_\epsilon(t)$, requires expanding the exponent in Eq. (6.41) to the second order. This leads to the diffusion equation in the *energy direction* for the distribution function $F_\epsilon(t)$,

$$\mathcal{D}\, \partial_\epsilon^2 F_\epsilon(t) - \partial_t F_\epsilon(t) = 0, \tag{14.80}$$

where the energy diffusion coefficient is $\mathcal{D} = D\langle(\partial_t \mathbf{A})^2\rangle = De^2\langle \mathbf{E}^2\rangle$ and the angular brackets denote averaging over the fast time-dependent oscillations of the external field.

The energy diffusion equation (14.80) implies that the characteristic energy of an electron, that is, temperature, in a system subject to an ac perturbation grows as $T \propto \sqrt{\mathcal{D}t}$. The total energy of the electron gas $\propto \nu T^2 \propto \nu \mathcal{D} t$ grows linearly in time, resulting in a Joule heating. Indeed, the change of the total energy, that is, the energy absorption, is calculated as

$$W = \frac{\mathrm{d}}{\mathrm{d}t} \nu \int \mathrm{d}\epsilon\, \epsilon\, \frac{1}{2}(1 - F_\epsilon(t)) = -\frac{\nu \mathcal{D}}{2} \int \mathrm{d}\epsilon\, \epsilon\, \partial_\epsilon^2 F_\epsilon(t) = \nu \mathcal{D}, \tag{14.81}$$

where we performed the energy integration by parts and used that $F_\infty = 1$ and $F_{-\infty} = -1$. For a harmonic ac electric field of the form $E(t) = E\cos\omega_0 t$, one finds $W = e^2 \nu D E^2/2$. This is the Joule heating of a metal with the conductivity $\sigma_{\mathrm{D}} = e^2 \nu D$, in agreement with $\mathbf{q} = 0$ limit of Eq. (14.61).

[9] In the time-reversal space the vector potential comes with the $\hat{\tau}_3$ matrix, cf Eq. (14.65). One may derive a similar action, where the time-dependent field comes with time-reversal $\hat{\tau}_0$ matrix structure [309, 310]. To this end one may start with, e.g. an ac scalar potential, which is random in space (a uniform-in-space scalar potential is a pure gauge, which does not induce any transitions). Upon averaging over the random spatial component of such scalar potential, one arrives at the action (14.77) with the ac field $\propto \hat{\tau}_0$.

Going beyond the stationary point treatment of the action (14.77), one may show [309, 310] that the energy diffusion coefficient $\mathcal{D}$ acquires weak localization corrections, similar in spirit to those discussed in Section 14.8. The one loop correction shows up in the case of $\propto \hat{\tau}_0$ ac field (see footnote 9). Since the problem is effectively one-dimensional (energy direction), it takes the form [310] (cf. the $d = 1$ case of Eq. (14.73))

$$W(t) = \nu\mathcal{D}\left(1 - \sqrt{\frac{t}{t_*}}\right), \tag{14.82}$$

where $t_* \propto \mathcal{D}/(\omega_0\delta)^2$, and δ is the mean level spacing. One may argue that such weak localization corrections are bound to lead to localization in the energy direction at $t \gtrsim t_*$. In this so-called *dynamical localization* scenario the initial diffusive growth of the effective temperature, $T(t) \propto \sqrt{\mathcal{D}t}$, is arrested at longer times and asymptotically $T(t \to \infty) \to \sqrt{\mathcal{D}t_*}$. The most famous example of the dynamical localization is provided by the quantum kicked rotor model [311, 312, 313, 314] (although it is not clear to what extent it may be described by the action like (14.77)).

14.10 Problems

14.10.1 Temperature Profile in a Mesoscopic Wire

Electronic distribution function in a short wire with an applied bias V is depicted in Fig. 14.2. This was obtained in an approximation that neglects any electron collisions and relaxation. Lets consider another limit of an intermediate-length wire, where a local electronic temperature, $T_e(x)$, and chemical potential, $\mu(x)$, are established due to electron–electron collisions. On the other hand, the wire is still short enough, so one can disregard electron–phonon collisions. As a result, the electronic temperature is different from the lattice one.

To find the temperature profile, consider the local electron energy $E_e(x)$ balance equation

$$\partial_t E_e = \nabla_x\big(\kappa(T_e)\nabla_x T_e\big) + \nu D\,(\nabla_x\mu)^2, \tag{14.83}$$

where the last term on the right-hand side represents locally deposited Joule heat and $\kappa(T_e)$ is the thermal conductance. In a metal the latter is given by $\kappa(T_e) = Dc_V(T_e) = (\pi^2/3)\nu D\, T_e$. (This statement is known as the Wiedemann–Franz law, stating $\kappa/\sigma_D = \mathcal{L}\, T_e$, where the Lorenz number is $\mathcal{L} = \pi^2/(3e^2)$; see Problem 18.8.1 for a derivation.) In a stationary state one puts $\partial_t E_e = 0$ and $\nabla_x\mu = eV/L$; the latter is a consequence of the charge continuity equation $\partial_t \rho_e = \nabla_x(\nu D\nabla_x\mu)$. Solve stationary equation (14.83) with the boundary conditions $T_e(0) = T_e(L) = 0$ to find the electronic temperature profile $T_e(x)$. What is the ratio of a maximal temperature in the middle of the wire and the applied voltage eV? Plot the corresponding distribution functions $n(\epsilon, x)$ for a few representative values of $x \in [0, L]$.

14.10.2 Electrodynamics of Disordered Metals

Consider the sigma-model action $S[\hat{Q}, \mathbf{A}, V]$, Eq. (14.48), with both classical and quantum vector and scalar potentials. Expand the action up to the second order in deviations from the metallic stationary point $\hat{Q} = \hat{\Lambda}$ and perform the Gaussian integration over the fluctuations $d^{\mathrm{cl}}, d^{\mathrm{q}}$. Find the effective *quadratic* action $S_{\mathrm{N}}[\mathbf{A}, V]$. Show that its quantum-classical (i.e. retarded) component has the form

$$S_{\mathrm{N}}^{\mathrm{q-cl}}[\mathbf{A}, V] = \sigma_{\mathrm{D}} \sum_{\mathbf{q},\omega} \begin{pmatrix} \mathbf{A}^{*\mathrm{q}}_{\mathbf{q},\omega}, & V^{*\mathrm{q}}_{\mathbf{q},\omega} \end{pmatrix} \begin{pmatrix} \frac{\omega^2}{Dq^2 - i\omega} & \frac{\mathbf{q}\omega}{Dq^2 - i\omega} \\ \frac{\omega \mathbf{q}}{Dq^2 - i\omega} & \frac{q^2}{Dq^2 - i\omega} \end{pmatrix} \begin{pmatrix} \mathbf{A}^{\mathrm{cl}}_{\mathbf{q},\omega} \\ V^{\mathrm{cl}}_{\mathbf{q},\omega} \end{pmatrix}, \tag{14.84}$$

where $\sigma_{\mathrm{D}} = e^2 \nu D$. The diagonal components of this matrix were already derived in Section 14.6 (cf. Eqs. (14.61) and (14.60)). Derive the advanced and Keldysh parts of this action. Show that the preceding action may be written in terms of the gauge-invariant electric field $\mathbf{E}_{\mathbf{q},\omega} = i\omega \mathbf{A}_{\mathbf{q},\omega} + i\mathbf{q} V_{\mathbf{q},\omega}$ as

$$S_{\mathrm{N}}^{\mathrm{q-cl}}[\mathbf{E}] = \sigma_{\mathrm{D}} \sum_{\mathbf{q},\omega} \frac{\mathbf{E}^{*\mathrm{q}}_{\mathbf{q},\omega} \mathbf{E}^{\mathrm{cl}}_{\mathbf{q},\omega}}{Dq^2 - i\omega}. \tag{14.85}$$

This action describes dynamics of the electric field propagation in a noninteracting disordered metal. We shall discuss its generalization for the interacting case in Chapter 16; see Problem 16.6.1.

The electron current and density are given by variations of this action over the quantum components of the vector and scalar potential correspondingly. One thus finds

$$\mathbf{j}_{\mathbf{q},\omega} = \sigma_{\mathrm{D}} \left[\frac{\omega^2 \mathbf{A}^{\mathrm{cl}}_{\mathbf{q},\omega}}{Dq^2 - i\omega} + \frac{\mathbf{q}\omega V^{\mathrm{cl}}_{\mathbf{q},\omega}}{Dq^2 - i\omega} \right] = \sigma_{\mathrm{D}} \frac{-i\omega\, \mathbf{E}^{\mathrm{cl}}_{\mathbf{q},\omega}}{Dq^2 - i\omega}; \tag{14.86}$$

$$\rho_{\mathbf{q},\omega} = \sigma_{\mathrm{D}} \left[\frac{\omega \mathbf{q} \cdot \mathbf{A}^{\mathrm{cl}}_{\mathbf{q},\omega}}{Dq^2 - i\omega} + \frac{q^2 V^{\mathrm{cl}}_{\mathbf{q},\omega}}{Dq^2 - i\omega} \right] = \sigma_{\mathrm{D}} \frac{-i\mathbf{q} \cdot \mathbf{E}^{\mathrm{cl}}_{\mathbf{q},\omega}}{Dq^2 - i\omega}. \tag{14.87}$$

From here, verify the continuity and derive the material relation:

$$i\mathbf{q} \cdot \mathbf{j}_{\mathbf{q},\omega} - i\omega \rho_{\mathbf{q},\omega} = 0; \qquad \mathbf{j}_{\mathbf{q},\omega} + Di\mathbf{q}\rho_{\mathbf{q},\omega} = \sigma_{\mathrm{D}} \mathbf{E}^{\mathrm{cl}}_{\mathbf{q},\omega}. \tag{14.88}$$

In real space-time these relations acquire the form

$$\nabla_{\mathbf{r}} \cdot \mathbf{j} + \partial_t \rho = 0; \qquad \mathbf{j} + D\nabla_{\mathbf{r}} \rho = \sigma_{\mathrm{D}} \mathbf{E}^{\mathrm{cl}}. \tag{14.89}$$

In equilibrium, one can rewrite the action (14.85) in imaginary time (i.e. Matsubara) formalism, according to the recipe of Section 3.3. Substituting there the material relation, one ends up with the imaginary time action for the density and current fluctuations in the noninteracting metal:

$$S_{\rm N}[\mathbf{j},\rho] = \frac{1}{2\sigma_{\rm D}} \iint \mathrm{d}^d\mathbf{r}\, \mathrm{d}^d\mathbf{r}' \iint_0^\beta \mathrm{d}\tau\, \mathrm{d}\tau'\, \mathbf{g}(\mathbf{r},\tau)\mathcal{D}(\mathbf{r}-\mathbf{r}',\tau-\tau')\mathbf{g}(\mathbf{r}',\tau'), \quad (14.90)$$

where $\mathbf{g}(\mathbf{r},\tau) = \mathbf{j}(\mathbf{r},\tau) + D\nabla_{\mathbf{r}}\rho(\mathbf{r},\tau)$ and $\mathcal{D}(\mathbf{q},\omega_m) = (Dq^2 + |\omega_m|)^{-1}$ is the imaginary time diffuson and $\omega_m = 2\pi m T$. This action was first derived by Levitov and Shytov [315]; see Problem 16.6.1.

14.10.3 Weak Localization Magneto-Conductance

Consider a disordered electronic system in the presence of a *static* magnetic field, characterized by a classical vector potential $\mathbf{A}(\mathbf{r})$. It enters the action (14.65) through the long derivative $\check{\partial}_{\mathbf{r}}\check{Q} = \nabla_{\mathbf{r}}\check{Q} + \mathrm{i}\mathbf{A}(\mathbf{r})[\hat{\gamma}^{cl}\hat{\tau}_3, \check{Q}]$. Using the parametrization (14.66), (14.67), show that the action of the diffuson modes (14.40) is not affected by the field. On the other hand, the Cooperon action (14.69) is affected and acquires the following form:

$$\mathrm{i}S\,[c_1,c_2] = -\frac{\pi\nu}{2}\sum_{\epsilon,\epsilon'}\int \mathrm{d}\mathbf{r}\, c_{1;\epsilon\epsilon'}(\mathbf{r})\left[-D(\nabla_{\mathbf{r}} - 2\mathrm{i}\mathbf{A}(\mathbf{r}))^2 - \mathrm{i}\epsilon - \mathrm{i}\epsilon'\right] c_{2;\epsilon'\epsilon}(\mathbf{r}). \quad (14.91)$$

The corresponding Cooperon propagator in real space, $\mathcal{C}(\mathbf{r},\mathbf{r}';\epsilon+\epsilon')$, is given by the inverse operator in the square brackets in this expression. Notice the factor of two in front of the vector potential, which shows that the Cooperon behaves as a charge- *two* particle in the magnetic field (while the diffuson has charge zero).

Let's focus on a uniform magnetic field $H = [\nabla_{\mathbf{r}} \times \mathbf{A}]$, perpendicular to a 2d metallic film. In this case the operator in (14.91) is the Schrödinger operator for the charge-two particle in the uniform field. Its spectrum is given by the familiar Landau levels $D\mathbf{q}^2 \to 4DH(n+1/2)$, where $n = 0, 1, 2, \ldots$. The weak localization correction (14.72) is $\delta\sigma(\omega) = -(2e^2D/\pi)\mathcal{C}(\mathbf{r},\mathbf{r};\omega)$. Show that it is given by

$$\delta\sigma(H) = -\frac{2e^2DH}{\pi^2}\sum_{n=0}^{\infty}\frac{1}{4DH(n+1/2)+1/\tau_\phi}, \quad (14.92)$$

where for the dc limit we substituted $-\mathrm{i}\omega \to 1/\tau_\phi$, frequency by the dephasing rate. The sum is logarithmically divergent at large n and should be cut off at $4DHn \approx 1/\tau_{\rm el}$. However, one can consider the magneto-conductance effect $\sigma(H) - \sigma(0)$, which is independent of the UV cutoff. Perform the summation and show that [316]

$$\sigma(H) - \sigma(0) = -\frac{e^2}{2\pi^2}\left[\log\left(\frac{H_\phi}{H}\right) - \Psi\left(\frac{1}{2} + \frac{H_\phi}{H}\right)\right], \tag{14.93}$$

where $H_\phi = (4D\tau_\phi)^{-1} = (2l_\phi)^{-2}$, and $\Psi(z)$ is the Euler digamma function. Plot the magneto-conductance in the range $0 < H < 50H_\phi$. Explain why the magneto-conductance is positive. For the phase-breaking length $l_\phi = 10^{-4}$cm, calculate the characteristic magnetic field H_ϕ in Gauss. The logarithmic increase of the magneto-conductance is terminated at $H \approx l_{\mathrm{el}}^{-2} \approx (D\tau_{\mathrm{el}})^{-1}$. See, for example, Ref. [307] for experimental manifestations of this phenomenon.

15
Mesoscopic Effects

The physical properties of small metallic devices exhibit sample-to-sample fluctuations due to random realizations of the disorder potential. We first study fluctuations of the electron spectrum and density of states, deriving Altshuler–Shklovskii and Wigner–Dyson statistics (the oscillatory part of the latter requires careful consideration of nonclassical stationary points of the sigma-model). We then characterize mesoscopic fluctuations of the current, including a universal result for small bias conductance, as well as large bias fluctuations of the current–voltage characteristics. Following Nazarov, we consider full counting statistics of quasi-1d diffusive wire and derive the Dorokhov distribution of the eigenvalues of its transmission matrix. Finally, we introduce a generic tunneling action, which is used to study full counting statistics and dynamics of charge transfer through an open chaotic cavity.

15.1 Spectral Statistics

Consider a disordered metallic grain of size L such that $L \gg l_{\rm el}$, where $l_{\rm el} = v_{\rm F}\tau_{\rm el}$ is the elastic mean free path. The spectrum of the Schrödinger equation consists of a discrete set of levels, ϵ_n, that may be characterized by the *sample-specific* density of states (DOS), $\nu(\epsilon) = \sum_n \delta(\epsilon - \epsilon_n)$. This quantity fluctuates strongly and usually cannot (and doesn't need to) be calculated analytically. One may average it over realizations of disorder to obtain a mean DOS: $\langle \nu(\epsilon)\rangle_{\rm dis}$. The latter is a smooth function of energy on the scale of the Fermi energy and thus may be taken as a constant $\langle \nu(\epsilon_{\rm F})\rangle_{\rm dis} \equiv \nu$. This is the DOS that was used in the previous chapter.[1]

One may wonder how to describe sample-to-sample (i.e. mesoscopic) fluctuations of the DOS $\nu(\epsilon)$ and, in particular, how much a given spectrum at one energy

[1] Actually, in this section we use the total DOS, not normalized per unit volume and thus having dimensionality of one over energy.

ϵ is correlated with itself at another energy ϵ'. To answer this question one may calculate the spectral correlation function

$$R(\epsilon_1, \epsilon_2) \equiv \big\langle (\nu(\epsilon_1) - \nu)(\nu(\epsilon_2) - \nu) \big\rangle_{\text{dis}} = \langle \nu(\epsilon_1)\nu(\epsilon_2) \rangle_{\text{dis}} - \nu^2. \tag{15.1}$$

This function was calculated by Altshuler and Shklovskii [317]. Here we discuss it using the Keldysh nonlinear sigma-model.

Since for a single quantum level $G^{\text{R}} - G^{\text{A}} = -2\pi \text{i}\delta(\epsilon - \epsilon_0)$, see Eqs. (10.31a) and (10.31b), DOS may be written as

$$\nu(\epsilon) = \frac{\text{i}}{2\pi}\text{Tr}\{G^{\text{R}}(\epsilon) - G^{\text{A}}(\epsilon)\} = \frac{1}{2\pi}\text{Tr}\{\langle \psi_1 \bar{\psi}_1 \rangle - \langle \psi_2 \bar{\psi}_2 \rangle\} = -\frac{1}{2\pi}\langle \vec{\bar{\Psi}} \hat{\sigma}_3 \vec{\Psi} \rangle,$$

where the angular brackets denote quantum (as opposed to disorder) averaging and the indices are in Keldysh space. To generate a DOS at any given energy one adds a source term

$$\text{i}S[J] = -\sum_{\epsilon} J_\epsilon \int \text{d}\mathbf{r}\, \vec{\bar{\Psi}}(\epsilon, \mathbf{r}) \hat{\sigma}_3 \vec{\Psi}(\epsilon, \mathbf{r}) = -\iint \text{d}t\text{d}t' \int \text{d}\mathbf{r}\, \vec{\bar{\Psi}}(\mathbf{r}, t) J_{t-t'} \hat{\sigma}_3 \vec{\Psi}(\mathbf{r}, t')$$

to the fermionic action (10.35). After averaging over disorder and changing to the $\hat{Q}$-matrix representation, the DOS source term is translated to

$$\text{i}S_{\text{DOS}}[J] = \pi\nu \int \frac{\text{d}\epsilon}{2\pi} J_\epsilon \int \text{d}\mathbf{r}\, \text{Tr}\{\hat{Q}_{\epsilon\epsilon}(\mathbf{r})\hat{\sigma}_3\}. \tag{15.2}$$

Then the DOS is generated by $\nu(\epsilon) = \delta\langle Z[J]\rangle_{\text{dis}}/\delta J_\epsilon$. It is now clear that $\langle \nu(\epsilon)\rangle_{\text{dis}} = \frac{1}{2}\nu\langle \text{Tr}\{\hat{Q}_{\epsilon\epsilon}\hat{\sigma}_3\}\rangle_Q$. Substituting $\hat{Q}_{\epsilon\epsilon} = \hat{\Lambda}_\epsilon$, Eq. (14.25), one finds $\langle \nu(\epsilon)\rangle_{\text{dis}} = \nu$, as it should be, of course. It is also easy to check that the fluctuations around $\hat{\Lambda}$ do not change the result (all the fluctuation diagrams cancel owing to the causality constraints). We are now in a position to calculate the correlation function (15.1),

$$R(\epsilon_1, \epsilon_2) \equiv \frac{\delta^2 \langle Z[J]\rangle_{\text{dis}}}{\delta J_{\epsilon_1}\delta J_{\epsilon_2}} - \nu^2 = \nu^2\left[\frac{1}{4}\Big\langle \text{Tr}\{\hat{Q}_{\epsilon_1\epsilon_1}\hat{\sigma}_3\}\text{Tr}\{\hat{Q}_{\epsilon_2\epsilon_2}\hat{\sigma}_3\}\Big\rangle_Q - 1\right]. \tag{15.3}$$

Employing the parametrization (14.38) of the $\hat{Q}$-matrix and expanding up to second order in the diffusive fluctuations $\hat{\mathcal{W}}$, one finds

$$\text{Tr}\{\hat{Q}_{\epsilon_1\epsilon_1}\hat{\sigma}_z\} = \frac{1}{2}\Big[4 - 2F_{\epsilon_1}d^{\text{q}}_{\epsilon_1\epsilon_1} - 2d^{\text{q}}_{\epsilon_1\epsilon_1}F_{\epsilon_1} + \sum_{\epsilon}\big(d^{\text{cl}}_{\epsilon_1\epsilon}d^{\text{q}}_{\epsilon\epsilon_1} + d^{\text{q}}_{\epsilon_1\epsilon}d^{\text{cl}}_{\epsilon\epsilon_1}\big)\Big]. \tag{15.4}$$

Since $\langle d^{\text{q}}d^{\text{q}}\rangle_{\mathcal{W}} = 0$, the only nonvanishing terms contributing to Eq. (15.3) are those with no d^{cl} and d^{q} at all (they cancel the $-\nu^2$ term in Eq. (15.1)) and those of the type $\langle d^{\text{cl}}d^{\text{q}}d^{\text{cl}}d^{\text{q}}\rangle_{\mathcal{W}}$. Collecting the latter terms, one finds

$$R(\epsilon_1\epsilon_2) = \frac{\nu^2}{16}\int \text{d}\mathbf{r} \sum_{\epsilon,\epsilon'}\Big\langle \Big(d^{\text{cl}}_{\epsilon_1\epsilon}d^{\text{q}}_{\epsilon\epsilon_1} + d^{\text{q}}_{\epsilon_1\epsilon}d^{\text{cl}}_{\epsilon\epsilon_1}\Big)\Big(d^{\text{cl}}_{\epsilon_2\epsilon'}d^{\text{q}}_{\epsilon'\epsilon_2} + d^{\text{q}}_{\epsilon_2\epsilon'}d^{\text{cl}}_{\epsilon'\epsilon_2}\Big)\Big\rangle_{\mathcal{W}}. \tag{15.5}$$

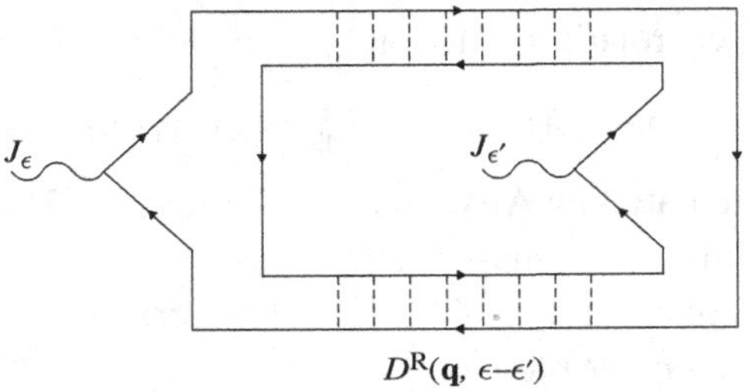

Figure 15.1 Diagram representing mesoscopic fluctuations of the density of states, $R(\epsilon_1,\epsilon_2)$, Eq. (15.3). It is generated from the Wick contraction $\langle d^{\mathrm{cl}}_{\epsilon_1\epsilon} d^{\mathrm{q}}_{\epsilon\epsilon_1} d^{\mathrm{q}}_{\epsilon_2\epsilon'} d^{\mathrm{cl}}_{\epsilon'\epsilon_2}\rangle_{\mathcal{W}} \to \langle d^{\mathrm{cl}}_{\epsilon_1\epsilon} d^{\mathrm{q}}_{\epsilon_2\epsilon'}\rangle_{\mathcal{W}} \langle d^{\mathrm{q}}_{\epsilon\epsilon_1} d^{\mathrm{cl}}_{\epsilon'\epsilon_2}\rangle_{\mathcal{W}} \propto [\mathcal{D}^{\mathrm{R}}(\mathbf{q},\epsilon_1-\epsilon_2)]^2 \delta_{\epsilon_1\epsilon'}\delta_{\epsilon_2\epsilon}$; see Eq. (15.5). There is also a similar diagram for the advanced diffusons.

Now one has to perform Wick's contractions, using the correlation function $\langle d^{\mathrm{cl}}_{\epsilon\epsilon'} d^{\mathrm{q}}_{\epsilon'\epsilon}\rangle_{\mathcal{W}} \propto \mathcal{D}^{\mathrm{R}}(\mathbf{q},\epsilon-\epsilon')$, which follows from Eq. (14.43), and also take into account that $\int d\epsilon\, [\mathcal{D}^{\mathrm{R(A)}}(\mathbf{q},\epsilon-\epsilon_1)]^2 = 0$, owing to the integration of a function that is analytic in the entire upper (lower) half-plane of ϵ. Due to the last observation, the only surviving Wick contractions are those where both energy summations in Eq. (15.5) are killed by the restrictions $\delta_{\epsilon,\epsilon_1}$ and $\delta_{\epsilon',\epsilon_2}$, or vice versa. As a result,

$$R(\epsilon_1,\epsilon_2) = \frac{1}{4\pi^2}\sum_{\mathbf{q}}\Big[\big(\mathcal{D}^{\mathrm{R}}(\mathbf{q},\epsilon_1-\epsilon_2)\big)^2 + \big(\mathcal{D}^{\mathrm{A}}(\mathbf{q},\epsilon_1-\epsilon_2)\big)^2\Big], \tag{15.6}$$

where the momentum summation stands for a summation over the discrete modes of the diffusion operator $D\nabla^2_{\mathbf{r}}$ with the zero current (zero derivative) boundary conditions at the edge of the metallic grain. This is the result of Altshuler and Shklovskii [317] for the unitary symmetry class.[2] Note that the correlation function $R(\epsilon_1,\epsilon_2)$ depends only on the energy difference $\omega = \epsilon_1 - \epsilon_2$. A diagrammatic representation of the $R(\epsilon_1,\epsilon_2)$ function is shown in Fig. 15.1. Adopting an explicit form of the diffusion propagator, we find the spectral correlation function in the form

$$R(\omega) = \frac{1}{2\pi^2}\,\mathrm{Re}\sum_n \frac{1}{\big(Dq_n^2 - i\omega\big)^2}, \tag{15.7}$$

[2] In the orthogonal symmetry class with unbroken time-reversal symmetry, Section 14.7, there are additional degrees of freedom – the Cooperon modes. They lead to the additional contribution to the correlation function (15.6), where $\mathcal{D}^{\mathrm{R(A)}}(\mathbf{q},\epsilon_1-\epsilon_2) \to \mathcal{C}^{\mathrm{R(A)}}(\mathbf{q},\epsilon_1-\epsilon_2)$; see Eq. (14.70). Since in the absence of the magnetic fields the diffuson $\mathcal{D}$ and the Cooperon $\mathcal{C}$ have the same analytic form, the correlation function $R(\epsilon_1,\epsilon_2)$ in the orthogonal case is twice as big as in the unitary one.

where, for example, for a rectangular grain, the wavenumbers of the diffusion operator are quantized as $q_n^2 = \sum_\mu \pi^2 n_\mu^2/L_\mu^2$, with $\mu = x, y, z$; $n_\mu = 0, 1, 2, \ldots$, and L_μ are the spatial dimensions of the mesoscopic sample.

For a small energy difference $\omega \lesssim E_{\mathrm{Th}} = D/L_\mu^2$, where E_{Th} is known as the Thouless energy, only the lowest homogenous mode, $q_n = 0$, of the diffusion operator (the so-called zero mode) may be retained: $R(\omega) = -1/(2\pi^2\omega^2)$. This result is universal, in the sense that it does not depend on any microscopic details of the metallic grain. The fact that the correlation function is negative means that the energy levels are *less* likely to be found at a small distance ω from each other. This is a manifestation of the energy level repulsion. Note that the correlations decay very slowly – as the inverse square of the energy distance. In fact this result is a part of the celebrated *random matrix* theory (RMT) statistics [318]. The random matrix prediction for the unitary symmetry class, known as Wigner–Dyson statistics, reads as

$$R_{\mathrm{RMT}}(\omega) = -\frac{1 - \cos(2\pi\omega/\delta)}{2\pi^2\omega^2}, \tag{15.8}$$

where $\delta = 1/\langle \nu \rangle_{\mathrm{dis}}$ is the mean level spacing. In addition to the $-1/(2\pi^2\omega^2)$ term obtained earlier, it contains the oscillatory function of the energy difference with period δ. These oscillations reflect the discreteness of the underlying energy spectrum. They also regularize the behavior of the correlation function at small energy difference $\omega \to 0$. The oscillatory term cannot be found in the framework of perturbation theory in small fluctuations around the stationary configuration $\hat{\Lambda}$. However, it may be recovered once additional stationary configurations are taken into account [319]. We discuss its derivation in the next section.

For larger energy difference $E_{\mathrm{Th}} \lesssim \omega \lesssim 1/\tau_{\mathrm{el}}$, one may substitute summation over the discrete spectrum of the diffusion operator in Eq. (15.7) by integration over $\mathbf{q}$. This way one finds $R(\omega) \propto -E_{\mathrm{Th}}^{-d/2}\,\omega^{d/2-2}$. To discuss this result it is convenient to look at the mean square fluctuation of the number of energy levels in the energy window E, defined as $\Sigma(E) = \iint_0^E d\epsilon_1 d\epsilon_2 R(\epsilon_1 - \epsilon_2)$. If the energy level sequence were completely random, that is, Poisson distributed, it would result in $\Sigma_{\mathrm{P}}(E) = E/\delta$, where the subscript stands for Poisson. The random matrix result (15.8) means $\Sigma_{\mathrm{RMT}}(E) \propto \ln(E/\delta) \ll \Sigma_{\mathrm{P}}(E)$, where $\delta \ll E \lesssim E_{\mathrm{Th}}$. This signifies a spectacular rigidity of the energy spectrum, due to the level repulsion at energies less than the Thouless energy. At larger energies, $E_{\mathrm{Th}} \lesssim \omega \lesssim 1/\tau_{\mathrm{el}}$, we found that $\Sigma_{\mathrm{AS}}(E) \propto (E/E_{\mathrm{Th}})^{d/2}$. The Altshuler–Shklovskii variance is much larger than the RMT prediction, but still much less than the Poisson limit. This shows that the spectral rigidity persists for beyond the Thouless energy, but takes a different (softer) form than the RMT prediction. Such a level statistics in the intermediate range of energies is essentially responsible for mesoscopic fluctuations of many physical quantities. Some examples are given in what follows; for reviews see [305, 268, 320].

15.2 Wigner–Dyson Statistics

Here, following Altland and Kamenev [321], we discuss how the oscillatory part of RMT Wigner–Dyson statistics (15.8) may be derived in the framework of the Keldysh nonlinear sigma-model. As explained, RMT appears as the $\mathbf{q} = 0$ limit of the generic spectral statistics. The corresponding action may be thus limited to spatially uniform configurations of the $\hat{Q}_{\epsilon\epsilon'}$-matrix field and takes the form (cf. Eq. (14.33))

$$iS[\hat{Q}] = -i\pi\nu \sum_{\epsilon} \mathrm{Tr}\{(\hat{\epsilon} + i\hat{J}_{\epsilon})\hat{Q}_{\epsilon\epsilon}\}, \tag{15.9}$$

where $\hat{J}_{\epsilon}$ is the source field, introduced in Eq. (15.2).[3] The perturbative calculation, which leads us to the smooth part of RMT statistics, was done by considering Gaussian fluctuations around the *classical* stationary point, Eq. (14.25), of the action (15.9). Here we show that the oscillatory part originates from taking into account other *nonclassical* stationary points.

Notice that the distribution function F_{ϵ} does not show up in the calculations of the previous section. This is natural, since we discuss the spectrum of the noninteracting system, which is completely independent of the occupation numbers. We can thus put $F_{\epsilon} = 0$ without loss of generality. It is also convenient to discretize the energy ϵ with the small step δ_{ϵ}, such that the relevant energy interval $-E_{\mathrm{Th}} < \epsilon < E_{\mathrm{Th}}$ is divided into $K = 2E_{\mathrm{Th}}/\delta_{\epsilon} \gg 1$ intervals. Then the $\hat{Q}$-matrix is understood as a $2K \times 2K$ matrix in the product of the energy and Keldysh spaces.

Variation of the action (15.9) with respect to $\hat{Q}$ under the nonlinear constraint $\hat{Q}^2 = 1$ leads to the stationary point equation $[\hat{\epsilon}, \hat{Q}] = 0$; see Section 14.4. It is solved by any diagonal matrix, but the nonlinear constraint $\hat{Q}^2 = 1$ restricts acceptable solutions to $\hat{Q} = \mathrm{diag}\{\pm 1, \ldots, \pm 1; \pm 1, \ldots, \pm 1\}$, in total 2^{2K} distinct configurations. The causality constraint, that is, the demand for $[\hat{G}^{-1} + i/2\tau_{\mathrm{el}}\hat{Q}]^{-1}$ to be $\mathrm{diag}\{G^{\mathrm{R}}, G^{\mathrm{A}}\}$ in Keldysh space, Section 14.3, selects the single *classical* configuration $\hat{\Lambda} = \mathrm{diag}\{1, \ldots, 1; -1, \ldots, -1\}$. It is clear that the action (15.9) of such a classical configuration is zero, $S[\hat{\Lambda}] = 0$ (in the absence of the source $\hat{J}_{\epsilon}$), in agreement with the general structure of the theory; see Eq. (2.53).

To calculate the contribution of Gaussian fluctuations around the classical stationary point, one parametrizes the $\hat{Q}$-matrix manifold as in Eqs. (14.38) and

[3] In Eq. (15.2) we have chosen it as $\hat{J}_{\epsilon} = J_{\epsilon}\hat{\sigma}_3$ to generate the $G^{\mathrm{R}}_{\epsilon} - G^{\mathrm{A}}_{\epsilon}$ combination. Here it is more convenient to keep it as a generic, but still diagonal in energy and Keldysh space matrix.

(14.39). Expanding to second order in the diffuson modes $d^{\rm cl}$ and $d^{\rm q}$, one finds for the zero-dimensional action (15.9) (cf. Eq. (14.42))

$$iS[d^{\rm cl}, d^{\rm q}] = -i\frac{\pi\nu}{2}\sum_{l,l'=1}^{K} d^{\rm q}_{ll'}(\epsilon_l^- - \epsilon_{l'}^+)d^{\rm cl}_{l'l}, \tag{15.10}$$

where $\epsilon_l^{\pm} = \epsilon_l \pm i0$ and we ignored the source part $\hat{J}_\epsilon$. Performing Gaussian integration, one finds for the corresponding contribution to the partition function

$$Z_0 \propto \prod_{l,l'} \frac{1}{\epsilon_l^- - \epsilon_{l'}^+} = \exp\left\{-\sum_{\epsilon,\epsilon'} \ln(\epsilon - \epsilon' - i0)\right\} = 1, \tag{15.11}$$

where we adopted the generic statement that the energy integral of a purely retarded or advanced function must be zero; see also the discussion after Eq. (2.52a). To show it explicitly, one has to involve a specific regularization of the high-energy behavior of the DOS, which renders convergence to the integral. It is possible to do, for example, for Gaussian RMT, where the average DOS takes the form of a semicircle [318]. The explicit calculation then shows [321] that indeed $Z_0 = 1$. It must be the case, however, for *any* physical regularization, because of the causality structure of the Keldysh correlation functions. Also, according to general principles, the *exact* partition function must be equal to unity, $Z = 1$. This means that all the other $2^{2K} - 1$ nonclassical stationary points do *not* contribute to the partition function. They (or rather one of them) still do contribute to the correlation function (15.3).

Consider a nonclassical stationary point, where the $\epsilon_{l'} = \epsilon_1$ position in the retarded block and $\epsilon_l = \epsilon_2$ position in the advanced block have the flipped signs, that is,

$$\hat{\hat{\Lambda}} = \mathrm{diag}\{\underbrace{1, 1, \ldots, 1, -1, 1, \ldots, 1}_{K \text{ retarded positions}};\ \underbrace{-1, -1, \ldots, -1, 1, -1, \ldots, -1}_{K \text{ advanced positions}}\}. \tag{15.12}$$

We also need to specify the diagonal source matrix, which is chosen to have J_1 in the retarded ϵ_1 and $-J_2$ in the advanced ϵ_2 positions, with all other positions being zero $\hat{J}_\epsilon = \mathrm{diag}\{0, \ldots, J_1, \ldots, 0; 0, \ldots, -J_2, \ldots, 0\}$. This source is tailored to produce $G^{\rm R}_{\epsilon_1} G^{\rm A}_{\epsilon_2} = \delta^2 Z[\hat{J}]/\delta J_1 \delta J_2$. The spectral correlation function (15.1) is then obtained as $R = -2\mathrm{Re}[G^{\rm R}_{\epsilon_1} G^{\rm A}_{\epsilon_2}]/(2\pi)^2$. We look now for the contribution of the nonclassical stationary point (15.12) to the generating function $Z[\hat{J}]$.

First, the action $S[\hat{\hat{\Lambda}}]$ is not zero any more. This is allowed, since the field configuration $\hat{\hat{\Lambda}}$ is not classical. Comparing the action $S[\hat{\hat{\Lambda}}]$ with $S[\hat{\Lambda}] = 0$, one notices that the former differs by $2\pi\nu\epsilon_1$, due to the flipped sign of the retarded ϵ_1 position, and by $-2\pi\nu\epsilon_2$, due to the flipped sign of the advanced ϵ_2 position. Finally

$S[\hat{\hat{\Lambda}}] = 2\pi\nu(\epsilon_1 - \epsilon_2)$. We look now for the contribution of the Gaussian fluctuations. Inspecting Eqs. (15.10) and (15.11), one understands that each pair of energies with the *opposite signs* of $\underline{\hat{Q}}_\epsilon$ brings in the factor $(\epsilon_l - \epsilon_{l'})^{-1}$. The pairs with the same signs of $\underline{\hat{Q}}_\epsilon$ do not contribute at all, since the corresponding $+\hat{1}$ and $-\hat{1}$ subblocks of $\underline{\hat{Q}}$ do not allow for any nontrivial rotations. Most of the energy pairs in $\hat{\hat{\Lambda}}$ are the same as in $\hat{\Lambda}$, and therefore the fluctuation factor is almost the same, that is, Z_0, Eq. (15.11). One has to correct, though, for the flipped signs of the retarded ϵ_1 and advanced ϵ_2 positions. As a result,

$$\begin{aligned}\tilde{Z}[\hat{J}] &= Z_0 \prod_{\epsilon_l \neq \epsilon_2} \frac{\epsilon_l^- - \epsilon_1^+}{\epsilon_l^- - \epsilon_2 + \mathrm{i}J_2} \prod_{\epsilon_{l'} \neq \epsilon_1} \frac{\epsilon_{l'}^+ - \epsilon_2^-}{\epsilon_{l'}^+ - \epsilon_1 - \mathrm{i}J_1} \mathrm{e}^{\mathrm{i}S[\hat{\hat{\Lambda}}]} \\ &= \prod_l \frac{\epsilon_l^- - \epsilon_1^+}{\epsilon_l^- - \epsilon_2 + \mathrm{i}J_2} \prod_{l'} \frac{\epsilon_{l'}^+ - \epsilon_2^-}{\epsilon_{l'}^+ - \epsilon_1 - \mathrm{i}J_1} \frac{\mathrm{i}J_2}{\epsilon_2^- - \epsilon_1^+} \frac{-\mathrm{i}J_1}{\epsilon_1^+ - \epsilon_2^-} \mathrm{e}^{\mathrm{i}S[\hat{\hat{\Lambda}}]} \\ &= -\frac{J_1 J_2}{(\epsilon_1 - \epsilon_2)^2} \mathrm{e}^{2\pi\mathrm{i}\nu(\epsilon_1 - \epsilon_2)} = -\frac{J_1 J_2}{\omega^2} \mathrm{e}^{2\pi\mathrm{i}\omega/\delta}. \end{aligned} \tag{15.13}$$

The ϵ_l-product in the first line runs over all advanced positions, safe for ϵ_2 and corrects for (i) the fact that one should not have included terms $(\epsilon_l^- - \epsilon_1^+)^{-1}$ in Z_0, since the retarded ϵ_1 position has now a -1 sign and (ii) instead one needs to include $(\epsilon_l^- - \epsilon_2 + \mathrm{i}J_2)^{-1}$ terms, since the advanced ϵ_2 position has now a $+1$ sign. We have also taken into account the source, $-\mathrm{i}J_2$, which shifts the ϵ_2 energy, see Eq. (15.9). Similarly, the $\epsilon_{l'}$-product runs over all retarded positions and achieves similar corrections. In the second line in Eq. (15.13) we took into account that $Z_0 = 1$ and completed the two products by one term each, to make them unrestricted. In the third line, we again used the fact that the unrestricted product may be written as $\prod_l \ldots = \exp\{\sum_\epsilon \ldots\} = 1$, see Eq. (15.11), due to the energy integration of purely retarded or advanced functions.

We now employ the fact that the corresponding contribution to the correlation function is given by $R = -(2\pi^2)^{-1}\mathrm{Re}\,\delta^2\tilde{Z}[\hat{J}]/\delta J_1 \delta J_2$ to find that the two stationary points $\hat{\Lambda}$ and $\hat{\hat{\Lambda}}$ lead exactly to the RMT Wigner–Dyson correlation function (15.8). Notice that $\tilde{Z}[\hat{J}] \propto J_1 J_2$, where J_1 and J_2 are sources at the two energies where the signs of the stationary point matrix were flipped. The first observation is that the nonclassical saddle point $\hat{\hat{\Lambda}}$ indeed does not contribute to the partition function $\tilde{Z}[0] = 0$. Another observation is that, if one flips any other sign, save the ϵ_1 and ϵ_2 positions, the corresponding contribution to the generating function is bound to be zero. Indeed, the source was chosen to be nonzero *only* for those two energies. In other words, it is the structure of the observable (and thus the source field) that selects a nonclassical stationary point that matters. As a corollary: different observables call for different nonclassical stationary configurations to be included.

The stationary point calculation requires a large action, which in our case means $\omega/\delta \gg 1$. Therefore, our method is only applicable for the energy window $\delta < \omega < E_{\mathrm{Th}}$. The fact that we were able to obtain the exact result (15.8), valid down to zero energy, is a peculiarity of the unitary symmetry class. The exact results down to zero energy may be obtained with the supersymmetry [290] and, in some cases, with the replica [287] technique. It is still an open problem to achieve this with the Keldysh technique. Another open question is a manifestation of the nonclassical stationary points in electron–electron interaction effects.

15.3 Universal Conductance Fluctuations

One of the most spectacular manifestations of mesoscopic fluctuations is sample to sample variations of electric conductance [305, 322]. In practice small changes of some external parameter, such as magnetic field, carrier density, impurity concentration, and others play the role of switching between the samples. They lead thus to random, but completely reproducible, fluctuations of the conductance around its mean value $g_{\mathrm{D}} = e^2\nu D/L$, where the DOS ν is understood as being integrated over the cross-section of the sample and L is its length along the current carrying direction. Our goal is to calculate the variance of such fluctuations, $\delta g = \sqrt{\langle (g - g_{\mathrm{D}})^2 \rangle_{\mathrm{dis}}}$. The remarkable observation, first made by Altshuler [323], and Lee and Stone [324], is that at low enough temperature the variance is universal,[4] $\delta g \propto e^2/\hbar$, where the proportionality coefficient weakly depends on geometry, but is independent of the average conductance g_{D} and other microscopic details. The effect is known as *universal conductance fluctuations* (UCF).

Here we follow the approach of Larkin and Khmel'nitskii [326], who calculated the second moment of the current $K(V_1, V_2) = \langle J(V_1)J(V_2)\rangle_{\mathrm{dis}}$ at two arbitrary voltages. The conductance variance is obtained as the small voltage limit $K = (\delta g)^2 V_1 V_2$. To perform this calculation it is convenient to introduce two replicas of the system which have the same realization as the disorder potential, but may have different external fields, such as, for example, applied voltage or vector potential. The generating function depends thus on two sets of the applied fields $Z[V_s, \mathbf{A}_s]$, where $s = 1, 2$ is the replica index. We use it to generate the current correlator by differentiation over the quantum components $\mathbf{A}_1^{\mathrm{q}}$ and $\mathbf{A}_2^{\mathrm{q}}$; see Eq. (14.55). Since both replicas have the same realization of the disorder, the averaging procedure of Section 14.2 mixes between them and the

[4] The higher moments and, more generally, the conductance distribution function also exhibit a certain degree of universality [325].

$\check{Q}$-matrix acquires non-diagonal structure in replica space. Yet the stationary point is replica-diagonal, $\check{\Lambda} = \text{diag}\{\hat{\Lambda}_1, \hat{\Lambda}_2\}$. In particular, its Keldysh block is diagonal, $2\hat{F} = \text{diag}\{2F(V_1), 2F(V_2)\}$, where $F(V_s)$ is the non-equilibrium distribution function (14.11), which solves the stationary point diffusion equation with the proper boundary conditions.

To account for UCF one needs to take into account replica-off-diagonal fluctuations of the $\check{Q}$-matrix around such a steady-state non-equilibrium stationary point. First we define the current–current correlation function as

$$K(V_1, V_2) = -\frac{e^2}{4}\frac{\delta^2 Z[V_s, \mathbf{A}_s]}{\delta\mathbf{A}_1^{\mathrm{q}}\delta\mathbf{A}_2^{\mathrm{q}}}\bigg|_{\mathbf{A}_s^{\mathrm{q}}=0} = -\frac{e^2}{4}\pi\nu D\Big\langle \mathrm{Tr}\{\hat{\gamma}^{\mathrm{q}}\check{Q}^{12}\hat{\gamma}^{\mathrm{q}}\check{Q}^{21}\}\Big\rangle_Q$$

$$+\frac{e^2}{4}(\pi\nu D)^2\Big\langle \mathrm{Tr}\{(\check{Q}\nabla_{\mathbf{r}}\check{Q})^{11}\hat{\gamma}^{\mathrm{q}}\}\,\mathrm{Tr}\{(\check{Q}\nabla_{\mathbf{r}}\check{Q})^{22}\hat{\gamma}^{\mathrm{q}}\}\Big\rangle_Q, \qquad (15.14)$$

where the superscripts are replica indices and traces run over the energy and Keldysh spaces. Here we have used Eqs. (14.52) and (14.54) to differentiate the action. We expand now each of the traces on the right-hand side to second order in the diffuson generators $\hat{d}^{\mathrm{cl}}$ and $\hat{d}^{\mathrm{q}}$, Eqs. (14.38) and (14.39), which are now 2×2 matrices in replica space.[5] Keeping only the terms that do not cancel (but rather double) between the diffuson and Cooperon channels, one finds

$$\mathrm{Tr}\{(\check{Q}\nabla_{\mathbf{r}}\check{Q})^{ss}\hat{\gamma}^{\mathrm{q}}\} \cong \sum_{\epsilon,\epsilon'}(\nabla_{\mathbf{r}}F_\epsilon^s)\Big[(d_{\epsilon\epsilon'}^{\mathrm{cl}})^{ss'}(d_{\epsilon'\epsilon}^{\mathrm{q}})^{s's} + (d_{\epsilon\epsilon'}^{\mathrm{q}})^{ss'}(d_{\epsilon'\epsilon}^{\mathrm{cl}})^{s's}\Big],$$

$$\mathrm{Tr}\{\hat{\gamma}^{\mathrm{q}}\check{Q}^{12}\hat{\gamma}^{\mathrm{q}}\check{Q}^{21}\} \cong \sum_{\epsilon,\epsilon'} 2(d_{\epsilon\epsilon'}^{\mathrm{cl}})^{12}(d_{\epsilon'\epsilon}^{\mathrm{cl}})^{21},$$

where we have omitted the $d^{\mathrm{q}}d^{\mathrm{q}}$ terms, since they vanish upon $\check{Q}$-integration. We now substitute these expressions into Eq. (15.14) and perform averaging over the fluctuating diffuson modes $d^{\mathrm{cl(q)}}$ according to Eqs. (14.43) and (14.46). In doing so, we employ that the gradient of the stationary non-equilibrium distribution (14.11) is a constant in space. Integrating then twice over the length of the wire L, the result may be written in the Fourier representation

[5] Since we calculate directly the dc current (unlike the linear response diagrammatics [323, 324], which deals with the low-frequency ac response), we should be careful not to include thermodynamic persistent currents. The latter, though almost zero in average, have mesoscopic fluctuations comparable to that of the dissipative currents [327] we are after. One way to exclude them is to work consistently in the time-reversal symmetric (orthogonal) ensemble, where the thermodynamic currents are absent. Then the Cooperon contributions are similar to the diffuson ones. One may show that the diffuson and Cooperon terms mutually cancel each other if one expands each of the two $\check{Q}$-matrices in $\mathrm{Tr}\{(\check{Q}\nabla_{\mathbf{r}}\check{Q})^{ss}\hat{\gamma}^{\mathrm{q}}\}$ to first order in $\check{\mathcal{W}}$, but that they double the result if one expands one of the $\check{Q}$-matrices to second order in $\check{\mathcal{W}}$, while putting the other one to be $\check{\Lambda}$. For the $\mathrm{Tr}\{\hat{\gamma}^{\mathrm{q}}\check{Q}^{12}\hat{\gamma}^{\mathrm{q}}\check{Q}^{21}\}$ term the situation is exactly opposite, i.e. only the term with each of the $\check{Q}$-matrices expanded to first order survives.

$$L^2K(V_1,V_2) = (2eD)^2\sum_{\epsilon,\epsilon';\mathbf{q}}\Big[\big|\mathcal{D}^{\mathrm{R}}(\mathbf{q},\omega)\big|^2 + \frac{1}{2}\,\mathrm{Re}\big(\mathcal{D}^{\mathrm{R}}(\mathbf{q},\omega)\big)^2\Big](\nabla_{\mathbf{r}}F^1_{\epsilon})(\nabla_{\mathbf{r}}F^2_{\epsilon'}), \tag{15.15}$$

where $\omega = \epsilon - \epsilon'$ and $\nabla_{\mathbf{r}}F^s_\epsilon = (F^{\mathrm{eq}}_{\epsilon-eV_s} - F^{\mathrm{eq}}_\epsilon)/L$ and we took into account the factor of two, due to the Cooperon modes.

Let us first analyze this result in the linear response regime, where the distribution function may be approximated as $\nabla_{\mathbf{r}}F^s_\epsilon \approx -\partial_\epsilon F^{\mathrm{eq}}_\epsilon eV_s/L$; see Eq. (14.11). Since the conductance is given by $g = J/V$, one finds for the conductance variance

$$\langle(\delta g)^2\rangle_{\mathrm{dis}} = K(V_1,V_2)/V_1V_2 = g_1^2 + \frac{1}{2}\,g_2^2, \tag{15.16}$$

where

$$g_1^2 = \left(\frac{2e^2D}{\pi\hbar L^2}\right)^2\int_{-\infty}^{+\infty}\frac{\mathrm{d}\omega}{2T}\,\mathcal{F}\left(\frac{\omega}{2T}\right)\sum_{\mathbf{q}}\frac{1}{(Dq^2)^2+\omega^2}, \tag{15.17a}$$

$$g_2^2 = \left(\frac{2e^2D}{\pi\hbar L^2}\right)^2\int_{-\infty}^{+\infty}\frac{\mathrm{d}\omega}{2T}\,\mathcal{F}\left(\frac{\omega}{2T}\right)\mathrm{Re}\sum_{\mathbf{q}}\frac{1}{(Dq^2-\mathrm{i}\omega)^2}, \tag{15.17b}$$

and $\mathcal{F}(x) = [x\coth(x)-1]/\sinh^2(x) = \frac{1}{4}\int \mathrm{d}y\partial_y\tanh(y+\frac{x}{2})\partial_y\tanh(y-\frac{x}{2})$. Notice that $\int\mathrm{d}x\,\mathcal{F}(x) = 1$ and thus at a low temperature $T \ll E_{\mathrm{Th}} = D/L^2$ the $\mathcal{F}(x)$ function may be regarded as a delta-function. As a result, at low temperature

$$g_1^2 = g_2^2 = \left(\frac{2e^2}{\pi\hbar}\right)^2\sum_{\mathbf{q}}\frac{1}{(Lq)^4}. \tag{15.18}$$

The $\mathbf{q}$-summation should be understood as a sum over eigenvalues of the diffusion operator within the metal, with zero boundary conditions at the metallic contacts (no diffuson fluctuations) and zero current boundary conditions at the contacts with an insulator. For example, for a quasi-one-dimensional wire of length L with metallic leads the quantization condition is $qL = \pi n$, where $n = 1,2,\ldots$ and thus $\sum_{\mathbf{q}}(Lq)^{-4} = 1/90$. Therefore we found that the conductance variance $\langle(\delta g)^2\rangle_{\mathrm{dis}} = c_d(e^2/2\pi\hbar)^2$, where c_d is a dimensionality- and geometry-dependent coefficient, for example, $c_1 = 4/15$ (notice that we deal here with the spinless carriers). The fact that this coefficient is insensitive to the strength of disorder constitutes the essence of *universality* of conductance fluctuations.

At a higher temperature $T \gg E_{\mathrm{Th}}$, one may put $\mathcal{F}(x) \approx \mathcal{F}(0) = 1/3$ and find

$$g_1^2 \approx \left(\frac{2e^2}{\pi\hbar}\right)^2\frac{E_{\mathrm{Th}}}{T}\,\frac{\pi}{6}\sum_{\mathbf{q}}^{\sqrt{T/D}}\frac{1}{(Lq)^2+(L/l_\phi)^2}, \tag{15.19}$$

while $g_2^2 \ll g_1^2$. Here we have introduced the dephasing length $l_\phi(T)$ as a long-distance cutoff, which exists, for example, due to electron–electron interactions [329, 330]. Therefore the conductance variance decreases with temperature as $(E_{\rm Th}/T)\min\{1, (E_{\rm Th}\tau_\phi)^{1-d/2}\}$, where $\tau_\phi = l_\phi^2/D$, in $d \le 2$. In the case $2 < d < 4$ the conductance variance scales as $(E_{\rm Th}/T)^{2-d/2}$. This is a reflection of the fact that the spectral statistics partially loses its rigidity at energies larger than the Thouless energy, Section 15.1.

For $V_s > E_{\rm Th}$ and $T = 0$ the energy integrals in Eq. (15.15) are restricted to the intervals $0 < \epsilon < eV_s$ due to the factors $\nabla_{\mathbf{r}} F_\epsilon^s = (F_{\epsilon - eV_s}^{\rm eq} - F_\epsilon^{\rm eq})/L$. One can then define the correlation function of the *differential* conductances as $K_g = \partial_{V_1}\partial_{V_2} K(V_1, V_2)$ [326]. Since the voltages enter only as the upper limits of integrations over ϵ and ϵ', the differentiation simply removes the energy integrals. As a result one finds for, for example, the $\left|\mathcal{D}^{\rm R}(\mathbf{q},\omega)\right|^2$ part of the correlation function

$$K_g = \left(\frac{2e^2}{\pi\hbar}\right)^2 \sum_{\mathbf{q}} \frac{1}{(Lq)^4 + \frac{e^2(V_1-V_2)^2}{E_{\rm Th}^2}} \propto \left(\frac{e^2}{\hbar}\right)^2 \left(\frac{E_{\rm Th}}{e|V_1 - V_2|}\right)^{2-d/2}, \tag{15.20}$$

where in the last approximate equality we assumed $e|V_1 - V_2| > E_{\rm Th}$ and substituted the eigenvalue summation by integration. The variance of the random part of the current–voltage characteristic is given by $\langle(\delta J(V))^2\rangle_{\rm dis} = \iint_0^V dV_1 dV_2\, K_g(V_1, V_2) \approx (e^2/\hbar)^2\big[V(E_{\rm Th}/e) + V^{d/2}(E_{\rm Th}/e)^{2-d/2}\big]$, where the first term on the right-hand side comes from the diagonal $e|V_1 - V_2| \lesssim E_{\rm Th}$, and the second one comes from the off-diagonal long-range correlations. For $eV \gg E_{\rm Th}$, the diagonal part dominates in dimensions $d < 2$, leading to $|\delta J(V)| \propto (e^2/\hbar)\sqrt{VE_{\rm Th}/e}$. This means that the current–voltage characteristics executes a random walk with the voltage "step" $E_{\rm Th}/e$ and the corresponding current "step" $eE_{\rm Th}/\hbar$ in a random direction. (There is, of course, a linear, nonrandom part of the current–voltage curve, $\langle J(V)\rangle_{\rm dis} = g_{\rm D} V$.) For dimensions $d > 2$ the long-range part dominates the correlator and one finds $|\delta J(V)| \propto (e^2/\hbar)V^{d/4}(E_{\rm Th}/e)^{1-d/4}$. Therefore, the random part is much more profound and the deviations from the linear current–voltage characteristic scale as $|\delta J(V)| \propto V^{3/4}$ in $d = 3$.

15.4 Full Counting Statistics in a Disordered Wire

In this section we discuss the statistics of *quantum* fluctuations of the current through a disordered mesoscopic wire. We have already encountered this problem in Section 13.3, where we dealt with the full counting statistics of a single quantum channel characterized by the tunneling probability $|\mathbf{t}|^2$. We arrived at the conclusion that the transmitted charge exhibits a *binomial* distribution, with the generating function $Z(\eta)$ given by Eq. (13.35). A disordered quasi-one-dimensional wire may be considered as a collection of quantum channels with transmission probabilities being randomly distributed, reflecting the randomness of the quenched disorder potential in the wire. Our goal here is to calculate the generating function of the current cumulants, which leads to shot noise, third moment, and so on. We shall find that it may be obtained from the single channel result (13.35) by averaging it over a certain distribution of transmissions $\mathcal{P}(|\mathbf{t}|^2)$, known as the Dorokhov distribution [331]. Both the distribution function itself and the way it emerges from the nonlinear sigma-model [332] are rather educating.

Consider two leads with the chemical potentials shifted by an externally applied voltage eV and connected to each other by a diffusive quasi-one-dimensional wire of length L. The wire conductance is $g_D = \sigma_\mathrm{D}/L$. Electron transport along the wire is fully described by the disorder-averaged generating function $Z[\eta] = \int \mathbf{D}[\hat{Q}] \exp(iS[\hat{Q}, A_\eta])$. The action is given by (14.48), while the auxiliary vector potential $\hat{A}_\eta$ enters the problem through the covariant derivative (14.49). We choose $\hat{A}_\eta$ to be purely quantum, that is, without the classical component, as

$$\hat{A}_\eta(t) = \frac{\hat{\gamma}^\mathrm{q}}{2L} \begin{cases} \eta, & 0 < t < t_0, \\ 0, & \text{otherwise.} \end{cases} \tag{15.21}$$

Here the quantum Keldysh matrix $\hat{\gamma}^\mathrm{q}$ is given by (10.40) and η is the already familiar, see Section 13.3, *counting field*. The action $S[\hat{Q}, A_\eta]$ is accompanied by the boundary conditions on the $\hat{Q}(x)$-matrix at the two ends of the wire:

$$\hat{Q}(0) = \begin{pmatrix} 1 & 2F_\epsilon \\ 0 & -1 \end{pmatrix}, \qquad \hat{Q}(L) = \begin{pmatrix} 1 & 2F_{\epsilon-eV} \\ 0 & -1 \end{pmatrix}. \tag{15.22}$$

Knowing $Z[\eta]$, one can find then any moment $\langle q^n \rangle$ of the charge transferred between reservoirs during the time of measurement t_0 via differentiation of $Z[\eta]$ with respect to the counting field η. The irreducible cumulants are defined as $\mathcal{C}_1 = \langle q \rangle = q_0$ and $\mathcal{C}_n = \langle (q - q_0)^n \rangle$ with $n = 2, 3, \ldots$, where $q = \int_0^{t_0} J(t) dt$ and $q_0 = t_0 g_\mathrm{D} V = t_0 \langle J \rangle$, where g_D is the average diffusive conductance. They may be found through the expansion of the logarithm of $Z[\eta]$ in powers of the counting field:

$$\ln Z[\eta] = \sum_{n=1}^{\infty} \frac{(i\eta)^n}{n!} \mathcal{C}_n. \tag{15.23}$$

We calculate $Z[\eta]$ in the stationary path approximation. The latter disregards both the localization effects and mesoscopic fluctuations of the counting statistics. That is, $\mathcal{C}_n$, which we are after, are *disorder-averaged* moments of the quantum fluctuations. To this end we look for a stationary configuration $\hat{Q}$, which extremizes the action $S[\hat{Q}, A_\eta]$, with the boundary conditions (15.22). The difficulty is that the action $S[\hat{Q}, A_\eta]$ depends explicitly on the quantum vector potential A_η and the solution of the corresponding stationary point equation is not known for an arbitrary A_η. This obstacle can be overcome by realizing that the spatially uniform vector potential (15.21) is a pure gauge and it can be gauged away from the action $S[\hat{Q}, A_\eta] \to S[\hat{Q}_\eta]$ by the gauge transformation

$$\hat{Q}(x\,; t, t') = \mathrm{e}^{\,\mathrm{i}x\hat{A}_\eta(t)}\, \hat{Q}_\eta(x\,; t, t')\, \mathrm{e}^{-\mathrm{i}x\hat{A}_\eta(t')}. \tag{15.24}$$

It comes with a price, though: the boundary conditions (15.22) change accordingly,

$$\hat{Q}_\eta(0) = \hat{Q}(0), \qquad \hat{Q}_\eta(L) = \mathrm{e}^{-\mathrm{i}\eta\hat{\gamma}^{\mathrm{q}}/2}\, \hat{Q}(L)\, \mathrm{e}^{\mathrm{i}\eta\hat{\gamma}^{\mathrm{q}}/2}. \tag{15.25}$$

The advantage of this transformation is that the stationary path equation for $\hat{Q}_\eta$, which is nothing else but the Usadel equation (14.34)

$$D\frac{\partial}{\partial x}\left(\hat{Q}_\eta\, \frac{\partial \hat{Q}_\eta}{\partial x}\right) = 0, \tag{15.26}$$

can be solved explicitly now. To this end notice that, according to Eq. (15.26), $\hat{Q}_\eta \partial_x \hat{Q}_\eta = -\partial_x \hat{Q}_\eta \hat{Q}_\eta = \hat{J}$ is a constant, that is, x-independent, matrix. Since $\hat{Q}_\eta^2 = \hat{1}$, $\hat{J}$ anti-commutes with $\hat{Q}_\eta$, that is, $\hat{Q}_\eta \hat{J} + \hat{J}\hat{Q}_\eta = 0$. As a result, one finds $\hat{Q}_\eta(x) = \hat{Q}_\eta(0)\exp\left(\hat{J}x\right)$. Putting $x = L$ and multiplying by $\hat{Q}_\eta(0)$ from the left, one expresses the yet unknown matrix $\hat{J}$ through the boundary conditions (15.25): $\hat{J} = L^{-1}\ln\left[\hat{Q}_\eta(0)\hat{Q}_\eta(L)\right]$.

Having determined the stationary configuration of the $\hat{Q}_\eta$ matrix, for a nonzero counting field η, one substitutes it back into the action $S[\hat{Q}_\eta]$ to find the generating function:

$$\ln Z[\eta] = \mathrm{i}S[\hat{Q}_\eta] = -\frac{\pi\nu D}{4}\,\mathrm{Tr}\{(\partial_x \hat{Q}_\eta)^2\} = \frac{\pi\nu D}{4}\,\mathrm{Tr}\{\hat{J}^2\},$$

where we used that $\partial_x \hat{Q}_\eta = -\hat{J}\hat{Q}_\eta = \hat{Q}_\eta \hat{J}$ and $\hat{Q}_\eta^2 = \hat{1}$. Calculating the time integrals, one goes to the Wigner transform $\iint \mathrm{d}t\mathrm{d}t' \to t_0 \int \frac{\mathrm{d}\epsilon}{2\pi}$, where t_0 emerges from the integral over the central time, and finds

$$\ln Z[\eta] = \frac{\pi g_{\mathrm{D}} t_0}{4}\int \frac{\mathrm{d}\epsilon}{2\pi}\,\mathrm{Tr}\ln^2\left[\hat{Q}(0)\ \mathrm{e}^{-\mathrm{i}\eta\hat{\gamma}^{\mathrm{q}}/2}\, \hat{Q}(L)\ \mathrm{e}^{\mathrm{i}\eta\hat{\gamma}^{\mathrm{q}}/2}\right]. \tag{15.27}$$

We analyze Eq. (15.27) in the zero-temperature limit, $T = 0$, where the distribution function $F_\epsilon = \tanh(\epsilon/2T) \to \mathrm{sign}(\epsilon)$. The algebra can be simplified by performing the rotation $\hat{\mathcal{Q}} = \hat{\mathcal{O}}^{-1}\hat{Q}\hat{\mathcal{O}}$ with the help of the matrix

$$\hat{\mathcal{O}} = \frac{1}{\sqrt{2}}\begin{pmatrix} 1 & -1 \\ 1 & 1 \end{pmatrix}. \tag{15.28}$$

One should note also that $\hat{\mathcal{O}}^{-1}\exp(\pm i\eta\hat{\gamma}^{q}/2)\hat{\mathcal{O}} = \exp(\pm i\eta\hat{\sigma}_3/2)$. It is not difficult to show that for $T = 0$ the only energy interval that contributes to the trace in Eq. (15.27) is $0 < \epsilon < eV$. Furthermore, in this interval the rotated Q-matrices are energy independent and given by

$$\hat{Q}(0) = \begin{pmatrix} -1 & -2 \\ 0 & 1 \end{pmatrix}, \qquad \hat{Q}(L) = \begin{pmatrix} 1 & 0 \\ -2 & -1 \end{pmatrix}. \tag{15.29}$$

As a result, the ϵ-integration in Eq. (15.27) gives a factor of eV and, inserting $\hat{Q}$ into $\ln Z[\eta]$, the latter reduces to

$$\ln Z[\eta] = \frac{g_{\mathrm{D}} eVt_0}{8}\,\mathrm{Tr}\ln^2\begin{pmatrix} -1+4e^{i\eta} & 2 \\ -2e^{i\eta} & -1 \end{pmatrix}. \tag{15.30}$$

Since the trace is invariant with respect to the choice of basis, it is convenient to evaluate it in the basis where the matrix under the logarithm in Eq. (15.30) is diagonal. Solving the eigenvalue problem and calculating the trace, one finds

$$\ln Z[\eta] = g_{\mathrm{D}} eVt_0 \ln^2\left[\sqrt{e^{i\eta}} + \sqrt{e^{i\eta}-1}\right]. \tag{15.31}$$

Knowing $\ln Z[\eta]$, one can extract the cumulants by expanding it in powers of η and employing Eq. (15.23). For example, $C_1 = q_0$, $C_2 = q_0/3$, while $C_3 = q_0/15$, and so on. Notice that to derive the full generating function, we had to consider the nonclassical stationary configuration of the $\hat{Q}$-matrix field. This configuration was selected by a quantum source A_η, which is necessary to generate the proper observable. The situation is thus rather similar to the one we encountered in calculation of the Wigner–Dyson statistics in Section 15.2.

The generating function (15.31) of the counting statistics in the disordered wire may be obtained from the single channel result $\ln Z[\eta, |\mathrm{t}|^2]$, Eq. (13.35), by averaging the latter over the distribution function of channel transparencies:

$$\ln Z[\eta] = \int_0^1 d|\mathrm{t}|^2\, \mathcal{P}(|\mathrm{t}|^2)\, \ln Z\left[\eta, |\mathrm{t}|^2\right]. \tag{15.32}$$

To find the distribution function $\mathcal{P}(|\mathrm{t}|^2)$ it is convenient to define $z = e^{i\eta} - 1$ and change the integration variable to $r = \sqrt{1-|\mathrm{t}|^2}$. Then one finds for the z-derivative of Eq. (15.32)

$$g_{\mathrm{D}} eVt_0 \frac{\ln(\sqrt{z+1}+\sqrt{z})}{\sqrt{z}\sqrt{z+1}} = \frac{eVt_0}{2\pi}\int_0^1 2r dr\, \mathcal{P}(r)\frac{1-r^2}{1+z-zr^2}.$$

It is now a simple matter to check that the only way this relation may be satisfied for any z is if $\mathcal{P}(r) = \pi g_{\mathrm{D}}/(1-r^2)r$. As a result, one finds the celebrated Dorokhov

[331] distribution of the transmission probabilities of individual quantum channels in a disordered wire:

$$\mathcal{P}(|\mathbf{t}|^2) = \frac{\pi g_D}{|\mathbf{t}|^2\sqrt{1-|\mathbf{t}|^2}}. \tag{15.33}$$

The surprising feature of this distribution is that it is *bimodal*, that is, it exhibits two peaks: a strong one near $|\mathbf{t}|^2 = 0$ and a weaker one near $|\mathbf{t}|^2 = 1$. This means that, while most of the channels are blocked by the disorder potential, there is always a fraction of them that are almost completely transparent. Notice that to make the distribution normalizable, one needs to modify it at exponentially small transmission probabilities. Such a modification, however, is inconsequential for calculation of any moment of $|\mathbf{t}|^2$, and thus for any reasonable observable. One may view, therefore, the normalization of the Dorokhov distribution (15.33) as a fitting parameter to reproduce correctly the average Landauer conductance of the diffusive wire, $g_D = \langle|\mathbf{t}|^2\rangle/(2\pi)$, where the angular brackets imply integration with the measure (15.33). Once the normalization is fixed, one may calculate other observables. For example, the average shot noise, according to Eq. (13.19), is given by [334, 262]

$$\langle\mathcal{S}(V)\rangle_{\text{dis}} = \frac{|2eV|}{2\pi}\left\langle|\mathbf{t}|^2(1-|\mathbf{t}|^2)\right\rangle = \frac{|2eV|g_D}{2}\int_0^1 d|\mathbf{t}|^2\,\sqrt{1-|\mathbf{t}|^2} = \frac{|2eV|g_D}{3}, \tag{15.34}$$

which is one-third of the Schottky value $|2eV|g_D = 2eJ$ (in agreement with the second cumulant quoted after Eq. (15.31)). We found that the binomial nature of the charge transport, along with the Dorokhov distribution of the transmissions, results in the shot noise suppression by the so-called Fano factor of $1/3$.[6]

15.5 Tunneling Action

Consider two disordered metallic leads separated by a tunneling barrier. The setup is described by the so-called tunneling action, which we already employed in Section 13.6:

$$S_T = \int_C dt \int_{\mathbf{r}\in L} d\mathbf{r} \int_{\mathbf{r}'\in R} d\mathbf{r}' \big[W_{\mathbf{rr}'}\bar{\psi}_L(\mathbf{r})\psi_R(\mathbf{r}') + W^*_{\mathbf{r}'\mathbf{r}}\bar{\psi}_R(\mathbf{r}')\psi_L(\mathbf{r})\big], \tag{15.35}$$

[6] An alternative way [335] to obtain the Fano factor of 1/3 is to look for the noise power in the form

$$\langle\mathcal{S}(V)\rangle_{\text{dis}} = -\frac{e^2}{2}\frac{\delta^2 Z[\mathbf{A}]}{\delta(\mathbf{A}^q)^2} = -\frac{e^2\pi\nu D}{2}\operatorname{Tr}\{\hat{\gamma}^q\hat{Q}\hat{\gamma}^q\hat{Q} - \hat{1}\} = g_D\int_0^L \frac{dx}{L}\int d\epsilon\,\left[1 - F_\epsilon^2(x)\right],$$

where we used Eq. (14.54) in the stationary point approximation $\hat{Q} = \hat{\Lambda}$. We now substitute the two-step zero temperature distribution function $F_\epsilon(x) = (1 - x/L)\text{sign}(\epsilon) + (x/L)\text{sign}(\epsilon - eV)$, Eq. (14.11), and perform the energy and coordinate integrations. This way, we find the same result as in Eq. (15.34).

where $\psi_{L(R)}$ and $\bar{\psi}_{L(R)}$ are fermionic fields describing annihilation and creation of electrons to the left (right) of the tunneling barrier. The $W_{\mathbf{rr}'}$ is a tunneling matrix restricted to the vicinity of the junction, since the overlap of electron wavefunctions decay exponentially away from it. Consider, for example, a pointlike (pinhole) tunneling, where $W_{\mathbf{rr}'} = W\delta(\mathbf{r}-\mathbf{r}')$. To find a relation between the matrix element W and the tunneling probability $|\mathbf{t}|^2$ through the pinhole one needs to sum over all partial amplitudes. For example, an electron may go directly from the left lead to the right lead with an amplitude $\propto W$. Alternatively it may go to the right lead, then back to the left and then finally back to the right one. The corresponding amplitude is $\propto W(\mathrm{i}\pi\nu)^2|W|^2$, where the two factors $(\mathrm{i}\pi\nu)$ appear due to summation of the perturbation theory energy denominators over intermediate virtual states $\sum_k(\epsilon-\xi_k)^{-1}$ (for simplicity we assume the same DOS in the left and right leads, $\nu_R = \nu_L = \nu$). Collecting all such "excursions" of going back and forth between the leads and summing the corresponding geometric series, one finds for the tunneling probability

$$|\mathbf{t}|^2 = (2\pi\nu)^2\left|\frac{W}{1+(\pi\nu)^2|W|^2}\right|^2. \tag{15.36}$$

For a more detailed derivation of this relation see appendix C of [336].

Since the tunneling action (15.35) is quadratic in fermion fields, the Gaussian integration over them is straightforward, leading to the disorder-averaged action of the form

$$\langle Z\rangle_{\mathrm{dis}} = \int \mathbf{D}[\hat{Q}_L,\hat{Q}_R]\,\exp\left(\mathrm{i}S[\hat{Q}_L,\hat{Q}_R]\right),$$
$$\mathrm{i}S = -\frac{\pi\nu}{4\tau_{\mathrm{el}}}\sum_{a=L,R}\mathrm{Tr}\{\hat{Q}_a^2\} + \mathrm{Tr}\ln\begin{pmatrix}\hat{G}_L^{-1}+\frac{i}{2\tau_{\mathrm{el}}}\hat{Q}_L & \hat{W}\\ \hat{W}^\dagger & \hat{G}_R^{-1}+\frac{\mathrm{i}}{2\tau_{\mathrm{el}}}\hat{Q}_R\end{pmatrix}. \tag{15.37}$$

Deriving Eq. (15.37), one has to introduce two $\hat{Q}$-matrices to decouple the disorder mediated four-fermion terms, Eq. (14.19), in each of the two leads independently. It was assumed for simplicity that both leads are characterized by equal elastic mean free times and electronic densities of states. Equation (15.37) exhibits the 2×2 matrix structure in the space of left–right leads. Note also that the tunneling matrix elements $\hat{W}$ form a unit matrix in the Keldysh subspace.

Introducing the notation $\hat{\mathbf{G}}_a^{-1} = \hat{G}_a^{-1} + \frac{i}{2\tau_{\mathrm{el}}}\hat{Q}_a$, one identically rewrites the last term of the action $\mathrm{i}S[\hat{Q}_L,\hat{Q}_R]$ in Eq. (15.37) as

$$\mathrm{Tr}\ln\begin{pmatrix}\hat{\mathbf{G}}_L^{-1} & \hat{W}\\ \hat{W}^\dagger & \hat{\mathbf{G}}_R^{-1}\end{pmatrix} = \mathrm{Tr}\ln\begin{pmatrix}\hat{\mathbf{G}}_L^{-1} & 0\\ 0 & \hat{\mathbf{G}}_R^{-1}\end{pmatrix} + \mathrm{Tr}\ln\left[\hat{1}+\begin{pmatrix}0 & \hat{\mathbf{G}}_L\hat{W}\\ \hat{\mathbf{G}}_R\hat{W}^\dagger & 0\end{pmatrix}\right].$$

Expanding now $\mathrm{Tr}\ln \hat{\mathbf{G}}_a^{-1}$ in gradients of the $\hat{Q}_a$ matrix around the saddle point $\hat{\Lambda}_a$, one obtains the sigma-model action (14.33) for each of the two leads independently. The coupling between them is provided by the second term on the right-hand side, which defines the tunneling action $\mathrm{i}S_\mathrm{T}[\hat{Q}_\mathrm{L}, \hat{Q}_\mathrm{R}]$. To proceed, let us expand the logarithm in powers of $\hat{W}$ and employ the local nature of the tunneling matrix elements, which allows us to substitute $\hat{\mathbf{G}}_a \to \hat{\mathbf{G}}_a(\mathbf{r},\mathbf{r}) = -\mathrm{i}\pi\nu\hat{Q}_a$, see Eq. (14.22). This way one finds for the tunneling part of the action

$$\mathrm{i}S_\mathrm{T}[\hat{Q}_\mathrm{L}, \hat{Q}_\mathrm{R}] = -\sum_{l=1}^{\infty}\frac{(\mathrm{i}\pi\nu)^{2l}}{l}\mathrm{Tr}(\hat{Q}_\mathrm{L}\hat{W}\hat{Q}_\mathrm{R}\hat{W}^\dagger)^l = \mathrm{Tr}\ln\left[\hat{1} + \pi^2\nu^2|W|^2\hat{Q}_\mathrm{L}\hat{Q}_\mathrm{R}\right].$$

Employing the commutativity under the trace operation, the last logarithm may be written as $\ln\left[\hat{1} + \pi^2\nu^2|W|^2\hat{Q}_\mathrm{R}\hat{Q}_\mathrm{L}\right]$. It is convenient to take the half sum of the two: $\frac{1}{2}\mathrm{Tr}\ln[1+(\pi\nu|W|)^4+(\pi\nu|W|)^2(\hat{Q}_\mathrm{L}\hat{Q}_\mathrm{R}+\hat{Q}_\mathrm{R}\hat{Q}_\mathrm{L})]$, where we used that $\hat{Q}_a^2 = \hat{1}$. Since, due to causality, $\mathrm{Tr}\{\hat{Q}_a^2\} = 0$, see the discussion after Eq. (14.26), $\mathrm{i}S_\mathrm{T}[\hat{Q}_\mathrm{L}, \hat{Q}_\mathrm{R}] = 0$, if $\hat{Q}_\mathrm{L} = \hat{Q}_\mathrm{R}$. To explicitly implement this constraint it is convenient to subtract a $\hat{Q}_a$-matrix-independent constant $\frac{1}{2}\mathrm{Tr}\ln[1 + (\pi\nu|W|)^2]^2$. Employing the relation (15.36) between the $\hat{W}$ matrix and channel transmissions, one obtains for the tunneling action [337, 338, 339]

$$\mathrm{i}S_\mathrm{T}[\hat{Q}_\mathrm{L}, \hat{Q}_\mathrm{R}] = \frac{1}{2}\sum_n \mathrm{Tr}\ln\left[\hat{1} - \frac{|\mathbf{t}_n|^2}{4}\left(\hat{Q}_\mathrm{L} - \hat{Q}_\mathrm{R}\right)^2\right], \tag{15.38}$$

where we used the channel basis, where the $\hat{W}$ matrix is diagonal. If all transmissions are small, $|\mathbf{t}_n|^2 \ll 1$, one may expand Eq. (15.38) to the leading order and find

$$\mathrm{i}S_\mathrm{T}[\hat{Q}_\mathrm{L}, \hat{Q}_\mathrm{R}] = -\frac{\pi g_\mathrm{T}}{4}\mathrm{Tr}\left\{(\hat{Q}_\mathrm{L} - \hat{Q}_\mathrm{R})^2\right\} \tag{15.39}$$

where $g_\mathrm{T} = \sum_n |\mathbf{t}_n|^2/(2\pi)$ is the Landauer tunneling conductance (13.10) in units of $e^2/\hbar$. This expression appears to be a direct generalization of the gradient term in the sigma-model action (14.33), with the substitution $g_\mathrm{D} = \hbar\nu D \to g_\mathrm{T}$. Equation (15.38) goes beyond the weak tunneling limit and allows us to treat the mesoscopic transport in an arbitrary two-terminal geometry. Its generalization for the multi-terminal case was developed by Nazarov *et al.* [337, 340, 341].

As the first application of Eq. (15.38) we obtain the Levitov formula for the full counting statistics, Section 13.3. As was explained previously, Eq. (15.25), the counting field η introduces the twisted boundary conditions, which in our present case read as

$$\hat{Q}_\mathrm{L} = \hat{\Lambda}_\mathrm{L}, \qquad \hat{Q}_\mathrm{R} = \mathrm{e}^{-\mathrm{i}\eta\hat{\gamma}^\mathrm{q}/2}\,\hat{\Lambda}_\mathrm{R}\,\mathrm{e}^{\mathrm{i}\eta\hat{\gamma}^\mathrm{q}/2}, \tag{15.40}$$

where we used that the $\hat{Q}_a$-matrices of the two metallic leads may be fixed to their stationary configurations $\hat{\Lambda}_a$. To evaluate the trace logarithm in Eq. (15.38) it is convenient to perform the rotation (15.28), which leads to

$$\hat{\mathcal{Q}}_{\mathrm{L}} - \hat{\mathcal{Q}}_{\mathrm{R}} = 2\begin{pmatrix} F_{\mathrm{L}} - F_{\mathrm{R}} & (1-F_{\mathrm{R}})\mathrm{e}^{-\mathrm{i}\eta} - (1-F_{\mathrm{L}}) \\ (1+F_{\mathrm{R}})\mathrm{e}^{\mathrm{i}\eta} - (1+F_{\mathrm{L}}) & F_{\mathrm{R}} - F_{\mathrm{L}} \end{pmatrix}_{\mathrm{K}},$$

The square of this matrix is proportional to the 2×2 unit matrix with the proportionality coefficient $(1+F_{\mathrm{R}})(1-F_{\mathrm{L}})(1-\mathrm{e}^{\mathrm{i}\eta}) + (1-F_{\mathrm{R}})(1+F_{\mathrm{L}})(1-\mathrm{e}^{-\mathrm{i}\eta})$. As a result, the generating function

$$Z[\eta] = \mathrm{e}^{\mathrm{i}S_{\mathrm{T}}[\hat{\mathcal{Q}}_{\mathrm{L}},\hat{\mathcal{Q}}_{\mathrm{R}}]} = \mathrm{e}^{\sum_{n,\epsilon} \ln[1+|\mathrm{t}_n|^2 n_{\mathrm{L}}(1-n_{\mathrm{R}})(\mathrm{e}^{\mathrm{i}\eta}-1)+|\mathrm{t}_n|^2 n_{\mathrm{R}}(1-n_{\mathrm{L}})(\mathrm{e}^{-\mathrm{i}\eta}-1)]} \qquad (15.41)$$

coincides with Eq. (13.31), which was derived in Section 13.3 assuming ballistic one-dimensional leads. We have used the fact here that $F_{\mathrm{L(R)}} = 1 - 2n_{\mathrm{L(R)}}$, where $n_{\mathrm{L}} = n_{\mathrm{F}}(\epsilon)$ and $n_{\mathrm{R}} = n_{\mathrm{F}}(\epsilon - eV)$ are the Fermi occupation functions. Our present derivation shows that as long as the leads are good metals, their nature does not affect the counting statistics of a mesoscopic device.

15.6 Chaotic Cavity

As another example, we consider a 0d metallic grain or cavity connected to the left and right metallic leads by ideal reflectionless (i.e. $|\mathrm{t}_n|^2 = 1$) contacts having N_{L} and N_{R} open channels, respectively. The zero dimensional nature of the cavity is ensured by the condition $eV, T \ll E_{\mathrm{Th}}$ and manifests itself in the fact that the cavity $\hat{Q}_{\mathrm{C}}$-matrix may be taken as a constant in space. Moreover, if $T \ll eV$ the $\hat{Q}_{\mathrm{C}}$-matrix is energy independent in the entire energy interval $0 < \epsilon < eV$, where V is the voltage applied between the left and right leads. The tunneling action of the cavity is given by $\mathrm{i}S_{\mathrm{C}}[\eta, \hat{Q}_{\mathrm{C}}] = \mathrm{i}S_{\mathrm{T}}[\hat{Q}_{\mathrm{L}}, \hat{Q}_{\mathrm{C}}] + \mathrm{i}S_{\mathrm{T}}[\hat{Q}_{\mathrm{C}}, \hat{Q}_{\mathrm{R}}]$, where the $\hat{Q}_{\mathrm{L(R)}}$-matrices of the two leads are the same as in Eq. (15.40). The 0d $\hat{Q}_{\mathrm{C}}$-matrix of the cavity may be parametrized as $\hat{Q}_{\mathrm{C}} = \mathrm{e}^{-\mathrm{i}\eta_{\mathrm{C}}\hat{\gamma}^{\mathrm{q}}/2}\,\hat{\Lambda}_{\mathrm{C}}\,\mathrm{e}^{\mathrm{i}\eta_{\mathrm{C}}\hat{\gamma}^{\mathrm{q}}/2}$, with yet unspecified quantum counting field η_{C} and classical distribution function $F_{\mathrm{C}} = 1 - 2n_{\mathrm{C}}$. Employing Eqs. (15.38) and (15.41) and putting $n_{\mathrm{L}} = 0$ and $n_{\mathrm{R}} = 1$, which is the case for $0 < \epsilon < eV$, while n_{C} is an energy-independent constant, one finds

$$\mathrm{i}S_{\mathrm{C}} = \frac{eVt_0}{2\pi}\left\{N_{\mathrm{L}} \ln\left[1 + (1-n_{\mathrm{C}})(\mathrm{e}^{\mathrm{i}\eta_{\mathrm{C}}} - 1)\right] + N_{\mathrm{R}} \ln\left[1 + n_{\mathrm{C}}(\mathrm{e}^{\mathrm{i}(\eta-\eta_{\mathrm{C}})} - 1)\right]\right\}. \qquad (15.42)$$

Since the $\hat{Q}_{\mathrm{C}}$-matrix is free to adjust to the external conditions, the generating function is given by $Z_{\mathrm{C}}[\eta] = \int \mathbf{D}\hat{Q}_{\mathrm{C}}\,\mathrm{e}^{\mathrm{i}S_{\mathrm{C}}}$. If $eVt_0N_{\mathrm{L(R)}} \gg 1$ the corresponding integration may be performed in the stationary point approximation. This implies that one has to look for stationary values of η_{C} and n_{C} that satisfy the equations of motion $\partial S_C/\partial\eta_C = 0$ and $\partial S_C/\partial n_C = 0$ and substitute them back into the action (15.42). This way, one finds $\ln Z_C[\eta] = iS_C[\eta]$. In general, the equations of motion are transcendental and their solution can't be found in an analytic form. The exception is the symmetric case $N_L = N_R = N$, where the equations of motion may be easily solved, leading to $\eta_C = \eta/2$ and $n_C = 1/2$. The corresponding generating function is [342]

$$\ln Z_C[\eta] = \frac{eVt_0}{2\pi}\, 2N \ln\left[1 + \frac{1}{2}(e^{i\eta/2} - 1)\right], \tag{15.43}$$

which is rather different from the 1d diffusive wire result (15.31). Indeed, employing Eqs. (13.35) and (15.32), one may check that the proper distribution function of transmission probabilities takes the form

$$\mathcal{P}(|\mathrm{t}|^2) = \frac{N}{\pi}\,\frac{1}{\sqrt{|\mathrm{t}|^2(1 - |\mathrm{t}|^2)}}. \tag{15.44}$$

Similarly to the Dorokhov distribution (15.33), the cavity distribution is bimodal. On the other hand, it is not so heavily tilted toward closed channels and is normalized as $\int_0^1 d|\mathrm{t}|^2\, \mathcal{P}(|\mathrm{t}|^2) = N$. One can extract now the transmitted charge cumulants by expanding in powers of $i\eta$, Eq. (15.23). The average transmitted charge is $C_1 = t_0 Ve^2N/(4\pi\hbar) = q_0$, which means that the average conductance of the system is $(N/2)(e^2/2\pi\hbar)$. This is natural since we have two ideal contacts in series, each having the Landauer conductance $N(e^2/2\pi\hbar)$.[7] The second cumulant is given by $C_2 = q_0/4$, leading to the Fano factor of 1/4. Thus, the shot noise suppression in a symmetric 0d chaotic cavity is rather different from that in a quasi-1d diffusive wire. (In the latter case the Fano factor is 1/3.)

Finally, the tunneling action allows us to describe not only the statistics of a stationary current flow, but also transient processes and dynamic correlation functions. To this end one has to include the dynamic action for $\hat{Q}$-matrices of metallic nodes. For the 0d cavity the corresponding dynamic action (14.33) is $iS[\hat{Q}_C] = \pi\nu \mathrm{Tr}\{\partial_t \hat{Q}_C\}$. For $T = 0$ and energy interval $0 < \epsilon < eV$ the $\hat{Q}_C$ matrix may be parametrized by an energy-independent quantum component η_C and classical occupation number n_C. In a dynamic setting they both may be slow functions of time. The dynamic action acquires the form

[7] Expanding the action (15.42) to first order in η and η_C, one finds

$$iS_C \approx (eVt_0/2\pi)\big[N_L(1 - n_C)i\eta_C + N_R n_C i(\eta - \eta_C)\big].$$

Optimizing it with respect to η_C and n_C results in the occupation $n_C = N_L/(N_L + N_R)$ and the counting field $\eta_C = N_R\eta/(N_L + N_R)$. Substituting it back to the action leads to $iS_C[\eta] = i\eta(eVt_0/2\pi)N_L N_R/(N_L + N_R)$. This is the series conductance of two contacts with conductances $N_L(e^2/2\pi\hbar)$ and $N_R(e^2/2\pi\hbar)$, as it should be. Expanding the action to higher orders in η and η_C, one may extract few higher cumulants for arbitrary N_L and N_R [342].

$$iS = \frac{\pi \nu eV}{2\pi} \int dt \, \mathrm{Tr}\left\{\left(\partial_t \, e^{-i\eta_C \hat{\gamma}^q/2}\right) \hat{\Lambda}_C \, e^{i\eta_C \hat{\gamma}^q/2}\right\} = -i\nu eV \int dt \, \eta_C \dot{n}_C, \quad (15.45)$$

where after the time differentiation acting on the leftmost function, we found for the trace $-i(\dot{\eta}_C/2)\,\mathrm{Tr}\{\hat{\gamma}^q\hat{\Lambda}_C\} = -i\dot{\eta}_C F_C$. Putting then $F_C = 1-2n_C$ and integrating by parts, one obtains Eq. (15.45). One can now define the "momentum" $p = i\eta_C$ and "coordinate" $n = eV\nu n_C$, where the latter has the meaning of the number of electrons brought to the cavity by an external voltage and the former is the corresponding quantum field. The total action, including dynamic and tunneling parts, then takes the form

$$iS[p,n] = -\int dt \, [p\dot{n} - H(p,n)] . \quad (15.46)$$

Compare Eqs. (4.20) and (4.79), where the (reaction) Hamiltonian is specified by the tunneling part of the action (15.42) and takes the form (e.g. for $\eta = 0$)

$$H(p,n) = \frac{eV}{2\pi}\left\{N_L \ln\left[1 + \left(1 - \frac{n}{eV\nu}\right)(e^p - 1)\right] + N_R \ln\left[1 + \frac{n}{eV\nu}(e^{-p} - 1)\right]\right\}. \quad (15.47)$$

One notices that this formulation is virtually identical to the treatment of reaction models of Section 4.11. This implies that the problem may be reformulated in terms of a certain Master equation for the probability distribution function $\mathcal{P}(n,t)$. For example, expanding the logarithms to first order in $e^{\pm p} - 1$ and comparing with Eq. (4.78), one concludes that the reaction rate for moving one particle from the left lead to the cavity is $(N_L/2\pi\nu)(eV\nu - n)$, while the rate of moving one particle from the cavity to the right lead is $(N_R/2\pi\nu)n$. The higher orders of expansion provide the rates of multi-particle processes. One can now employ the machinery of classical stochastic systems and reaction models, Chapter 4, to describe the dynamics of the charge transport through the chaotic cavity. For more details see [268, 343].

15.7 Problems

15.7.1 Aharonov–Bohm Conductance Oscillations

Consider a quasi-1d mesoscopic ring of length L with an Aharonov–Bohm magnetic flux Φ through it. Calculate dc conductivity of the ring, $\sigma(\Phi)$. To this end, consider the weak-localization correction (14.72), which exhibits sensitivity to the flux through the Cooperon propagator, $\delta\sigma(\Phi) = -(2e^2D/\pi)\sum_{\mathbf{q}} C(\mathbf{q};\Phi)$. In the presence of the magnetic field, the latter is governed by the action (14.91) with $\Phi = \oint dx\, A_x$. In the ring geometry the Aharonov–Bohm flux dictates quantization condition of the Cooperon momenta $q_n^x = 2\pi(n - 2\varphi)/L$, where $\varphi = \Phi/\Phi_0$ with $\Phi_0 = 2\pi\hbar/e$ is the flux quantum, and n is an integer. Show that

$$\delta\sigma(\Phi) = -\frac{2e^2D}{\pi L}\sum_{n=-\infty}^{\infty}\frac{1}{E_{\text{Th}}(2\pi)^2(n-2\varphi)^2+\frac{1}{\tau_\phi}} = -\frac{e^2 l_\phi}{\pi^2}\sum_{m=1}^{\infty} e^{-mL/l_\phi}\cos(4\pi m\varphi), \tag{15.48}$$

where $E_{\text{Th}} = D/L^2$ is the Thouless energy and $l_\phi = \sqrt{D\tau_\phi}$ is the phase coherence length. Perform the last summation to find a closed expression for $\delta\sigma(\Phi)$.

One finds the periodic oscillations of the conductivity with the period $\Phi_0/2$. Indeed, only *even* harmonics are present in the sum. The odd harmonics, while allowed by symmetries, are absent. This is a result of the disorder averaging. Indeed, one may quite generally write $\delta\sigma(\varphi) = \sum_m \sigma_m \cos(2\pi m\varphi)$. We found that $\langle\sigma_{2m+1}\rangle = 0$, while $\langle\sigma_{2m}\rangle \propto l_\phi\, e^{-mL/l_\phi}$, where $\langle\ldots\rangle$ denotes ensemble averaging. However, in any specific realization of the ring the odd harmonics may appear as well. To estimate them one may look for, for example, $\langle\sigma_1^2\rangle$. To this end, one needs to evaluate the $\cos(2\pi(\varphi \pm \varphi'))$ harmonics of $\langle\delta\sigma(\varphi)\delta\sigma(\varphi')\rangle$. This may be done following Section 15.3, where in addition to the diffuson propagators (which are subject to the flux $\varphi-\varphi'$), one uses Cooperons (which are subject to the flux $\varphi+\varphi'$). Employ Eqs. (15.16)–(15.18) with the flux-shifted momentum quantization, as earlier, to calculate $\sqrt{\langle\sigma_1^2\rangle}$. Compare it with $\langle\sigma_2\rangle$, calculated previously.

Calculate flux-dependent conductivity of a 2d cylinder with the Aharonov–Bohm flux through it. To this end, perform 2d momentum summation of the Cooperon propagator as $\sum_{\mathbf{q}} C(\mathbf{q};\Phi) = L^{-1}\sum_{q_n^x}\int dq^z/(2\pi)C(q_n^x, q^z;\Phi)$ with the quantization condition for q_n^x explained earlier. The $\Phi_0/2$ periodic conductance oscillations in the cylinder were predicted by Altshuler, Aronov, and Spivak [345] and soon after observed by Sharvin and Sharvin [346]. If the length of the cylinder is longer than the dephasing length, $L_z \gg l_\phi$, it results in an effective ensemble averaging, $\langle\delta\sigma(\varphi)\rangle$. In this case, there is no need to evaluate mesoscopic fluctuations.

16
Electron–Electron Interactions in Disordered Metals

We discuss singlet channel electron–electron interactions in disordered systems with broken time-reversal symmetry. In particular we derive the zero-bias anomaly, Altshuler–Aronov corrections to the conductivity, the kinetic equation, and the energy relaxation rate.

16.1 $\mathcal{K}$-Gauge

In Section 14.6 we obtained the diffusive density response as a result of Gaussian fluctuations of the $\hat{Q}$-matrix around its stationary value $\hat{\Lambda}$. These Gaussian fluctuations are induced by linear coupling between, say, scalar potentials V^{α} and the diffuson modes specified by Eqs. (14.58). While there is nothing wrong with these tactics, one can substantially simplify the theory by fine-tuning the stationary $\hat{Q}$-matrix configuration. To this end one needs to acknowledge that the scalar V and vector $\mathbf{A}$ potentials can distort the stationary matrix $\hat{\Lambda}$ and deflect it to another point on the soft manifold $\hat{Q}^2 = \hat{1}$. The way to approach this latter true saddle point is to request that the linear coupling, mentioned previously, vanishes.

To implement this program one may use the *gauge invariance* of the nonlinear sigma-model. It implies that the gauge transformation of the $\hat{Q}$-matrix

$$\hat{Q}_{\mathcal{K}}(\mathbf{r};t,t') = \mathrm{e}^{-\mathrm{i}\hat{\mathcal{K}}(\mathbf{r},t)}\,\hat{Q}_{tt'}(\mathbf{r})\,\mathrm{e}^{\mathrm{i}\hat{\mathcal{K}}(\mathbf{r},t')}, \tag{16.1}$$

along with the gauge transformation of the electromagnetic potentials

$$\hat{V}_{\mathcal{K}}(\mathbf{r},t) = \hat{V}(\mathbf{r},t) + \partial_t\hat{\mathcal{K}}(\mathbf{r},t)\,; \qquad \hat{\mathbf{A}}_{\mathcal{K}}(\mathbf{r},t) = \hat{\mathbf{A}}(\mathbf{r},t) + \nabla_{\mathbf{r}}\hat{\mathcal{K}}(\mathbf{r},t) \tag{16.2}$$

preserves the form of the sigma-model action (14.48). That is, the action may be written[1] with $\hat{Q}, \mathbf{A}, V \to \hat{Q}_{\mathcal{K}}, \mathbf{A}_{\mathcal{K}}, V_{\mathcal{K}}$. Notice also that the observables, such as, for

[1] One needs a bit more extra care with the first term on the right-hand side of Eq. (14.48). To show its invariance one goes back to Eq. (14.47) and performs the gauge transformations (16.1), (16.2) under the Tr ln sign, [297, 24].

example, current (14.55) are invariant under the transformations (16.1) and (16.2).

The matrix $\hat{\mathcal{K}}(\mathbf{r},t) = \mathcal{K}^\alpha(\mathbf{r},t)\hat{\gamma}^\alpha$ is defined through two scalar fields $\mathcal{K}^\alpha(\mathbf{r},t)$ with $\alpha = (\mathrm{cl},\mathrm{q})$, which are specified in what follows. One may use now the freedom of choosing $\mathcal{K}^\alpha$ to make the stationary configuration of the *gauged transformed* $\hat{Q}_{\mathcal{K}}$-matrix to be maximally close to $\hat{\Lambda}$, Eq. (14.25), *in the presence* of electromagnetic potentials. To this end we substitute $\hat{Q}_{\mathcal{K}} = \hat{\Lambda} - [\hat{\mathcal{W}}, \hat{\Lambda}]/2$ (cf. derivation of the Usadel equation (14.34)) in the gauge-transformed action (14.48) and demand that there are no terms linear in the rotation generators $\hat{\mathcal{W}}_{\epsilon_1\epsilon_2}(\mathbf{r})$ *times* the *first power* of the electromagnetic potentials $\mathbf{A}_{\mathcal{K}}, V_{\mathcal{K}}$. Notice that, since $\hat{\Lambda}_\epsilon$ is the proper stationary point in the absence of electromagnetic potentials, there are no linear terms in the deviations $\hat{\mathcal{W}}$ themselves. This leads to the following condition:

$$\left(\hat{\Lambda}_{\epsilon_+}\hat{\gamma}^\alpha - \hat{\gamma}^\alpha\hat{\Lambda}_{\epsilon_-}\right)V^\alpha_{\mathcal{K}}(\mathbf{r},\omega) - \left(\hat{\gamma}^\alpha - \hat{\Lambda}_{\epsilon_+}\hat{\gamma}^\alpha\hat{\Lambda}_{\epsilon_-}\right)D\,\mathrm{div}\mathbf{A}^\alpha_{\mathcal{K}}(\mathbf{r},\omega) = 0, \tag{16.3}$$

where $\epsilon_\pm = \epsilon \pm \omega/2$. It is in general impossible to satisfy this matrix condition for any ϵ and ω by a choice of two fields $\mathcal{K}^\alpha(\mathbf{r},\omega)$. In thermal equilibrium, however, there is a "magic" fact that

$$\frac{1 - F^{\mathrm{eq}}_{\epsilon_+}F^{\mathrm{eq}}_{\epsilon_-}}{F^{\mathrm{eq}}_{\epsilon_+} - F^{\mathrm{eq}}_{\epsilon_-}} = \coth\frac{\omega}{2T} \equiv B_\omega, \tag{16.4}$$

which depends on ω only, but *not* on ϵ. This allows for the condition (16.3) to be satisfied for all ϵ's, if the following relation holds between the gauge transformed potentials (16.2):

$$\vec{V}_{\mathcal{K}}(\mathbf{r},\omega) = \begin{pmatrix} 1 & 2B_\omega \\ 0 & -1 \end{pmatrix} D\,\mathrm{div}\vec{\mathbf{A}}_{\mathcal{K}}(\mathbf{r},\omega), \tag{16.5}$$

where we employed vector notations for, for example, $\vec{V}_{\mathcal{K}} = (V^{\mathrm{cl}}_{\mathcal{K}}, V^{\mathrm{q}}_{\mathcal{K}})^{\mathrm{T}}$. This equation specifies the special gauge, which we call the $\mathcal{K}$-gauge for both classical and quantum components of the electromagnetic potentials.

The advantage of the $\mathcal{K}$-gauge is that the action does not contain terms linear in the deviations of the $\hat{Q}_{\mathcal{K}}$ matrix from its stationary point $\hat{\Lambda}$ *and linear* in the electromagnetic potentials. Note that there are still terms that are linear in $\hat{\mathcal{W}}$ and quadratic in the electromagnetic potentials. This means that, strictly speaking, $\hat{\Lambda}$ is not the exact stationary point on the $\hat{Q}_{\mathcal{K}}$ manifold for any realization of the electromagnetic potentials.[2] However, since the deviations from the true saddle point are pushed to second order in potentials, the $\mathcal{K}$-gauge substantially simplifies the structure of the perturbation theory. Moreover, this state of affairs holds

[2] Notice that the root of all complications is an attempt to accommodate the *quantum* components of the electromagnetic potentials into the "deflected" stationary point. If only classical components are present, one may look for the stationary solution having the causality structure (14.25). The problem is then reduced to the solution of the diffusion equation (14.36) in the presence of external (classical) potentials.

only in equilibrium. For out-of-equilibrium situations condition (16.3) cannot be identically satisfied. Therefore terms linear in $\hat{\mathcal{W}}$ and electromagnetic fields unavoidably appear in the action. As we shall explain, it is precisely these terms that are responsible for the collision integral in the kinetic equation.

With the help of Eq. (16.2) the definition of the $\mathcal{K}$-gauge, Eq. (16.5), may be viewed as an explicit relation determining the gauge fields $\mathcal{K}^\alpha$ through the electromagnetic potentials V^α and $\mathbf{A}^\alpha$. Taking $\hat{\mathbf{A}} = 0$ for simplicity, one finds for the classical and quantum components of the gauge field

$$(D\nabla_{\mathbf{r}}^2 + \mathrm{i}\omega)\,\mathcal{K}^{\mathrm{cl}}(\mathbf{r},\omega) + 2B_\omega D\nabla_{\mathbf{r}}^2\,\mathcal{K}^{\mathrm{q}}(\mathbf{r},\omega) = V^{\mathrm{cl}}(\mathbf{r},\omega); \tag{16.6a}$$

$$(D\nabla_{\mathbf{r}}^2 - \mathrm{i}\omega)\,\mathcal{K}^{\mathrm{q}}(\mathbf{r},\omega) = -V^{\mathrm{q}}(\mathbf{r},\omega). \tag{16.6b}$$

By going to Fourier space, one may resolve these equations and write the result in the vector form

$$\vec{\mathcal{K}}(\mathbf{q},\omega) = -\hat{\mathcal{D}}(\mathbf{q},\omega)\hat{\mathcal{B}}_\omega^{-1}\vec{V}(\mathbf{q},\omega), \tag{16.7}$$

where we have introduced bosonic matrices in Keldysh space

$$\hat{\mathcal{D}}(\mathbf{q},\omega) = \begin{pmatrix} \mathcal{D}^{\mathrm{K}}(\mathbf{q},\omega) & \mathcal{D}^{\mathrm{R}}(\mathbf{q},\omega) \\ \mathcal{D}^{\mathrm{A}}(\mathbf{q},\omega) & 0 \end{pmatrix}; \qquad \hat{\mathcal{B}}_\omega = \begin{pmatrix} 2B_\omega & 1_\omega^{\mathrm{R}} \\ -1_\omega^{\mathrm{A}} & 0 \end{pmatrix}, \tag{16.8}$$

with the components (cf. Eq. (14.44))

$$\mathcal{D}^{\mathrm{R(A)}}(\mathbf{q},\omega) = \left(Dq^2 \mp \mathrm{i}\omega\right)^{-1}, \qquad \mathcal{D}^{\mathrm{K}}(\mathbf{q},\omega) = B_\omega\left[\mathcal{D}^{\mathrm{R}}(\mathbf{q},\omega) - \mathcal{D}^{\mathrm{A}}(\mathbf{q},\omega)\right]. \tag{16.9}$$

Equation (16.7) provides an explicit *linear* relation between the scalar potential and the gauge fields $\mathcal{K}^\alpha$. It thus defines the $\hat{Q}_{\mathcal{K}}$-matrix in the $\mathcal{K}$-gauge according to Eq. (16.1). As was explained, the latter possesses a stationary configuration which is close to the metallic stationary point $\hat{\Lambda}$ (with deviations being quadratic in electromagnetic potentials).

Let us evaluate now the generating function $\langle Z[V]\rangle_{\mathrm{dis}}$ (we put $\mathbf{A}=0$ for simplicity) in the *stationary point approximation* for the $\hat{Q}_{\mathcal{K}}$-matrix in the $\mathcal{K}$-gauge, that is, we put $\hat{Q}_{\mathcal{K}} = \hat{\Lambda}$. It is given by $\langle Z[V]\rangle_{\mathrm{dis}} = \mathrm{e}^{\mathrm{i}S[\hat{\Lambda},V_{\mathcal{K}}]}$, with the action given by Eq. (14.48),

$$\mathrm{i}S[\hat{\Lambda},V_{\mathcal{K}}] = \frac{\mathrm{i}\nu}{2}\,\mathrm{Tr}\left\{\hat{V}_{\mathcal{K}}\hat{\sigma}_1\hat{V}_{\mathcal{K}}\right\} + \frac{\pi\nu D}{4}\,\mathrm{Tr}\left\{[\nabla_{\mathbf{r}}\hat{\mathcal{K}},\hat{\Lambda}]^2\right\}. \tag{16.10}$$

Notice that, while we disregarded the bare vector potential, there is still the gauge vector potential $\hat{\mathbf{A}}_{\mathcal{K}} = \nabla_{\mathbf{r}}\hat{\mathcal{K}}$, Eq. (16.2). We use now the explicit relation between the scalar potential V^{α} and the gauge potentials $\mathcal{K}^{\alpha}$, Eq. (16.7), to express the action (16.10) in terms of the former. The straightforward algebra yields[3]

$$iS[\hat{\Lambda}, V] = i \sum_{\mathbf{q},\omega} \vec{V}^{\mathrm{T}}(-\mathbf{q}, -\omega)\, \hat{\Pi}(\mathbf{q},\omega)\vec{V}(\mathbf{q},\omega), \tag{16.13}$$

where, in agreement with the general definition (10.49), the polarization matrix has the typical causality structure of a quadratic bosonic matrix action in Keldysh space,

$$\hat{\Pi}(\mathbf{q},\omega) = \begin{pmatrix} 0 & \Pi^{\mathrm{A}}(\mathbf{q},\omega) \\ \Pi^{\mathrm{R}}(\mathbf{q},\omega) & \Pi^{\mathrm{K}}(\mathbf{q},\omega) \end{pmatrix}, \tag{16.14}$$

with the components

$$\Pi^{\mathrm{R(A)}}(\mathbf{q},\omega) = \frac{\nu D q^2}{Dq^2 \mp i\omega}, \qquad \Pi^{\mathrm{K}}(\mathbf{q},\omega) = B_{\omega}\big[\Pi^{\mathrm{R}}(\mathbf{q},\omega) - \Pi^{\mathrm{A}}(\mathbf{q},\omega)\big], \tag{16.15}$$

as expected from our discussion in Section 14.6; see Eq. (14.60). We have thus rederived the diffusive form of the density response. Notice, however, that this time we did not employ Gaussian fluctuations of the $\hat{Q}$-matrix around its stationary point. Instead the gauge-transformed $\hat{Q}_{\mathcal{K}}$ matrix was fixed right at its stationary point $\hat{\Lambda}$. The diffusive propagators (16.9) came from the choice of the gauge and not from the diffuson propagators of the $\hat{Q}$-matrix fluctuations. The choice of the $\mathcal{K}$-gauge substantially simplifies the theory of interacting electrons in the presence of the disorder potential.

16.2 Nonlinear Sigma-Model for Interacting Systems

We turn now to the discussion of electron–electron interactions in disordered systems. In this chapter we limit ourselves to the effects of the *singlet* channel interactions only; see Eq. (10.64). The Cooper channel interactions are considered in Chapter 17. The singlet part of the interaction action takes the form (Eq. (10.64))

$$S_{\text{s-int}} = -\frac{1}{2}\int_{C} dt \iint d\mathbf{r}d\mathbf{r}'\, \rho(\mathbf{r},t) U_{\mathrm{s}}(\mathbf{r}-\mathbf{r}')\rho(\mathbf{r}',t), \tag{16.16}$$

[3] To evaluate the last trace in Eq. (16.10), it is useful to employ the following relation:

$$\int_{-\infty}^{+\infty} d\epsilon\, \mathrm{Tr}\left\{\hat{\gamma}^{\alpha}\hat{\gamma}^{\beta} - \hat{\gamma}^{\alpha}\hat{\Lambda}_{\epsilon_+}\hat{\gamma}^{\beta}\hat{\Lambda}_{\epsilon_-}\right\} = 4\omega(\hat{\mathcal{B}}_{\omega}^{-1})^{\alpha\beta}, \tag{16.11}$$

where $\epsilon_{\pm} = \epsilon \pm \omega/2$, and the matrix $\hat{\mathcal{B}}$ is defined by Eq. (16.8). Equation (16.11) is a consequence of the following integral relations between bosonic and fermionic distribution functions:

$$\int_{-\infty}^{+\infty} d\epsilon\, (F_{\epsilon_+} - F_{\epsilon_-}) = 2\omega, \qquad \int_{-\infty}^{+\infty} d\epsilon\, (1 - F_{\epsilon_+}F_{\epsilon_-}) = 2\omega B_{\omega}. \tag{16.12}$$

where electron density $\rho(\mathbf{r}, t)$ is defined by Eq. (10.63) and the singlet interaction potential is given by $U_s(\mathbf{q}) = U(\mathbf{q}) - U_t$, where $U(\mathbf{q})$ is the bare (e.g. Coulomb) interaction and U_t is the triplet interaction constant. We decouple now the singlet interaction action with the help of the real bosonic Hubbard–Stratonovich field $\varphi(\mathbf{r}, t)$, exactly as we did in Eq. (10.65). We then split the bosonic field φ into forward and backward branches and perform a Keldysh rotation, as in Eq. (10.66). As a result of this procedure we arrive at *non-interacting* disordered electrons subject to a vector of dynamic fluctuating fields $\vec{\varphi} = (\varphi^{\mathrm{cl}}, \varphi^{\mathrm{q}})^{\mathrm{T}}$. The Hubbard–Stratonovich weight of the vector bosonic field is $\vec{\varphi}^{\mathrm{T}} U_{\mathrm{s}}^{-1}(\mathbf{q})\hat{\sigma}_1\vec{\varphi}$ (cf. Eq. (10.66)).

The fluctuating auxiliary field $\vec{\varphi}$ plays exactly the same role as the source scalar potential $\vec{V}$ in the nonlinear sigma-model action (14.48). It is convenient to perform the gauge transformation to the $\mathcal{K}$-gauge, with $\vec{\mathcal{K}} = -\hat{\mathcal{D}}\hat{\mathcal{B}}^{-1}\vec{\varphi}$ (cf. Eq. (16.7)). We then obtain the disorder-averaged partition function in the following form:

$$\langle Z\rangle_{\mathrm{dis}} = \int \mathbf{D}[\varphi]\, \mathrm{e}^{\mathrm{i}\mathrm{Tr}\{\vec{\varphi}^{\mathrm{T}}\hat{U}_{\mathrm{RPA}}^{-1}\vec{\varphi}\}} \int \mathbf{D}[\hat{Q}_{\mathcal{K}}]\, \mathrm{e}^{\mathrm{i}S_0[\hat{Q}_{\mathcal{K}}]+\mathrm{i}S_1[\hat{Q}_{\mathcal{K}},\nabla_{\mathbf{r}}\mathcal{K}]+\mathrm{i}S_2[\hat{Q}_{\mathcal{K}},\nabla_{\mathbf{r}}\mathcal{K}]}, \quad (16.17)$$

where S_l with $l = 0, 1, 2$ contain the lth power of the electromagnetic potentials and are given by

$$\mathrm{i}S_0[\hat{Q}_{\mathcal{K}}] = -\frac{\pi\nu}{4}\,\mathrm{Tr}\left\{D(\nabla_{\mathbf{r}}\hat{Q}_{\mathcal{K}})^2 - 4\partial_t\hat{Q}_{\mathcal{K}}\right\}, \quad (16.18a)$$

$$\mathrm{i}S_1[\hat{Q}_{\mathcal{K}}, \nabla_{\mathbf{r}}\mathcal{K}] = \mathrm{i}\pi\nu\,\mathrm{Tr}\left\{\hat{\varphi}_{\mathcal{K}}\hat{Q}_{\mathcal{K}} - D(\nabla_{\mathbf{r}}\hat{\mathcal{K}})\hat{Q}_{\mathcal{K}}(\nabla_{\mathbf{r}}\hat{Q}_{\mathcal{K}})\right\}, \quad (16.18b)$$

$$\mathrm{i}S_2[\hat{Q}_{\mathcal{K}}, \nabla_{\mathbf{r}}\mathcal{K}] = \frac{\pi\nu D}{4}\,\mathrm{Tr}\left\{[\nabla_{\mathbf{r}}\hat{\mathcal{K}}, \hat{Q}_{\mathcal{K}}]^2 - [\nabla_{\mathbf{r}}\hat{\mathcal{K}}, \hat{\Lambda}]^2\right\}; \quad (16.18c)$$

see also Eqs. (14.52) and (14.54). In the last term $\mathrm{i}S_2$ we have added and subtracted the contribution of the stationary point $\hat{Q}_{\mathcal{K}} = \hat{\Lambda}$. The latter contributes to the part of the action quadratic in $\vec{\varphi}$ fields and is given by Eqs. (16.10) and (16.13) (where we put $\vec{V} \to \vec{\varphi}$). Together with the bare Hubbard–Stratonovich term $\vec{\varphi}^{\mathrm{T}} U_{\mathrm{s}}^{-1}(\mathbf{q})\hat{\sigma}_1\vec{\varphi}$ it goes into the definition of the random phase approximation (RPA) inverse interaction matrix

$$\hat{U}_{\mathrm{RPA}}^{-1}(\mathbf{q}, \omega) = U_{\mathrm{s}}^{-1}(\mathbf{q})\hat{\sigma}_1 + \hat{\Pi}(\mathbf{q}, \omega), \quad (16.19)$$

where the diffusive polarization matrix is given by Eqs. (16.14) and (16.15).

Equations (16.17)–(16.19) constitute an effective nonlinear sigma-model for an interacting disordered Fermi liquid. The model consists of two interacting fields: the matrix field $\hat{Q}_{\mathcal{K}}$, obeying the nonlinear constraint $\hat{Q}_{\mathcal{K}}^2 = \hat{1}$, and the bosonic longitudinal field $\nabla_{\mathbf{r}}\hat{\mathcal{K}}$ (or equivalently $\hat{\varphi}$). As will be apparent later, the $\hat{Q}_{\mathcal{K}}$-field describes fluctuations of the quasiparticle distribution function, whereas $\hat{\varphi}$ (or $\hat{\mathcal{K}}$) represents propagation of electromagnetic modes through the media.

Since the fields $\vec{\varphi}$ and $\vec{\mathcal{K}}$ are linearly related through $\vec{\mathcal{K}} = -\hat{\mathcal{D}}\hat{\mathcal{B}}^{-1}\vec{\varphi}$, Eq. (16.7), one can change the φ-integration in Eq. (16.17) to a $\mathcal{K}$-integration. To this end we need to evaluate the following correlation function:

$$\mathcal{V}^{\alpha\beta}(x-x') \equiv -2\mathrm{i}\langle \mathcal{K}^{\alpha}(x)\mathcal{K}^{\beta}(x')\rangle = -2\mathrm{i}\int \mathbf{D}[\vec{\varphi}]\, \mathcal{K}^{\alpha}(x)\mathcal{K}^{\beta}(x')\, \mathrm{e}^{\mathrm{i}\mathrm{Tr}\{\vec{\varphi}^{T}\hat{U}_{\mathrm{RPA}}^{-1}\vec{\varphi}\}}, \tag{16.20}$$

where $x = \mathbf{r}, t$ and the factor $-2\mathrm{i}$ is used for convenience. Employing the linear relation between $\hat{\varphi}$ and $\hat{\mathcal{K}}$ along with the fact that $\langle \varphi^{\alpha}\varphi^{\beta}\rangle = (\mathrm{i}/2)U_{\mathrm{RPA}}^{\alpha\beta}$, one finds for the gauge field correlation function

$$\hat{\mathcal{V}}(\mathbf{q},\omega) = \hat{\mathcal{D}}(\mathbf{q},\omega)\hat{\mathcal{B}}_{\omega}^{-1}\hat{U}_{\mathrm{RPA}}(\mathbf{q},\omega)\big(\hat{\mathcal{B}}_{-\omega}^{-1}\big)^{\mathrm{T}}\hat{\mathcal{D}}^{\mathrm{T}}(-\mathbf{q},-\omega). \tag{16.21}$$

Employing the explicit form of the $\hat{\mathcal{D}}$ and $\hat{\mathcal{B}}$ matrices, Eq. (16.8), one finds that the bosonic correlation matrix $\hat{\mathcal{V}}(\mathbf{q},\omega)$ has the standard Keldysh structure

$$\hat{\mathcal{V}}(\mathbf{q},\omega) = \begin{pmatrix} \mathcal{V}^{\mathrm{K}}(\mathbf{q},\omega) & \mathcal{V}^{\mathrm{R}}(\mathbf{q},\omega) \\ \mathcal{V}^{\mathrm{A}}(\mathbf{q},\omega) & 0 \end{pmatrix}, \tag{16.22}$$

with the elements

$$\mathcal{V}^{\mathrm{R(A)}}(\mathbf{q},\omega) = -\frac{1}{(Dq^2 \mp \mathrm{i}\omega)^2}\left(U_{\mathrm{s}}^{-1}(\mathbf{q}) + \frac{\nu Dq^2}{Dq^2 \mp \mathrm{i}\omega}\right)^{-1}; \tag{16.23a}$$

$$\mathcal{V}^{\mathrm{K}}(\mathbf{q},\omega) = B_{\omega}\big[\mathcal{V}^{\mathrm{R}}(\mathbf{q},\omega) - \mathcal{V}^{\mathrm{A}}(\mathbf{q},\omega)\big]. \tag{16.23b}$$

This propagator corresponds to the RPA screened dynamic Coulomb interaction, dressed by the two diffusons at the vertices, Fig. 16.1(a). Therefore, the role of the gauge field $\mathcal{K}$ is to automatically take into account both the RPA screening of the interactions, Fig. 16.1(b), and its vertex renormalization by the diffusons. Owing to the linear dependence between $\hat{\varphi}$ and $\hat{\mathcal{K}}$, we use integration over the $\hat{\varphi}$ or $\hat{\mathcal{K}}$ fields interchangeably. The essence is that the correlation function of the two $\hat{\mathcal{K}}^{\alpha}$ fields is given by Eqs. (16.20)–(16.23).

16.3 Zero Bias Anomaly

We discuss now modifications of the single-particle density of states (DOS) $\nu(\epsilon)$ of the disordered electron gas due to Coulomb interactions. As was briefly mentioned in Section 14.8, the disorder potential itself does not alter the *average* DOS. This is not the case any more when interactions are present. As we shall see, interactions lead to an energy-dependent DOS, singularly suppressed at the Fermi energy. The effect is known as the zero bias anomaly and is directly measurable in tunneling experiments.

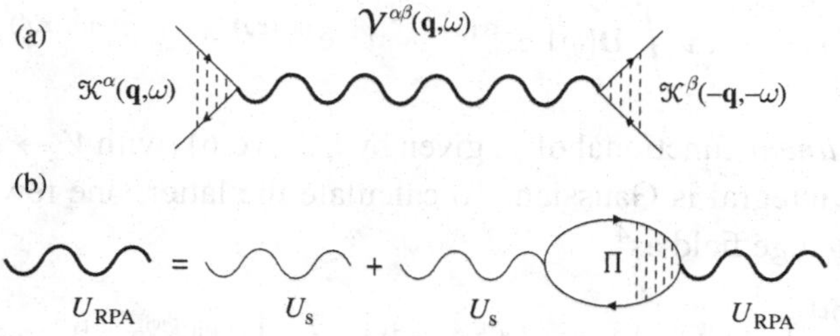

Figure 16.1 (a) Diagrammatic representation of the gauge field propagator $\hat{\mathcal{V}}(\mathbf{q},\omega)$: the wavy line represents the Coulomb interaction. Vertices dressed by the diffusons are shown by the ladders of dashed lines. (b) Screened Coulomb interaction in RPA, $\hat{U}_{\mathrm{RPA}}(\mathbf{q},\omega)$. Bold and thin wavy lines represent screened and bare interactions, respectively, and the loop represents the diffusive polarization matrix $\hat{\Pi}$, Eq. (16.14).

We are interested in the disorder-averaged single-particle Green function at coinciding spatial points, defined as

$$\mathcal{G}^{ab}(t-t') = -\mathrm{i}\langle\langle \psi_a(\mathbf{r},t)\bar{\psi}_b(\mathbf{r},t')\rangle\rangle = \left\langle \left[\hat{G}^{-1} + \frac{\mathrm{i}}{2\tau_{\mathrm{el}}}\hat{Q} - \hat{\varphi} \right]^{-1}_{\mathbf{rr};tt';ab} \right\rangle_{\varphi,\hat{Q}_{\mathcal{K}}}, \qquad (16.24)$$

where $a, b = 1, 2$ and $\langle\langle \ldots \rangle\rangle$ denotes both quantum and disorder averaging. To obtain the second equality one introduces a source term, directly coupled to the bilinear combination of Grassmann fields, into the fermionic action. Following the same algebra as in Section 14.5, performing Keldysh rotation and disorder averaging, one finds that the source term enters the trace logarithm in Eq. (14.21). Differentiating the latter with respect to the source and putting it to zero, one obtains Eq. (16.24), where $\langle \ldots \rangle_{\varphi,\hat{Q}_{\mathcal{K}}}$ denotes averaging with the weight given by Eqs. (16.17) and (16.18). It is important to mention that the Green function $\hat{\mathcal{G}}(t-t')$ is *not* gauge-invariant and is evaluated in a particular gauge, the one where the chemical potential is set to be at $\epsilon = 0$. It is fixed by the setup of the tunneling experiment. It is thus the bare $\hat{Q}$-matrix and *not* the gauged-transformed one, $\hat{Q}_{\mathcal{K}}$, which appears in the last equality in Eq. (16.24).

We evaluate the integral over the $\hat{Q}_{\mathcal{K}}$-matrix in Eq. (16.24) in the stationary point approximation, neglecting both the massive and massless fluctuations around the stationary point. According to the stationary point condition (14.22), $\left[\hat{G}^{-1} + (\mathrm{i}/2\tau_{\mathrm{el}})\hat{Q} - \hat{\varphi}\right]^{-1}_{\mathbf{rr};tt'} = -\mathrm{i}\pi\nu\hat{Q}_{tt'}(\mathbf{r})$. Moreover, according to Eq. (16.1), $\hat{Q}_{tt'}(\mathbf{r}) = \mathrm{e}^{\mathrm{i}\hat{\mathcal{K}}(\mathbf{r},t)}\,\hat{\Lambda}_{t-t'}\mathrm{e}^{-\mathrm{i}\hat{\mathcal{K}}(\mathbf{r},t')}$, where we adopt that in the stationary point approximation $\hat{Q}_{\mathcal{K}}(\mathbf{r};t,t') = \hat{\Lambda}_{t-t'}$. Since $S_l[\hat{\Lambda},\mathcal{K}] = 0$ for $l = 0, 1, 2$, Eq. (16.18), we find for the Green function

$$\hat{\mathcal{G}}(t-t') = -\mathrm{i}\pi\nu \int \mathbf{D}[\varphi]\, \mathrm{e}^{\mathrm{i}\mathrm{Tr}\{\vec{\varphi}^{\mathrm{T}}\hat{U}_{\mathrm{RPA}}^{-1}\vec{\varphi}\}}\, \mathrm{e}^{\mathrm{i}\hat{\mathcal{K}}(\mathbf{r},t)}\, \hat{\Lambda}_{t-t'}\, \mathrm{e}^{-\mathrm{i}\hat{\mathcal{K}}(\mathbf{r},t')}. \tag{16.25}$$

Since $\hat{\mathcal{K}}$ is the *linear* functional of $\hat{\varphi}$, given by Eq. (16.6) (with $\hat{V} \to \hat{\varphi}$), the remaining functional integral is Gaussian. To calculate the latter, one rewrites the phase factors of the gauge field as[4]

$$\mathrm{e}^{\pm\mathrm{i}\mathcal{K}^{\alpha}\hat{\gamma}^{\alpha}} = \frac{\hat{\gamma}^{\mathrm{cl}}}{2}\left[\mathrm{e}^{\pm\mathrm{i}(\mathcal{K}^{\mathrm{cl}}+\mathcal{K}^{\mathrm{q}})} + \mathrm{e}^{\pm\mathrm{i}(\mathcal{K}^{\mathrm{cl}}-\mathcal{K}^{\mathrm{q}})}\right] + \frac{\hat{\gamma}^{\mathrm{q}}}{2}\left[\mathrm{e}^{\pm\mathrm{i}(\mathcal{K}^{\mathrm{cl}}+\mathcal{K}^{\mathrm{q}})} - \mathrm{e}^{\pm\mathrm{i}(\mathcal{K}^{\mathrm{cl}}-\mathcal{K}^{\mathrm{q}})}\right].$$

Then, for example,

$$\left\langle \mathrm{e}^{\mathrm{i}[\mathcal{K}_t^{\mathrm{cl}}+\mathcal{K}_t^{\mathrm{q}}-\mathcal{K}_{t'}^{\mathrm{cl}}+\mathcal{K}_{t'}^{\mathrm{q}}]}\right\rangle_{\varphi} = \left\langle \mathrm{e}^{\mathrm{i}\int \mathrm{d}t_1\, \mathcal{K}_{t_1}^{\alpha} J_{t_1}^{\alpha}}\right\rangle_{\varphi} = \mathrm{e}^{-(\mathrm{i}/4)\int \mathrm{d}t_1 \mathrm{d}t_2\, J_{t_1}^{\alpha} \mathcal{V}_{t_1-t_2}^{\alpha\beta} J_{t_2}^{\beta}},$$

where $J_{t_1}^{\mathrm{cl}} = \delta(t_1-t) - \delta(t_1-t')$, $J_{t_1}^{\mathrm{q}} = \delta(t_1-t) + \delta(t_1-t')$, and the correlator, $\langle \mathcal{K}_{t_1}^{\alpha}\mathcal{K}_{t_2}^{\beta}\rangle_{\varphi} = (\mathrm{i}/2)\mathcal{V}_{t_1-t_2}^{\alpha\beta}$; see Eq. (16.20). Employing the causality structure of $\mathcal{V}^{\alpha\beta}$, Eq. (16.22), along with $\mathcal{V}_{-t}^{\mathrm{K}} = \mathcal{V}_t^{\mathrm{K}}$ and $\mathcal{V}_{-t}^{\mathrm{R}} = \mathcal{V}_t^{\mathrm{A}}$, we find for this particular contribution $\exp\{(\mathrm{i}/2)[\mathcal{V}_{t-t'}^{\mathrm{K}} - \mathcal{V}_0^{\mathrm{K}} - \mathcal{V}_{t-t'}^{\mathrm{R}} + \mathcal{V}_{t-t'}^{\mathrm{A}}]\}$. Collecting all contributions and employing that $\mathcal{V}_0^{\mathrm{R}} + \mathcal{V}_0^{\mathrm{A}} = 0$ and $\mathcal{V}_{t-t'}^{\mathrm{R}} = 0$ for $t < t'$, while $\mathcal{V}_{t-t'}^{\mathrm{A}} = 0$ for $t > t'$, one finds for the Green function

$$\hat{\mathcal{G}}(t) = -\mathrm{i}\pi\nu \sum_{\alpha\beta} \left(\hat{\gamma}^{\alpha}\, \hat{\Lambda}_t\, \hat{\gamma}^{\beta}\right) \mathbb{B}^{\alpha\beta}(t), \tag{16.26}$$

where the auxiliary propagator $\mathbb{B}^{\alpha\beta}(t)$ has the standard bosonic structure, as, for example, in Eq. (16.22), with

$$\mathbb{B}^{\mathrm{R(A)}}(t) = \mathrm{i}\exp\left(\mathrm{i}[\mathcal{V}^{\mathrm{K}}(t) - \mathcal{V}^{\mathrm{K}}(0)]/2\right)\sin\left(\mathcal{V}^{\mathrm{R(A)}}(t)/2\right), \tag{16.27a}$$

$$\mathbb{B}^{\mathrm{K}}(t) = \exp\left(\mathrm{i}[\mathcal{V}^{\mathrm{K}}(t) - \mathcal{V}^{\mathrm{K}}(0)]/2\right)\cos\left([\mathcal{V}^{\mathrm{R}}(t) - \mathcal{V}^{\mathrm{A}}(t)]/2\right). \tag{16.27b}$$

The gauge field propagator, $\hat{\mathcal{V}}(\mathbf{r},t)$, defined by Eqs. (16.22) and (16.23), enters Eq. (16.27) at the coinciding spatial points

$$\hat{\mathcal{V}}(t) = \int \frac{\mathrm{d}\omega}{2\pi}\, \mathrm{e}^{-\mathrm{i}\omega t} \sum_{\mathbf{q}} \hat{\mathcal{V}}(\mathbf{q},\omega). \tag{16.28}$$

Before proceeding with the calculations of the DOS, it is instructive to check that the stationary point approximation employed to derive Eq. (16.26) respects the

[4] This equation is based on the following property: consider an arbitrary function of a linear superposition of Pauli matrices $f(a + \mathbf{b}\hat{\sigma})$, where a is a number and $\mathbf{b}$ is a vector. The observation is that $f(a + \mathbf{b}\hat{\sigma}) = A + \mathbf{B}\hat{\sigma}$, where A is some new number and $\mathbf{B}$ a new vector. To see this, let us choose the z-axis along the direction of the $\mathbf{b}$ vector. Then the eigenvalues of the operator $a + \mathbf{b}\hat{\sigma}$ are $a \pm b$, and the corresponding eigenvalues of the operator $f(a + \mathbf{b}\hat{\sigma})$ are $f(a \pm b)$. Thus, one concludes that $A = \frac{1}{2}[f(a+b) + f(a-b)]$ and $\mathbf{B} = \frac{\mathbf{b}}{2b}[f(a+b) - f(a-b)]$.

fermionic FDT. For this purpose it is convenient to rewrite identically Eq. (16.26) in the following form (cf. Eqs. (2.41)):

$$\mathcal{G}^{>(<)}(t) = -\mathrm{i}\pi\nu\Lambda_t^{>(<)}\mathbb{B}^{>(<)}(t), \tag{16.29}$$

where

$$\mathbb{B}^{\mathrm{R}}(t) - \mathbb{B}^{\mathrm{A}}(t) = \mathbb{B}^{>}(t) - \mathbb{B}^{<}(t)\,; \qquad \mathbb{B}^{\mathrm{K}}(t) = \mathbb{B}^{>}(t) + \mathbb{B}^{<}(t) \tag{16.30}$$

and the same relations hold for the components of the fermionic Green functions $\hat{\Lambda}$ and $\hat{\mathcal{G}}$. Employing Eqs. (16.27) along with the bosonic FDT relation for the components of $\hat{\mathcal{V}}(\mathbf{q},\omega)$, one obtains

$$\begin{aligned}\mathbb{B}^{>(<)}(t) &= \frac{\mathrm{e}^{-\frac{\mathrm{i}}{2}\mathcal{V}^{\mathrm{K}}(0)}}{2}\exp\left\{\frac{\mathrm{i}}{2}\sum_{\mathbf{q},\omega}\mathrm{e}^{-\mathrm{i}\omega t}(\mathcal{V}^{\mathrm{R}}(\mathbf{q},\omega)-\mathcal{V}^{\mathrm{A}}(\mathbf{q},\omega))\left(\coth\frac{\omega}{2T}\pm 1\right)\right\}\\ &= \frac{1}{2}\exp\left\{\int\frac{d\omega}{2\pi}\,\mathrm{Im}\sum_{q}\mathcal{V}^{\mathrm{R}}(\mathbf{q},\omega)\left[\coth\frac{\omega}{2T}(1-\cos\omega t)\pm\mathrm{i}\sin\omega t\right]\right\}.\end{aligned} \tag{16.31}$$

According to FDT the equilibrium bosonic and fermionic Green functions in the frequency representation must satisfy the following relations:

$$\mathbb{B}^{>}(\omega) = \mathrm{e}^{\omega/T}\,\mathbb{B}^{<}(\omega)\,; \qquad \mathcal{G}^{>}(\epsilon) = -\mathrm{e}^{\epsilon/T}\,\mathcal{G}^{<}(\epsilon). \tag{16.32}$$

Since $\coth(\omega/2T)+1 = \mathrm{e}^{\omega/T}(\coth(\omega/2T)-1)$, the exponent in Eq. (16.31) satisfies the bosonic FDT. It is then not difficult to see that Eqs. (16.32) indeed hold.[5]

Having found the Green function, Eq. (16.26), we obtain the single-particle (or tunneling) DOS according to the standard definition,

$$\nu(\epsilon) \equiv \frac{\mathrm{i}}{2\pi}\left[\mathcal{G}^{\mathrm{R}}(\epsilon)-\mathcal{G}^{\mathrm{A}}(\epsilon)\right] = \frac{\nu}{\tanh(\epsilon/2T)}\int dt\; \mathrm{e}^{\mathrm{i}\epsilon t}\,F_t\left[\mathbb{B}^{>}(t)+\mathbb{B}^{<}(t)\right], \tag{16.33}$$

where in the last equality we made use of the fermionic and bosonic FDT relations (16.30) and (16.32) and the fact that $\Lambda_t^{\mathrm{K}} = 2F_t$, Eq. (14.25). We restrict ourselves to the analysis of this result only at zero temperature. Noting that for $T = 0$ the Fourier

[5] If a pair of bosonic functions $b^{>}(t)$ and $b^{<}(t)$ (being transformed to the frequency representation) satisfies Eq. (16.32), then for any analytic function $f(z)$ the pair $f^{>}(t) \equiv f(b^{>}(t))$ and $f^{<}(t) \equiv f(b^{<}(t))$ also satisfies it. Indeed,

$$f^{>(<)}(\omega) = \int dt\,\mathrm{e}^{\mathrm{i}\omega t} f\left(\int\frac{d\omega'}{2\pi}\,b^{>(<)}(\omega')\,\mathrm{e}^{-\mathrm{i}\omega' t}\right).$$

Expanding f on the right-hand side in a Taylor series and performing the t integration, we see that in each order of the expansion $f^{>}(\omega) = \exp(\omega/T)f^{<}(\omega)$. Taking $b^{>(<)}(t)$ to be the exponent in Eq. (16.31) and f to be the exponential function proves that $\mathbb{B}^{>}$ and $\mathbb{B}^{<}$ are related through FDT. It is then a simple matter to check that if the fermionic $\Lambda^{>(<)}$ and bosonic $\mathbb{B}^{>(<)}$ satisfy FDT, Eq. (16.32), then so do the fermionic functions $\mathcal{G}^{>(<)}$ given by Eq. (16.29).

transform of the equilibrium fermionic distribution function is $F_t = (i\pi t)^{-1}$, one obtains

$$\nu(\epsilon) = \frac{\nu}{\pi} \int dt \, \frac{\sin|\epsilon|t}{t} \, \exp\left\{ \int_0^\infty \frac{d\omega}{\pi} \, \mathrm{Im} \sum_{\mathbf{q}} \mathcal{V}^{\mathrm{R}}(\mathbf{q},\omega)(1-\cos\omega t) \right\}$$
$$\times \cos\left\{ \int_0^\infty \frac{d\omega}{\pi} \, \mathrm{Im} \sum_{\mathbf{q}} \mathcal{V}^{\mathrm{R}}(\mathbf{q},\omega) \sin\omega t \right\}. \tag{16.34}$$

Let us focus on the two-dimensional case, where the bare Coulomb interaction is given by $U(\mathbf{q}) = 2\pi e^2/q$, with $\mathbf{q}$ being the 2d wavenumber. Employing Eq. (16.23a) with such interactions and performing 2d momentum as well as energy integrations leads to

$$\int_0^{+\infty} \frac{d\omega}{\pi} \sum_{\mathbf{q}} \mathrm{Im}[\mathcal{V}^{\mathrm{R}}(\mathbf{q},\omega)] \begin{pmatrix} \cos\omega t - 1 \\ \sin\omega t \end{pmatrix} = \frac{1}{8\pi^2 g} \begin{cases} \ln\frac{t}{\tau_{\mathrm{el}}} \ln(t\tau_{\mathrm{el}}\omega_0^2) + 2\mathbb{C}\ln(t\omega_0) \\ -\pi\ln(t\omega_0) \end{cases}, \tag{16.35}$$

where $g = \hbar\nu D$ is the dimensionless conductance per square, $\omega_0 = D\kappa^{-2}$, $\kappa = (2\pi e^2\nu)^{-1}$ is the Thomas–Fermi 2d screening radius and $\mathbb{C} = 0.577\ldots$ is the Euler constant. The remaining time integral in Eq. (16.34) may be performed in the saddle-point approximation, resulting in

$$\nu(\epsilon) = \nu \exp\left\{ -\frac{1}{8\pi^2 g} \ln(|\epsilon|\tau_{\mathrm{el}})^{-1} \ln(\tau_{\mathrm{el}}\omega_0^2/|\epsilon|) \right\} \tag{16.36}$$

for $|\epsilon| \ll 1/\tau_{\mathrm{el}}$ and $\nu(\epsilon) \to \nu$ in the opposite limit. We see thus that the interactions suppress the single-particle DOS near the Fermi energy (i.e. $\epsilon = 0$) in a singular way. This effect, observed frequently in tunneling experiments, is known as the zero bias anomaly. Its physics may be traced back to the fluctuating phase (i.e. gauge factor) of the Green function; see Eq. (16.25). Upon averaging over the fluctuating scalar potential $\vec{\varphi}$ this phase produces the "Debye–Waller factor," which exponentially suppresses the tunneling amplitude.

Does the DOS really go to zero at $\epsilon = 0$, as Eq. (16.36) suggests? We can not answer this question. Indeed, we have adopted a stationary point approximation for the $\hat{Q}_{\mathcal{K}}$-integral, disregarding diffuson fluctuations. Being taken into account, the latter give rise to perturbative corrections $\sim g^{-1}\ln(|\epsilon|\tau_{\mathrm{el}})$; see Section 16.4. Therefore one can trust Eq. (16.36) only if $|\ln(|\epsilon|\tau_{\mathrm{el}})| \lesssim g$. Notice, however, that since the exponent in Eq. (16.36) contains an additional large logarithm $\ln(\tau_{\mathrm{el}}\omega_0^2/|\epsilon|) > 1$, it may be large (justifying keeping the result in the exponential form, as opposed to expanding it to the lowest nontrivial order). In other words, the averaging of the fluctuating phase factors sums up the leading terms in powers of g^{-1} times the logarithm *square*, while the omitted diffuson fluctuations bring only subleading terms with powers of g^{-1} times the logarithm. The perturbative result for the zero bias anomaly (most relevant for experiment) was first derived by Altshuler and Aronov [347, 348] and Altshuler, Aronov and Lee [349]. The exponentiated version, Eq. (16.36), was derived in a number of ways [282, 350, 351, 352, 353]. Here we followed the sigma-model calculation of [297].

16.4 Altshuler–Aronov Effect

The zero-bias anomaly is entirely a consequence of the fluctuating phase factors $e^{\pm i\hat{\mathcal{K}}}$, Eq. (16.25). In other words it is essentially due to the *non-gauge-invariance* of the single-particle Green function. On the other hand, observables such as the current, Eq. (14.55), are gauge invariant. This statement manifests itself in the identical cancelation of the phase factors under the trace sign in Eq. (14.55). Still, there are interaction corrections to the current, known as the Altshuler–Aronov effect [348].

Since the current is gauge-invariant, one may evaluate it in any gauge and the $\mathcal{K}$-gauge is by far the most convenient. Focusing on the part of the dc ($\omega = 0$) current (14.55) that is due to the presence of the fluctuating vector potential in the covariant derivative operator $\hat{\partial}_{\mathbf{r}}$, one finds

$$\delta\mathbf{J}_{\mathrm{dc}} = \frac{e\pi\nu D}{2}\,\mathrm{i}\Big\langle \sum_{\epsilon,\epsilon',\epsilon''} \mathrm{Tr}\{\hat{\gamma}^{\mathrm{q}}\hat{Q}_{\mathcal{K};\epsilon\epsilon'}\hat{\gamma}^{\alpha}\mathbf{A}^{\alpha}_{\mathcal{K};\epsilon'-\epsilon''}\hat{Q}_{\mathcal{K};\epsilon''\epsilon}\}\Big\rangle_{\varphi,\hat{Q}_{\mathcal{K}}}, \tag{16.37}$$

where we wrote the trace in the energy representation. In the absence of an external classical vector potential, $\mathbf{A}^{\alpha}_{\mathcal{K}}$ is a pure gauge field, Eq. (16.2), $\mathbf{A}^{\alpha}_{\mathcal{K}} = \nabla_{\mathbf{r}}\mathcal{K}^{\alpha}$, having both classical and quantum components. We now parametrize the $\hat{Q}_{\mathcal{K}}$-matrices as in Eqs. (14.38) and (14.39) and expand to first order in the fluctuating diffuson fields d^{α}. This way, we obtain for the first-order fluctuation correction to the dc current

$$\delta\mathbf{J}_{\mathrm{dc}} = e\pi\nu D\,\mathrm{i}\Big\langle\sum_{\epsilon,\epsilon'}(F_{\epsilon}+F_{\epsilon'})\Big[-d^{\mathrm{cl}}_{\epsilon\epsilon'}\mathbf{A}^{\mathrm{q}}_{\mathcal{K};\epsilon'-\epsilon} + d^{\mathrm{q}}_{\epsilon\epsilon'}\big(\mathbf{A}^{\mathrm{cl}}_{\mathcal{K};\epsilon'-\epsilon}(F_{\epsilon'}-F_{\epsilon}) + \mathbf{A}^{\mathrm{q}}_{\mathcal{K};\epsilon'-\epsilon}(1-F_{\epsilon'}F_{\epsilon})\big)\Big]\Big\rangle_{\varphi,\hat{Q}_{\mathcal{K}}}. \tag{16.38}$$

At a first glance the averaging over the $\hat{Q}_{\mathcal{K}}$-matrix fluctuations (i.e. over $d^{\alpha}_{\epsilon\epsilon'}$ fields) of this expression vanishes. Indeed, the $\mathcal{K}$-gauge was constructed in such a way as to eliminate the terms linear in d^{α} and $\mathbf{A}^{\alpha}_{\mathcal{K}}$ in the action. If so, the products $d^{\alpha}\mathbf{A}^{\beta}_{\mathcal{K}}$ in Eq. (16.38) should average out to zero. This is indeed the case in equilibrium: no current is flowing. However, we are interested in a non-equilibrium situation, where an external bias is applied across the system. Neglecting for a moment collision processes (which are certainly present in the interacting system), one may think of the distribution function F_ϵ as of the two-step function given by Eq. (14.11), Fig. 14.2(b). Since this is a non-equilibrium function, the procedure of Section 16.1 can't completely eliminate the term $\propto d^\alpha\, \mathbf{A}^\beta_{\mathcal{K}}$ in the action. It is this term, remaining due to the non-equilibrium character of the distribution function, that makes the average fluctuating current $\delta\mathbf{J}_{\rm dc}$ nonzero.

In particular we focus on the $\mathrm{i}S_1 = -\mathrm{i}\pi\nu D\mathrm{Tr}\{\hat{\mathbf{A}}_{\mathcal{K}}\hat{Q}_{\mathcal{K}}\nabla_{\mathbf{r}}\hat{Q}_{\mathcal{K}}\}$ part of the action, Eq. (16.18b), where the gradient operator acts on the stationary distribution function $F_\epsilon(\mathbf{r})$, rather than on the diffuson fields $d^\alpha_{\epsilon\epsilon'}(\mathbf{r})$ (the latter was exactly the strategy that led us to Eq. (16.3) and eventually to the definition of the $\mathcal{K}$-gauge; the former option is absent in equilibrium). Once again, employing the parametrizations of the $\hat{Q}_{\mathcal{K}}$-matrices through Eqs. (14.38) and (14.39), expanding to first order in the fluctuating diffuson fields d^α and focusing only on the terms with $\nabla_{\mathbf{r}}F$, we find for the corresponding part of S_1 action

$$\delta S_1 = -2\pi\nu D\sum_{\epsilon,\epsilon'} d^{\rm q}_{\epsilon'\epsilon}\Big[\mathbf{A}^{\rm cl}_{\mathcal{K};\epsilon-\epsilon'}\nabla_{\mathbf{r}}(F_\epsilon - F_{\epsilon'}) - \mathbf{A}^{\rm q}_{\mathcal{K};\epsilon-\epsilon'}\nabla_{\mathbf{r}}(1 - F_\epsilon F_{\epsilon'})\Big]. \quad (16.39)$$

We expand now $\mathrm{e}^{\mathrm{i}\delta S_1}$ to first order and perform the functional averaging of $\langle\delta\mathbf{J}_{\rm dc}\mathrm{i}\delta S_1\rangle_{\varphi,\hat{Q}_{\mathcal{K}}}$ with the action, which includes the $\vec{\varphi}^{\rm T}U^{-1}_{\rm RPA}\vec{\varphi}$ part of the fluctuating scalar potential, Eq. (16.17), along with the Gaussian diffuson action (14.42). In doing so we must understand that $\mathbf{A}^\alpha_{\mathcal{K}} = \nabla_{\mathbf{r}}\mathcal{K}^\alpha$ is a pure gauge field. Since $\langle d^{\rm q}d^{\rm q}\rangle = 0$, only the first term on the right-hand side of Eq. (16.38) contributes. Moreover, since $\langle\mathbf{A}^{\rm q}_{\mathcal{K}}\mathbf{A}^{\rm q}_{\mathcal{K}}\rangle = 0$, only the first term on the right-hand side of Eq. (16.39) remains. Finally, employing that $\langle\mathcal{K}^{\rm cl}_{\epsilon-\epsilon'}(\mathbf{r}')\mathcal{K}^{\rm q}_{\epsilon'-\epsilon}(\mathbf{r})\rangle = (\mathrm{i}/2)\mathcal{V}^{\rm R}(\mathbf{r}'-\mathbf{r},\epsilon-\epsilon')$, Eq. (16.20), while the propagator of the diffusons is given by $\langle d^{\rm cl}_{\epsilon\epsilon'}(\mathbf{r})d^{\rm q}_{\epsilon'\epsilon}(\mathbf{r}')\rangle = -2\mathcal{D}^{\rm R}(\mathbf{r}-\mathbf{r}',\epsilon-\epsilon')/\pi\nu$, Eq. (14.43), we find for the interaction correction to the dc current

$$\delta\mathbf{J}_{\rm dc} = 2\pi\mathrm{i}e\nu D^2\int d\mathbf{r}'\sum_{\epsilon,\epsilon'}\mathcal{D}^{\rm R}(\mathbf{r}-\mathbf{r}',\epsilon-\epsilon')\,\nabla_{\mathbf{r}}\nabla_{\mathbf{r}'}\mathcal{V}^{\rm R}(\mathbf{r}'-\mathbf{r},\epsilon-\epsilon')$$
$$\times\big(F_\epsilon(\mathbf{r}) + F_{\epsilon'}(\mathbf{r})\big)\big(\nabla_{\mathbf{r}'}F_{\epsilon'}(\mathbf{r}') - \nabla_{\mathbf{r}'}F_\epsilon(\mathbf{r}')\big). \quad (16.40)$$

We shall restrict ourselves to the analysis of this expression in the linear response regime only. Then $\nabla_{\mathbf{r}'}F_\epsilon = (F_{\epsilon-eV} - F_\epsilon)/L = -\partial_\epsilon F_\epsilon eV/L$ (cf. Eq. (14.11)), where V/L is the electric field across the sample and all distribution functions are now understood as the equilibrium ones. We can now change integration variables as $(\epsilon+\epsilon')/2 \to \epsilon$ and $\epsilon - \epsilon' \to \omega$ and perform the explicit integration over ϵ. We then find for the Altshuler–Aronov [348] correction to the dc conductivity, given by $\delta\sigma_{\rm AA} = \delta\mathbf{J}_{\rm dc}L/V$,

$$\delta\sigma_{\rm AA} = -{\rm i}\,\frac{2\sigma_D}{\pi d}\int {\rm d}\omega\,\frac{\partial}{\partial\omega}\left[\omega\coth\frac{\omega}{2T}\right]\sum_{\mathbf{q}}\mathcal{D}^R(\mathbf{q},\omega)\,D\mathbf{q}^2\mathcal{V}^R(\mathbf{q},\omega)$$
$$\approx {\rm i}\,\frac{2e^2D}{\pi d}\int {\rm d}\omega\,\frac{\partial}{\partial\omega}\left[\omega\coth\frac{\omega}{2T}\right]\sum_{\mathbf{q}}\frac{1}{(Dq^2-i\omega)^2}, \qquad (16.41)$$

where $\sigma_D = e^2\nu D$. In the second line here we employed the limit of strong Coulomb interactions, $U_{\rm s}^{-1}(\mathbf{q}) \ll \Pi^{\rm R}$, and thus made the approximation that $D\mathbf{q}^2\mathcal{V}^{\rm R}(\mathbf{q},\omega) \approx -[\nu(D\,\mathbf{q}^2 - {\rm i}\omega)]^{-1}$; see Eqs. (16.23a). Focusing on $d = 2$ and performing momentum integration, one finds

$$\delta\sigma_{\rm AA} = -\frac{e^2}{2\pi^2}\ln(T\tau_{\rm el})^{-1}, \qquad (16.42)$$

where the elastic scattering rate $\tau_{\rm el}^{-1}$ enters as the upper cutoff in the logarithmic integral over the frequency ω. We found thus a negative singular correction with the relative value $\delta\sigma_{\rm AA}/\sigma_D \propto g^{-1}\ln(1/T\tau_{\rm el})$, where $g = \hbar\nu D$ is the dimensionless conductance per square. Notice that the correction is smaller (in the large logarithm) than the non-gauge-invariant zero bias effect considered in Section 16.3. Such a relative smallness of the $\hat{Q}_{\mathcal{K}}$-fluctuation corrections justifies the non-perturbative approach to the zero bias anomaly, which treats $\hat{Q}_{\mathcal{K}}$ in the stationary point approximation. One may trust the conductivity correction as long as $T \gg \tau_{\rm el}^{-1}{\rm e}^{-g}$. To go beyond this limit the RG treatment of the interacting sigma-model was developed in [282, 283, 354]. Reviews of the perturbative interaction corrections in disordered conductors can be found in [348, 356, 357].[6] In $d = 1, 3$ one also finds the singular suppression of the conductivity at a low temperature, which scales as $\delta\sigma_{\rm AA} \sim (d-2)T^{d/2-1}$.

16.5 Kinetic Equation

The aim of this section is to show how the kinetic equation for the distribution function $F_\epsilon(\mathbf{r},t)$ appears in the framework of the Keldysh formulation. In Section 14.4 it was demonstrated that the diffusive kinetic equation for noninteracting fermions is nothing but the stationary point equation for the effective action of the $\hat{Q}$-matrix. In the case of interacting electrons it is obtained from the action $S[\hat{Q}_{\mathcal{K}}, \nabla_{\mathbf{r}}\mathcal{K}]$, Eqs. (16.18), by first integrating out the fast degrees of freedom: both diffuson fluctuations, d^α, and electromagnetic fields, $\mathcal{K}^\alpha$ (or equivalently φ^α).

Let us outline the logic of the procedure, which leads from the partition function (16.17) to the kinetic equation. As the first step we separate slow and fast

[6] To appreciate the simplifications offered by the sigma-model treatment, one may notice that the original derivation includes consideration of eight diagrams, five of which cancel each other. The calculation of the remaining three involves an elaborated analytical continuation procedure.

degrees of freedom in the action, $S_l[\hat{Q}_{\mathcal{K}}, \partial_{\mathbf{r}}\mathcal{K}]$, where $l = 0, 1, 2$ (see Eqs. (16.18)). The former are encoded in the distribution function $F_{tt'}(\mathbf{r})$, which after a Wigner transformation acquires the form $F_\epsilon(\mathbf{r}, t)$. The fast degrees of freedom are represented by the diffuson modes $d^\alpha_{tt'}(\mathbf{r})$ and the electromagnetic modes $\mathcal{K}^\alpha(\mathbf{r}, t)$. This separation is achieved by an appropriate parametrization of the $\hat{Q}_{\mathcal{K}}$-matrix. A convenient choice is $\hat{Q}_{\mathcal{K}} = \hat{\mathcal{U}}_{\mathrm{s}}^{-1} \circ \hat{Q}_{\mathrm{fast}} \circ \hat{\mathcal{U}}_{\mathrm{s}}$ (cf. Eq. (14.75)), where the slow rotation matrices are

$$\hat{\mathcal{U}}_{\mathrm{s}} = \mathrm{e}^{\hat{\sigma}_- Z} \circ \mathrm{e}^{\hat{\sigma}_+ F} = \begin{pmatrix} 1 & F \\ Z & 1 + Z \circ F \end{pmatrix}, \qquad \hat{\mathcal{U}}_{\mathrm{s}}^{-1} = \mathrm{e}^{-\hat{\sigma}_+ F} \circ \mathrm{e}^{-\hat{\sigma}_- Z} \tag{16.43}$$

and the fast part of the $\hat{Q}_{\mathcal{K}}$-matrix is parametrized by the fluctuating diffuson fields $\hat{Q}_{\mathrm{fast}} = \exp\{-\hat{\mathcal{W}}_{\mathrm{f}}/2\} \circ \hat{\sigma}_3 \circ \exp\{\hat{\mathcal{W}}_{\mathrm{f}}/2\}$ (compare this parametrization with that given by Eq. (14.38)). In the last equation $Z_{tt'}(\mathbf{r})$ (not to be confused with the partition function) may be thought of as the *quantum* component of the distribution function $F_{tt'}(\mathbf{r})$. It is very similar in spirit to the fields $\delta_{\tilde{G}}$ and $\delta_{\tilde{\Sigma}}$, used in Eq. (12.45). Although $Z_{tt'}(\mathbf{r})$ is put to zero at the end of the calculations, it was emphasized in [357, 358] that $Z_{tt'}(\mathbf{r})$ must be kept explicitly in the $\hat{Q}$-matrix parametrization to obtain the proper form of the collision integral in the kinetic equation.

As the second step, one performs integrations over φ^α (or equivalently $\mathcal{K}^\alpha$, since the relation between them is fixed by Eq. (16.7)), and over the $\hat{\mathcal{W}}_{\mathrm{f}}$ fields in the partition function (16.17), to arrive at the effective action

$$\mathrm{e}^{\mathrm{i}S^{\mathrm{eff}}[F,Z]} = \int \mathbf{D}[\hat{\mathcal{W}}_{\mathrm{f}}, \varphi]\, \mathrm{e}^{\mathrm{i}S[F,Z,\hat{\mathcal{W}}_{\mathrm{f}},\nabla_{\mathbf{r}}\mathcal{K}]}. \tag{16.44}$$

The resulting effective action S^{eff} depends on F and its quantum component Z, and possibly the classical external fields, such as, for example, scalar or vector potentials. One then looks for the stationary point equation for the distribution function F as

$$\left.\frac{\delta S^{\mathrm{eff}}[F,Z]}{\delta Z}\right|_{Z=0} = 0, \tag{16.45}$$

which is the sought kinetic equation.

Proceeding along these lines, one expands the action (16.18) up to second order in the fast fields $\hat{\mathcal{W}}_{\mathrm{f}}$, $\hat{\varphi}$ and $\hat{\mathcal{K}}$ and up to first order in the auxiliary slow field Z.[7] The dependence on the slow distribution function F is to be kept exact. For the slow part of the S_0 action (16.18a) one finds after standard Wigner transformation that $\mathrm{i}S_0^{\mathrm{eff}} = 2\pi\nu\,\mathrm{Tr}\left\{\left[D\nabla_{\mathbf{r}}^2 F_\epsilon(\mathbf{r},t) - \partial_t F_\epsilon(\mathbf{r},t)\right] Z_\epsilon(\mathbf{r},t)\right\}$. Upon variation, Eq. (16.45),

[7] By keeping terms quadratic in Z and splitting them subsequently with a Hubbard–Stratonovich field, one arrives at the stochastic Boltzmann–Langevin terms [235, 236] in the kinetic equation, as explained in Section 12.4. Such terms allow us to find the shot noise and higher cumulants in the framework of the kinetic equation, see also [359, 360, 361].

with respect to Z, one recovers the noninteracting kinetic equation (14.36). The fast part of S_0 is, of course, the diffuson action (14.40).

Turning to the S_1 part of the action, Eq. (16.18b), one notices that, since it is linear in the fast field $\mathcal{K}$, one needs to keep only the terms linear in the fast diffuson modes $\hat{\mathcal{W}}_{\mathrm{f}}$. In equilibrium such terms were eliminated by the choice of the $\mathcal{K}$-gauge, which dictated the relation (16.7) between $\vec{\mathcal{K}}$ and $\vec{\varphi} \leftrightarrow \vec{V}$. Away from equilibrium, it is still convenient to keep the relation (16.7) intact, where the bosonic distribution function B_ω is understood as

$$B_\omega(\mathbf{r},t) = \frac{1}{2\omega}\int_{-\infty}^{+\infty} d\epsilon' \, [1 - F_{\epsilon'}(\mathbf{r},t)F_{\epsilon'-\omega}(\mathbf{r},t)] \,. \tag{16.46}$$

Notice that if $F_\epsilon = \tanh(\epsilon/2T)$, then $B_\omega = \coth(\omega/2T)$; compare Eq. (16.4). After some algebra one finds the term $\sim \mathcal{I}[F]\nabla^2_{\mathbf{r}}\mathcal{K}^{\mathrm{cl}}d^{\mathrm{q}}$ in the S_1 action, where the functional

$$\mathcal{I}[F] \equiv 1 - F_{\epsilon-\omega}(\mathbf{r},t)F_\epsilon(\mathbf{r},t) - B_\omega(\mathbf{r},t)\big[F_\epsilon(\mathbf{r},t) - F_{\epsilon-\omega}(\mathbf{r},t)\big] \tag{16.47}$$

is obviously zero in equilibrium.[8] There is also a term linear in the auxiliary field Z of the form $\sim Z\nabla^2_{\mathbf{r}}\mathcal{K}^{\mathrm{q}}d^{\mathrm{cl}}$. Expanding e^{iS_1} to second order and focusing on the product of these two terms, one finds after Gaussian integration over $\mathcal{K}^\alpha$ and d^α (see Eq. (8.43) and see [24] for details)

$$iS_1^{\mathrm{eff}} = -4i\pi\nu\,\mathrm{Tr}\left\{\left(Dq^2\right)^2\left[\mathcal{D}^{\mathrm{A}}(\mathbf{q},\omega)\mathcal{V}^{\mathrm{R}}(\mathbf{q},\omega) - \mathcal{D}^{\mathrm{R}}(\mathbf{q},\omega)\mathcal{V}^{\mathrm{A}}(\mathbf{q},\omega)\right]\mathcal{I}[F]Z\right\}. \tag{16.48}$$

The remaining S_2 part of the action (16.18c) is already quadratic in the fast fields $\hat{\mathcal{K}}$ and therefore may be taken at $\hat{\mathcal{W}}_{\mathrm{f}} = 0$. Typical terms linear in Z have the structure of $\mathrm{Tr}\{(\partial_{\mathbf{r}}\mathcal{K}^{\mathrm{q}})F(\partial_{\mathbf{r}}\mathcal{K}^{\mathrm{cl}})FZ\}$ and $\mathrm{Tr}\{(\partial_{\mathbf{r}}\mathcal{K}^{\mathrm{cl}})(\partial_{\mathbf{r}}\mathcal{K}^{\mathrm{cl}})FZ\}$. Keeping all of them and performing Gaussian integrations over $\mathcal{K}^\alpha$ [24], one finds

$$iS_2^{\mathrm{eff}} = 2i\pi\nu\,\mathrm{Tr}\left\{Dq^2[\mathcal{V}^{\mathrm{R}}(\mathbf{q},\omega) - \mathcal{V}^{\mathrm{A}}(\mathbf{q},\omega)]\mathcal{I}[F]Z\right\}, \tag{16.49}$$

where we took into account that according to Eqs. (16.7) and (16.21) there is an FDT-like relation $\mathcal{V}^{\mathrm{K}}(\mathbf{r},\mathbf{r},t) = B_\omega(\mathbf{r},t)\sum_{\mathbf{q}}\left[\mathcal{V}^{\mathrm{R}}(\mathbf{q},\omega) - \mathcal{V}^{\mathrm{A}}(\mathbf{q},\omega)\right]$ for the Keldysh component of the propagator at the coinciding spatial arguments, which includes the effective bosonic distribution (16.46).

Finally, combining $S_0^{\mathrm{eff}}[F,Z]$ together with $S_{1,2}^{\mathrm{eff}}[F,Z]$, Eqs. (16.48) and (16.49), and employing Eq. (16.45), one arrives at the kinetic equation

$$\partial_t F_\epsilon(\mathbf{r},t) - D\partial^2_{\mathbf{r}}F_\epsilon(\mathbf{r},t) = I^{\mathrm{coll}}[F], \tag{16.50}$$

[8] Here we disregarded terms with $\nabla_{\mathbf{r}}F$, which are crucial to derive, for example, the Altshuler–Aronov correction; see Eq. (16.39). In the context of the kinetic equation they lead to Altshuler–Aronov renormalization of the diffusion constant in the kinetic term. On the other hand, terms $\sim \mathcal{I}[F]$ lead to the collision integral.

where the collision integral is given by

$$I^{\mathrm{coll}}[F]=\sum_{\mathbf{q},\omega}\mathcal{M}(\mathbf{q},\omega)\Big[1-F_{\epsilon-\omega}(\mathbf{r},t)F_{\epsilon}(\mathbf{r},t)-B_{\omega}(\mathbf{r},t)[F_{\epsilon}(\mathbf{r},t)-F_{\epsilon-\omega}(\mathbf{r},t)]\Big]. \tag{16.51}$$

The interaction kernel here stands for

$$\begin{aligned}\mathcal{M}(\mathbf{q},\omega)&=-iDq^2\Big\{2Dq^2\big[\mathcal{D}^{\mathrm{A}}\mathcal{V}^{\mathrm{R}}-\mathcal{D}^{\mathrm{R}}\mathcal{V}^{\mathrm{A}}\big]-\big[\mathcal{V}^{\mathrm{R}}-\mathcal{V}^{\mathrm{A}}\big]\Big\}\\&=-2\,\mathrm{Re}[\mathcal{D}^{\mathrm{R}}(\mathbf{q},\omega)]\,\mathrm{Im}[U^{\mathrm{R}}_{\mathrm{RPA}}(\mathbf{q},\omega)],\end{aligned} \tag{16.52}$$

where in the last line we used $\mathcal{V}^{\mathrm{R(A)}}(\mathbf{q},\omega)=-\big[\mathcal{D}^{\mathrm{R(A)}}(\mathbf{q},\omega)\big]^2U^{\mathrm{R(A)}}_{\mathrm{RPA}}(\mathbf{q},\omega)$, which is a direct consequence of Eq. (16.19a), along with $2Dq^2=(\mathcal{D}^{\mathrm{R}})^{-1}+(\mathcal{D}^{\mathrm{A}})^{-1}$. This collision integral is to be compared with the one, Eq. (11.11), for the clean systems. They both contain the same combination of the distribution functions, which renders cancelation of the collision integral by the equilibrium Fermi-Dirac distribution. The difference is that in the clean case the distribution function depends on the momentum $\mathbf{k}$. In the disordered case the elastic impurity scattering quickly equilibrates over the momenta directions, while the relatively slow inelastic electron–electron scattering provides the energy equilibration. Thus in the latter case the distribution function depends only on the energy $\epsilon=\xi_{\mathbf{k}}$. The main difference, however, is the singular scattering kernel $\sum_{\mathbf{q}}\mathcal{M}(\mathbf{q},\omega)$. The physical reason for such an enhanced scattering cross-section is the absence of momentum conservation in the disordered system and thus a larger phase space available for the scattering.

Recalling the connection between the fermion distribution function and the occupation number $n_\epsilon(\mathbf{r},t)=(1-F_\epsilon(\mathbf{r},t))/2$ and employing Eqs. (16.12) and (16.46), one identically rewrites the right-hand side of Eq. (16.50) as [362, 363]

$$\begin{aligned}I^{\mathrm{coll}}&=\sum_{\mathbf{q},\omega,\epsilon'}\frac{8\pi\mathcal{M}(\mathbf{q},\omega)}{\omega}\\&\times\Big[n_\epsilon n_{\epsilon'-\omega}(1-n_{\epsilon'})(1-n_{\epsilon-\omega})-n_{\epsilon'}n_{\epsilon-\omega}(1-n_\epsilon)(1-n_{\epsilon'-\omega})\Big].\end{aligned} \tag{16.53}$$

This form clearly shows the "out" minus "in" structure of the collision integral (the seemingly wrong sign is associated with the fact that the left-hand side of Eq. (16.50) acts on $F_\epsilon=1-2n_\epsilon$). The present discussion can be generalized to include the spin triplet interaction channel. The corresponding kinetic equation and the collision integral were discussed in [364, 365].

Following the strategy of Section 11.1 we focus now on the "out" relaxation rate [363, 348, 366] for an electron of energy ϵ. It enters the spatially uniform kinetic equation $\partial_t n_\epsilon=-n_\epsilon/\tau_{ee}(\epsilon)$ and is given by

$$\frac{1}{\tau_{\mathrm{ee}}(\epsilon)}=\int d\omega d\epsilon'\frac{\sum_{\mathbf{q}}\mathcal{M}(\mathbf{q},\omega)}{\pi\omega}\,n_{\mathrm{F}}(\epsilon'-\omega)[1-n_{\mathrm{F}}(\epsilon')][1-n_{\mathrm{F}}(\epsilon-\omega)], \tag{16.54}$$

where all occupation numbers were substituted by the Fermi functions. This is appropriate if one is interested in small (linear) deviations of n_ϵ from its equilibrium value $n_F(\epsilon)$. Equation (16.54) simplifies considerably at zero temperature, $T = 0$. Indeed, the Fermi functions limit the energy integrations to the range $0 < \omega < \epsilon$, while $0 < \epsilon' < \omega$, where the product of all occupation numbers is unity. Integration over ϵ' brings in the factor of ω. In the universal limit of strong interactions, $U_s^{-1} \ll \Pi^R$, the kernel $\mathcal{M}(\mathbf{q}, \omega)$ acquires a simple form and one finds for the "out" relaxation rate

$$\frac{1}{\tau_{ee}(\epsilon)} = \frac{2}{\pi \nu} \int_0^\epsilon d\omega\, \omega \sum_{\mathbf{q}} \frac{1}{(Dq^2)^2 + \omega^2} = \frac{|\epsilon|}{4\pi g}, \tag{16.55}$$

where the last equality is given for $d = 2$, with $g = \hbar\nu D \gg 1$ being the dimensionless conductance per square. For $d = 1, 3$ the "out" relaxation rate scales with the energy as $\tau_{ee}^{-1}(\epsilon) \propto (1/\nu_d)(|\epsilon|/D)^{d/2}$, here $|\epsilon| \lesssim 1/\tau_{el}$, see [348] for further details. One notices that in all cases the small energy electron–electron relaxation rate appears to be much faster than in the case of clean systems, Eq. (11.12). This is a consequence of the aforementioned increase in the number of available states, due to the absence of the integral of motion (momentum) in the disordered case. Still $\tau_{ee}^{-1} \ll |\epsilon|$, justifying the quasiparticle picture employed to write the kinetic equation. Since $\omega \sim \epsilon$, this fact also justifies the division into the slow and fast degrees of freedom adopted to derive Eqs. (16.48) and (16.49). Indeed, the slow ones have characteristic frequencies of order $1/\tau_{ee}$, while the fast ones are of order ω.

16.6 Problems

16.6.1 Semiclassical Treatment of the Zero Bias Anomaly

Here we discuss a complimentary view on the zero bias anomaly of Section 16.3 developed by Levitov and Shytov [315]. It treats the zero bias suppression of the density of states as an under-barrier tunneling phenomenon. Since corresponding classical trajectories (i.e. solutions of the classical equations of motion) are forbidden, one looks for imaginary time trajectories. The barrier for an incoming electron is created by the Coulomb repulsion from electrons in the metal. The tunneling is associated with rearrangements of electron density and current. There is an imaginary time action for such a process, which suppresses its amplitude.

The imaginary time action for the current and density fluctuations in noninteracting metal is given by Eq. (14.90). Coulomb interaction, $U(\mathbf{r}-\mathbf{r}')$, adds an extra term:

$$S_N[\mathbf{j}, \rho] = \frac{1}{2}\iint d^{d+1}x\, d^{d+1}x' \left[\frac{1}{\sigma_D}\mathbf{g}(x)\mathcal{D}(x-x')\mathbf{g}(x') + \rho(x)U(\mathbf{r}-\mathbf{r}')\delta(\tau-\tau')\rho(x')\right], \tag{16.56}$$

where $x = (\mathbf{r}, \tau)$, $\mathbf{g}(x) = \mathbf{j}(x) + D\nabla_\mathbf{r}\rho(x)$, and $\mathcal{D}(\mathbf{q}, \omega) = (Dq^2 + |\omega|)^{-1}$ is the imaginary time diffuson. The current and density fluctuations are restricted by the (imaginary time) continuity equation $\nabla_\mathbf{r} \cdot \mathbf{j}(\mathbf{r}, \omega) + |\omega|\rho(\mathbf{r}, \omega) = \mathcal{J}(\mathbf{r}, \omega)$, where the source term $\mathcal{J}(x)$ is specified in what follows. The restriction enters the action through the Lagrange multiplier, $\lambda(x)$, as

$$S[\mathbf{j}, \rho, \lambda] = S_\mathrm{N}[\mathbf{j}, \rho] + \int \mathrm{d}^{d+1}x\, \lambda(x)\big[\nabla_\mathbf{r} \cdot \mathbf{j}(x) + |\omega|\rho(x) - \mathcal{J}(x)\big]. \tag{16.57}$$

Perform variations of this action with respect to $\mathbf{j}(x)$, $\rho(x)$, and $\lambda(x)$. Show that, in addition to the continuity relation, upon elimination of $\lambda(x)$ it leads to the following equations of motion:

$$\mathbf{j} + D\nabla_\mathbf{r}\rho = \sigma_\mathrm{D}\mathbf{E}; \qquad \mathbf{E}(\mathbf{r}, \tau) = -\nabla_\mathbf{r} \int \mathrm{d}\mathbf{r}'\, U(\mathbf{r}-\mathbf{r}')\rho(\mathbf{r}', \tau). \tag{16.58}$$

Solve these equations to show that $\rho_{\mathbf{q},\omega} = \mathcal{J}_{\mathbf{q},\omega}/(Dq^2 + |\omega| + \sigma_\mathrm{D} q^2 U_\mathbf{q})$ and $\mathbf{g}_{\mathbf{q},\omega} = (-i\mathbf{q})\sigma_\mathrm{D} U_\mathbf{q} \rho_{\mathbf{q},\omega}$. Substitute it into the action (16.56) and show that

$$S_\mathrm{N} = \frac{1}{2}\sum_{\mathbf{q},\omega} \frac{U_\mathbf{q}\, |\mathcal{J}_{\mathbf{q},\omega}|^2}{(Dq^2 + |\omega|)(Dq^2 + |\omega| + \sigma_\mathrm{D} q^2 U_\mathbf{q})}. \tag{16.59}$$

Notice that the action is proportional to the interaction potential $U_\mathbf{q}$.

An electron insertion and extraction after the imaginary time $2\tau_0$ in a certain location may be described by the source $\mathcal{J}(x) = e\delta(\mathbf{r})\big(\delta(\tau + \tau_0) - \delta(\tau - \tau_0)\big)$ and therefore $\mathcal{J}_{\mathbf{q},\omega} = 2ie \sin \omega\tau_0$. In 2d the Coulomb potential is $U_\mathbf{q} = 2\pi/q$. Calculate the action (16.59) and show that

$$S_\mathrm{N}(\tau_0) = \frac{e^2}{8\pi^2\sigma_\mathrm{D}} \log\frac{\tau_0}{\tau_\mathrm{el}} \log(\tau_0\tau_\mathrm{el}\omega_0^2), \tag{16.60}$$

where $\omega_0 = (2\pi\sigma_\mathrm{D})^2/D$. Finally the density of states is $\nu(\epsilon) = \nu \int \mathrm{d}\tau_0\, \mathrm{e}^{-S_\mathrm{N}(\tau_0)+2\tau_0\epsilon}$. Perform the integral in the stationary point approximation and compare your result with Eq. (16.36).

16.6.2 Near-Field Heat Transfer in Disordered Metals

The near-field heat transfer in clean systems is considered in Problem 12.6.2. Using kinetic equation in the disordered limit (16.51), show that the expression (12.72) still holds [367, 368] with the polarization operators given by Eq. (14.60). Perform the frequency and wavenumber integrals and find the near-field heat flux $J_h^\mathrm{NF}(T)$ in the disordered limit. Compare your result with the radiative Stefan–Boltzmann result and the clean limit of the near-field mechanism, Problem 12.6.2.

Consider the linear response regime, where $\delta T = T_1 - T_2 \ll T_2$, then $J_h = \partial_T J_h^{\rm NF}(T)\delta T$. On the other hand, $J_h = \mathrm{d}E_e/\mathrm{d}t = (\mathrm{d}E_e/\mathrm{d}T)(\mathrm{d}T/\mathrm{d}t) = -C(T)\,\mathrm{d}\delta T/\mathrm{d}t$, where E_e is the energy of the electron gas in one layer and $C(T) \propto \nu T$ is the heat capacity of the electron gas. From here one finds $\mathrm{d}\delta T/\mathrm{d}t = -\,\delta T/\tau_T$, where the relaxation rate is $1/\tau_T = \partial_T J_h^{\rm NF}(T)/C(T)$. Show that $1/\tau_T \propto T/\mathrm{g}$, which is very different from the Fermi liquid relaxation rate $\propto T^2/\epsilon_{\rm F}$, (11.12), but is consistent with the relaxation rate (16.55) in the disordered limit.

17

Dynamics of Disordered Superconductors

We generalize the nonlinear sigma-model to include interactions in the Cooper channel. As its stationary point approximation we derive coupled Usadel, self-consistency and Poisson equations. We use them to derive the Josephson current of an SNS junction, as well as the dispersion of the collective Carlson–Goldman mode of the superconductors. In the gapless case one may explicitly integrate out fermionic degrees of freedom, obtaining the time-dependent Ginzburg–Landau action. The latter is used to derive the Aslamazov–Larkin fluctuation correction to the normal state conductivity.

17.1 Cooper Channel Interactions

So far we have been mostly discussing the effects of singlet channel electron–electron interactions. We turn now to the Cooper channel interactions; see Eqs. (10.62)–(10.64). As was realized by Bardeen, Cooper, and Schrieffer (BCS) [369], exchange of virtual phonons with large wavenumbers $q \sim k_F$ mediates effective attractive interactions in the Cooper channel. The latter lead to formation of two-electron bound states – Cooper pairs, which form a condensate below a certain critical temperature T_c. The superfluid current of such a condensate is charged and results in a phenomenon known as superconductivity. In this chapter we restrict ourselves to the theory of disordered superconductors, that is, those where $T_c \ll \hbar/\tau_{el}$. This condition is indeed fulfilled for many conventional superconductors (but not for cold atom realizations). It actually considerably simplifies the theory by ensuring the spatially local character of the correlation functions.[1]

We write the local Cooper channel interaction action with attractive interactions as (cf. Eq. (10.64))

[1] Indeed, the characteristic length scale, the coherence length, $\xi_0 = \sqrt{D/T_c}$, appears to be much larger than the mean free path $l_{el} \propto \sqrt{D\tau_{el}}$, which is the scale where the nonlocal correlations decay.

$$S_{\mathrm{BCS}} = \frac{\lambda}{\nu}\int_C \mathrm{d}t \int \mathrm{d}\mathbf{r}\, \bar{\Phi}(\mathbf{r},t)\Phi(\mathbf{r},t), \tag{17.1}$$

where the time integral runs along the closed time contour, and the complex Cooper pair density is given by $\Phi(\mathbf{r},t) = \psi_\downarrow(\mathbf{r},t)\psi_\uparrow(\mathbf{r},t)$, while its conjugate is $\bar{\Phi}(\mathbf{r},t) = \bar{\psi}_\uparrow(\mathbf{r},t)\bar{\psi}_\downarrow(\mathbf{r},t)$. Notice the opposite sign with respect to Eq. (10.64), which stresses the attractive nature of the interactions, $\lambda > 0$. Actually the phonon-mediated interactions should be described by a time-nonlocal kernel $\lambda(t-t')$, or its Fourier image $\lambda(\omega)$. Following BCS, we approximate it as being a constant $\lambda(\omega) \approx \lambda$ for energy transfer below the Debye frequency, $|\omega| < \omega_{\mathrm{D}}$, and zero otherwise. This is justified by the fact that observables depend on the cutoff ω_{D} only logarithmically (see the following discussion).

This four-fermion interaction term may be decoupled via the Hubbard–Stratonovich transformation by introducing an auxiliary *complex* bosonic field $\Delta(\mathbf{r},t)$:

$$\begin{aligned} \mathrm{e}^{\,iS_{\mathrm{BCS}}} &= \int \mathbf{D}[\Delta]\; \mathrm{e}^{\,\mathrm{i}\int \mathrm{d}x\,[-\frac{\nu}{\lambda}|\Delta(x)|^2 + \Delta(x)\bar{\Phi}(x) + \Delta^*(x)\Phi(x)]} \\ &= \int \mathbf{D}[\Delta]\; \mathrm{e}^{\,\mathrm{i}\int \mathrm{d}x\,[-\frac{\nu}{\lambda}|\Delta(x)|^2 + \Delta(x)\bar{\psi}_\uparrow(x)\bar{\psi}_\downarrow(x) + \Delta^*(x)\psi_\downarrow(x)\psi_\uparrow(x)]}, \end{aligned} \tag{17.2}$$

here $x = (\mathbf{r},t)$ and $\int \mathrm{d}x = \int_C \mathrm{d}t \int \mathrm{d}\mathbf{r}$. After the Hubbard–Stratonovich transformation (17.2) the fermionic action becomes quadratic in Grassmann fields and may be written in the matrix form

$$S = \frac{1}{2}\int \mathrm{d}x\, (\bar{\psi}_\uparrow, -\psi_\downarrow) \begin{pmatrix} \mathrm{i}\partial_t + \frac{(\nabla_{\mathbf{r}}+\mathrm{i}\mathbf{A})^2}{2m} + V_{\mathrm{dis}} & \Delta \\ -\Delta^* & -\mathrm{i}\partial_t + \frac{(\nabla_{\mathbf{r}}-\mathrm{i}\mathbf{A})^2}{2m} + V_{\mathrm{dis}} \end{pmatrix} \begin{pmatrix} \psi_\uparrow \\ \bar{\psi}_\downarrow \end{pmatrix}, \tag{17.3}$$

along with the identical (in the absence of the Zeeman term) action for the $(\psi_\downarrow, -\bar{\psi}_\uparrow)^{\mathrm{T}}$ spinor. To make further notation compact, it is convenient to introduce fermionic bispinors

$$\bar{\Psi} = \frac{1}{\sqrt{2}}\left(\bar{\psi}_\uparrow, -\psi_\downarrow, \bar{\psi}_\downarrow, \psi_\uparrow\right); \qquad \Psi = \frac{1}{\sqrt{2}}\left(\psi_\uparrow, \bar{\psi}_\downarrow, \psi_\downarrow, -\bar{\psi}_\uparrow\right)^{\mathrm{T}}, \tag{17.4}$$

defined in the four-dimensional space Ω, which can be viewed as the direct product N$\otimes$S of the Gor'kov–Nambu [370, 371] (N) $(\psi_\uparrow, \bar{\psi}_\downarrow)$ and spin (S) spaces $(\psi_\uparrow, \psi_\downarrow)$. In principle, the choice of the bispinors is not unique. One can rearrange components of the bispinors in a different manner, separating explicitly the time-reversal (TR) $(\psi, \bar{\psi})$ and spin spaces. Finally, one may equally think of Ψ as representing the direct product of the Nambu and time-reversal subspaces. These representations are equivalent, that is, $\Omega = \mathrm{N}\otimes\mathrm{S} \propto \mathrm{S}\otimes\mathrm{TR} \propto \mathrm{N}\otimes\mathrm{TR}$, and the choice between them is dictated by convenience in calculations for a particular problem at hand. We shall use the N $\otimes$ S choice and omit the S part, since the theory is diagonal in the latter subspace,[2] as illustrated by Eq. (17.3). The vectors $\bar{\Psi}$ and Ψ are not independent and are related to each other, $\bar{\Psi} = (\check{C}\Psi)^{\mathrm{T}}$, by the charge-conjugation matrix $\check{C} \equiv \mathrm{i}\hat{\tau}_2 \otimes \hat{s}_1$, where $\hat{\tau}_i$ and $\hat{s}_i$, for $i = 0, 1, 2, 3$, are Pauli matrices acting in the Nambu and spin subspaces, respectively; the $\hat{\sigma}_i$ matrices, as before, act in the bosonic Keldysh subspace, while $\hat{\gamma}^{\mathrm{cl(q)}}$ act in the fermionic Keldysh subspace. To avoid confusion, we shall specify, where appropriate, Keldysh and Nambu subspaces by subscripts K and N, respectively.

We now employ the same strategy as outlined in Chapters 10–16. First we split the fermionic fields into forward and backward branches and perform the Keldysh–Larkin–Ovchinnikov rotation; see Section 10.3. This way, an additional Keldysh (K) subspace is introduced. We then perform disorder averaging of the partition function and then split the emerging four-fermion term with the matrix field $\check{Q}_{tt'}(\mathbf{r})$, which acts in the $\Omega \otimes$ K $\otimes$ time space; see Chapter 14. At this point the action is quadratic in fermionic spinors and they may be integrated out, resulting in the following disorder-averaged partition function:

$$\langle Z\rangle_{\mathrm{dis}} = \int \mathbf{D}[\varphi, \Delta]\, \mathrm{e}^{\frac{\mathrm{i}}{2}\mathrm{Tr}\{\check{\varphi} U_{\mathrm{s}}^{-1} \check{\Sigma}_1 \check{\varphi}\} - \frac{\mathrm{i}\nu}{2\lambda}\mathrm{Tr}\{\check{\Delta}^{\dagger}\check{\Sigma}_1\check{\Delta}\}} \int \mathbf{D}[\check{Q}]\; \mathrm{e}^{\mathrm{i}S[\check{Q},\Delta,\mathbf{A},\varphi]};$$

$$\mathrm{i}S[\check{Q}, \Delta, \mathbf{A}, \varphi] = -\frac{\pi\nu}{8\tau_{\mathrm{el}}}\mathrm{Tr}\{\check{Q}^2\} + \mathrm{Tr}\ln\left[\check{G}^{-1} + \frac{\mathrm{i}}{2\tau_{\mathrm{el}}}\check{Q} - \check{\varphi} - \mathbf{v}_{\mathrm{F}}\check{\mathbf{A}} + \check{\Delta}\right], \quad (17.5)$$

which generalizes Eq. (14.47). We have also included the scalar potential φ, originating from the singlet (i.e. Coulomb) interaction potential $U_{\mathrm{s}}(\mathbf{q})$, Eq. (10.65). Hereafter, we use the check symbol to denote 4×4 matrices acting in the K $\otimes$ N space and the hat symbol for the 2×2 matrices acting in either Nambu or Keldysh subspaces. The bare inverse Green function in K $\otimes$ N space is given by $\check{G}^{-1} = \mathrm{i}\check{\mathcal{T}}_3 \partial_t + \nabla_{\mathbf{r}}^2/2m + \mu$, where $\check{\mathcal{T}}_3 = \hat{\gamma}^{\mathrm{cl}} \otimes \hat{\tau}_3$. Equation (17.5) also contains the matrix $\check{\Sigma}_1 = \hat{\sigma}_1 \otimes \hat{\tau}_0$ along with the bosonic matrix fields, defined as

$$\begin{aligned} \check{\varphi}(\mathbf{r}, t) &= \varphi^{\alpha}(\mathbf{r}, t)\, \hat{\gamma}^{\alpha} \otimes \hat{\tau}_0, \qquad \check{\mathbf{A}}(\mathbf{r}, t) = \mathbf{A}^{\alpha}(\mathbf{r}, t)\, \hat{\gamma}^{\alpha} \otimes \hat{\tau}_3, \\ \check{\Delta}(\mathbf{r}, t) &= \Delta^{\alpha}(\mathbf{r}, t)\, \hat{\gamma}^{\alpha} \otimes \hat{\tau}_+ - \Delta^{*\alpha}(\mathbf{r}, t)\, \hat{\gamma}^{\alpha} \otimes \hat{\tau}_-, \end{aligned} \quad (17.6)$$

with $\hat{\tau}_{\pm} = (\hat{\tau}_1 \pm \mathrm{i}\hat{\tau}_2)/2$. The $\check{Q}$-matrix also has a 4×4 structure in Keldysh and Nambu spaces along with the matrix structure in the time domain.

We use now the fact that the elastic scattering rate $1/\tau_{\mathrm{el}}$ is the largest energy scale in the problem (indeed we focus on disordered superconductors with $T_{\mathrm{c}}\tau_{\mathrm{el}} \ll 1$).

[2] The omission of this subspace excludes rotations in, e.g. the $(\psi_{\uparrow}, \bar{\psi}_{\uparrow})$ time-reversal subspace. These rotations are exactly the soft modes "normal Cooperons" which are responsible for the weak localization effects; see Section 14.7. We also miss the triplet diffuson soft modes associated with the rotations in $(\psi_{\uparrow}, \psi_{\downarrow})$ subspace. As long as localization and spin-scattering effects are not considered, such an omission is indeed justified. I am indebted to M. Skvortsov for discussing this issue.

Following the discussion of Section 14.3, it allows one to separate soft modes in $\check{Q}$-matrix space. The latter obey the nonlinear constraint $\check{Q} \circ \check{Q} = \check{1}$. Focusing on this soft-mode manifold, one may expand the logarithm under the trace operation in Eq. (17.5) in spatial and temporal gradients of the $\check{Q}$-matrix as well as in the relatively weak fluctuating bosonic fields φ and Δ (similar to the calculations presented in Section 14.5). As a result, one obtains the action of disordered superconductors in the following form (cf. Eq. (14.65)):

$$\mathrm{i}S[\check{Q}, \Delta, \mathbf{A}, \varphi] = \frac{\mathrm{i}\nu}{4}\mathrm{Tr}\{\check{\varphi}\check{\Sigma}_1\check{\varphi}\} - \frac{\pi\nu}{8}\mathrm{Tr}\{D(\hat{\partial}_{\mathbf{r}}\check{Q})^2 - 4\check{T}_3\partial_t\check{Q} - 4\mathrm{i}\check{\varphi}\check{Q} + 4\mathrm{i}\check{\Delta}\check{Q}\}. \tag{17.7}$$

The first (anomalous) term on the right-hand side is the static polarizability of the electronic band, as explained in Eq. (14.54). The last term is the Keldysh nonlinear sigma-model action, generalized for disordered superconductors, first obtained by Feigel'man, Larkin, and Skvortsov [372, 373] (for the replica version see, for example, [374]). The covariant derivative $\hat{\partial}_{\mathbf{r}}$ is defined as in Eq. (14.49):

$$\hat{\partial}_{\mathbf{r}}\check{Q} = \nabla_{\mathbf{r}}\check{Q} + \mathrm{i}[\mathbf{A}\check{T}_3, \check{Q}]. \tag{17.8}$$

Notice the $\hat{\tau}_3$ Nambu matrix in the definition of the matrix vector potential (17.6), which signifies the pair-breaking nature of the magnetic field. Also notice that $\nu = \nu_\uparrow + \nu_\downarrow$ stands for the density of states for two spin components.

17.2 Usadel Equation

Averaging over the strong quenched disorder and taking care of the large energy scale $1/\tau_{\mathrm{el}}$ results in the nonlinear sigma-model action (17.7), supplemented by the nonlinear constraint $\check{Q}^2 = \check{1}$. We focus now on this soft-mode manifold and incorporate smaller energy scales associated with the superconducting correlations (and with the screened Coulomb interactions). To this end we look for a stationary matrix $\underline{\check{Q}}_{t_1t_2}(\mathbf{r})$ on the soft-mode Goldstone manifold defined by the nonlinear constraint $\check{Q}^2 = \check{1}$. This is achieved with the help of parametrizations $\check{Q} = \exp(-\check{\mathcal{W}}/2) \circ \underline{\check{Q}} \circ \exp(\check{\mathcal{W}}/2)$ and expansion to first order in the rotation generators $\check{\mathcal{W}}$ as $\check{Q} \approx \underline{\check{Q}} - [\check{\mathcal{W}}\overset{\circ}{,} \underline{\check{Q}}]/2$. Substituting this form into the action (17.7) and demanding that the first order in $\check{\mathcal{W}}$ vanishes, one finds the stationary point condition

$$\hat{\partial}_{\mathbf{r}}(D\underline{\check{Q}}\,\hat{\partial}_{\mathbf{r}}\underline{\check{Q}}) - \{\check{T}_3\partial_t, \underline{\check{Q}}\}_+ + \mathrm{i}[\check{\Delta} - \check{\varphi}, \underline{\check{Q}}\,] = 0, \tag{17.9}$$

which is known as the Usadel equation [299] (cf. the normal state Eq. (14.34)).

We shall look for solutions of this equation in the subspace of *classical* configurations, which exhibit the causality structure and thus have the form

$$\check{Q} = \begin{pmatrix} \hat{Q}^{\mathrm{R}} & \hat{Q}^{\mathrm{K}} \\ 0 & \hat{Q}^{\mathrm{A}} \end{pmatrix}_{\mathrm{K}}, \tag{17.10}$$

with retarded, advanced, and Keldysh components being matrices in the Nambu subspace. Hereafter, we omit underlining the stationary configuration of the sigma-model action. In the subspace of classical configurations the nonlinear constraint $\check{Q}^2 = \check{1}$ is resolved by the following conditions:

$$\hat{Q}^{\mathrm{R}}\hat{Q}^{\mathrm{R}} = \hat{Q}^{\mathrm{A}}\hat{Q}^{\mathrm{A}} = \hat{1}, \qquad \hat{Q}^{\mathrm{R}}\hat{Q}^{\mathrm{K}} + \hat{Q}^{\mathrm{K}}\hat{Q}^{\mathrm{A}} = 0. \tag{17.11}$$

Substituting the classical ansatz (17.10) into the Usadel equation (17.9), and assuming that the fields $\Delta(\mathbf{r}, t)$ and $\varphi(\mathbf{r}, t)$ are pure *classical*, one finds a closed equation for the retarded (advanced) component of the stationary $\check{Q}$-matrix:

$$\hat{\partial}_{\mathbf{r}}\big(D\,\hat{Q}^{\mathrm{R(A)}}\hat{\partial}_{\mathbf{r}}\hat{Q}^{\mathrm{R(A)}}\big) + \mathrm{i}\big[\epsilon\hat{\tau}_3, \hat{Q}^{\mathrm{R(A)}}\big] + \mathrm{i}\big[\hat{\Delta} - \hat{\varphi}, \hat{Q}^{\mathrm{R(A)}}\big] = 0, \tag{17.12}$$

where we employed a Fourier representation for the operator ∂_t. As we shall see, this equation describes the quasiparticle spectrum of the superconductor.

First we look for a spatially uniform and time translationally invariant solution $\hat{\Lambda}_\epsilon^{\mathrm{R(A)}}$ that satisfies

$$\mathrm{i}\epsilon\big[\hat{\tau}_3, \hat{\Lambda}_\epsilon^{\mathrm{R(A)}}\big] + \mathrm{i}\big[\hat{\Delta}, \hat{\Lambda}_\epsilon^{\mathrm{R(A)}}\big] = 0. \tag{17.13}$$

While writing this equation, we have assumed that the order parameter field $\Delta(\mathbf{r}, t)$ has developed a nonzero expectation value (condensate) Δ, which is time- and space-independent.[3] Its value will be determined below from the self-consistency condition. Without loss of generality we assume that Δ is real, thus $\hat{\Delta} = \mathrm{i}\Delta\hat{\tau}_2$, see Eq. (17.6). Taking into account that $\big(\hat{Q}^{\mathrm{R(A)}}\big)^2 = \hat{1}$, one finds the solution of Eq. (17.13) as

$$\hat{\Lambda}_\epsilon^{\mathrm{R}} = \frac{\operatorname{sign}\epsilon}{\sqrt{(\epsilon + \mathrm{i}0)^2 - \Delta^2}} \begin{pmatrix} \epsilon & \Delta \\ -\Delta & -\epsilon \end{pmatrix}_{\mathrm{N}} = \hat{\tau}_3 u_\epsilon + \mathrm{i}\hat{\tau}_2 v_\epsilon, \tag{17.14}$$

where

$$u_\epsilon = \frac{\epsilon \operatorname{sign}\epsilon}{\sqrt{(\epsilon + \mathrm{i}0)^2 - \Delta^2}}, \qquad v_\epsilon = \frac{\Delta \operatorname{sign}\epsilon}{\sqrt{(\epsilon + \mathrm{i}0)^2 - \Delta^2}}, \qquad u_\epsilon^2 - v_\epsilon^2 = 1 \tag{17.15}$$

and $\hat{\Lambda}_\epsilon^{\mathrm{A}} = -\hat{\tau}_3 u_\epsilon^* - \mathrm{i}\hat{\tau}_2 v_\epsilon^*$. The infinitesimal imaginary increments are chosen to have the retarded (advanced) component analytic in the entire upper (lower) half-plane of the complex energy ϵ. The $\operatorname{sign}\epsilon$ factor ensures that $\hat{\Lambda}_\epsilon^{\mathrm{R(A)}} \to \pm\hat{\tau}_3$ for

[3] Since the scalar potential φ does not have a finite expectation value, we drop it from the equation for the equilibrium matrix $\hat{\Lambda}_\epsilon^{\mathrm{R}}$.

$|\epsilon| \gg \Delta$, as expected for a normal metal; see Eq. (14.25). Notice also that u_ϵ and v_ϵ are real for $|\epsilon| > \Delta$ and pure imaginary for $|\epsilon| < \Delta$. In the latter case the sign ϵ factor also ensures continuity of these functions at $\epsilon = 0$.

We turn now to the determination of the order parameter Δ. Varying the action (17.5) and (17.7) with respect to the quantum component $\Delta^{*\mathrm{q}}(\mathbf{r}, t)$ of the order parameter field, one finds the stationary point equation for $\Delta^{\mathrm{cl}}(\mathbf{r}, t)$:

$$\Delta^{\mathrm{cl}}(\mathbf{r}, t) = \frac{\pi\lambda}{2} \mathrm{Tr}\{(\hat{\gamma}^{\mathrm{q}} \otimes \hat{\tau}_-)\check{Q}\} = \frac{\pi\lambda}{2} \mathrm{Tr}\{\hat{\tau}_- \hat{Q}^{\mathrm{K}}\}, \tag{17.16}$$

where in the last equality we restricted ourselves to the classical ansatz (17.10). Variation with respect to $\Delta^{\mathrm{q}}(\mathbf{r}, t)$ leads to the complex conjugate equation for $\Delta^{*\mathrm{cl}}$. These equations provide the self-consistency conditions for the complex order parameter Δ. Looking for a constant real solution of the form $\Delta^{\mathrm{cl}}(\mathbf{r}, t) = \Delta$, one obtains $\Delta = (\pi\lambda/2)\int(\mathrm{d}\epsilon/2\pi)(\hat{\Lambda}^{\mathrm{K}}_\epsilon)_{12}$. According to FDT, $\hat{\Lambda}^{\mathrm{K}} = \tanh(\epsilon/2T) \times(\hat{\Lambda}^{\mathrm{R}}_\epsilon - \hat{\Lambda}^{\mathrm{A}}_\epsilon)$, and thus $\hat{\Lambda}^{\mathrm{K}}_\epsilon = 0$ for $|\epsilon| < \Delta$. As a result, one finds the self-consistency condition for $\Delta = \Delta(T)$:[4]

$$\Delta = \lambda\Delta \int_{|\Delta|}^{\omega_{\mathrm{D}}} \mathrm{d}\epsilon \, \frac{\tanh \epsilon/2T}{\sqrt{\epsilon^2 - |\Delta|^2}}. \tag{17.17}$$

This equation admits a nonzero solution for $T < T_{\mathrm{c}}$, where the critical temperature is determined from the condition $1 = \lambda\int_0^{\omega_{\mathrm{D}}}(\mathrm{d}\epsilon/\epsilon)\tanh\epsilon/2T_{\mathrm{c}}$. On the other hand, at $T = 0$ the order parameter $\Delta(0)$ is given by the solution of $1 = \lambda\int_{\Delta(0)}^{\omega_{\mathrm{D}}} \mathrm{d}\epsilon/\sqrt{\epsilon^2 - \Delta^2(0)}$. This leads to the simple relation between the two [375] $\Delta(0) = 1.76T_{\mathrm{c}}$, which is reasonably well satisfied for many conventional superconductors.

Having determined $\hat{\Lambda}^{\mathrm{R}}_\epsilon$ and the order parameter $\Delta(T)$, one may discuss the single-particle density of states. The latter is defined as $\nu(\epsilon) = -\mathrm{Im}\, \mathcal{G}^{\mathrm{R}}(\epsilon)/\pi$, where $\mathcal{G}^{\mathrm{R}}$ is the retarded single-particle Green function (traced over the spin indices) discussed in Section 16.3. Following the same root (cf. Eq. (16.25)), one may show that in the stationary point approximation the density of states is given by

$$\nu(\epsilon) = \frac{\nu}{2} \mathrm{Re}\,\mathrm{Tr}\{\hat{\tau}_3 \hat{\Lambda}^{\mathrm{R}}_\epsilon\} = \nu\,\theta(|\epsilon| - \Delta)\frac{|\epsilon|}{\sqrt{\epsilon^2 - \Delta^2}}. \tag{17.18}$$

We find thus that the quasiparticle spectrum has a gap, that is, $\nu(\epsilon) = 0$ for $|\epsilon| < \Delta(T)$, as long as $T < T_{\mathrm{c}}$. Immediately outside the gap the density of states exhibits an integrable square root singularity, which has profound consequences for quasiparticle dynamics; see Section 17.5. It is instructive to check that $\int \mathrm{d}\epsilon(\nu(\epsilon) - \nu) = 0$. This observation implies that the total number of single-particle

[4] To bring it to textbook form, e.g. [375], one changes variables as $\xi = \sqrt{\epsilon^2 - |\Delta|^2}$ to write $\epsilon = \sqrt{|\Delta|^2 + \xi^2}$, and finds $1 = \lambda\int_0^{\omega_{\mathrm{D}}}\left(\mathrm{d}\xi/\sqrt{|\Delta|^2 + \xi^2}\right)\tanh\sqrt{|\Delta|^2 + \xi^2}/2T$.

states is not altered by the gap opening. Instead, the states are "pushed" from the gap interval into above-the-gap peaks.

Let us now focus on the Keldysh component of the Usadel equation (17.9), given by the (1, 2) element in K space. It provides another equation, which takes the form

$$\partial_{\mathbf{r}}(D\,\hat{Q}^{\mathrm{R}}\partial_{\mathbf{r}}\hat{Q}^{\mathrm{K}} + D\,\hat{Q}^{\mathrm{K}}\partial_{\mathbf{r}}\hat{Q}^{\mathrm{A}}) + \mathrm{i}[\epsilon\hat{\tau}_3, \hat{Q}^{\mathrm{K}}] + \mathrm{i}[\hat{\Delta} - \hat{\varphi}, \hat{Q}^{\mathrm{K}}] = 0. \tag{17.19}$$

The last of the conditions (17.11) may be explicitly resolved by the standard parametrization of the Keldysh component as

$$\hat{Q}^{\mathrm{K}} = \hat{Q}^{\mathrm{R}} \circ \hat{F} - \hat{F} \circ \hat{Q}^{\mathrm{A}}, \tag{17.20}$$

where $\hat{F}$ is a 2×2 matrix in Nambu space (provided $(\hat{Q}^{\mathrm{R}})^2 = (\hat{Q}^{\mathrm{A}})^2 = \hat{1}$), which may be thought of as a matrix distribution function. Using the fact that the $\check{Q}_{t,t'}(\mathbf{r})$ matrix has the symmetries of the $\Psi(t,\mathbf{r})\bar{\Psi}(t',\mathbf{r})$ bilinear combination, one may show [376] that among the four components of $\hat{F}_{t,t'}(\mathbf{r})$ only two are linearly independent. Following Schmid–Schön [377] and Larkin–Ovchinnikov [28], we choose them as

$$\hat{F} = \begin{pmatrix} F_{\mathrm{L}} + F_{\mathrm{T}} & 0 \\ 0 & F_{\mathrm{L}} - F_{\mathrm{T}} \end{pmatrix}_{\mathrm{N}} = F_{\mathrm{L};tt'}(\mathbf{r})\hat{\tau}_0 + F_{\mathrm{T};tt'}(\mathbf{r})\hat{\tau}_3, \tag{17.21}$$

where the subscripts L and T refer to the *longitudinal* and *transversal* components of the distribution function. The meaning of this notation will be clarified at the end of Section 17.5. The fact that $\check{Q}_{t,t'}(\mathbf{r})$ has the symmetries of $\Psi(t,\mathbf{r})\bar{\Psi}(t',\mathbf{r})$ translates into the statement that $F_{\mathrm{L};tt'}$ is odd while $F_{\mathrm{T};tt'}$ is even under the interchange of $t \leftrightarrow t'$. Correspondingly their Wigner transforms $F_{\mathrm{L}}(\epsilon,\mathbf{r},t)$ and $F_{\mathrm{T}}(\epsilon,\mathbf{r},t)$ are odd and even functions of energy ϵ, respectively. Notice that there are *not* two different quasiparticle distribution functions. Instead, the odd and even components of the one and only quasiparticle distribution function appear in the $\hat{F}$-matrix with the $\hat{\tau}_0$ and $\hat{\tau}_3$ Nambu matrices. In equilibrium $F_{\mathrm{L}}(\epsilon) = \tanh\epsilon/2T$ and $F_{\mathrm{T}}(\epsilon) = 0$. Being even, $F_{\mathrm{T}}(\epsilon)$ is responsible for the charge current and density, while the odd $F_{\mathrm{L}}(\epsilon)$ provides the energy current and density.

It is sometimes useful to subtract from the Keldysh Usadel equation (17.19) the retarded one (17.12) convoluted with $\hat{F}$ and add $\hat{F}$ convoluted with the advanced equation (17.12), schematically $\mathrm{K} - \mathrm{R} \circ \mathrm{F} + \mathrm{F} \circ \mathrm{A}$. As a result, one obtains the equation for the distribution matrix $\hat{F}$:

$$\begin{aligned} &D\left(\partial_{\mathbf{r}}^2\hat{F} - \partial_{\mathbf{r}}(\hat{Q}^{\mathrm{R}}\partial_{\mathbf{r}}\hat{F}\hat{Q}^{\mathrm{A}}) + \hat{Q}^{\mathrm{R}}\partial_{\mathbf{r}}\hat{Q}^{\mathrm{R}}\partial_{\mathbf{r}}\hat{F} - \partial_{\mathbf{r}}\hat{F}\hat{Q}^{\mathrm{A}}\partial_{\mathbf{r}}\hat{Q}^{\mathrm{A}}\right) \\ &\quad + \mathrm{i}\hat{Q}^{\mathrm{R}}[\epsilon\hat{\tau}_3, \hat{F}] - \mathrm{i}[\epsilon\hat{\tau}_3, \hat{F}]\hat{Q}^{\mathrm{A}} + \mathrm{i}\hat{Q}^{\mathrm{R}}[\hat{\Delta} - \hat{\varphi}, \hat{F}] - \mathrm{i}[\hat{\Delta} - \hat{\varphi}, \hat{F}]\hat{Q}^{\mathrm{A}} = 0. \end{aligned} \tag{17.22}$$

As mentioned previously, only two components of this equation are independent. One can thus (i) take the Nambu trace of this equation and (ii) multiply it by $\hat{\tau}_3$ and then take the Nambu trace. This way, one finds two scalar equations for the two unknown functions $F_{\rm L}$ and $F_{\rm T}$. Being written in the Wigner transformed representation, they are known as Larkin–Ovchinnikov equations [28]. As we shall see in Section 17.5, the Wigner transformation may be dangerous due to the singularities in the density of states. If this is the issue, it is more convenient to work directly with Eq. (17.19) written in the energy representation than with Eq. (17.22).

Finally, varying the action (17.5) and (17.7) with respect to the quantum component of the scalar potential $\varphi^{\rm q}(\mathbf{r},t)$ and adopting the classical ansatz (17.10), one obtains the Poisson equation for the scalar potential $\varphi^{\rm cl}=\varphi$:

$$\nabla_{\mathbf{r}}^2\varphi = 4\pi e^2\Big[\nu\varphi + \frac{\pi\nu}{4}\,{\rm Tr}\{\hat\tau_0\,(\hat Q^{\rm K}-\hat\Lambda^{\rm K})\}\Big], \tag{17.23}$$

where the first term on the right-hand side originates from the variation of the first (anomalous) term in the action (17.7), which represents the static polarizability of the electronic band. In the last term we have subtracted the equilibrium matrix $\hat\Lambda^{\rm K}$, which is supposed to carry the same charge as the positive background. The Poisson equation together with the self-consistency condition (17.16) and the Usadel equations for the spectrum (17.12) and the distribution function (17.22) (or equivalently (17.19)) constitute the closed system of equations governing the dynamics of the superconductor. In the following section we consider some applications of these equations.

17.3 Stationary Superconductivity

In the presence of boundaries or proximity to a normal metal one faces the problem of spatially nonuniform superconductivity. In this case, both Δ and $\hat Q^{\rm R(A)}$ acquire a coordinate dependence, and one should look for solutions of Eqs. (17.9) and (17.16). In doing so, we will assume that $\check Q_{tt'}$ is static, that is, independent of the central time. As a result, $\check Q_\epsilon$ is diagonal in the energy representation. The nonlinear constraints $(\hat Q^{\rm R(A)})^2=\hat 1$, Eq. (17.11), may be explicitly resolved by the angular parametrization [378] for the retarded and advanced blocks of the $\check Q$-matrix:

$$\hat Q^{\rm R}(\mathbf{r},\epsilon)=\begin{pmatrix}\cosh\vartheta & \sinh\vartheta\,\exp(i\chi)\\ -\sinh\vartheta\,\exp(-i\chi) & -\cosh\vartheta\end{pmatrix}_{\rm N}, \tag{17.24a}$$

$$\hat Q^{\rm A}(\mathbf{r},\epsilon)=-\hat\tau_3\big[\hat Q^{\rm R}\big]^\dagger\hat\tau_3=\begin{pmatrix}-\cosh\vartheta^* & -\sinh\vartheta^*\exp(i\chi^*)\\ \sinh\vartheta^*\exp(-i\chi^*) & \cosh\vartheta^*\end{pmatrix}_{\rm N}, \tag{17.24b}$$

where $\vartheta(\mathbf{r},\epsilon)$ and $\chi(\mathbf{r},\epsilon)$ are *complex* coordinate- and energy-dependent angles. Substituting $\hat Q^{\rm R}$ in the form of (17.24) into Eq. (17.12), one finds from the diagonal elements of the corresponding matrix equation

$$D\,\hat{\partial}_{\mathbf{r}}\big(\sinh^2\vartheta\;\hat{\partial}_{\mathbf{r}}\chi\big) = 2\mathrm{i}|\Delta|\sinh\vartheta\,\sin(\theta-\chi), \tag{17.25}$$

where the order parameter is parametrized as $\Delta(\mathbf{r}) = |\Delta(\mathbf{r})|\exp\{\mathrm{i}\theta(\mathbf{r})\}$. From the off-diagonal block of the matrix equation (17.12), using Eq. (17.25), one obtains

$$D\,\hat{\partial}_{\mathbf{r}}^2\vartheta + 2\mathrm{i}\epsilon\sinh\vartheta - 2\mathrm{i}|\Delta|\cosh\vartheta\,\cos(\theta-\chi) = \frac{D}{2}\big(\hat{\partial}_{\mathbf{r}}\chi\big)^2\sinh 2\vartheta. \tag{17.26}$$

We proceed with Eq. (17.22) for the distribution matrix $\hat{F}$, parametrized as in Eq. (17.21). (i) Taking the Nambu trace of the matrix equation (17.22) and (ii) multiplying it by $\hat{\tau}_3$ and then taking the trace, one finds two coupled kinetic equations for the stationary non-equilibrium distribution functions $F_{\mathrm{L(T)}}(\mathbf{r},\epsilon)$, which can be written in the following form [379]:

$$\hat{\partial}_{\mathbf{r}}\big(\mathcal{D}_{\mathrm{L}}\hat{\partial}_{\mathbf{r}}F_{\mathrm{L}} - D\mathcal{Y}\hat{\partial}_{\mathbf{r}}F_{\mathrm{T}}\big) + D\mathcal{J}_{\mathrm{S}}\hat{\partial}_{\mathbf{r}}F_{\mathrm{T}} = \mathcal{I}^{\mathrm{a}}, \tag{17.27a}$$

$$\hat{\partial}_{\mathbf{r}}(\mathcal{D}_{\mathrm{T}}\hat{\partial}_{\mathbf{r}}F_{\mathrm{T}} + D\mathcal{Y}\hat{\partial}_{\mathbf{r}}F_{\mathrm{L}}) + D\mathcal{J}_{\mathrm{S}}\hat{\partial}_{\mathbf{r}}F_{\mathrm{L}} = \mathcal{I}^{\mathrm{b}}. \tag{17.27b}$$

Here we have introduced energy- and coordinate-dependent diffusion coefficients

$$\mathcal{D}_{\mathrm{L}} = \frac{D}{4}\mathrm{Tr}\left\{\hat{\tau}_0 - \hat{Q}^{\mathrm{R}}\hat{Q}^{\mathrm{A}}\right\} = \frac{D}{2}\left[1 + |\cosh\vartheta|^2 - |\sinh\vartheta|^2\cosh\left(2\mathrm{Im}[\chi]\right)\right], \tag{17.28a}$$

$$\mathcal{D}_{\mathrm{T}} = \frac{D}{4}\mathrm{Tr}\left\{\hat{\tau}_0 - \hat{\tau}_3\hat{Q}^{\mathrm{R}}\hat{\tau}_3\hat{Q}^{\mathrm{A}}\right\} = \frac{D}{2}\left[1 + |\cosh\vartheta|^2 + |\sinh\vartheta|^2\cosh\left(2\mathrm{Im}[\chi]\right)\right], \tag{17.28b}$$

as well as the density of the supercurrent carrying states $\mathcal{J}_{\mathrm{S}}(\mathbf{r},\epsilon)$ and the spectral density $\mathcal{Y}(\mathbf{r},\epsilon)$, defined as

$$\mathcal{J}_{\mathrm{S}} = \frac{1}{4}\mathrm{Tr}\left\{\hat{\tau}_3(\hat{Q}^{\mathrm{R}}\hat{\partial}_{\mathbf{r}}\hat{Q}^{\mathrm{R}} - \hat{Q}^{\mathrm{A}}\hat{\partial}_{\mathbf{r}}\hat{Q}^{\mathrm{A}})\right\} = -\mathrm{Im}\left(\sinh^2\vartheta\;\hat{\partial}_{\mathbf{r}}\chi\right), \tag{17.29}$$

$$\mathcal{Y} = \frac{1}{4}\mathrm{Tr}\left\{\hat{Q}^{\mathrm{R}}\hat{\tau}_3\hat{Q}^{\mathrm{A}}\right\} = \frac{1}{2}|\sinh\vartheta|^2\sinh\left(2\mathrm{Im}[\chi]\right). \tag{17.30}$$

Finally, the right-hand sides of Eqs. (17.27) contain the two "collision integrals," which are due to Andreev-like interactions of non-equilibrium quasiparticles with the order parameter. Both of them are proportional to the transversal component of the distribution function:

$$\mathcal{I}^{\mathrm{a}} = \frac{F_{\mathrm{T}}}{2}\,\mathrm{Tr}\left\{\hat{\tau}_3(\hat{Q}^{\mathrm{R}}\hat{\Delta} + \hat{\Delta}\hat{Q}^{\mathrm{A}})\right\} = 2F_{\mathrm{T}}|\Delta|\mathrm{Re}\left[\sinh\vartheta\,\sin(\theta-\chi)\right], \tag{17.31a}$$

$$\mathcal{I}^{\mathrm{b}} = \frac{F_{\mathrm{T}}}{2}\,\mathrm{Tr}\left\{\hat{Q}^{\mathrm{R}}\hat{\Delta} + \hat{\Delta}\hat{Q}^{\mathrm{A}}\right\} = -2F_{\mathrm{T}}|\Delta|\mathrm{Im}\left[\sinh\vartheta\,\cos(\theta-\chi)\right]. \tag{17.31b}$$

The actual collision integrals associated with the inelastic and electron–phonon interactions are discussed in Section 18.5; see also Kopnin [376]. Equations (17.25), (17.26), and (17.27), together with the spectral quantities (17.28)–(17.31), represent a complete set of equations for disordered superconductors applicable within the static approximation. These equations are accompanied by the self-consistency relation (see Eq. (17.16))

$$\Delta(\mathbf{r}) = \frac{\lambda}{4}\int d\epsilon \left\{\left[\sinh\vartheta\, e^{i\chi} + \sinh\vartheta^* e^{i\chi^*}\right] F_L - \left[\sinh\vartheta\, e^{i\chi} - \sinh\vartheta^* e^{i\chi^*}\right] F_T\right\}, \tag{17.32}$$

resulting in a coordinate-dependent order parameter field $\Delta(\mathbf{r})$.

One should supplement these equations with the boundary conditions that express the current continuity [380–382]. For a tunneling junction with conductance g_T, sandwiched between two metals, the continuity dictates that at the junction

$$\nu_L D_L \check{Q}_L \hat{\partial}_{\mathbf{r}}^{\perp} \check{Q}_L = g_T[\check{Q}_L, \check{Q}_R] = \nu_R D_R \check{Q}_R \hat{\partial}_{\mathbf{r}}^{\perp} \check{Q}_R, \tag{17.33}$$

where L/R denote left/right of the junction, respectively, and $\hat{\partial}_{\mathbf{r}}^{\perp}$ stands for the derivative normal to the junction.

An analytic solution of the system of kinetic equations (17.25)–(17.27) is rarely possible. In general, one has to rely on numerical methods. To find a solution for a given transport problem, one may proceed as follows [378].

(i) Start with a certain $\Delta(\mathbf{r})$. Usually one takes $\Delta = \text{const}$ everywhere in the superconductors and $\Delta = 0$ in the normal metals.
(ii) Solve the Usadel equations (17.25)–(17.26) for the retarded Green function, thus determining the spectral angles $\vartheta(\mathbf{r}, \epsilon)$ and $\chi(\mathbf{r}, \epsilon)$.
(iii) Use these solutions to calculate the spectral kinetic quantities $\mathcal{D}_{L,T}(\mathbf{r}, \epsilon)$, $\mathcal{J}_S(\mathbf{r}, \epsilon)$ and $\mathcal{Y}(\mathbf{r}, \epsilon)$.
(iv) Solve the kinetic equations (17.27) for $F_{L/T}(\mathbf{r}, \epsilon)$.
(v) Calculate a new $\Delta(\mathbf{r})$ from Eq. (17.32) and iterate this procedure until self-consistency is achieved.

Having solved the kinetic equations, one may determine the physical quantities of interest. For example, the electric current is given by the derivative of the action (17.7) with respect to the quantum component of the vector potential (cf. Eq. (14.55.) This way, one finds

$$\begin{aligned}\mathbf{J} &= \frac{e\pi\nu D}{4}\,\mathrm{Tr}\{\hat{\gamma}^q \otimes \hat{\tau}_3 \check{Q}\,\hat{\partial}_{\mathbf{r}}\check{Q}\} = \frac{e\pi\nu D}{4}\,\mathrm{Tr}\{\hat{\tau}_3(\hat{Q}^R\,\hat{\partial}_{\mathbf{r}}\hat{Q}^K + \hat{Q}^K\,\hat{\partial}_{\mathbf{r}}\hat{Q}^A)\} \\ &= \frac{e\pi\nu D}{4}\,\mathrm{Tr}\{\hat{\tau}_3(\hat{Q}^R\,\hat{\partial}_{\mathbf{r}}\hat{Q}^R\hat{F} - \hat{F}\hat{Q}^A\,\hat{\partial}_{\mathbf{r}}\hat{Q}^A + \hat{\partial}_{\mathbf{r}}\hat{F} - \hat{Q}^R\hat{\partial}_{\mathbf{r}}\hat{F}\hat{Q}^A)\}.\end{aligned} \tag{17.34}$$

The first two terms associated with the gradients of the quasiparticle spectrum give the supercurrent $\mathbf{J}_{\mathrm{S}}$. The last two terms proportional to the gradients of the distribution function are the normal component of the current $\mathbf{J}_{\mathrm{N}}$. In terms of the quantities introduced previously, one finds the following for the super and normal current components, respectively:

$$\mathbf{J}_{\mathrm{S}} = \frac{e\nu D}{2} \int \mathrm{d}\epsilon\, F_{\mathrm{L}}(\mathbf{r},\epsilon)\, \mathcal{J}_{\mathrm{S}}(\mathbf{r},\epsilon); \qquad \mathbf{J}_{\mathrm{N}} = \frac{e\nu}{2} \int \mathrm{d}\epsilon\, \mathcal{D}_{\mathrm{T}}(\mathbf{r},\epsilon)\, \partial_{\mathbf{r}} F_{\mathrm{T}}(\mathbf{r},\epsilon), \tag{17.35}$$

where we used that $\mathrm{Tr}\{\hat{Q}^{\mathrm{R}}\partial_{\mathbf{r}}\hat{Q}^{\mathrm{R}}\} = \mathrm{Tr}\{\hat{\tau}_3\hat{Q}^{\mathrm{R}}\hat{Q}^{\mathrm{A}}\} = 0$. Notice that the normal component is the same as Eq. (14.55) with the substitution $D \to \mathcal{D}_{\mathrm{T}}$. Indeed, it is only the component of the distribution function that is even in energy, that is, F_{T}, which contributes to the normal current (14.55). As opposed to the normal current, the supercurrent is given by the odd component of the distribution function F_{L} and therefore may exist even in equilibrium, where $F_{\mathrm{L}} = \tanh \epsilon/2T$.

17.4 SNS System

As an example of an application of the general formalism developed in the previous section, we consider a superconductor–normal metal–superconductor (SNS) system. We shall assume that the normal part (N) is represented by a quasi-one-dimensional nano-wire with cross-section diameter less than the superconducting coherence length $\xi_0 = \sqrt{D/\Delta(0)}$. This allows us to treat the problem as 1d, that is, the spatial dependence is restricted to the x-coordinate only. The superconducting leads (S) are assumed to be bulk and 3d, which allows us to fix their order parameter to be $\Delta = \mathrm{const}$ and avoid solving the self-consistency condition. That is, the normal nano-wire does not damage the superconductivity in bulk leads. In the normal wire there is no Cooper channel attraction, that is, $\lambda = 0$ and thus, according to the self-consistency equation (17.16), the order parameter is absent, $\Delta = 0$. Nevertheless, the leads induce superconductivity inside the normal region, through the coordinate-dependent $\check{Q}$-matrix. This phenomenon is known as the *proximity* effect.

We start from the case where both superconducting leads have the same phase of the order parameter, which we choose as $\theta = 0$. The $\hat{Q}^{\mathrm{R}}(x,\epsilon)$-matrix inside the wire, that is, for $|x| < L/2$, may be parametrized as in Eq. (17.24a) with $\chi = 0$ and thus $\partial_x \chi = 0$. As a result, the Usadel equation (17.26) simplifies considerably and reads as

$$D\,\partial_x^2 \vartheta(x,\epsilon) + 2\mathrm{i}\epsilon \sinh \vartheta(x,\epsilon) = 0. \tag{17.36}$$

Indeed, $\Delta(x) = 0$ for $|x| < L/2$. At the interfaces with the superconducting leads at $x = \pm L/2$, this equation is supplemented by the boundary conditions

$\vartheta(\pm L/2, \epsilon) = \vartheta_{BCS}(\epsilon)$, where $\tanh\vartheta_{BCS}(\epsilon) = \Delta/\epsilon$; see Eqs. (17.15) and (17.24). Having solved Eq. (17.36) for the angle $\vartheta(x, \epsilon)$, one finds the local density of states as $\nu(x, \epsilon) = \nu\mathrm{Re}\big[\cosh\vartheta(x, \epsilon)\big]$; see Eq. (17.18).

It is convenient to perform rotation $\vartheta(x, \epsilon) = \mathrm{i}\pi/2 - \zeta(x, \epsilon)$ such that Eq. (17.36) becomes real and allows the straightforward integration

$$\sqrt{\frac{\epsilon}{E_{\mathrm{Th}}}} = \int_{\zeta_{\mathrm{BCS}}}^{\zeta_0} \frac{d\zeta}{\sqrt{\sinh\zeta_0 - \sinh\zeta}} \equiv K(\zeta_0, \epsilon), \tag{17.37}$$

where $E_{\mathrm{Th}} = D/L^2$ is the Thouless energy of the wire, $\zeta_0 = \zeta(0, \epsilon)$, and $\sinh\zeta_{\mathrm{BCS}} = \epsilon/\sqrt{\Delta^2 - \epsilon^2}$. Equation (17.37) defines ζ_0 as a function of the energy ϵ. Knowing $\zeta_0(\epsilon)$ one determines the density of states in the middle of the wire as $\nu(0, \epsilon) = \nu\mathrm{Im}[\sinh\zeta_0(\epsilon)]$.

In the limit of a long wire, $L \gg \xi_0$, the density of states is distorted from its normal value ν in the deep sub-gap limit, $\epsilon \ll \Delta$. One may approximate thus $\zeta_{\mathrm{BCS}} \approx 0$ and the function on the right-hand side of Eq. (17.37) is essentially energy independent, $K(\zeta_0, \epsilon) \approx K(\zeta_0, 0)$. It exhibits the maximum $K_{\max} = K(\zeta_0^{\max}) \approx 1.75$ at $\zeta_0^{\max} \approx 1.5$. On the other hand, the left-hand side of Eq. (17.37) exceeds $K_{\max}$ for $\epsilon > (K_{\max})^2 E_{\mathrm{Th}} = \epsilon_g$. Thus, for all energies $\epsilon < \epsilon_g$, Eq. (17.37) has only a real solution for ζ_0 and therefore $\nu(0, \epsilon) \equiv 0$, since $\nu(0, \epsilon) \propto \mathrm{Im}\big[\sinh\zeta_0\big]$. For $\epsilon > \epsilon_g$ the function ζ_0 becomes complex and gives a finite density of states. Right above the gap, $0 < \epsilon - \epsilon_g \ll \epsilon_g$, one finds with the help of Eq. (17.37)

$$\nu(\epsilon) \approx 3.7\delta^{-1}\sqrt{\frac{\epsilon}{\epsilon_g} - 1}, \tag{17.38}$$

where $\nu(\epsilon) = \int dx\, \nu(x, \epsilon)$ is the global density of states, integrated over the volume of the wire, and $\delta = 1/(\nu L)$ is its level spacing. We found thus an induced mini-gap inside the long normal wire. The density of states exhibits square root nonanalytic behavior above the mini-gap. Note that since $\epsilon_g \propto E_{\mathrm{Th}} \ll \Delta$, the approximation $\zeta_{\mathrm{BCS}}(\epsilon_g) \approx 0$ is well justified.

In the opposite limit of a short wire, $L \ll \xi_0$, or equivalently $E_{\mathrm{Th}} \gg \Delta$, Eq. (17.37) is still applicable. However, one must keep the full energy dependence of $\zeta_{\mathrm{BCS}}(\epsilon)$. One may show that the energy gap is given by $\epsilon_g = \Delta - \Delta^3/8E_{\mathrm{Th}}^2$ and is only slightly smaller than the bulk gap Δ. This is natural, since the proximity effect for the short wire is expected to be strong. Immediately above the induced gap, the density of states again exhibits square root nonanalyticity, Fig. 17.1. The coefficient in front of it, however, is large, $\nu(\epsilon) \sim \delta^{-1}(E_{\mathrm{Th}}/\Delta)^2\sqrt{\epsilon/\epsilon_g - 1}$, [383].

We turn now to the case of a finite phase difference θ between the order parameters of the two S leads. This leads to the boundary condition $\chi(L/2, \epsilon) - \chi(-L/2, \epsilon) = \theta$. Our goal is to calculate the Josephson supercurrent $J_S(\theta)$, which flows through the normal wire under these conditions. For the step function order

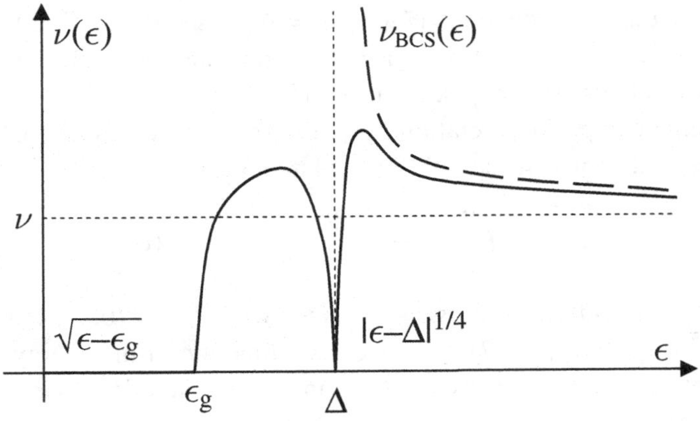

Figure 17.1 Integrated density of states in a short SNS system, after [383]. Above the proximity gap ϵ_g it exhibits square-root nonanalyticity, while at large energy it approaches the BCS result (17.18). One also finds a soft gap $\nu(\epsilon) \sim |\epsilon - \Delta|^{1/4}$ around the bulk order parameter Δ.

parameter, $\Delta(x) = \Delta$ for $|x| > L/2$ and $\Delta = 0$ for $|x| < L/2$, Eqs. (17.25) and (17.26) acquire the form

$$D\,\partial_x\big(\sinh^2\vartheta\,\partial_x\chi\big) = 0, \tag{17.39a}$$

$$D\,\partial_x^2\vartheta + 2i\epsilon\sinh\vartheta = \frac{D}{2}(\partial_x\chi)^2\sinh 2\vartheta. \tag{17.39b}$$

The boundary conditions for ϑ are $\vartheta(\pm L/2,\epsilon) = \vartheta_{\mathrm{BCS}}(\epsilon)$, while for the phase $\chi(L/2,\epsilon) - \chi(-L/2,\epsilon) = \theta$. For the short wire, $L \ll \xi_0$, the second term on the left-hand side of (17.39b) is smaller than the gradient term by $\epsilon/E_{\mathrm{Th}} \ll 1$ and thus may be neglected. Equation (17.39a) allows for the first integral of motion $\sinh^2\vartheta\,\partial_x\chi = -\mathcal{J}$ (cf. Eq. (17.29)). One may thus eliminate $\partial_x\chi$ from Eq. (17.39b) and find $\partial_x^2\vartheta = \mathcal{J}^2\cosh\vartheta/\sinh^3\vartheta = -(\mathcal{J}^2/2)\,\partial_\vartheta\sinh^{-2}\vartheta$. This "Newtonian" equation may be exactly integrated:

$$\cosh\vartheta(x,\epsilon) = \cosh\vartheta_0\cosh\left(\frac{\mathcal{J}x}{\sinh\vartheta_0}\right), \tag{17.40}$$

where $\vartheta_0 = \vartheta(0,\epsilon)$. Knowing $\vartheta(x,\epsilon)$, one inserts it back into the first integral of (17.39a), $\theta = \int_{-L/2}^{L/2}\mathrm{d}x\,\partial_x\chi = -\mathcal{J}\int_{-L/2}^{L/2}\mathrm{d}x/\sinh^2\vartheta(x,\epsilon)$, to find

$$\tan(\theta/2) = -\frac{1}{\sinh\vartheta_0}\tanh\left(\frac{\mathcal{J}}{2\sinh\vartheta_0}\right). \tag{17.41}$$

This equation along with Eq. (17.40) taken at the NS interfaces, $x = \pm L/2$, constitute the system of two algebraic equations for two unknown quantities: $\mathcal{J}$

and ϑ_0. Such an algebraic problem may be straightforwardly solved, resulting in $\mathcal{J}(\epsilon,\theta) = -(2/L)\sinh\vartheta_0 \operatorname{arctanh}\big[\sinh\vartheta_0 \tan(\theta/2)\big]$, whereas the angle at $x = 0$ is $\sinh\vartheta_0 = \sinh\vartheta_{\mathrm{BCS}}/\sqrt{1+\tan^2(\theta/2)\cosh^2\vartheta_{\mathrm{BCS}}}$, where the boundary angle is $\cosh\vartheta_{\mathrm{BCS}} = \epsilon/\sqrt{\epsilon^2-\Delta^2}$. Knowing $\mathcal{J}(\epsilon,\theta)$, one finds the Josephson supercurrent with the help of Eq. (17.35) as

$$J_{\mathrm{S}}(\theta) = \frac{e\nu D}{2}\int_0^\infty d\epsilon \, \tanh\left(\frac{\epsilon}{2T}\right) \operatorname{Im}\mathcal{J}(\epsilon,\theta). \tag{17.42}$$

Using the obtained solution for $\mathcal{J}(\epsilon,\theta)$, one finds that

$$\operatorname{Im}\mathcal{J}(\epsilon,\theta) = \frac{1}{L}\frac{\pi\Delta\cos(\theta/2)}{\sqrt{\epsilon^2-\Delta^2\cos^2(\theta/2)}} \tag{17.43}$$

for $\Delta\cos(\theta/2) < \epsilon < \Delta$, and $\operatorname{Im}\mathcal{J}(\epsilon,\theta) = 0$ otherwise. Employing Eqs. (17.42) and (17.43), one arrives at the result derived by Kulik and Omelyanchuk [384] for the zero-temperature Josephson current of the short diffusive SNS junction:

$$J_{\mathrm{S}}(\theta) = \frac{\pi g_{\mathrm{D}}\Delta}{e}\cos(\theta/2)\operatorname{arctanh}\big[\sin(\theta/2)\big], \tag{17.44}$$

where $g_{\mathrm{D}} = e^2\nu D/L$ is the conductance of the normal wire. The phase dependence is 2π-periodic with $J_{\mathrm{S}}(0) = J_{\mathrm{S}}(\pi) = 0$ and has a logarithmical divergent derivative at $\theta = \pi$.

17.5 Collective Modes of Disordered Superconductors

We now consider the spectrum of collective modes in disordered superconductors. They are combined fluctuations of the complex order parameter Δ, the $\check{Q}$-matrix (which encompasses both the quasiparticle spectrum and their distribution function) and, for charged superconductors, the scalar potential φ. The problem is somewhat similar to the collective modes of the collisionless plasma considered in Chapter 7 and described by the combined kinetic and Poisson equations. However, in the case of the superconductor one has to add the dynamics of the order parameter, governed by the self-consistency condition (17.16), and the associated change in the quasiparticle spectrum, governed by the retarded (or advanced) component of the Usadel equation. To find the spectrum of a small oscillation we linearize this system of equations with respect to small deviations, that is, put $\check{Q} = \check{\Lambda} + \delta\check{Q}$, while $\Delta = \Delta + \delta\Delta$, and restrict ourselves to the linear order in $\delta\check{Q}$, $\delta\Delta$, and φ.

Taking into account that $[\epsilon\hat{\tau}_3, \hat{\Lambda}^{\mathrm{K}}] + [\hat{\Delta}, \hat{\Lambda}^{\mathrm{K}}] = 0$, the linearized Keldysh component of the Usadel equation (17.19) for $\hat{Q}^{\mathrm{K}}_{\epsilon\epsilon'} = \hat{\Lambda}^{\mathrm{K}}_{\epsilon}\delta_{\epsilon\epsilon'} + \delta\hat{Q}^{\mathrm{K}}_{\epsilon\epsilon'}$ takes the form

$$\begin{aligned} &D\hat{\Lambda}^{\mathrm{R}}\nabla_{\mathbf{r}}^{2}\delta\hat{Q}^{\mathrm{K}}+\mathrm{i}[\epsilon\hat{\tau}_{3},\delta\hat{Q}^{\mathrm{K}}]+\mathrm{i}[\hat{\Delta},\delta\hat{Q}^{\mathrm{K}}] \\ &\quad=-\mathrm{i}[\delta\hat{\Delta},\hat{\Lambda}^{\mathrm{K}}]+\mathrm{i}[\hat{\varphi},\hat{\Lambda}^{\mathrm{K}}]-D\hat{\Lambda}^{\mathrm{K}}\nabla_{\mathbf{r}}^{2}\delta\hat{Q}^{\mathrm{A}}, \end{aligned} \tag{17.45}$$

where $\delta\hat{\Delta}(\mathbf{r},t)=\delta\Delta(\mathbf{r},t)\hat{\tau}_{+}-\delta\Delta^{*}(\mathbf{r},t)\hat{\tau}_{-}$. It should be supplemented with the linearized self-consistency condition; see Eq. (17.16),

$$\delta\Delta=\frac{\pi\lambda}{2}\,\mathrm{Tr}\{\hat{\tau}_{-}\,\delta\hat{Q}^{\mathrm{K}}\} \tag{17.46}$$

and, for charged Cooper pairs, with the linearized Poisson equation; see Eq. (17.23),

$$(4\pi e^{2})^{-1}\nabla_{\mathbf{r}}^{2}\varphi=\nu\varphi+\frac{\pi\nu}{4}\,\mathrm{Tr}\{\hat{\tau}_{0}\,\delta\hat{Q}^{\mathrm{K}}\}. \tag{17.47}$$

To solve the linearized Keldysh Usadel equation, one needs to find first the advanced component $\hat{Q}^{\mathrm{A}}=\hat{\Lambda}^{\mathrm{A}}+\delta\hat{Q}^{\mathrm{A}}$ of the $\hat{Q}$-matrix. Since the latter appears in Eq. (17.45) with the gradient square, one may omit gradient terms in the corresponding advanced component of the Usadel equation. (We are seeking for the lowest nonvanishing term in $D\mathbf{q}^{2}$.) It thus takes the form (cf. Eq. (17.12))

$$\mathrm{i}[\epsilon\hat{\tau}_{3},\delta\hat{Q}^{\mathrm{A}}]+\mathrm{i}[\hat{\Delta},\delta\hat{Q}^{\mathrm{A}}]=-\mathrm{i}[\delta\hat{\Delta},\hat{\Lambda}^{\mathrm{A}}]+\mathrm{i}[\hat{\varphi},\hat{\Lambda}^{\mathrm{A}}]. \tag{17.48}$$

One may distinguish between the longitudinal and transverse variations of the order parameter. The former changes the amplitude of the order parameter $\delta\Delta^{\mathrm{L}}=\delta|\Delta|$, while the latter changes its phase $\delta\Delta^{\mathrm{T}}=|\Delta|\delta\theta$. Here $\delta\Delta^{\mathrm{L(T)}}(\mathbf{r},t)$ are two *real* functions in space and time. In the adopted convention where $\hat{\Delta}=\mathrm{i}\hat{\tau}_{2}\Delta$, the matrix forms of the corresponding variations are $\delta\hat{\Delta}^{\mathrm{L}}=\mathrm{i}\hat{\tau}_{2}\delta\Delta^{\mathrm{L}}(\mathbf{r},t)$, while $\delta\hat{\Delta}^{\mathrm{T}}=\mathrm{i}\hat{\tau}_{1}\delta\Delta^{\mathrm{T}}(\mathbf{r},t)$. Since the self-consistency relation (17.17) fixes the modulus of the order parameter, one expects (and indeed finds) that the mode of the longitudinal fluctuations has a gap. The corresponding gap turns out to be $2|\Delta|$. On the other hand, the phase θ of the order parameter is not fixed by the self-consistency equation. As a result, the corresponding transversal mode may appear to be gapless. Yet, in the charged system, gradients of the phase are associated with charged supercurrents, which lead to a charge buildup and thus cost the Coulomb energy. As a result, at $T=0$ the corresponding mode acquires a plasmon gap; see Chapter 7. The situation is more interesting at $T\lesssim T_{\mathrm{c}}$, where $\Delta\ll T$. In this case there is a substantial thermal population of quasiparticle states. As a result, the induced supercurrents may be counterbalanced by the normal currents of the thermally excited quasiparticles, allowing for charge neutrality to be preserved. This leads to the gapless Carlson–Goldman [386] collective mode, which we consider in what follows.[5]

[5] At the NS interface the neutralizing normal current flows in the N region and thus is spatially separated from the superconductor. As a result the gapless (phason) mode exists even at $T=0$, [387].

As a warmup exercise we consider a neutral superconductor, that is, $\varphi = 0$, where the transverse mode is expected to be gapless at all temperatures. If the order parameter variations are restricted to the transverse form, that is, $\delta\hat{\Delta} = \mathrm{i}\hat{\tau}_1\delta\Delta^{\mathrm{T}}(\mathbf{r},t)$, one may seek for a solution of the advanced equation (17.48) in the following form:

$$\delta\hat{Q}^{\mathrm{A}}_{\epsilon\epsilon'}(\mathbf{r}) = \hat{\tau}_0 r^{\mathrm{A}}_{\epsilon\epsilon'}(\mathbf{r}) + \hat{\tau}_1 s^{\mathrm{A}}_{\epsilon\epsilon'}(\mathbf{r}). \tag{17.49}$$

We take into account that, $\mathrm{i}[\epsilon\hat{\tau}_3, \hat{\tau}_1 s^{\mathrm{A}}] = -\hat{\tau}_2\{\epsilon, s^{\mathrm{A}}\}_+ = -\hat{\tau}_2(\epsilon+\epsilon')s^{\mathrm{A}}_{\epsilon\epsilon'}$, along with the fact that $\Lambda^{\mathrm{A}}_\epsilon = -\hat{\tau}_3 u^*_\epsilon - \mathrm{i}\hat{\tau}_2 v^*_\epsilon$ and thus the right-hand side of Eq. (17.48) is $-\mathrm{i}[\delta\hat{\Delta}, \hat{\Lambda}^{\mathrm{A}}] = \delta\Delta^{\mathrm{T}}_{\epsilon-\epsilon'}\big[\,\mathrm{i}\hat{\tau}_2(u^*_\epsilon + u^*_{\epsilon'}) + \hat{\tau}_3(v^*_\epsilon + v^*_{\epsilon'})\big]$. Collecting coefficients in front of the $\mathrm{i}\hat{\tau}_2$ and $\hat{\tau}_3$ Nambu matrices, one obtains two linear algebraic equations for $s^{\mathrm{A}}_{\epsilon\epsilon'}$ and $r^{\mathrm{A}}_{\epsilon\epsilon'}$:

$$\mathrm{i}(\epsilon+\epsilon')s^{\mathrm{A}}_{\epsilon\epsilon'} = \delta\Delta^{\mathrm{T}}_\omega(u^*_\epsilon + u^*_{\epsilon'});$$
$$\mathrm{i}(\epsilon-\epsilon')r^{\mathrm{A}}_{\epsilon\epsilon'} + 2\mathrm{i}\Delta s^{\mathrm{A}}_{\epsilon\epsilon'} = \delta\Delta^{\mathrm{T}}_\omega(v^*_\epsilon + v^*_{\epsilon'}),$$

where $\omega = \epsilon - \epsilon'$. Their solution is

$$s^{\mathrm{A}}_{\epsilon\epsilon'} = -\mathrm{i}\delta\Delta^{\mathrm{T}}_\omega\,\frac{u^*_\epsilon + u^*_{\epsilon'}}{\epsilon+\epsilon'}; \qquad r^{\mathrm{A}}_{\epsilon\epsilon'} = \mathrm{i}\delta\Delta^{\mathrm{T}}_\omega\,\frac{v^*_\epsilon - v^*_{\epsilon'}}{\epsilon+\epsilon'}, \tag{17.50}$$

where we used that $\epsilon v^*_\epsilon = \Delta u^*_\epsilon$ (cf. Eq. (17.15)).[6]

Having found $\delta\hat{Q}^{\mathrm{A}}$, we turn now to the Keldysh Usadel component (17.45). Its solution may be sought for in the same form as (17.49) with $s^{\mathrm{A}}_{\epsilon\epsilon'}, r^{\mathrm{A}}_{\epsilon\epsilon'} \to s^{\mathrm{K}}_{\epsilon\epsilon'}, r^{\mathrm{K}}_{\epsilon\epsilon'}$. Substituting it in Eq. (17.45) and taking into account that, due to FDT, $\Lambda^{\mathrm{K}}_\epsilon = F^{\mathrm{eq}}_\epsilon(\Lambda^{\mathrm{R}}_\epsilon - \Lambda^{\mathrm{A}}_\epsilon) = F^{\mathrm{eq}}_\epsilon\big[\hat{\tau}_3(u_\epsilon + u^*_\epsilon) + \mathrm{i}\hat{\tau}_2(v_\epsilon + v^*_\epsilon)\big]$, one obtains two linear equations for $s^{\mathrm{K}}_{\epsilon\epsilon'}$ and $r^{\mathrm{K}}_{\epsilon\epsilon'}$:

$$\begin{aligned}
&\big[D\mathbf{q}^2 u_\epsilon - \mathrm{i}(\epsilon+\epsilon')\big]\,s^{\mathrm{K}}_{\epsilon\epsilon'} + D\mathbf{q}^2 v_\epsilon r^{\mathrm{K}}_{\epsilon\epsilon'} = \delta\Delta^{\mathrm{T}}_\omega\big[F^{\mathrm{eq}}_\epsilon(u_\epsilon + u^*_\epsilon) + F^{\mathrm{eq}}_{\epsilon'}(u_{\epsilon'} + u^*_{\epsilon'})\big]\\
&\qquad\qquad - D\mathbf{q}^2 F^{\mathrm{eq}}_\epsilon\big[(u_\epsilon + u^*_\epsilon)s^{\mathrm{A}}_{\epsilon\epsilon'} + (v_\epsilon + v^*_\epsilon)r^{\mathrm{A}}_{\epsilon\epsilon'}\big];\\
&\big[D\mathbf{q}^2 v_\epsilon - 2\mathrm{i}\Delta\big]\,s^{\mathrm{K}}_{\epsilon\epsilon'} + \big[D\mathbf{q}^2 u_\epsilon - \mathrm{i}(\epsilon-\epsilon')\big]\,r^{\mathrm{K}}_{\epsilon\epsilon'}\\
&= \delta\Delta^{\mathrm{T}}_\omega\big[F^{\mathrm{eq}}_\epsilon(v_\epsilon + v^*_\epsilon) + F^{\mathrm{eq}}_{\epsilon'}(v_{\epsilon'} + v^*_{\epsilon'})\big] - D\mathbf{q}^2 F^{\mathrm{eq}}_\epsilon\big[(u_\epsilon + u^*_\epsilon)r^{\mathrm{A}}_{\epsilon\epsilon'} + (v_\epsilon + v^*_\epsilon)s^{\mathrm{A}}_{\epsilon\epsilon'}\big].
\end{aligned}$$

Notice that only the $s^{\mathrm{K}}_{\epsilon\epsilon'}$ component enters the self-consistency condition (17.46). (The $r^{\mathrm{K}}_{\epsilon\epsilon'}$ component enters the Poisson equation (17.47) and is necessary for the consideration of charged superconductors.) It is convenient to solve this system by first excluding the spatial gradients. One then finds for $r^{\mathrm{K}}_{\epsilon\epsilon'}$

$$r^{\mathrm{K}}_{\epsilon\epsilon'}(\mathbf{q}) = -\frac{\mathrm{i}\delta\Delta^{\mathrm{T}}_\omega(\mathbf{q})}{\epsilon+\epsilon'}\big[F^{\mathrm{eq}}_\epsilon(v_\epsilon + v^*_\epsilon) - F^{\mathrm{eq}}_{\epsilon'}(v_{\epsilon'} + v^*_{\epsilon'})\big]. \tag{17.51}$$

[6] It is instructive to verify that $\hat{\Lambda}^{\mathrm{A}}_\epsilon\delta\hat{Q}^{\mathrm{A}}_{\epsilon\epsilon'} + \delta\hat{Q}^{\mathrm{A}}_{\epsilon\epsilon'}\hat{\Lambda}^{\mathrm{A}}_{\epsilon'} = 0$ and therefore the nonlinear constraint $(\hat{Q}^{\mathrm{A}})^2 = \hat{1}$, Eq. (17.11), is satisfied to linear order in $\delta\Delta^{\mathrm{T}}$.

Keeping now the terms to first order in $D\mathbf{q}^2$, one can solve for $s^{\mathrm{K}}_{\epsilon\epsilon'}(\mathbf{q})$ and find

$$s^{\mathrm{K}}_{\epsilon\epsilon'}(\mathbf{q}) = \frac{\mathrm{i}\delta\Delta^{\mathrm{T}}_{\omega}(\mathbf{q})}{\epsilon+\epsilon'}\Big[F^{\mathrm{eq}}_{\epsilon}(u_\epsilon + u^*_\epsilon) + F^{\mathrm{eq}}_{\epsilon'}(u_{\epsilon'} + u^*_{\epsilon'}) + \frac{\mathrm{i}D\mathbf{q}^2}{\epsilon+\epsilon'}$$
$$\times\left\{F^{\mathrm{eq}}_{\epsilon}(u_\epsilon + u^*_\epsilon)u^*_{\epsilon'} - F^{\mathrm{eq}}_{\epsilon'}(u_{\epsilon'} + u^*_{\epsilon'})u_\epsilon + F^{\mathrm{eq}}_{\epsilon}(v_\epsilon + v^*_\epsilon)v^*_{\epsilon'} - F^{\mathrm{eq}}_{\epsilon'}(v_{\epsilon'} + v^*_{\epsilon'})v_\epsilon\right\}\Big], \tag{17.52}$$

where we used $u_\epsilon^2 - v_\epsilon^2 = (u^*_\epsilon)^2 - (v^*_\epsilon)^2 = 1$.

Finally, one substitutes $\delta\hat{Q}^{\mathrm{K}}_{\epsilon\epsilon'} = \hat{\tau}_0 r^{\mathrm{K}}_{\epsilon\epsilon'} + \hat{\tau}_1 s^{\mathrm{K}}_{\epsilon\epsilon'}$ into the self-consistency condition (17.46). The latter takes the form $\mathrm{i}\delta\Delta^{\mathrm{T}}_{\omega} = (\lambda/4)\int d\epsilon\, s^{\mathrm{K}}_{\epsilon,\epsilon-\omega}$. Since both parts of this relation are $\sim \delta\Delta^{\mathrm{T}}_{\omega}(\mathbf{q})$, its compatibility dictates the dispersion relation $\omega(\mathbf{q})$ of the collective oscillation mode. Putting first $\mathbf{q} = 0$ and $\omega = 0$ (i.e. $\epsilon = \epsilon'$) in Eq. (17.52) and comparing it with Eq. (17.17), one observes that the self-consistency relation is satisfied identically. This is a manifestation of the fact that a static, spatially uniform variation of the phase $\delta\theta = \delta\Delta^{\mathrm{T}}/\Delta$ of the order parameter is consistent with the gap equation (17.17). The remaining finite-frequency and finite-wavenumber part of the self-consistency relation takes the form

$$\left[\omega^2\, a(\omega) - D\mathbf{q}^2\, b(\omega)\right]\delta\Delta^{\mathrm{T}}_{\omega}(\mathbf{q}) = 0, \tag{17.53}$$

where $\omega^2 a(\omega)$ is the ϵ-integral of the first line in Eq. (17.52) with the subtracted static part

$$a(\omega) = \frac{1}{\omega^2}\int_{\Delta}^{\infty}\frac{d\epsilon\, F^{\mathrm{eq}}_{\epsilon}\,\epsilon}{\sqrt{\epsilon^2-\Delta^2}}\left[\frac{1}{2\epsilon-\omega} + \frac{1}{2\epsilon+\omega} - \frac{1}{\epsilon}\right] = \int_{\Delta}^{\infty} d\epsilon\,\frac{\tanh\epsilon/2T}{\sqrt{\epsilon^2-\Delta^2}}\frac{1}{4\epsilon^2-\omega^2} \tag{17.54}$$

and $-D\mathbf{q}^2\, b(\omega)$ is the ϵ-integral of the second line in Eq. (17.52). This latter integral requires some care. In particular, one should pay attention to the interval $\Delta < \epsilon < \Delta + \omega$, where $(u_{\epsilon'} + u^*_{\epsilon'})u_\epsilon = 0$ (remember $\epsilon' = \epsilon - \omega$), but $(u_\epsilon + u^*_\epsilon)u^*_{\epsilon'} = 2\mathrm{i}\epsilon\epsilon'/(\sqrt{\epsilon^2-\Delta^2}\sqrt{\Delta^2-\epsilon'^2})$ is pure imaginary! In the same way $(v_{\epsilon'} + v^*_{\epsilon'})v_\epsilon = 0$, while $(v_\epsilon + v^*_\epsilon)v^*_{\epsilon'} = 2\mathrm{i}\Delta^2/(\sqrt{\epsilon^2-\Delta^2}\sqrt{\Delta^2-\epsilon'^2})$ is pure imaginary. This interval (and similarly $-\Delta < \epsilon < -\Delta + \omega$) provides thus a real part to $b(\omega)$ (together with an imaginary i in front of $D\mathbf{q}^2$ in Eq. (17.52)). In the limit of small frequency $\omega \ll 2\Delta$ one finds

$$\operatorname{Re} b(0) = \frac{1}{2}\int_{\Delta}^{\Delta+\omega} d\epsilon\,\frac{F^{\mathrm{eq}}_{\epsilon}}{4\epsilon^2}\,\frac{2\epsilon^2 + 2\Delta^2}{\sqrt{\epsilon^2-\Delta^2}\sqrt{\Delta^2-(\epsilon-\omega)^2}} \approx \frac{F^{\mathrm{eq}}_{\Delta}}{4\Delta}\int_{\Delta}^{\Delta+\omega}\frac{d\epsilon}{\sqrt{\epsilon-\Delta}\sqrt{\Delta+\omega-\epsilon}}. \tag{17.55}$$

The amazing observation is that, despite being restricted to the interval of length ω, the last integral is simply equal to π. As a result, one finds that $\operatorname{Re} b(0) = (\pi/4\Delta)$

$\times \tanh(\Delta/2T)$. Notice that if we worked in the Wigner transformed representation, this part of the integral would be missing. Instead, one is left with ϵ-integrals that are divergent at $|\epsilon| = \Delta$. Such inapplicability of the Wigner transformation is due to the singular nature of the density of states. It may be mitigated by magnetic impurities or by taking $T > T_c$. Contribution to $b(\omega)$ from the outside of the interval $\Delta < \epsilon < \Delta + \omega$ is imaginary and is given by

$$\operatorname{Im} b(0) = -\frac{1}{2}\int_{\Delta+\omega}^{\infty} d\epsilon \, \frac{F_\epsilon^{\rm eq} - F_{\epsilon-\omega}^{\rm eq}}{4\epsilon^2} \, \frac{2\epsilon^2 + 2\Delta^2}{\sqrt{\epsilon^2 - \Delta^2}\sqrt{(\epsilon-\omega)^2 - \Delta^2}} . \tag{17.56}$$

One may evaluate now $a(0)$ and $\operatorname{Im} b(0)$ in the two limiting cases: (i) $T = 0$ where $F_\epsilon^{\rm eq} = \operatorname{sign}\epsilon$ and (ii) $T_c - T \ll T_c$ where $\Delta(T) \ll T$ and thus in the significant region of energy $F_\epsilon^{\rm eq} \approx \epsilon/2T$.

For $T = 0$ one finds $a(0) = 1/4\Delta^2$, while $\operatorname{Im} b(0) = 0$ and $\operatorname{Re} b(0) \approx \pi/4\Delta$. Employing Eq. (17.53), one finds the acoustic mode with the dispersion relation [388]

$$\omega = \omega_B(\mathbf{q}) = \sqrt{\pi\Delta(0)D}\, q. \tag{17.57}$$

This is essentially the Bogoliubov mode of neutral superfluid discussed in Chapter 8. At $T = 0$ there are no thermally excited quasiparticles and thus no Landau damping of the collective mode. For a finite $T \ll \Delta$ one finds $\operatorname{Im} b(0) \sim -\omega e^{-\Delta/T}$ and thus the Landau damping is exponentially suppressed. The speed of sound appears to be dependent on both the diffusion constant and the amplitude of the order parameter $c = \sqrt{\pi\Delta(0)D}$.

For $T_c - T \ll T_c$ one finds $a(0) = \pi/16T\Delta$ and $\operatorname{Im} b(0) = -(\omega/8T\Delta)\ln(\Delta/\omega)$, where the logarithmic factor originates due to the near divergence of the integral in Eq. (17.56) at the lower boundary. Taking into account that in this limit $\operatorname{Re} b(0) \approx \pi/8T$, the collective mode dispersion is found as

$$\omega_B(\mathbf{q}) = \sqrt{2\Delta(T)D}\, q - i\,\frac{Dq^2}{2\pi}\ln\frac{\Delta}{Dq^2}. \tag{17.58}$$

As long as $Dq^2 \ll \Delta(T)$ the Landau damping is relatively ineffective and the mode is underdamped. In this limit one also finds $Dq^2 \ll \omega_B(\mathbf{q}) \ll \Delta$, which justifies the approximations made previously. Notice that the acoustic mode exists even in the disordered system, if, of course, the wavelength is greater than the mean free path $q \ll 1/l_{\rm el}$. The latter condition is indeed fulfilled, since we work in the regime where $\Delta \ll 1/\tau_{\rm el}$.

We turn now to the case of a superconductor with charged Cooper pairs (such as a metal). In this case one needs to involve the Poisson equation (17.47) along with the self-consistency relation and keep the scalar potential φ on the right-hand side of the linearized Usadel equations (17.45) and (17.48). One can still look for a solution of the Usadel equations in the form of a linear superposition of the $\hat{\tau}_0$ and $\hat{\tau}_1$ Pauli matrices in Nambu space; see Eq. (17.49). The corresponding coefficients $r^{\mathrm{A(K)}}_{\epsilon\epsilon'}$ and $s^{\mathrm{A(K)}}_{\epsilon\epsilon'}$ are now linear superpositions of terms proportional to $\delta\Delta^{\mathrm{T}}_{\omega}(\mathbf{q})$ and $\varphi_\omega(\mathbf{q})$. For example,

$$s^{\mathrm{K}}_{\epsilon\epsilon'} = \frac{\mathrm{i}\delta\Delta^{\mathrm{T}}_{\omega}}{\epsilon+\epsilon'}\left[F^{\mathrm{eq}}_{\epsilon}(u_\epsilon + u^*_\epsilon) + (\epsilon \leftrightarrow \epsilon')\right] - \frac{\varphi_\omega}{\epsilon+\epsilon'}\left[F^{\mathrm{eq}}_{\epsilon}(v_\epsilon + v^*_\epsilon) - (\epsilon \leftrightarrow \epsilon')\right]; \tag{17.59a}$$

$$r^{\mathrm{K}}_{\epsilon\epsilon'} = -\frac{\varphi_\omega}{\omega}\left[F^{\mathrm{eq}}_{\epsilon}(u_\epsilon + u^*_\epsilon) - (\epsilon \leftrightarrow \epsilon')\right] - \frac{\mathrm{i}\omega\delta\Delta^{\mathrm{T}}_{\omega} - 2\Delta\varphi_\omega}{(\epsilon+\epsilon')\,\omega}\left[F^{\mathrm{eq}}_{\epsilon}(v_\epsilon + v^*_\epsilon) - (\epsilon \leftrightarrow \epsilon')\right], \tag{17.59b}$$

where we have omitted gradient terms $\sim D\mathbf{q}^2$ for brevity. We now substitute these expressions into the self-consistency condition (17.46) and the Poisson equation (17.47), which acquire the form

$$\mathrm{i}\delta\Delta^{\mathrm{T}}_{\omega} = \frac{\pi\lambda}{2}\int\frac{d\epsilon}{2\pi}\, s^{\mathrm{K}}_{\epsilon\epsilon-\omega}; \qquad -\frac{\mathbf{q}^2}{4\pi e^2}\,\varphi_\omega = \nu\varphi_\omega + \frac{\pi\nu}{2}\int\frac{d\epsilon}{2\pi}\, r^{\mathrm{K}}_{\epsilon\epsilon-\omega}. \tag{17.60}$$

Performing the energy integrations as explained previously, one finds two linear homogenous equations for $\delta\Delta^{\mathrm{T}}_{\omega}$ and φ_ω, which may be written as follows [388]:

$$\left[\begin{pmatrix} \omega^2 & 2\mathrm{i}\omega\Delta \\ -2\mathrm{i}\omega\Delta & 4\Delta^2 \end{pmatrix} a(\omega) - \begin{pmatrix} D\mathbf{q}^2 b(\omega) & 0 \\ 0 & D\mathbf{q}^2 c(\omega) - \frac{\mathbf{q}^2}{4\pi e^2 \nu} \end{pmatrix}\right]\begin{pmatrix} \delta\Delta^{\mathrm{T}}_{\omega} \\ \varphi_\omega \end{pmatrix} = 0\,. \tag{17.61}$$

The $(1,1)$ element here is the already familiar condition (17.53), which (in the absence of φ) leads to the dispersion relation of the Bogoliubov mode. The matrix in front of $a(\omega)$ follows from the energy integration of Eqs. (17.59). Notice that the integral of the first term on the right-hand side of Eq. (17.59b) exactly cancels the anomalous term $\nu\phi_\omega$ in the Poisson equation. The fact that the determinant of this matrix is zero means that at $\mathbf{q} = 0$ the physical observables depend only on the gauge-invariant combination, $\mathrm{i}\omega\delta\Delta^{\mathrm{T}}_{\omega} - 2\Delta\varphi_\omega \sim \partial_t\theta + 2\varphi$, see Eq. (17.59b), rather than on the phase of the order parameter θ and the scalar potential φ separately. The only[7] term here we haven't derived explicitly for the lack of space is $c(\omega)$. However, its limiting behavior at $T = 0$ and $T_c - T \ll T_c$ may be understood without any algebra. Indeed, since at $T = 0$ the electronic spectrum has a gap and there are no thermal quasiparticles above the gap, the system is incompressible and does not respond to the gradient of the scalar potential, that is,

[7] The off-diagonal terms in the last matrix in Eq. (17.61) are small in the parameter $D\mathbf{q}^2/\omega$ and thus omitted.

$c(\omega) = 0$. The technical reason is the same which led to $\mathrm{Im}\, b(0) = 0$ at $T = 0$; see Eq. (17.56). On the other hand, at $T_c - T \ll T_c$ there are plenty of excited quasiparticles well above the superconducting gap $\Delta(T)$. Their response to the scalar potential is basically the same as for normal metals. (Superconducting correlations bring only small corrections $\sim \Delta(T)/T \ll 1$.) One thus concludes that $-D\mathbf{q}^2 c(\omega) = D\mathbf{q}^2/(D\mathbf{q}^2 - \mathrm{i}\omega) \approx \mathrm{i}D\mathbf{q}^2/\omega$, see Eq. (14.60), and thus $c(\omega) \approx -\mathrm{i}/\omega$.

We are prepared now to discuss collective modes of charged superconductors. Their dispersion relation is obtained from the requirement that the determinant of the matrix in Eq. (17.61) is zero. At $T = 0$ one has $a = 1/4\Delta^2$, $b = \pi/4\Delta$, and $c = 0$. Demanding a zero determinant, one finds, for example, as $\mathbf{q} \to 0$,

$$\omega(0) = \omega_{\mathrm{p}} = \sqrt{4\pi^2 e^2 \nu \Delta(0) D}. \tag{17.62}$$

This frequency may be called the superconducting plasmon frequency (see Section (7.2)), if one identifies $n_{\mathrm{s}} = \pi\nu\Delta D m$ as the density of superconducting electrons. The mode clearly has a gap, due to the Coulomb energy cost associated with the charge redistribution by supercurrents. This result is quantitatively applicable only if $\omega_{\mathrm{p}} \ll 2\Delta(0)$, which requires a strongly disordered substance.

At T close to T_c there is a large population of quasiparticles, which allow charge neutrality to be maintained. We can thus work in the limit of $e^2 \to \infty$. Recalling that in this case $a = \pi/16T\Delta$, $b = [1 - \mathrm{i}(\omega/\pi\Delta)\ln(\Delta/\omega)]\pi/8T$, and $c = -\mathrm{i}/\omega$, one finds from the condition of the vanishing determinant

$$\omega^2 - \left[2\Delta D\mathbf{q}^2 - \mathrm{i}\frac{\pi\Delta^2\omega}{2T}\right]\left(1 - \mathrm{i}\frac{\omega}{\pi\Delta}\ln\frac{\Delta}{\omega}\right) = 0. \tag{17.63}$$

Assuming for a moment that both imaginary terms are relatively small, one finds for the dispersion relation of the collective mode, first observed by Carlson and Goldman (CG) [386],

$$\omega_{\mathrm{CG}}(\mathbf{q}) = \sqrt{2\Delta(T)D}\,q - \mathrm{i}\frac{\pi\Delta^2(T)}{4T} - \mathrm{i}\frac{Dq^2}{2\pi}\ln\frac{\Delta(T)}{Dq^2}. \tag{17.64}$$

The two imaginary terms here have very different natures. The last one is the Landau damping due to excitation of thermal quasiparticles, already discussed in the context of the Bogoliubov mode (17.58). It makes the CG mode overdamped for $\omega \gtrsim \Delta(T)$. The first one originates from the product of the $D\mathbf{q}^2 b(\omega)$ and $4\Delta^2 a(\omega)$ terms of the matrix (17.61). The former describes spatial variations of the order parameter $\delta\Delta^{\mathrm{T}}$, while the latter is responsible for the quasiparticle number nonconservation in the presence of Δ. Such a nonconservation may be interpreted as the Andreev reflection processes, which convert quasiparticles into quasiholes and back. The corresponding damping term is therefore associated with the Andreev reflection of quasiparticles on the spatial fluctuations of the order parameter. It makes the CG mode overdamped for $\omega \lesssim \Delta^2(T)/T$. In between the Andreev and Landau damped regimes, for $\Delta^2(T)/T \lesssim \omega \lesssim \Delta(T)$, there is a propagating acoustic mode $\omega_{\mathrm{CG}}(\mathbf{q}) = cq$, where the speed of sound

$$c = \sqrt{2\Delta(T)D} \tag{17.65}$$

coincides with the speed of the Bogoliubov mode (17.58). This result was first obtained by Schmid and Schön [389]. For clean superconductors with $T_{\mathrm{c}}\tau_{\mathrm{el}} \gg 1$ the corresponding velocity is given by $c \propto v_{\mathrm{F}}\sqrt{\Delta(T)/T}$, [390]. It is clear that the propagating mode may exist only at a temperature very close to T_{c}, where $\Delta(T) \ll T$.

For the transverse fluctuations of the order parameter $\delta\hat{\Delta} \sim i\hat{\tau}_1$ we were able to find a solution of the Usadel equation for $\delta\hat{Q}^{\mathrm{K}}$ as a superposition of the $\hat{\tau}_0$ and $\hat{\tau}_1$ matrices. The corresponding deviation of the distribution matrix $\hat{F}$, Eq. (17.21), is therefore restricted to the $\hat{\tau}_3$ direction, $\delta\hat{F} \sim \hat{\tau}_3$.[8] This is the reason the corresponding (even in energy) component $F_{\mathrm{T}}(\mathbf{r}, \epsilon)$ is called the transverse distribution. For the longitudinal variations of the order parameter $\delta\hat{\Delta} \sim i\hat{\tau}_2$, the Usadel equation is solved by a linear superposition of the $\hat{\tau}_2$ and $\hat{\tau}_3$ matrices. The corresponding mode, known as the Schmid or longitudinal one, may be derived in exactly the same way as the Bogoliubov mode. It appears to have a gap $\omega_{\mathrm{L}}(0) = 2\Delta$. The corresponding deviation of the distribution matrix $\hat{F}$, Eq. (17.21), is restricted to the $\hat{\tau}_0$ direction, $\delta\hat{F} \sim \hat{\tau}_0$, allowing us to identify the odd in energy part $F_{\mathrm{L}}(\mathbf{r}, \epsilon)$ as the longitudinal one.

17.6 Time-Dependent Ginzburg–Landau Theory

As we saw in Section 17.5, solution of the Usadel equation at $T < T_{\mathrm{c}}$ requires careful account of the coherence factors u_ϵ and v_ϵ. The latter carry information about the singular BCS density of states, which is crucial for the derivation of the transverse (Goldstone) mode; see the discussion after Eq. (17.55). This difficulty is not there for $T > T_{\mathrm{c}}$, or in the presence of magnetic impurities, which allow for superconductivity without the gap in the density of states. In these cases one may considerably simplify the theory by explicitly integrating out the fermionic $\check{Q}$-matrix degrees of freedom. This may be achieved in the Gaussian approximation, which accounts for the diffusive nature of electron motion in metals, but disregards localization effects. As a result of this procedure one obtains the theory of coupled bosonic fields $\Delta(\mathbf{r}, t)$ and $\mathbf{A}(\mathbf{r}, t)$, known as the time-dependent Ginzburg–Landau (TDGL) equation.

We shall restrict ourselves to the $T > T_{\mathrm{c}}$ situation, where the order parameter field $\Delta(\mathbf{r}, t)$ does *not* have a nontrivial expectation value. Yet it exhibits thermal

[8] Indeed, $\delta\hat{Q}^K \sim \Lambda^R\delta\hat{F} - \delta\hat{F}\Lambda^A$. Since $\Lambda^{R(A)}$ has only $\hat{\tau}_3$ and $\hat{\tau}_2$ components, $\delta\hat{F}$ must be restricted to $\hat{\tau}_3$ components only.

fluctuations that manifest themselves in the singular behavior of various observables at $T \to T_c$. While TDGL is suitable for treating non-equilibrium dynamics of the order parameter field, it is based on the assumption that the underlying electronic degrees of freedom (encoded in the $\check{Q}$-matrix) are in a state of local thermal equilibrium. This is possible due to scale separation between slow and long wavelength fields $\Delta(\mathbf{r},t)$, $\mathbf{A}(\mathbf{r},t)$ and the relatively fast electronic $\check{Q}$-matrix field. Indeed, the latter is governed by electron–electron or electron–phonon relaxation times, which are finite at $T \gtrsim T_c$. On the other hand, the dynamics of the former is associated with the Ginzburg–Landau time $\tau_{\mathrm{GL}} \propto (T - T_c)^{-1}$ and the coherence (or correlation) length $\xi \propto \sqrt{D\tau_{\mathrm{GL}}}$, divergent as $T \to T_c$.

The program of integrating out the Gaussian $\check{Q}$-matrix fluctuations is greatly simplified by the choice of a convenient gauge [391]. Here we employ the gauge invariance of the theory, which manifests itself in the invariance of the action (17.7) with respect to the following set of transformations (cf. Eqs. (16.1) and (16.2)):

$$\begin{aligned}
&\check{Q}_{\mathcal{K},tt'}(\mathbf{r}) = \mathrm{e}^{-\mathrm{i}\check{\mathcal{K}}(\mathbf{r},t)}\, \check{Q}_{tt'}(\mathbf{r})\, \mathrm{e}^{\mathrm{i}\check{\mathcal{K}}(\mathbf{r},t')}; \qquad &&\check{\Delta}_{\mathcal{K}}(\mathbf{r},t) = \mathrm{e}^{-\mathrm{i}\check{\mathcal{K}}(\mathbf{r},t)}\check{\Delta}(\mathbf{r},t)\,\mathrm{e}^{\mathrm{i}\check{\mathcal{K}}(\mathbf{r},t)};\\
&\check{\varphi}_{\mathcal{K}}(\mathbf{r},t) = \check{\varphi}(\mathbf{r},t) + \check{\mathcal{T}}_3\partial_t\check{\mathcal{K}}(\mathbf{r},t); &&\check{\mathbf{A}}_{\mathcal{K}}(\mathbf{r},t) = \check{\mathbf{A}}(\mathbf{r},t) + \nabla_{\mathbf{r}}\check{\mathcal{K}}(\mathbf{r},t),
\end{aligned} \tag{17.66}$$

where $\check{\mathcal{K}} = \mathcal{K}^\alpha\hat{\gamma}^\alpha \otimes \hat{\tau}_3$ is a matrix in $\mathrm{K}\otimes\mathrm{N}$ space, characterized by two scalar fields $\mathcal{K}^{\mathrm{cl,q}}(\mathbf{r},t)$. It is convenient to choose them such that the gauge transformed electromagnetic potentials $\varphi^\alpha_{\mathcal{K}}$ and $\mathbf{A}^\alpha_{\mathcal{K}}$ satisfy the $\mathcal{K}$-gauge condition (16.5). (The explicit form of the corresponding gauge fields $\mathcal{K}^\alpha$ is given by Eq. (16.7) with $V \to \varphi$.) As explained in Section 16.1, the advantage of the $\mathcal{K}$-gauge is that (in equilibrium) the action does not contain terms linear in the electromagnetic potentials *and* deviations of the $\check{Q}_{\mathcal{K}}$-matrix from the *metallic* saddle point $\check{\Lambda} = \hat{\Lambda}\otimes\hat{\tau}_3$, where $\hat{\Lambda}_\epsilon$ is given by Eq. (14.25).

Following Eqs. (14.37)–(14.39), it is convenient to parametrize the $\check{Q}_{\mathcal{K}}$-matrix manifold specified by the nonlinear constraint $\check{Q}^2_{\mathcal{K}} = \check{1}$, as [372, 391]

$$\check{Q}_{\mathcal{K}} = \check{\mathcal{U}}^{-1}\circ \mathrm{e}^{-\check{\mathcal{W}}/2}\circ(\hat{\sigma}_3\otimes\hat{\tau}_3)\circ\mathrm{e}^{\check{\mathcal{W}}/2}\circ\check{\mathcal{U}}, \tag{17.67}$$

with the following choice of rotation generators:[9]

$$\check{\mathcal{W}}_{tt'}(\mathbf{r}) = \begin{pmatrix} c_{tt'}(\mathbf{r})\hat{\tau}_+ - c^*_{tt'}(\mathbf{r})\hat{\tau}_- & d^{\mathrm{cl}}_{tt'}(\mathbf{r})\hat{\tau}_0 + \tilde{d}^{\mathrm{cl}}_{tt'}(\mathbf{r})\hat{\tau}_3 \\ d^{\mathrm{q}}_{tt'}(\mathbf{r})\hat{\tau}_0 + \tilde{d}^{\mathrm{q}}_{tt'}(\mathbf{r})\hat{\tau}_3 & \tilde{c}_{tt'}(\mathbf{r})\hat{\tau}_+ - \tilde{c}^*_{tt'}(\mathbf{r})\hat{\tau}_- \end{pmatrix}_{\mathrm{K}}, \qquad \check{\mathcal{U}} = \hat{\mathcal{U}}\otimes\hat{\tau}_0\,. \tag{17.68}$$

[9] Notice that the structure is somewhat different from the weak-localization Cooper channel parametrizations (14.66) and (14.67), which may be traced back to the fact that in that case the saddle point is $\check{\Lambda} = \mathrm{diag}\{\hat{\Lambda}, \hat{\Lambda}^{\mathrm{T}}\}_{\mathrm{TR}}$, while in the present case it is $\check{\Lambda} = \hat{\Lambda}\otimes\hat{\tau}_3$, where $\hat{\tau}_3$ operates in the Nambu space N. The difference is due to the presence of the symplectic symmetry (14.64) in the time-reversal space TR, but not in the Nambu space.

As compared with Eq. (14.39), $\check{\mathcal{W}}$ contains twice as many diffuson modes, which are described by four Hermitian matrices in time space: $\{d^{\mathrm{cl}}, \tilde{d}^{\mathrm{cl}}\}$, representing fluctuations of the longitudinal and transversal components of the distribution matrix (17.21), respectively, and their quantum counterparts $\{d^{\mathrm{q}}, \tilde{d}^{\mathrm{q}}\}$. It also contains the Cooperon modes described by two independent *complex* matrix fields $\{c, \tilde{c}\}$. The Nambu structure of the components of the Keldysh matrix (17.68) is chosen in a way to assure anticommutativity $\{(\hat{\sigma}_3 \otimes \hat{\tau}_3), \check{\mathcal{W}}\}_+ = 0$. One expands now the action (17.7) in powers of the $\check{\mathcal{W}}$ fluctuations and performs a Gaussian integration over the set of independent diffuson and Cooperon modes $\{d^\alpha, \tilde{d}^\alpha, c, \tilde{c}\}$. Some details of this procedure are discussed toward the end of this section. As a result of it one finds

$$\int \mathbf{D}[\check{Q}_{\mathcal{K}}]\; \mathrm{e}^{\,\mathrm{i}S[\check{Q}_{\mathcal{K}}, \Delta_{\mathcal{K}}, \mathbf{A}_{\mathcal{K}}]} = \mathrm{e}^{\,\mathrm{i}S_{\mathrm{eff}}[\Delta_{\mathcal{K}}, \mathbf{A}_{\mathcal{K}}]}, \tag{17.69}$$

where the effective bosonic action consists of three contributions,

$$S_{\mathrm{eff}} = S_{\mathrm{GL}} + S_{\mathrm{S}} + S_{\mathrm{N}}, \tag{17.70}$$

with the subscripts denoting Ginzburg–Landau (GL), the supercurrent (S), and the normal current (N) parts, respectively. We now discuss them one by one.

The time-dependent Ginzburg–Landau part of the action $S_{\mathrm{GL}}[\Delta_{\mathcal{K}}, \mathbf{A}_{\mathcal{K}}]$ originates from pairing two $\check{\mathcal{W}}\check{\Delta}_{\mathcal{K}}$ vertices, Fig. 17.2(a), as well as from pairing them with the $\check{\mathcal{W}}\check{A}_{\mathcal{K}}\check{\mathcal{W}}$ and $\check{\mathcal{W}}\check{A}_{\mathcal{K}}\check{\mathcal{W}}\check{A}_{\mathcal{K}}$ vertices; see the derivation that follows. This leads to[10]

$$S_{\mathrm{GL}} = 2\nu \mathrm{Tr}\left\{\vec{\Delta}^{\dagger}_{\mathcal{K}}(\mathbf{r}, t)\hat{L}^{-1}\vec{\Delta}_{\mathcal{K}}(\mathbf{r}, t)\right\}, \tag{17.71}$$

where $\vec{\Delta}_{\mathcal{K}} = (\Delta^{\mathrm{cl}}_{\mathcal{K}}, \Delta^{\mathrm{q}}_{\mathcal{K}})^{\mathrm{T}}$. The matrix $\hat{L}^{-1}$ has the typical bosonic causality structure in Keldysh space,

$$\hat{L}^{-1} = \begin{pmatrix} 0 & (L^{-1})^{\mathrm{A}} \\ (L^{-1})^{\mathrm{R}} & (L^{-1})^{\mathrm{K}} \end{pmatrix}_{\mathrm{K}}, \tag{17.72}$$

with components given by[11]

$$(L^{-1})^{\mathrm{R(A)}} = \frac{\pi}{8T}\left[\mp\partial_t + D\big(\nabla_{\mathbf{r}} + 2\mathrm{i}e\mathbf{A}^{\mathrm{cl}}_{\mathcal{K}}\big)^2 - \tau_{\mathrm{GL}}^{-1} - \frac{7\zeta(3)}{\pi^3 T_{\mathrm{c}}}|\Delta^{\mathrm{cl}}_{\mathcal{K}}|^2\right], \tag{17.73a}$$

[10] The nonlinear term $\sim \Delta^{*\mathrm{q}}_{\mathcal{K}}\Delta^{\mathrm{cl}}_{\mathcal{K}}|\Delta^{\mathrm{cl}}_{\mathcal{K}}|^2$ originates from expanding the term $\check{Q}_{\mathcal{K}}\check{\Delta}_{\mathcal{K}}$ to *third* order in $\check{\mathcal{W}}$ and pairing it with the three vertices $\check{\mathcal{W}}\check{\Delta}_{\mathcal{K}}$ [391].

[11] It is instructive to compare the GL action with the Gross–Pitaevskii one, Eq. (8.15). While the latter describes inertial dissipationless dynamics, the former prescribes the diffusive relaxation. The inertial dynamics and the corresponding acoustic mode are only partially recovered at $T < T_{\mathrm{c}}$, thanks to the singular nature of the BCS density of states (DOS), Section 17.5. The difference may be traced back to the fact that the superconducting order parameter Δ interacts with the quasiparticles having a finite DOS, ν, at $T > T_{\mathrm{c}}$. On the other hand, the Bose–Einstein condensate Φ_0 interacts with the quasiparticles whose DOS goes to zero at small energy (e.g. as ϵ^2 in 3d).

$$(L^{-1})^{\mathrm{K}} = B_\omega \left[(L^{-1})^{\mathrm{R}}(\omega) - (L^{-1})^{\mathrm{A}}(\omega)\right] \approx \frac{\mathrm{i}\pi}{2}, \tag{17.73b}$$

where in equilibrium $B_\omega = \coth \omega/2T$ and we used $\omega \ll T \approx T_\mathrm{c}$. The Ginzburg–Landau relaxation time is defined as $\tau_{\mathrm{GL}} = \pi/8(T - T_\mathrm{c})$. Comparing the action (17.71)–(17.73) with Eq. (9.5), one notices that it describes the classical second-order phase transition with the complex nonconservative order parameter (in the absence of the electromagnetic fields). In the Hohenberg–Halperin classification, Section 9.5, it belongs to the model A class. The corresponding Landau free energy (9.2) is given by

$$\mathcal{F}_{\mathrm{GL}}[\Delta, \mathbf{A}] = \frac{\pi\nu}{8T} \int d\mathbf{r} \left[D\left|\partial_\mathbf{r}\Delta\right|^2 + \tau_{\mathrm{GL}}^{-1}\left|\Delta\right|^2 + \frac{g}{2}\left|\Delta\right|^4 \right], \tag{17.74}$$

with the critical parameter $\tau_{\mathrm{GL}}^{-1} \propto T - T_\mathrm{c}$ and nonlinearity $g = 7\zeta(3)/\pi^3 T_\mathrm{c}$. The complex order parameter field Δ happens to be charged with the charge $2e$ and therefore interacts with the vector potential through the covariant derivative $\nabla_\mathbf{r} \to \partial_\mathbf{r} = \nabla_\mathbf{r} + 2ie\mathbf{A}$. The useful feature of the $\mathcal{K}$-gauge is the absence of interactions with the scalar electromagnetic potential (i.e. the time derivative does not acquire the covariant form in this specific gauge). The vector potential dynamics is governed by the usual Maxwell equations. The media provides the right-hand side of the Maxwell equations, which follow from the two additional terms in the effective action (17.70): $S_\mathrm{S} + S_\mathrm{N}$.

The supercurrent part of the action $S_\mathrm{S}[\Delta_\mathcal{K}, \mathbf{A}_\mathcal{K}]$ originates from the diagrams of Fig. 17.2 (c),(d). It is given by[12]

$$S_\mathrm{S} = -\frac{\pi e\nu D}{T} \operatorname{Tr}\left\{\mathbf{A}_\mathcal{K}^{\mathrm{q}} \operatorname{Im}\left[\Delta_\mathcal{K}^{*\mathrm{cl}}(\nabla_\mathbf{r} + 2ie\mathbf{A}_\mathcal{K}^{\mathrm{cl}})\Delta_\mathcal{K}^{\mathrm{cl}}\right]\right\}. \tag{17.75}$$

Being differentiated with respect to $\mathbf{A}^\mathrm{q}$, it provides an expression for the supercurrent that coincides with the conventional Ginzburg–Landau definition [375]: $\mathbf{J}_\mathrm{S} = -\delta S_\mathrm{S}/2\delta\mathbf{A}_\mathcal{K}^\mathrm{q} = \delta\mathcal{F}_{\mathrm{GL}}/\delta\mathbf{A}$, Eq. (17.74).

The main part of the normal action $S_\mathrm{N}[\Delta_\mathcal{K}, \mathbf{A}_\mathcal{K}]$ is obtained by substituting the normal metal values $\check{Q}_\mathcal{K} = \check{\Lambda}$ and $\check{\Delta}_\mathcal{K} = 0$ into the action $S[\check{Q}_\mathcal{K}, \Delta_\mathcal{K}, \mathbf{A}_\mathcal{K}, \varphi_\mathcal{K}]$, Eq. (17.7). The calculation is virtually identical to the one leading to Eq. (16.17), but this time we choose to express the result through the vector potential instead of the scalar one. (The two are related through the $\mathcal{K}$-gauge condition (16.5).) This way, one finds

$$S_\mathrm{N} = e^2\nu D \operatorname{Tr}\left\{ \vec{\mathbf{A}}_\mathcal{K}^{\mathrm{T}} \begin{pmatrix} 0 & D\nabla_\mathbf{r}^2 + \partial_t \\ D\nabla_\mathbf{r}^2 - \partial_t & 4iT \end{pmatrix}_\mathrm{K} \vec{\mathbf{A}}_\mathcal{K} \right\}, \tag{17.76}$$

12 Note that in Eq. (17.71) and throughout the rest of this section we have restored the electron charge e accompanying source fields $\mathbf{A} \to e\mathbf{A}$ and $\varphi \to e\varphi$, such that $\mathbf{A}$ and φ are now actual electromagnetic potentials.

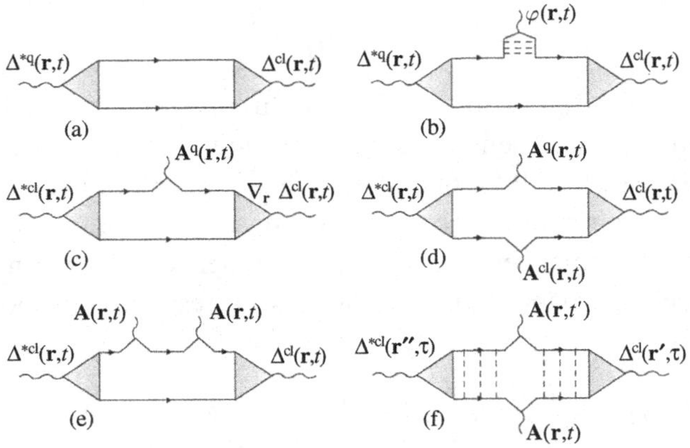

Figure 17.2 Diagrammatic representation of the effective action S_{eff}. (a) Ginzburg–Landau part S_{GL}, Eq. (17.71). (b) Anomalous Gor'kov–Eliashberg coupling of the scalar potential with the order parameter (see Eq. (17.83) and discussions in what follows). (c) Paramagnetic and (d) diamagnetic parts of the supercurrent action S_{S}. (e) DOS and (f) Maki–Thompson corrections to the normal current action S_{N}. In the case of diagrams (e) and (f) there are two possible choices for the vector potentials: *classical–quantum*, which is a part of the current, and *quantum–quantum*, which is its FDT counterpart.

where $\vec{\mathbf{A}}_{\mathcal{K}} = (\mathbf{A}^{\mathrm{cl}}_{\mathcal{K}}, \mathbf{A}^{\mathrm{q}}_{\mathcal{K}})^{\mathrm{T}}$. As we shall see, S_{N} describes the normal dissipative current, as well as the corresponding FDT-related Johnson–Nyquist noise; see Eq. (13.17). The superconducting fluctuations lead to the renormalization of the density of states ν and the diffusion coefficient D in Eq. (17.76) by terms proportional to $|\Delta^{\mathrm{cl}}_{\mathcal{K}}|^2$. They are known as density of states (DOS) [392] and Maki–Thompson (MT) [393] corrections, respectively, Fig. 17.2(e), (f). For example, the DOS correction calls for the renormalization $\nu \to \nu + \delta\nu(\mathbf{r},t)$, where $\delta\nu(\mathbf{r},t) = -\nu(7\zeta(3)/4\pi^2)|\Delta^{\mathrm{cl}}_{\mathcal{K}}(\mathbf{r},t)|^2/T^2$. For detailed discussion of the corresponding parts of the action see [391].

One can use the effective action (17.70) to derive the *stochastic* time-dependent Ginzburg–Landau equation. To this end one needs to get rid of terms quadratic in quantum components of the fields: $|\Delta^{\mathrm{q}}_{\mathcal{K}}|^2$ in S_{GL}, and $(\mathbf{A}^{\mathrm{q}}_{\mathcal{K}})^2$ in S_{N}. For the former this is achieved through the Hubbard–Stratonovich transformation with the help of an auxiliary complex field ξ_Δ:

$$\mathrm{e}^{-\pi\nu\,\mathrm{Tr}\left\{|\Delta^{\mathrm{q}}_{\mathcal{K}}|^2\right\}} = \int \mathbf{D}[\xi_\Delta]\, \mathrm{e}^{-\frac{\pi\nu}{4T}\mathrm{Tr}\left\{\frac{|\xi_\Delta|^2}{4T} - i\xi^*_\Delta\Delta^{\mathrm{q}}_{\mathcal{K}} - i\xi_\Delta\Delta^{*\mathrm{q}}_{\mathcal{K}}\right\}}. \qquad (17.77)$$

As a result, the action S_{GL}, Eq. (17.71), acquires a linear form in the quantum component of the order parameter $\Delta^{\mathrm{q}}_{\mathcal{K}}$. Integration over the latter leads to the functional delta-function, imposing the stochastic TDGL equation

$$\left[\partial_t + \tau_{\mathrm{GL}}^{-1} - D\big[\nabla_{\mathbf{r}} + 2ie\mathbf{A}_{\mathcal{K}}^{\mathrm{cl}}(\mathbf{r},t)\big]^2 + \frac{7\zeta(3)}{\pi^3 T}|\Delta_{\mathcal{K}}^{\mathrm{cl}}(\mathbf{r},t)|^2\right]\Delta_{\mathcal{K}}^{\mathrm{cl}}(\mathbf{r},t) = \xi_\Delta(\mathbf{r},t), \tag{17.78}$$

where, in agreement with FDT, the complex Gaussian white noise $\xi_\Delta(\mathbf{r},t)$ has the following correlator:

$$\langle \xi_\Delta(\mathbf{r},t)\xi_\Delta^*(\mathbf{r}',t')\rangle = 2T\,\frac{8T}{\pi\nu}\,\delta(\mathbf{r}-\mathbf{r}')\delta(t-t'). \tag{17.79}$$

Indeed, the stationary solution of the corresponding Fokker–Planck equation (see Eq. (9.11)) is $e^{-\mathcal{F}_{\mathrm{GL}}/T}$.

In a similar way, one decouples the term quadratic in $\mathbf{A}_{\mathcal{K}}^{\mathrm{q}}$ in the action (17.76) by introducing a vector Hubbard–Stratonovich field $\vec{\xi}_{\mathbf{j}}(\mathbf{r},t)$:

$$\mathrm{e}^{-4T\sigma\,\mathrm{Tr}\left\{[\mathbf{A}_{\mathcal{K}}^{\mathrm{q}}]^2\right\}} = \int \mathbf{D}[\vec{\xi}_{\mathbf{j}}]\;\mathrm{e}^{-\mathrm{Tr}\left\{\frac{\vec{\xi}_{\mathbf{j}}^{\,2}}{4T\sigma} + 2i\mathbf{A}_{\mathcal{K}}^{\mathrm{q}}\vec{\xi}_{\mathbf{j}}\right\}}, \tag{17.80}$$

where $\sigma = e^2\nu D$ is the conductivity with DOS and MT renormalizations. The resulting action is now linear in $\mathbf{A}_{\mathcal{K}}^{\mathrm{q}}$, allowing us to define the charge density $\rho(\mathbf{r},t) = -\,\delta S_{\mathrm{eff}}/2\delta\varphi^{\mathrm{q}}(\mathbf{r},t)$ and the current density $\mathbf{J}(\mathbf{r},t) = -\,\delta S_{\mathrm{eff}}/2\delta\mathbf{A}^{\mathrm{q}}(\mathbf{r},t)$. It is important to emphasize that the differentiation here is performed over the bare electromagnetic potentials $\{\mathbf{A},\varphi\}$, while the action S_{eff} (17.70) is written in terms of the gauged ones $\{\mathbf{A}_{\mathcal{K}},\varphi_{\mathcal{K}}\}$. The connection between the two $\{\mathbf{A},\varphi\} \rightleftarrows \{\mathbf{A}_{\mathcal{K}},\varphi_{\mathcal{K}}\}$ is provided by the functional $\mathcal{K}[\mathbf{A},\varphi]$, which is defined by Eq. (16.5). A simple algebra then leads to a set of the continuity relation $\partial_t\rho(\mathbf{r},t) + \mathrm{div}\,\mathbf{J}(\mathbf{r},t) = 0$ and the expression for the current density entering the Maxwell equation $\nabla_{\mathbf{r}} \times \mathbf{H} - \dot{\mathbf{E}} = \mathbf{J}$,

$$\mathbf{J} = \sigma\mathbf{E} - D\nabla_{\mathbf{r}}\rho + \frac{\pi e\nu D}{2T}\,\mathrm{Im}\left\{\Delta_{\mathcal{K}}^{*\mathrm{cl}}\big[\nabla_{\mathbf{r}} + 2ie\mathbf{A}_{\mathcal{K}}^{\mathrm{cl}}\big]\Delta_{\mathcal{K}}^{\mathrm{cl}}\right\} + \vec{\xi}_{\mathbf{j}}(\mathbf{r},t), \tag{17.81}$$

where $\mathbf{E}(\mathbf{r},t) = \partial_t\mathbf{A}_{\mathcal{K}} - \nabla_{\mathbf{r}}\varphi_{\mathcal{K}}$ is the electric field. The current fluctuations are induced by the vector Gaussian white noise with the correlator

$$\left\langle \xi_{\mathbf{j}}^{\mu}(\mathbf{r},t)\,\xi_{\mathbf{j}}^{\nu}(\mathbf{r}',t')\right\rangle = 2T\sigma\,\delta_{\mu\nu}\delta(\mathbf{r}-\mathbf{r}')\delta(t-t'), \tag{17.82}$$

ensuring validity of the FDT. Equations (17.78) and (17.81) along with the continuity relation, supplemented by the Maxwell equations, constitute the complete set describing dynamics of the superconductors at $T \gtrsim T_{\mathrm{c}}$.

It is instructive to rewrite the TDGL equation (17.78) back in the original gauge. This is achieved by the substitution of the gauged order parameter $\Delta_{\mathcal{K}}^{\mathrm{cl}} = \Delta^{\mathrm{cl}}\exp\big(-2ie\mathcal{K}^{\mathrm{cl}}\big)$ and the vector potential, Eq. (17.66), into Eq. (17.78). This way, one finds the following equation for the bare order parameter Δ^{cl}:

$$[\partial_t - 2ie\partial_t \mathcal{K}^{\mathrm{cl}}]\Delta^{\mathrm{cl}} = \Big[D\left[\nabla_{\mathbf{r}} + 2ie\mathbf{A}^{\mathrm{cl}}\right]^2 - \tau_{\mathrm{GL}}^{-1} - \frac{7\zeta(3)}{\pi^3 T}|\Delta^{\mathrm{cl}}|^2\Big]\Delta^{\mathrm{cl}} + \xi_\Delta, \tag{17.83}$$

where we have redefined the complex noise as $\xi_\Delta \to \xi_\Delta \exp\left(2ie\mathcal{K}^{\mathrm{cl}}\right)$, which, however, does not change its correlation function (17.79). Unlike the TDGL equation frequently found in the literature, the left-hand side of Eq. (17.83) contains the Gor'kov–Eliashberg [395] anomalous term $\partial_t \mathcal{K}^{\mathrm{cl}}(\mathbf{r}, t)$ instead of the scalar potential $\varphi^{\mathrm{cl}}(\mathbf{r}, t)$; see Fig. 17.2(b). In a generic case $\mathcal{K}^{\mathrm{cl}}(\mathbf{r}, t)$ is a nonlocal functional of the scalar and longitudinal vector potentials given by Eq. (16.7), which for the classical component reads as

$$\left(D\nabla_{\mathbf{r}}^2 - \partial_t\right)\mathcal{K}^{\mathrm{cl}}(\mathbf{r}, t) = \varphi^{\mathrm{cl}}(\mathbf{r}, t) - D\,\mathrm{div}\mathbf{A}^{\mathrm{cl}}(\mathbf{r}, t). \tag{17.84}$$

The equivalence $\varphi^{\mathrm{cl}} = -\partial_t \mathcal{K}^{\mathrm{cl}}$ holds for spatially uniform potentials; however, in general the two are distinct. The standard motivation behind writing the scalar potential $\varphi^{\mathrm{cl}}(\mathbf{r}, t)$ on the left-hand side of the TDGL equation is gauge invariance. Note, however, that the local gauge transformation

$$\Delta \to \Delta\, \mathrm{e}^{-2ie\chi}, \qquad \varphi \to \varphi + \partial_t \chi, \qquad \mathbf{A} \to \mathbf{A} + \nabla_{\mathbf{r}}\chi, \qquad \mathcal{K} \to \mathcal{K} - \chi$$

leaves Eq. (17.83) unchanged, and therefore this form of the TDGL equation is perfectly gauge invariant. The last expression here is an immediate consequence of Eq. (17.84) and the rules of the gauge transformation for φ and $\mathbf{A}$. In the $\mathcal{K}$-gauge, Eq. (17.66), specified by $\chi = \mathcal{K}^{\mathrm{cl}}$, the anomalous Gor'kov–Eliashberg term disappears from the left-hand side of the TDGL equation (17.83), and one recovers Eq. (17.78).

17.6.0.1 Derivation of the Effective Action

Substituting the parametrizations (17.67) and (17.68) into the action (17.7) and expanding to second order in the complex Cooperon modes c and $\bar{c}$, one finds the following quadratic actions (throughout this section we omit the gauge-specifying subscript $\mathcal{K}$ for brevity):

$$iS^{c}[c, \Delta] = -\frac{\pi\nu}{2}\,\mathrm{Tr}\Big\{c^*_{\epsilon\epsilon'}[Dq^2 - i(\epsilon + \epsilon')]c_{\epsilon'\epsilon} + 2i\Delta^{c}_{\epsilon\epsilon'}c^*_{\epsilon'\epsilon} - 2i\Delta^{*c}_{\epsilon\epsilon'}c_{\epsilon'\epsilon}\Big\}, \tag{17.85a}$$

$$iS^{\bar{c}}[\bar{c}, \Delta] = -\frac{\pi\nu}{2}\,\mathrm{Tr}\Big\{\bar{c}^*_{\epsilon\epsilon'}[Dq^2 + i(\epsilon + \epsilon')]\bar{c}_{\epsilon'\epsilon} - 2i\Delta^{\bar{c}}_{\epsilon\epsilon'}\bar{c}^*_{\epsilon'\epsilon} + 2i\Delta^{*\bar{c}}_{\epsilon\epsilon'}\bar{c}_{\epsilon'\epsilon}\Big\}, \tag{17.85b}$$

where the following form factors were introduced:

$$\Delta^{c}_{\epsilon\epsilon'} = \Delta^{\mathrm{cl}}(\epsilon - \epsilon') + F^{\mathrm{eq}}_{\epsilon}\Delta^{\mathrm{q}}(\epsilon - \epsilon'), \qquad \Delta^{\bar{c}}_{\epsilon\epsilon'} = \Delta^{\mathrm{cl}}(\epsilon - \epsilon') - F^{\mathrm{eq}}_{\epsilon'}\Delta^{\mathrm{q}}(\epsilon - \epsilon'). \tag{17.86}$$

It is important to emphasize that the diffuson modes $\{d^\alpha, \tilde{d}^\alpha\}$ couple to Δ only starting from the quadratic order in $\check{\mathcal{W}}$. These terms produce nonlocal and nonlinear interaction vertices for the order parameter and will not be considered here; see [391] for more details. We have also utilized the advantage of the $\mathcal{K}$-gauge to avoid terms linear in $\check{\mathcal{W}}$ and the electromagnetic potentials, see Section 16.1. To perform Gaussian integrations over the Cooperon modes c and $\tilde{c}$ it is convenient to find stationary configurations of the quadratic forms in Eqs. (17.85), which are given by

$$c_{\epsilon\epsilon'}(\mathbf{q}) = \frac{-2\mathrm{i}\Delta^{c}_{\epsilon\epsilon'}(\mathbf{q})}{Dq^2 - \mathrm{i}(\epsilon+\epsilon')}, \qquad \tilde{c}_{\epsilon\epsilon'}(\mathbf{q}) = \frac{2\mathrm{i}\Delta^{\tilde{c}}_{\epsilon\epsilon'}(\mathbf{q})}{Dq^2 + \mathrm{i}(\epsilon+\epsilon')}. \tag{17.87}$$

The Gaussian integral is calculated by substituting these stationary configurations back into the quadratic action (17.85), which leads to

$$\mathrm{i}S_{\mathrm{GL}}[\Delta] = 4\pi\nu \sum_{\mathbf{q}} \iint \frac{d\epsilon d\omega}{4\pi^2} \frac{[\Delta^{*\mathrm{cl}} + F^{\mathrm{eq}}_{\epsilon-\omega/2}\Delta^{*\mathrm{q}}][\Delta^{\mathrm{cl}} + F^{\mathrm{eq}}_{\epsilon+\omega/2}\Delta^{\mathrm{q}}]}{Dq^2 - 2\mathrm{i}\epsilon}, \tag{17.88}$$

where $\Delta^{\mathrm{cl(q)}} = \Delta^{\mathrm{cl(q)}}(\mathbf{q},\omega)$ with $\omega = \epsilon - \epsilon'$ and $\epsilon = (\epsilon+\epsilon')/2$. We have also employed the fact that $F^{\mathrm{eq}}_{\epsilon}$ is an odd function to change variables as $\epsilon \to -\epsilon$ in the contribution coming from the $\tilde{c}$ modes. The contribution to $\mathrm{i}S_{\mathrm{GL}}[\Delta]$ with the two classical components of the order parameter $\sim\Delta^{*\mathrm{cl}}\Delta^{\mathrm{cl}}$ vanishes identically after the ϵ-integration, being an integral of the purely retarded function. This is a manifestation of the normalization condition for the Keldysh-type action (see Section 2.7 for discussions). The action therefore acquires the standard Keldysh causality structure, Eqs. (17.71) and (17.72). Combining it with the bare Hubbard–Stratonovich term $\sim\Delta^{*\mathrm{q}}\Delta^{\mathrm{cl}}/\lambda$, Eq. (17.5), one obtains for its retarded and advanced components

$$(L^{-1})^{\mathrm{R(A)}}(\mathbf{q},\omega) = -\frac{1}{\lambda} - \mathrm{i}\int d\epsilon \, \frac{F^{\mathrm{eq}}_{\epsilon\mp\omega/2}}{Dq^2 - 2\mathrm{i}\epsilon}. \tag{17.89}$$

This expression can be reduced to a more familiar form. Indeed, adding and subtracting the right-hand side of Eq. (17.89) taken at zero frequency and momentum, one writes for, for example, the retarded component

$$(L^{-1})^{\mathrm{R}}(\mathbf{q},\omega) = -\frac{1}{\lambda} + \int_{-\omega_{\mathrm{D}}}^{+\omega_{\mathrm{D}}} d\epsilon \, \frac{F^{\mathrm{eq}}_{\epsilon}}{2\epsilon} - \mathrm{i}\int d\epsilon \left[\frac{F^{\mathrm{eq}}_{\epsilon}}{Dq^2 - \mathrm{i}\omega - 2\mathrm{i}\epsilon} + \frac{F^{\mathrm{eq}}_{\epsilon}}{2\mathrm{i}\epsilon}\right]. \tag{17.90}$$

The first integral on the right-hand side is logarithmically divergent and is to be cut off at the Debye frequency ω_{D}. Recalling the definition of the critical temperature T_{c}, Eq. (17.17), one notices that the first two terms here combine to give $\ln T_{\mathrm{c}}/T \approx (T_{\mathrm{c}} - T)/T$. The last integral is convergent and may be evaluated exactly, leading to

$$
\begin{aligned}
(L^{-1})^{\mathrm{R}}(\mathbf{q},\omega) &= \ln\frac{T_{\mathrm{c}}}{T} - \psi\left(\frac{Dq^2 - \mathrm{i}\omega}{4\pi T} + \frac{1}{2}\right) + \psi\left(\frac{1}{2}\right) \\
&\approx \frac{\pi}{8T}\Big(\mathrm{i}\omega - Dq^2 - \tau_{\mathrm{GL}}^{-1}\Big),
\end{aligned} \tag{17.91}
$$

where $\psi(x)$ is the digamma function and $\tau_{\mathrm{GL}}^{-1} = 8(T - T_{\mathrm{c}})/\pi$. Since according to the last expression $Dq^2 \sim \omega \sim \tau_{\mathrm{GL}}^{-1} \ll T$, the expansion of the digamma function is justified. Transforming back to the space-time representation using $\mathrm{i}\omega \to -\partial_t$ and $q^2 \to -\nabla_{\mathbf{r}}^2 \to -\big(\nabla_{\mathbf{r}} + 2\mathrm{i}e\mathbf{A}^{\mathrm{cl}}\big)^2$, where we also included the vector potential, one arrives at Eq. (17.73a). To derive the Keldysh component of the $\hat{L}^{-1}$ matrix, one adds to Eq. (17.88) zero in the form $-4\pi\nu\,\mathrm{Tr}\{\Delta^{*\mathrm{q}}\Delta^{\mathrm{q}}/[Dq^2 - 2\mathrm{i}\epsilon]\} = 0$, which vanishes upon ϵ integration by causality. Employing the familiar identity $F^{\mathrm{eq}}_{\epsilon-\omega/2}F^{\mathrm{eq}}_{\epsilon+\omega/2} - 1 = B_\omega(F^{\mathrm{eq}}_{\epsilon-\omega/2} - F^{\mathrm{eq}}_{\epsilon+\omega/2})$, one obtains Eq. (17.73b).

The supercurrent part of the action S_{S} emerges from the $\mathrm{Tr}\{[\check{\mathbf{A}},\check{Q}]\hat{\partial}_{\mathbf{r}}\check{Q}\}$ term upon second-order expansion over the Cooperon modes, which leads to

$$
S_{\mathrm{S}} = \frac{\mathrm{i}\pi\nu}{4}\,\mathrm{Tr}\Big\{c^*_{tt'}(\mathbf{r})\mathcal{N}_{tt'}c_{t't}(\mathbf{r}) + \tilde{c}^*_{tt'}(\mathbf{r})\mathcal{N}_{tt'}\tilde{c}_{t't}(\mathbf{r})\Big\}, \tag{17.92}
$$

where

$$
\mathcal{N}_{tt'} = \delta(t - t')\frac{2eD}{T}\left[\frac{1}{2}\,\mathrm{div}\mathbf{A}^{\mathrm{q}}(\mathbf{r},t) + \mathbf{A}^{\mathrm{q}}(\mathbf{r},t)\big[\nabla_{\mathbf{r}} + 2\mathrm{i}e\mathbf{A}^{\mathrm{cl}}(\mathbf{r},t)\big]\right]. \tag{17.93}
$$

Deriving $\mathcal{N}_{tt'}$, one uses an approximation for the equilibrium Fermi function in the time representation

$$
F^{\mathrm{eq}}_{t-t'} = -\frac{\mathrm{i}T}{\sinh(\pi T(t - t'))} \overset{t-t' \gg 1/T}{\longrightarrow} \frac{\mathrm{i}}{2T}\,\delta'(t - t'), \tag{17.94}
$$

which is applicable for slowly varying external fields. To perform the Gaussian integration over the Cooperon modes, one substitutes the stationary configurations (17.87) into Eq. (17.92). To this end it is convenient to rewrite them in the space-time representation. This is achieved by noticing that, since $\epsilon \approx T$ and $Dq^2 \approx \tau_{\mathrm{GL}}^{-1}$, $Dq^2 \ll \epsilon$. Therefore one may think of the denominators in Eq. (17.87) as $0 \mp \mathrm{i}(\epsilon + \epsilon')$, which greatly simplifies the Fourier transformation to the time domain. This way, one finds

$$
c_{tt'}(\mathbf{r}) \approx -\mathrm{i}\theta(t - t')\Delta^{\mathrm{cl}}\left(\mathbf{r}, \frac{t + t'}{2}\right) + \chi(t - t')\Delta^{\mathrm{q}}\left(\mathbf{r}, \frac{t + t'}{2}\right), \tag{17.95a}
$$

$$
\tilde{c}_{tt'}(\mathbf{r}) \approx \mathrm{i}\theta(t - t')\Delta^{\mathrm{cl}}\left(\mathbf{r}, \frac{t + t'}{2}\right) - \chi(t - t')\Delta^{\mathrm{q}}\left(\mathbf{r}, \frac{t + t'}{2}\right), \tag{17.95b}
$$

$$
\chi(t) = \int_{-\infty}^{+\infty} \frac{d\epsilon}{2\pi}\tanh\left(\frac{\epsilon}{2T}\right)\frac{\mathrm{e}^{-\mathrm{i}\epsilon t}}{\epsilon + \mathrm{i}0} = \frac{2}{\pi}\,\mathrm{arctanh}\left(\mathrm{e}^{-\pi T|t|}\right), \tag{17.95c}
$$

where the step function $\theta(t-t')$ is understood as being smeared by about $1/T$ and in particular $\theta(0) = 1/2$. Now one can perform integration over t' in Eq. (17.92) with the help of the delta-function from Eq. (17.93). Keeping only the classical component of the order parameter, since $\mathcal{N}$ is already proportional to the quantum component $\mathbf{A}^{\mathrm{q}}$, one obtains S_{S}, Eq. (17.75).

17.7 Fluctuating Superconductivity

Although for temperatures above T_c the expectation value of the order parameter is zero, the expectation value of $|\Delta^{\mathrm{cl}}|^2$ is finite. Indeed, since the TDGL equation (17.78) or (17.83) contains the noise ξ_Δ, it forces the order parameter to fluctuate around zero, creating a finite correlation function of the form $\langle \Delta^{\mathrm{cl}}(\mathbf{r})\Delta^{*\mathrm{cl}}(\mathbf{r}')\rangle = \mathrm{i}L^{\mathrm{K}}(\mathbf{r}-\mathbf{r}')/(2\nu)$; see Eq. (17.71). Thermodynamic and transport observables in the metal are affected by the fluctuating order parameter. For example, the current (17.81) contains the term $\sim \mathrm{Im}\,\Delta^{*\mathrm{cl}}\partial_{\mathbf{r}}\Delta^{\mathrm{cl}}$. Upon averaging over the fluctuations there is a finite residual effect of such a sensitivity on the instantaneous value of the order parameter. This fact leads to a scope of fluctuation corrections to magnetization, density of states, conductivity, and so on, which usually exhibit singular behavior at $T \to T_c$; see Larkin and Varlamov [396] for a review.

Here we focus on fluctuation corrections to the conductivity, which originate from the expression (17.81) for the electric current. The latter consist of the normal (the first two terms on the right-hand side of Eq. (17.81)) and the supercurrent (the third term), coming from S_{N} and S_{S}, respectively (the noise $\vec{\xi}_{\mathbf{j}}$, while important for the current noise spectrum, does not contribute to the average current). The normal part includes the Δ^{cl}-dependent density of states and diffusion constant renormalization. Upon averaging over the order parameter fluctuations they lead to DOS [392] and Maki–Thompson [393] corrections to the conductivity. We shall not derive them here, referring the reader to [396, 391, 24]. Instead we concentrate on the supercurrent part, which leads to a correction to the conductivity discovered by Aslamazov and Larkin [397]. Averaging the supercurrent part of Eq. (17.81) over the fluctuations of the order parameter, one obtains

$$\delta\mathbf{J}_{\mathrm{S}}(\mathbf{r}) = \frac{\pi e D}{4T}\left[\frac{1}{2}(\nabla_{\mathbf{r}} - \nabla_{\mathbf{r}'}) + 2\mathrm{i}e\mathbf{A}\right]L^{\mathrm{K}}(\mathbf{r},\mathbf{r}')\Big|_{\mathbf{r}'=\mathbf{r}}, \tag{17.96}$$

where $L^{\mathrm{K}}(\mathbf{r},\mathbf{r}')$ is the Keldysh component of the full *dressed* correlation matrix. In the absence of external fields the latter is even in $\mathbf{r}-\mathbf{r}'$ and thus $\delta\mathbf{J}_{\mathrm{S}} = 0$. One needs therefore to evaluate how $L^{\mathrm{K}}(\mathbf{r},\mathbf{r}')$ is affected by applied fields. This task leads to the Dyson equation for the correlation matrix $\hat{L}$. We restrict ourselves to the linear response on a spatially uniform ac vector potential $\mathbf{A}(\Omega)$. The latter enters the bosonic operator $\hat{L}^{-1}$ in Eq. (17.71) as $(\pi D/2T)\hat{\sigma}_1 e\mathbf{A}\mathrm{i}\nabla_{\mathbf{r}}$ and we invert the operator to first order in such a perturbation (neglecting the nonlinear terms in TDGL). Going to the Fourier representation, one finds for the ac current at frequency Ω

$$\delta \mathbf{J}_S = \frac{i\pi eD}{4T} \sum_{\mathbf{q},\omega} \left[\mathbf{q} \left(\hat{L}(\mathbf{q}, \omega + \Omega) \frac{\pi D}{2T} \hat{\sigma}_1 e(\mathbf{A}\mathbf{q}) \hat{L}(\mathbf{q}, \omega) \right)^{\mathrm{K}} + 2e\mathbf{A} L^{\mathrm{K}}(\mathbf{q}, \omega) \right], \tag{17.97}$$

where all correlators $\hat{L}$ are understood hereafter as not containing the external vector potential, and the superscript K means that the Keldysh, that is, (cl, cl), component of the correlation matrix is taken. Performing the matrix multiplication and employing the bosonic FDT relation $L^{\mathrm{K}} = B_\omega(L^{\mathrm{R}} - L^{\mathrm{A}})$, one finds for the first term in the square brackets on the right-hand side of Eq. (17.97)

$$\frac{\pi eD}{2T} \mathbf{q}(\mathbf{A}\mathbf{q}) \Big[B_\omega L^{\mathrm{R}}_{\omega+\Omega} L^{\mathrm{R}}_\omega - B_{\omega+\Omega} L^{\mathrm{A}}_{\omega+\Omega} L^{\mathrm{A}}_\omega + (B_{\Omega+\omega} - B_\omega) L^{\mathrm{R}}_{\omega+\Omega} L^{\mathrm{A}}_\omega \Big]. \tag{17.98}$$

Let us look first at the $\Omega = 0$ limit. The last term on the right-hand side vanishes, while the first two along with the last (paramagnetic) terms on the right-hand side of Eq. (17.97) combine to yield the following expression:

$$2eB_\omega \nabla_{\mathbf{q}} \big[(\mathbf{A}\,\mathbf{q})(L^{\mathrm{R}}_\omega(\mathbf{q}) - L^{\mathrm{R}}_\omega(\mathbf{q})) \big],$$

where we employed the fact that $\nabla_{\mathbf{q}} L^{\mathrm{R(A)}}_\omega(\mathbf{q}) = (\pi D\mathbf{q}/4T)(L^{\mathrm{R(A)}}_\omega)^2$; see Eq. (17.73a). Being the full gradient, this expression vanishes upon summation over momenta $\mathbf{q}$. As expected, a vector potential that is constant in space and time does not produce any current.

We look now for the first order in the external frequency Ω. The last term on the right-hand side of Eq. (17.98) yields $\Omega(\partial_\omega B_\omega) L^{\mathrm{R}}_\omega L^{\mathrm{A}}_\omega$. Shifting the integration variable $\omega + \Omega \to \omega$ in the second term, one finds for the first and second terms:

$$\begin{aligned} B_\omega L^{\mathrm{R}}_{\omega+\Omega} L^{\mathrm{R}}_\omega - B_\omega L^{\mathrm{A}}_\omega L^{\mathrm{A}}_{\omega-\Omega} &\to \Omega B_\omega \big[(\partial_\omega L^{\mathrm{R}}_\omega) L^{\mathrm{R}}_\omega + L^{\mathrm{A}}_\omega (\partial_\omega L^{\mathrm{A}}_\omega) \big] \\ = \frac{\Omega}{2} B_\omega \big[\partial_\omega (L^{\mathrm{R}}_\omega)^2 + \partial_\omega (L^{\mathrm{A}}_\omega)^2 \big] &\to -\frac{\Omega}{2} (\partial_\omega B_\omega) \big[(L^{\mathrm{R}}_\omega)^2 + (L^{\mathrm{A}}_\omega)^2 \big], \end{aligned}$$

where in the last instance we involved ω-integration by parts. As a result, one finds for the current

$$\begin{aligned} \delta \mathbf{J}_S &= -\left(\frac{\pi eD}{4T} \right)^2 i\Omega \sum_{\mathbf{q},\omega} \mathbf{q}(\mathbf{A}\mathbf{q}) (\partial_\omega B_\omega) \Big[(L^{\mathrm{R}}_\omega)^2 + (L^{\mathrm{A}}_\omega)^2 - 2L^{\mathrm{R}}_\omega L^{\mathrm{A}}_\omega \Big] \\ &= \left(\frac{\pi eD}{4T} \right)^2 \int \frac{d\omega}{2\pi} (\partial_\omega B_\omega) \sum_{\mathbf{q}} \mathbf{q}\,(\mathbf{E}\,\mathbf{q}) \left[L^{\mathrm{R}}_\omega(\mathbf{q}) - L^{\mathrm{A}}_\omega(\mathbf{q}) \right]^2, \end{aligned} \tag{17.99}$$

where the electric field is given by $\mathbf{E} = \partial_t \mathbf{A} = -i\Omega\mathbf{A}$. The Aslamazov–Larkin (AL) correction to the conductivity is given by $\delta\sigma_{\mathrm{AL}} = \delta \mathbf{J}_S / \mathbf{E}$. Since for

$T - T_c \ll T_c$, $\omega \sim \tau_{GL}^{-1} \ll T$, one may approximate the bosonic distribution function as $B_\omega = \coth \omega/2T \approx 2T_c/\omega$. Recalling that the order parameter propagators are $L^{R(A)} = (8T/\pi)[\pm i\omega - Dq^2 - \tau_{GL}^{-1}]^{-1}$, Eq. (17.73a), and performing the frequency integration in Eq. (17.99), one finds

$$\delta\sigma_{AL} = \frac{8T_c e^2 D}{d} \sum_{\mathbf{q}} \frac{Dq^2}{(Dq^2 + \tau_{GL}^{-1})^3}. \tag{17.100}$$

Further analysis depends on the effective dimensionality d of the system. In particular, for a film of thickness less than the superconducting coherence length $\xi = \sqrt{D\tau_{GL}}$, which is effectively two-dimensional, $d = 2$, one obtains $\delta\sigma_{AL} = e^2 T_c \tau_{GL}/2\pi$. Recalling that $\tau_{GL} = \pi/8(T - T_c)$, one finally obtains [397]

$$\delta\sigma_{AL} = \frac{e^2}{16\hbar} \frac{T_c}{T - T_c}. \tag{17.101}$$

As a precursor of the superconducting transition, the conductivity is singularly enhanced as $T \to T_c$. Moreover, the singular correction depends only on the reduced temperature $T/T_c - 1$ and is insensitive to the bare conductivity and other microscopic details of the film. For an arbitrary dimensionality $d < 4$, Eq. (17.100) yields $\delta\sigma_{AL} \propto \tau_{GL}^{2-d/2}$. This leads in $d = 1$ to the fluctuation correction, which scales as $\delta\sigma_{AL} \propto (T - T_c)^{-3/2}$, while in $d = 3$ it goes as $\delta\sigma_{AL} \propto (T - T_c)^{-1/2}$.

17.8 Problems

17.8.1 Gapless Superconductivity

Consider a superconductor in a presence of a static, spatially uniform perturbation, which breaks the time-reversal symmetry. This may be, for example, a small concentration of magnetic impurities, or an in-plane magnetic field, $H_\parallel$, applied to a thin film of thickness $d < \xi$, where $\xi = \sqrt{D/\Delta}$ is the coherence length. In the latter case one introduces an (x, y)-independent vector potential $A_x = H_\parallel z$, and integrates the action (17.7) over the z-coordinate, $|z| < d/2$. This leads to 2d spatially uniform action

$$iS = -\frac{\pi\nu d}{8} \operatorname{Tr}\Big\{ -\frac{\gamma}{2} [\check{\mathcal{T}}_3, \check{Q}]^2 + 4i\epsilon \check{\mathcal{T}}_3 \check{Q} + 4i\check{\Delta}\check{Q} \Big\}, \tag{17.102}$$

where $\gamma = \frac{1}{6} D(H_\parallel d)^2$ is the energy scale of the time-reversal symmetry breaking perturbation.

Perform variation with respect to $\check{Q}$ and derive the Usadel equation for the stationary matrix $\hat{\Lambda}_\epsilon^R$ (cf. Eq. (17.13)). Parametrize $\hat{\Lambda}_\epsilon^R$ with the help of the *complex* rotation angle in the Nambu space, $\vartheta(\epsilon)$, as in Eq. (17.24), and show that the latter satisfy the following equation:

$$\epsilon = \Delta \coth \vartheta - i\gamma \cosh \vartheta . \tag{17.103}$$

Solution of this equation determines, among other things, the density of states:

$$\nu(\epsilon) = \frac{\nu}{4}\,\mathrm{Tr}\{\hat{\tau}^3 \hat{\Lambda}^R - \hat{\Lambda}^A \hat{\tau}^3\} = \nu\, \mathrm{Re}\,[\cosh \vartheta(\epsilon)] . \tag{17.104}$$

Within the energy gap, $|\epsilon| < \epsilon_g$, where $0 \leq \epsilon_g \leq \Delta$, the density of states is zero, that is, $\mathrm{Re}\,[\cosh \vartheta] = 0$, and thus the angle is $\vartheta = -i\pi/2 + \theta$, with real θ. This brings $\epsilon = \Delta \tanh \theta - \gamma \sinh \theta$. By investigating the right-hand side of this equation, show that it may be indeed satisfied only in the limited range of energies, $|\epsilon| < \epsilon_g$, where the gap ϵ_g is given by the *astroid* condition [398]

$$\epsilon_g^{2/3} + \gamma^{2/3} = \Delta^{2/3} . \tag{17.105}$$

Investigate behavior of the density of states, $\nu(\epsilon)$, immediately above the gap, that is, $\epsilon \gtrsim \epsilon_g$. Notice that according to the astroid condition, the gap exists as long as $\gamma < \Delta$.

Let's look now at the self-consistency relation (17.32) at $T = 0$, that is, with $F_{\mathrm{L}}(\epsilon) = \mathrm{sign}(\epsilon)$ and $F_{\mathrm{T}}(\epsilon) = 0$. It takes the form (notice, $\vartheta(-\epsilon) = -\vartheta^*(\epsilon)$):

$$1 = \frac{\lambda}{\Delta}\,\mathrm{Re}\int_0^{\omega_{\mathrm{D}}} d\epsilon \sinh \vartheta = \frac{\lambda}{\Delta}\,\mathrm{Re}\int d\tau\, \frac{d\epsilon}{d\tau}\,\tau = -\lambda\, \mathrm{Re}\int \frac{d\tau}{\sqrt{1+\tau^2}} \left[\frac{1}{\tau} + i\frac{\gamma}{\Delta}\tau^2\right], \tag{17.106}$$

where $\tau = \sinh \vartheta$ and $d\epsilon/d\tau$ is calculated using Eq. (17.103). One can now eliminate the interaction strength, λ, by dividing this equation by that at $\gamma = 0$, which defines the bare order parameter Δ_0. Think carefully about the limits of the τ-integral (they are different for $\gamma < \Delta$ vs. $\gamma > \Delta$) and perform the integration explicitly.[13] This leads to the relation $\log(\Delta_0/\Delta) = g(\gamma/\Delta)$, where $g(x) = \pi x/4$, if $x < 1$, and $g(x) = \ln(x + \sqrt{x^2 - 1}) - \frac{1}{2x}\sqrt{x^2 - 1} + \frac{x}{2} \arcsin x^{-1}$, if $x > 1$.

If you look for a point where $\Delta \to 0$, this spells $x \to \infty$ and $g(x) \to \log(2x)$. This means that the self-consistency loses a nonzero solution once $\gamma \geq \Delta_0/2$. This is the *quantum phase transition* point, where the superconductivity is destroyed by the time-reversal-symmetry breaking perturbation. On the other hand, the gap disappears if $\gamma \geq \Delta$, that is, at $x \geq 1$, or $\log(\Delta_0/\Delta) \geq \pi/4$. Therefore, within the interval $e^{-\pi/4} = 0.456 < \gamma/\Delta_0 < 1/2$ *below* the quantum phase transition, the self-consistency condition leads to a nonzero order parameter $0 < \Delta < \gamma$, while the gap in the spectrum is already fully suppressed. This is the phenomenon of the *gapless superconductivity* [398]. Solve Eq. (17.103) numerically and plot the density of states (17.104) as a function of energy for $\gamma < \Delta$, $\gamma = \Delta$, and $\gamma > \Delta$ [414].

[13] Use the fact that for $\omega_{\mathrm{D}} \gg \Delta_0$ the upper limit of the τ-integral is $\Delta/\omega_{\mathrm{D}} \ll 1$, and $\Delta_0/\omega_{\mathrm{D}} \ll 1$ for the normalization integral. There is thus a narrow interval, where $\int_{\Delta/\omega_{\mathrm{D}}}^{\Delta_0/\omega_{\mathrm{D}}} d\tau/\tau = \log(\Delta_0/\Delta)$.

17.8.2 Microwave Absorption in Superconductors

Consider a superconducting thin film irradiated with a microwave electromagnetic field with frequency ω. In a spatially uniform case this leads to the action $iS_{\text{coll}}[\check{Q}] = \frac{\pi\nu D}{8}\,\text{Tr}\left\{[\check{\mathcal{T}}_3\mathbf{A},\check{Q}]^2\right\}$ (cf. the first term in Eq. (17.102)), where $\mathbf{A}(t)$ is the ac vector potential, associated with the electromagnetic wave. This term leads to the collision integral, given by

$$\frac{\nu(\epsilon)}{\nu}\,\partial_t F_{L,\epsilon} = I_{\text{coll}}[F_{\text{L},\epsilon}] = \frac{1}{\pi\nu}\,\text{Tr}_{\text{N}}\left\{\hat{\tau}_0\left(\frac{\delta iS_{\text{coll}}}{\delta\check{Q}_{\epsilon\epsilon}}\right)^{(1,2)}_{\text{K}}\right\}, \tag{17.107}$$

where $\nu(\epsilon)$ is the superconducting density of states (17.18) and we have restricted ourselves to the longitudinal component of the distribution function and thus taken the Nambu trace $\text{Tr}_{\text{N}}\{\hat{\tau}_0\ldots\}$. Calculate the variation, take its $(1,2)$ Keldysh block, and put the $\check{Q}$-field to its stationary point value, $\check{Q}=\check{\Lambda}_\epsilon$, which is diagonal in the energy space, with $\hat{\Lambda}^{\text{K}}_\epsilon = F_{L,\epsilon}(\hat{\Lambda}^{\text{R}}_\epsilon - \hat{\Lambda}^{\text{A}}_\epsilon)$. Finally take the trace over the Nambu subspace. This way you'll find for the collision integral:

$$I_{\text{coll}}[F_{\text{L},\epsilon}] = \sum_{\pm}\mathcal{M}_{\epsilon,\epsilon'}(F_{\text{L},\epsilon'} - F_{\text{L},\epsilon}); \qquad \mathcal{M}_{\epsilon,\epsilon'} = D(\mathbf{A})^2\left[\text{Re}\,u_\epsilon\text{Re}\,u_{\epsilon'} + \text{Re}\,v_\epsilon\text{Re}\,v_{\epsilon'}\right], \tag{17.108}$$

where the summation stands for absorption and emission processes $\epsilon' = \epsilon\pm\omega$. It's worth noticing that for an ultrasound stimulation, the expression is similar, but with $\mathcal{M}_{\epsilon,\epsilon'} \propto [\text{Re}\,u_\epsilon\text{Re}\,u_{\epsilon'} - \text{Re}\,v_\epsilon\text{Re}\,v_{\epsilon'}]$ instead, see Eq. (18.30). The difference may be traced back to the presence of the time-reversal symmetry breaking matrices $\hat{\tau}_3$, multiplying the vector potential, but not the phonon field.

Calculate the energy absorption, by evaluating change of the total electron energy $\partial_t E_e = \partial_t\int d\epsilon\,\epsilon\,\nu(\epsilon)\frac{1}{2}(1-F_{\text{L},\epsilon}) = -\frac{\nu}{2}\int d\epsilon\,\epsilon\,I_{\text{coll}}[F_{\text{L},\epsilon}] = \text{Re}[\sigma(\omega)](\omega\mathbf{A})^2$, where the last equality expresses the Joule heating as a product of the electric field $e\omega\mathbf{A}$ and the in-phase current $\text{Re}[\sigma(\omega)](e\omega\mathbf{A})$. Substitute the collision integral (17.108) and rearrange the energy integrals to find for, for example, $0<\omega<2\Delta$:

$$\text{Re}[\sigma(\omega)] = \frac{\sigma_{\text{D}}}{\omega}\int_\Delta^\infty d\epsilon\,(F_{\text{L},\epsilon+\omega} - F_{\text{L},\epsilon})\frac{\epsilon(\epsilon+\omega)+\Delta^2}{\sqrt{\epsilon^2-\Delta^2}\sqrt{(\epsilon+\omega)^2-\Delta^2}}, \tag{17.109}$$

where $\sigma_{\text{D}} = e^2\nu D$ is the Drude conductivity of the normal metal. This is the Mattis–Bardeen formula [399] for the microwave absorption in superconductors. Derive the corresponding expression for $\omega > 2\Delta$. In the linear response regime one may take the equilibrium value for the distribution function $F_{\text{L},\epsilon} = \tanh(\epsilon/2T)$.

Show that in the limit $\omega \ll T \ll \Delta$ the Mattis–Bardeen formula leads to

$$\mathrm{Re}[\sigma(\omega)] = \sigma_\mathrm{D}\frac{2\Delta}{T}\,\mathrm{e}^{-\Delta/T}\left[\log\left(\frac{T}{\omega}\right) - \gamma_\mathrm{E}\right], \tag{17.110}$$

where γ_E is the Euler gamma; while in the limit $\omega \ll 2\Delta \ll T \lesssim T_c$ to

$$\mathrm{Re}[\sigma(\omega)] = \sigma_\mathrm{D}\left[1 + \frac{\Delta}{2T}\log\left(\frac{2\Delta}{\omega}\right)\right]. \tag{17.111}$$

What is the Kramers–Kronig counterpart of the logarithmic singularity at small frequency in the $\mathrm{Im}[\sigma(\omega)]$? The singularity is a consequence of the BCS density of states.[14] Moreover, the coefficient in front of $\log(1/\omega)$ is proportional to $\partial_\epsilon F_{\mathrm{L},\epsilon}|_{\epsilon=\Delta}$. Both of these features are particularly sensitive to deviations from the linear response, which tend to smear the singularity. Such deviations come from (i) the BCS density of states being rounded up by $\gamma \propto D\mathbf{A}^2$, as discussed in Problem 17.8.1, and (ii) a non-equilibrium distribution function, which is a subject of Problem 17.8.3.

17.8.3 Stimulated Superconductivity

Here we discuss the kinetic equation (17.107), (17.108) beyond the linear response regime. For BCS superconductors, $\mathrm{Re}\,u_\epsilon = \mathrm{Re}\,v_\epsilon = 0$ if $|\epsilon| < \Delta$. This implies that for $\omega < 2\Delta$ both initial and final energies $\epsilon, \epsilon' > \Delta$ (or $\epsilon, \epsilon' < -\Delta$). These processes conserve the number of thermal quasiparticles. (The pair-breaking processes with, e.g. $\epsilon < -\Delta$ and $\epsilon' > \Delta$, which create extra quasiparticles, are only possible for $\omega > 2\Delta$.) Therefore radiation with $\omega < 2\Delta$ excites the quasiparticles to higher energies, without changing their total number. This means *less* quasiparticles at $\epsilon \approx \Delta$ in an irradiated film than without the radiation. According to the self-consistency condition (17.16), it is $F_{\mathrm{L},\Delta}$ that is primarily responsible for the value of Δ. Thus, heating the quasiparticles with $\omega < 2\Delta$ radiation *increases* the order parameter. Compare it with the usual phonon-mediated heating, which, of course, supresses the order parameter. The key difference is that while the former conserves quasiparticle number, the latter creates more quasiparticles. Enhancement of the order parameter by radiation was seen experimentally [402, 403] and soon explained by Eliashberg [400]; for a review see [401].

Let us focus on $T \lesssim T_c$, where $\omega \ll 2\Delta \ll T$. Expand the right-hand side of Eq. (17.108) to the second powers in ω to arrive at the energy diffusion equation

$$\partial_t\big(\nu(\epsilon)F_{L,\epsilon}\big) = \nu\,\partial_\epsilon\big(\omega^2\mathcal{M}_{\epsilon,\epsilon}\,\partial_\epsilon F_{L,\epsilon}\big). \tag{17.112}$$

[14] It is interesting to note, however, that the logarithmic singularity is *not* present in the *ultrasound* absorption by quasiparticles, cf. Eq. (18.48), due to the different combination of u_ϵ and v_ϵ coherence factors, mentioned previously.

This is a generalization of the energy diffusion in normal metals, discussed in Section 14.9. It shows that in the superconductors the diffusion coefficient is energy-dependent:

$$\mathcal{D}(\epsilon) = \omega^2 \mathcal{M}_{\epsilon,\epsilon} = D(\omega \mathbf{A})^2 \frac{\epsilon^2 + \Delta^2}{\epsilon^2 - \Delta^2} \theta(|\epsilon| - \Delta); \tag{17.113}$$

compare Eq. (14.80). According to Eq. (17.112), the total number of quasiparticles, $\int_\Delta^\infty d\epsilon\, \nu(\epsilon) \frac{1}{2}(1 - F_{L,\epsilon})$, is conserved.

The only stationary solution with no energy current is $F_{L,\epsilon} = \text{const}$, which cannot satisfy the quasiparticle conservation. This indicates that one has to invoke some relaxation mechanism, for example, phonons, to find a stationary distribution. Assuming some (energy-dependent) relaxation time $\tau_{\text{rel}}(\epsilon)$, the characteristic diffusion distance in the energy direction is found from $\epsilon^* = \sqrt{\mathcal{D}(\epsilon^*)\tau_{\text{rel}}(\epsilon^*)}$. The distribution function may be approximated as $F_{L,\epsilon} = F^* = \text{const}$ for $\epsilon < \epsilon^*$ and $F_{L,\epsilon} = \tanh(\epsilon/2T)$ for $\epsilon > \epsilon^*$, where F^* is found from the quasiparticle number conservation. Substitute such $F_{L,\epsilon}$ in the self-consistency relation and show that $\Delta \approx \epsilon^*$ for $\Delta(T) < \epsilon^* < \Delta(0)$ and $\Delta \to \Delta(0)$ for $\Delta(0) < \epsilon^*$. Here $\Delta(T) \ll \Delta(0)$ is the equilibrium order parameter.

18

Electron–Phonon Interactions

In this chapter we consider interactions of electrons with acoustic phonons. We derive the corresponding sigma-model and show how it yields kinetic theory for electron and phonon distributions. We obtain electron–phonon thermalization times along with the ultrasound attenuation rate in normal metals and superconductors.

18.1 Phonon Action and Propagators

Lattice displacement at a point $(\mathbf{r}, t)$ is described by a real vector field $\mathbf{u}(\mathbf{r}, t)$ with the Cartesian components $u^{\mu}(\mathbf{r}, t)$, where $\mu = x, y, z$. It is convenient to pass right away to the Fourier basis, $\mathbf{r} \to \mathbf{q}$, $t \to \omega$ and consider $\mathbf{u}_{\mathbf{q},\omega}$ as phonon degrees of freedom. Notice that the latter are formally complex variables, obeying $\bar{\mathbf{u}}_{\mathbf{q},\omega} = \mathbf{u}_{-\mathbf{q},-\omega}$. In accordance with our considerations of real boson fields in Section 6.4, their Keldysh action is given by

$$S[\mathbf{u}] = \frac{\rho_m}{2} \sum_{\mathbf{q},\omega,j} \bar{u}^{\mu,\alpha}_{\mathbf{q},\omega} \left[\omega^2 - \left(\omega^{(j)}_{\mathbf{q}} \right)^2 \right] \eta^{(j)}_{\mu\nu}(\mathbf{q})\, \hat{\sigma}_1^{\alpha\beta}\, u^{\nu,\beta}_{\mathbf{q},\omega}, \tag{18.1}$$

where ρ_m is the material mass density, $\alpha, \beta = cl, q$ and $\hat{\sigma}_1$ is the Pauli matrix in the Keldysh space.[1] Index j is used to label phonon modes whose dispersion relations are denoted as $\omega^{(j)}_{\mathbf{q}}$. We shall restrict ourselves to the low-frequency acoustic modes, but the optical ones may be treated in exactly the same way. The acoustic modes come with longitudinal, $j = l$, and transversal, $j = t$, polarizations. Their long-wavelength dispersion relations are $\omega^{(j)}_{\mathbf{q}} = v_j q$, where $v_{l,t}$ are the speeds of longitudinal and transversal sound correspondingly. In the action (18.1) these two polarizations are encoded by the two projectors

[1] Notice dimensions: $[\mathbf{u}(\mathbf{r}, t)] = L$, thus $[\mathbf{u}_{\mathbf{q},\omega}] = L^{d+1}T$, while $[\rho_m] = M/L^d$ and $[\sum_{\mathbf{q},\omega}] = L^{-d}T^{-1}$. Therefore $[S[\mathbf{u}]] = ML^2/T = [\hbar]$.

$$\eta^{(l)}_{\mu\nu}(\mathbf{q}) = \frac{q_\mu q_\nu}{q^2}; \qquad \eta^{(t)}_{\mu\nu}(\mathbf{q}) = \delta_{\mu\nu} - \frac{q_\mu q_\nu}{q^2}, \tag{18.2}$$

with matrix multiplication properties: $\eta^{(j)}\eta^{(j)} = \eta^{(j)}$; $\eta^{(t)}\eta^{(l)} = 0$, $\eta^{(t)} + \eta^{(l)} = 1$, and $\operatorname{tr}\eta^{(t)} = d - 1$; $\operatorname{tr}\eta^{(l)} = 1$, showing that there are $d - 1$ transversal and one longitudinal mode. The longitudinal part of the action is expressed in terms of $(q_\nu u^\nu) \sim \operatorname{div}\mathbf{u}$, representing lattice compression or dilution. On the contrary, the transversal displacements are not associated with local variations of the lattice density.

The phonon propagators,

$$D^{\alpha\beta}_{\nu\mu}(\mathbf{q}, \omega) = -\mathrm{i}\langle u^{\nu,\alpha}_{\mathbf{q},\omega}\, \bar{u}^{\mu,\beta}_{\mathbf{q},\omega}\rangle, \tag{18.3}$$

may be straightforwardly read out from the Gaussian action (18.1). In particular, the corresponding retarded (advanced) components are given by

$$D^{R(A)}_{\nu\mu}(\mathbf{q}, \omega) = \frac{1}{\rho_m} \sum_{j=l,t} \frac{\eta^{(j)}_{\nu\mu}(\mathbf{q})}{(\omega \pm \mathrm{i}0)^2 - \left(\omega^{(j)}_{\mathbf{q}}\right)^2}. \tag{18.4}$$

Their imaginary parts may be written as

$$\operatorname{Im} D^{R(A)}_{\nu\mu}(\mathbf{q}, \omega) = \pm\frac{\pi}{\rho_m} \sum_{j=l,t} \frac{\eta^{(j)}_{\nu\mu}(\mathbf{q})}{2\,\omega} \left[\delta\left(\omega - \omega^{(j)}_{\mathbf{q}}\right) + \delta\left(\omega + \omega^{(j)}_{\mathbf{q}}\right)\right]. \tag{18.5}$$

The Keldysh component is given by the matrix $D^K = D^R B - B D^A$, where B is the matrix of phonon distribution functions. We will not discriminate between $d - 1$ transversal modes and assign them the same scalar distribution function, $B^{(t)}$, while longitudinal ones may have a different one, $B^{(l)}$. Under this assumption the distribution matrix may be written as $B = \eta^{(l)} B^{(l)} + \eta^{(t)} B^{(t)}$. The Wigner transform of these two-point objects is $B^{(j)}(\mathbf{r}, t; \mathbf{q}, \omega)$. Due to the fact that the distribution functions appear in product with $\operatorname{Im} D^R$, which contains the delta function, one may put $\omega = \omega^{(j)}_{\mathbf{q}}$ and consider simpler objects $B^{(j)}_{\mathbf{q}}(\mathbf{r}, t) = B^{(j)}(\mathbf{r}, t; \mathbf{q}, \omega^{(j)}_{\mathbf{q}})$. Occasionally we will restrict ourselves to distributions isotropic in the phonon momenta space, where the distribution functions do not depend on momentum direction, but only on its absolute value. In this case, one can consider $B_{\omega^{(j)}_{\mathbf{q}}}(\mathbf{r}, t)$, or for brevity $B^{(j)}_{\omega}(\mathbf{r}, t)$. In thermal equilibrium they are given by $B^{(l)}_{\omega} = B^{(t)}_{\omega} = \coth(\omega/2T)$ in accordance with the fluctuation-dissipation relation. Away from equilibrium $B^{(j)}_{\mathbf{q}}(\mathbf{r}, t)$ should be found from kinetic equations, discussed in Section 18.6.

18.2 Fröhlich and Pippard Electron–Phonon Interactions

The most obvious mechanism of electron–phonon interactions in metals is due to the Coulomb attraction. Indeed, since the conduction band electrons are delocalized and separated from the underlying crystal lattice, the latter carries a positive charge density. It is important that, due to the global charge neutrality, the lattice charge density is exactly equal (and opposite) to the average electronic charge density, ρ_0, which may be written as

$$\rho_0 = \int^{k_F} \frac{d^d\mathbf{k}}{(2\pi)^d} = \int_0^{\epsilon_F} d\epsilon\ \nu(\epsilon) = \frac{k_F v_F \nu}{d}, \tag{18.6}$$

where we assumed a parabolic band with the density of states $\nu(\epsilon) = \nu(\epsilon/\epsilon_F)^{d/2-1}$ and $\nu = \nu(\epsilon_F)$, where $\epsilon_F = k_F v_F/2$. The phonon displacement, $\mathbf{u}(\mathbf{r}, t)$, creates thus a nonuniform lattice charge density $\rho_{\text{lat}}(\mathbf{r}, t) = \rho_0 \text{div}\, \mathbf{u}(\mathbf{r}, t) - \rho_0$. Such lattice charge interacts with the electronic charge density, $\rho_e(\mathbf{r}, t) = \bar{\psi}(\mathbf{r}, t)\psi(\mathbf{r}, t)$, via the Coulomb interactions.

The Coulomb interactions are screened by the fast electronic degrees of freedom, while slow phonons do not effectively participate in the screening. One should thus use RPA interactions $U^{-1}_{\text{RPA}}(\mathbf{q}, \omega) = (4\pi e^2/q^2)^{-1} + \Pi_0(\mathbf{q}, \omega)$, discussed in Section 10.7. Here $\Pi_0(\mathbf{q}, \omega)$ is the electronic polarization operator, Sections 10.5 and 10.6. Since phonon frequencies are typically much smaller than the electronic ones, one is justified considering the static limit of the polarization operator, $\Pi_0(\mathbf{q}, 0) \approx \nu$; see Eq. (10.56) in the clean case, or Eq. (14.60) in the disordered case. Moreover, for long-wavelength acoustic phonons the characteristic wavenumbers are much less than inverse screening radius, $q \ll \sqrt{4\pi e^2 \nu}$. This leads to the approximately frequency- and momentum-independent interactions $U^{-1}_{\text{RPA}}(\mathbf{q}, \omega) \approx \nu$. In the space-time representation those are local and instantaneous interactions between the lattice distortion, $\rho_0 \text{div}\, \mathbf{u}(\mathbf{r}, t)$, and the electronic density, $\bar{\psi}(\mathbf{r}, t)\psi(\mathbf{r}, t)$, of the form

$$\begin{aligned} S^{(\text{F})}[\mathbf{u}, \psi] &= \frac{1}{\nu} \int_C dt \int d\,\mathbf{r}\ \rho_0 \text{div}\, \mathbf{u}(\mathbf{r}, t)\, \bar{\psi}(\mathbf{r}, t)\psi(\mathbf{r}, t) \\ &= \int_C dt \sum_{\mathbf{k},\mathbf{q}} \bar{\psi}\left(\mathbf{k} + \frac{\mathbf{q}}{2}, t\right) \Gamma^{(\text{F})}_{\mu\nu} i q^\mu u^\nu_{\mathbf{q},t} \psi\left(\mathbf{k} - \frac{\mathbf{q}}{2}, t\right), \end{aligned} \tag{18.7}$$

where

$$\Gamma^{(\text{F})}_{\mu\nu} = \frac{\rho_0}{\nu}\, \delta_{\mu\nu} = \frac{k_F v_F}{d}\, \delta_{\mu\nu}; \tag{18.8}$$

compare Eq. (18.6). This is the celebrated Fröhlich electron–phonon coupling [404] (denoted thus with the superscript (F)). Notice that it provides electron interactions only with the *longitudinal* phonons. Indeed, the transversal ones do not lead to the lattice charge density distortion and thus do not contribute to the Coulomb energy. It is also important to notice that the coupling constant, $k_F v_F/d$, is independent of lattice properties as well as details of the interactions. It depends only on the Fermi surface parameters. This is due to the assumed perfect screening of the Coulomb interactions by the conduction electrons. This is a valid approximation in good metals and superconductors, where the Fermi velocity is much larger than the speed of sound, $v_F \gg v_l$. If this condition is not satisfied, there may be important deviations from the Fröhlich form, Eqs. (18.7), (18.8).

While legitimate in the clean case, the Fröhlich coupling misses an important piece of the physics in the "dirty" limit, $ql_{\text{el}} \ll 1$, where q is phonon wavenumber and $l_{\text{el}} = v_F \tau_{\text{el}}$ is electron elastic mean free path. As was first realized by Pippard [406], phonons not only deform the lattice but also displace impurities, transforming formerly static disorder potential $V_{\text{dis}}(\mathbf{r})$ into the dynamic one, $V_{\text{dis}}(\mathbf{r} + \mathbf{u}(\mathbf{r}, t))$. Colloquially, this leads to the electron density being dragged along with the lattice displacement and providing a compensation for the induced lattice charge $e\rho_0 \operatorname{div} \mathbf{u}$. In other words, the displaced impurity potential provides fast elastic relaxation of the electron distribution around the Fermi surface locally deformed by phonons. These ideas were put on the quantitative basis by Tsuneto [407] and Schmid [408], who showed that in the limit $ql_{\text{el}} \ll 1$ the Fröhlich coupling, Eq. (18.7), should be modified as

$$S^{(\text{P})}[\mathbf{u}, \psi] = \int_C dt \sum_{\mathbf{k},\mathbf{q}} \bar{\psi}\left(\mathbf{k} + \frac{\mathbf{q}}{2}, t\right) \Gamma^{(\text{P})}_{\mu\nu}(\mathbf{k}) \mathrm{i} q^{\mu} u^{\nu}_{\mathbf{q},t} \psi\left(\mathbf{k} - \frac{\mathbf{q}}{2}, t\right), \qquad (18.9)$$

where the Pippard (P) vertex is given by

$$\Gamma^{(\text{P})}_{\mu\nu}(\mathbf{k}) = \frac{k_F v_F}{d} \delta_{\mu\nu} - k_\mu v_\nu, \qquad (18.10)$$

where $v_\nu = \partial \epsilon_{\mathbf{k}}/\partial k_\nu = k_\nu/m$ for the parabolic dispersion $\epsilon_{\mathbf{k}} = k^2/2m$. The first term on the right-hand side of Eq. (18.10) is already familiar Fröhlich coupling. The second one originates from the impurities motion.

The remarkable feature of the Pippard coupling is that its integral over the Fermi surface vanishes

$$\int d\Omega_{\mathbf{k}}\, \Gamma^{(\text{P})}_{\mu\nu}(\mathbf{k}) = 0, \qquad (18.11)$$

where $\int d\Omega_{\mathbf{k}}$ denotes angular integration in the reciprocal space. Therefore in the zeroth approximation the impurities motion cancels the Coulomb effect and eliminates the scalar Fröhlich coupling between longitudinal phonons and the electron density. The remaining coupling is of the quadrupole nature, described by the traceless symmetric tensor (18.10). This greatly *reduces* electron–phonon interactions in the "dirty" limit, $ql_{\text{el}} \ll 1$, in comparison with the scalar Fröhlich coupling.

On the bright side, it predicts almost equal coupling of electrons with the long-wavelength longitudinal and transversal phonons (see the following discussion) – the feature that is routinely observed experimentally. The precise cancellation of the scalar coupling, expressed in Eq. (18.11), is not a law of nature. It is a consequence of the assumed perfect screening and the impurities rigidly following the lattice displacements. If either of these assumptions is not fulfilled, some degree of the scalar Fröhlich coupling remains, significantly enhancing interactions with the longitudinal phonons.

The most straightforward way to derive Eqs. (18.9), (18.10) [407, 409] is by performing a unitary transformation, which yields a Hamiltonian in the co-moving reference frame, where the impurity potential is static. Alternative ways maybe found in Refs. [410, 411, 412]. We shall not repeat these derivations here. Instead, we accept Eqs. (18.9), (18.10) as a phenomenological starting point and derive an effective nonlinear sigma model which incorporates electron–phonon interaction in the Pippard–Tsuneto–Schmid form. Problem 18.8.4 includes impurities motion in the sigma-model from the first principles and shows that it leads to the same result. This provides an indirect justification for the microscopic vertex in the form of Eqs. (18.9), (18.10).

18.3 Nonlinear Sigma–Model for Electron–Phonon Interactions

Following Section 14.2 we now perform the averaging over the static disorder and introduce the time nonlocal matrix field $\hat{Q}_{t,t'}(\mathbf{r})$ to split emerging four-fermion term. The resulting action, including electron–phonon coupling, Eqs. (18.9), (18.10), is now quadratic in the fermionic fields, which may be integrated out in the usual way, leading to (cf. Eq. (14.21)):

$$
iS[\hat{Q},\hat{\mathbf{u}}] = -\frac{\pi\nu}{4\tau}\,\mathrm{Tr}\{\hat{Q}^2\} + \mathrm{Tr}\ln\left[\hat{G}^{-1} + \frac{i}{2\tau_{\mathrm{el}}}\hat{Q} + \Gamma^{(\mathrm{P})}_{\mu\nu}\partial^{\mu}\hat{u}^{\nu}\right], \tag{18.12}
$$

where $\hat{G}^{-1}$ is the free fermion inverse Green function (10.36), $\hat{G}^{-1} = i\partial_t + \nabla^2_{\mathbf{r}}/2m + \mu \approx i\partial_t + i\mathbf{v}_F\nabla_{\mathbf{r}}$. The Keldysh matrix of lattice displacements has the standard form $\hat{\mathbf{u}} = \mathbf{u}^{\alpha}\hat{\gamma}^{\alpha}$, where $\alpha = cl, q$ and $\hat{\gamma}^{cl} = \hat{\sigma}_0, \hat{\gamma}^{q} = \hat{\sigma}_1$ are the two vertex matrices in the fermionic Keldysh space.

The soft diffusive modes of the action are described by the manifold $\hat{Q}^2 = \hat{1}$ and therefore one can write $\hat{Q} = \hat{\mathcal{R}}^{-1} \circ \hat{\Lambda} \circ \hat{\mathcal{R}}$, Eq. (14.27), where $\hat{\Lambda}$ is the fermionic Green function at coinciding spatial points, Eq. (14.25). One then introduces impurity-dressed Green function $\hat{\mathcal{G}}^{-1} = \hat{G}^{-1} + \frac{i}{2\tau_{\mathrm{el}}}\hat{\Lambda}$ and rewrites the action (18.12) as (cf. Eq. (14.32))

$$
iS[\hat{Q},\hat{\mathbf{u}}] = \mathrm{Tr}\ln\Big[\hat{1} + i\hat{\mathcal{G}}\hat{\mathcal{R}}\partial_t\hat{\mathcal{R}}^{-1} + i\hat{\mathcal{G}}\hat{\mathcal{R}}\mathbf{v}_F\nabla_{\mathbf{r}}\hat{\mathcal{R}}^{-1} + \hat{\mathcal{G}}\,\hat{\mathcal{R}}\,\Gamma^{(\mathrm{P})}_{\mu\nu}\partial^{\mu}\hat{u}^{\nu}\,\hat{\mathcal{R}}^{-1}\Big]. \tag{18.13}
$$

Finally, one expands the logarithm here to the lowest nonvanishing orders. First, neglecting the electron–phonon Γ-term, one obtains the already familiar nonlinear sigma-model action, $iS[\hat{Q}]$, Eq. (14.33). Focusing now on the phonon-induced term, it is easy to see that the first order in Γ term vanishes upon integration over the fast fermionic momentum, $\mathbf{k}$, due to the relation (18.11), see footnote 5 in Chapter 14 following Eq. (14.32). It is this point, where the Pippard–Tsuneto–Schmid coupling, Eqs. (18.9), (18.10), is qualitatively different from the Fröhlich one (the latter would bring the first order $\mathrm{Tr}\{(\rho_0 \mathrm{div}\, \hat{\mathbf{u}})\hat{Q}\}$ term). Going to the second order in $\Gamma^{(\mathrm{P})}$, one finds

$$iS[\hat{Q},\hat{\mathbf{u}}] = iS[\hat{Q}] - \frac{1}{2}\mathrm{Tr}\big[\hat{\mathcal{G}}\hat{\mathcal{R}}\Gamma^{(\mathrm{P})}_{\mu\nu}\partial^{\mu}\hat{u}^{\nu}\hat{\mathcal{R}}^{-1}\hat{\mathcal{G}}\hat{\mathcal{R}}\Gamma^{(\mathrm{P})}_{\eta\lambda}\partial^{\eta}\hat{u}^{\lambda}\hat{\mathcal{R}}^{-1}\big].$$

We employ now Eq. (14.30) along with

$$\sum_{\mathbf{k}}\mathcal{G}^R(\mathbf{k},\epsilon)\mathcal{G}^A(\mathbf{k},\epsilon) = 2\pi\nu\tau_{\mathrm{el}}; \quad \sum_{\mathbf{k}}\mathcal{G}^R(\mathbf{k},\epsilon)k_{\mu}v_{\nu}\,\mathcal{G}^A(\mathbf{k},\epsilon) = \frac{2\pi\nu k_F v_F\tau_{\mathrm{el}}}{d}\,\delta_{\mu\nu};$$

$$\sum_{\mathbf{k}}\mathcal{G}^R(\mathbf{k},\epsilon)k_{\mu}v_{\nu}\,\mathcal{G}^A(\mathbf{k},\epsilon)k_{\eta}v_{\lambda} = \frac{2\pi\nu k_F^2 v_F^2\tau_{\mathrm{el}}}{d(d+2)}\left(\delta_{\mu\nu}\delta_{\eta\lambda}+\delta_{\mu\eta}\delta_{\nu\lambda}+\delta_{\mu\lambda}\delta_{\nu\eta}\right),$$

to finally find

$$iS[\hat{Q},\hat{\mathbf{u}}] = iS[\hat{Q}] + \frac{\pi\nu k_F^2 D}{4}\,\mathrm{Tr}\Big\{[\hat{Q},\partial^{\mu}\hat{u}^{\nu}][\hat{Q},\partial^{\eta}\hat{u}^{\lambda}]\Big\}\,\Upsilon_{\mu\nu,\eta\lambda}, \tag{18.14}$$

where $[\hat{Q},\partial^{\mu}\hat{u}^{\nu}] = \hat{Q}_{tt'}(\mathbf{r})\partial^{\mu}\hat{u}^{\nu}_{\mathbf{r},t'} - \partial^{\mu}\hat{u}^{\nu}_{\mathbf{r},t}\hat{Q}_{tt'}(\mathbf{r})$ and $D = v_F^2\tau_{\mathrm{el}}/d$ is the diffusion constant,

$$\Upsilon_{\mu\nu,\eta\lambda} \equiv \frac{1}{d+2}\left[\delta_{\mu\eta}\delta_{\nu\lambda}+\delta_{\mu\lambda}\delta_{\nu\eta}-\frac{2}{d}\,\delta_{\mu\nu}\delta_{\eta\lambda}\right]. \tag{18.15}$$

Notice that, due to the symmetry of this tensor with respect to $\mu\leftrightarrow\nu$ and $\eta\leftrightarrow\lambda$ permutations, one can rewrite the electron–phonon vertex (18.14) in terms of the linear strain tensor $\hat{\varepsilon}^{\mu\nu}_{\mathbf{r},t} = (\partial^{\mu}\hat{u}^{\nu}_{\mathbf{r},t}+\partial^{\nu}\hat{u}^{\mu}_{\mathbf{r},t})/2$:

$$iS[\hat{Q},\hat{\varepsilon}] = iS[\hat{Q}] + \frac{\pi\nu k_F^2 D}{2(d+2)}\left[\mathrm{Tr}\Big\{[\hat{Q},\hat{\varepsilon}^{\mu\nu}][\hat{Q},\hat{\varepsilon}^{\nu\mu}]\Big\} - \frac{1}{d}\,\mathrm{Tr}\Big\{[\hat{Q},\hat{\varepsilon}^{\mu\mu}][\hat{Q},\hat{\varepsilon}^{\nu\nu}]\Big\}\right]. \tag{18.16}$$

The spatially local, second-order in the lattice displacement, $\mathbf{u}$, vertex (18.14), or (18.16) is the leading term describing interaction of phonons with the electronic degrees of freedom in disordered metals, in the $ql_{\mathrm{el}} \ll 1$ limit.

The present derivation relies on the phenomenological electron–phonon coupling (18.9), (18.10). Problem 18.8.4 addresses how it may be systematically deduced from the phonon-displaced disorder potential $V_{\mathrm{dis}}(\mathbf{r}+\mathbf{u}(\mathbf{r},t))$. The scalar Fröhlich vertex, Eq. (18.8), leads to the linear in displacement vertex

$S \propto \rho_0 \mathrm{Tr}\{\mathrm{div}\,\hat{\mathbf{u}}\,\hat{Q}\}$. Such a term is analogous to the scalar potential term in Eq. (14.48). Therefore, the electron–phonon kinetics with scalar Fröhlich interactions is very similar to the theory of electron–electron interactions, considered in Chapter 16. The only difference is that the electron–electron effective interaction $U_{\mathrm{RPA}}(\mathbf{q}, \omega)$ is substituted by the longitudinal phonon propagator $q^{\mu} q^{\nu} D_{\nu\mu}(\mathbf{q}, \omega)$, Eq. (18.5). Such a theory was developed, for example, in Ref. [413]. It leads to the disorder-enhanced electron–phonon interaction effects, akin to the disorder-enhanced electron–electron interactions phenomena of Chapter 16. In the presence of the perfect screening the scalar term is absent, however, due to the traceless form of the Pippard–Tsuneto–Schmid electron–phonon vertex (18.10). In this case one has to resort to the quadratic quadrupole vertex (18.14), first derived in Ref. [414]. It leads to disorder-*suppressed* electron–phonon interactions effects, considered in Refs. [407, 408, 410, 411, 412, 414]. The sigma-model action (18.14) provides a compact way of deriving these results, which will be demonstrated in what follows.

18.4 Electron Kinetics in Normal Metals

To derive electron collision integral due to the electron–phonon interactions one first averages the action (18.14) over the fluctuations of the lattice displacements, $\mathbf{u}(\mathbf{r}, t)$. With the help of Eq. (18.3) one obtains the collision action from Eq. (18.14):

$$\mathrm{i}\langle S[\hat{Q}, \hat{\mathbf{u}}]\rangle_{\mathbf{u}} = \mathrm{i}S[\hat{Q}] + \mathrm{i}S_{\mathrm{coll}}[\hat{Q}], \tag{18.17}$$

where

$$\mathrm{i}S_{\mathrm{coll}}[\hat{Q}] = \frac{\pi \nu k_F^2 D}{4} \sum_{\mathbf{q},\omega} \mathrm{Tr}\big\{\hat{Q}_{\epsilon-\omega,\epsilon'-\omega}\hat{\gamma}^{\alpha}\hat{Q}_{\epsilon',\epsilon}\hat{\gamma}^{\beta}\big\}\, D^{\alpha\beta}_{\nu\lambda}(\mathbf{q}, \omega) q^{\mu} q^{\eta} \Upsilon_{\mu\nu,\eta\lambda}, \tag{18.18}$$

where the trace, Tr, operation involves Keldysh matrix structure along with the summation over ϵ and ϵ'. We now adopt the strategy of Sections 14.4 and 17.2, where the kinetic (Usadel) equation is identified as the stationary point condition for the $\hat{Q}$-field, which stabilizes the action. We thus look for the matrix stationary point equation $\delta\langle S[\hat{Q}, \hat{\mathbf{u}}]\rangle_{\mathbf{u}}/\delta\hat{Q}_{\epsilon\epsilon}(\mathbf{r}) = 0$. As explained in Section 14.4, its Keldysh $(1, 2)$ component constitutes the kinetic equation for the distribution function F_{ϵ}. This way one finds

$$\partial_t F_{\epsilon} - \nabla_{\mathbf{r}}\,[D \nabla_{\mathbf{r}} F_{\epsilon}] = I_{\mathrm{coll}}[F_{\epsilon}(\mathbf{r}, t)], \tag{18.19}$$

where the electron–phonon collision integral is given by

$$I_{\mathrm{coll}}[F_{\epsilon}(\mathbf{r}, t)] = \frac{1}{\pi \nu} \left(\frac{\delta \mathrm{i} S_{\mathrm{coll}}}{\delta \hat{Q}_{\epsilon\epsilon}(\mathbf{r})} \right) \Bigg|^{(1,2)}_{\hat{Q}=\hat{\Lambda}}. \tag{18.20}$$

The variational derivative here ought to be restricted to the sigma-model target space, $\hat{Q}^2 = 1$. A simple way to ensure this is to use the $\hat{Q}$-matrix parameterization $\hat{Q} = e^{-\hat{\mathcal{W}}/2}\hat{\Lambda}e^{\hat{\mathcal{W}}/2} \approx \hat{\Lambda} + \frac{1}{2}[\hat{\Lambda}, \hat{\mathcal{W}}]$ and expand the action (18.18) to the linear order in $\hat{\mathcal{W}}_{\epsilon\epsilon}$. Here $\hat{\mathcal{W}}$'s are infinitesimal generators of the symmetry transformations. This way one obtains

$$I_{\text{coll}}[F_\epsilon] = \frac{k_F^2 D}{4\mathrm{i}} \sum_{\mathbf{q},\epsilon'} \Big[\hat{\gamma}^\beta \hat{\Lambda}_{\epsilon'} \hat{\gamma}^\alpha \hat{\Lambda}_\epsilon - \hat{\Lambda}_\epsilon \hat{\gamma}^\beta \hat{\Lambda}_{\epsilon'} \hat{\gamma}^\alpha\Big]^{(1,2)} \times D^{\alpha\beta}_{\nu\lambda}(\mathbf{q}, \epsilon - \epsilon') q^\mu q^\eta \Upsilon_{\mu\nu,\eta\lambda}. \tag{18.21}$$

We now employ the classical form of $\hat{\Lambda}_\epsilon$, Eq. (14.25), along with the vertex matrices, $\hat{\gamma}^{cl,q}$ to evaluate the Keldysh (1, 2) component (cf. with Eq. (16.51) for electron–electron interactions)[2]

$$I_{\text{coll}}[F_\epsilon] = \sum_{\epsilon'} \mathcal{M}(\epsilon - \epsilon')\Big[1 - F_\epsilon F_{\epsilon'} - B_{\epsilon-\epsilon'}(F_\epsilon - F_{\epsilon'})\Big], \tag{18.22}$$

where the dimensionless phonon matrix element is given by

$$\mathcal{M}(\omega) = 2k_F^2 D \sum_{\mathbf{q}} \mathrm{Im}[D^R_{\nu\lambda}(\mathbf{q}, \omega)]\, q^\mu q^\eta \Upsilon_{\mu\nu,\eta\lambda} = \mathcal{M}_l(\omega) + (d-1)\mathcal{M}_t(\omega), \tag{18.23}$$

with the superscripts $j = l, t$ denoting the one longitudinal and $(d-1)$ transversal modes. Employing Eqs. (18.3) and (18.15), one finds:[3]

$$\mathcal{M}_j(\omega) = \frac{b_j \pi k_F^2 D}{\rho_m} \sum_{(\mp),\mathbf{q}} \frac{q^2}{\omega} \delta(\omega \mp v_j q) = \frac{b_j \pi \Omega_d}{(2\pi)^d} \frac{k_F^2 D|\omega|^d \operatorname{sign}\omega}{\rho_m v_j^{d+2}} \equiv \frac{|\omega|^d \operatorname{sign}\omega}{\Theta_j^d}, \tag{18.24}$$

where $\Omega_d = 2\pi^{d/2}/\Gamma(d/2)$ is the area of S_{d-1} unit sphere and the coefficients are given $b_t = 1/(d+2)$; $b_l = b_t(2d-2)/d$. The two characteristic energy scales $\Theta_{l,t}$ are defined through the last equality in (18.24).

Consider now a situation where the electrons are heated by an external radiation or a current. Due to the fast electron–electron relaxation they establish an almost thermal distribution with an elevated temperature, T_e, which may be significantly larger than the lattice temperature, T. To discuss the heat transfer between electron and phonon subsystems, define electronic energy density as $E_e = \int d\epsilon\, \nu(\epsilon)\epsilon\, n_\epsilon$, where the fermionic occupation number, n_ϵ, is found from $F_\epsilon = 1 - 2n_\epsilon$. In a spatially uniform case one finds from the kinetic equation

[2] Notice that the 1 in the square brackets on the right-hand side is kept for convenience. It is actually given by $(\Lambda^R_\epsilon - \Lambda^A_\epsilon)\sum_{\epsilon'}[\Lambda^R_{\epsilon'} D^R(\epsilon' - \epsilon) + \Lambda^A_{\epsilon'} D^A(\epsilon' - \epsilon)] = 0$, since it involves the energy integration of the sum of the retarded and advanced correlators, Eqs. (2.45) and (10.29). However, recalling that $\Lambda^{R(A)} = \pm 1$, one can formally write this expression as $\propto \sum_{\epsilon'}[D^R(\epsilon' - \epsilon) - D^A(\epsilon' - \epsilon)] \propto \sum_{\epsilon'} \mathrm{Im}[D^R(\epsilon' - \epsilon)]$, resulting in the first term in the square brackets in Eq. (18.22).

[3] Here we employed $q^\mu q^\eta \Upsilon_{\mu\nu,\eta\lambda} = q^2[b_t \eta^{(t)}_{\nu\lambda} + b_l \eta^{(l)}_{\nu\lambda}]$ with $b_t = 1/(d+2)$; $b_l = b_t(2d-2)/d$ along with $\mathrm{tr}\,\eta^{(t)} = d - 1$ and $\mathrm{tr}\,\eta^{(l)} = 1$.

(18.19), $\partial_t E_e = -(\nu/2)\int d\epsilon\, \epsilon\, I_{\text{coll}}$, where we put $\nu(\epsilon) \approx \nu$, anticipating that the integral is localized near the Fermi energy (see the following discussion). We now use that in a state of partial equilibrium $1 - F_\epsilon F_{\epsilon'} = (F_\epsilon - F_{\epsilon'})\coth \omega/2T_e$, where $\omega = \epsilon - \epsilon'$ and $B_{\epsilon-\epsilon'} = \coth \omega/2T$. Noticing also that $\int d\epsilon\, \epsilon(F_\epsilon - F_{\epsilon-\omega}) = \omega^2$, one finds, from Eqs. (18.22) and (18.24), [410, 411]

$$\partial_t E_e = \nu \int_0^\infty \frac{d\omega}{2\pi} \mathcal{M}(\omega)\,\omega^2 \left[n_B\left(\frac{\omega}{T}\right) - n_B\left(\frac{\omega}{T_e}\right)\right] = \left(T^{d+3} - T_e^{d+3}\right)\alpha_d \nu \sum_{j=1}^{d} \frac{1}{\Theta_j^d}, \tag{18.25}$$

where the bosonic occupation number is $n_B = (B_\omega - 1)/2 = (e^{\omega/T} - 1)^{-1}$ and $\alpha_3 = 4\pi^5/63$, $\alpha_2 = 12\zeta(5)/\pi \approx 3.96$, $\alpha_1 = \pi^3/30$. The j summation hereafter runs over one longitudinal mode and $d - 1$ transversal ones.

Dynamics of electronic temperature relaxation is found as $\partial_t E_e = C(T_e)\partial_t T_e$, where the electronic specific heat is $C(T_e) = \partial E_e/\partial T_e = \frac{\pi^2}{3}\nu T_e$. This way, one obtains

$$\partial_t T_e = \left(T^{d+3} - T_e^{d+3}\right)\frac{3\alpha_d}{\hbar\pi^2 T_e}\sum_{j=1}^{d}\frac{1}{\Theta_j^d} \approx -T_e^{d+2}\frac{3\alpha_d}{\hbar\pi^2}\sum_{j=1}^{d}\frac{1}{\Theta_j^d}, \tag{18.26}$$

where in the last equality we took the limit $T_e \gg T$ and restored $\hbar$ to have proper dimensions. This leads to a characteristic relaxation time $\tau \propto \hbar\Theta_j^d/T_e^{d+1}$ and a slow power-law cooling $T_e(t) \sim t^{-1/(d+1)}$ at $t > \tau$. In the opposite limit when $\delta T = T_e - T \ll T$ one finds exponential equilibration, $\delta T(t) \propto e^{-t/\tau_E}$, with the energy relaxation rate [408]

$$\frac{\hbar}{\tau_E} = \frac{3\alpha_d(d+3)}{\pi^2}\left(\frac{1}{\Theta_l^d} + \frac{d-1}{\Theta_t^d}\right) T^{d+1}. \tag{18.27}$$

We found thus that the cooling rate of electrons in disordered metals decreases extremely fast at temperatures $T < \Theta_t < \Theta_l$. This can lead to spectacular effects due to an effective decoupling of electrons and the lattice, resulting in overheating of the electronic subsystem by an applied current [415].

To give an order of magnitude estimate consider aluminum with $\rho_m = 2.7\,\text{g/cm}^3$, $\epsilon_F = 11.7\,\text{eV}$ and $v_F = 1.7 \times 10^8\,\text{cm/s}$, $v_t = 3 \times 10^5\,\text{cm/s}$, $v_l = 6.4 \times 10^5\,\text{cm/s}$. Taking the elastic mean free path as $l_{\text{el}} = 100\,\text{nm}$, one obtains diffusion constant $D = 570\,\text{cm}^2/\text{s}$, and $\Theta_t = 10\,\text{K}$, while $\Theta_l = 35\,\text{K}$. The energy relaxation time is given by $\tau_E = 3(1\text{K}/T)^4\,\mu\text{s}$. This may be compared with the electron–electron relaxation time, evaluated in Section 16.5. The latter scales as $\tau_{\text{ee}} = \hbar\nu(\hbar D/T)^{d/2}$. For our example of aluminum one finds $\tau_{\text{ee}} = 2.5(1\text{K}/T)^{3/2}\,\mu\text{s}$. Thus, for $T \gtrsim 1\,\text{K}$ the electron–phonon time is shorter, implying fast equilibration between electrons and the lattice. For smaller temperatures aluminum is superconducting, of course, requiring a separate consideration.

18.5 Kinetics of Quasiparticles in Superconductors

Quasiparticles in superconductors may be either thermal or non-equilibrium, excited by external circuit fluctuations, RF signal, or injected through, for example, a tunneling contact. They occupy states above the energy gap, $|\epsilon| > \Delta$. Such quasiparticles constitute a major source of decoherence in superconducting qubits [416, 417]. Understanding their thermalization rates is thus of primary importance for improving performance of prototypical quantum computers. At small temperature, $T \ll \Delta$, the quasiparticle concentration is small and therefore their mutual interactions are usually inconsequential. In this situation the dominant mechanism of the quasiparticles' thermalization is their interactions with the acoustic phonons.[4]

The beauty of the sigma-model description, presented in Section 18.3 is that it is immediately applicable to superconductors. One should only enlarge the dimensionality of the Q-matrix from the Keldysh to Keldysh$\otimes$Nambu space, $\hat{Q} \to \check{Q}$ and introduce a factor of $1/2$ in front of the action (18.12) to acknowledge Nambu doubling of the degrees of freedom. With these simple modifications Eqs. (18.14) and (18.18) hold, and one can employ them to describe kinetics of quasiparticle–phonon interactions. Following Section 17.2, one looks for the stationary point of the action (18.18), in particular its $(1,2)$ Keldysh component. The resulting equation is still a matrix in the Nambu space, which needs to be projected onto Nambu $\hat{\tau}_0$ and $\hat{\tau}_3$ components. This way, one obtains kinetic equations for the longitudinal (energy), F_L, and transversal (charge), F_T, components of the distribution function; see Sections 17.2 and 17.3. Without phonons the corresponding kinetic equations are given by Eqs. (17.27). Interactions with the acoustic phonons provide the collision terms. The latter directly follow from Eq. (18.21) by substituting the BCS form of the $\check{\Lambda}$ matrix, Eqs. (17.14) and (17.15), multiplying it by $\hat{\tau}_0$ and $\hat{\tau}_3$ matrices and taking the trace over the Nambu space.

We use now the BCS Green functions, which are matrices in the Nambu space, Eq. (17.14), $\hat{\Lambda}^{\mathrm{R}}_\epsilon = \hat{\tau}_3 u_\epsilon + i\hat{\tau}_2 v_\epsilon$ and $\hat{\Lambda}^{\mathrm{A}}_\epsilon = -\hat{\tau}_3 u^*_\epsilon - i\hat{\tau}_2 v^*_\epsilon$, where the coherence factors are given by Eq. (17.15) and obey $u_\epsilon^2 - v_\epsilon^2 = 1$. The Keldysh component is given by $\hat{\Lambda}^K = \hat{\Lambda}^R\hat{F} - \hat{F}\hat{\Lambda}^A$, where $\hat{F}$ is the distribution matrix in the Nambu space, which may be written as $\hat{F} = F_{L,\epsilon}(\mathbf{r},t)\hat{\tau}_0 + F_{T,\epsilon}(\mathbf{r},t)\hat{\tau}_3$, Eq. (17.21).[5] To keep the presentation compact we first assume that there is no charge imbalance, that is, $F_T = 0$ and derive the kinetic equation for the longitudinal component. To this end we project the $(1,2)$ Keldysh component of the Usadel equation onto the $\hat{\tau}_0$ Nambu component: $\mathrm{Tr_N}\left[\hat{\tau}_0 \delta\langle S[\check{Q},\mathbf{u}]\rangle_{\mathbf{u}}/\delta\check{Q}_{\epsilon\epsilon}\Big|^{(1,2)}_{\mathrm{K}}\right] = 0$. This way, we find:

[4] Indeed, the optical phonon frequency, being of the order of the Debye frequency, ω_D, is much larger than the superconducting energy gap, Δ, making them ineffective for relaxation of low-energy quasiparticles.

[5] In equilibrium $F_{L,\epsilon} = \tanh(\epsilon/2T)$ and $F_{T,\epsilon} = 0$.

$$\frac{\nu(\epsilon)}{\nu}\,\partial_t F_{L,\epsilon} - \nabla_{\mathbf{r}}\left[D\nabla_{\mathbf{r}}F_{L,\epsilon}\right] = I_{\text{coll}}[F_L], \tag{18.28}$$

where $\nu(\epsilon) = \nu\,\text{Re}\,u_\epsilon$ is the BCS density of states, Eq. (17.18). The collision integral is given by Eq. (18.21), where the superconducting $\check{\Lambda}$ is used and the trace is taken over the Nambu space. A straightforward calculation yields

$$I_{\text{coll}}[F_L] = \sum_{\epsilon'} \mathcal{M}(\epsilon,\epsilon')\Big[1 - F_{L,\epsilon}F_{L,\epsilon'} - B_{\epsilon-\epsilon'}(F_{L,\epsilon} - F_{L,\epsilon'})\Big], \tag{18.29}$$

where the matrix element is, [410, 414],

$$\mathcal{M}(\epsilon,\epsilon') = \mathcal{M}(\epsilon-\epsilon')$$
$$[\text{Re}\,u_\epsilon \text{Re}\,u_{\epsilon'} - \text{Re}\,v_\epsilon \text{Re}\,v_{\epsilon'}] = \mathcal{M}(\epsilon-\epsilon')\frac{|\epsilon\epsilon'| - \Delta^2\text{sign}(\epsilon\epsilon')}{\sqrt{\epsilon^2-\Delta^2}\sqrt{\epsilon'^2-\Delta^2}}, \tag{18.30}$$

and $\mathcal{M}(\epsilon-\epsilon')$ is the normal state matrix element, Eq. (18.23). For BCS superconductors $\text{Re}\,u_{\epsilon'} = \text{Re}\,v_{\epsilon'} = 0$ for $|\epsilon'| < \Delta$, Eq. (17.15). This implies that the collision integral exists only if both initial and final state energies exceed the gap, Δ, that is, $|\epsilon|, |\epsilon'| > \Delta$. This is to be expected, since there are no quasiparticle states below energy Δ.

Consider relaxation of a quasiparticle initially residing at some energy $\epsilon > \Delta$. After absorbing or emitting an acoustic phonon, it acquires a final energy ϵ' with the rate given by Eq. (18.29). It is convenient to split the energy integration in this equation into the two regions $\epsilon' > \Delta$ and $\epsilon' < -\Delta$. The former processes represent scattering of quasiparticles without changing their total number. The latter leads to the quasiparticle recombination, resulting in decrease of their population. As shown in what follows, the recombination process is very slow, allowing it to reach a state of partial thermal equilibrium within the band of quasiparticles, $\epsilon > \Delta$, and quasiholes, $\epsilon < -\Delta$, without equilibrating between the two bands. To discuss such partially equilibrated state, notice that, if both $\epsilon, \epsilon' > \Delta$, the collision integral (18.29) is nullified by $F_{L,\epsilon>\Delta} = \tanh(\epsilon - \mu^*)/2T$, where T is the lattice temperature and $\mu^* < \Delta$ is an effective chemical potential residing below the bottom of the quasiparticle band. Similarly, if $\epsilon, \epsilon' < -\Delta$, the collision integral (18.29) is nullified by $F_{L,\epsilon<-\Delta} = \tanh(\epsilon + \mu^*)/2T$, rendering $F_{L,\epsilon}$ to be the odd function of energy, as it should.

The effective chemical potential, μ^*, is determined by the quasiparticle density, given by $\rho_{\text{qp}} = \int_\Delta^\infty d\epsilon\,\nu(\epsilon)n_{\text{qp}}(\epsilon)$, where the quasiparticle occupation number is found from $F_{L,\epsilon} = 1 - 2n_{\text{qp}}(\epsilon)$. In the partially equilibrated state with $T \ll \Delta - \mu^*$, one finds

$$\rho_{\text{qp}} \approx \nu \int_\Delta^\infty d\epsilon\,\frac{\epsilon}{\sqrt{\epsilon^2-\Delta^2}}\,e^{-(\epsilon-\mu^*)/T} = \nu\sqrt{\frac{\pi\Delta T}{2}}\,e^{-(\Delta-\mu^*)/T}. \tag{18.31}$$

The effective chemical potential, $\mu^*(t)$, is a slow function of time, which reaches zero in true equilibrium. To describe dynamics of the quasiparticle density, one substitutes it into the kinetic equation (18.28), where the final energy integration is restricted to $\epsilon' < -\Delta$:

$$\partial_t \rho_{\rm qp} = \nu \int_{\Delta}^{\infty} d\epsilon \int_{-\infty}^{-\Delta} \frac{d\epsilon'}{2\pi} \mathcal{M}(\epsilon, \epsilon')(F_{L,\epsilon} - F_{L,\epsilon'}) \left[n_B\left(\frac{\epsilon - \epsilon'}{T}\right) - n_B\left(\frac{\epsilon - \epsilon' - 2\mu^*}{T}\right)\right],$$

where we have used the fact, that in the state of partial equilibrium with $\epsilon > \Delta$ and $\epsilon' < -\Delta$, $1 - F_{L,\epsilon}F_{L,\epsilon'} = (F_{L,\epsilon} - F_{L,\epsilon'})\coth((\epsilon - \epsilon' - 2\mu^*)/2T)$. For $T \ll \Delta - \mu^*$ the integrals here are dominated by the near vicinities of $\pm\Delta$, correspondingly, where $F_{L,\epsilon} \approx 1$ and $F_{L,\epsilon'} \approx -1$. This way one finds

$$\partial_t \rho_{\rm qp} = \nu\, \mathcal{M}(2\Delta)\Delta T \left[e^{-2\Delta/T} - e^{-2(\Delta - \mu^*)/T} \right] \approx -\rho_{\rm qp}^2 \frac{2}{\pi\nu} \sum_{j=1}^{d} \left(\frac{2\Delta}{\Theta_j}\right)^d, \tag{18.32}$$

where in the last approximate equality we assumed $T \ll \mu^* < \Delta$ and used Eqs. (18.24) and (18.31). The fact that $\partial_t \rho_{\rm qp} \propto -\rho_{\rm qp}^2$ reflects the binary collision nature of the recombination process. Indeed, a quasiparticle needs to "collide" (through the phonon exchange) with a quasihole for the recombination to occur. The recombination rate is thus proportional to the product of their densities. Equation (18.32) predicts that after a dynamical perturbation, creating excess quasiparticles, the quasiparticle concentration is given by the universal law,

$$\rho_{\rm qp}(t) = \left(\frac{\Theta_t}{2\Delta}\right)^d \frac{\pi\hbar\nu}{2(d-1)t} = \frac{\rho_0}{d(d-1)} \left(\frac{\Theta_t}{2\Delta}\right)^d \frac{\pi\hbar}{\epsilon_F t}, \tag{18.33}$$

where ρ_0 is the total electron density and we employed Eq. (18.6). This is a remarkable result, which shows that the concentration of quasiparticles does not depend either on lattice temperature (provided $T \ll \Delta$) or on an initial quasiparticle concentration. It only depends on material parameters and time, t, the system is allowed to equilibrate after a kick. The t^{-1} algebraic decay persists up until an exponentially long (in $e^{\Delta/T}$) time, when the quasiparticle concentration approaches its equilibrium value $\rho_{\rm qp} \sim \rho_0 \sqrt{\Delta T/\epsilon_F^2}\, e^{-\Delta/T}$; compare Eq. (18.31) with $\mu^* = 0$.

Indeed, the presence of non-equilibrium quasiparticles has been detected in qubits [418, 419], where time t is limited by the coherence time and can't be extended indefinitely. To eliminate these undesirable long-lived quasiparticles, the qubit designs employ magnetic vortex [419, 420], or normal metal [421] traps and gap engineering by variation of the film thickness [422]. The basic idea is to incorporate regions with a reduced superconducting gap near the qubit. The non-equilibrium quasiparticles diffusing to these regions and partially equilibrating down to the local gap, appear to be trapped and do not influence the qubit operation. To evaluate efficiency of, for example, a vortex trap, one needs to extend the electron–phonon collision integral treatment to superconductors with the broken time reversal symmetry and suppressed gap [414], but we do not go into details here.

Before concluding this section, we briefly discuss relaxation of the charge imbalance, described by the even in energy, $F_{T,\epsilon}$, component of the distribution function. Projecting the $(1,2)$ Keldysh component of the Usadel equation onto the $\hat{\tau}_3$ Nambu component, $\mathrm{Tr_N}\left[\hat{\tau}_3\delta\langle S[\check{Q},\mathbf{u}]\rangle_{\mathbf{u}}/\delta\check{Q}_{\epsilon\epsilon}\Big|_{\mathrm{K}}^{(1,2)}\right]=0$, one finds

$$\frac{\nu(\epsilon)}{\nu}\,\partial_t F_{T,\epsilon}-\nabla_{\mathbf{r}}\left[\mathcal{D}_T(\epsilon)\nabla_{\mathbf{r}}F_{T,\epsilon}\right]=\tilde{I}_{\mathrm{coll}}[F_T,F_L], \tag{18.34}$$

where $\mathcal{D}_T(\epsilon)=D\epsilon^2/(\epsilon^2-\Delta^2)$, Eq. (17.28), is the charge diffusion coefficient. The corresponding collision integral is given by

$$\tilde{I}_{\mathrm{coll}}[F_T,F_L]=\sum_{\epsilon'}\tilde{\mathcal{M}}(\epsilon,\epsilon')\Big[-F_{L,\epsilon}F_{T,\epsilon'}-F_{T,\epsilon}F_{L,\epsilon'}-B_{\epsilon-\epsilon'}(F_{T,\epsilon}-F_{T,\epsilon'})\Big], \tag{18.35}$$

where $|\epsilon|,|\epsilon'|>\Delta$ and the matrix element is

$$\tilde{\mathcal{M}}(\epsilon,\epsilon')=\mathcal{M}(\epsilon-\epsilon')\mathrm{Re}\,u_\epsilon\mathrm{Re}\,u_{\epsilon'}=\mathcal{M}(\epsilon-\epsilon')\,\frac{\nu(\epsilon)}{\nu}\,\frac{\nu(\epsilon')}{\nu}, \tag{18.36}$$

where $\mathcal{M}(\epsilon-\epsilon')$ is the normal state matrix element, Eq. (18.23). Using the fact that $F_{L,\epsilon}$, B_ω are odd and $\tilde{\mathcal{M}}(\epsilon,\epsilon')$ is odd under permutation of the two arguments, one can show conservation of the total imbalance $\partial_t\mathcal{Q}=0$, where $\mathcal{Q}\equiv\int d\epsilon\;\nu(\epsilon)F_{T,\epsilon}$. This means that the electron–phonon collisions do *not* induce a gap in the spectrum of Carlson–Goldman transverse collective mode, Section 17.5. Indeed, conservation of the $\mathbf{q}=0$ component of the imbalance implies $\omega_{\mathrm{CG}}(0)=0$, Eq. (17.64). Moreover, one can check that if both $F_{L,\epsilon}$ and B_ω are given by their equilibrium values, then $F_{T,\epsilon}\propto\cosh^{-2}(\epsilon/2T)$ nullifies the collision integral (18.35), [377]. It represents, therefore, the (quasi)stationary form of the transverse distribution function. It may be associated with a shift of the quasiparticle chemical potential with respect to the one of the condensate, $F_{T,\epsilon}=\delta\mu\,\partial_\mu F_{L,\epsilon}|_{\mu=0}$. Such chemical potential mismatch serves as a voltage and, according to the Josephson relation, leads to a steady-state rotation of the phase of the order parameter. One can see an indication of this looking at the self-consistency relation, Eq. (17.16), where F_T induces an imaginary correction (i.e. a phase) to an otherwise real Δ. Such steady-state phase rotation due to the conserved charge imbalance exists only in a completely isolated piece of a superconductor. Any coupling to leads or external circuitry results in equilibration of the imbalance and dampening of the phase rotation [377].

18.6 Phonon Kinetics

So far we have been tacitly assuming that the phonon modes are in thermal equilibrium at the lattice temperature T and considered kinetics of the electronic (and quasiparticle) subsystem. Here we broaden the scope, allowing for the phonon modes to be away from equilibrium, describing them with the distribution functions $B_{\mathbf{q}}^{(j)}$. Our aim thus is to derive a kinetic equation for real bosonic fields, discussed in Section 6.7.

Its collision integral is expressed through the components of the phonon self-energy matrix $\hat{\Sigma}(\mathbf{q},\omega)$. The latter may be directly read out from the electron–phonon interaction vertex (18.14), (18.15). To this end we disregard fluctuations of the $\hat{Q}_{t,t'}$-matrix degrees of freedom and set them to the metallic saddle point, $\hat{Q}_{t,t'} \to \hat{\Lambda}_{t-t'} \to \hat{\Lambda}_{\epsilon}$, where we use time translation invariance and, in the last transition, the Fourier transform. With this substitution the action (18.14) acquires the following form:

$$
\mathrm{i}S[\hat{\Lambda},\hat{\mathbf{u}}] = \frac{\pi\nu k_F^2 D}{2}\sum_{\mathbf{q},\omega,j} \bar{u}^{\nu,\alpha}_{\mathbf{q},\omega}\,\mathrm{Tr}\Big\{\hat{\Lambda}_{\epsilon-\omega}\hat{\gamma}^{\alpha}\hat{\Lambda}_{\epsilon}\hat{\gamma}^{\beta} - \hat{\gamma}^{\alpha}\hat{\gamma}^{\beta}\Big\}\, b_j q^2 \eta^{(j)}_{\nu\lambda}\, u^{\lambda,\beta}_{\mathbf{q},\omega}, \qquad (18.37)
$$

where we used $q^{\mu}q^{\eta}\Upsilon_{\mu\nu,\eta\lambda} = q^2[b_t\eta^{(t)}_{\nu\lambda} + b_l\eta^{(l)}_{\nu\lambda}]$, with the coefficients b_j defined after Eq. (18.24). The Tr operation implies trace over the Keldysh (and possibly Nambu) matrix structure along with summation over ϵ. This action is quadratic in lattice displacements, $\hat{\mathbf{u}}_{\mathbf{q},\omega}$, and therefore constitutes the self-energy correction to the bare phonon action (18.1). Employing the normal metal form of the $\hat{\Lambda}$ matrices, Eq. (14.25), one finds for the retarded, advanced, and Keldysh components of the phonon self-energy. (Also notice that $\alpha=\beta=cl$ component is zero, as it must.)

$$
\Sigma_j^{R(A)}(\mathbf{q},\omega) = \mp\mathrm{i}\,\frac{2\pi\nu k_F^2 D}{\rho_m}\, b_j q^2 \sum_{\epsilon}(F_{\epsilon} - F_{\epsilon-\omega}); \qquad (18.38a)
$$

$$
\Sigma_j^{K}(\mathbf{q},\omega) = -2\mathrm{i}\,\frac{2\pi\nu k_F^2 D}{\rho_m}\, b_j q^2 \sum_{\epsilon}(1 - F_{\epsilon}F_{\epsilon-\omega}), \qquad (18.38b)
$$

where we normalized the self-energy by material mass density, ρ_m, to have the latter as a common factor in front of the phonon propagator (18.3). Employing now Eqs. (6.54) and (6.55), one obtains the following kinetic equation (mind the change in notations relative to Eq. (6.54); we now call boson distribution B instead of F, reserving F for the fermion distribution):

$$
\partial_t B_{\mathbf{q}}^{(j)} + \mathbf{v}_{j\mathbf{q}}\nabla_{\mathbf{r}} B_{\mathbf{q}}^{(j)} = \nu\,\frac{b_j\pi\, k_F^2 D}{\rho_m}\,\frac{q^2}{\omega_{\mathbf{q}}^{(j)}}\sum_{\epsilon}\Big[1 - F_{\epsilon}F_{\epsilon-\omega_{\mathbf{q}}^{(j)}} - B_{\mathbf{q}}^{(j)}\big(F_{\epsilon} - F_{\epsilon-\omega_{\mathbf{q}}^{(j)}}\big)\Big], \qquad (18.39)
$$

where $\mathbf{v}_{j\mathbf{q}} = \nabla_{\mathbf{q}}\omega_{\mathbf{q}}^{(j)} = v_j\mathbf{q}/q$. For non-equilibrium phonons with the distribution function depending on the quantum numbers j and $\mathbf{q}$, instead of the energy $\omega = \omega_{\mathbf{q}}^{(j)}$, one should modify the electronic kinetic equation (18.19), (18.22) as

$$\partial_t F_\epsilon - \nabla_{\mathbf{r}}\left[D\nabla_{\mathbf{r}}F_\epsilon\right] = \frac{k_F^2 D}{2\rho_m}\sum_{j=1}^{d} b_j \sum_{(\mp),\mathbf{q}} \frac{\pm q^2}{\omega_{\mathbf{q}}^{(j)}}\left[1 - F_\epsilon F_{\epsilon\mp\omega_{\mathbf{q}}^{(j)}} - B_{\mathbf{q}}^{(j)}(F_\epsilon - F_{\epsilon\mp\omega_{\mathbf{q}}^{(j)}})\right]. \tag{18.40}$$

Notice that the electronic distribution function, F_ϵ, still depends only on energy, because of the "dirty" limit constraint, that is $\omega \ll v_j/l_{\text{el}} = v_j/v_F\tau_{\text{el}} \ll 1/\tau_{\text{el}}$. This means that electrons equilibrate fast over momentum directions due to the elastic scattering. This leaves their energy as the only relevant parameter. The closely related observation is that, in our treatment, phonons propagate ballistically, while electrons do so diffusively. The coupled kinetic equations (18.39) and (18.40) should be solved together to find electron and phonon distribution functions in a truly non-equilibrium setup.

Here we restrict ourselves to a sanity check demonstrating that the total energy of electron and phonon subsystems is conserved. For simplicity we focus on a spatially uniform and momentum isotropic case, where the phonon distribution function depends on $\omega_{\mathbf{q}}^{(j)} = \omega$ and the coupled kinetic equations take the following form:

$$\partial_t B_\omega^{(j)} = \nu\,\frac{b_j\pi\, k_F^2 D}{\rho_m v_j^2}\,\omega \sum_\epsilon \left[1 - F_\epsilon F_{\epsilon-\omega} - B_\omega^{(j)}(F_\epsilon - F_{\epsilon-\omega})\right]; \tag{18.41}$$

$$\partial_t F_\epsilon = \sum_{j=1}^{d}\sum_\omega \mathcal{M}_j(\omega)\left[1 - F_\epsilon F_{\epsilon-\omega} - B_\omega^{(j)}(F_\epsilon - F_{\epsilon-\omega})\right], \tag{18.42}$$

where odd in energy electron–phonon matrix elements, $\mathcal{M}_j(\omega)$, are given by Eq. (18.24). The electron energy was already discussed in Section 18.4, Eq. (18.25). It is given by $E_e = \int d\epsilon\, \nu(\epsilon)\epsilon\, n_\epsilon$, where the fermionic occupation number, n_ϵ, is found from $F_\epsilon = 1 - 2n_\epsilon$. The phonon energy density is given by

$$E_{\text{ph}} = \sum_{j=1}^{d}\sum_{\mathbf{q}} \omega_{\mathbf{q}}^{(j)} n_B^{(j)}(\omega_{\mathbf{q}}^{(j)}) = \frac{1}{2}\sum_{j=1}^{d}\int_0^\infty d\omega\, \omega N_j(\omega)\,(B_\omega^{(j)} - 1), \tag{18.43}$$

where $N_j(\omega) = \sum_{\mathbf{q}}\delta(\omega - \omega_{\mathbf{q}}^{(j)}) = |\omega|^{d-1}\Omega_d/(2\pi v_j)^d$ is the density of states of transversal and longitudinal acoustic phonons and $n_B^{(j)}(\omega) = (B_\omega^{(j)} - 1)/2$ is the boson occupation number. One thus finds

$$\partial_t E_{\text{ph}} = \frac{1}{4}\int d\omega\, \omega \sum_{j=1}^{d} N_j(\omega)\partial_t B_\omega^{(j)}; \qquad \partial_t E_e = -\frac{\nu}{2}\int d\epsilon\, \epsilon\, \partial_t F_\epsilon, \tag{18.44}$$

where we employed that $B_\omega^{(j)}$ is an odd function of ω to extent the integration to negative infinity. One now uses the kinetic equations (18.41) and (18.42) along with the matrix elements (18.24) to observe[6] that the total energy is indeed conserved: $\partial_t E_{\rm ph} + \partial_t E_e = 0$.

18.7 Ultrasound Attenuation

Probably the simplest and most experimentally relevant problem in phonon kinetics is attenuation of an ultrasound beam by its interactions with the electrons. Those lead to conversion of the sound energy to electron–hole pairs. For a high-intensity acoustic beam one may use kinetic equation (18.39) to calculate evolution of phonon occupation numbers along the propagation direction. Here we restrict ourselves with the linear response approach, where the damping of a certain phonon mode labeled by j and $\mathbf{q}$ is given by the imaginary part of its retarded self-energy, $\Sigma_j^R(\mathbf{q},\omega)$.

For normal metals one can immediately use Eq. (18.38a). To this end notice that the energy sum there $\sum_\epsilon (F_\epsilon - F_{\epsilon-\omega}) = \omega/\pi$, independently on a specific form of the electronic distribution. One thus finds

$$\Sigma_j^R(\mathbf{q},\omega) = -\mathrm{i}\,\frac{2\nu\, k_F^2 D}{\rho_m}\, b_j q^2 \omega \equiv -\mathrm{i}2\gamma_q^{(j)}\,\omega, \tag{18.45}$$

where the phonon scattering rate is

$$\gamma_q^{(j)} = b_j \frac{\nu\, k_F^2 D q^2}{\rho_m} \sim \frac{m}{M} D q^2, \tag{18.46}$$

where m/M is the ratio of electron to ion masses and D is electron diffusion coefficient. This form of the self-energy implies that the $\omega^2 - (\omega_\mathbf{q}^{(j)})^2$ in the bare phonon action (18.1) is replaced by $\omega^2 + \mathrm{i}\gamma_q^{(j)}\,\omega - (\omega_\mathbf{q}^{(j)})^2$. This means that on the semiclassical level the phonon displacements obey the Newtonian equation with friction[7]

$$\ddot{\mathbf{u}} = -(\omega_\mathbf{q}^{(j)})^2 \mathbf{u} - 2\gamma_q^{(j)}\,\dot{\mathbf{u}}. \tag{18.47}$$

[6] To this end we change the integration variable from ω to $\epsilon' = \epsilon - \omega$ and use the expression for the phonon density of states along with Eq. (18.24) to find

$$\partial_t E_{\rm ph} = \frac{\nu}{4}\sum_{j=1}^{d} \iint \frac{d\epsilon d\epsilon'}{2\pi}(\epsilon-\epsilon')\mathcal{M}_j(\epsilon-\epsilon')\Big[1 - F_\epsilon F_{\epsilon'} - B_{\epsilon-\epsilon'}^{(j)}(F_\epsilon - F_{\epsilon'})\Big].$$

Using the fact that $\mathcal{M}_j(\epsilon-\epsilon')$ is odd, while the expression in the square brackets is even upon exchange of ϵ and ϵ', one can replace $(\epsilon-\epsilon')$ by 2ϵ. This brings it exactly to the form $-\partial_t E_e$.

[7] The Keldysh component of the self-energy leads to the stochastic Langevin force acting on the oscillator. If electrons are in the thermal state, the latter obeys the fluctuation-dissipation relation.

For long wavelength, that is, small q, the oscillators are underdamped, $\gamma_q^{(j)} \ll \omega_{\mathbf{q}}^{(j)}$. Thus the ultrasound propagates in metals with the characteristic inverse decay length $\xi_q^{-1} = \gamma_q/v_j \sim v_j\tau_{\rm el}q^2$, where we used that $v_F^2 \sim v_j^2 M/m$. Notice that in more disordered materials (i.e. with shorter electronic $\tau_{\rm el}$) the sound propagates *better* (longer ξ_q). This is, of course, due to the Pippard effect of almost perfect compensation of the induced lattice charge by electrons dragged along with moving impurities. Also notice that the sound attenuation length is independent of the lattice temperature. As we will see, the situation is very different in superconductors, where the attenuation length grows rapidly as temperature is decreased.

We turn now to BCS superconductors. Due to the gap in the quasiparticle spectrum, one expects suppressed sound absorption for $\omega < 2\Delta$, that is, $q < 2\Delta/v_j$. To quantify this effect we go back to the effective phonon action (18.37) and use the superconducting saddle point for the electronic degrees of freedom, $\hat{\Lambda} \to \check{\Lambda}$ (along with an extra overall factor $1/2$ to compensate for Nambu doubling). Performing the trace over electronic Keldysh subspace we find for the retarded, that is, $\alpha = q$; $\beta = cl$, component of the phonon self-energy: $\mathrm{Tr}\{\check{\Lambda}_{\epsilon-\omega}\hat{\gamma}^q\check{\Lambda}_\epsilon\hat{\gamma}^{cl}\} = \mathrm{Tr}_N\{\hat{\Lambda}^K_{\epsilon-\omega}\hat{\Lambda}^R_\epsilon + \hat{\Lambda}^A_{\epsilon-\omega}\hat{\Lambda}^K_\epsilon\}$. We substitute now the BCS Green functions in the Nambu space, Eq. (17.14), $\hat{\Lambda}^{\rm R}_\epsilon = \hat{\tau}_3 u_\epsilon + \mathrm{i}\hat{\tau}_2 v_\epsilon$, $\hat{\Lambda}^{\rm A}_\epsilon = -\hat{\tau}_3 u^*_\epsilon - \mathrm{i}\hat{\tau}_2 v^*_\epsilon$, and $\hat{\Lambda}^K = \hat{\Lambda}^R\hat{F} - \hat{F}\hat{\Lambda}^A$, where $\hat{F} = F_{L,\epsilon}\hat{\tau}_0 + F_{T,\epsilon}\hat{\tau}_3$, Eq. (17.21), to find [407][8]

$$\begin{aligned}\mathrm{Im}\Sigma_j^R(\mathbf{q},\omega) &= -\frac{2\pi\nu\, k_F^2 D}{\rho_m}\, b_j q^2 \sum_\epsilon (F_{L,\epsilon} - F_{L,\epsilon-\omega})\,[\mathrm{Re}\, u_\epsilon \mathrm{Re}\, u_{\epsilon-\omega} - \mathrm{Re}\, v_\epsilon \mathrm{Re}\, v_{\epsilon-\omega}] \\ &= -\frac{\gamma_{\mathbf{q}}^{(j)}}{2}\int d\epsilon\,(F_{L,\epsilon} - F_{L,\epsilon-\omega})\,\frac{|\epsilon(\epsilon-\omega)| - \Delta^2 \mathrm{sign}\,\epsilon(\epsilon-\omega)}{\sqrt{\epsilon^2-\Delta^2}\sqrt{(\epsilon-\omega)^2-\Delta^2}}, \qquad (18.48)\end{aligned}$$

where the energy integral runs over the region $|\epsilon| > \Delta$ and $|\epsilon - \omega| > \Delta$. Notice that the charge imbalance component of the distribution, F_T, does not show up in this expression.

The integral in Eq. (18.48) is most readily analyzed in the limit $\omega \ll T, \Delta$. One may expand then to linear order in ω and use the fact that the fraction tends to 1, as $\omega \to 0$, to find $-\mathrm{Im}\Sigma_j^R = 2\gamma_q^{(j)}\int_\Delta^\infty d\epsilon\, \partial_\epsilon F_{L,\epsilon}\,\omega = 2\gamma_q^{(j)}(1 - F_{L,\Delta})\,\omega$. Finally one obtains for the low-frequency ultrasound attenuation rate in superconductors:

$$\gamma_q^{(j)\mathrm{SC}}(T) = 2\gamma_q^{(j)}\, n_\Delta, \qquad (18.49)$$

where $\gamma_q^{(j)}$ is the corresponding normal state temperature-independent attenuation and $n_\Delta = (1 - F_{L,\Delta})/2$ is the quasiparticle occupation number at the superconducting gap energy Δ. In equilibrium it is given by the Fermi function, $n_\Delta = (e^{\Delta/T} + 1)^{-1}$, where $\Delta = \Delta(T)$ is the temperature-dependent BCS

[8] Notice that this one-line sigma-model exercise takes 10 pages of heroic diagrammatic calculations in Ref. [407].

energy gap. Equation (18.49) predicts a sharp decrease in ultrasound attenuation at $T < T_c$, and exponentially small, $\sim e^{-\Delta/T}$, attenuation in the limit $T \ll \Delta$. This was indeed measured in numerous experiments. According to Eq. (18.48), the ultrasound attenuation is a sensitive tool to measure a non-equilibrium quasiparticle occupation n_Δ.

18.8 Problems

18.8.1 Temperature Profile in a Conductor

Find a temperature profile in a quasi-1D conductor of length L with an applied voltage V. Assume a state of local equilibrium, characterized by a coordinate-dependent electronic temperature and chemical potential. Put lattice temperature to zero, for simplicity.

Solution: If electrons are locally equilibrated (by electron–electron collisions) their distribution function acquire the form $F_\epsilon(x) = \tanh(\epsilon - \mu(x))/2T_e(x)$, where $\mu(x)$ is a local chemical potential, with the boundary conditions $\mu(\pm L/2) = \pm eV/2$, and $T_e(x)$ is a local electron temperature with $T_e(\pm L/2) = 0$. To find $\mu(x)$ and $T_e(x)$ we write equations for electron density, $\rho_e(x) = \nu \int d\epsilon\, n_\epsilon(x)$, and energy density, $E_e(x) = \nu \int d\epsilon\, (\epsilon - \mu(x))n_\epsilon(x)$, where $n_\epsilon(x) = (1 - F_\epsilon(x))/2$ is the local occupation number. Employing (18.19), calculating $\nabla_x^2 F_\epsilon$, and evaluating the energy integrals, one finds

$$\partial_t \rho_e - \nabla_x(\nu D \nabla_x \mu) = 0; \tag{18.50a}$$

$$\partial_t E_e - \nabla_x\left(\frac{\pi^2}{3}\nu D\, T_e \nabla_x T_e\right) = \nu D\, (\nabla_x \mu)^2 + (T^6 - T_e^6)\,\nu \sum_{j=1}^{3} \frac{\alpha_3}{\Theta_j^3}, \tag{18.50b}$$

where we employed Eq. (18.25) and $\int ds\, s^2/\cosh^2 s = \pi^2/6$. Here the dimensionality of the phonon bath is $d = 3$. Zero on the right-hand side of (18.50a) reflects conservation of particle number by the electron–phonon collisions. The term $\nu D\, (\nabla_x \mu)^2$ represents Joule heating of the electron gas by an applied bias. Here one identifies charge current, $J = e\nu D \nabla_x \mu$, and the heat current, $J_h = \frac{\pi^2}{3}\nu D T_e \nabla_x T_e$. The corresponding charge conductivity $\sigma_D = e^2 \nu D$ and the heat conductivity $\kappa = \frac{\pi^2}{3}\nu D T_e$ obey the Wiedemann–Franz law:

$$\frac{\kappa}{\sigma_D} = \mathfrak{L}\, T_e, \tag{18.51}$$

where the Lorenz number is $\mathfrak{L} = \pi^2/(3e^2)$.

In the stationary case one finds $\nabla_x^2 \mu = 0$ and thus $\mu(x) = eVx/L$. The temperature profile is then found from the following equation:

$$\frac{\pi^2}{6}\nabla_x^2 T_e^2 = -\left(\frac{eV}{L}\right)^2 + T_e^6 \frac{1}{D}\sum_{j=1}^{3}\frac{\alpha_3}{\Theta_j^3}, \tag{18.52}$$

where we assume a negligible lattice temperature, $T \ll T_e$. We only analyze it in the limiting cases of long and short wire. If the wire is long one may neglect the heat diffusion and find an almost uniform electron temperature (apart from the boundary layers) given by

$$T_e \sim \hbar\left(\frac{eV}{L}\right)^{1/3}\left(\frac{\rho_m v_t^5}{\hbar k_F^2}\right)^{1/6}. \tag{18.53}$$

Notice that, despite being a result of the balance between Joule heating and phonon cooling, the stationary T_e is independent of the wire's conductivity, νD. It only depends on the material parameters (mass density, speed of sound, and Fermi momentum) along with the local electric field, eV/L. Taking the numbers for aluminum, Section 18.4, one finds $T_e = 4\mathrm{K}(eV/L[\mathrm{cm}])^{1/3}$.

If the wire is short, one may disregard phonon cooling and consider the balance of the heat diffusion and Joule heating. Solving equation (18.52) with the boundary conditions $T_e(\pm L/2) = 0$, one finds a semicircular profile for the temperature

$$T_e(x) = eV\frac{\sqrt{3}}{2\pi}\sqrt{1-(2x/L)^2} \tag{18.54}$$

with the maximal temperature in the middle $T_e(0) \approx 0.27eV$. The short and long wire limits are distinguished by comparing this value with Eq. (18.53). This brings the characteristic length scale $L_V \propto (eV)^{-2}$, such that the short wire limit is $L \lesssim L_V$.

18.8.2 Phonon Diffusion

In 3D at large distances phonons propagate diffusively due to their scattering by electrons. Using the approach of Section 14.1, derive the corresponding diffusion equation for the phonon distribution function (assume locally equilibrated electrons with temperature $T_e(\mathbf{r}, t)$). Assuming a locally equilibrated lattice with temperature $T(\mathbf{r}, t)$, derive a flow equation for the lattice energy density and phonon heat conductivity.

Solution: Following Section 14.1 we assume almost isotropic boson distribution function: $B_{\mathbf{q}} = B_0(\omega_{\mathbf{q}}, \mathbf{r}, t) + \mathbf{n}_{\mathbf{q}} \cdot \mathbf{B}_1(\omega_{\mathbf{q}}, \mathbf{r}, t)$, where $\mathbf{n}_{\mathbf{q}}$ is a unit vector with the direction of $\mathbf{q}$ and we have suppressed polarization index j. Substituting this form in the phonon kinetic equation (18.39) and (i) integrating over directions of $\mathbf{n}_{\mathbf{q}}$; (ii) multiplying by $\mathbf{n}_{\mathbf{q}}$ and then integrating over directions, one finds

$$\partial_t B_0 + \frac{v_j}{3}\nabla_{\mathbf{r}} \cdot \mathbf{B}_1 = \gamma_q \left(\coth \frac{\omega_{\mathbf{q}}}{2T_e} - B_0 \right); \tag{18.55a}$$

$$\partial_t \mathbf{B}_1 + v_j \nabla_{\mathbf{r}} B_0 = -\gamma_q \mathbf{B}_1 \tag{18.55b}$$

where we used $1 - F_\epsilon F_{\epsilon-\omega} = (F_\epsilon - F_{\epsilon-\omega}) \coth \omega/2T_e$ along with $\sum_\epsilon (F_\epsilon - F_{\epsilon-\omega}) = \omega/\pi$ and Eq. (18.46). Neglecting $\partial_t \mathbf{B}_1$ in comparison with $\gamma_q \mathbf{B}_1$, we finally obtain the damped-diffusion equation for the isotropic part of the phonon distribution, $B_0(\omega_{\mathbf{q}}, \mathbf{r}, t)$:

$$\partial_t B_0 - D_{\text{ph}} \nabla_{\mathbf{r}}^2 B_0 = \gamma_q \left(\coth \frac{\omega_{\mathbf{q}}}{2T_e} - B_0 \right), \tag{18.56}$$

where the phonon diffusion coefficient,[9] $D_{\text{ph}}(\omega_{\mathbf{q}}) = v_j^2/(3\gamma_q) \propto \frac{v_j^4}{D\omega_{\mathbf{q}}^2} \frac{M}{m}$.

Assuming local equilibrium, $B_0(\omega_{\mathbf{q}}, \mathbf{r}, t) = \coth \omega_{\mathbf{q}}/2T(\mathbf{r}, t)$, and using the expression for the phonon energy (18.43), one finds

$$\partial_t E_{\text{ph}} - \nabla_{\mathbf{r}} \left(\sum_{j=1}^{3} \frac{v_j \rho_m}{18 b_j \nu D k_F^2} T \nabla_{\mathbf{r}} T \right) = (T_e^6 - T^6)\, \nu \sum_{j=1}^{3} \frac{\alpha_3}{\Theta_j^3}, \tag{18.57}$$

here $b_t = 1/5$ and $b_l = 4/15$. Comparing this with Eq. (18.50b) for the electron heat current, one notices that the lattice heat conductivity, $\kappa_{\text{ph}} \approx \frac{v_t \rho_m}{\nu D k_F^2} T$, scales linearly with temperature, in agreement with the Wiedemann–Franz law (18.51). Unlike the electron one, it is inversely proportional to the charge conductivity, νD. This leads to the upward renormalization of the Lorenz number of the form $\delta\mathcal{L} \sim (10/k_F l_{\text{el}})^2$, where we put $(M/m)^{1/4} \approx 10$. It may be noticeable in highly disordered heavy metal compounds.

18.8.3 Quantum Correction to Electron–Phonon Vertex

Evaluate the first quantum correction to the electron–phonon vertex (18.16).

Remarks: Since the electron–phonon vertex is proportional to the diffusion constant, D, it is tempting to assume that it is suppressed by the weak-localization correction, similar to Eq. (14.72). Indeed the electron–phonon vertex (18.16), $\sim\text{Tr}\{[\hat{Q}, \hat{\varepsilon}][\hat{Q}, \hat{\varepsilon}]\}$, where $\hat{\varepsilon}$ is the linear strain tensor, looks similar to vector potential vertex $\sim\text{Tr}\{[\check{Q}, \hat{\mathbf{A}}\hat{\tau}_3][\check{Q}, \hat{\mathbf{A}}\hat{\tau}_3]\}$, considered in Section 14.8 to evaluate the weak-localization correction. The difference is that the strain field does not break the time-reversal symmetry and thus comes without a $\hat{\tau}_3$ matrix in the time-reversal space. The calculation of the weak-localization correction proceeds exactly as in

[9] The scattering rate should also include phonon scattering by phonon nonlinearities and lattice imperfections. Those typically scale with higher powers of q, making scattering by electrons dominant at small temperatures.

Section 14.8, but the absence of $\hat{\tau}_3$ matrices results in the *opposite* sign of the quantum correction, as compared to conductivity. This can be seen even easier using the diagram in Fig. 14.3, where the external wavy lines represent now the strain tensor $\varepsilon^{\mu\nu}$, coupled to the fermion vertex $\Gamma^{(P)}_{\nu\mu}$, Eq. (18.10). The latter is quadratic in electron velocity. Thus, the fact that the two fermion vertices have opposite velocities does not lead to the negative sign, unlike the case of the weak localization correction to conductivity, Section 14.8, where the vertices are linear in the velocity. As a result, the first quantum correction *enhances* electron–phonon coupling [423] in Eq. (18.16)

$$\nu k_F^2 D \to \nu k_F^2 D \left(1 + \frac{1}{\pi\nu}\sum_{\mathbf{q}} \frac{1}{Dq^2 - \mathrm{i}\omega}\right),$$

where ω is a frequency of the phonon field $\varepsilon^{\mu\nu}(\omega)$.

18.8.4 Sigma-Model with Moving Impurities

Derive a sigma-model for the case where the impurity potential is dynamical due to slow phonon deformations of the lattice: $V_{\text{dis}}(\mathbf{r}) \to V_{\text{dis}}(\mathbf{r} + \mathbf{u}(\mathbf{r}, t))$.

Solution: It is convenient to shift $\mathbf{r}$ to write the interaction of the electron density $\rho_e(\mathbf{r}, t) = \bar{\psi}(\mathbf{r}, t)\psi(\mathbf{r}, t) - \rho_0$ with the disorder potential as

$$H_{\text{dis}} = \sum_{\mathbf{r}} V_{\text{dis}}(\mathbf{r})\rho_e(\mathbf{r} - \mathbf{u}(\mathbf{r}, t), t).$$

Performing averaging over the Gaussian distribution of short-ranged disorder, one finds the following action

$$\mathrm{i}S_{\text{dis}} = -\frac{1}{4\pi\nu\tau_{\text{el}}}\iint dt dt' \sum_{\mathbf{r}} \bar{\psi}_{\mathbf{r}-\mathbf{u},t}\psi_{\mathbf{r}-\mathbf{u},t}\bar{\psi}_{\mathbf{r}-\mathbf{u}',t'}\psi_{\mathbf{r}-\mathbf{u}',t'},$$

where $\mathbf{u} = \mathbf{u}(\mathbf{r}, t)$ and $\mathbf{u}' = \mathbf{u}(\mathbf{r}, t')$. One can now rearrange the fermionic fields and decouple the four-fermion action with the help of the nonlocal in time field $Q_{t,t'}(\mathbf{r})$. This leads to the following term:

$$\bar{\psi}_{\mathbf{r}-\mathbf{u},t} Q_{t,t'}(\mathbf{r})\psi_{\mathbf{r}-\mathbf{u}',t'} \approx \bar{\psi}_{\mathbf{r},t}\left[Q_{t,t'}(\mathbf{r}) - \hat{\mathcal{L}}_1 + \frac{1}{2}\hat{\mathcal{L}}_2\right]\psi_{\mathbf{r},t'}. \tag{18.58}$$

We have expanded fermionic fields to the second order in the displacement $\mathbf{u}$, which brings the two operators:

$$\hat{\mathcal{L}}_1 = \overleftarrow{\nabla}\cdot\mathbf{u}\, Q_{t,t'}(\mathbf{r}) + Q_{t,t'}(\mathbf{r})\mathbf{u}'\cdot\overrightarrow{\nabla}; \tag{18.59}$$
$$\hat{\mathcal{L}}_2 = \overleftarrow{\nabla}\overleftarrow{\nabla}\cdot\cdot\,\mathbf{u}\mathbf{u}\, Q + \overleftarrow{\nabla}\cdot\mathbf{u}\, 2Q\mathbf{u}'\cdot\overrightarrow{\nabla} + Q\,\mathbf{u}'\mathbf{u}'\cdot\cdot\,\overrightarrow{\nabla}\overrightarrow{\nabla},$$

where the arrows above the gradient operators show the direction of the differentiation in the context of Eq. (18.58). The action is now quadratic in the unshifted fermionic fields, which may be integrated out in the standard way, leading to the determinant:

$$\mathrm{Tr}\log\left\{G^{-1}-\varphi+\frac{\mathrm{i}}{2\tau_{\mathrm{el}}}\left[Q-\hat{\mathcal{L}}_1+\frac{1}{2}\hat{\mathcal{L}}_2\right]\right\}, \tag{18.60}$$

where $\varphi = \varphi_{\mathbf{r},t}$ is the scalar potential coming from the Hubbard–Stratonovich decoupling of the Coulomb interactions.

From this point on, one proceeds along the standard root of deriving Keldysh nonlinear sigma-model, Chapter 14. To this end one passes to the Keldysh 2×2 structure, by splitting the contour on forward and backward branches and performing the Keldysh rotation. One then realizes that the soft diffusive modes of the action are described by the manifold $\hat{Q}^2 = 1$ and therefore one can write $\hat{Q} = \hat{\mathcal{R}}^{-1}\hat{\Lambda}\hat{\mathcal{R}}$, where $\hat{\Lambda}$ is the Green function in coinciding spatial points, Eq. (14.25). This way, Eq. (18.60) may be rewritten as

$$\mathrm{Tr}\log\left\{1+\hat{\mathcal{G}}\hat{\mathcal{R}}[G^{-1},\hat{\mathcal{R}}^{-1}]-\hat{\mathcal{G}}\hat{\mathcal{R}}\left[\hat{\varphi}+\frac{\mathrm{i}}{2\tau_{\mathrm{el}}}\hat{\mathcal{L}}_1-\frac{\mathrm{i}}{4\tau_{\mathrm{el}}}\hat{\mathcal{L}}_2\right]\hat{\mathcal{R}}^{-1}\right\}, \tag{18.61}$$

where $\hat{\mathcal{G}}^{-1} = G^{-1}+\mathrm{i}\hat{\Lambda}/2\tau_{\mathrm{el}}$. Finally, one expands the logarithm here to the lowest nonvanishing orders. First neglecting $\hat{\mathcal{L}}_{1,2}$ terms, one obtains the standard nonlinear sigma-model action, cf. Eq. (14.48):

$$\mathrm{i}S_0 = \frac{\mathrm{i}\nu}{2}\mathrm{Tr}\{\hat{\varphi}\hat{\sigma}^1\hat{\varphi}\} - \frac{\pi\nu}{4}\mathrm{Tr}\{D(\partial_{\mathbf{r}}\hat{Q})^2 - 4\partial_t\hat{Q} - 4\mathrm{i}\hat{\varphi}\hat{Q}\}. \tag{18.62}$$

The first term on the right-hand side here represents static polarizability (i.e. screening) of the electronic band. It comes from the so-called retarded-retarded and advanced-advanced loops. The dynamic screening is encoded in $\pi\nu\mathrm{Tr}\{\hat{\varphi}\hat{Q}\}$ term along with fluctuations of the $\hat{Q}$ field around its stationary point $\hat{\Lambda}$.

We focus now on the phonon-induced $\hat{\mathcal{L}}_{1,2}$ terms, which originate from the motion of the impurities relative to the electronic liquid. It is easy to see that the first order in $\hat{\mathcal{L}}_1$ vanishes. One is thus left with the three terms: (i) first order in $\hat{\mathcal{L}}_1$ and in $\mathrm{i}\hat{\mathcal{R}}\mathbf{v}_F\cdot\overrightarrow{\nabla}\hat{\mathcal{R}}^{-1}$; (ii) first order in $\hat{\mathcal{L}}_2$; and (iii) second order in $\hat{\mathcal{L}}_1$. A straightforward, but somewhat lengthy evaluation of these three terms results:

$$\mathrm{i}S_{(\mathrm{i})} = -\mathrm{i}\pi\nu\frac{v_F k_F}{d}\mathrm{Tr}\{\hat{\mathbf{u}}\cdot\nabla\hat{Q}\} = \mathrm{i}\pi\rho_0\mathrm{Tr}\{\mathrm{div}\hat{\mathbf{u}}\,\hat{Q}\}; \tag{18.63a}$$

$$\mathrm{i}S_{(\mathrm{ii})} = -\mathrm{i}\frac{\pi\nu}{2\tau_{\mathrm{el}}}\frac{k_F^2}{d}\mathrm{Tr}\{\hat{\mathbf{u}}\cdot\hat{\mathbf{u}} - \hat{\mathbf{u}}\hat{Q}\hat{\mathbf{u}}\hat{Q}\}; \tag{18.63b}$$

$$\mathrm{i}S_{(\mathrm{iii})} = -\mathrm{i}S_{(\mathrm{ii})} + \frac{\pi\nu D k_F^2}{4}\mathrm{Tr}\big\{[\hat{Q},\partial^\mu\hat{u}^\nu][\hat{Q},\partial^\eta\hat{u}^\lambda]\big\}\,\Upsilon_{\mu\nu,\eta\lambda}, \tag{18.63c}$$

where $\Upsilon_{\mu\nu,\eta\lambda}$ is given by Eq. (18.15). Notice that the leading orders in $S_{\text{(iii)}}$ is exactly cancelled against $S_{\text{(ii)}}$. The second sub-leading term in Eq. (18.63c) originates from the gradient operators in $\hat{\mathcal{L}}_1$ acting on displacements $\hat{\mathbf{u}}$, as opposed to the Green functions, $\hat{\mathcal{G}}$.

The scalar linear coupling $S_{\text{(i)}}$ may be combined with the potential term in Eq. (18.62) by shifting the potential $\hat\varphi \to \hat\phi = \hat\varphi + \frac{\rho_0}{\nu}\text{div}\hat{\mathbf{u}}$. To deal with it we need to bring the Fröhlich electron–phonon coupling via the Coulomb interaction of the electron density and lattice compression. The corresponding action is

$$S_C = \int dt \left[\frac{1}{2} \sum_{\mathbf{q}} \varphi_{\mathbf{q},t} U_C^{-1} \varphi_{-\mathbf{q},t} + \sum_{\mathbf{r}} \varphi_{\mathbf{r},t} (\rho_0 \text{div}\,\mathbf{u} - \rho_e)_{\mathbf{r},t} \right], \tag{18.64}$$

where $U_C = 4\pi e^2/q^2$ is the bare Coulomb interaction. In the limit of the strong Coulomb interactions, $U_C \to \infty$, one is left with the static screening $\frac{\nu}{2}\text{Tr}\{\hat\varphi\hat\sigma^{1}\hat\varphi\}$ in Eq. (18.62) along with the Fröhlich interaction term $\hat\varphi\hat\sigma^{1}\rho_0 \text{div}\,\hat{\mathbf{u}}$ in Eq. (18.64) (the $-\varphi_{\mathbf{r},t}\rho_e(\mathbf{r},t)$ term is already fully included in (18.62)). Upon the aforementioned shift, these terms result in $\frac{\nu}{2}\text{Tr}\{\hat\phi\hat\sigma^{1}\hat\phi\} - \frac{\rho_0^2}{2\nu}\text{div}\hat{\mathbf{u}}\,\hat\sigma^{1}\text{div}\hat{\mathbf{u}}$. The first term here stands for the screened electron–electron interactions, unaffected by lattice compression, while the second one serves to renormalize upward the longitudinal sound velocity. This latter effect is already accommodated by using the correct value of v_l and thus no other effects of the scalar electron–phonon coupling, $S_{\text{(i)}}$, remain.

The only remaining term thus is the second – the quadrupole term in Eq. (18.63c), which coincides with Eq. (18.14). The latter was derived using phenomenological Pippard–Tsuneto–Schmid form, Eq. (18.9), of the electron–phonon coupling. The present first-principles derivation thus provides an independent justification for the Schmid theory [408].

References

[1] L. P. Kadanoff and G. Baym, *Quantum Statistical Mechanics* (Benjamin, 1962).
[2] A. A. Abrikosov, L. P. Gor'kov, and I. E. Dzyaloshinski, *Methods of Quantum Field Theory in Statistical Physics* (Dover, 1963).
[3] A. L. Fetter and J. D. Walecka, *Quantum Theory of Many-Particle Systems* (Dover, 1971, 2003).
[4] G. D. Mahan, *Many-Particle Physics* (Plenum Publishers, 1981, 1990, 2000).
[5] C. Itzykson and J.-B. Zuber, *Quantum Field Theory* (McGraw-Hill, 1980).
[6] J. W. Negele and H. Orland, *Quantum Many-Particle Systems* (Addison-Wesley, 1988).
[7] E. Fradkin, *Field Theories of Condensed Matter Systems* (Perseus Books, 1991).
[8] A. M. Tsvelik, *Quantum Field Theory in Condensed Matter Physics* (Cambridge University Press, 1995, 2003).
[9] A. Altland and B. D. Simons, *Condensed Matter Field Theory* (Cambridge University Press, 2006, 2010).
[10] N. Nagaosa, *Quantum Field Theory in Condensed Matter Physics* (Springer, 2010).
[11] T. Matsubara, "A new approach to quantum-statistical mechanics," *Prog. Theor. Phys.*, **14**, 351 (1955).
[12] J. Schwinger, "The special canonical group," *PNAS*, **46**, 1401 (1960); "Brownian motion of a quantum oscillator," *J. Math. Phys.*, **2**, 407 (1961).
[13] O. V. Konstantinov and V. I. Perel, "A graphical technique for computation of kinetic quantities," *Zh. Eksp. Teor. Fiz.*, **39**, 197 (1960); [*Sov. Phys. JETP*, **12**, 142 (1961)].
[14] L. V. Keldysh, "Diagram technique for nonequilibrium processes," *Zh. Eksp. Teor. Fiz.*, **47**, 1515 (1964); [*Sov. Phys. JETP*, **20**, 1018 (1965)].
[15] E. M. Lifshitz and L. P. Pitaevskii, *Physical Kinetics* (Butterworth-Heinemann, 1981).
[16] H. Smith and H. H. Jensen, *Transport Phenomena* (Clarendon Press, 1989).
[17] H. Haug and A.-P. Jauho, *Quantum Kinetics in Transport and Optics of Semiconductors* (Springer Series in Solid-State Sciences) (Springer, 2010).
[18] J. Rammer, *Quantum Field Theory of Non-equilibrium States* (Cambridge University Press, 2007).
[19] J. Rammer and H. Smith, "Quantum field-theoretical methods in transport theory of metals," *Rev. Mod. Phys.*, **58**, 323 (1986).
[20] P. Schwab and R. Raimondi, "Quasiclassical theory of charge transport in disordered interacting electron systems," *Ann. Phys. (Leipzig)*, **12**, 471 (2003).

[21] V. Spicka, B. Velicky, and A. Kalvova, "Long and short time quantum dynamics: I. Between Green's functions and transport equations," *Phys. E-Low-Dim. Syst. and Nanostruct.*, **29**, 154 (2005).
[22] V. Spicka, B. Velicky and A. Kalvova, "Long and short time quantum dynamics: II. Kinetic regime," *Phys. E-Low-Dim. Syst. and Nanostruct.*, **29**, 175 (2005).
[23] A. Kamenev, "Many-body theory of non-equilibrium systems," in *Nanophysics: Coherence and Transport*, edited by H. Bouchiat et al. (Elsevier, 2005), p. 177.
[24] A. Kamenev and A. Levchenko, "Keldysh technique and non-linear – model: Basic principles and applications," *Adv. Phys.*, **58**, 197 (2009).
[25] K. T. Mahanthappa, "Multiple production of photons in quantum electrodynamics," *Phys. Rev.*, **126**, 329 (1962).
[26] P. M. Bakshi and K. T. Mahanthappa, "Expectation value formalism in quantum field theory," *J. Math. Phys.*, **4**, 1 (1963).
[27] L. V. Keldysh, "Real-time nonequilibrium Green's functions," in *Progress in Nonequilibrium Green's Functions II*, edited by M. Bonitz and D. Semkat (World Scientific, 2003).
[28] A. I. Larkin and Yu. N. Ovchinnikov, "Nonlinear conductivity of superconductors in the mixed state," *Zh. Eksp. Teor. Fiz.*, **68**, 1915 (1975); [*Sov. Phys. JETP*, **41**, 960 (1975)].
[29] R. P. Feynman and F. L. Vernon Jr., "The theory of a general quantum system interacting with a linear dissipative system," *Ann. Phys.*, **24**, 118 (1963).
[30] H. W. Wyld, "Formulation of the theory of turbulence in an incompressible fluid," *Ann. Phys.*, **14**, 143 (1961).
[31] P. C. Martin, E. D. Siggia, and H. A. Rose, "Statistical dynamics of classical systems," *Phys. Rev., A*, **8**, 423 (1973).
[32] J. DeDominicis, "Techniques of field renormalization and dynamics of critical phenomena," *J. Physique (Paris)*, **37**, C1 (1976).
[33] R. P. Feynman and A. R. Hibbs, *Quantum Mechanics and Path Integrals* (McGraw-Hill, 1965).
[34] A. O. Caldeira and A. J. Leggett, "Quantum tunnelling in a dissipative system," *Ann. Phys. (NY)*, **149**, 374 (1983).
[35] L. D. Landau and E. M. Lifshitz, *Mechanics* (Butterworth-Heinemann, 1976).
[36] L. D. Landau and E. M. Lifshitz, *Quantum Mechanics Non-Relativistic Theory* (Butterworth-Heinemann, 1981).
[37] B. I. Ivlev and V. I. Melnikov, "Stimulation of tunneling by a high-frequency field: Decay of the zero-voltage state in Josephson junctions," *Pisma Zh. Eksp. Teor. Fiz.*, **41**, 116 (1985); [*JETP Lett.*, **41**, 142 (1985)]; *Phys. Rev. Lett.*, **55**, 1614 (1985).
[38] B. I. Ivlev and V. I. Melnikov, "Effect of resonant pumping on activated decay rates," *Phys. Lett. A*, **116**, 427–428 (1986).
[39] A. J. Leggett, S. Chakravarty, A. T. Dorsey, M. P. A. Fisher, A. Garg, and W. Zwerger, "Dynamics of the dissipative two-state system," *Rev. Mod. Phys.*, **59**, 1 (1987).
[40] V. Ambegaokar, U. Eckern, and G. Schön, "Quantum dynamics of tunneling between superconductors," *Phys. Rev. Lett.*, **48**, 1745 (1982).
[41] S. E. Korshunov, "Coherent and incoherent tunnelling in a Josephson junction with a 'periodic' dissipation," *Pisma Zh. Eksp. Teor. Fiz.*, **45**, (1987), 342 [*JETP Lett.* **45**, 434 (1987)].
[42] X. Wang and H. Grabert, "Coulomb charging at large conduction," *Phys. Rev. B*, **53**, 12621 (1996).

[43] Yu. V. Nazarov, "Coulomb blockade without tunnel junctions," *Phys. Rev. Lett.*, **82**, 1245 (1999).

[44] A. Altland, L. I. Glazman, A. Kamenev, and J. S. Meyer, "Inelastic electron transport in granular arrays," *Ann. Phys.*, **321**, 2566 (2006).

[45] M. Titov and D. Gutman, "Korshunov instantons out of equilibrium," *Phys. Rev. B*, **93**, 155428 (2016).

[46] A. S. Dotdaev, Ya. I. Rodionov, and K. S. Tikhonov, "Instantons in the out-of-equilibrium Coulomb blockade," *Phys. Lett. A*, **419**, 127736 (2021).

[47] H. K. Janssen, "On a Lagrangean for classical field dynamics and renormalization group calculations of dynamical critical properties," *Z. Physik B*, **23**, 377 (1976).

[48] N. G. Van Kampen, *Stochastic Processes in Physics and Chemistry*, 3rd ed. (North-Holland, 2007).

[49] C. W. Gardiner, *Handbook of Stochastic Methods for Physics, Chemistry, and the Natural Sciences* (Springer-Verlag, 1997).

[50] H. Risken, *The Fokker–Planck Equation: Methods of Solution and Applications* (Springer-Verlag, 1989).

[51] J. Zinn-Justin, *Quantum Field Theory and Critical Phenomena* (Oxford University Press, 2002).

[52] M. A. Shifman, In *ITEP Lectures on Particle Physics and Field Theory*, chapter IV (World Scientific, 1999).

[53] P. Hänggi, P. Talkner, and M. Borkovec, "Reaction-rate theory: Fifty years after Kramers," *Rev. Mod. Phys.*, **62**, 251 (1990).

[54] H. A. Kramers, "Brownian motion in a field of force and the diffusion model of chemical reactions," *Physica (Utrecht)*, **7**, 284 (1940).

[55] V. I. Melnikov, "Activated tunneling decay of metastable state: Solution of the Kramers problem," *Physica*, **130A**, 606 (1985).

[56] M. D. Donsker and S. R. S. Varadhan, "Asymptotic evaluation of certain Markov process expectations for large time," *Commun. Pure Appl. Math.*, **28**, 1 (1975).

[57] M. D. Donsker and S. R. S. Varadhan, "Asymptotic evaluation of certain Markov process expectations for large time, II," *Commun. Pure Appl. Math.* **28**, 279 (1975).

[58] M. D. Donsker and S. R. S. Varadhan, "Asymptotic evaluation of certain Markov process expectations for large time, III," *Commun. Pure Appl. Math.* **29**, 389 (1976).

[59] M. D. Donsker and S. R. S. Varadhan, "Asymptotic evaluation of certain Markov process expectations for large time, IV," *Commun. Pure Appl. Math.* **36**, 183 (1983).

[60] J. Hoppenau, D. Nickelsen, and A. Engel, "Level 2 and level 2.5 large deviation functionals for systems with and without detailed balance," *New J. Phys.*, **18**, 083010 (2016).

[61] N. R. Smith, "Anomalous scaling and first-order dynamical phase transition in large deviations of the Ornstein–Uhlenbeck process," *Phys. Rev. E*, **105**, 014120 (2022). arXiv:2109.14972.

[62] D. Nickelsen and H. Touchette, "Anomalous scaling of dynamical large deviations," *Phys. Rev. Lett.*, **121**, 090602 (2018).

[63] C. Jarzynski, "Nonequilibrium equality for free energy differences," *Phys. Rev. Lett.*, **78**, 2690 (1997).

[64] C. Jarzynski, "Equilibrium free-energy differences from nonequilibrium measurements: A master-equation approach," *Phys. Rev. E*, **56**, 5018 (1997).

[65] G. E. Crooks, "Nonequilibrium measurements of free energy differences for microscopically reversible Markovian systems," *J. Stat. Phys.*, **90**, 1481 (1998).

[66] G. E. Crooks, "Path-ensemble averages in systems driven far from equilibrium," *Phys. Rev. E*, **61**, 2361 (1999).

[67] A. Imparato and L. Peliti, "Work probability distribution in systems driven out of equilibrium," *Phys. Rev. E*, **72**, 046114 (2005).
[68] M. I. Dykman, E. Mori, J. Ross, and P. M. Hunt, "Large fluctuations and optimal paths in chemical kinetics," *J. Chem. Phys.*, **100**, 5735 (1994).
[69] V. Elgart and A. Kamenev, "Rare event statistics in reaction-diffusion systems," *Phys. Rev. E*, **70**, 041106 (2004).
[70] B. Meerson and P. V. Sasorov, "Noise-driven unlimited population growth," *Phys. Rev. E*, **78**, 060103(R) (2008).
[71] C. Escudero and A. Kamenev, "Switching rates of multistep reactions," *Phys. Rev. E*, **79**, 041149 (2009).
[72] M. Assaf and B. Meerson, "Extinction of metastable stochastic populations," *Phys. Rev. E*, **81**, 021116 (2010).
[73] V. N. Smelyanskiy, M. I. Dykman, and B. Golding, "Time oscillations of escape rates in periodically driven systems," *Phys. Rev. Lett.*, **82**, 3193 (1999).
[74] M. I. Dykman, B. Golding, L. I. McCann et al., "Activated escape of periodically driven systems," *Chaos*, **11**, 587 (2001).
[75] V. Melnikov, "On the stability of a center for time-periodic perturbations," *Trans. Mosc. Math. Soc.*, **12**, 3 (1963).
[76] J. Guckenheimer and P. Holmes, *Non-linear Oscillations, Dynamical Systems, and Bifurcations of Vector Fields* (Springer-Verlag, 1986).
[77] C. Escudero and J. A. Rodriguez, "Persistence of instanton connections in chemical reactions with time-dependent rates," *Phys. Rev. E*, **77**, 011130 (2008).
[78] M. Assaf, A. Kamenev, and B. Meerson, "Population extinction in a time-modulated environment," *Phys. Rev. E*, **78**, 041123 (2008).
[79] M. Assaf, A. Kamenev, and B. Meerson, "Population extinction risk in the aftermath of a catastrophic event," *Phys. Rev. E*, **79**, 011127 (2009).
[80] R. Graham and T. Tél, "Existence of a potential for dissipative dynamical systems," *Phys. Rev. Lett.*, **52**, 9 (1984).
[81] R. Graham and T. Tél, "On the weak-noise limit of Fokker-Planck models," *J. Stat. Phys.*, **35**, 729 (1984).
[82] R. Graham and T. Tél, "Weak-noise limit of Fokker-Planck models and nondifferentiable potentials for dissipative dynamical systems," *Phys. Rev. A*, **31**, 1109 (1985).
[83] R. Graham, "Macroscopic potentials, bifurcations and nose in dissipative system," in *Noise in Non-linear Dynamical Systems*, edited by F. Moss and P. V. E. McClintock (Cambridge University Press, 1989), p. 225.
[84] M. I. Dykman, M. M. Millonas, and V. N. Smelyanskiy, "Observable and hidden singular features of large fuctuations in nonequilibrium systems,"*Phys. Lett. A*, **195**, 53 (1994).
[85] V. N. Smelyanskiy, M. I. Dykman, and R. S. Maier, "Topological features of large fluctuations to the interior of a limit cycle," *Phys. Rev. E*, **55**, 2369 (1997).
[86] R. S. Maier and D. L. Stein, "Transition-rate theory for nongradient drift fields," *Phys. Rev. Lett.*, **69**, 3691 (1992).
[87] R. S. Maier and D. L. Stein, "Escape problem for irreversible systems," *Phys. Rev. E*, **48**, 931 (1993).
[88] R. S. Maier and D. L. Stein, "Effect of focusing and caustics on exit phenomena in systems lacking detailed balance," *Phys. Rev. Lett.*, **71**, 1783 (1993).
[89] M. I. Freidlin and A. D. Ventzel, *Random Perturbations in Dynamical Systems* (Springer-Verlag, 1984).

[90] A. Dembo and O. Zeitouni, *Large Deviations Techniques and Applications* (Springer-Verlag, 1998).
[91] M. Parker, A. Kamenev, and B. Meerson, "Noise-induced stabilization in population dynamics," *Phys. Rev. Lett.*, **107**, 180603 (2011).
[92] M. Parker and A. Kamenev, "Extinction in the Lotka–Volterra model," *Phys. Rev. E*, **80**, 021129 (2009).
[93] M. I. Dykman and M. A. Krivoglaz, "Spectral distribution of nonlinear oscillators with nonlinear friction due to a medium," *Physica Status Solidi (b)*, **68**, 111 (1975).
[94] G. Lindblad, "On the generators of quantum dynamical semigroups," *Commun. Math. Phys.*, **48**, 119130 (1976).
[95] V. Gorini, A. Kossakowski, and E. C. G. Sudarshan, "Completely positive dynamical semigroups of N-level systems," *J. Math. Phys.*, **17**, 821825 (1976).
[96] H.-P. Breuer and F. Petruccione, *The Theory of Open Quantum Systems* (Oxford University Press, 2002).
[97] L. M. Sieberer, S. D. Huber, E. Altman, and S. Diehl, "Nonequilibrium functional renormalization for driven-dissipative Bose-Einstein condensation," *Phys. Rev. B*, **89**, 134310 (2014).
[98] L. M. Sieberer, M. Buchhold, and S. Diehl, "Keldysh field theory for driven open quantum systems," *Rep. Prog. Phys.*, **79**, 096001 (2016).
[99] S. M. Barnett and P. M. Radmore, *Methods in Theoretical Quantum Optics* (Oxford Series in Optical and Imaging Sciences **15**) (Clarendon Press, 1997).
[100] M. I. Dykman and V. N. Smelyanskiy, "Quantum theory of transitions between stable states of a nonlinear oscillator interacting with a medium in a resonant field," *Zh. Eksp. Teor. Fiz.*, **94**, 61 (1988); [*Soviet Physics - JETP*, **67**, 1769 (1988)].
[101] M. Marthaler and M. I. Dykman, "Switching via quantum activation: A parametrically modulated oscillator," *Phys. Rev. A*, **73**, 042108 (2006).
[102] M. I. Dykman, "Periodically modulated quantum nonlinear oscillators," in *Fluctuating Nonlinear Oscillators. From Nanomechanics to Quantum Superconducting Circuits*, edited by M. I. Dykman (Oxford University Press 2012), pp. 165–197.
[103] M. Marthaler and M. I. Dykman, "Quantum interference in the classically forbidden region: A parametric oscillator," *Phys. Rev. A*, **76**, 010102(R) (2007).
[104] S. Diehl, E. Rico, M. A. Baranov, and P. Zoller, "Topology by dissipation in atomic quantum wires," *Nat. Phys*, **7**, 971977 (2011).
[105] C-E. Bardyn, M. A. Baranov, C. V. Kraus, E. Rico, A. Imamoglu, P. Zoller, and S. Diehl, "Topology by dissipation," *New J. Phys.*, **15**, 085001 (2013).
[106] Berislav Buča and Tomaž Prosen, "A note on symmetry reductions of the Lindblad equation: transport in constrained open spin chains," *New Journal of Physics*, **14**, 073007 (2012).
[107] V. Albert and L. Jiang, "Symmetries and conserved quantities in Lindblad master equations," *Phys. Rev. A*, **89**, 022118 (2014).
[108] S. Lieu, R. Belyansky, J. T. Young, R. Lundgren, V. Albert, A. V. Gorshkov, "Symmetry breaking and error correction in open quantum systems," *Phys. Rev. Lett.*, **125**, 240405 (2020).
[109] R. A. Santos, F. Iemini, A. Kamenev, and Y. Gefen, "A possible route towards dissipation-protected qubits using a multidimensional dark space and its symmetries," *Nature Communications*, **11**, 5899 (2020).
[110] F. Thompson and A. Kamenev, *Qubit Decoherence and Symmetry Restoration through Real-Time Instantons, Phys. Rev. Research*, **4**, 023020 (2022).

[111] M. I. Dykman, "Heating and cooling of local and quasilocal vibrations by a nonresonance field," *Fiz. Tverd. Tela*, **20**, 2264 (1978); [*Soviet Physics – Solid State*, **20**, 1306 (1978)].

[112] M. I. Dykman and M. A. Krivoglaz, "Theory of nonlinear oscillator interacting with a medium," in *Soviet Physics Reviews*, edited by I. M. Khalatnikov, **5**, 265 (1984).

[113] I. Wilson-Rae, N. Nooshi, W. Zwerger, and T. J. Kippenberg, "Theory of ground state cooling of a mechanical oscillator using dynamical backaction," *Phys. Rev. Lett.*, **99**, 093901 (2007).

[114] F. Marquardt, J. P. Chen, A. A. Clerk, and S. M. Girvin, "Quantum theory of cavity-assisted sideband cooling of mechanical motion," *Phys. Rev. Lett.*, **99**, 093902 (2007).

[115] A. J. Leggett, "Bose-Einstein condensation in the alkali gases: Some fundamental concepts," *Rev. Mod. Phys.*, **73**, 307 (2001).

[116] J. M. Luttinger, "An exactly soluble model of a many-fermion system," *J. Math. Phys.*, **4**, 1154 (1963).

[117] A. A. Vlasov, "About the vibrational properties of an electron gas," *Zh. Eksp. Teor. Fiz.*, **8**, 291 (1938).

[118] A. A. Vlasov, "On the kinetic theory of an assembly of particles with collective interaction," *J. Phys.*, **9**, 25 (1945).

[119] L. D. Landau, "On the vibrations of the electronic plasma," *Zh. Eksp. Teor. Fiz.*, **16**, 574 (1946); [*Sov. Phys. JETP*, **10**, 25 (1946)].

[120] H. Schamel, "Electron holes, ion holes and double layers: Electrostatic phase space structures in theory and experiment," *Phys. Rep.*, **140**, 161 (1986).

[121] W. E. Drummond and D. Pines, "Non-linear stability of plasma oscillations," *Nucl. Fusion Suppl. Pt. 3*, 1049 (1962).

[122] A. A. Vedenov, E. P. Velikhov, and R. Z. Sagdeev, "Quasilinear theory of plasma oscillations," *Nucl. Fusion Suppl., Pt. 2*, 465 (1962).

[123] L. A. Artsimovich and R. Z. Sagdeev, *Plasma Physics for Physicists* (Atomizdat, 1979).

[124] V. E. Zakharov, V. S. L'vov, and G. Falkovich, *Kolmogorov Spectra of Turbulence: Wave Turbulence* (Springer-Verlag, 1992).

[125] E. H. Lieb and W. Liniger, "Exact analysis of an interacting Bose gas. I. The general solution and the ground state," *Phys. Rev.*, **130**, 1605 (1963).

[126] E. H. Lieb, Exact analysis of an interacting Bose gas. II. The excitation spectrum, *Phys. Rev.*, **130**, 1616 (1963).

[127] A. Griffin, T. Nikuni, and E. Zaremba, *Bose-Condensed Gases at Finite Temperatures* (Cambridge University Press, 2009).

[128] V. N. Popov, *Functional Integrals and Collective Excitations* (Cambridge University Press, 1991).

[129] S. T. Beliaev, "Application of the method of quantum field theory to a system of bosons," *Sov. Phys. JETP*, **34**, 289 (1958).

[130] S. T. Beliaev, "Energy-spectrum of a non-ideal Bose gas," *Sov. Phys. JETP*, **34**, 433 (1958).

[131] U. Eckern, "Relaxation processes in a condensed Bose gas," *J. Low Temp. Phys.*, **54**, 333 (1984).

[132] M. J. Bijlsma, E. Zaremba, and H. T. C. Stoof, "Condensate growth in trapped Bose gases," *Phys. Rev. A*, **62**, 063609 (2000).

[133] H. T. C. Stoof, "Coherent versus incoherent dynamics during Bose-Einstein condensation in atomic gases," *J. Low Temp. Phys.*, **114**, 11 (1999).

[134] G. Baym and C. Ebner, "Phonon-quasiparticle interactions in dilute solutions of He3 in Superfluid He4: I. Phonon thermal conductivity and ultrasonic attenuation," *Phys. Rev.* **164**, 235 (1967).
[135] L. D. Landau and I. M. Khalatnikov, "The theory of viscosity of helium II. Calculation of the viscosity coefficient," *Zh. Eksp. Teor. Fiz.*, **19**, 709 (1949).
[136] D. M. Gangardt and A. Kamenev, "Bloch oscillations in a one-dimensional spinor gas," *Phys. Rev. Lett.*, **102**, 070402 (2009).
[137] C. N. Yang, "Some exact results for the many-body problem in one dimension with repulsive delta-function interaction," *Phys. Rev. Lett.*, **19** (1967), 1312; M. Gaudin, "Un systeme a une dimension de fermions en interaction," *Phys. Lett.*, **24A**, 55 (1967).
[138] S. Stringari, "Collective excitations of a trapped bose-condensed gas," *Phys. Rev. Lett.*, **77**, 2360 (1996).
[139] P. P. Kulish, S. V. Manakov, and L. D. Faddeev, "Comparison of the exact quantum and quasiclassical results for a nonlinear Schrödinger equation," *Theor. Math. Phys.*, **28**, 615620 (1976).
[140] J. S. Langer, "Theory of nucleation rates," *Phys. Rev. Lett.*, **21**, 973 (1968).
[141] K. G. Wilson, "The renormalization group: Critical phenomena and the Kondo problem," *Rev. Mod. Phys.*, **47**, 773 (1975).
[142] J. Cardy, *Scaling and Renormalization in Statistical Physics* (Cambridge University Press, 1996).
[143] N. Goldenfeld, *Lectures on Phase Transitions and the Renormalization Group* (Westview Press, 1992).
[144] L. Kadanoff, *Statistical Physics: Statics, Dynamics and Renormalization* (World Scientific, 2000).
[145] P. C. Hohenberg and B. I. Halperin, "Theory of dynamic critical phenomena," *Rev. Mod. Phys.*, **49**, 435 (1977).
[146] S. Sachdev, *Quantum Phase Transitions* (Cambridge University Press, 2001).
[147] A. Schmid, "Diffusion and localization in a dissipative quantum system," *Phys. Rev. Lett.*, **51**, 1506 (1983).
[148] S. A. Bulgadaev, "Phase diagram of a dissipative quantum system," *Pis'ma Zh. Eksp. Theor. Fiz.*, **39**, 264 (1984); [*JETP Lett.*, **39**, 315 (1984)].
[149] M. Doi, "Stochastic theory of diffusion-controlled reaction," *J. Phys. A*, **9**, 1479 (1976).
[150] L. Peliti, "Path integral approach to birth-death processes on a lattice," *J. de Physique*, **46**, 1469 (1984).
[151] H. D. Abarbanel, J. D. Bronzan, R. L. Sugar, and A. R. White, "Reggeon field theory: Formulation and use," *Phys. Rep.*, **21**, 119 (1975).
[152] H. K. Janssen, "On the nonequilibrium phase transition in reaction-diffusion systems with an absorbing stationary state," *Z. Phys. B*, **42**, 151 (1981).
[153] P. Grassberger, "On phase transitions in Schlögl's second model," *Z. Phys. B*, **47**, 365 (1982).
[154] J. L. Cardy and P. Grassberger, "Epidemic models and percolation," *J. Phys. A*, **18**, L267 (1985).
[155] J. L. Cardy and R. L. Sugar, "Directed percolation and Reggeon field theory," *J. Phys. A: Math. Gen.*, **13**, L423 (1980).
[156] H. Hinrichsen, "Non-equilibrium critical phenomena and phase transitions into absorbing states," *Adv. Phys.*, **49**, 815 (2000).

[157] G. Odor, *Universality in Nonequilibrium Lattice Systems: Theoretical Foundations* (World Scientific, 2008).
[158] J. Marro and R. Dickman, *Nonequilibrium Phase Transitions in Lattice Models* (Cambridge University Press, 1999).
[159] V. Elgart and A. Kamenev, "Classification of phase transitions in reaction-diffusion models," *Phys. Rev. E*, **74**, 041101 (2006).
[160] J. Cardy and U. C. Täuber, "Theory of branching and annihilating random walks," *Phys. Rev. Lett.*, **77**, 4780 (1996).
[161] J. Cardy and U. C. Täuber, "Field theory of branching and annihilating random walks," *J. Stat. Phys.*, **90**, 1 (1998).
[162] O. Al Hammal, J. A. Bonachela, and M. A. Munoz, "Absorbing state phase transitions with a non-accessible vacuum," *J. Stat. Mech.*, **12**, P12007 (2006).
[163] H. K. Janssen, F. van Wijland, O. Deloubriere, and U. C. Täuber, "Pair contact process with diffusion: Failure of master equation field theory," *Phys. Rev. E*, **70**, 056114 (2004).
[164] M. Kardar, G. Parisi, and Y. Zhang, "Dynamic scaling of growing interfaces," *Phys. Rev. Lett.*, **56**, 889 (1986).
[165] D. Forster, D. R. Nelson, and M. J. Stephen, "Large-distance and long-time properties of a randomly stirred fluid," *Phys. Rev. A*, **16**, 732 (1977).
[166] E. Frey and U. C. Täuber, "Two-loop renormalization-group analysis of the Burgers-Kardar-Parisi-Zhang equation," *Phys. Rev. E*, **50**, 1024 (1994); *Phys. Rev. E*, **51**, 6319 (1995).
[167] C. A. Doty and J. M. Kosterlitz, "Exact dynamical exponent at the Kardar- Parisi-Zhang roughening transition," *Phys. Rev. Lett.*, **69**, 1979 (1992).
[168] E. Frey, U. C. Täuber, and H. K. Janssen, "Scaling regimes and critical dimensions in the Kardar-Parisi-Zhang problem," *Europhys. Lett.*, **47**, 14 (1999).
[169] H. C. Fogedby, "Soliton approach to the noisy Burgers equation: Steepest descent method," *Phys. Rev. E*, **57**, 4943 (1998).
[170] H. C. Fogedby, "Canonical phase-space approach to the noisy Burgers equation: Probability distributions," *Phys. Rev. E*, **59**, 5065 (1999).
[171] L. Bertini, A. De Sole, D. Gabrielli, G. Jona-Lasinio, and C. Landim, "Fluctuations in stationary nonequilibrium states of irreversible processes," *Phys. Rev. Lett.*, **87**, 040601 (2001).
[172] L. Bertini, A. De Sole, D. Gabrielli, G. Jona-Lasinio, and C. Landim, "Macroscopic fluctuation theory for stationary non-equilibrium states," *J. Stat. Phys.*, **107**, 635 (2002).
[173] L. Bertini, A. De Sole, D. Gabrielli, G. Jona-Lasinio, and C. Landim, "Macroscopic fluctuation theory," *Rev. Mod. Phys.*, 87, 593 (2015).
[174] J. Tailleur, J. Kurchan, and V. Lecomte, "Mapping out-of-equilibrium into equilibrium in one-dimensional transport models," *J. Phys. A*, **41**, 505001 (2008).
[175] D. A. Huse, C. L. Henley, and D. S. Fisher, "Huse, Henley, and Fisher respond," *Phys. Rev. Lett.*, **55**, 2924 (1985).
[176] I. V. Kolokolov and S. E. Korshunov, "Optimal fluctuation approach to a directed polymer in a random medium," *Phys. Rev. B*, **75**, 140201(R) (2007).
[177] I. V. Kolokolov and S. E. Korshunov, "Universal and nonuniversal tails of distribution functions in the directed polymer and Kardar-Parisi-Zhang problems," *Phys. Rev. B*, **78**, 024206 (2008).
[178] I. V. Kolokolov and S. E. Korshunov, "Explicit solution of the optimal fluctuation problem for an elastic string in a random medium," *Phys. Rev. B*, **80**, 031107 (2009).

[179] B. Meerson, E. Katzav, and A. Vilenkin, "Large deviations of surface height in the Kardar-Parisi-Zhang equation," *Phys. Rev. Lett.*, **116**, 070601 (2016).
[180] A. Kamenev, B. Meerson, and P. V. Sasorov, "Short-time height distribution in 1d KPZ equation: Starting from a parabola," *Phys. Rev. E*, **94**, 032108 (2016).
[181] P. Le Doussal, S. N. Majumdar, A. Rosso, and G. Schehr, "Exact short-time height distribution in the one-dimensional Kardar-Parisi-Zhang equation and edge fermions at high temperature," *Phys. Rev. Lett.*, **117**, 070403 (2016).
[182] M. Janas, A. Kamenev, and B. Meerson, "Dynamical phase transition in large deviation statistics of the Kardar-Parisi-Zhang equation," *Phys. Rev. E*, **94**, 032133 (2016).
[183] G. Falkovich, I. Kolokolov, V. Lebedev, and A. Migdal, "Instantons and intermittency," *Phys. Rev. E*, **54**, 4896 (1996).
[184] G. Falkovich, K. Gawedzki, and M. Vergassola, "Particles and fields in fluid turbulence," *Rev. Mod. Phys.*, **73**, 913 (2001).
[185] T. Grafke, R. Grauer, and T. Schäfer, "The instanton method and its numerical implementation in fluid mechanics," *J. Phys. A*, **48**, 333001 (2015).
[186] C. Kipnis, C. Marchioro, and E. Presutti, "Heat flow in an exactly solvable model," *J. Stat. Phys.*, **27**, 65 (1982).
[187] V. Lecomte, A. Imparato, and F. van Wijland, "Current fluctuations in systems with diffusive dynamics, in and out of equilibrium," *Prog. Theor. Phys. Suppl.*, **184**, 276 (2010).
[188] P. L. Krapivsky and B. Meerson, "Fluctuations of current in nonstationary diffusive lattice gases," *Phys. Rev. E*, **86**, 031106 (2012).
[189] V. Dotsenko, "Bethe ansatz derivation of the Tracy-Widom distribution for one-dimensional directed polymers," *Europhys. Lett.*, **90**, 20003 (2010).
[190] P. Calabrese and P. Le Doussal, "Exact solution for the Kardar-Parisi-Zhang equation with flat initial conditions," *Phys. Rev. Lett.*, **106**, 250603 (2011).
[191] P. Calabrese and P. Le Doussal, "The KPZ equation with flat initial condition and the directed polymer with one free end," *J. Stat. Mech.*, **P06001** (2012).
[192] T. Imamura and T. Sasamoto, "Exact solution for the stationary Kardar-Parisi-Zhang equation," *Phys. Rev. Lett.* **108**, 190603 (2012).
[193] T. Imamura and T. Sasamoto, "Stationary correlations for the 1D KPZ equation," *J. Stat. Phys.*, **150**, 908 (2013).
[194] A. Borodin, I. Corwin, P. L. Ferrari, and B. Vetö, "Height fluctuations for the stationary KPZ equation," *Math. Phys. Anal. Geom.* **18**, 1 (2015).
[195] C. A. Tracy and H. Widom, "Level-spacing distributions and the Airy kernel," *Comm. Math. Phys.*, **159**, 174 (1994).
[196] J. Baik and E. M. Rains, "Limiting distributions for a polynuclear growth model with external sources," *J. Stat. Phys.*, **100**, 523 (2000).
[197] A. I. Chernykh and M. G. Stepanov, "Large negative velocity gradients in Burgers turbulence," *Phys. Rev. E*, **64**, 026306 (2001).
[198] M. J. Ablowitz, D. J. Kaup, A. C. Newell, and H. Segur, "The inverse scattering transform-Fourier analysis for nonlinear problems," *Studies in Appl. Math.*, **53**, 249315 (1974).
[199] A. Krajenbrink and P. Le Doussal, "The inverse scattering of the Zakharov-Shabat system solves the weak noise theory of the Kardar-Parisi-Zhang equation," *Phys. Rev. Lett.* **127**, 064101 (2021).
[200] A. Krajenbrink and P. Le Doussal, "Inverse scattering solution of the weak noise theory of the Kardar-Parisi-Zhang equation with flat and Brownian initial conditions," Phys. Rev. E **105**, 054142 (2022).

[201] E. Bettelheim, N. R. Smith, and B. Meerson, "Inverse scattering method solves the problem of full statistics of nonstationary heat transfer in the Kipnis-Marchioro-Presutti model," Phys. Rev. Lett. **128**, 130602 (2022).
[202] N. R. Smith, A. Kamenev, and B. Meerson, "Landau theory of the short-time dynamical phase transitions of the Kardar-Parisi-Zhang interface," *Phys. Rev. E*, **97**, 042130 (2018).
[203] A. Krajenbrink and P. Le Doussal, "Simple derivation of the $-H^{5/2}$ tail for the 1D KPZ equation," *J. Stat. Mech.*, **063210** (2018).
[204] Yier Lin and Li-Cheng Tsai, "Short time large deviations of the KPZ equation," *Commun. Math. Phys.*, **386**, 359393 (2021).
[205] A. Krajenbrink and P. Le Doussal, "Exact short-time height distribution in 1D KPZ equation with Brownian initial condition," *Phys. Rev. E*, **96**, 020102 (2017).
[206] T. Iwatsuka, Y. T. Fukai, and K. A. Takeuchi, "Direct evidence for universal statistics of stationary Kardar-Parisi-Zhang interfaces," *Phys. Rev. Lett.*, **124**, 250602 (2020).
[207] A. K. Hartmann, B. Meerson, and P. Sasorov, "Observing symmetry-broken optimal paths of stationary Kardar-Parisi-Zhang interface via a large-deviation sampling of directed polymers in random media," *Phys. Rev. E*, **104**, 054125 (2021).
[208] S. F. Edwards and D. R. Wilkinson, "The surface statistics of a granular aggregate," *Proceedings of the Royal Society of London. Series A, Mathematical and Physical Sciences*, **381**, pp. 17–31 (1982).
[209] T. Nattermann and Lei-Han Tang, "Kinetic surface roughening. I. The Kardar-Parisi-Zhang equation in the weak-coupling regime," *Phys. Rev. A* **45**, 7156 (1992).
[210] R. Hirota, "Exact solution of the Kortewegde Vries equation for multiple collisions of solitons," *Phys. Rev. Lett.*, **27**, 1192 (1971).
[211] J. Hietarinta, "Hirotas bilinear method and soliton solutions," *Physics AUC*, **15**, 31 (2005).
[212] F. A. Berezin, *The Method of Second Quantization* (Academic Press, 1966).
[213] J. A. Hertz, "Quantum critical phenomena," *Phys. Rev. B*, **14**, 1165 (1976).
[214] A. J. Millis, "Effect of a nonzero temperature on quantum critical points in itinerant fermion systems," *Phys. Rev. B*, **48**, 7183 (1993).
[215] D. Belitz, T. R. Kirkpatrick, and T. Vojta, "How generic scale invariance influences quantum and classical phase transitions," *Rev. Mod. Phys.*, **77**, 579 (2005).
[216] D. L. Maslov and A. V. Chubukov, "Nonanalytic paramagnetic response of itinerant fermions away and near a ferromagnetic quantum phase transition," *Phys. Rev. B*, **79**, 075112 (2009).
[217] S. Sachdev, "Quantum phase transitions and conserved charges," *Z. Phys. B*, **94**, 469 (1994).
[218] P. A. Lee, "Gauge field, Aharonov-Bohm flux, and high-T_c superconductivity," *Phys. Rev. Lett.*, **63**, 680 (1989).
[219] A. A. Abrikosov and I. M. Khalatnikov, "The theory of a Fermi liquid," *Rep. Prog. Phys.*, **22**, 329 (1959).
[220] J. C. Slonczewski, "Current-driven excitation of magnetic multilayers," *J. Magn. Magn. Mater.*, **286**, L1 (1996).
[221] L. Berger, "Emission of spin waves by a magnetic multilayer traversed by a current," *Phys. Rev. B*, **54**, 9353 (1996).
[222] R. N. Gurzhi, "Hydrodynamic effects in solids at low temperature," *Usp. Fiz. Nauk*, **94**, 689 (1968); [*Sov. Phys. Usp.*, **11**, 255 (1968)].

[223] L. Levitov and G. Falkovich, "Electron viscosity, current vortices and negative nonlocal resistance in graphene," *Nat. Phys.* **12**, 672 (2016).
[224] D. Xiao, M.-C. Chang, and Q. Niu, "Berry phase effects on electronic properties," *Rev. Mod. Phys.*, **82**, 1959 (2010).
[225] E. Bettelheim, "Derivation of one-particle semiclassical kinetic theory in the presence of non-Abelian Berry curvature," *J. of Phys. A: Mathematical and Theoretical*, **50**, 415303 (2017).
[226] L. Faddeev and R. Jackiw, "Hamiltonian reduction of unconstrained and constrained systems," *Phys. Rev. Lett.*, **60**, 1692 (1988).
[227] C. Duval, Z. Horvath, P. A. Horvathy, L. Martina, and P. Stichel, "Berry phase correction to electron density in solids and exotic dynamics," *Mod. Phys. Lett. B*, **20**, 373 (2006).
[228] S. Li, A. V. Andreev, and B. Z. Spivak, "Anomalous transport phenomena in $p_x + ip_y$ superconductors," *Phys. Rev. B*, **92**, 100506(R) (2015).
[229] N. A. Sinitsyn, A. H. MacDonald, T. Jungwirth, V. K. Dugaev, and Jairo Sinova, "Anomalous Hall effect in a two-dimensional Dirac band: The link between the Kubo-Streda formula and the semiclassical Boltzmann equation approach," *Phys. Rev. B*, **75**, 045315 (2007).
[230] K. I. Seetharam, C-E. Bardyn, N. H. Lindner, M. S. Rudner, and G. Refael, "Controlled population of Floquet-Bloch states via coupling to Bose and Fermi baths," *Phys. Rev. X*, **5**, 041050 (2015).
[231] K. I. Seetharam, C-E. Bardyn, N. H. Lindner, M. S. Rudner, and G. Refael, "Steady states of interacting Floquet insulators," *Phys. Rev. B*, **99**, 014307 (2019).
[232] N. Tsuji, T. Oka, and H. Aoki, "Correlated electron systems periodically driven out of equilibrium: Floquet + DMFT formalism," *Phys. Rev. B*, **78**, 235124 (2008).
[233] M. Genske and A. Rosch, "Floquet-Boltzmann equation for periodically driven Fermi systems," *Phys. Rev. A*, **92**, 062108 (2015).
[234] T. Bilitewski and N. R. Cooper, "Scattering theory for Floquet-Bloch states," *Phys. Rev. A*, **91**, 033601 (2015).
[235] Sh. M. Kogan and A. Ya. Shulman, "Theory of fluctuations in a nonequilibrium electron gas," *Zh. Eksp. Teor. Fiz.*, **56**, 862 (1969); [*Sov. Phys. JETP* **29**, 467 (1969)].
[236] Sh. M. Kogan, "Equations for the correlation functions using a generalized Keldysh technique," *Phys. Rev. A*, **44**, 8072 (1991).
[237] K. E. Nagaev, "The Boltzmann Langevin approach: A simple quantum-mechanical derivation," *Physica E*, **74**, 461 (2015).
[238] A. I. Larkin and Y. N. Ovchinnikov, "Quasiclassical method in the theory of superconductivity," *JETP*, **28**, 960 (1969).
[239] A. Kitaev, *Hidden Correlations in the Hawking Radiation and Thermal Noise*, talk given at Fundamental Physics Prize Symposium, Nov. 10, 2014. Stanford SITP seminars, Nov. 11 and Dec. 18, 2014.
[240] J. Maldacena, S. H. Shenker, and D. A. Stanford, "Bound on chaos," *J. High Energ. Phys.*, **106** (2016).
[241] I. L. Aleiner, L. Faoro, and L. B. Ioffe, "Microscopic model of quantum butterfly effect: Out-of-time-order correlators and traveling combustion waves," *Annals of Physics*, **375**, 378–406 (2016).
[242] R. A. Fisher, "The wave of advance of advantageous genes," *Ann. Eugenics*, **7**, 355 (1937).
[243] A. Kolmogoroff, I. Petrovsky, and N. Piscounoff, *Study of the Diffusion Equation with Growth of the Quantity of Matter and Its Application to a Biology Problem*,

Bulletin de luniversité détat á Moscou, Ser. int., Section A, Vol. 1 (1937); [translated in P. Pelcé, *Dynamics of Curved Fronts* (Academic Press, San Diego, 1988)].

[244] D. Panja, "Effects of fluctuations on propagating fronts," *Physics Reports*, **393**, 87–174 (2004).

[245] G. A. Brooker and J. Sykes, "Transport properties of a Fermi liquid," *Phys. Rev. Lett.* **21**, 279 (1968).

[246] V. V. Kabanov and A. S. Alexandrov, "Electron relaxation in metals: Theory and exact analytical solutions," *Phys. Rev. B* **78**, 174514 (2008).

[247] J. B. Pendry, "Radiative exchange of heat between nanostructures," *J. Phys.: Condens. Matter*, **11**, 6621 (1999).

[248] K. Kim, B. Song, V. Fernndez-Hurtado, W. Lee, W. Jeong, L. Cui, D. Thompson, J. Feist, M. T. H. Reid, F. J. Garca-Vidal, J. C. Cuevas, E. Meyhofer, and P. Reddy, "Radiative heat transfer in the extreme near field," *Nature*, **528**, 387 (2015).

[249] R. St-Gelais, L. Zhu, S. Fan, and M. Lipson, "Near-field radiative heat transfer between nanostructures in the deep sub-wavelength regime," *Nature Nanotechnology*, **11**, 515 (2016).

[250] G. Bimonte, T. Emig, M. Kardar, and M. Krüger, "Nonequilibrium fluctuational quantum electrodynamics: Heat radiation, heat transfer, and force," *Annu. Rev. Condens. Matter Phys.*, **8**, 119 (2017).

[251] G. D. Mahan, "Tunneling of heat between metals," *Phys. Rev. B*, **95**, 115427 (2017).

[252] J-H. Jiang and J-S. Wang, "Caroli formalism in near-field heat transfer between parallel graphene sheets," *Phys. Rev. B*, **96**, 155437 (2017).

[253] M. Büttiker, "Role of scattering amplitudes in frequency-dependent current fluctuations in small conductors," *Phys. Rev. B*, **45**, 3807 (1992).

[254] R. Landauer, "Spatial variation of currents and fields due to localized scatterers in metallic conduction," *IBM J. Res. Dev.*, **1**, 233 (1957).

[255] R. Landauer, "Transport as a consequence of incident carrier flux," in *Localization, Interaction, and Transport Phenomena*, edited by G. Bergmann and Y. Bruynseraede (Springer-Verlag, 1985), p. 38.

[256] M. Büttiker, Y. Imry, R. Landauer, and S. Pinhas, "Generalized many-channel conductance formula with application to small rings," *Phys. Rev. B*, **31**, 6207 (1985).

[257] Y. Imry, "Physics of mesoscopic systems," in *Perspectives on Condensed Matter Physics*, edited by G. Grinstein and E. Mazenko (Singapore: World Scientific, 1986).

[258] Y. Imry, *Introduction to Mesoscopic Physics* (Oxford University Press, 1997).

[259] S. Datta, *Electronic Transport in Mesoscopic Systems* (Cambridge University Press, 1995).

[260] G. B. Lesovik, "Excess quantum noise in 2D ballistic point contacts," *JETP Lett.*, **49**, 592 (1989).

[261] Sh. M. Kogan, *Electronic Noise and Fluctuations in Solids* (Cambridge University Press, 1996).

[262] Ya. M. Blanter and M. Büttiker, "Shot noise in mesoscopic conductors," *Phys. Rep.*, **336**, 1 (2000).

[263] T. Martin, "Noise in mesoscopic physics," in *Nanophysics: Coherence and Transport*, edited by H. Bouchiat et al. (Elsevier, 2005), p. 283.

[264] L. S. Levitov and G. B. Lesovik, "Charge distribution in quantum shot noise," *Pis'ma Zh. Eksp. Teor. Fiz.*, **58**, 225 (1993); [*JETP Lett.*, **58**, 230 (1993)].

[265] G. B. Lesovik and L. S. Levitov, "Noise in an ac biased junction: Nonstationary Aharonov-Bohm effect," *Phys. Rev. Lett.*, **72**, 538 (1994).

[266] H. Lee and L. S. Levitov, "Current fluctuations in a single tunnel junction," *Phys. Rev. B*, **53**, 7383 (1996).

[267] L. S. Levitov, "The statistical theory of mesoscopic noise," in *Quantum Noise in Mesoscopic Physics*, edited by Yu. V. Nazarov and Ya. M. Blanter (Kluwer, 2003), p. 373.

[268] Yu. V. Nazarov and Ya. M. Blanter, *Quantum Transport: Introduction to Nanoscience* (Cambridge University Press, 2009).

[269] A. V. Andreev and A. Kamenev, "Counting statistics of an adiabatic pump," *Phys. Rev. Lett.*, **85**, 1294 (2000).

[270] J. Tobiska and Yu. V. Nazarov, "Inelastic interaction corrections and universal relations for full counting statistics in a quantum contact," *Phys. Rev. B*, **72**, 235328 (2005).

[271] P. W. Brouwer, "Scattering approach to parametric pumping," *Phys. Rev. B*, **58**, 10135 (1998).

[272] T. Holstein and H. Primakoff, "Field dependence of the intrinsic domain magnetization of a ferromagnet," *Phys. Rev.*, **58**, 1098 (1940).

[273] A. L. Chudnovskiy, J. Swiebodzinski, and A. Kamenev, "Spin-torque shot noise in magnetic tunnel junctions," *Phys. Rev. Lett.*, **101**, 066601 (2008).

[274] W. F. Brown, Jr, "Thermal fluctuations of a single-domain particle," *Phys. Rev.*, **130**, 1677 (1963).

[275] Y. Meir and N. S. Wingreen, "Landauer formula for the current through an interacting electron region," *Phys. Rev. Lett.*, **68**, 2512 (1992).

[276] D. A. Ivanov, H. W. Lee, and L. S. Levitov, "Coherent states of alternating current," *Phys. Rev. B*, **56**, 6839 (1997).

[277] D. C. Glattli and P. S. Roulleau, "Levitons for electron quantum optics," *Physica Status Solidi (b)*, **254**, 1600650 (2017).

[278] F. Pierre, H. Pothier, D. Esteve, and M. H. Devoret, "Energy redistribution between quasiparticles in mesoscopic silver wires," *J. Low Temp. Phys.*, **118**, 437 (2000).

[279] S. F. Edwards, "A second quantization formulation for electrons in disordered systems," *J. Phys. C*, **8**, 1660 (1975).

[280] F. J. Wegner, "The mobility edge problem: Continuous symmetry and a conjecture," *Z. Phys. B*, **35**, 207 (1979).

[281] K. B. Efetov, A. I. Larkin, and D. E. Khmelnitskii, "Interaction of diffusion modes in the theory of localization," *Zh. Eksp. Teor. Fiz.*, **79**, 1120 (1980); [*Sov. Phys. JETP*, **52**, 568 (1980)].

[282] A. M. Finkel'stein, "Electron liquid in disordered conductors," edited by I. M. Khalatnikov, *Soviet Scientific Reviews* **14** (Harwood, 1990).

[283] D. Belitz and T. R. Kirkpatrick, "The Anderson-Mott transition," *Rev. Mod. Phys.*, **66**, 261 (1994).

[284] M. Mézard, G. Parisi, and M. Virasoro, *Spin Glass Theory and Beyond*, (World Scientific, 1987).

[285] A. Kamenev and M. Mézard, "Wigner-Dyson statistics from the replica method," *J. Phys. A*, **32**, 4373 (1999).

[286] A. Kamenev and M. Mezard, "Level correlations in disordered metals: The replica," *Phys. Rev. B*, **60**, 3944 (1999).

[287] E. Kanzieper, "Replica field theories, Painleve transcendents, and exact correlation functions," *Phys. Rev. Lett.*, **89**, 250201 (2002).

[288] E. Kanzieper, "Replica approach in random matrix theory," in *Oxford Handbook of Random Matrix Theory*, edited by G. Akemann, J. Baik, and P. Di Francesco (Oxford University Press, 2011).
[289] K. B. Efetov, "Supersymmetry and theory of disordered metals," *Adv. Phys.*, **32**, 53 (1983).
[290] K. B. Efetov, *Supersymmetry in Disorder and Chaos* (Cambridge University Press, 1997).
[291] A. D. Mirlin, "Statistics of energy levels and eigenfunctions in disordered systems," *Phys. Rep.*, **326**, 259 (2000).
[292] H. Sompolinsky, "Time-dependent order parameters in spin-glasses," *Phys. Rev. Lett.*, **47**, 935 (1981).
[293] H. Sompolinsky and A. Zippelius, "Relaxational dynamics of the Edwards-Anderson model and the mean-field theory of spin-glasses," *Phys. Rev. B*, **25**, 6860 (1982).
[294] V. S. Dotsenko, M. V. Feigelman, and L. B. Ioffe, "Spin glasses and related problems," in *Soviet Scientific Reviews*, **15** (Harwood Academic, 1990), pp. 1–250.
[295] L. F. Cugliandolo, "Lecture notes in slow relaxation and non-equilibrium dynamics in condensed matter," in *Les Houches Session 77 July 2002*, edited by J-L. Barrat, J. Dalibard, J. Kurchan, and M. V. Feigel'man (Springer-Verlag, 2003), pp. 367–521.
[296] M. L. Horbach and G. Schön, "Dynamic nonlinear sigma-model of electron localization," *Ann. Phys. (Leipzig)*, **505**, 51 (1993).
[297] A. Kamenev and A. V. Andreev, "Electron-electron interactions in disordered metals: Keldysh formalism," *Phys. Rev. B*, **60**, 2218 (1999).
[298] C. Chamon, A. W. Ludwig, and C. Nayak, "Schwinger-Keldysh approach to disordered and interacting electron systems: Derivation of Finkelsteins renormalization-group equations," *Phys. Rev. B*, **60**, 2239 (1999).
[299] K. Usadel, "Generalized diffusion equation for superconducting alloys," *Phys. Rev. Lett.*, **25**, 507 (1970).
[300] D. B. Gutman and Y. Gefen, "Shot noise in disordered junctions: Interaction corrections," *Phys. Rev. B*, **64**, 205317 (2001).
[301] P. W. Anderson, "Absence of diffusion in certain random lattices," *Phys. Rev.*, **109**, 1492 (1958).
[302] L. P. Gor'kov, A. I. Larkin, and D. E. Khmelnitskii, "Particle conductivity in a two-dimensional random potential," *Pis'ma Zh. Eksp. Teor. Fiz.*, **30**, 248 (1979); [*Sov. Phys. JETP. Lett.*, **30**, 228 (1979)].
[303] B. L. Altshuler, A. G. Aronov, D. E. Khmelnitskii, and A. I. Larkin, "Coherent effects in disordered conductors," in *Quantum Theory of Solids*, edited by I. M. Lifshits (MIR Publishers, 1983), pp. 130–237.
[304] P. A. Lee and T. V. Ramakrishnan, "Disordered electronic systems," *Rev. Mod. Phys.*, **57**, 287 (1985).
[305] A. L. Altshuler, P. A. Lee, and R. A. Webb, eds. *Mesoscopic Phenomena in Solids* (North Holland, 1991).
[306] S. Hikami, "Anderson localization in a nonlinear-σ-model representation," *Phys. Rev. B*, **24**, 2671 (1981).
[307] G. Bergmann, "Weak localization in thin films a time-of-flight experiment with conduction electrons," *Phys. Rep.*, **107**, 1 (1984).
[308] E. Abrahams, P. W. Anderson, D. C. Licciardello, and T. V. Ramakrishnan, "Scaling theory of localization: Absence of quantum diffusion in two dimensions," *Phys. Rev. Lett.*, **42**, 673 (1979).

[309] M. A. Skvortsov, "Quantum correction to the Kubo formula in closed mesoscopic systems," *Phys. Rev. B*, **68**, 041306(R) (2003).
[310] D. M. Basko, M. A. Skvortsov, and V. E. Kravtsov, "Dynamic localization in quantum dots: Analytical theory," *Phys. Rev. Lett.*, **90**, 096801 (2003).
[311] B. V. Chirikov, F. M. Izrailev, and D. L. Shepelyansky, "Dynamical stochasticity in classical and quantum mechanics," *Sov. Sci. Rev. Sec.*, C2, 209 (1981).
[312] D. R. Grempel, R. E. Prange, and S. Fishman, "Quantum dynamics of a non-integrable system," *Phys. Rev. A*, **29**, 1639 (1984).
[313] C. Tian, A. Kamenev, and A. I. Larkin, "Weak dynamical localization in periodically kicked cold atomic gases," *Phys. Rev. Lett.*, **93**, 124101 (2004).
[314] F. Haake. *Quantum Signatures of Chaos, second edition* (Springer, New York, 2000).
[315] L. S. Levitov and A. V. Shytov, "Semiclassical theory of the Coulomb anomaly," *JETP Lett.*, **66**, 214221 (1997).
[316] B. L. Altshuler, D. Khmel'nitzkii, A. I. Larkin, and P. A. Lee, "Magnetoresistance and Hall effect in a disordered two-dimensional electron gas," *Phys. Rev. B*, **22**, 5142 (1980).
[317] B. L. Altshuler and B. I. Shklovskii, "Repulsion of energy levels and conductivity of small metal samples," *Zh. Eksp. Teor. Fiz.*, **91**, 220 (1986); [*Sov. Phys. JETP*, **64**, 127 (1986)].
[318] M. L. Mehta, *Random Matrices* (Academic, 1991).
[319] A. V. Andreev and B. L. Altshuler, "Spectral statistics beyond random matrix theory," *Phys. Rev. Lett.*, **75**, 902 (1995).
[320] E. Akkermans and G. Montambaux, *Mesoscopic Physics of Electrons and Photons* (Cambridge University Press, 2007).
[321] A. Altland and A. Kamenev, "Wigner-Dyson statistics from the Keldysh σ-model," *Phys. Rev. Lett.*, **85**, 5615 (2000).
[322] C. P. Umbach, S. Washburn, R. B. Laibowitz, and R. A. Webb, "Magnetoresistance of small, quasi-one-dimensional, normal-metal rings and lines," *Phys. Rev. B*, **30**, 4048 (1984).
[323] B. L. Altshuler, "Fluctuations in extrinsic conductivity of disordered conductors," *Pis'ma Zh. Eksp. Teor. Fiz.*, **41**, 530 (1985).
[324] P. A. Lee and A. D. Stone, "Universal conductance fluctuations in metals," *Phys. Rev. Lett.*, **55**, 1622 (1985).
[325] B. L. Altshuler, V. E. Kravtsov, and I. V. Lerner, "Distribution of mesoscopic fluctuations and relaxation processes in disordered conductors," in *Mesoscopic Phenomena in Solids*, edited by B. L. Altshuler, P. A. Lee, and R. A. Webb (North-Holland, 1991), p. 449.
[326] A. I. Larkin and D. E. Khmelnitskii, "Mesoscopic fluctuations of current-voltage characteristics," *Zh. Eksp. Teor. Fiz.*, **91**, 1815 (1986); [*Sov. Phys. JETP*, **64**, 1075 (1986)].
[327] H. F. Cheung, E. K. Riedel, and Y. Gefen, "Persistent currents in mesoscopic rings and cylinders," *Phys. Rev. Lett.*, **62**, 587 (1989).
[328] E. K. Riedel and F. von Oppen, "Mesoscopic persistent current in small rings," *Phys. Rev. B*, **47**, 15449 (1993).
[329] B. L. Altshuler, A. G. Aronov, and D. E. Khmelnitskii, "Effects of electron-electron collisions with small energy transfers on quantum localisation," *J. Phys. C*, **15**, 7367 (1982).
[330] I. L. Aleiner and Ya. M. Blanter, "Inelastic scattering time for conductance fluctuations," *Phys. Rev. B*, **65**, 115317 (2002).

[331] O. N. Dorokhov, "On the coexistence of localized and extended electronic states in the metallic phase," *Solid State Commun.*, **51**, 381 (1984).
[332] Yu. V. Nazarov, "Limits of universality in disordered conductors," *Phys. Rev. Lett.*, **73**, 134 (1994).
[333] Yu. V. Nazarov, "Universalities of weak localization," *Ann. Phys. (Leipzig)*, **8**, 507 (1999).
[334] C. W. J. Beenakker and M. Büttiker, "Suppression of shot noise in metallic diffusive conductors," *Phys. Rev. B*, **46**, 1889 (1992).
[335] K. E. Nagaev, "On the shot noise in dirty metal contacts," *Phys. Lett. A*, **169**, 103 (1992).
[336] I. L. Aleiner, P. W. Brouwer, and L. I. Glazman, "Quantum effects in Coulomb blockade," *Phys. Rep.*, **358**, 309 (2002).
[337] Yu. V. Nazarov, "Novel circuit theory of Andreev reflection," *Superlattices and Microstruct.*, **25**, 1221 (1999).
[338] W. Belzig and Yu. V. Nazarov, "Full current statistics in diffusive normal-superconductor structures," *Phys. Rev. Lett.*, **87**, 067006 (2001).
[339] M. V. Feigelman, A. Kamenev, A. I. Larkin, and M. A. Skvortsov, "Weak charge quantization on a superconducting island," *Phys. Rev. B*, **66**, 054502 (2002).
[340] Yu. V. Nazarov, "Coulomb blockade without tunnel junctions," *Phys. Rev. Lett.*, **82**, 1245 (1999).
[341] I. Snyman and Yu. V. Nazarov, "Keldysh action of a multiterminal time-dependent scatterer," *Phys. Rev. B*, **77**, 165118 (2008).
[342] Ya. M. Blanter, H. Schomerus, and C. W. J. Beenakker, "Effect of dephasing on charge-counting statistics in chaotic cavities," *Physica E*, **11**, 1 (2001).
[343] S. Pilgram, A. N. Jordan, E. V. Sukhorukov, and M. Büttiker, "Stochastic path integral formulation of full counting statistics," *Phys. Rev. Lett.*, **90**, 206801 (2003).
[344] A. N. Jordan, E. V. Sukhorukov, and S. Pilgram, "Fluctuation statistics in networks: A stochastic path integral approach," *J. Math. Phys.*, **45**, 4386 (2004).
[345] B. L. Altshuler, A. G. Aronov, and B. Z. Spivak, "The Aharonov-Bohm effect in disordered conductors," *Pis'ma Zh. Eksp. Teor. Fiz.*, **33**, 101 (1981); [*JETP Lett.*, **33**, 94 (1981)].
[346] D. Yu. Sharvin and Yu. V. Sharvin, "Magnetic-flux quantization in a cylindrical film of a normal metal," *Pis'ma Zh. Eksp. Teor. Fiz.*, **34**, 285 (1981); [*JETP Lett.*, **34**, 272 (1981)].
[347] B. L. Altshuler and A. G. Aronov, "Contribution to the theory of disordered metals in strongly doped semiconductors," *Zh. Eksp. Teor. Fiz.*, **77** 2028 (1979); [*Sov. Phys. JETP*, **50**, 968 (1979)].
[348] B. L. Altshuler and A. G. Aronov. in *Electron–Electron Interaction in Disordered Systems*, edited by A. J. Efros and M. Pollak (Elsevier, 1985), p. 1.
[349] B. L. Altshuler, A. G. Aronov, and P. A. Lee, "Interaction effects in disordered Fermi systems in two dimensions," *Phys. Rev. Lett.*, **44**, 1288 (1980).
[350] Yu. V. Nazarov, "Anomalous current-voltage characteristics of tunnel junctions," *Zh. Eksp. Teor. Fiz.*, **96**, 975 (1989); [*Sov. Phys. JETP*, **68**, 561 (1989)].
[351] A. M. Finkel'stein, "Suppression of superconductivity in homogeneously disordered systems," *Physica B*, **197**, 636 (1994).
[352] L. S. Levitov and A. V. Shytov, "Semiclassical theory of the Coulomb anomaly," *Pis'ma Zh. Eksp. Teor. Fiz.*, **66**, 200 (1997); [*JETP Lett.*, **66**, 214 (1997)].
[353] P. Kopietz, "Coulomb gap in the density of states of disordered metals in two dimensions," *Phys. Rev. Lett.*, **81**, 2120 (1998).

[354] A. M. Finkel'stein, "Disordered electron liquid with interactions," in *50 Years of Anderson Localization*, edited by E. Abrahams (Singapore: World Scientific, 2010), p. 385.
[355] A. M. Finkel'stein, "Disordered electron liquid with interactions," *Int. J. Mod. Phys. B*, **24**, 1855 (2010).
[356] H. Fukuyama, "Interaction effects in the weakly localized regime of two- and three-dimensional disordered system," in *Electron–Electron Interactions in Disordered Systems*, edited by A. J. Efros and M. Pollak (Elsevier, 1985).
[357] G. Zala, B. N. Narozhny, and I. L. Aleiner, "Interaction corrections at intermediate temperatures: Longitudinal conductivity and kinetic equation," *Phys. Rev. B*, **64**, 214204 (2001).
[358] G. Catelani and I. L. Aleiner, "Interaction corrections to thermal transport coefficients in disordered metals: The quantum kinetic equation approach," *Zh. Eksp. Teor. Fiz.*, **127**, 372 (2005); [*Sov. Phys. JETP*, **100**, 331 (2005)].
[359] K. E. Nagaev, "Cascade Boltzmann-Langevin approach to higher-order current correlations in diffusive metal contacts," *Phys. Rev. B*, **66**, 075334 (2002).
[360] D. B. Gutman, A. D. Mirlin, and Y. Gefen, "Kinetic theory of fluctuations in conducting systems," *Phys. Rev. B*, **71**, 085118 (2005).
[361] D. A. Bagrets, "Full current statistics of incoherent cold electrons," *Phys. Rev. Lett.*, **93**, 236803 (2004).
[362] B. L. Altshuler, "Temperature dependence of impurity conductivity of metals at low temperatures," *Zh. Eksp. Teor. Fiz.*, **75**, 1330 (1978); [*Sov. Phys. JETP*, **48**, 670 (1978)].
[363] B. L. Altshuler and A. G. Aronov, "Damping of one-electron excitations in metals," *Pis'ma Zh. Eksp. Teor. Fiz.*, **30**, 514 (1979); [*Sov. Phys. JETP Lett.*, **30**, 482 (1979)].
[364] O. V. Dimitrova and V. E. Kravtsov, "Infrared catastrophe in two-quasiparticle collision integral," *JETP Lett.*, **86**, 670 (2007).
[365] N. M. Chtchelkatchev and I. S. Burmistrov, "Energy relaxation in the spin-polarized disordered electron liquid," *Phys. Rev. Lett.*, **100**, 206804 (2008).
[366] A. Schmid, "On the dynamics of electrons in an impure metal," *Z. Phys.*, **271**, 251 (1974).
[367] A. Kamenev, "Near-field heat transfer between disordered conductors," arXiv:1811.10187.
[368] J. L. Wise, D. M. Basko, and F. W. J. Hekking, "Role of disorder in plasmon-assisted near-field heat transfer between two-dimensional metals," *Phys. Rev. B*, **101**, 205411 (2020).
[369] J. Bardeen, L. N. Cooper, and J. R. Schrieffer, "Theory of superconductivity," *Phys. Rev.*, **108**, 1175 (1957).
[370] L. P. Gor'kov, "On the energy spectrum of superconductors," *Zh. Eksp. Teor. Fiz.*, **34** (1958), 735; [*Sov. Phys. JETP*, **7**, 505 (1958)].
[371] Y. Nambu, "Quasi-particles and gauge invariance in the theory of superconductivity," *Phys. Rev.*, **117**, 648 (1960).
[372] M. V. Feigel'man, A. I. Larkin, and M. A. Skvortsov, "Keldysh action for disordered superconductors," *Phys. Rev. B*, **61**, 12361 (2000).
[373] M. A. Skvortsov, A. I. Larkin, and M. V. Feigel'man, "Superconductive proximity effect in interacting disordered conductors," *Phys. Rev. B*, **63**, 134507 (2001).
[374] I. V. Lerner, "Nonlinear sigma model for normal and superconducting systems: A pedestrian approach," in *Quantum Phenomena in Mesoscopic Systems*, edited by B. L. Altshuler, et al. (IOS Press, 2003), p. 271.

[375] M. Tinkham. *Introduction to Superconductivity* (McGraw Hill, 1996).
[376] N. B. Kopnin, *Theory of Nonequilibrium Superconductivity* (Oxford University Press, 2001).
[377] A. Schmid and G. Schön, "Linearized kinetic equations and relaxation processes of a superconductor near T_c," *J. Low Temp. Phys.*, **20**, 207 (1975).
[378] W. Belzig, F. K. Wilhelm, C. Bruder, G. Schön, and A. D. Zaikin, "Quasi-classical Green's function approach to mesoscopic superconductivity," *Superlatt. Microstruct.*, **25**, 1251 (1999).
[379] A. I. Larkin and Yu. N. Ovchinnikov, "Nonlinear effects during the motion of vortices in superconductors," *Zh. Eksp. Teor. Fiz.*, **73**, 299 (1977); [*Sov. Phys. JETP*, **46**, 155 (1977)].
[380] A. V. Zaitsev, "Quasiclassical equations of the theory of superconductivity for continuous metals and the properties of constricted microcontacts," *Zh. Eksp. Teor. Fiz.*, **86**, 1742 (1984); [*Sov. Phys. JETP*, **59**, 1015 (1984)].
[381] M. Yu. Kuprianov and V. F. Lukichev, "Influence of boundary transparency on the critical current of 'dirty' $SS'S$ structures," *Zh. Eksp. Teor. Fiz.*, **94**, 139 (1988); [*Sov. Phys. JETP*, **67**, 1163 (1988)].
[382] C. J. Lambert, R. Raimondi, V. Sweeney, and A. F. Volkov, "Boundary conditions for quasiclassical equations in the theory of superconductivity," *Phys. Rev. B*, **55**, 6015 (1997).
[383] A. Levchenko, "Crossover in the local density of states of mesoscopic superconductor/normal-metal/superconductor junctions," *Phys. Rev. B*, **77**, 180503(R) (2008).
[384] I. O. Kulik and A. N. Omelyanchuk, "Properties of superconducting microbridges in the pure limit," *Sov. J. Low Temp. Phys.*, **3**, 459 (1977).
[385] I. O. Kulik and A. N. Omelyanchuk, "Josephson effect in superconducting bridges – Microscopic theory," *Sov. J. Low Temp. Phys.*, **4**, 142 (1978).
[386] R. V. Carlson and A. M. Goldman, "Propagating order-parameter collective modes in superconducting films" *Phys. Rev. Lett.*, **34**, 11 (1975).
[387] B. N. Narozhny, I. L. Aleiner, and B. L. Altshuler, "Theory of interaction effects in normal-metal-superconductor junctions out of equilibrium," *Phys. Rev. B*, **60**, 7213 (1999).
[388] I. O. Kulik, O. Entin-Wohlman, and R. Orbach, "Pair susceptibility and mode propagation in superconductors: A microscopic approach," *J. Low Temp. Phys.*, **43**, 591 (1981).
[389] A. Schmid and G. Schön, "Collective oscillations in a dirty superconductor," *Phys. Rev. Lett.*, **34**, 942 (1975).
[390] S. N. Artemenko and A. F. Volkov, "Collisionless relaxation of the energy gap in superconductors," *Zh. Eksp. Teor. Fiz.*, **65**, 2038 (1973); [*Sov. Phys. JETP*, **38**, 1018 (1974)].
[391] A. Levchenko and A. Kamenev, "Keldysh Ginzburg-Landau action of fluctuating superconductors," *Phys. Rev. B*, **76**, 094518 (2007).
[392] J. P. Hurault and K. Maki, "Breakdown of the mean field theory in the superconducting transition region," *Phys. Rev. B*, **2**, 2560 (1970).
[393] K. Maki, "The critical fluctuation of the order parameter in type-II superconductors," *Prog. Teor. Phys.*, **39**, 897 (1968).
[394] R. S. Thompson, "Microwave, flux flow, and fluctuation resistance of dirty type-II superconductors," *Phys. Rev. B*, **1**, 327 (1970).
[395] L. P. Gor'kov and G. M. Eliashberg, "Generalization of Ginzburg-Landau equations for non-stationary problems in the case of alloys with paramagnetic

impurities," *Zh. Eksp. Teor. Fiz.*, **54**, 612 (1968); [*Sov. Phys. JETP*, **27**, 328 (1969)].
[396] A. I. Larkin and A. Varlamov. *Theory of Fluctuations in Superconductors* (Oxford University Press, 2005).
[397] L. G. Aslamazov and A. I. Larkin, "Effect of fluctuations on the properties of a superconductor above the critical temperature," *Fiz. Tverd. Tela.*, **10**, 1104 (1968); [*Sov. Phys. Solid State*, **10**, 875 (1968)].
[398] A. I. Larkin, "Superconductor of small dimensions in a strong magnetic field," *Zh. Eksp. Teor. Fiz.*, **48**, 232 (1965); [*Sov. Phys. JETP*, **21**, 153 (1965)].
[399] D. C. Mattis and J. Bardeen, "Theory of the anomalous skin effect in normal and superconducting metals," *Phys. Rev.*, **111**, 412 (1958).
[400] G. M. Eliashberg, "Inelastic electron collisions and nonequilibrium stationary states in superconductors," *JETP Letters*, **11**, 114 (1970); [*Sov. Phys. JETP*, **34**, 668 (1972)].
[401] J. A. Pals, K. Weiss, P. M. T. M. van Attekum, R. E. Horstman, and J. Wolter, "Non-equilibrium superconductivity in homogeneous thin films," *Physics Reports*, **89**, 323–390 (1982).
[402] A. F. G. Wyatt, V. M. Dmitriey, W. S. Moore, and F. W. Sheard, "Microwave-enhanced critical supercurrents in constricted tin films," *Phys. Rev. Lett.*, **16**, 1166 (1966).
[403] A. H. Dayem and J. J. Wiegand, "Behavior of thin-film superconducting bridges in a microwave field," *Phys. Rev.*, **155**, 419 (1967).
[404] H. Fröhlich, "Theory of the superconducting state. I. The ground state at the absolute zero of temperature," *Phys. Rev.*, **79**, 845 (1950).
[405] H. Fröhlich, "Electrons in lattice fields," *Adv. Phys.*, **3**, 325 (1954).
[406] A. B. Pippard, "Ultrasonic attenuation in metals," *Phil. Mag.*, **46**, 1104 (1955).
[407] T. Tsuneto, "Ultrasonic attenuation in superconductors," *Phys. Rev.*, **121**, 402 (1961).
[408] A. Schmid, "Electron-phonon interaction in an impure metal," *Z. Phys.*, **259**, 421 (1973).
[409] A. Shtyk, "Kinetics of electron-phonon fluctuations in disordered superconductors," Ph.D. thesis submitted to Landau Institute (Moscow 2016, in Russian).
[410] M. Yu. Reizer and A. V. Sergeyev, "Electron-phonon interaction in impure metals and superconductors," *Zh. Eksp. Teor. Fiz.* **90**, 1056 (1986); [*Sov. Phys. JETP*, **63**, 616 (1986)].
[411] V. I. Yudson and V. E. Kravtsov, "Electron kinetics in isolated mesoscopic rings driven out of equilibrium," *Phys. Rev B*, **67**, 155310 (2003).
[412] A. Shtyk and M. Feigel'man, "Ultrasonic attenuation via energy diffusion channel in disordered conductors," *Phys. Rev. B*, **92**, 195101 (2015).
[413] B. L. Altshuler, "Temperature dependence of impurity conductivity of metals at low temperatures," *Zh. Eksp. Teor. Fiz.*, **75**, 1330 (1978); [*Sov. Phys. JETP*, **48**, 670 (1978)].
[414] Y. Savich, L. Glazman, and A. Kamenev, "Quasiparticle relaxation in superconducting nanostructures," *Phys. Rev. B*, **96**, 104510 (2017).
[415] M. Ovadia, B. Sacp, and D. Shahar, "Electron-phonon decoupling in disordered insulators," *Phys. Rev. Lett.*, **102**, 176802 (2009).
[416] M. H. Devoret and R. S. Schoelkopf, "Superconducting circuits for quantum information: An outlook," *Science*, **339**, 1169 (2013).

[417] T. Karzig, C. Knapp, R. M. Lutchyn et al., "Scalable designs for quasiparticle-poisoning-protected topological quantum computation with Majorana zero modes," *Phys. Rev. B*, **95**, 235305 (2017).

[418] U. Vool, I. M. Pop, K. Sliwa et al., "Non-Poissonian quantum jumps of a fluxonium qubit due to quasiparticle excitations," *Phys. Rev. Lett.*, **113**, 247001 (2014).

[419] C. Wang, Y. Y. Gao, et al., "Measurement and control of quasiparticle dynamics in a superconducting qubit," *Nat. Commun.*, **5**, 5836 (2014).

[420] M. Taupin, I. M. Khaymovich, M. Meschke, A. S. Mel'nikov, and J. P. Pekola, "Quasiparticle trapping in Meissner and vortex states of mesoscopic superconductors," *Nat. Commun.* **7**, 10977 (2016).

[421] R.-P. Riwar, A. Hosseinkhani, L. D. Burkhart et al., "Normal-metal quasiparticle traps for superconducting qubits," *Phys. Rev. B*, **94**, 104516 (2016).

[422] R-P. Riwar and G. Catelani, "Efficient quasiparticle traps with low dissipation through gap engineering," *Phys. Rev. B*, **100**, 144514 (2019).

[423] M. V. Feigel'man and V. E. Kravtsov, "Electron-phonon cooling power in Anderson insulators," *Phys. Rev. B*, **99**, 125415 (2019).

Index

Printed in the United States
by Baker & Taylor Publisher Services

Printed in the United States
by Baker & Taylor Publisher Services